Physical Geology

THE STRUCTURE AND PROCESSES OF THE EARTH

B. CLARK BURCHFIEL
Massachusetts Institute of Technology

ROBERT J. FOSTER
San Jose State University

EDWARD A. KELLER
University of California, Santa Barbara

WILTON N. MELHORN
Purdue University

DOUGLAS G. BROOKINS
University of New Mexico

LEIGH W. MINTZ
California State University, Hayward

HAROLD V. THURMAN
Mt. San Antonio College

CHARLES E. MERRILL PUBLISHING CO.
A Bell & Howell Company
Columbus Toronto London Sydney

Cover Photo: "The Bowl." Muddy Mountains.
Kaibab thrust over recumbent syncline.
Clark County, Nevada, May, 1950.
Courtesy of U.S Geological Survey.

Published by Charles E. Merrill Publishing Co.
A Bell & Howell Company
Columbus, Ohio 43216

This book was set in Trump and Benguiat
Text Design and Production Coordination: Ann Mirels
Copy Editor: Linda M. Johnstone

Library of Congress Catalog Card Number: 81-84623
International Standard Book Number: 0-675-09913-7
Printed in the United States of America
1 2 3 4 5 6 7 8 9 10 – 86 85 84 83 82

THE AUTHORS

B. CLARK BURCHFIEL, *Professor of Geology, Massachusetts Institute of Technology, Cambridge, Massachusetts.* Dr. Burchfiel earned his Ph.D. at Yale University. His area of special interest is structural geology. He has been consultant to eight major oil companies and has participated in numerous professional organizations. He currently serves on the National Academy of Sciences Geological Sciences Board and is chairman of the Committee on Publications of the Geological Society of America. Dr. Burchfiel has published extensively. Both American and international projects have been completed under his direction.

ROBERT J. FOSTER, *Professor of Physical Science, San Jose State University, San Jose, California.* Dr. Foster received his Ph.D. from the University of Washington. The area of physical geology is his specialty. He has participated in geologic conferences and has done consulting work in the western states. In addition, Dr. Foster is a member of numerous professional organizations, including the American Association of Petroleum Geologists. He served as consulting editor for the Earth Science Series published by Charles E. Merrill.

EDWARD A. KELLER, *Associate Professor of Environmental Studies, University of California, Santa Barbara.* Dr. Keller received his Ph.D. from Purdue University and has special interest in geomorphology. He has been an environmental consultant in five states and is active in research and public service projects. His latest grant from the U.S. Geological Survey is for "Tectonic Geomorphology and Possible Future Seismic Activity of the Central Ventura Basin, California." Dr. Keller is the author of *Environmental Geology,* now in its third edition.

WILTON N. MELHORN, *Professor of Geosciences and Civil Engineering, Purdue University, Lafayette, Indiana. Also, Research Geologist, Laboratory for Application of Remote Sensing (LARS), Purdue University. In addition, Adjunct Professor, Mackay School of Mines, University of Nevada, Reno.* Dr. Melhorn received his Ph.D.. from the University of Michigan and specializes in geomorphology. He has been a consultant to numerous corporations concerning geologic problems, has been involved in many public service projects, and is active in many professional societies. He recently completed two contracts with the Environmental Protection Agency concerning groundwater in Indiana.

DOUGLAS G. BROOKINS, *Professor of Geology, University of New Mexico, Albuquerque, New Mexico.* Dr. Brookins received his Ph.D. from Massachusetts Institute of Technology where he specialized in the area of earth resources. He has presented numerous papers and has given symposia both nationally and internationally. Dr. Brookins is the author of the text, *Earth Resources, Energy, and the Environment.*

LEIGH W. MINTZ, *Associate Vice President, Academic Programs and Professor of Geological Sciences, California State University, Hayward.* Dr. Mintz earned his Ph.D. from the University of California at Berkeley. His postdoctoral studies in the area of earth history have taken him to London, Paris, and Bonn. Dr. Mintz is active in a number of professional organizations and in many community service projects. He is the author of *Historical Geology,* third edition.

HAROLD V. THURMAN, *Instructor, Department of Physical Sciences, Mt. San Antonio College, Walnut, California.* Mr. Thurman received his master's degree from California State University at Los Angeles and has special interest in the area of oceanography. He is the author of *Introductory Oceanography,* third edition, and is currently working on two other texts. He has also served as an exploration and development geologist in the oil industry.

PREFACE

- Mount St. Helens erupts, ejecting several cubic kilometers of volcanic ash into the atmosphere and downing approximately 800 million board feet of timber through the force of the blast.
- *Voyagers 1* and *2* send back photographs of volcanoes erupting on Jupiter's moon Io.
- A house, several automobiles, part of a community swimming pool, and portions of two businesses are swallowed up by a large sinkhole which developed quickly in Winter Park, Florida.
- Plate tectonic theory explains the creation of the Hawaiian Islands in terms of plate movement and hot spots.
- The animal and plant life in and around lakes in the Adirondacks of Upstate New York decreases as the acid rain formed because of air pollution in the fossil-fuel-burning industrial Midwest falls into the watershed of the area, increasing the lakes' acidity and the speed of weathering.
- The fringes of the Sahara Desert expand, accompanied by droughts in the Sahel which subject new land areas to wind erosion, drifting sand, and other desert processes.

GEOLOGISTS ARE INTERESTED IN ALL OF THESE EVENTS AND processes. The warm-water vents and Mount St. Helens contribute to the knowledge of plate tectonics, the exciting theory which explains the structure of the earth. Information about the volcanoes of Io, as well as helping scientists understand the structure of the Jovian moon, contributes to information about volcanic processes on earth. Geologists are concerned with natural geological processes and how human action affects those processes. Sinkholes and earthquakes are two examples of the many natural earth processes geologists are seeking to better understand. Acid rain and the desertification of the Sahel are obvious examples of the human impact. By burning fossil fuels and other industrial products, and by overgrazing, removal of trees, intensive cropping of agricultural marginal lands, and numerous other activities, humans continuously interact with geologic processes.

A wealth of data has recently emerged concerning global and regional geologic processes operating today and over geologic time. As more information is added to the vast store of knowledge about the planet earth, it becomes increasingly difficult for one person to write an introductory survey of the field. Because of this, we at Merrill decided to consolidate the efforts of several experts in the subfields of geology, with the final product a modern, comprehensive, introductory textbook. We provided these experts with support, thorough market research, and reviews; and they in turn provided the knowledge and expertise to write *Physical Geology.*

Bob Foster, Project Editor, has written the introductory chapter and the discussions of earth materials. Clark Burchfiel explains the structure of the earth's interior and presents the theory of plate tectonics and mountain building. Leigh Mintz provides information on geologic time, along with a brief introduction to earth history. Ed Keller, Bill Melhorn, and Hal Thurman have written a thorough survey of geomorphology, and Doug Brookins has contributed an introduction to earth resources.

The authors and editors have designed a text based on market research. *Physical Geology* has been written in a logical order, but is also flexible enough to be adapted to individual course organizations. In Chapter 1, readers are introduced to the field of geology and to many basic geological principles. The last half of the chapter includes information about the moon and other planets in the solar system, giving a good idea of the earth's place within the universe. Minerals and the three rock groups are covered in Chapters 2, 3, 5, and 6. This organization allows students quickly to become familiar with the terminology and concepts they will encounter in the laboratory. Chapter 4, on weathering, precedes the discussion of sedimentary rocks so that students will understand in advance the processes that produce the raw materials of the sedimentary part of the rock cycle. The interior processes of the earth are covered in Chapters 7, 8, and 9. Chapter 7 covers some of the geophysical properties of the earth, and earthquakes. Chapters 8 and 9 demonstrate how those properties fit into the plate tectonic model, and how this model explains mountains and other earth structures. Chapter 10 is divided into two parts. In the first half, the concept of geologic time, and its relevance to the study of the earth, are examined. The second half serves as a brief overview of earth history and is related to plate tectonics as presented in Chapter 8. Chapters 11–16 deal with the processes that wear down and shape the surface of the earth. These six chapters give a comprehensive and modern introduction to these important traditional topics. Chapter 17 provides students with an introduction to earth resources. Information on resource availability and the importance of resources in the world today is included.

Each chapter of *Physical Geology* includes several pedagogical aids. For example, a list of *key words* is found at the beginning of each chapter, giving students an overview of the information to be presented, as well as helping them to become conversant with some of the more important terminology. Numerous *line drawings* help students clarify geologic concepts, and the many *photographs* enhance their understanding of the geologic processes operating on earth. At the end of each chapter is a *summary,* which provides students with a brief review of the material they should now understand. A list of *review exercises* helps them to test their knowledge, and a detailed list of *suggested readings* gives those wishing further information a head start in finding it. The *appendixes* at the end of the book cover rock and mineral identification, maps, unit measurement systems, and rock textures, and will be useful in laboratory courses or for further explanation in lecture. The comprehensive *glossary* will prove to be a valuable tool in helping students learn the vocabulary of geologists.

All prefaces include thank yous, and a project of this size has a long list. Many people have worked on *Physical Geology* to help make it a useful text for both instructors and students. First, thanks are owed to Thomas Henyey, University of Southern California; Donald Parker, Baylor University; Jesse O. Snowden, University of New Orleans; and Charles P. Walters, Kansas State University, all of whom reviewed the entire manuscript. Their comments were a very helpful guide for both authors and editors as the manuscript evolved to its final form. In addition, Leigh Royden reviewed Chapters 7, 8, and 9, and provided assistance in editing and improving that section of the manuscript. A debt of thanks is also owed to the many geologists who completed our initial survey of the physical geology course. It would be impossible to name everyone who responded, but together you helped to shape the book and your contribution is appreciated.

Special thanks go to Ann Mirels, who not only designed the book but also skillfully oversaw the project from beginning to end; James H. Hubbard and David Germon, who coordinated the artwork and photos; Linda Johnstone, who copyedited the text; Nancy Hall, who was an immeasurable help as the book moved through production; Joan Foster, who compiled the index; Chris Conty, who conducted the market survey used to shape the organization of the text; and Greg Spatz, who conceived of the project and persuaded the authors that it was needed and worth doing. A thanks also to all the other individuals who contributed to the publication of *Physical Geology.*

Pamela S. Cooper
Editor, College Division
Charles E. Merrill

CONTENTS

1
PLANET EARTH: AN OVERVIEW

2
MINERALOGY

3
IGNEOUS ROCKS

4 WEATHERING AND SOILS

5 SEDIMENTARY ROCKS

6 METAMORPHIC ROCKS

7 THE EARTH'S INTERIOR

8 PLATE TECTONICS

9 THE EARTH'S CRUST AND MOUNTAIN BUILDING

10
GEOLOGIC TIME AND EARTH HISTORY

11
MASS WASTING

12
GROUNDWATER AND RELATED PHENOMENA

13
RIVERS AND LANDFORMS

14

GLACIERS AND GLACIATION

15

DESERTS AND ARID ZONES

16

OCEANS AND SHORELINES

17

EARTH RESOURCES, ENERGY, AND THE ENVIRONMENT

APPENDIXES

GLOSSARY 473

INDEX 493

- geology
- geologic time
- igneous rock
- metamorphic rock
- sedimentary rock
- relative date
- absolute date
- radioactive decay

1

- fossil
- uniformitarianism
- geologic column (time scale)
- continental drift
- sea-floor spreading
- mid-ocean ridge
- volcanic island arc
- plate tectonics
- convergent plate boundary
- transform fault
- universe
- solar system
- galaxy
- red shift
- planet
- satellite
- comet
- meteor
- asteroid
- mare
- crater
- seismograph
- scarp
- polar ice cap
- hydrocarbon
- internal energy

Planet Earth: An Overview

GEOLOGY IS THE STUDY OF THE EARTH. THAT WE LIVE ON THE earth is reason enough to study it. The more we know about our planet, especially its environment and resources, the better we can understand, use, and appreciate it. To us, the earth is the most important body in the universe.

In the broader view, however, the importance of the earth shrinks. At least a half dozen stars are known to have faint, unseen companions, but it is not known whether these companions are very faint stars or massive planets revolving about the observed star. Thus, the earth appears to be an average planet. However, at least in the solar system, the earth is unique in having abundant water, exemplified by the oceans, and an atmosphere that can support life. The earth's surface temperature, controlled by its distance from the sun, makes these features possible, and these features, in turn, make life possible on the earth. The development and history of life are important aspects of historical geology. The space program has also revealed that the earth is unique among the planets so far studied in having a magnetic field. The earth's magnetic field, for reasons that will be discussed later, is believed to be caused by a liquid iron core that may also store the energy which causes the formation of such surface features as mountain ranges. These features, which, as far as we know, are unique to the earth, are central to the processes of erosion and deformation of the earth's surface that are the main aspects of physical geology. Thus, geology is in large part the study of the consequences of the earth's unique features.

Geology has contributed a great deal to civilization, both intellectually and economically. Among the great concepts gained from geological studies are an understanding of the great age of the earth, and the development of an absolute time scale. Geology differs from most other sciences in that it is concerned with **absolute time**. Time appears in the equations of physics and chemistry, but these sciences are generally concerned with rates of change, and the time is relative, not absolute. **Geologic time** extends back almost 5 billion years to when the earth formed. Thus, geology is concerned with immense lengths of time when measured against human experience. Although it is difficult to comprehend the lengths of time involved in geologic processes, this must be done to appreciate geology fully.

The fundamental principle that underlies most of geology is simply that the present processes occurring on the earth have occurred throughout geologic time. Therefore, ancient rocks can be interpreted in terms of present processes.

Another important point learned from geology is that constant change, both biological and physical, has been and is occurring on the earth.

The origin and development of life is part of geology. It is closely related to the history and development of the earth's surface and thus cannot be separated from the physical history of the earth. Geology shows that, in the broadest sense, all life is related. Biologists share this concern with life, but much of the evidence is geologic in nature. Thus, geology and biology overlap in part.

The economic contributions of geology to civilization show that in many ways, too, geology is a very practical science. Geologic knowledge is used to locate and to exploit our mineral resources. Except for water and soil, all mineral resources, such as sand and gravel, petroleum, coal, and metals, are nonrenewable. Once mined, they are gone, and new deposits must be found. Geologists have discovered the deposits of metal and energy-producing minerals on which our civilization is based. We take these things for granted now; but a hundred years ago, when the West was opening up and the industrial revolution was occurring, these mineral deposits were being discovered at a rapid rate and geologists were the most influential scientists of the day. At this time, too, the principles of geology were being formulated.

Today, the study of mineral resources is of utmost importance, but geologists are also concerned with other economic problems such as urbanization. The development of large cities has resulted in the building of large structures, such as tall buildings and dams. Geology helps in designing foundations for these structures. Examples of both large and small structures that have failed through neglect of simple geologic principles, easily understood by elementary students, are common (Fig. 1–1). Dams fail because they are

FIGURE 1–1
Landsliding at Point Firmin, California. Geologic studies can prevent losses such as this. (Spence air photo)

built near active faults or on porous foundations. During their first rainy season, new highways are washed out or blocked by landslides. Homes built on hillsides are destroyed by landslides and mudflows. Geologists have also recognized the need for earthquake-resistant structures in some areas and have helped in their design.

SCOPE OF GEOLOGY

Geology is concerned both with the processes operating in and on the earth, and with the history of the earth, including the history of life. In the broadest sense, geology includes the study of the continents, the oceans, the atmosphere, and the earth's magnetic and radiation fields. Clearly, this scope is too broad for any one scientist, so geologists generally, but not exclusively, limit themselves to the solid earth that can be studied directly. Other specialties have developed to investigate the other aspects of the earth. Geophysicists study the deep parts of the earth and its fields, mainly by indirect methods; oceanographers are concerned with the hydrosphere; and meteorologists study the atmosphere.

Even with this restriction, geology is a very broad field. Most geologists specialize in one or more facets of geology, much as engineers specialize in various fields of physical science such as electronics or construction. However, geology is even broader than engineering because it encompasses both physical and biological science.

Mention of a few of the specialties in geology will illustrate. Those who study minerals and rocks need specialized training in chemistry and physics, as does the geochemist, who is concerned with chemical processes in the earth. Those who study fossils must be trained in biology of plants and animals, both vertebrate and invertebrate. Those who study deformed rocks must know mechanics. Groundwater and petroleum geologists must be familiar with hydrodynamics. A complete listing would be very long, but these examples will illustrate the point. All of these specialties overlap somewhat, and, no matter what the specialty, a geologist must be familiar with all facets of geology.

THE ROCK CYCLE

The materials that form the earth's surface are called *rocks*. Rocks are classified into three types according to their origin. Two of these types are formed by processes deep in the earth and therefore tell us something about conditions within the crust. They are:

1. **Igneous rocks**, which solidify from a melt
2. **Metamorphic rocks**, which are rocks that have been changed, generally by high temperature and pressure within the crust

The third type, which records the conditions at the surface, is:

3. **Sedimentary rocks**, which are composed of the weathered fragments of older rocks that are deposited in layers near the earth's surface by water, wind, and ice

Much of geology is concerned with the interactions among the forces that produce these three rock types. The relationships are quite involved, but can be illustrated by the rock cycle (Fig. 1–2).

GEOLOGIC TIME

Rocks can be dated both relatively and absolutely. **Relative dates** are of two types. The simpler of these is the determination of the local sequence based on relationships such as the fact that the oldest beds are at the bottom of undisturbed sedimentary rocks. Determining these relationships by mapping the rocks—that is, plotting on an accurate map the positions of rock bodies—is the main method of geologic study. The second way of dating rocks is by referring them to a geologic time scale based on the study of the fossils in the rocks. **Absolute dating** is a more recent development and is based on radioactive decay of elements in rocks; this method has been used to calibrate the relative geologic time scale developed from the study of fossils.

The recognition of the immensity of geologic time and the development of methods for subdividing geologic time are among the great intellectual accomplishments of humans. It is difficult to discuss the development of ideas in generalities, but a brief outline is presented here. Discoveries made near the end of the eighteenth century led to rapid progress, although there had been earlier studies. The culmination of this early work was the theory now known as **uniformitarianism**, published by the Englishman

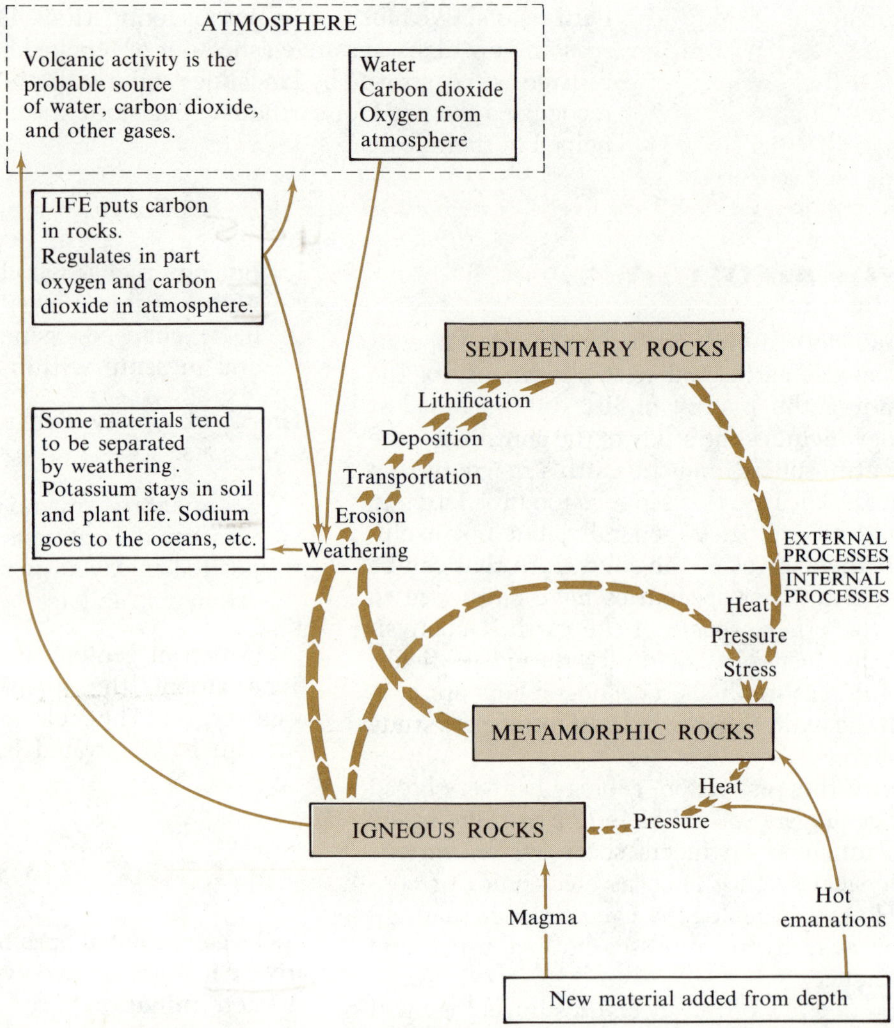

FIGURE 1–2
The middle of this diagram, showing the relationships among igneous, sedimentary, and metamorphic rocks, is what is generally considered the rock cycle. The upper and lower parts of the figure show how material is added to and subtracted from the rock cycle.

James Hutton in 1785 and 1795. Uniformitarianism simply means that the present is the key to the past; that is, the history of old rocks can be interpreted by noting how similar rocks are being formed today. Hutton's book was poorly written, and the theory did not gain wide acceptance until it was popularized by his colleague Playfair in 1802. Uniformitarianism, although a simple idea, is the keystone of modern geology. A simple example will show the power of this theory.

Many varieties of shell-bearing animals can be seen in the mud along the seashore at low tide. A similar mudstone with similar shells encountered in a canyon wall is interpreted, using uniformitarianism, as a former seabottom that has been lithified and uplifted. If this seems too elementary, it need only be remembered that just a few years before Hutton's time, only a small number of people recognized that fossil shells are evidence of once-living organisms.

Recently, uniformitarianism, as just presented, has been criticized as being oversimplified. The objection is that geological processes have modified the earth; therefore, at least in some cases, these processes do not operate now in the same ways that they did in the past. For example, after the origin of life, organisms changed the composition of the atmosphere, making further creation of life impossible. Thus, though uniformitarianism in its simplified version can explain many geologic phenomena, the doctrine has limitations that must be taken into account.

Geologic history is based largely on the study of sedimentary rocks and their contained fossils. Such studies have led to the development of a geologic time scale showing the relative ages of rocks. In recent years, it has been possible, using measurements of the radioactive decay of certain elements in rocks, to determine the absolute ages of some rocks. This ability to date igneous and metamorphic rocks has expanded the range of geologic history.

After several false starts, historical geology developed in the last part of the eighteenth century and the first half of the nineteenth. In 1782, the great

French chemist Antoine Lavoisier demonstrated that, near Paris, the quarries dug for pottery and porcelain clay all exposed the same sequence of sedimentary rocks. Georges Cuvier and Alexandre Brongniart published maps in 1810 and 1822 showing the distribution of the various rock types around Paris, and in 1815 William Smith published a geologic map of England. Cuvier and Brongniart studied the fossils in each of the sedimentary layers they mapped and discovered, as did Smith, that each layer contained a different group of fossils. They discovered that a sedimentary bed can be identified by its fossils. That work set the scene for the next 20 years, for during this period the geologic time scale was developed.

The **geologic time scale**, or **column**, was the result of individual work by a number of people in western Europe. These men studied the rocks and their fossils at well-exposed places, generally near their homes, and described these rocks and their fossils in books and papers. They called these sequences of rocks **systems** and named each of the systems. As the studies proceeded in this haphazard manner, the general sequence of older to younger was recognized, and the gaps were filled in. It was also shown that rocks in similar stratigraphic position, although far removed from the type area, contained the same or very similar fossils. Thus, it took work by many people spread over a wide area to demonstrate that fossils can be used to date rocks, not only in local areas, but all over Europe. In this way, a geologic column, or time scale, was developed as a standard of reference; fossils from other areas could be correlated or dated by comparison with fossils on which the time scale was based. The geologic time scale in current use is shown in Fig. 1–3. Almost all of the systems or periods were originally defined in the first half of the nineteenth century.

The changes in marine invertebrates during geologic history are obvious to even a casual observer;

Era	Approx. Age in Millions of Years (Radioactivity)	Period or System *Period refers to a time measure; system refers to the rocks deposited during a period.*	
Cenozoic		Recent (Holocene)	Neogene
		Pleistocene	
	7	Tertiary: Pliocene	
	26	Miocene	
	37–38	Oligocene	Paleogene
	53–54	Eocene	
		Paleocene	
	65		
Mesozoic		Cretaceous	
	136	Jurassic	
	190–195	Triassic	
	225		
Paleozoic		Permian	
	280	Pennsylvanian	Carboniferous
	310	Mississippian	
	345	Devonian	
	395	Silurian	
	430–440	Ordovician	
	500	Cambrian	
	570		
Precambrian	700	*First multicelled organisms*	
	3500	*First one-celled organisms*	
	4000	*Approximate age of oldest rocks discovered*	
	4500	*Approximate age of meteorites*	

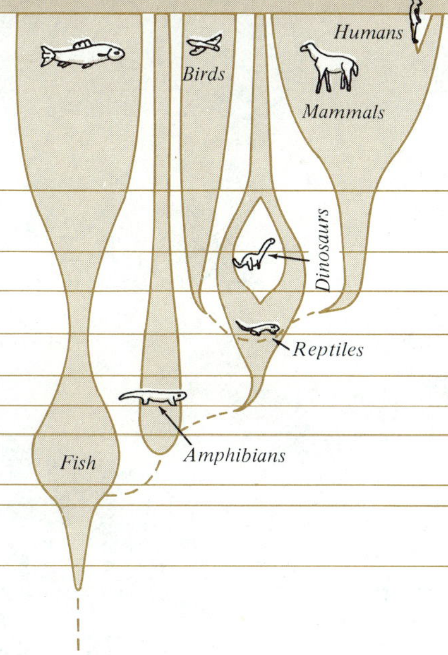

FIGURE 1–3
The geologic time scale. Shown to the right is a very simplified diagram showing the development of life. Not included on the diagram are many types of invertebrate fossils such as clams, brachiopods, corals, sponges, snails, and so forth, which first appeared in the Cambrian or Ordovician and have continued to the present.

for this reason, such fossils can be used to date the enclosing rocks (see Fig. 1–4). The changes between periods are not as great but are just as obvious to a trained observer.

The geologic column or time scale provided only a relative time scale to which rocks could be referred. It also showed the general development of life and so set the stage for Darwin, who published his theory of evolution in 1859.

Although the idea of evolution was not new to Darwin, his book started a controversy that is still continuing in nonscientific circles. Until his time, no controversy existed between geology and religion, because even the geologists interpreted geologic history in biblical terms. Biblical scholars thought that the earth was about 6000 years old, based on their interpretation of the Old Testament. However, in the hundred years before Darwin, some geologists began to recognize that a

A.

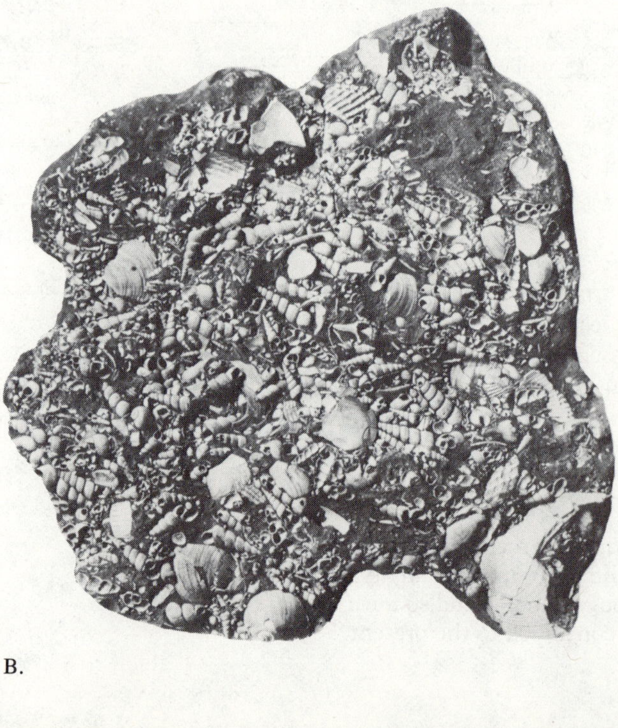

B.

FIGURE 1–4
Brief comparison of these two fossiliferous slabs of rock show the great changes in life that have occurred during geologic time. A. Pennsylvanian fossils from Kansas. Bryozoa and brachiopods are most abundant. (Photo from Ward's Natural Science Establishment, Inc., Rochester, N. Y.) B. Miocene fossils from Virginia. Snails and clams are most abundant. (Photo by W. T. Lee, U.S. Geological Survey)

much longer time was necessary for the deposition of the sedimentary rocks and other geologic events than students of the Bible would allow. Before evolution was accepted, the changes in fossil life throughout the geologic column were explained by a catastrophic dying out of all life and a new creation of the animals found as fossils in the succeeding bed. This idea had its roots in the biblical Noachian Flood and the widely held theory of Werner, a late eighteenth century German, who believed that all surface rocks, including igneous rocks, were precipitated in a universal sea. Darwin had a pronounced influence on the development of geologic thought, but his theory also needed the long duration of geologic time envisioned by the uniformitarianists.

Thus, the central point became the age of the earth. The geologists and the evolutionists needed a very old earth. In 1899, Joly calculated the age at 90 million years by dividing the total amount of salt in the ocean by the amount of salt annually brought into the ocean by the rivers. Estimates of the age of the earth based on the time necessary to deposit the rocks of the geologic time scale were made by many geologists in the latter part of the nineteenth century; and although they varied between 3 million and 1.584 billion years, about 100 million years was an average. The lack of precision and the subjective nature of the assumptions made these estimates suspect. Lord Kelvin, one of the most influential physicists of the day, entered the fray and attempted to calculate the age of the earth from thermodynamics. He assumed that the earth began as a melted body and has been cooling ever since. He calculated that 20–40 million years had passed from the time when the earth had cooled enough for life to exist to the present. About the turn of the century, when Kelvin was an old man but still refining his calculations, radioactivity was discovered. Radioactivity is a source of heat within the earth, and so Kelvin's heat-flow calculations were based on an incorrect assumption.

Radioactivity was discovered by Becquerel in 1896. In 1902, Rutherford and Soddy showed that by radioactive decay, an element is changed in part to another element. This sounded much like the quest of the alchemists, who tried to change lead into gold, and was not immediately accepted by many scientists. However, in 1905 Boltwood noted the association of lead and uranium in uranium minerals and that the ratio of lead to uranium was higher in the older minerals. Boltwood used this information to calculate the age of these minerals, and one of them had a computed age of over 2 billion years. This greatly extended the possible age of the earth and was quickly accepted by a few geologists. Present methods of radioactive dating show the earth to be about 5 billion years old. These methods will be described later in this chapter. Thus, in a single stroke, radioactivity provided the energy source not taken into account by Kelvin and a means to find the age of the earth.

THE MOBILE EARTH

The earth is a dynamic planet whose surface is constantly undergoing change. Given the length of geologic time, one can comprehend that the hills and mountains are not eternal features, but that eventually they will be worn flat by erosion. Earthquakes and volcanoes reveal a restless interior. Deformed rocks that have been bent into folds and broken also bear mute testimony to the power of the earth's internal energy.

CONTINENTAL DRIFT

The idea that the continents have moved around on the earth's surface is an old one. Almost every school child notices that the continents fit together like the pieces of a jigsaw puzzle. The fit is not perfect (Fig. 1–5), but erosion and deformation could have distorted the pieces. The fit of the continents is more than just shape; the rock layers and types also match. Again, the fit is not perfect because of erosion and deformation. The ages of the rocks and their structures also are similar.

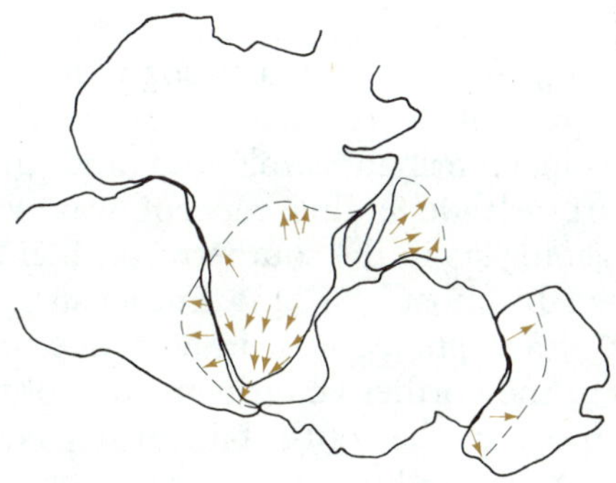

FIGURE 1–5
Reconstruction of the ancient continent of the Southern Hemisphere. The glaciers and their directions of movement are indicated.

In the nineteenth century, geologic studies in the southern continents revealed other kinds of evidence. Large areas of glacial deposits and glacially eroded bedrock were discovered in India, South America, Africa, and Australia. The glaciers were continental glaciers such as those in the northern continents during the much more recent Ice Age. The glacial deposits on the southern continents are all about 250 million years old. These glacial rocks are now found on both sides of

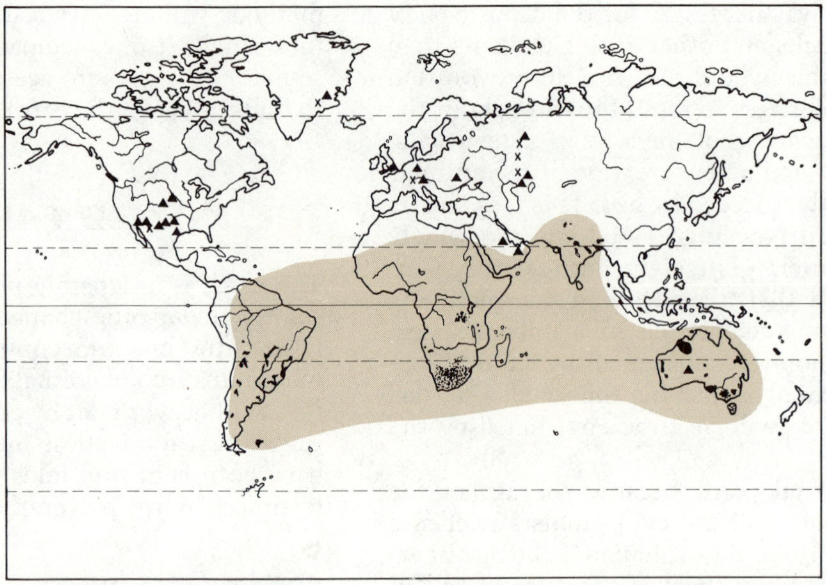

FIGURE 1–6
Map showing the present distribution of the ancient glaciation of the Southern Hemisphere. The area involved in the hypothetical southern continent existing before drifting is stippled. The areas of glacial deposits are indicated in black and extend from north of the equator to near the South Pole. Reefs (crosses) and evaporites (triangles) far to the north indicate much warmer climates. (Reefs and evaporites from Opdyke, 1962)

the equator in areas of warm climates (Fig. 1–6). Even if the South Pole is moved to the most favorable location among these ancient glaciers, some are still near the equator. However, if the continents are joined as in Fig. 1–5, the glacial areas and the directions of ice movement recorded in the rocks all match well. The rocks associated with the glacial deposits also are similar on all the southern continents. On each of these continents, the rocks contain similar fossils of both plants and reptiles. These organisms could not have migrated across the seas that presently separate these continents.

Therefore, the evidence is strong that at least the southern continents were once a single continent, and that this supercontinent broke up and the pieces drifted apart. Although this concept was widely accepted by geologists in the southern continents, it was not believed by most U.S. and European geologists. Their lack of acceptance was, in part, due to the fact that rocks in the northern continents do not match so clearly as those in the south, but mostly because no mechanism that could cause continental drift was known.

SEA-FLOOR SPREADING

Sea-floor spreading is a much newer idea than continental drift. Although it had been proposed much earlier, it was finally accepted in the 1960s when the evidence for it became overwhelming. All of the oceans have ridges or rises running through them, called **mid-ocean ridges**. These are areas of volcanic activity and shallow earthquakes. New ocean floor is created at the ridges by the volcanic activity (Fig. 1–7). This is shown by the age and thickness of the rocks on the ocean floors. These rocks on the ocean floor have been sampled by drilling. The age of the volcanic rocks that comprise the bedrock of the oceans is progressively older as one moves away from the ridges in both directions. In the same way, the sediments above the bedrock become thicker, and the deepest sediments older, with distance from the mid-ocean ridges. The conclusion seems inescapable that new ocean floors are being created at the mid-ocean ridges, and that the ocean floors are moving away from the ridges. The rate of movement is only a few centimeters per year, but given the earth's great age, the total movement is very large.

If new ocean floor is created at the mid-ocean ridges, then the ocean floors must be consumed somewhere; otherwise, the earth must expand. The **volcanic island arcs** are places where the ocean floor apparently returns to the interior of the earth (Fig. 1–8). The volcanic island arcs are so called because they are islands formed by volcanoes that appear as curves or arcs on a map. They are geologically active areas, with deep earthquakes occurring as well as the volcanic activity.

It is believed that the ocean floor returns to the interior of the earth at the volcanic island arcs because the earthquakes occur on a plane sloping under the island arc (Fig. 1–8). Movement of a slab of ocean floor would cause the earthquakes. The volcanoes could be the result of partial melting of the slab of ocean floor. Such a process is suggested by the rock cycle (Fig. 1–2). Downward movement of the ocean floor is also suggested by the deep submarine trenches in front of volcanic island arcs (Fig. 1–8).

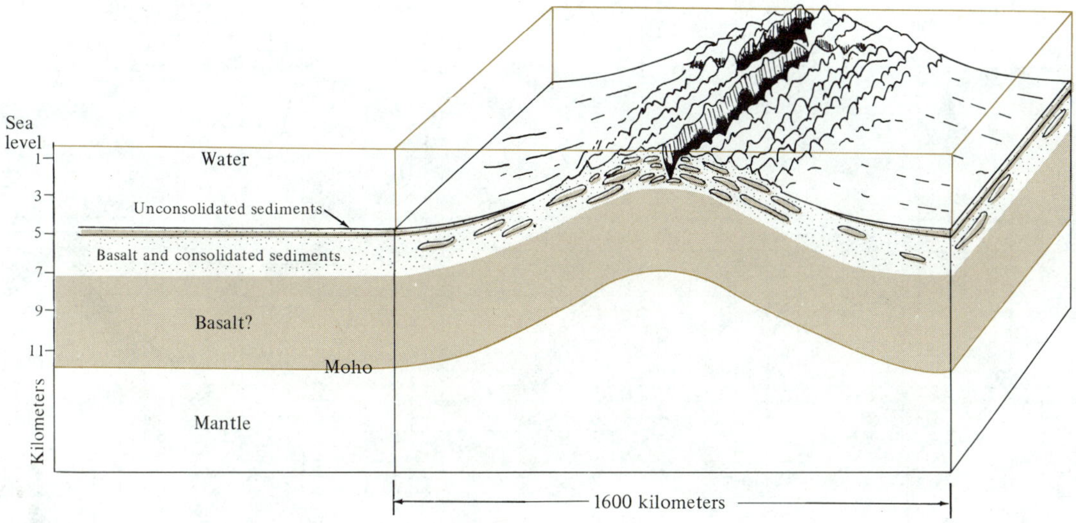

FIGURE 1–7
Volcanic activity at mid-ocean ridges creates the rocks that underlie the oceans.

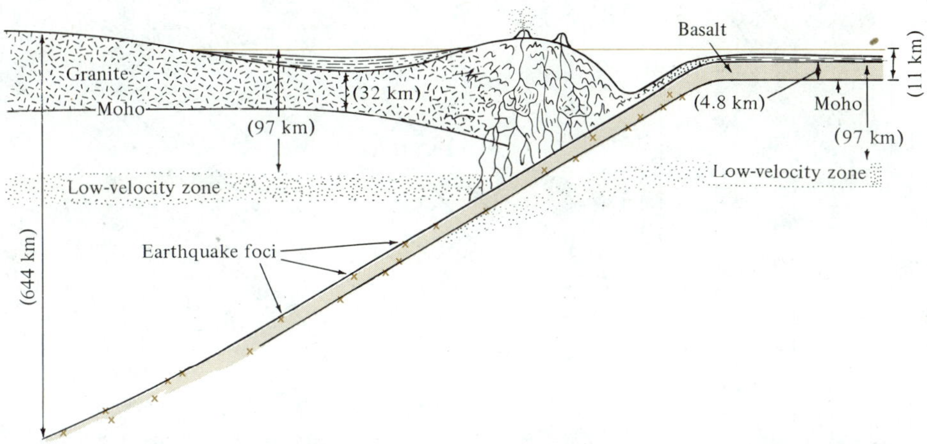

FIGURE 1–8
Volcanic island arc, showing location of earthquake foci. Because of the distances involved, this figure is not drawn to scale.

In sea-floor spreading, new ocean floor is being created at the mid-ocean ridges. The ocean floor moves slowly toward the volcanic island arcs, where it is consumed. It seems reasonable that this movement of the surface rocks must be matched by an opposite movement in the earth's interior.

PLATE TECTONICS

Continental drift and sea-floor spreading come together as parts of **plate tectonics**. In plate tectonics, the surface of the earth is believed to be composed of a number of thin slabs, or plates, that move across the surface (Fig. 1–9). The plates are outlined by the locations of earthquakes as shown in Fig. 1–9. The plates are formed at the mid-ocean ridges, and they return to the earth's interior at **convergent plate boundaries** such as the volcanic island arcs. Where the plates pass each other, earthquakes are also caused by the motion, and these plate boundaries are called **transform faults**.

The continents are made of granite, a rock that is less dense than the ocean-floor rocks. The continents ride atop the ocean-floor basalt, and so as the plates move, they carry the continents with them (Fig. 1–10). In this way, the same mechanism that causes sea-floor spreading also is the cause of continental drift. Movements of the continents through geologic time are shown in Fig. 1–11.

As the plates move across the earth's surface, they collide from time to time. Apparently, new mid-ocean ridges sometimes form. The type of collision is determined by the composition of the plates at the point of collision. If ocean floor collides with continent, the ocean floor is consumed at the resulting convergent plate boundary (Fig. 1–12). If ocean floor meets ocean

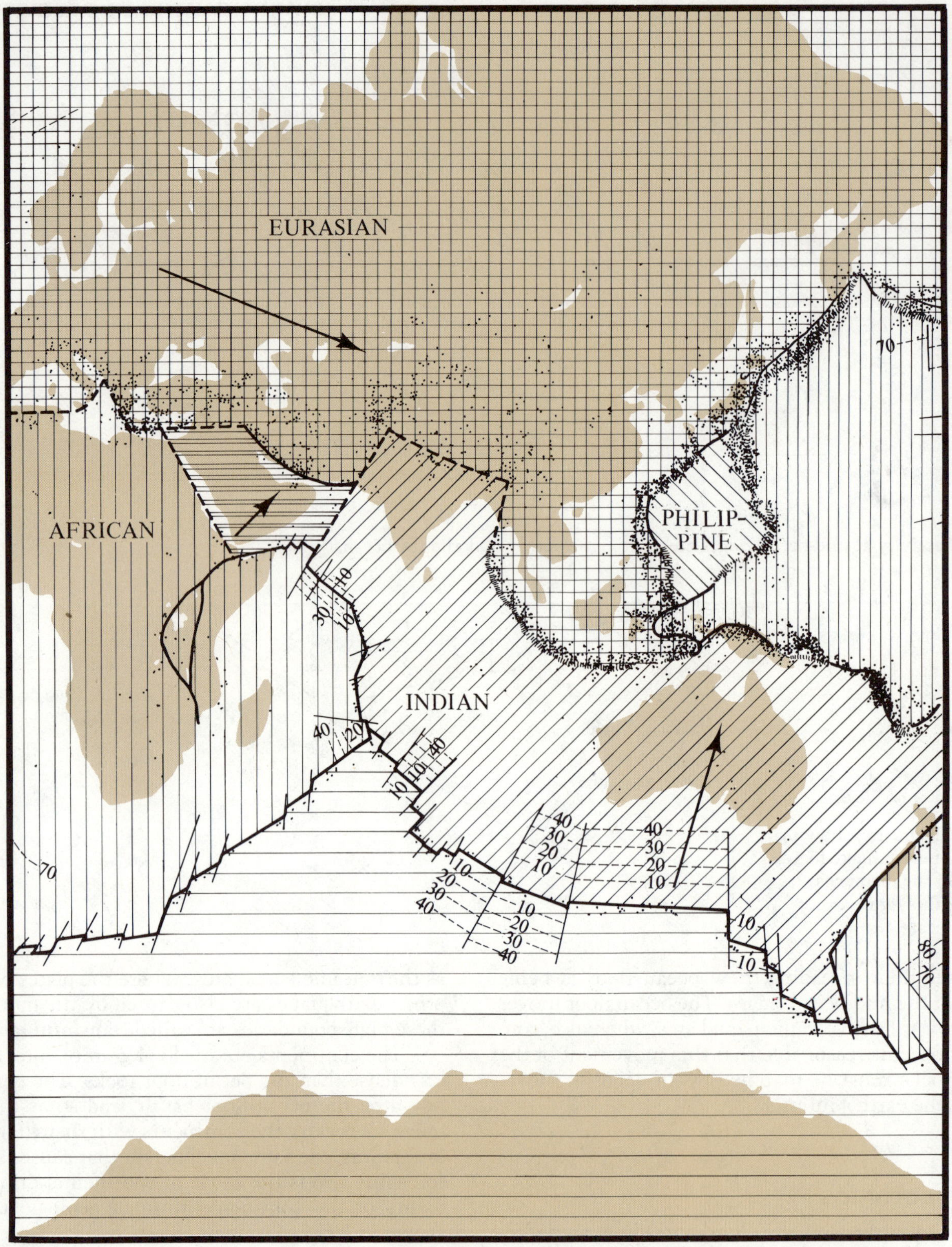

FIGURE 1–9
The crustal plates. In general, each plate begins at a ridge (heavy line) and moves toward a trench (hatching). Earthquake epicenters (dots) tend to outline the plates. Transform faults are shown by thin lines. The dashed lines show the ages (in millions of years), based on magnetic data. The arrows designate the general direction of movement

floor, one plate is consumed and a volcanic island arc forms (Fig. 1–13). If continent collides with continent, a mountain range is formed (Fig. 1–14). In all of these collisions, the rocks are deformed by the collision. The rocks, especially the sedimentary rocks that form in layers, are bent and folded, and they are broken and

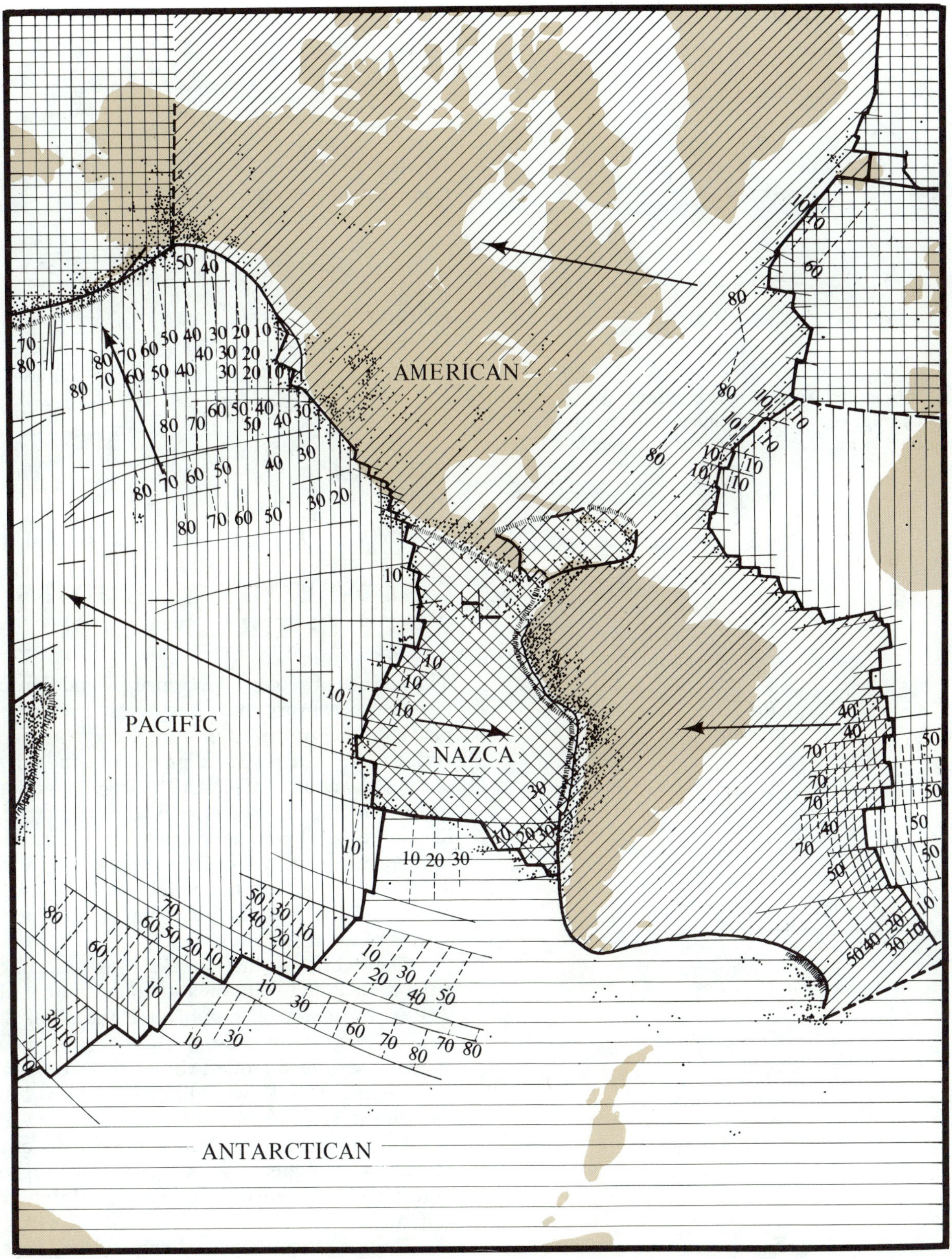

FIGURE 1–9 (cont.)
of the plates. (Information and interpretation from many sources, mainly workers at Lamont-Doherty Geological Observatory)

faulted. Igneous rocks are also formed where plates are consumed, and the rising igneous rocks cause metamorphism and deformation. This, then, is the origin of the folded and faulted rocks that form the earth's mountain ranges. In our exploration of other planets, no such structures yet have been found.

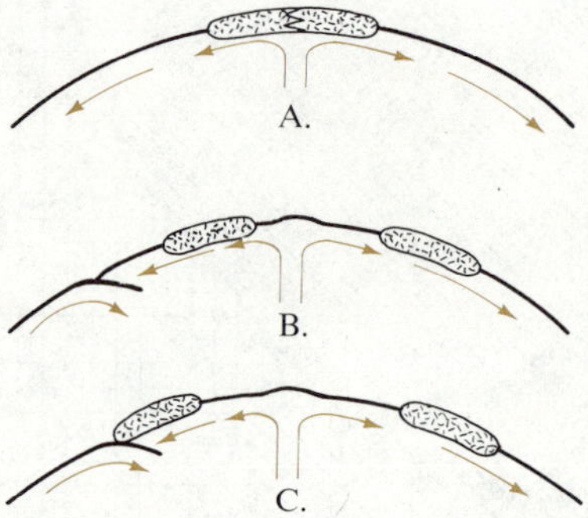

FIGURE 1–10
Possible role of sea-floor spreading in continental drift. Formation of a ridge beneath a continent could result in splitting of the continent and movement to a trench.

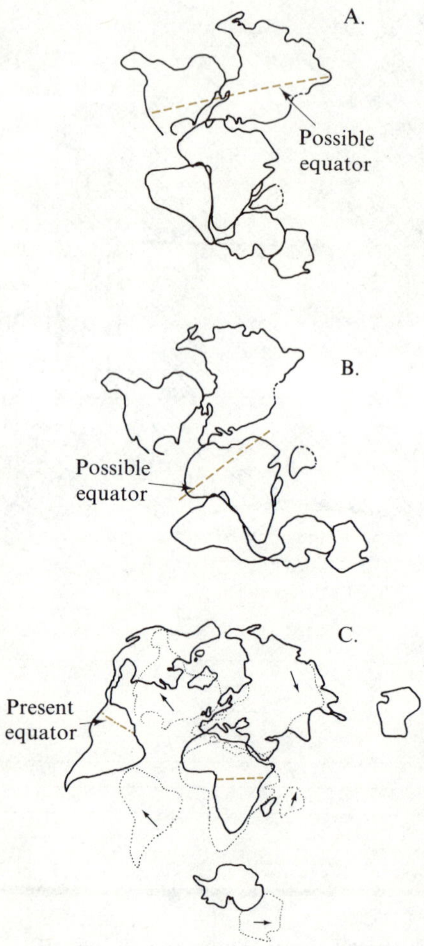

FIGURE 1–11
Movement of continents. A. Late Paleozoic time, 250 million years ago. B. End of Jurassic period, 136 million years ago. C. End of Cretaceous period, 65 million years ago, is shown dotted, and present positions are shown by the solid lines. (Based on R. S. Deitz and J. C. Holden, 1970)

FIGURE 1–12
Continent-ocean plate collision.

FIGURE 1–13
Ocean-ocean plate collision.

FIGURE 1–14
Continent-continent plate collision.

DISCOVERING THE EARTH'S PLACE IN THE UNIVERSE

The science of geology developed around the end of the eighteenth century, but long before then, the earth's place in the universe was explored. In many early civilizations, the priests kept track of the movements of the sun, moon, and planets relative to the stars. This was of practical value, as well as religious, because it led to the development of calendars and prediction of such events as eclipses. The Greeks went further and developed a system to explain the motions of the heavens. This was the first big step in finding the earth's place in the universe.

The work of the Greeks was summarized in A.D. 140 by Ptolemy. In the Ptolemaic system, the earth was at the center of the universe and all of the other objects in the solar system—the sun, moon, and planets—revolved around the earth. Although not all of the earlier Greeks would have agreed with this system, the great advantage of Ptolemy's work was that it could be used to predict future movements of the sky. Thus, it was of great use in astrology and navigation. Columbus and the early explorers used the Ptolemaic system.

The first important challenge to the idea that the earth and human beings' place was at the center of the universe came at the Renaissance about 1400 years after Ptolemy. Putting the sun at the center and having the planets revolve around the sun made a much simpler system in the sense that fewer motions were necessary. This system was published posthumously in 1543 by Nicholas Copernicus. Moving humans away from the center created a controversy that raged for a hundred years because proof of the earth's daily rotation and yearly revolution came much later. The main advantage of the Copernican system over the Ptolemaic system was that it was simpler. Both had many inaccuracies in prediction. In the early part of the seventeenth century, Johannes Kepler, using data collected by Tycho Brahe, was able to describe the motions of planets and showed that Copernicus was basically correct.

Galileo Galilei, an Italian, was a contemporary of Brahe and Kepler. Although he worked in many fields, only parts of his studies of astronomy and mechanics will be mentioned here. In 1609, Galileo built one of the first telescopes and used it to observe the heavens. He saw things no one had ever seen before, and many of his observations could not be explained by the Ptolemaic system. He announced his discoveries in two books but was later forced to recant by the Inquisition. As a result, he had to spend the last years of his life in virtual house arrest. It was during this period that he carried out many of his experiments in mechanics, experiments which later helped Newton.

Isaac Newton, an English physicist, using the work of Kepler and Galileo, discovered the laws of mechanics and of gravity. He worked out these laws about 1664 and published them in 1686. It was then possible to understand why the planets orbit the sun. This discovery profoundly influenced eighteenth century thought. In fact, it was widely believed that almost everything could be explained by Newton's laws. It was not until the early part of the twentieth century that Einstein showed that Newton's laws do not apply in all situations.

Thus, by Newton's time the earth's place in the sun's family was established. The distance to the sun, 150 million km, was measured in 1672. But where is the sun in the universe? New instruments, especially big telescopes, had to be built before this question could be answered and our place in the universe known, if indeed it ever can be known fully.

The first step out of the solar system was to measure the distance to some nearby stars. This was not done until 1838, when Bessel measured the very small change in angle to a nearby star, as the earth moves from one side of its orbit to the other. This geometric method can be used only to an angle of 0.008 second of arc, which corresponds to a distance of about 400 light-years. Beyond this distance, the angle is too small to measure accurately. The closest star is about 4 light-years away.

It was not until the early twentieth century that greater distances could be measured. During much of the nineteenth century, spectroscopic analysis was developed and applied to starlight. In 1911, Ejner Hertzsprung in Holland and H. N. Russell in the United States noted that there is a relationship between spectral type and absolute brightness of a star. The absolute brightness can be calculated from the apparent or measured brightness for those stars whose distance is known. It then became possible to calculate the distance from the apparent brightness of any star whose spectral type was known, because from the spectral type one knows the absolute brightness.

The size of the known universe was growing rapidly, but for the most part the distances measured were to stars in our galaxy, the Milky Way. (All of the stars seen with the naked eye are within our galaxy, although some more distant objects such as the Andromeda galaxy can be seen without a telescope.) During the nineteenth century, astronomers noted many objects whose light was soft and fuzzy. By 1850, several spiral galaxies had been noted, but their nature was in doubt. In 1908, some stars were resolved in some of these objects, suggesting that they might be vast groups of stars like the Milky Way. It was not until 1925, when the new 100-in. (250-cm) telescope at Mount Wilson became available, that Edwin Hubble was able to report that these objects are composed of many stars. Objects the size of the Milky Way, so distant that they appear to be stars except in the largest telescopes, reveal a universe whose size is beyond comprehension. The diameter of the Milky Way itself is 80,000 light-years.

In 1912, Henrietta Leavitt made a discovery that enabled measurement of the distance to a galaxy. She observed that there is a relationship between the brightness and the period of a type of variable star called a Cepheid. Using this, as well as other methods, the size of the known universe quickly expanded to over 1 billion light-years. The total number of galaxies may be as great as 100 billion.

Perhaps the most astonishing thing revealed was Hubble's discovery in 1929 that all of these galaxies are moving away from us—that is, the universe is expanding. The further away a galaxy, the faster it is receding from us. This motion is measured from the **red shift**, a movement of the lines in a galaxy's spectrum toward the red. This change in frequency is similar to the change in pitch of a train whistle as the train moves away from the listener.

Thus, in about 500 years our perception of the earth has changed from its being the center of the universe to its being a small planet orbiting an average star in one of perhaps 100 billion galaxies. (See Fig. 1–15.) Research continues in astronomy, and new, strange objects (quasistellar objects—QSOs) have been discovered. Some of these QSOs emit radiowaves, and some have very large red shifts, suggesting very high

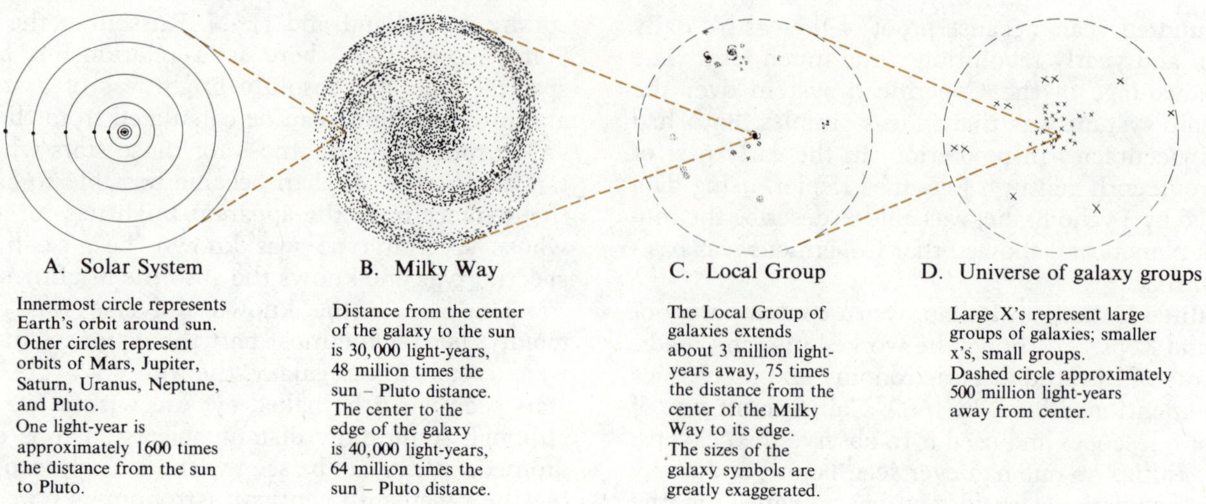

FIGURE 1–15
The earth's place in the universe.

velocities and perhaps great distances. These objects, or others yet to be discovered, may again change our ideas on the earth's place in the universe.

THE SOLAR SYSTEM

The **solar system** is composed of the sun and the planets that orbit around the sun, the satellites of the planets, and comets, meteors, and asteroids. The nine planets all revolve in the same direction around the sun. The orbits are all elliptical, very nearly circular, and lie close to a single plane, except Pluto, whose orbit is inclined (see Fig. 1–16). Seen from Earth, the sun and planets follow the same path across the sky. The inner planets—Mercury, Venus, Earth, and Mars—form a group in which the individual members are more or less similar in size. The next four planets—Jupiter, Saturn, Uranus, and Neptune—comprise a similar grouping of much larger planets. Pluto is much smaller than the other outer planets and has a tilted, eccentric orbit, all of which suggest it may not be a true planet. It may once have been a satellite of Neptune and escaped into its own orbit. Between Mars and Jupiter are several thousand small bodies, the asteroids, that are probably the origin of meteorites.

THE MOON

The moon orbits Earth and rotates on its axis at the same rate, and so only one side of the moon is visible from Earth. Our only knowledge of the far side of the moon comes from space probes that have photographed this region. They have revealed a cratered surface with almost no smooth areas (Fig. 1–17).

The visible face of the moon has large, dark, smooth areas and lighter colored, more rugged areas. The early astronomers called the dark regions seas

FIGURE 1–16
The orbits of the planets. Because Pluto is almost 40 times further away from the sun than is Earth, it is not practical to draw this diagram to scale. The inner planets (circles) are drawn to scale, and the distances among the great planets are drawn to a different scale. Eccentricity of the orbits is neglected. The edge view shows that, with the exception of Pluto, the orbits lie nearly in the same plane (ecliptic).

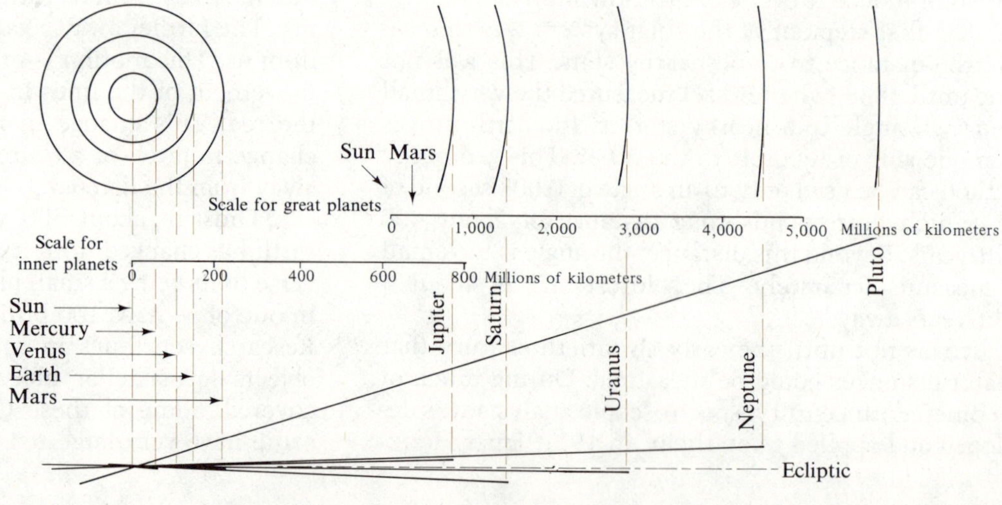

FIGURE 1–17
Hidden side of the moon photographed by *Lunar Orbiter III* from an altitude of about 1450 km, looking south from a point about 400 km south of the lunar equator. The prominent crater is about 240 km in diameter, and its rim shows evidence of inward slumping like many of the moon's craters. (Photo courtesy NASA)

[*maria* (pl.), *mare* (sing.)] and the lighter areas, lands, in a fanciful analogy to Earth. Both have craters, but the lighter uplands have more craters than the lower maria. The craters range from hundreds of kilometers across down to a few centimeters in diameter. Most of these craters probably were formed by meteorite im-

FIGURE 1–18
Oblique view of the crater Kepler taken by *Orbiter III* from a height of 52.6 km. Kepler has rays extending radially in all directions. The rays are believed to be the material splashed out by the impact and are visible from Earth only near full moon. The light-colored streaks in the lower right may be ray material, especially as the associated smaller craters form lines radial from Kepler. Many domes are visible; one in the foreground has a crater at its summit, suggesting that the domes may be volcanic. The line in the foreground may be a fault. (Photo courtesy NASA)

FIGURE 1–19
An array of domes near the older impact crater Marius. The domes are from 3 to 16 km in diameter and from 300 to 450 m high. Many have pits on the summit and so resemble volcanoes. The wrinkle ridges may be volcanic or subsidence features. (Photo courtesy NASA)

pacts. The moon, unlike Earth, is not protected by an atmosphere, so more meteorites hit the moon than Earth. The relative ages of various parts of the moon's surface can be determined by the density of the craters, the older surfaces having more craters.

Evidence of volcanic activity on the moon is abundant. Some of the craters may be volcanic in origin, and volcanic activity probably modifies many of the craters. A number of domes with pits or craters at the top, strongly suggesting volcanoes, are shown in Figs. 1–18 and 1–19.

Other probable volcanic features are the wrinkle ridges shown in Fig. 1–19. Ridges like these are tens of meters high and extend for hundreds of kilometers. They are almost entirely confined to the maria. Their origin is unknown, and it has been suggested that they may be caused by downsinking of the maria, or they may be the location of feeders of the surface flows.

The nature of the moon's surface layer and surface processes is very different from Earth's because of the lack of water and atmosphere. Downslope movements occur on steep crater walls (Fig. 1–20). Some very slow

FIGURE 1–20
Closer view of Copernicus from *Orbiter II.* The hummocky terrain in the foreground is ejecta from the crater. Details of the sliding of the rim can be seen. The central peaks within the crater are about 300 m high. The low area in the left foreground is apparently a fault trough filled by younger volcanic rocks. (Photo courtesy NASA)

erosion does occur, because the older craters are more rounded than younger ones. This rounding is believed to be caused by the impact of micrometeorites and atomic particles from the sun.

The moon has a rugged surface with many mountains; however, these mountains have apparently been formed by impacts and volcanic activity. Thus, they are very different from the compressive folded and faulted mountains of Earth.

Carbon compounds have been found in the moon rocks, but no trace of life has been discovered.

The seismographs left on the moon reveal that most of the moon's "earthquakes" are caused by tidal attraction and occur when the moon is closest to Earth. In comparison to Earth, the moon is an almost dead planet. Most of the moonquakes have very low magnitudes, and most appear to originate from two active zones about 950 km long, one running north-south and the other northeast-southwest. Meteorite falls have also been recorded by the seismographs. Seismic studies reveal that the moon is layered.

The history of the moon begins with its formation, probably about 4.6 billion years ago. This is the same date as the origin of Earth and the rest of the solar system. The outer part of the moon probably melted at this time. The infall of material and the resulting compaction would cause heating. This melting would produce the crust of the moon, which crystallized about 4 billion years ago. After the crust formed, the interior of the moon would begin to heat. This heating would be caused by radioactive elements. It would first melt the layer below the crust, and could be the origin of the mare volcanics that formed between 3.9 and 3.1 billion years ago. Melting would then progress downward as the upper layers cooled. This process could produce the partially melted core predicted by the seismic data. After 3.1 billion years, the only process acting on the surface has been meteorite impacts.

MERCURY

Mercury is too small and too close to the sun for detailed telescope study, even though it has no atmosphere to obscure observation. Space probes have revealed cratered highland areas and presumably younger, uncratered basin areas. In this, Mercury is similar to Mars and the moon. The basins are much like those on the moon but have many more cracks and wrinkle ridges (Fig. 1–21). The rocks filling these basins appear to be volcanic, like the similar rocks on the moon;

FIGURE 1–21
Part of Caloris Basin is shown on the left side of this mosaic of *Mariner 10* images. Caloris Basin is the largest feature so far discovered on Mercury and is 1300 km in diameter. The floor of the basin has many cracks and ridges. (Photo courtesy Jet Propulsion Laboratory)

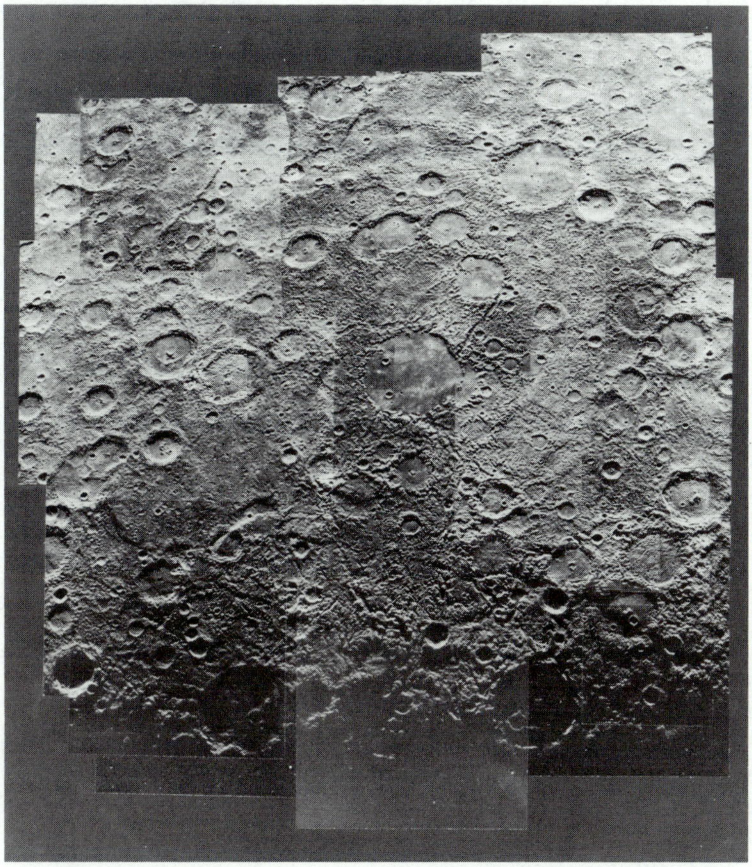

FIGURE 1–22
This unusual terrain, seen only on Mercury, is termed weird terrain. The hills and ridges cut across both smooth and cratered areas. This image is on the opposite side of Mercury from Caloris Basin, shown in Fig. 1–21. (Photo courtesy Jet Propulsion Laboratory)

however, unlike Mars and the moon, none of the photos of Mercury show any overt volcanic features such as volcanoes. As on Mars, the distribution of cratered and basin areas is not uniform.

The big basins on Mercury are apparently impact craters caused by huge meteorites. At points on the opposite side of the planet from such basins are what is called **weird terrain** (Fig. 1–22). It is believed that the shock effects of great impacts are concentrated at such places. The moon has similar big basins, and somewhat similar weird terrain is found there also.

Mercury's unique feature is many scarps over 485 km long and 3 km high (Fig. 1–23). These scarps cut both smooth and cratered areas and may have been formed by faulting. They may be compressional features, and, if so, they are the only compressional features seen on any planet except Earth.

Mercury's magnetic field is about 1 percent that of Earth's, but even this weak field is much greater than that of Venus or Mars. This suggests that Mercury may have or may once have had a liquid iron core, or that the present field is caused by remanent magnetism in the rocks. Mercury's specific gravity is almost as high as Earth's, further suggesting that it may have an iron core. All of this suggests that Mercury melted, at least partially, early in its history.

FIGURE 1–23
The scarps running from upper left to lower right in this image of Mercury are about 300 km long. They may be compressional features. (Photo courtesy Jet Propulsion Laboratory)

MARS

Mars has a reddish color and so was named for a god of war. The surface is covered with fine material and boulders that appear to be weathered basalt. It is farther from the sun than is Earth, and so has lower temperatures. Because its rotational axis is tilted with respect to its orbital plane, it has seasons like Earth. Telescopic observation reveals that polar ice caps form and disappear each year, and the color of the planet changes from bluish green in spring to brown in the fall, suggesting seasonal growth to some observers (Fig. 1–24). *Mariner 9* photographs reveal dust clouds, so the observed color changes may be caused by dust storms and clouds. Mars's atmosphere is composed mainly of carbon dioxide and a little water vapor. Mariner spacecraft revealed that this inhospitable atmosphere is less than 1 percent as dense as Earth's. The average temperature at the equator at noon is probably 4°C, and the temperature may fall to −130°C at night. Thus, the conditions are not favorable for life on Mars, but some may exist. The pictures returned by Mariner space probes show evidence of both internal and surface activity.

Dust storms have been observed on Mars. As a storm subsides, first the higher elevations come into view, and finally the lowlands. In spite of the thin atmosphere, winds up to 480 km/hour are possible, and 275-km/hour winds are not uncommon. The dust, probably the size of sand, may result from meteorite impacts or other processes. The wind may move the dust to lower elevations, and in this way create the featureless plains of Mars. The pits and basins of Fig. 1–25 could also be caused by wind erosion. The cliff,

FIGURE 1–24
The northern hemisphere of Mars. The shrinking north polar ice cap is at the top, and the huge volcanoes are near the bottom. (Photo courtesy NASA)

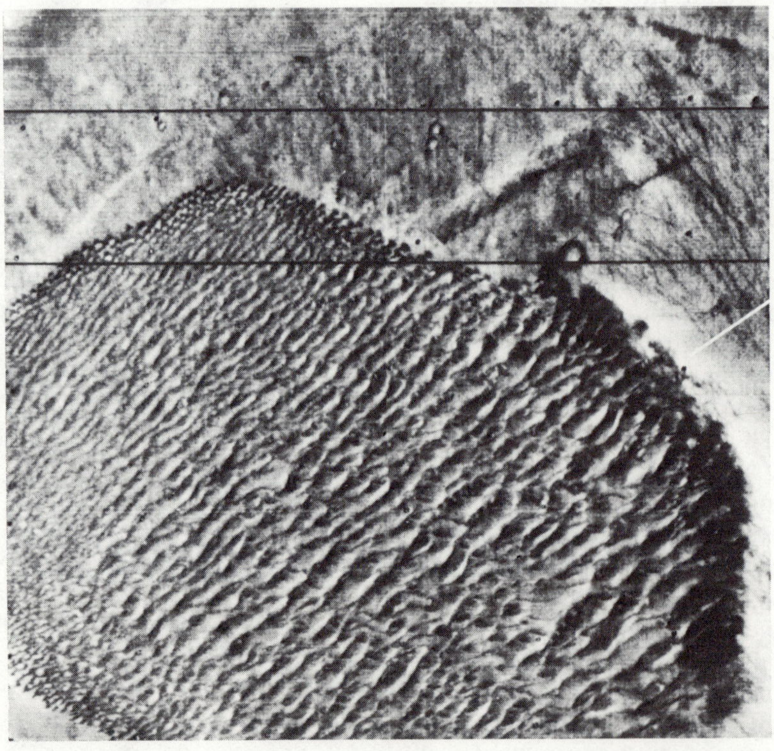

FIGURE 1–25
A dune field within a 150-km-wide crater on Mars. (Photo courtesy NASA)

up to 1.6 km high, that surrounds Olympus Mons (see Fig. 1–26) may also have been produced by wind erosion.

The volcanic area has Olympus Mons (see Fig. 1–26), the largest volcano ever seen. It is about 8 km high and 480 km across at the base; it has a broad complex crater, and its flanks are covered with what appears to be flow material. Nearby are three big volcanoes in a row, one of which is over 24 km high.

The equatorial region near the volcanic area is a plateau cut by what appear to be faults (Fig. 1–27). The valleys shown in Fig. 1–27A are part of a parallel system that extends more than 4800 km. They are interpreted as tensional features, perhaps associated with uplift. Many years ago, some astronomers reported a network of faint lines which were incorrectly called canals. The report of these "canals" caused speculation as to the existence of intelligent life on Mars. Perhaps these valleys are the "canals" seen by some astronomers. Another valley that also may be of fault origin is shown in Fig. 1–27B. Note the lines of craters that parallel the main valley.

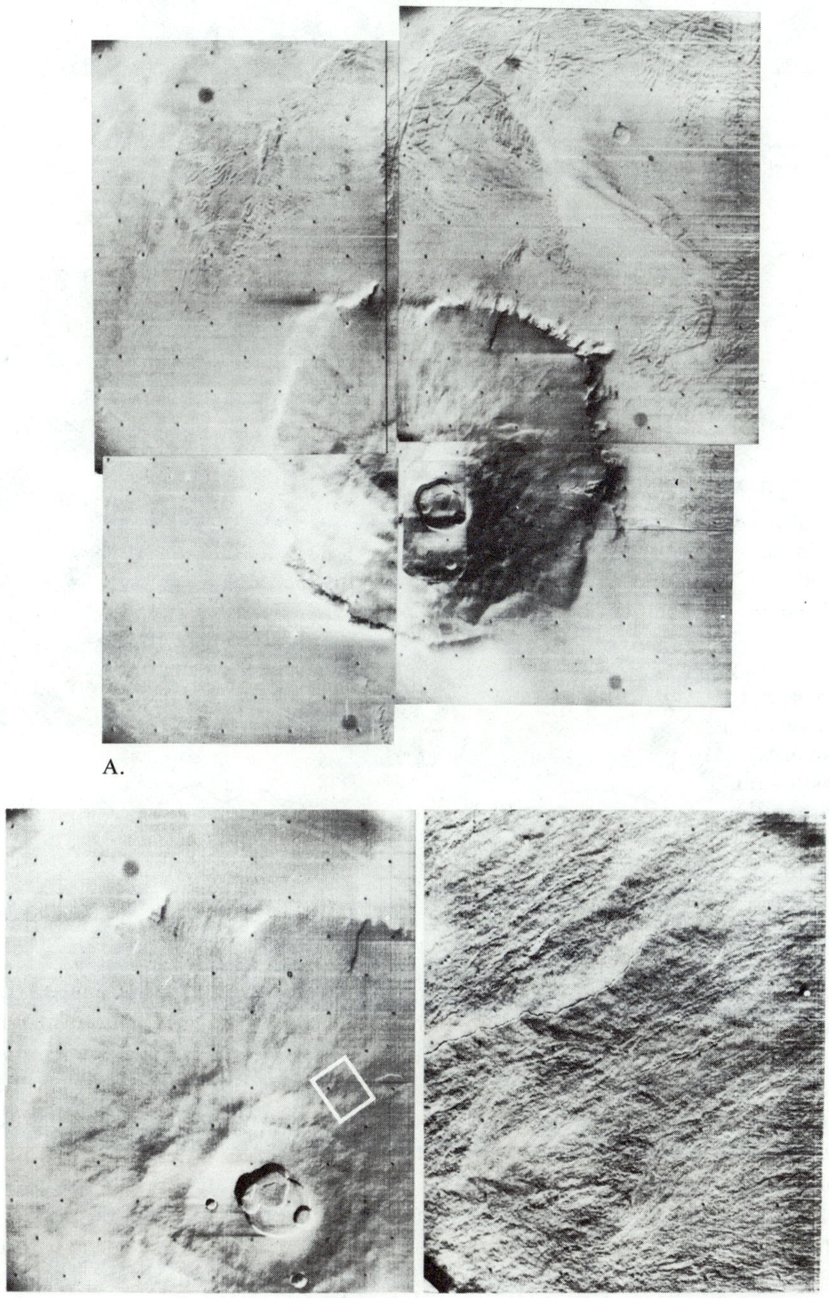

A. B. C.

FIGURE 1–26
Olympus Mons, a gigantic volcanic mountain on Mars. A. The mountain is 500 km in diameter across the base. The complex crater is 65 km in diameter. B. Closer view of the summit. C. Telephoto view of the flank showing features that resemble flow material on Earth. The area shown is 43 × 56 km. (Photos courtesy NASA)

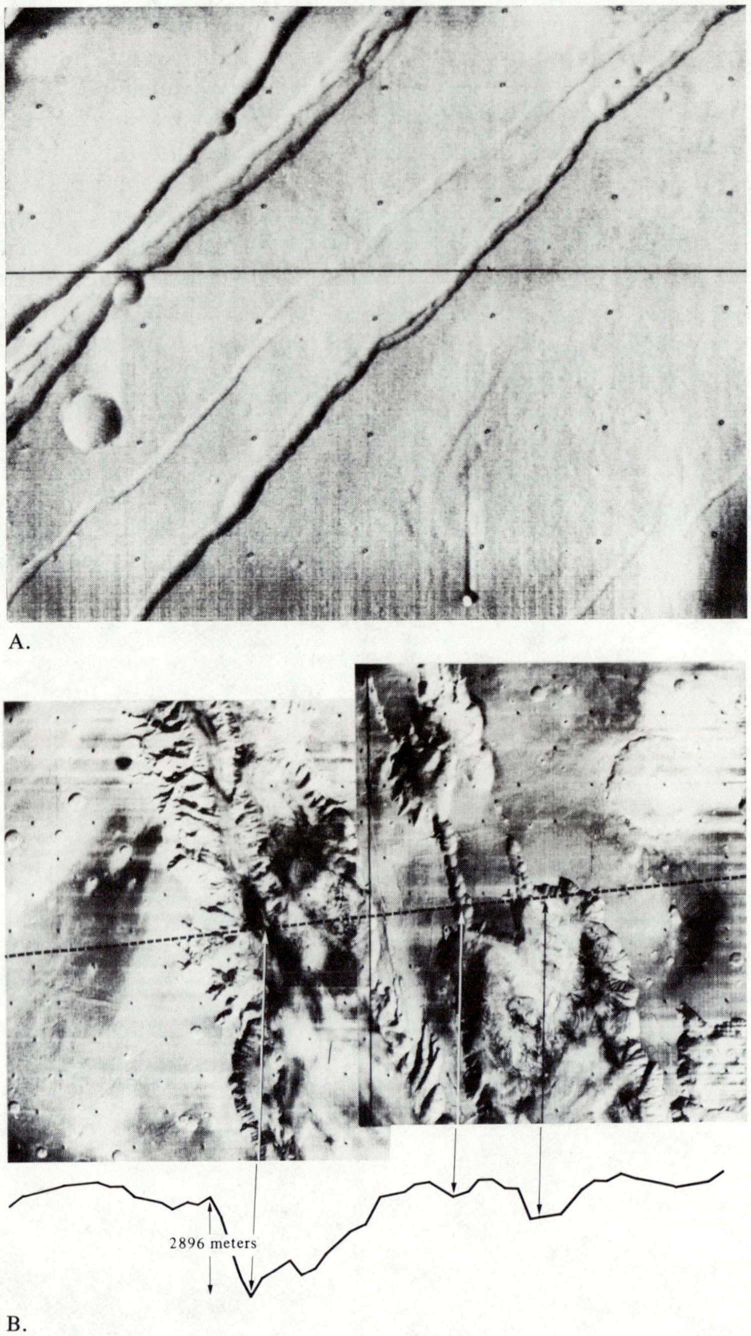

FIGURE 1–27
Valleys of probable fault origin on Mars. A. These structures are part of a system of parallel fissures extending more than 4800 km. The widest valley is about 1.6 km across and has a smaller valley inside it. The photo covers an area 34 × 42 km. B. A valley on Mars much deeper than the Grand Canyon, approximately 120 km across. (Photos courtesy NASA)

Mars also has features that suggest erosion by running water, although very little water is present in the atmosphere. An example of a long, sinuous gully resembling a river course is shown in Fig. 1–28. The water might have come from melting part of the polar caps; that will be discussed later. It is also possible that water in the form of ice is present in the "soil." The pits and basins shown in Fig. 1–25 are near the south pole, and so could have formed by the melting of subsurface ice. The chaotic terrain of short ridges and small valleys shown in Fig. 1–29 is similar to landslide and collapsed areas on Earth caused by melting subsurface ice. Clouds on some of the big volcanoes suggest the possibility that they may be emitting gases, possibly water vapor, that might at times be abundant enough to cause rain, and so erosion.

Craters are common in many areas, and, although some are of volcanic origin, many are caused by meteorite impact. This is not unexpected, because Mars is near the asteroid belt and has a thin atmosphere. The effects of meteorite impacts are clearly shown on Mars's moon Phobos (Fig. 1–30). Mars's equator almost separates cratered from uncratered areas. Why such a hemispherical division occurs is not known.

The polar ice caps appear to be dual ice caps. Much of the ice is frozen carbon dioxide, and this is the part that disappears and re-forms seasonally. The rest is composed of water ice. The laminated structure near the ice caps suggests that windblown dust accumulates in the ice at times. These deposits seem to be at least 100 m thick.

VENUS

Venus is named for the goddess of love because it is visible only near sunrise and sunset. It is closer to the sun than is Earth, and thus is hotter. It has a thick atmosphere, composed almost entirely of carbon dioxide. Carbon dioxide causes a greenhouse effect, which accounts for Venus's high temperature. Also present in minor amounts are hydrogen, water vapor, carbon monoxide, and hydrofluoric and hydrochloric acids.

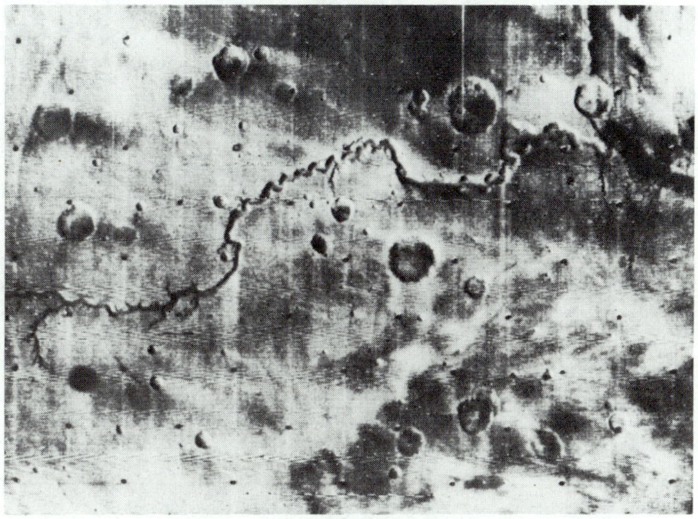

A.

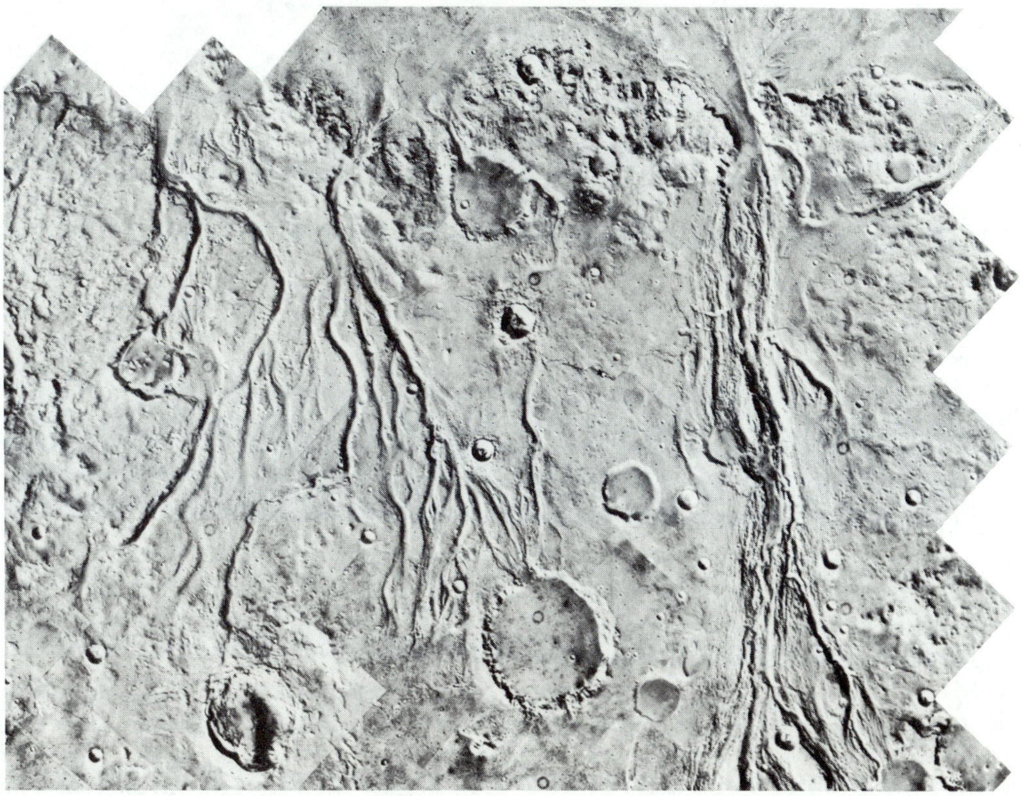

B.

FIGURE 1–28
Erosional features on Mars. A. Riverlike gully suggests erosion by running water. B. The pattern of these gullies suggests erosion by floodwater.

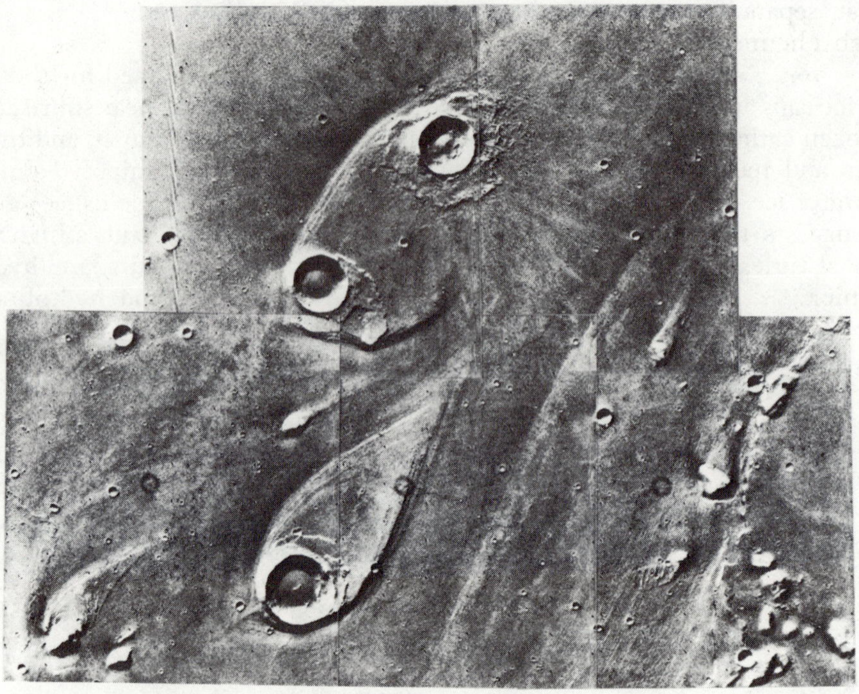

C.

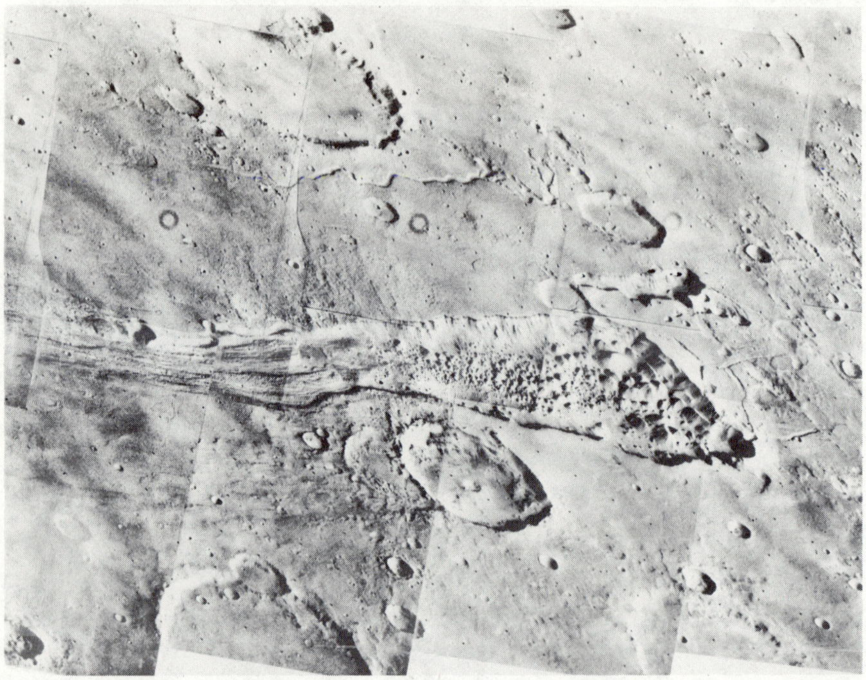

D.

FIGURE 1–28 (cont.)
C. These teardrop-shaped islands are also suggestive of erosion by floodwater.
D. The chaotic terrain at the head of this gully suggests that melting of subsurface ice may have been the source of the fluid that eroded the gully. (Photos courtesy NASA)

The surface temperature is near 465°C, making life a very remote possibility. The atmosphere has a solid cloud cover 72–97 km thick, so the surface is not visible. In spite of Venus's slow rotation, the surface temperature does not change as much as expected from day to night, suggesting a circulation in the atmosphere that transfers the heat. Venus is the only planet that rotates on its axis in a clockwise direction. No magnetic field has been detected, suggesting that Venus, too, lacks a liquid core, although its rate of rotation may be too slow to develop a magnetic field. In 1975, two Russian soft-landers operated for about an hour each before the high surface temperatures incapacitated them. The photographs they sent back to Earth showed angular

FIGURE 1–29
Chaotic terrain on Mars is shown on the right. The chaotic terrain is a jumble of ridges, each of which is 2.5–3.2 km wide and 1.6–8 km long. (*Mariner 6* photo courtesy NASA)

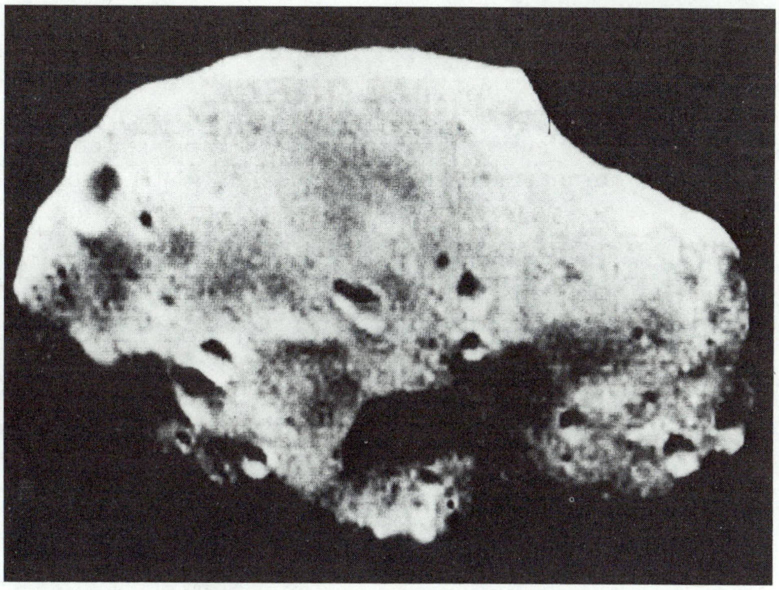

FIGURE 1–30
Mariner 9 photograph of Mars's moon Phobos. Phobos is about 21 × 26 km. The age and strength of Mars's innermost moon is shown by the impact crater. (Photo courtesy NASA)

rocks and no sand or dust. Radar observations suggest that Venus may have mountain ranges and craters, but little can be said about their type or origin.

LARGER MOONS OF JUPITER AND SATURN

The larger moons of Jupiter and Saturn range in size from slightly smaller than our moon to somewhat larger than Mercury. Jupiter and Saturn have at least 15 satellites each, but only 4 of Jupiter's and 1 of Saturn's are similar in size to the inner planets. Only one other moon, Neptune's Triton, is of similar size, but very little is known about it.

The most interesting of these bodies is Jupiter's Io, which is slightly larger (3640-km diameter) than our moon and has about the same density. Spacecraft images have revealed eight active volcanoes (Fig. 1–31). The volcanic eruptions are caused by tidal flexing by the other nearby large satellites, especially Europa. The gravitational attraction of the other satellites deforms

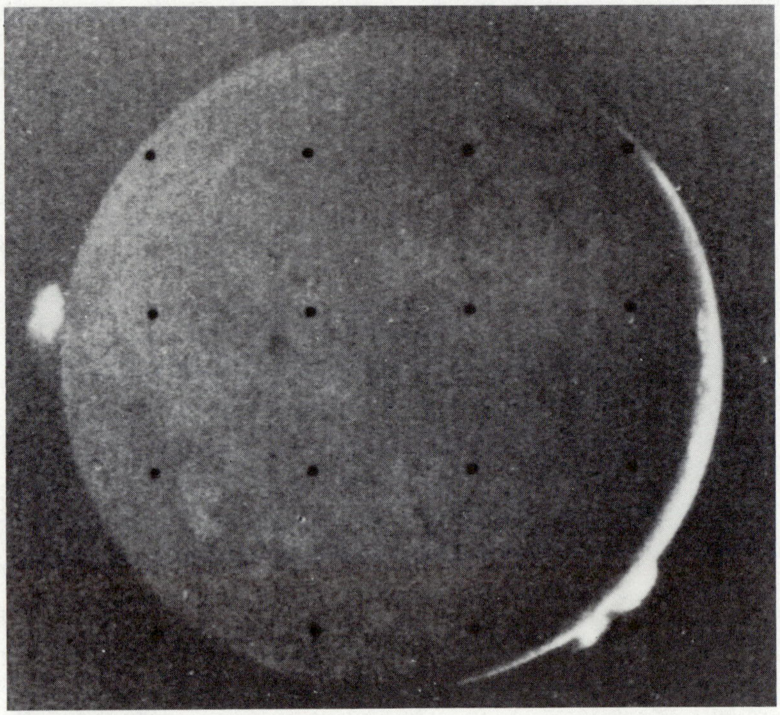

FIGURE 1–31
Io. Three erupting volcanoes can be seen in this view taken from 1.2 million km. (Photo courtesy NASA)

Io, heating it by friction. The material erupted by the volcanoes is sulfur and sulfur dioxide, and these materials are responsible for Io's colors—red, orange, black, and white. Io and the other large satellites are obviously much different from the inner planets.

The other large satellites of Jupiter are composed of rock and water. Europa (3050-km diameter) has a density of 3.03 g/cm^3, which is somewhat less than our moon's. It is believed to be composed of rock with a layer of ice about 100 km thick at the surface. The surface shows a tangle of lines, perhaps cracks in the ice (Fig. 1–32). Ganymede (5270-km diameter) has a density of 1.93 g/cm^3 and is believed to be about half rock and half water. Callisto (5000-km diameter) is less

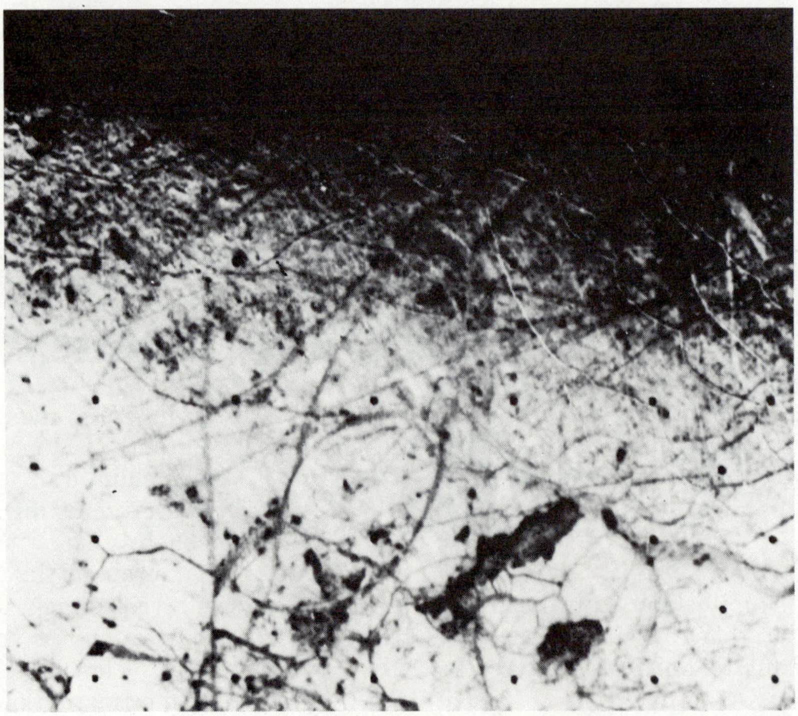

FIGURE 1–32
Europa. The surface of this satellite of Jupiter shows a tangle of lines that may be cracks in its icy surface. (Photo courtesy NASA)

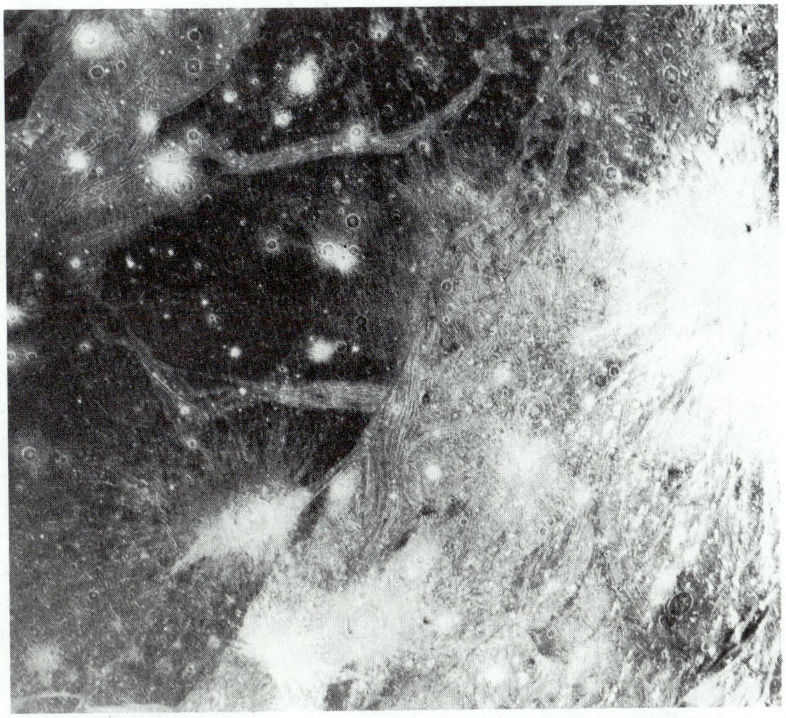

FIGURE 1–33
Ganymede. Light and dark grooved areas and craters are shown in this view.

dense and contains more water than rock. Ganymede's surface has dark cratered areas and lighter grooved, perhaps faulted, regions (Fig. 1–33). Callisto's surface is covered with craters (Fig. 1–34).

Saturn has only one large moon, Titan, 5000 km in diameter. Because Titan has a thick atmosphere, little is known about its surface. Although only slightly larger than Mercury, Titan is able to retain an atmosphere because it is so cold. The atmosphere, like Earth's, is composed largely of nitrogen. The other constituents of Titan's atmosphere are hydrocarbons, especially methane. The surface temperature is estimated to be about −184°C, and liquid nitrogen may be present on the surface. The action of the sun's energy

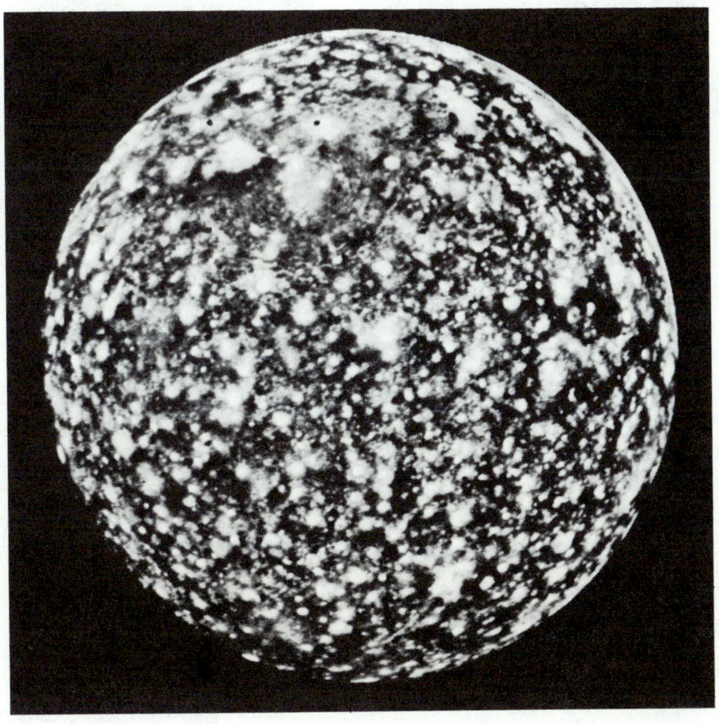

FIGURE 1–34
Callisto. This satellite of Jupiter has a cratered surface. (Photo courtesy NASA)

on this atmosphere has produced brown gases that are probably similar to photochemical smog on Earth, which forms as the result of the sun's actions on hydrocarbons in our atmosphere.

COMPARATIVE GEOLOGY OF THE INNER PLANETS AND THE MOON

The inner planets have been studied using spacecraft, and their structures and processes appear somewhat like Earth's. Samples have been returned from the moon, and close-up images of it and the other inner planets have been obtained. Images have also been returned of some of the moons of Jupiter and Saturn, and, although similar in size, they are different from the inner planets.

The most obvious difference between Earth, and the moon and inner planets, is the absence or greatly reduced amount of erosion on the latter, because, with the exception of Venus, they have tenuous or no atmospheres. This aids our interpretation of their internal processes because surface forms are not destroyed by erosion. All of the inner planets and their moons have been bombarded by meteorites, and the resulting craters are conspicuous on all except Earth, where erosion has destroyed most.

Sea-floor spreading and plate tectonics are the main internal processes on Earth, and somewhat similar processes appear to operate on some of the other planets. The energy for these internal processes must come from within the planet. The amount of internal energy apparently is proportional to the size of the planet. The other planets are smaller than Earth, and they show some aspects of plate tectonics. We will begin with the smallest, our moon (3475-km diameter), and proceed to Mercury (4828-km diameter), Mars (6758-km diameter), and end with Venus (6228-km diameter).

The similarities among the moon and the inner planets so far studied by space probe are striking. The similarity of surface features strongly suggests similar histories for the moon and these planets, Mars and Mercury. Rocks have been returned only from the moon, so the dating of events on Mars and Mercury depends on correlation with the moon on degree of cratering. Their histories begin with their formation about 4.6 billion years ago. From then until about 4 billion years ago, cratering occurred at a rapid rate, forming the cratered highlands. Apparently, the inner planets experienced a high influx of meteorites during this interval. (Erosion has largely obliterated evidence of this event on Earth.) From 4 billion to about 3.1 billion years ago, the number of impacts diminished rapidly, before the less cratered volcanic lowlands formed.

The moon's cratered uplands differ somewhat in composition from the volcanic maria. This is seen from Earth as a difference in reflectivity. Similar differences in reflectivity are apparent on Mars and Mercury. This suggests that they, like Earth, underwent some melting early in their histories. The resulting layering is revealed by seismic studies on Earth and the moon. The energy released by crystallization of Earth's fluid core may be the energy that causes deformation of the crust. The early crystallization of the cores of the smaller planets may account for their lack of compressional features and for the great age of their surface features. The larger of the small planets, Mars and Mercury, have some surface deformation. Mars has great fractures, and Mercury has long scarps, suggesting again that size and internal energy are related.

The smaller planets also differ from Earth in that their surface features are not evenly distributed. Mars's equator is almost the dividing line between a cratered hemisphere and a smooth one. Mercury also has cratered and smooth halves. The side of the moon opposite Earth is more cratered than the side facing us. Why such a distribution of upland and lowland exists is not clear, although at present there is more continent in Earth's northern hemisphere than in the southern. At times in the geologic past, Earth's continents are believed to have been joined into a single large continent. Thus, had Earth run out of internal energy earlier, as the smaller planets did, a land hemisphere and an ocean hemisphere might have resulted.

SUMMARY

Geology is the study of the earth. The earth is unique in having abundant water, and an atmosphere and temperature that can support life. The fundamental principle of geology is uniformitarianism, which is simply that the present processes occurring on the earth have always gone on. Geologic studies have revealed that constant change, both physical and biological, has been and is occurring on the earth.

The rocks of the crust are classified by origin into three types: igneous (solidified from a melt or magma), metamorphic (changed from another rock, generally by high temperature and pressure), and sedimentary (deposited in layers on the earth's surface by water, wind, and ice). The rock cycle shows the relationships among the three rock types.

Geology differs from most sciences in that it is concerned with absolute time. Absolute dates are based on radioactive decay of elements in rocks. Relative dates are based on local physical relationships, or on the fossils in the rocks. The geologic time scale is a relative time scale based on fossils.

Continental drift is shown by the shapes of the continents and the match of their rocks and fossils. Sea-floor spreading is the theory which suggests that new sea floor is created at the mid-ocean ridges, and that the sea floors move slowly away from the ridges and eventually are consumed at volcanic island arcs.

Plate tectonics assumes that the earth's surface is composed of a number of slabs, or plates, that move across the surface. Some of the plates carry continents. The plates form at mid-ocean ridges and are consumed at convergent boundaries, such as volcanic island arcs. Where one plate passes another, the boundary is called a transform fault.

When plates collide, they form convergent boundaries, and the surface rocks are deformed. If a plate composed of ocean floor collides with a similar plate, one of the plates is consumed at the resulting convergent boundary and a volcanic island arc forms. If ocean floor collides with a continental boundary, the ocean floor is consumed at the resulting convergent plate boundary. When continent collides with continent, a mountain range is formed.

The solar system is composed of the sun and the planets, comets, meteors, asteroids, and the satellites of the planets.

The inner planets, Mercury, Venus, Earth, and Mars, are smaller and denser than the more distant great planets.

Because the moon orbits Earth and rotates on its axis at the same rate, only one side is visible to us. The dark, smooth areas are called maria (mare, sing.), and the lighter regions are uplands.

Most moonquakes are low magnitude and are caused by tidal effects. They originate from two zones and reveal a layered moon.

The moon probably formed about 4.6 billion years ago and probably melted soon after this. The oldest rocks so far found are from the highlands and are about 4 billion years old. The maria basalts were extruded between 3.9 and 3.1 billion years ago. Since then, the moon has been quiet, with only meteorite impacts and perhaps some limited local volcanic activity.

Mercury has cratered upland areas and uncratered basins, probably of impact origin. Scarps up to 3 km high and 480 km long suggest faulting. Mars has dust storms, seasonal ice caps, huge volcanoes, evidence of faulting, apparent erosion by running water, and impact craters, as well as featureless and chaotic terrains. Venus has a thick atmosphere and is very hot. None of these inner planets have mountain ranges like Earth's.

The larger moons of Jupiter and Saturn range in size from slightly smaller than our moon to somewhat larger than Mercury. Io, a moon of Jupiter, has eight known active volcanoes. These volcanoes, which emit sulfur and sulfur dioxide, have as their energy source gravitation flexing caused by the other nearby large moons. The other large moons of Jupiter are believed to have crusts composed mainly of water ice. They have grooves, cracks, and other surface features of unknown origin. Saturn's only large moon is Titan, which has an atmosphere composed mainly of nitrogen. Titan is the only satellite known to have an atmosphere.

An obvious difference between Earth, and the moon and inner planets, is the almost complete lack of erosion on the latter.

All of the inner planets and their moons have been bombarded by meteorites; the resulting craters are conspicuous on all except Earth, where erosion has destroyed most of the craters.

The internal energy of a planet is apparently proportional to its size.

Our moon (3475-km diameter) shows evidence of ancient volcanic activity.

Mercury (4828-km diameter) has some scarps, suggesting faulting.

Mars (6758-km diameter) has huge volcanoes and long valleys that probably indicate faulting.

The much larger Earth (12,750-km diameter) has enough internal energy to have plate tectonics, with its associated volcanism, faulting, and deformation of the crust.

REVIEW EXERCISES

1 In what ways is the earth unique?
2 What are some of the economic uses of geology?
3 What are some of the ways geology can aid in planning the growth of an area?
4 Diagram the main elements of the rock cycle.
5 Briefly define *igneous rocks. Sedimentary rocks. Metamorphic rocks.*
6 What are the differences between relative dates and absolute dates in geology?
7 What is meant by the term *uniformitarianism*?
8 How was the geologic column or time scale established?
9 Outline some of the ways used to try to determine the age of the earth before radioactivity was discovered.
10 What is the evidence for continental drift?
11 Outline the theory of sea-floor spreading.
12 What three features are the boundaries of plates?
13 Describe the three types of convergent plate boundaries.
14 List the components of the solar system.
15 Briefly outline the history of the moon.
16 Why is Io unique so far as we know in the solar system?
17 Which is the only satellite with an atmosphere?
18 List the differences between Earth and Mercury. Earth and Mars. Earth and Venus.

SUGGESTED READINGS

GEOLOGY

ADAMS, F. D. *The Birth and Development of the Geological Sciences.* New York: Dover Publications, 1954, 506 pp. (paperback). Originally published in 1938.

ALBRITTON, C. C., ed. *The Fabric of Geology.* Reading Mass.: Addison-Wesley Publishing Co., 1963, 372 pp.

AMERICAN GEOLOGICAL INSTITUTE. *Geology: Science and Profession.* Washington, D.C.: American Geological Institute, 1965, 28 pp.

GEIKIE, A. *The Founders of Geology.* New York: Dover Publications, 1962, 486 pp. (paperback). Originally published in 1905 by Macmillan and Co.

GILLISPIE, C. C. *Genesis and Geology.* New York: Harper and Row Publishers, 1951, 306 pp. (paperback).

MURRAY, R. C., and TEDROW, J. C. F. *Forensic Geology: Earth Sciences and Criminal Investigations.* New Brunswick, N. J.: Rutgers University Press, 1975, 217 pp.

ROY, C. J. *The Sphere of the Geological Scientist.* Washington, D.C.: American Geological Institute, 1962, 28 pp.

SHORT, N. M., AND OTHERS. *Mission to Earth: Landsat Views the World.* NASA SP-360. Washington, D.C.: U.S. Government Printing Office, 1977, 459 pp.

U.S. GEOLOGICAL SURVEY. *Earth Science in the Public Service.* Professional Paper 921. Washington, D.C.: U.S. Government Printing Office, 1974, 73 pp.

GEOLOGIC TIME

EICHER, D. L. *Geologic Time,* 2nd ed. Englewood Cliffs, N.J.: Prentice-Hall, 1976, 152 pp. (paperback).

HAY, E. A. "Uniformitarianism Reconsidered." *Journal of Geological Education* (1967), Vol. 15, No. 1, pp. 11–12.

McALESTER, A. L. *The History of Life,* 2nd ed. Englewood Cliffs, N.J.: Prentice-Hall, 1977, 160 pp. (paperback).

MINTZ, L. W. *Historical Geology,* 3rd ed. Columbus, Ohio: Charles E. Merrill Publishing Co., 1981, 611 pp.

NICHOLS, R. L. "The Comprehension of Geologic Time." *Journal of Geological Education* (1974), Vol. 22, No. 2, pp. 65–68.

MOBILE EARTH

DEWEY, J. F. "Plate Tectonics." *Scientific American* (1972), Vol. 226, No. 5, pp. 56–68.

DICKINSON, W. R. "Plate Tectonics in Geologic History." *Science* (1971), Vol. 174, No. 4005, pp. 107–13.

HALLAM, A. "Continental Drift and the Fossil Record." *Scientific American* (1972), Vol. 227, No. 5, pp. 56–66.

HEIRTZLER, J. R. "Sea-Floor Spreading." *Scientific American* (1968), Vol. 219, No. 6, pp. 60–70.

KURTÉN, B. "Continental Drift and Evolution." *Scientific American* (1969), Vol. 220, No. 3, pp. 54–64.

VINE, F. J. "Sea-Floor Spreading: New Evidence." *Journal of Geological Education* (1969), Vol. 17, No. 1, pp. 6–16.

WILSON, J. T. "Some Aspects of the Current Revolution in the Earth Sciences." *Journal of Geological Education* (1969), Vol. 17, No. 4, pp. 145–50.

WYLLIE, P. J. *The Dynamic Earth: Textbook in Geosciences.* New York: John Wiley and Sons, 1971, 416 pp.

PLANETS

EBBIGHAUSEN, E. G. *Astronomy,* 4th ed. Columbus, Ohio: Charles E. Merrill Publishing Co., 1980, 159 pp. (paperback).

HEAD, J. W., AND OTHERS. "Geologic Evolution of the Terrestrial Planets." *American Scientist* (1977), Vol. 65, No. 1, pp. 21–29.

KING, E. A. *Space Geology: An Introduction.* New York: John Wiley and Sons, 1976, 400 pp.

LEWIS, J. S. "The Chemistry of the Solar System." *Scientific American* (1974), Vol. 230, No. 3, pp. 51–65.

SAGAN, C. "The Solar System." *Scientific American* (1975), Vol. 233, No. 3, pp. 22–31. This entire issue is devoted to the solar system.

MARS

ARVIDSON, R. E.; BINDER, A. B.; and JONES, K. L. "The Surface of Mars." *Scientific American* (1978), Vol. 238, No. 3, pp. 76–89.

CARR, M. H. "The Volcanoes of Mars." *Scientific American* (1976), Vol. 234, No. 1, pp. 32–43.

HOROWITZ, N. H. "The Search for Life on Mars." *Scientific American* (1977), Vol. 237, No. 5, pp. 52–61.

MURRAY, B. C. "Mars from *Mariner 9.*" *Scientific American* (1973), Vol. 228, No. 1, pp. 48–69.

VENUS

JAMES, J. N. "The Voyage of *Mariner II.*" *Scientific American* (1963), Vol. 209, No. 1, pp. 70–84.

MOON

DYAL, P., and PARKIN, C. W. "The Magnetism of the Moon." *Scientific American* (1971), Vol. 225, No. 2, pp. 63–73.

GOLDREICH, P. "Tides and the Earth-Moon System." *Scientific American* (1972), Vol. 226, No. 4, pp. 43–52.

HAMMOND, A. L. "Lunar Science: Analyzing the Apollo Legacy." *Science* (1973), Vol. 179, No. 4080, pp. 1313–15.

HARTMANN, W. K. "Cratering in the Solar System." *Scientific American* (1977), Vol. 236, No. 1, pp. 84–99.

MASON, B. "The Lunar Rocks." *Scientific American* (1971), Vol. 225, No. 4, pp. 48–58.

RUNCORN, S. K. "Corals as Paleontological Clocks." *Scientific American* (1966), Vol. 215, No. 4, pp. 26–33.

SCHMITT, H. H. "*Apollo 17* Report on the Valley of Taurus-Littrow." *Science* (1973), Vol. 182, No. 4113, pp. 681–90. A report on the geology of the moon, written by a geologist who was there.

SCIENCE. (1970), Vol. 167, No. 3918. *Apollo 11* issue.

WOOD, J. A. "The Geology of Mars." *American Scientist* (1980), Vol. 68, No. 6, pp. 626–35.

———. "The Lunar Soil." *Scientific American* (1970), Vol. 223, No. 2, pp. 14–23.

- element
- atom
- electron
- neutron
- proton
- mineral
- crystalline
- glass
- daughter element
- electron capture
- radiation halo
- fission track
- cosmic ray
- Van Allen radiation belt
- aurora
- fission

2

- crystal
- rock
- ion
- SiO_4 tetrahedron
- nucleus
- atomic number
- isotope
- atomic mass number
- electron shell
- ionic bond
- covalent bond
- metallic bond
- radioactivity
- alpha particle
- beta particle
- gamma ray
- half-life
- fusion
- polymorph
- hardness
- specific gravity
- conchoidal fracture
- cleavage
- streak
- silicate
- plagioclase
- potassium feldspar
- perthite
- olivine
- pyroxene
- augite
- amphibole
- hornblende
- mica

GEOLOGIC TIME

EICHER, D. L. *Geologic Time,* 2nd ed. Englewood Cliffs, N.J.: Prentice-Hall, 1976, 152 pp. (paperback).

HAY, E. A. "Uniformitarianism Reconsidered." *Journal of Geological Education* (1967), Vol. 15, No. 1, pp. 11–12.

McALESTER, A. L. *The History of Life,* 2nd ed. Englewood Cliffs, N.J.: Prentice-Hall, 1977, 160 pp. (paperback).

MINTZ, L. W. *Historical Geology,* 3rd ed. Columbus, Ohio: Charles E. Merrill Publishing Co., 1981, 611 pp.

NICHOLS, R. L. "The Comprehension of Geologic Time." *Journal of Geological Education* (1974), Vol. 22, No. 2, pp. 65–68.

MOBILE EARTH

DEWEY, J. F. "Plate Tectonics." *Scientific American* (1972), Vol. 226, No. 5, pp. 56–68.

DICKINSON, W. R. "Plate Tectonics in Geologic History." *Science* (1971), Vol. 174, No. 4005, pp. 107–13.

HALLAM, A. "Continental Drift and the Fossil Record." *Scientific American* (1972), Vol. 227, No. 5, pp. 56–66.

HEIRTZLER, J. R. "Sea-Floor Spreading." *Scientific American* (1968), Vol. 219, No. 6, pp. 60–70.

KURTÉN, B. "Continental Drift and Evolution." *Scientific American* (1969), Vol. 220, No. 3, pp. 54–64.

VINE, F. J. "Sea-Floor Spreading: New Evidence." *Journal of Geological Education* (1969), Vol. 17, No. 1, pp. 6–16.

WILSON, J. T. "Some Aspects of the Current Revolution in the Earth Sciences." *Journal of Geological Education* (1969), Vol. 17, No. 4, pp. 145–50.

WYLLIE, P. J. *The Dynamic Earth: Textbook in Geosciences.* New York: John Wiley and Sons, 1971, 416 pp.

PLANETS

EBBIGHAUSEN, E. G. *Astronomy,* 4th ed. Columbus, Ohio: Charles E. Merrill Publishing Co., 1980, 159 pp. (paperback).

HEAD, J. W., AND OTHERS. "Geologic Evolution of the Terrestrial Planets." *American Scientist* (1977), Vol. 65, No. 1, pp. 21–29.

KING, E. A. *Space Geology: An Introduction.* New York: John Wiley and Sons, 1976, 400 pp.

LEWIS, J. S. "The Chemistry of the Solar System." *Scientific American* (1974), Vol. 230, No. 3, pp. 51–65.

SAGAN, C. "The Solar System." *Scientific American* (1975), Vol. 233, No. 3, pp. 22–31. This entire issue is devoted to the solar system.

MARS

ARVIDSON, R. E.; BINDER, A. B.; and JONES, K. L. "The Surface of Mars." *Scientific American* (1978), Vol. 238, No. 3, pp. 76–89.

CARR, M. H. "The Volcanoes of Mars." *Scientific American* (1976), Vol. 234, No. 1, pp. 32–43.

HOROWITZ, N. H. "The Search for Life on Mars." *Scientific American* (1977), Vol. 237, No. 5, pp. 52–61.

MURRAY, B. C. "Mars from *Mariner 9.*" *Scientific American* (1973), Vol. 228, No. 1, pp. 48–69.

VENUS

JAMES, J. N. "The Voyage of *Mariner II.*" *Scientific American* (1963), Vol. 209, No. 1, pp. 70–84.

MOON

DYAL, P., and PARKIN, C. W. "The Magnetism of the Moon." *Scientific American* (1971), Vol. 225, No. 2, pp. 63–73.

GOLDREICH, P. "Tides and the Earth-Moon System." *Scientific American* (1972), Vol. 226, No. 4, pp. 43–52.

HAMMOND, A. L. "Lunar Science: Analyzing the Apollo Legacy." *Science* (1973), Vol. 179, No. 4080, pp. 1313–15.

HARTMANN, W. K. "Cratering in the Solar System." *Scientific American* (1977), Vol. 236, No. 1, pp. 84–99.

MASON, B. "The Lunar Rocks." *Scientific American* (1971), Vol. 225, No. 4, pp. 48–58.

RUNCORN, S. K. "Corals as Paleontological Clocks." *Scientific American* (1966), Vol. 215, No. 4, pp. 26–33.

SCHMITT, H. H. "*Apollo 17* Report on the Valley of Taurus-Littrow." *Science* (1973), Vol. 182, No. 4113, pp. 681–90. A report on the geology of the moon, written by a geologist who was there.

SCIENCE. (1970), Vol. 167, No. 3918. *Apollo 11* issue.

WOOD, J. A. "The Geology of Mars." *American Scientist* (1980), Vol. 68, No. 6, pp. 626–35.

———. "The Lunar Soil." *Scientific American* (1970), Vol. 223, No. 2, pp. 14–23.

Mineralogy

MINERALOGY, WHICH IS NOW ONLY A BRANCH OF GEOLOGY, WAS probably its parent. The first geological books were mineralogies, and they bear little resemblance to modern mineralogy books. They made no distinction between minerals and fossils, and included myths and folklore associated with the minerals.

Mineralogy came out of the Middle Ages with the writings of Georg Bauer (1494–1555), who, following the custom of the time, wrote in Latin under the Latinized name Agricola. Like many of the early people in geology, Agricola was a physician, as this was the only scientific training available. He was a Saxon German and lived in Joachimstal, near the famous silver mines. His hobby was the study of minerals and fossils, and he wrote several books on these subjects. His most important book was *De Re Metallica,* published in 1556. This was the first book on mining and metallurgy. It was very popular and was one of the first books translated into local languages that could be read by the common people. This work was published at the right time, about 100 years after the invention of printing, enabling it to be widely circulated; and it was on a practical subject that was well received. It is interesting to note that this took place during the age of such men as Da Vinci, Copernicus, and Luther.

Another of Agricola's books, *De Natura Fossilum,* was on minerals and fossils. It was published in 1546 and classified minerals on the basis of physical properties.

Nicolas Steno (1638–1687) was about 100 years ahead of his time in understanding geology. He will be introduced here for his work in mineralogy, and mentioned later when stratigraphy is considered. He was born a Dane, Niels Steensen, but he also Latinized his name because he wrote in Latin. In 1668, he published a book in which he recognized the true nature of fossils. He also noted from his study of quartz crystals that the angles between the faces are always the same. This was the beginning of the science of crystallography, one of the offshoots of mineralogy.

Abraham Gottlob Werner (1749–1817) did much for mineralogy but probably held back geology for 100 years. He studied at Freiberg from 1769 to 1771 and then went on to the University of Leipzig, where he wrote a mineralogy book, *Von der äusserlichen Kennzeichen der Fossilien,* in 1774. On the basis of this book, Werner was invited back to Freiberg the same year and became a professor at the age of 25. He was the chief proponent of the so-called mixed system of mineralogy as the most practical at the time. He knew that this was not the best system and strived for the ideal system based on composition. He constantly improved his system during the 40 years he stayed at the Academy at Freiberg. In his last revision in 1817, he recognized 317 minerals in classes such as earthy minerals, saline minerals, combustible minerals, and metallic minerals.

Werner also founded the neptunist theory of the origin of the earth. This latter work was not at all in the same class with his mineralogy. This theory was based on his idea that all rocks were precipitates from a universal ocean. Although Werner wrote very little, even when his theory of the earth was under attack, he was extremely popular, and his fame, based in part on his mineralogy, was spread far and wide by his students.

During Werner's lifetime, further advances were made in crystallography. Abbé René-Just Haüy (1743–1822), a French botanist, became interested in mineralogy after hearing a lecture at the Royal Gardens in Paris. The consistency of form of plants and minerals intrigued him. After accidentally dropping a calcite crystal, he noted that no matter what the shape of the original crystal, the cleavage fragments all have the same (rhombohedral) shape. From this, he deduced that all minerals are made of tiny building blocks, thus beginning our modern ideas of unit cells. He also developed the six crystal systems.

Mineralogy developed very rapidly in the nineteenth century. Crystallography advanced with the development of instruments to measure accurately the angles between crystal faces. The most important advance, however, was the development of chemical analysis so that the composition of minerals could be determined.

The new mineralogy based on chemistry came from the great Swedish chemist Johann Jacob Berzellius (1779–1848). Berzellius discovered selenium and thorium, and was the first to isolate calcium, barium, strontium, silicon, and zirconium. He also devised the system of writing chemical formulas and equations that is still used. In 1814, he published in Stockholm *A New System of Mineralogy.* In this book, he recognized silicates, bringing order out of chaos.

With these chemical and crystallographic methods available, mineralogists studied many minerals and much new data was produced. This in turn led to the publication of books that described and classified minerals. Two outstanding books of this era that are still in use (in much more modern editions) are a German text by Groth, published in Germany, and James Dwight Dana's *System of Mineralogy,* published in the United States. Dana (1813–1895) was a professor at Yale who made important advances in structural geology.

In the twentieth century, the discovery of X rays enabled probing the internal structure of minerals. This was first done in 1912 by Max von Laue and his students W. Friedrich and P. Knipping. Their discovery proved that the atoms in a crystal have an orderly arrangement. Very soon after this, Sir William Bragg and his son, Laurence, used X rays to determine the internal structure of halite (rock salt), and a new era in mineralogy began. Since then, X-ray methods have been refined, and the structures of many minerals have been determined.

ABUNDANCE OF ELEMENTS

On casual examination, the surface of the earth appears to be quite complex. Many different-appearing rocks and soils are to be seen. The complexity is much more apparent than real, and close study reveals a very simple composition. The chemical composition of the crust of the earth will be considered first.

Over 100 elements are known; of these, 80 are stable. The crust of the earth is composed almost entirely of only 8 elements, with all of the other elements making up less than 1 percent, as shown in Table 2–1. Thus, in spite of its heterogeneous aspect, the crust of the earth has a very simple composition.

Notice that many of our common, cheap metals, such as chrome, nickel, zinc, copper, lead, and tin, are really rather rare and form only about 0.04 percent of the crust. How these elements are concentrated into mineable deposits will be discussed in later chapters.

Before discussing the important implications of these data, their origin should be considered. All abundance data are calculated from estimates of the amounts of the different rock types and their chemical analyses. The main problem is estimating the amounts of the rock types; however, geologic maps showing the distribution of rock types have been made for many parts of the earth's surface. In spite of this, such estimates are difficult, and different workers have made different estimates. The amounts shown in the tables are subject to some revision, but they do show the general composition of the crust. Estimates of the composition of the whole earth or of the solar system are quite different.

DEFINITIONS

So far, we have considered **elements**, which are materials that cannot be subdivided by ordinary chemical methods. The smallest unit of an element is the **atom**. The atom can be broken down into smaller particles such as electrons, neutrons, protons, and so forth, but

TABLE 2–1
The average composition of continental crustal rocks. (A few very rare elements and short-lived radioactive elements are omitted.)

Symbol	Name	Weight (%)	Symbol	Name	Weight (%)	Symbol	Name	Weight (%)
O	Oxygen	46.4	Nd	Neodymium	0.0028	Mo	Molybdenum	0.00015
Si	Silicon	28.15	La	Lanthanum	0.0025	Ho	Holmium	0.00015
Al	Aluminum	8.23	Co	Cobalt	0.0025	Eu	Europium	0.00012
Fe	Iron	5.63	Sc	Scandium	0.0022	Tb	Terbium	0.00011
Ca	Calcium	4.15	N	Nitrogen	0.0020	Lu	Lutetium	0.00008
Na	Sodium	2.36	Li	Lithium	0.0020	I	Iodine	0.00005
Mg	Magnesium	2.33	Nb	Niobium	0.0020	Tl	Thallium	0.000045
K	Potassium	2.09	Ga	Gallium	0.0015	Tm	Thulium	0.000025
Ti	Titanium	0.57	Pb	Lead	0.00125	Sb	Antimony	0.00002
H	Hydrogen	0.14	B	Boron	0.0010	Cd	Cadmium	0.00002
P	Phosphorus	0.105	Th	Thorium	0.00096	Bi	Bismuth	0.000017
Mn	Manganese	0.095	Sm	Samarium	0.00073	In	Indium	0.00001
F	Fluorine	0.0625	Gd	Gadolinium	0.00073	Hg	Mercury	0.000008
Ba	Barium	0.0425	Pr	Praseodymium	0.00065	Ag	Silver	0.000007
Sr	Strontium	0.0375	Dy	Dysprosium	0.00052	Se	Selenium	0.000005
S	Sulphur	0.026	Yb	Ytterbium	0.0003	A(r)	Argon	0.000004
C	Carbon	0.020	Hf	Hafnium	0.0003	Pd	Palladium	0.000001
Zr	Zirconium	0.0165	Cs	Cesium	0.0003	Pt	Platinum	0.000001
V	Vanadium	0.0135	Er	Erbium	0.00028	Te	Tellurium	0.000001
Cl	Chlorine	0.013	Be	Beryllium	0.00028	Ru	Ruthenium	0.000001
Cr	Chrome	0.010	U	Uranium	0.00027	Rh	Rhodium	0.0000005
Rb	Rubidium	0.009	Br	Bromine	0.00025	Os	Osmium	0.0000005
Ni	Nickel	0.0075	Ta	Tantalum	0.0002	Au	Gold	0.0000004
Zn	Zinc	0.0070	Sn	Tin	0.0002	He	Helium	0.0000003
Ce	Cerium	0.0067	As	Arsenic	0.00018	Re	Rhenium	0.0000001
Cu	Copper	0.0055	Ge	Germanium	0.00015	Ir	Iridium	0.0000001
Y	Yttrium	0.0033	W	Tungsten	0.00015			

SOURCE: After B. Mason, 1966; and K. Krauskopf, 1968.

except in the case of radioactive elements, this requires the large amounts of energy available in atom-smashing machines such as the cyclotron.

In nature, elements are combined to form **minerals**, which can be defined as:

1. naturally occurring, crystalline,
2. inorganic substances with
3. a definite small range in chemical composition and physical properties.

All three conditions must be met by a mineral.

Minerals are **crystalline** substances; that is, they have an orderly internal structure (arrangement of atoms), as opposed to such things as **glasses**, which are supercooled liquids and have no internal order. A **crystal** is a solid form bounded by smooth planes which give an outward manifestation of the orderly internal structure (see Fig. 2–18). Note the distinction between *crystal* and *crystalline.* Although all minerals are crystalline, they do not necessarily occur as geometric crystals.

The minerals group themselves naturally to form **rocks**, which can be defined as:

1. a natural aggregate of one or more minerals, or
2. any essential and appreciable part of the solid portion of the earth (or any other part of the solar system).

It should be noted that this definition does not require all rocks to be composed entirely of minerals. Some noncrystalline material is present in many quickly chilled volcanic rocks.

OXYGEN AND SILICON, THE BUILDING BLOCKS OF THE ROCK-FORMING MINERALS

Because the rocks of the crust are composed predominantly of eight elements, it would be suspected at first glance that these eight elements, which must form most minerals, would occur in many different combinations. However, most rocks are composed of only a few combinations of elements—or minerals. The reason for this happy situation, which means a beginner needs to learn only a few minerals to identify most rocks, can be understood from the data in Table 2–2. Rocks are composed of over 90 percent oxygen by volume, because oxygen atoms are abundant and relatively large. Therefore, the number of possible combinations is controlled by the ways in which oxygen atoms can be arranged to build minerals. This is especially true because relatively few other atoms are readily available to form minerals. Therefore, although about 2000 minerals have been discovered, only 20 are common, and fewer than 10 form well over 90 percent of all rocks.

TABLE 2–2
Main elements in the continental crust.

Element	Weight (%)	Atom (%)	Volume (%)
Oxygen	46.40	62.17	94.05
Silicon	28.15	21.51	0.88
Aluminum	8.23	6.54	0.48
Iron	5.63	2.16	0.48
Calcium	4.15	2.22	1.19
Sodium	2.36	2.20	1.11
Magnesium	2.33	2.05	0.32
Potassium	2.09	1.15	1.49
Total	99.34	100.00	100.00

Thus, the key to understanding the structure of minerals lies in how the atoms (more correctly called **ions**, because the atoms have gained or lost one or more electrons) in a mineral are arranged, that is, how the other atoms are fitted among the oxygen atoms. Clearly, the size of the atoms in question, compared to the size of the oxygen atom, will determine the structure of any mineral. This relationship is shown in two dimensions in Fig. 2–1. The number of oxygen ions that surround an element can be calculated from simple geometric relationships.

Small atom fits in space between 3 large atoms　　Larger atom fits in space between 4 large atoms

FIGURE 2–1
The relative sizes of atoms determine how they can fit together. Shown here in two dimensions.

As can be predicted from the abundance data, most of the rock-forming minerals are composed largely of oxygen and silicon, generally with aluminum and at least one more of the "big eight." The sizes of oxygen and silicon are such that the small silicon ion fits inside four oxygen ions. This unit, which is the building block of the silicate minerals, is called the **SiO_4 tetrahedron**, because the four oxygen ions can be considered to be situated on the four corners of a tetrahedron with the silicon ion at the center (Fig. 2–2).

The structures of the silicate minerals are determined by the way these SiO_4 tetrahedra are arranged. Two of these tetrahedra, if joined at a corner, can share

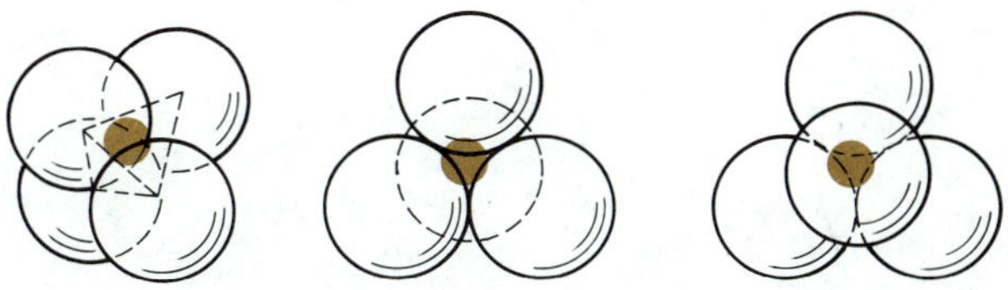

FIGURE 2–2
Three views of an SiO_4 tetrahedron. The large spheres are oxygen, and the small spheres are silicon.

a single oxygen ion. This forms a strong bond. With this type of linkage, through sharing of oxygen, the structure of the silicate minerals is formed. The possible structures are shown in Fig. 2–3.

ATOMIC STRUCTURE

The sizes of ions determine the internal geometry of minerals and other crystalline substances; however, some features of atomic structure must be learned in order to understand how minerals are held together. This study also will provide an understanding of natural radioactivity, which is an important source of energy in the earth, and which will be discussed in later chapters.

Atoms are composed of smaller particles, and although physicists have discovered many such particles, only a few need be discussed here. Most of the mass of an atom is in its **nucleus**. The nucleus is composed of **protons** and **neutrons** (except in the case of the hydrogen nucleus, which is composed of only a single proton). **Protons** are heavy and have a positive electrical charge. The nucleus is surrounded by a cloud of the much lighter, negatively charged electrons (Fig. 2–4). An atom is mostly empty space because the diameter of the outer electron orbits, which is the size of the atom and averages nearly 0.00000002 cm, is between 20,000 and 200,000 times the diameter of the nucleus. The number of electrons equals the number of protons; hence, an atom is electrically neutral. The

		Formula of Silicon-Oxygen Unit	*Number of Oxygen Shared*	*Example*
1. Single tetrahedron		(SiO_4)	0	Olivine
2. Double tetrahedron		(Si_2O_7)	1	Epidote
3. Ring	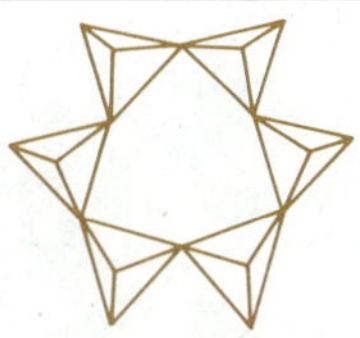	(Si_6O_{18})	2	Tourmaline
4. Chains	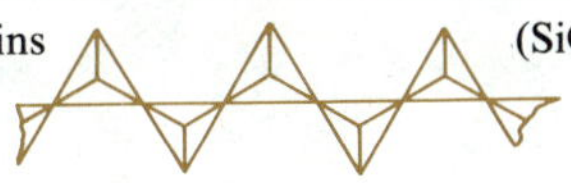	(SiO_3)	2	Pyroxene (augite)
5. Double chains	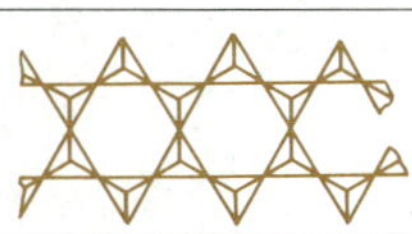	(Si_4O_{11})	2 & 3	Amphibole (hornblende)
6. Sheets	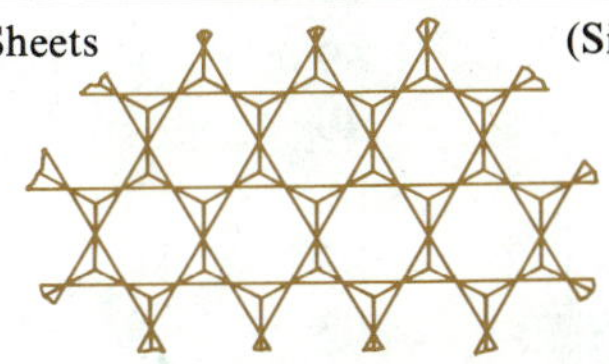	(Si_2O_5)	3	Micas (muscovite, biotite)
7. Three-dimensional networks	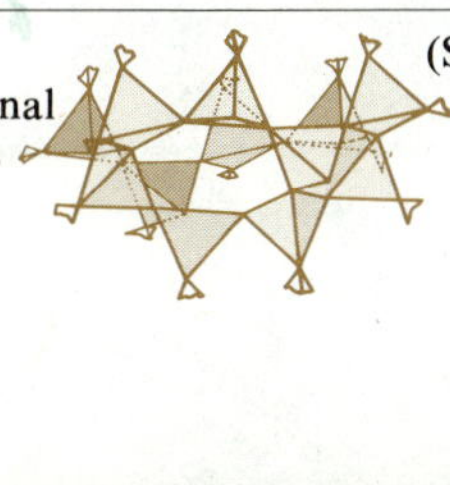	(SiO_2)	4	Quartz, feldspars

FIGURE 2–3
Structures of silicate minerals.

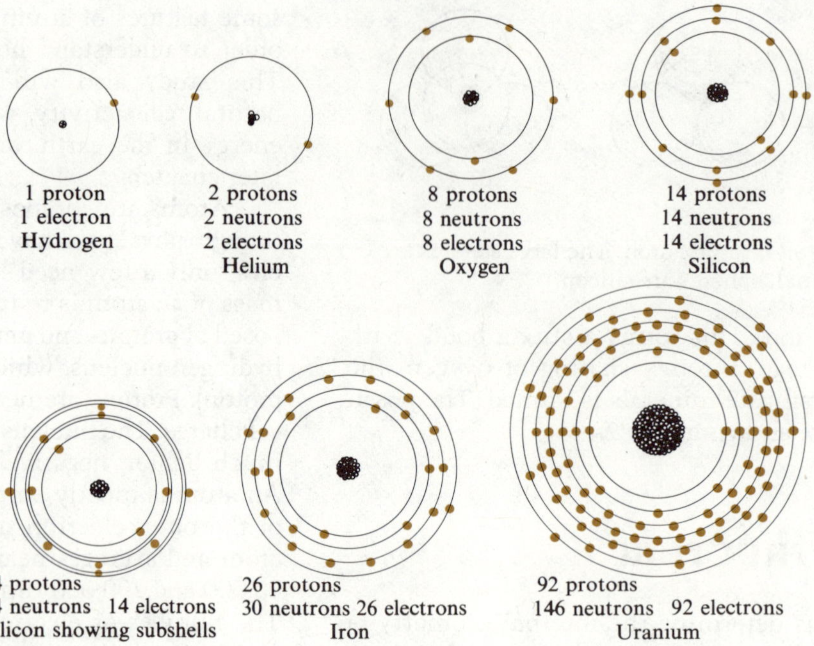

FIGURE 2–4
Diagrammatic representation of some common atoms. Although the electron shells are shown in only two dimensions here, the electrons actually move in spherical shells. The electron shells are complicated by subshells, shown only for silicon. Although the nuclei contain almost all of the mass of the atoms, the atomic radii are between 20,000 and 200,000 times the radii of the nuclei.

number of protons and electrons determines the chemical properties, and therefore the identity of an element. The number of protons (or electrons) is called the **atomic number** of an element. In addition, an atom may have **neutrons** in its nucleus. Neutrons have a mass slightly greater than that of a proton plus an electron, and are electrically neutral. The number of neutrons in the atom of an element may vary, giving rise to **isotopes**. Changing the number of neutrons in the atom of an element does not change the chemical properties, but it does change the mass of the atom. The **atomic mass number** is the number of neutrons and protons in the nucleus. Most elements have naturally occurring isotopes, as shown in Table 2–3.

TABLE 2–3
Examples of naturally occurring isotopes.

Element	Symbol*	Atomic Number (= number of protons)	Nucleus		Atomic Mass Number (= number of protons and neutrons)	Electrons (= number of protons)	Remarks
			Protons	Neutrons			
Hydrogen	${}_{1}H^{1}$	1	1	0	1	1	Common form
	${}_{1}H^{2}$	1	1	1	2	1	Deuterium
Helium	${}_{2}He^{4}$	2	2	2	4	2	Common form
	${}_{2}He^{3}$	2	2	1	3	2	
Oxygen	${}_{8}O^{16}$	8	8	8	16	8	Common form
	${}_{8}O^{18}$	8	8	10	18	8	
Carbon	${}_{6}C^{12}$	6	6	6	12	6	Common form
	${}_{6}C^{14}$	6	6	8	14	6	
Potassium	${}_{19}K^{39}$	19	19	20	39	19	Common form
	${}_{19}K^{40}$	19	19	21	40	19	
Lead	${}_{82}Pb^{206}$	82	82	124	206	82	
	${}_{82}Pb^{207}$	82	82	125	207	82	
	${}_{82}Pb^{208}$	82	82	126	208	82	
Uranium	${}_{92}U^{235}$	92	92	143	235	92	
	${}_{92}U^{238}$	92	92	146	238	92	

*The subscript is the atomic number, and the superscript is the atomic mass number. Sometimes only the superscript is used because only the mass number can vary for a given element.

BONDING

The bonding of atoms to form minerals is done in several ways, most of which involve the electrons surrounding the nucleus. The electrons are not randomly distributed around the nucleus, but move in distinct orbits. Each orbit, or **shell**, contains a fixed number of electrons, as shown diagrammatically in two dimensions in Fig. 2–4. Most of the electron shells have subshells, as indicated in Fig. 2–4 and Table 2–4. Up to seven shells are possible to accommodate the number of electrons necessary to balance the number of protons in the nucleus. As the number of protons in the nucleus increases, the electron shells are filled. The first shell contains 2 electrons when filled, and the second, 8. The third shell can contain up to 18 electrons; the fourth, 32; the fifth, 32; the sixth, 9; and the seventh, 2. The electrons are distributed among the outer shells of any element in such a manner that the outer shell always contains 8 or fewer electrons. Larger numbers can be accommodated only in interior shells.

The main principle involved in mineral bonding is to achieve a stable configuration with the outer shell filled. This can be accomplished in different ways. One possibility is for the atom to gain or lose electrons. An example of this is an atom with one electron in its outer shell losing this electron to another atom with seven electrons in its outer shell. The resulting ions have opposite charges and so are mutually attracted (Fig. 2–5). This is called **ionic bonding**, and many minerals have ionic bonds.

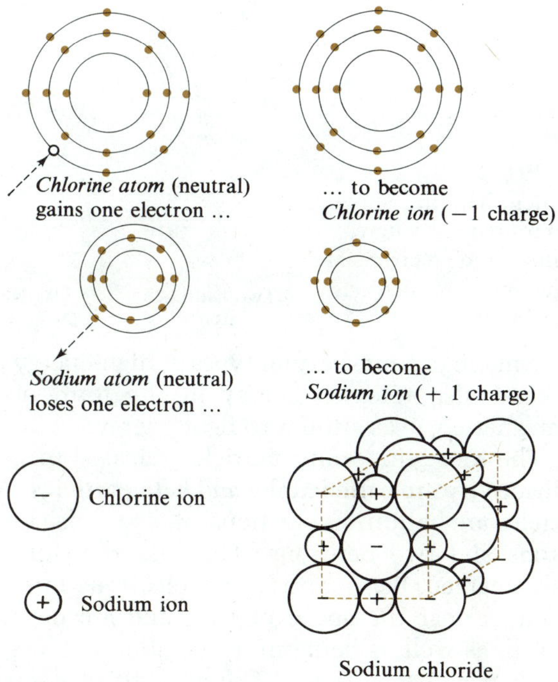

FIGURE 2–5
Ionic bonding. Ions are formed by the completion of an electron ring by the gain or loss of electrons. The resulting ions are held together (bonded) by the attraction of unlike electric charges.

Another way that the outer shell of an atom can be filled is by two or more atoms sharing electrons. In this way, the outer shells of the atoms are filled even though the total number of electrons is not enough to satisfy both atoms individually. This bonding, called **covalent bonding**, is shown in Fig. 2–6. Diamond is one of the few minerals with covalent bonding, and this accounts for the hardness of the diamond. Bonding of the silicate minerals is midway between ionic and covalent.

A third common type of bonding is the **metallic bond**. In a metal, the atoms are closely packed, so the electrons are not "owned" by any particular nucleus. This allows the electrons to move rather freely through the metal, accounting for the good thermal and electrical conductivity of metals. In minerals, metallic bonds are found in some sulfides and in native elements, such as gold.

The size of atoms or ions is important in determining the structure of minerals. Ions within about 10 percent of the same size may mutually substitute in many minerals. Some examples of ion pairs that can mutually substitute are silicon and aluminum, sodium and calcium, iron and magnesium, and gold and silver. See Table 2–4 and Fig. 2–7. The type of bond also affects possible substitution, and although potassium and gold are nearly the same size, they do not substitute for each other. Potassium forms ionic bonds, and gold forms covalent bonds with oxygen.

Present knowledge of the internal structure of minerals comes from X-ray studies. X rays have wavelengths in the same size range as atoms and so are deflected by the atoms in a crystal. This deflection occurs because the atoms in a crystal are packed in an orderly way. If the packing were not orderly, the X rays would be deflected randomly. Instead, the regular pattern of the deflected X rays can be used to determine the internal structure of a mineral (Fig. 2–8).

NATURAL RADIOACTIVITY

The material in this section is not directly related to mineralogy, the main topic of this chapter. The geological application of this section appears in later chapters. It is included here because it is directly related to atomic structure.

Not all nuclei are stable. **Radioactive nuclei** disintegrate spontaneously, releasing energy in the process. Understanding why and how this takes place was one of the great advances in physics during the first half of the twentieth century. In natural radioactivity, the un-

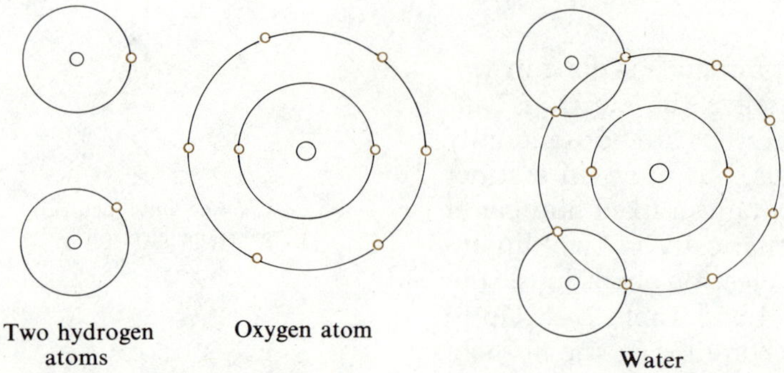

FIGURE 2–6
Covalent bonding, shown diagrammatically here, results when electrons are shared by more than one atom.

stable nucleus emits several types of high-energy particles and also releases energy in the form of electromagnetic waves similar to light energy.

The more important particles released in natural radioactivity are called **alpha** and **beta particles**. Alpha particles are helium nuclei (ions) and so consist of two protons and two neutrons. They are therefore positively charged. Beta particles are electrons moving at velocities near the speed of light and are negatively charged, as well as being much smaller and less massive than alpha particles. The amount of energy released by emission of a particle depends on the mass and the velocity of the particle.

The other type of radiation in natural radioactivity is gamma radiation. **Gamma rays** are energy in the form of electromagnetic waves. Visible light, radiowaves, and X rays are also energy in the form of electromagnetic waves and differ only in their frequency or wavelength. Gamma rays differ from X rays only in frequency; gamma rays have a somewhat higher frequency.

In radioactive decay, an atom is changed to an atom of another element. This means that the number of atoms of a radioactive element decreases with the passage of time. If the rate of disintegration is known, then by measuring the amounts of the parent and daughter elements, the age of crystallization of the mineral containing the parent element can be found. This is the principle of radioactive dating and works only because the rate of disintegration is not affected by temperature, pressure, and so on.

The rate of radioactive decay is expressed in terms of the **half-life**. In one half-life, one-half of the radioactive atoms present will decay to the **daughter element**. In the next half-life, one-half of the remaining radioactive atoms will decay, and so on (Fig. 2–9).

The products of radioactive decay can be understood rather easily from the foregoing descriptions of nuclei and radiation particles. Every element has a certain equal number of protons and electrons, and a variable number of neutrons that determines the isotope of that element. If the element decays by **alpha emission**,

TABLE 2–4
Atomic and ionic radii of some elements. Note the general relationship between gain or loss of electrons and the difference between the atomic and ionic radii.

Element	Atomic Number	Number of Electrons per Shell and Subshell						Ion Formed	Number of Electrons Gained or Lost to Form Ion	Radii (angstroms)	
		1st	2nd	3rd	4th	5th	6th			Ionic	Atomic
Oxygen	8	2	2•4					O^{-2}	2 gained	1.40	0.60
Sodium	11	2	2•6	1				Na^{+}	1 lost	0.97	1.86
Magnesium	12	2	2•6	2				Mg^{+2}	2 lost	0.66	1.60
Aluminum	13	2	2•6	2•1				Al^{+3}	3 lost	0.51	1.43
Silicon	14	2	2•6	2•2				Si^{+4}	4 lost	0.42	1.17
Sulfur	16	2	2•6	2•4				S^{-2}	2 gained	1.85	1.04
Chlorine	17	2	2•6	2•5				Cl^{-}	1 gained	1.81	1.07
Potassium	19	2	2•6	2•6	1			K^{+}	1 lost	1.33	2.31
Calcium	20	2	2•6	2•6•2	2			Ca^{+2}	2 lost	0.99	1.96
Titanium	22	2	2•6		2			Ti^{+3}	3 lost	0.76	1.46
								Ti^{+4}	4 lost	0.68	1.46
Iron	26	2	2•6	2•6•6	2			Fe^{+2}	2 lost	0.74	1.24
								Fe^{+3}	3 lost	0.64	1.24
Silver	47	2	2•6	2•6•10	2•6•10	1		Ag^{+}	1 lost	1.26	1.44
Gold	79	2	2•6	2•6•10	2•6•10	2•6•9	1	Au^{+}	1 lost	1.37	1.44

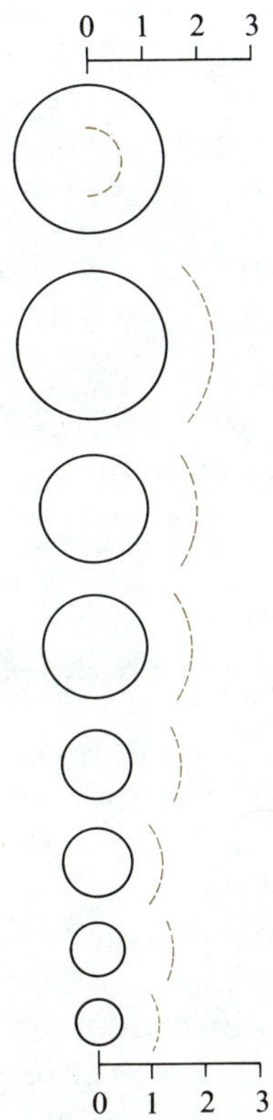

FIGURE 2–7
Comparison of ionic sizes of some common elements. Atomic sizes are indicated by the dashed lines. The scale is in angstroms (1 angstrom is 1.0×10^{-7} mm).

it loses the two protons and two neutrons that make up the alpha particle. Thus, the atomic number is reduced by 2, and the atomic mass is reduced by 4, as shown in the following equation:

$$_{62}Sm^{147} \rightarrow {}_{60}Nd^{143} + {}_{2}\alpha^{4}$$

Alpha particles can be seen on a luminous-dial watch. In a dark room, the flash of light produced each time an alpha particle hits the zinc sulfide on the dial of the watch can be seen with a magnifying lens.

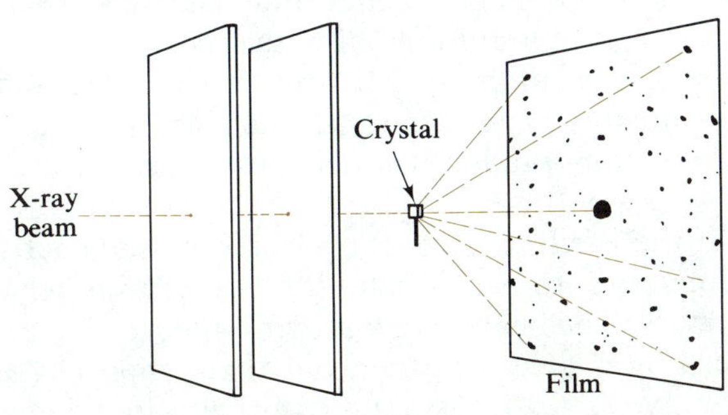

FIGURE 2–8
A beam of X rays is deflected by the atoms in a crystal into a regular pattern, thus revealing the internal structure of the crystal.

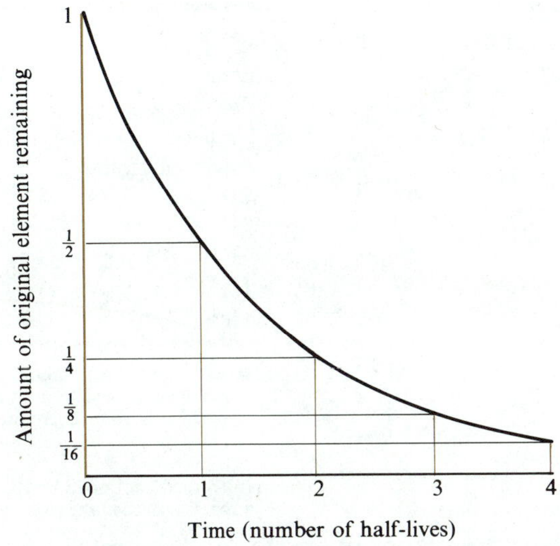

FIGURE 2–9
Rate of radioactive decay. During each half-life, one-half of the remaining amount of the radioactive element decays to its daughter element.

In **beta decay**, an electron is lost. This can be thought of as occurring because a neutron separates into a proton and an electron; the electron is emitted at high velocity, and the proton is retained by the nucleus. Thus, the atomic number is increased by 1, and the atomic mass remains the same. The additional proton in the nucleus is balanced by gaining an electron for the outer shell. Free electrons generally are available and can be captured by the atom. In equation form,

$$_{37}Rb^{87} \rightarrow {}_{38}Sr^{87} + \beta^{-}$$

The third geologically important type of radioactivity occurs by **electron capture**. In this case, the nucleus takes an electron from one of the inner shells. This electron unites with a proton to form a neutron. Thus the atomic number is reduced by 1, and the mass remains the same, as in

$$_{19}K^{40} + e^{-} \rightarrow {}_{18}A^{40}$$

The only changes in rate of radioactive decay occur in decay by electron capture, in which the rate changes slightly as the abundance of nearby electrons changes.

In most radioactivity, gamma rays are also emitted. Emission of a particle may leave the nucleus in an excited state, and emission of gamma rays can allow the atom to settle down. In any case, mass must be conserved in a radioactive event, and this can explain why gamma rays are emitted. Note in Table 2–5 that the mass of a neutron is slightly greater than the combined masses of a proton and an electron. Using beta decay as an example, the question is, Where did the excess mass go when the neutron changed into a proton and an electron? The answer is that the mass was

TABLE 2–5
Charge and mass of some particles and atoms.

	Charge	Mass		Charge	Mass
Electron	−1	0.000549	$_{8}O^{16}$	0	16.000000 (standard)
Proton	+1	1.007593	$_{8}O^{18}$	0	18.004855
Neutron	0	1.008982	$_{82}Pb^{204}$	0	204.0361
Alpha particle	+2	4.002675	$_{82}Pb^{206}$	0	206.0386
$_{1}H^{1}$	0	1.008142	$_{82}Pb^{207}$	0	207.0403
$_{1}H^{2}$	0	2.014735	$_{82}Pb^{208}$	0	208.0414
$_{1}H^{3}$	0	3.016997	$_{90}Th^{232}$	0	232.11034
$_{2}He^{3}$	0	3.016977	$_{92}U^{235}$	0	235.11704
$_{2}He^{4}$	0	4.003873	$_{92}U^{238}$	0	238.12493

NOTE: Masses are in atomic mass units (amu) based on oxygen atom = 16.0000 amu (old unit) [1 amu = 1/16 of the weight of an oxygen atom. A molecular weight of oxygen is 16 g, and there are 6.02×10^{23} atoms (Avogadro's number) in a molecular weight. Therefore,

$$1 \text{ amu} = \frac{1}{16} \times \frac{16 \text{ g}}{6.02 \times 10^{23}} = 1.66 \times 10^{-24} \text{ g}$$

(10^{23} means 1 followed by 23 zeros; 10^{-24} means 1/1 followed by 24 zeros)]. Note that the mass of an electron plus a proton is less than the mass of a neutron.

changed into energy in the form of gamma rays. The equivalence of mass and energy was postulated by Einstein, and experiments have so far shown this to be correct. His famous equation is

$$E = mc^2$$

or, in words, energy equals mass times the speed of light squared. However, in beta decay the energy of the gamma rays is less than the amount of mass lost. This led Enrico Fermi to postulate the emission of a small neutral particle he called the **neutrino**. Many years later, the existence of neutrinos was proved. The neutrino is only one of a few dozen tiny particles found or postulated by atomic scientists. How they are formed, and whether they exist in the nucleus as particles or in the form of energy, are not known. With this in mind, it is clear that the preceding descriptions of how radioactive decay takes place are only convenient fictions to account for the main results.

RADIOACTIVITY IN GEOLOGY

The naturally radioactive isotopes are shown in Table 2–6 and Fig. 2–10. Note that there are three radioactive series that go through many steps before a stable isotope is formed, as well as a number of isotopes that decay to a stable product in a single step. Also, some isotopes, such as potassium 40, decay by two processes to two different daughter products. In addition to the three series shown is the neptunium series; however, the half-life of this series is much shorter than that of the others, so only very small amounts of some of the members of this series are found in nature.

Only a few of the natural radioactive isotopes are of geologic importance. Some of these are useful in age determination, and others are sources of radioactive heating of the earth. Those that have proved most useful in dating are carbon 14, potassium 40, rubidium 87, uranium 235, uranium 238, and thorium 232. The requirements for dating are a reasonable rate of decay (half-life), the retention of daughter isotopes, and the existence of common minerals containing the parent element. Those most important as a source of radioactive heating can be determined from consideration of abundance data and amount of energy released, both of which are shown in the tables. They are uranium 235, uranium 238, thorium 232, and potassium 40.

Calculation of the amount of energy released by radioactivity is simple. One needs to know only the masses of the parent and daughter elements and the masses of the particles released (see Table 2–5). These data and Einstein's famous equation $E = mc^2$ (energy equals loss of mass multiplied by the speed of light squared) can be used to determine the amount of energy released by the decay of one atom of the parent isotope. The rate of decay (half-life) is then used to determine how often an atom of the parent decays. The data for the main heat-producing isotopes are shown in Table 2–7. Notice that because all of these isotopes have been decaying throughout geologic time, more of them were present in the early history of the earth and so produced more heat at that time. Note also that potassium 40 and uranium 235 both have much shorter half-lives than uranium 238 and thorium 232 and so are assumed to have been much more abundant at that time and thus produced much more heat during the early stages of the earth.

In the earth, the energy released by radioactivity goes mainly into heat. Each kind of radiation behaves differently. The heavy, slow-moving alpha particles are stopped by a sheet of paper, but their impact is very destructive. They impart some of their kinetic energy to electrons and, being heavy, may disrupt whole atoms. Beta particles are much lighter and faster. They are about 100 times more penetrating than alpha particles but do about 100 times less damage. Because they are too light to affect nuclei, they lose their kinetic energy by interaction with electrons. Gamma rays move with the velocity of light and have no mass. They are about 100 times more penetrating than beta particles but do much less damage. Like beta rays, gamma rays only disrupt electrons. Heat is molecular

TABLE 2–6
Summary of naturally occurring radioactive isotopes.

A. Single-Step Decay

Isotope	Decay	Half-Life (years, unless noted)	Percentage of Total Amount of Element in Crust
Neutron	$^1n \rightarrow {}^1H^+ + \beta^-$	12 min	—
Tritium ($_1H^3$)	$_1H^3 \rightarrow He^3 + \beta^-$	12.26	0.0001–0.00001
Carbon 14	$_6C^{14} \rightarrow {}_7N^{14} + \beta^-$	5660 ± 30	(in biologic material)
Potassium 40	$_{19}K^{40}$ → (89%) $_{20}Ca^{40} + \beta^-$; → (11%) $_{18}A^{40}$ (electron capture)	1.3×10^9	0.0118
Rubidium 87	$_{37}Rb^{87} \rightarrow {}_{38}Sr^{87} + \beta^-$	4.7×10^{10}	27.85
Indium 115	$_{49}In^{115} \rightarrow {}_{50}Sn^{115} + \beta^-$	6×10^{14}	95.72
Lanthanum 138	$_{57}La^{138}$ → (85%) $_{58}Ce^{138} + \beta^-$; → (15%) $_{56}Ba^{138}$ (electron capture)	1.1×10^{11}	0.089
Samarium 147	$_{62}Sm^{147} \rightarrow {}_{60}Nd^{143} + {}_2\alpha^4$	1.06×10^{11}	15.1
Lutetium 176	$_{71}Lu^{176}$ → (33%) $_{72}Hf^{176} + \beta^-$; → (67%) $_{70}Yb^{176}$ (electron capture)	2.1×10^{10}	2.59
Rhenium 187	$_{75}Re^{187} \rightarrow {}_{76}Os^{187} + \beta^-$	4×10^{10}	62.93

B. Multiple-Step Decay
All isotopes with atomic number greater than 83 are radioactive, and all of these that occur naturally are members of the three radioactive families listed below. Some isotopes with atomic number less than 83 also occur in these families. The complete decay of these families is shown in Fig. 2–10.

Isotope	Decay	Half-Life (years, unless noted)	Percentage of Total Amount of Element in Crust
Uranium 235	$_{92}U^{235} \rightarrow {}_{82}Pb^{207} + 7{}_2\alpha^4 + 4\beta$	7.1×10^8	0.72
Uranium 238	$_{92}U^{238} \rightarrow {}_{82}Pb^{206} + 8{}_2\alpha^4 + 6\beta$	4.5×10^9	99.28
Thorium 232	$_{90}Th^{232} \rightarrow {}_{82}Pb^{208} + 6{}_2\alpha^4 + 4\beta$	1.4×10^{10}	100

NOTE: A few isotopes with very long half-lives and of very small abundances are not included.

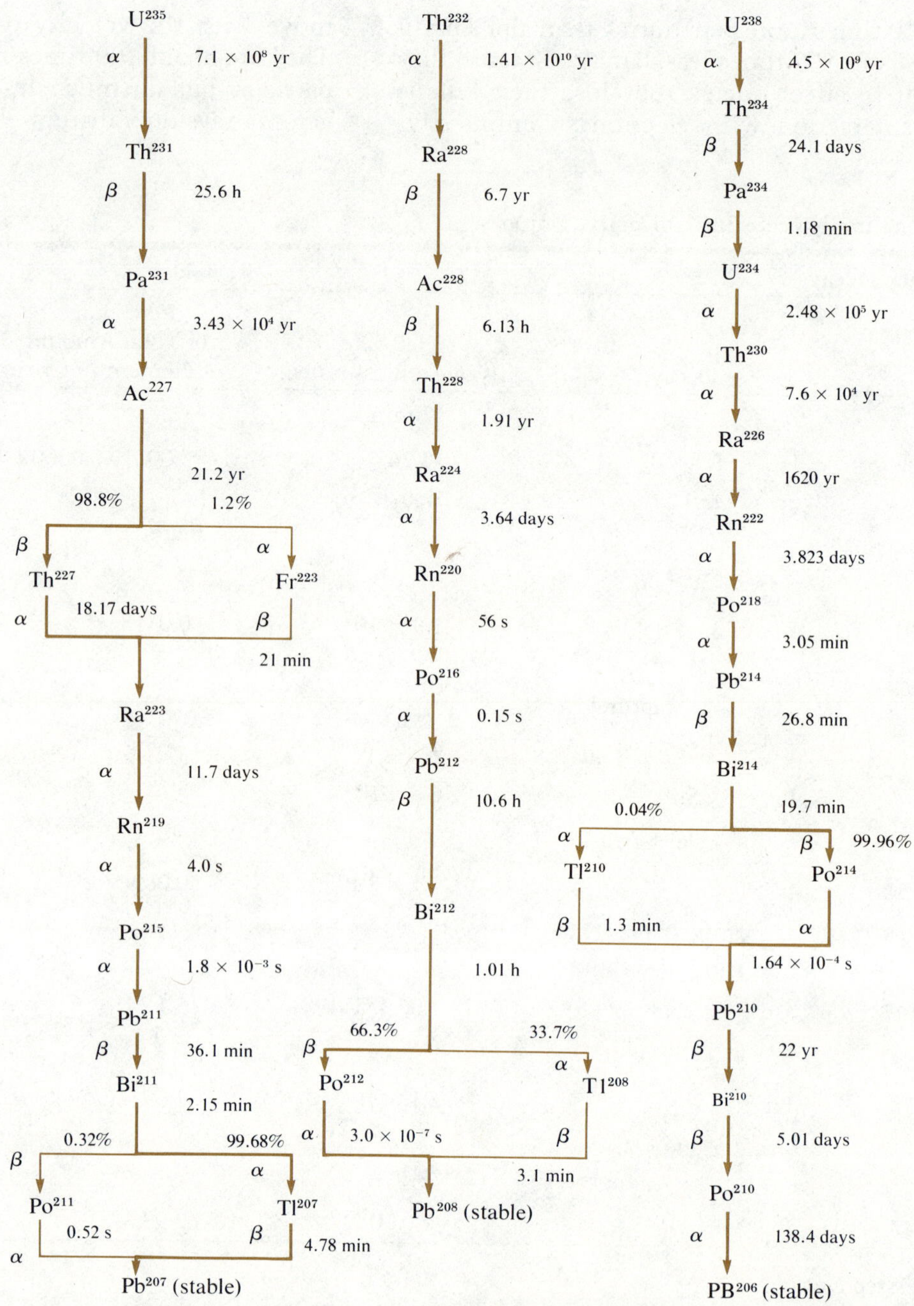

FIGURE 2–10
Decay schemes for uranium 235, uranium 238, and thorium 232. Type of decay, branching ratios, and half-lives are indicated.

motion, so the main effect of radioactivity is to heat the surroundings by imparting motion to electrons and atoms. Other effects are damage to a crystal lattice and increase of potential energy by moving electrons from one shell to another. These less important features are discussed below.

It should also be clear that only gamma rays have high energy and can penetrate very far, so most radiation detectors work mainly on gamma rays. Alpha rays are absorbed by about 10 cm of air, and beta rays by about a meter of air, but gamma rays can travel over a kilometer in air. Less than 2 cm of rock will absorb alpha and beta radiation, but gamma rays can penetrate over 30 cm of rock. These points should be taken into account when prospecting for radioactive mineral deposits. One means of detecting radioactivity is shown in Fig. 2–11.

The crystal structure of minerals containing alpha-emitting elements can be damaged or destroyed by the alpha particles, especially if the mineral has weak bonding. In other cases, alpha radiation produces discoloration, and this can cause gray (smoky) quartz, violet fluorite, and blue or yellow halite. Beta particles can also cause discoloration in some minerals. Another

TABLE 2–7
Radioactive heat production in rocks.

Radioactive Isotope	Granitic Rocks: Present	Granitic Rocks: 4.5 Billion Years Ago	Basalt: Present	Basalt: 4.5 Billion Years Ago
Uranium 238	10.60	22.2	1.34	2.68
Uranium 235	0.46	37.5	0.06	4.74
Thorium 232	11.47	14.4	1.67	2.10
Potassium 40	3.30	40.1	0.73	8.89
Total	25.83	114.2	3.80	18.41

SOURCE: J. W. Winchester, "Terrestrial Heat Flow, Radioactivity, and the Chemical Composition of the Earth's Interior," *Journal of Geological Education* (1966), Vol. 14, No. 5, pp. 200–204.
NOTE: At the time the earth formed, 4.6 billion years ago, the shorter-lived radioactive elements were more abundant and so produced more heat than at present. The units used are 10^{-14} cal/s per gram of rock.

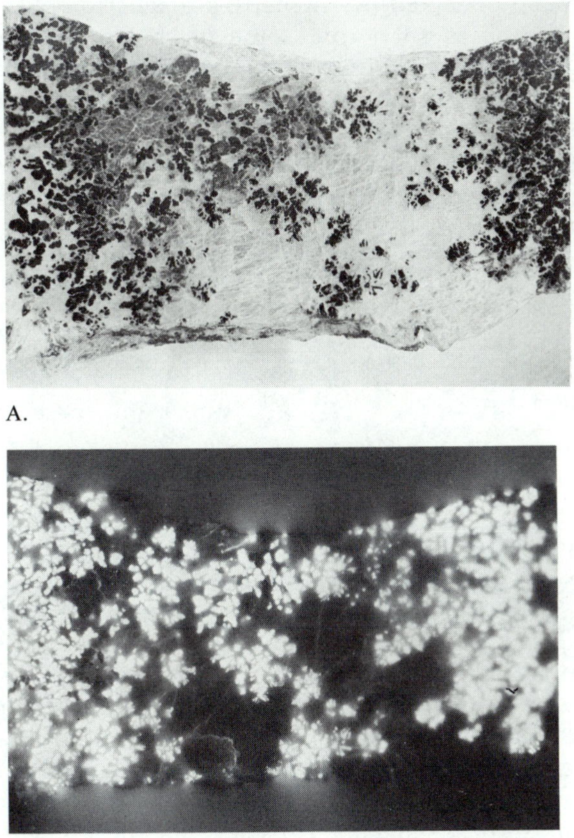

A.

B.

FIGURE 2–11
The mineral uraninite. A. An ordinary photograph. B. An autoradiograph made by placing photographic film next to the mineral in a darkroom. The radiation from the uraninite exposes the film even though no visible light reaches the film. This can be done without a darkroom by wrapping the film in opaque paper and laying the flat mineral on the film. Exposure times are generally a few weeks. (Photo from Ward's Natural Science Establishment, Inc., Rochester, N. Y.)

common example of alpha damage is the **halos** or rings that commonly surround tiny crystals of radioactive minerals. These small crystals generally occur as inclusions in later-formed minerals, and dark halos are common in biotite. Uranium and thorium minerals emit alpha particles with various amounts of energy, and because the alpha particles do most of their damage near the end of their movement, traveling a distance determined by the energy of each particle, they form a series of concentric halos (Fig. 2–12). The amount of damage done to a mineral (the thickness and darkness of the halo) can be used to calculate the age of the mineral, but so far such dating methods have met with only limited success. Uranium 238 also decays by spontaneous fission at a very much slower rate than it does by alpha decay. The high-energy fission particles released in this way do much damage to the enclosing crystal. Under high magnification, the paths of such particles can be seen in some minerals. The path is a small tube, not unlike a bullet hole (Fig. 2–13). By counting the tubes, the number of fission disintegrations that have occurred can be determined. The number of remaining uranium atoms, the source of the particles, can be determined by instrumental analysis, and thus the age of the mineral can be calculated.

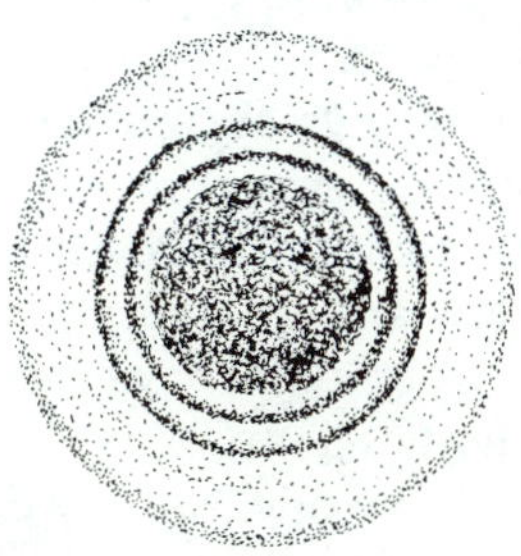

FIGURE 2–12
Radiation halo. Such halos are commonly seen in biotite crystals where they surround tiny crystals of radioactive minerals such as uraninite. Typical size is less than 0.1 mm.

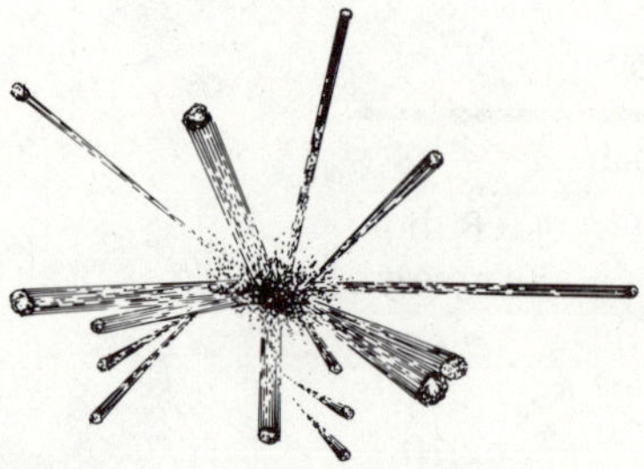

FIGURE 2–13
Fossil fission tracks. The tubes are caused by high-energy radioactive particles escaping from the mineral at the center. The tubes are much like bullet holes. The longer tubes are about 0.01 mm in length.

Radiation can increase the energy of an atom by moving its electrons into different orbital shells and by trapping within the atom other electrons from beta particles or from other atoms. Remember that each electron belongs in a certain shell. This energy can be released by freeing the electrons so that they can escape or move back to the original shell. One way to do this is to excite the atom by heating it. The energy is released in the form of light. This phenomenon can be seen easily by heating a mineral such as fluorite in a frying pan in a dark room. The complete explanation of this phenomenon would require much deeper knowledge of atomic structure. Because the amount of energy stored depends on the length of time a mineral has been exposed to radiation, the age of some minerals can be determined by measuring the amount of stored energy. So far, this has not proved to be an accurate method of dating minerals.

NEUTRONS—COSMIC RAYS

To complete our understanding of natural radiation, it is necessary to consider another particle. That particle is the neutron, and it is particularly interesting because of its central role in atomic and nuclear bombs. The main source of natural neutrons is **cosmic rays**, and only a few of the possible types of neutron reactions will be considered here.

Primary cosmic rays are nuclei (atoms that have lost their electrons) moving at nearly the speed of light. Although they are mainly hydrogen and helium, most other elements are also present in small quantities. That they come from outer space is shown by their increase with height. The number does not change appreciably from night to day, so they do not come from the sun. The atoms are believed to come from exploding stars (supernova), and are accelerated to very high energy levels by the associated magnetic fields.

The primary cosmic rays collide with the atoms of gas that form the earth's atmosphere, producing secondary cosmic rays. These secondary cosmic rays are particles knocked out of the atoms; they consist of neutrons, protons, and other particles. Neutrons outside a nucleus are not stable, but decay, with a half-life of 12.8 min, to a proton and a beta particle. Each primary cosmic ray can produce several secondary cosmic rays, each of the secondaries can produce more cosmic rays, and so on. Each successive generation has less energy; however, primary cosmic rays and certain of the early-formed particles have great penetrating power. Neutrons have great penetrating power because they have no charge, and thus are not deflected by protons or electrons.

Some of the less energetic nuclei of the primary cosmic rays form the **Van Allen radiation belts** around the earth. Because these nuclei have electric charges, they are deflected by the earth's magnetic field. The interaction is such that some of the less energetic charged particles move in a spiral course around lines of force of the earth's magnetic field. Following such a path, these particles move from pole to pole and back (Fig. 2–14). The more energetic particles are able to penetrate the magnetic field; these are the primary cosmic rays discussed previously.

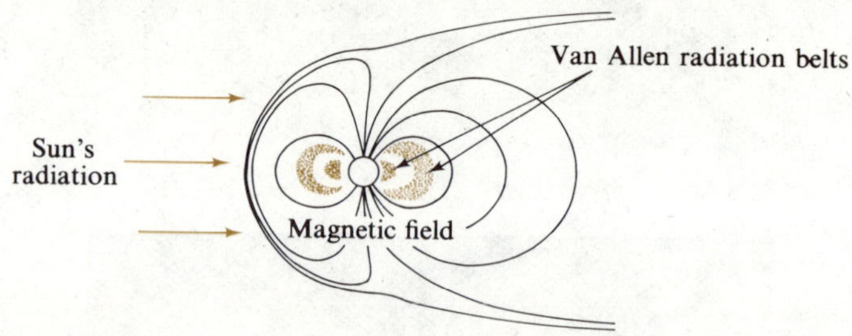

FIGURE 2–14
The earth's magnetic field and radiation belts.

In addition to cosmic rays, there are also particles from the sun that affect the earth. These particles are electrons and protons, and though they are constantly expelled from the sun, at times many more are expelled. These periods of intense solar activity correspond with intense sunspot activity and solar flares. Because these charged particles disrupt the earth's magnetic field, these events are called **magnetic storms**. Such changes, of course, affect the earth's radiation belts, as well as radio communications. These particles also follow the lines of force of the earth's magnetic field and so fall on the earth near the magnetic poles. It is the interaction between these particles and the atoms of oxygen and nitrogen in the upper atmosphere that causes the aurora, or northern and southern lights. The processes may be likened to the effect of an electrical current on neon gas that produces the red light in a neon sign.

Of the cosmic rays, only the neutron is important geologically. A neutron colliding with a nitrogen atom knocks a proton out, converting the nitrogen to carbon 14:

$$_7N^{14} + n^1 \rightarrow {}_6C^{14} + {}_1H^1$$

Carbon 14 is radioactive and decays by beta emission to nitrogen 14:

$$_6C^{14} \rightarrow {}_7N^{14} + \beta^-$$

These reactions are utilized in the carbon-14 method of dating.

Cosmic rays also make possible the dating of meteorites. When cosmic rays hit iron atoms, helium 3 is produced. Thus, the amount of helium 3 reveals how long the iron meteorite has undergone cosmic-ray bombardment. This effect was first noticed when attempts were made to date meteorites by measuring the amount of helium produced by alpha decay. More helium than expected was found, and analysis of the helium revealed both helium 4 from alpha activity and helium 3 from cosmic-ray activity. It was also discovered that the helium 3 concentration is greatest near the surface of a meteorite and becomes less dense in the interior, as would be expected from cosmic-ray activity. Similar effects were noted with other isotopes formed by cosmic-ray bombardment. Thus, ordinary radioactive dating can tell the time of crystallization of a meteorite, and cosmic-ray dating tells when the body broke up to form a meteorite.

The principle involved in these cosmic-ray transformations is used commonly by nuclear physicists. By bombarding atoms with high-speed particles, alpha particles and electrons, as well as neutrons, they have succeeded in making many isotopes. Radioactive isotopes of almost every element have been made artificially.

The energy used in geologic processes ultimately comes from the atom. Neutrons are what causes the splitting, or **fission**, of atoms. The large, unstable uranium 235 atom releases neutrons when it undergoes fission, making a continuous or chain reaction possible. Natural uranium consists of only 0.7 percent uranium 235. The 99.3 percent uranium 238 absorbs neutrons, preventing natural chain reactions. Thus, to produce an atomic bomb, uranium 235 must be concentrated in a large enough mass (the critical mass) that the neutrons cannot escape.

The opposite of fission is **fusion**, the making of a more massive atom from lighter atoms. This is the process that releases energy in the sun and other stars. An example is

$$_1H^1 + {}_1H^1 \rightarrow {}_1H^2$$
$$_1H^2 + {}_1H^1 \rightarrow {}_2He^3$$
$$_2He^3 + {}_2He^3 \rightarrow {}_2He^4 + 2\ {}_1H^1$$

or, in summary, the net reaction is

$$4\ {}_1H^1 \rightarrow {}_2He^4 + 2e^-$$

These reactions take place at very high temperatures of the order of several billion degrees. Such temperatures occur in uranium atomic bombs. The hydrogen bomb, which is a fusion device, is made by adding deuterium ($_1H^2$) and tritium ($_1H^3$) to a uranium bomb, producing what is called a thermonuclear device.

IDENTIFICATION OF MINERALS—PHYSICAL PROPERTIES

The definition of mineral suggests that minerals can be identified in several ways. The fact that minerals have reasonably fixed chemical compositions suggests that ordinary methods of chemical analysis can be used, and, historically, this approach was important in the development of both mineralogy and chemistry. The drawbacks to its current use are that a laboratory, as well as much skill and knowledge, is needed, and the relatively recent discovery that certain minerals having identical chemical compositions have very different internal (crystal) structures. In this latter case, the **polymorphs** ("many forms") generally have very different characteristics, such as graphite and diamond, which are both composed of pure carbon. Proof that certain minerals are indeed polymorphs had to wait until the use of X rays to probe crystal structure was perfected. Another problem in the use of chemical analysis is that analysis of the all-important silicate minerals is an extremely difficult task, even in the best-equipped laboratories. Fortunately, this problem can be bypassed by using the easily determined physical properties to identify minerals, and this can be supplemented where necessary with spot chemical tests.

The **physical properties** include such things as color, luster, hardness, weight (specific gravity), crystal form, fluorescence, taste, solubility, cleavage, magnetism, radioactivity, and many more. The use of the more important physical properties follows.

HARDNESS

The **hardness** of a mineral is its resistance to scratching. As might be guessed, it is a difficult property to measure exactly because the amount of force, the shape of the scratches, and the relation of the surface tested to the crystal structure will all affect the measured hardness. In spite of this, the gross hardness is easily measured and is of great help in identification. Because surface alteration may change the apparent hardness, fresh surfaces only must be tested.

The hardness scale (Mohs) is based on 10 minerals. The steps are not of equal value but are arbitrarily defined. They are reasonably equally spaced, except that the step between corundum and diamond is very large:

1. Talc (softest mineral)
2. Gypsum
3. Calcite
4. Fluorite
5. Apatite
6. Orthoclase
7. Quartz
8. Topaz
9. Corundum
10. Diamond (hardest mineral)

A useful hardness scale is:

2.
—Fingernail
3.
—Copper cent (A bright, shiny one should be used, or only the hardness of the tarnish will be tested.)
4.
5.
—Knife blade or glass plate
6. —File
7. —Quartz
8.

To determine the hardness of a mineral, you must find the softest mineral on the hardness scale that will scratch it. With a little practice, only a knife blade and your fingernail will be necessary to estimate the hardness. Note that two substances of the same hardness can scratch each other. A pitfall for the neophyte is accepting any mark on the mineral as a scratch, because just as soft chalk leaves a mark on a hard blackboard, a soft mineral may leave a mark on a hard one. You can check this by trying to remove the mark with your moistened finger, by feeling the mark with your fingernail (even a tiny depression can be detected), and by looking very closely at the mark with a lens. The final test should always be to reverse the test and see whether the mineral can scratch the test mineral.

FRACTURE AND CLEAVAGE

Fracture and cleavage describe the way a mineral breaks. Any irregular break is termed a **fracture**. Many terms have been coined to describe fracture, but most, such as *even, fibrous, splintery* or *hackly,* are self-explanatory. A common type of fracture, called **conchoidal**, is the hollow, rounded type of break that occurs in glass (see Fig. 3–32).

A mineral has a **cleavage** if it has a direction of weakness that, when broken, produces a smooth plane that reflects light (Figs. 2–15 and 2–16). The common minerals may have up to six such directions. Note that a cleavage is a direction; therefore, one cleavage direction will, in general, produce two cleavage surfaces. Cleavages can be recognized by their smoothness and by their tendency to form pairs or steps, as suggested in the sketches. A series of steps is more common in minerals with poor cleavage, that is, minerals in which the cleavage direction is only slightly weaker than any other direction in the mineral. Even in this case, each of the cleavage surfaces will be smooth and parallel and so will reflect light as a unit.

The angle between cleavage directions can be important in the identification of some minerals, as noted in Appendix A.

The planes of weakness that are the cleavage directions in a mineral are caused by the atomic structure of the mineral. Cleavages form along directions of weak bonding as shown in Fig. 2–17.

CRYSTAL FORM

All crystalline substances crystallize in one of six crystal systems (Figs. 2–18 through 2–20). If the mineral

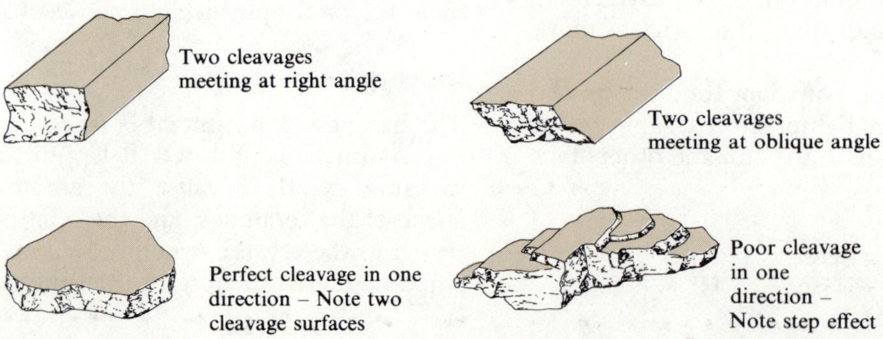

FIGURE 2–15
Cleavage.

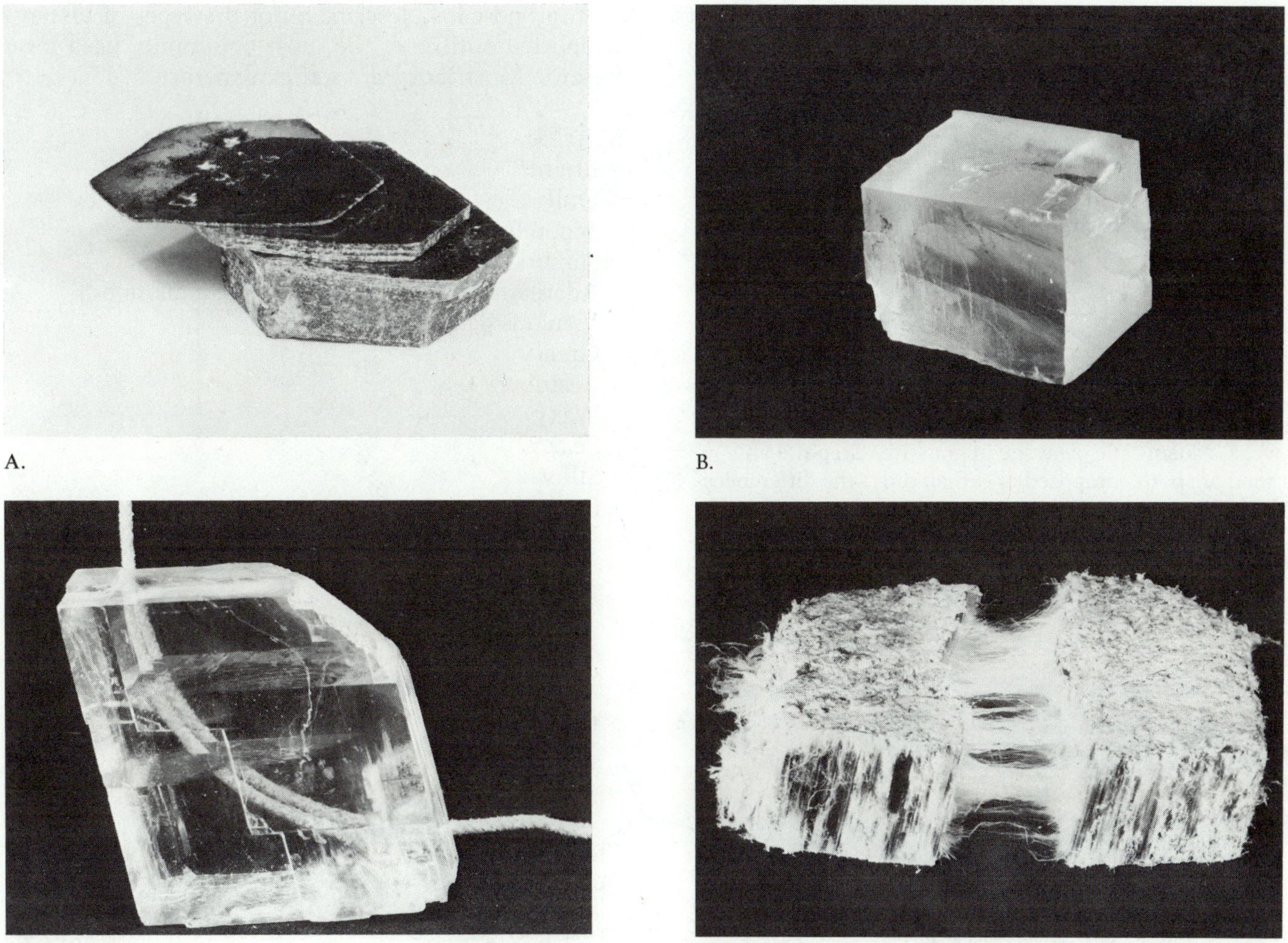

A. B. C. D.

FIGURE 2–16
Minerals showing cleavage. A. Mica crystal with one cleavage. B. Galena, cubic cleavage. C. Calcite. Note the double refraction of the string behind the crystal. D. Serpentine asbestos with fibrous cleavage. (Photos from Ward's Natural Science Establishment, Inc., Rochester, N.Y.)

grows in unrestricted space, it develops the external shape of its crystal form; if it cannot grow its external shape, its crystalline nature can be determined only under the microscope or by X-ray analysis. Many external forms are possible in each of the systems; however, the system can be determined by the symmetry of the crystal. Crystallography is a fascinating subject, and more information can be found in the mineralogy references listed at the end of this chapter.

COLOR

Strictly speaking, **color** means the color of a fresh, unaltered surface, although in some cases the tarnished or weathered color may help in identification. For some minerals, color is diagnostic, but many, such as quartz, may have almost any color, due to slight inpurities.

STREAK

Streak is the color of the powdered mineral. To see the streak, the mineral is rubbed on a piece of unglazed porcelain called a streak plate. The color of the streak in many minerals is a more constant property than the color. The streak may be a very different color from the mineral color, and this, too, is of help in indentification. Care should be taken, especially when working with small disseminated crystals, that only the mineral, and not the matrix, is rubbed on the streak plate. Many powdered minerals are white and are said to have no streak. Minerals with a hardness greater than the streak plate will, of course, have no streak.

LUSTER

Luster is the way a mineral reflects light. The two main types of luster are *metallic* and *nonmetallic.* The distinction is difficult to describe in words; either it looks like metal, or it does not. A few fall in between and can be called *submetallic*. Dozens of terms have been proposed to describe all of the types of luster. A simplified outline of the most commonly used terms follows. Most terms are self-explanatory.

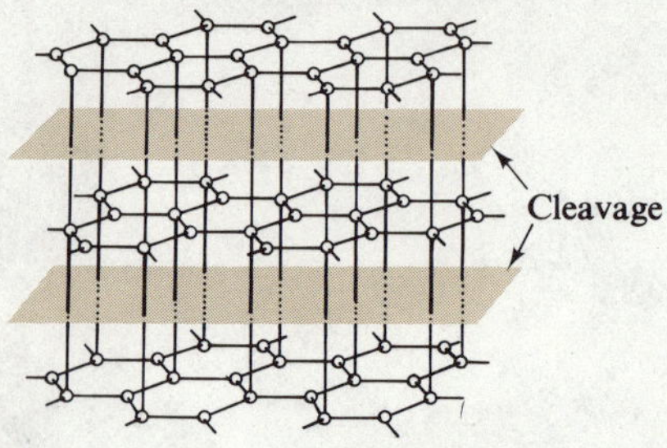

FIGURE 2–17
Structure of graphite. The bonds between the carbon sheets are weak, causing the cleavage in graphite. Graphite and diamond are both composed of carbon only; the differences between them are a function of the internal structure.

- Metallic
  - Bright
  - Dull
- Submetallic (best to avoid if possible)
- Nonmetallic
  - Adamantine (brilliant luster, like a diamond)
  - Vitreous (glassy)
  - Greasy
  - Resinous
  - Waxy
  - Pearly
  - Silky
  - Dull or earthy

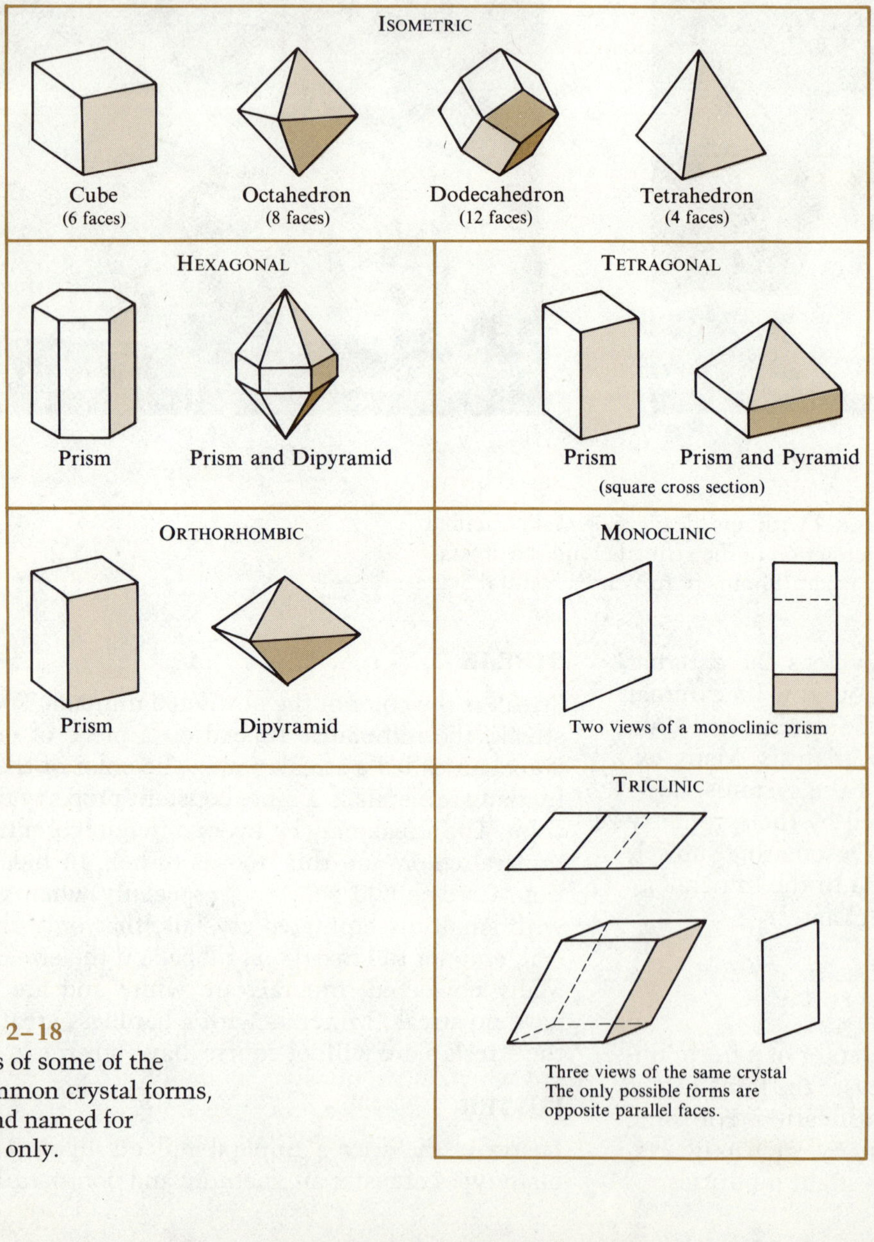

FIGURE 2–18
Examples of some of the more common crystal forms, shown and named for reference only.

FIGURE 2–19
Mineral specimens showing crystal forms. A. Quartz. B. Galena. C. Orthoclase (potassium feldspar). D. Pyrite. (Parts A and B from Ward's Natural Science Establishment, Inc., Rochester, N.Y.; parts C and D from Smithsonian Institution)

SPECIFIC GRAVITY

Specific gravity is a measure of the relative weight of a substance. It is the ratio of the mass of a substance to the mass of an equal volume of water and is measured by weighing the substance in air and in water:

$$\text{Specific gravity} = \frac{\text{weight in air}}{\text{weight in air} - \text{weight in water}}$$

This is a difficult measurement to make with most specimens. Estimations of specific gravity can be made fairly accurately by one who is experienced in handling minerals. It is of help in identification, especially with the very heavy minerals.

The **density** of a substance is its mass per unit volume; when expressed in grams per cubic centimeter, density is numerically equal to specific gravity.

TASTE AND SOLUBILITY

Taste and solubility can be determined by touching the specimen with the tongue. A soluble mineral will have the feel of a lump of sugar, and an insoluble mineral may feel like glass.

Reaction with dilute hydrochloric acid is another solubility test. Calcite will react, producing effervescence. Dolomite (see Tables A–1 and A–2 in Appendix A) must be powdered to increase the surface area before effervescence occurs. Dolomite will also react with hot dilute acid or with strong acid, both of which also increase the rate of chemical reaction.

Other physical properties may require special equipment to test. Examples are radioactivity (Geiger counter), magnetism (magnet), fluorescence (ultraviolet light).

ROCK-FORMING MINERALS

The main rock-forming minerals are **silicates**, minerals in which elements such as iron, magnesium, sodium, calcium, and potassium are combined with aluminum, silicon, and oxygen. Their identification will be men-

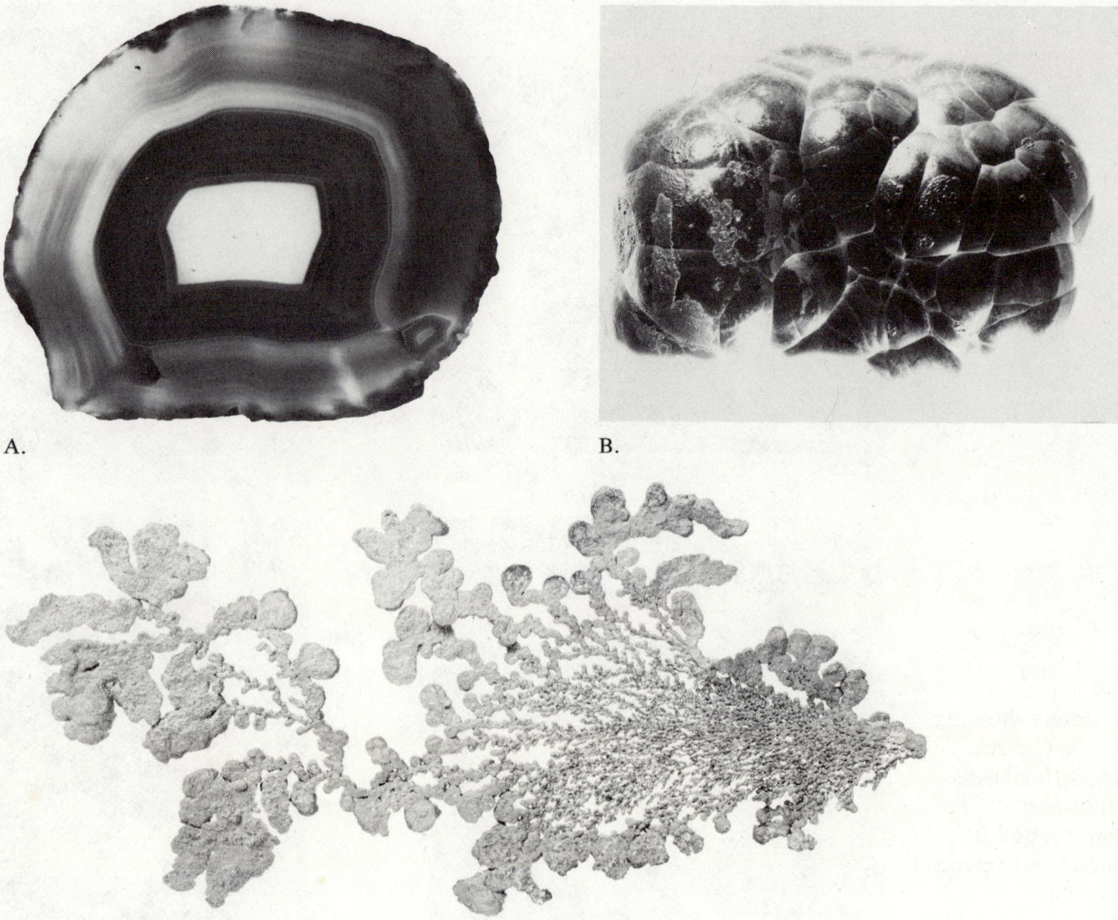

A. B. C.

FIGURE 2–20
Minerals whose crystalline nature is not readily apparent. In most of the specimens shown, the individual crystals are too small to be seen with the naked eye. A. Agate. B. Hematite. C. Native copper. (Part A from Ward's Natural Science Establishment, Inc., Rochester, N.Y.; parts B and C from Smithsonian Institution)

tioned here and summarized in Appendix A. Their structures are shown in Fig. 2–3. A common classification of minerals is given in Table 2–8.

FELDSPARS

Feldspars are the most abundant minerals (see Fig. 2–21). They are composed of sodium, calcium, and potassium, combined with silicon, aluminum, and oxygen. There are two main types, and they are complicated by mixing.

Plagioclase feldspar is an example of a continuous mineral series. The high-temperature plagioclase is *calcic,* $CaAl_2Si_2O_8$; at low temperature, *sodic plagioclase,* $NaAlSi_3O_8$, forms. A continuous series of intermediate plagioclase containing both sodium and calcium can form, depending on the temperature and composition of the melt. This is possible because sodium and calcium ions are about the same size. Aluminum and silicon, which are also nearly the same size, also mutually substitute. The microscope must be used to determine the relative amounts of sodium and calcium in plagioclase.

Potassium feldspar, $KAlSi_3O_8$, is more complicated. A number of varieties have been recognized, but their identification requires microscopic or X-ray techniques. The variety that generally forms in granitic rocks is commonly, although generally incorrectly, called *orthoclase;* and the variety in volcanic rocks, which crystallize at higher temperatures than granitic rocks, is called *sanidine.*

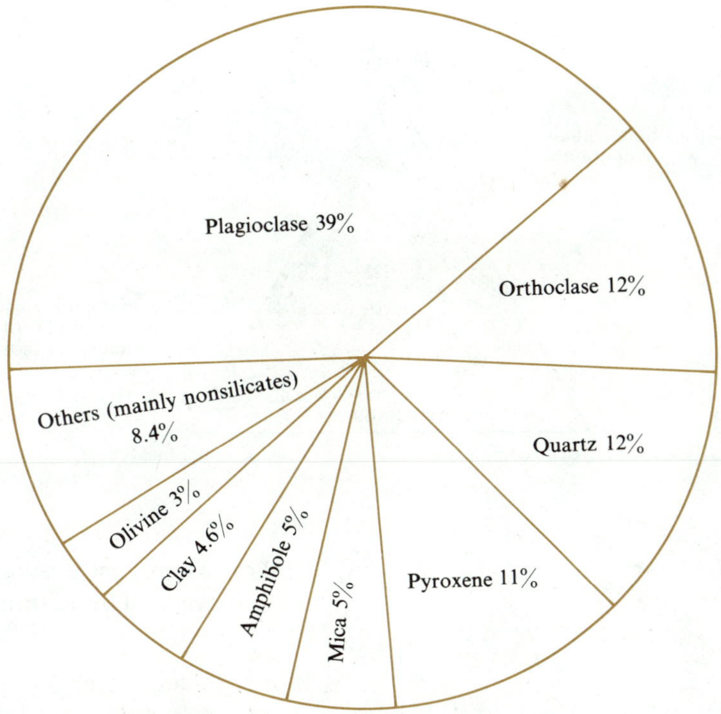

FIGURE 2–21
Relative abundances of minerals in the earth's crust. (Data from Ronov and Yaroshevsky, 1969)

TABLE 2–8
Common minerals and their classification.

Native elements:
- gold; graphite, C

Sulfides (and sulphosalts):
- pyrite, FeS_2; galena, PbS

Oxides and hydroxides:
- hematite, Fe_2O_3; magnetite, Fe_3O_4; limonite, $FeO(OH) \cdot n\,H_2O$

Halides:
- halite, NaCl; fluorite, CaF_2

Carbonates, nitrates, borates, and iodates:
- calcite, $CaCO_3$; dolomite, $CaMg(CO_3)_2$

Sulfates, chromates, molybdates, tungstates:
- gypsum, $CaSO_4 \cdot 2\,H_2O$; anhydrite, $CaSO_4$

Phosphates, arsenates, and vanadates:
- apatite, $Ca_5(F,Cl,OH)\,(PO_4)_3$

Silicates

Limited amounts of sodium can be accommodated in potassium feldspar in spite of the differences in ionic sizes. This limited mixing is further complicated in that the temperature of formation further controls the amount of mixing (Fig. 2–22).

At high temperatures, such as in volcanic rocks, all mixtures of sodium and potassium are possible and can be preserved by rapid chilling. Slow cooling will allow the potassium feldspar and the sodic plagioclase to unmix. In some rocks, the unmixing is complete and two distinct crystals form; but in many, the two feldspar crystals are intergrown and form *perthite.*

Feldspars are identified by their light color, hardness (6), and two cleavages at nearly right angles. Striations on one cleavage direction will distinguish plagioclase (Fig. 2–23). Perthite will appear as mottled, irregular veining of different color (Fig. 2–24).

OLIVINE

Olivine, $(Mg,Fe)_2SiO_4$, is the simplest of the ferromagnesian or dark-colored minerals. The parentheses in the formula mean that magnesium and iron may be present in varying amounts because they are nearly the same size. Magnesium olivine forms at higher temperatures than the iron-rich types. Olivine generally forms small, rounded crystals that, because of alteration, may give deceptively low hardness tests. It can be recognized by its distinctive olive green color.

PYROXENE

Pyroxene is the family name of a very large and complex group of minerals. Here, only *augite,* $(Ca,Na)(Mg,Fe^{II},Fe^{III},Al)(Si,Al)_2O_6$, the main rock-forming mineral of the family, will be considered. The possible range in composition is suggested by its complex formula. Augite is generally recognized by its dark color, hardness (5–6), and two cleavages that meet at nearly right angles.

FIGURE 2–22
Triangular composition diagram showing the possible mixtures of the three feldspars and their temperature stability. This diagram is based on experimental data. Compositions on the left side are common in the granitic rocks discussed in the next chapter. A feldspar composed of 50 percent sodic plagioclase and 50 percent potassium feldspar would fall on point A on the diagram.

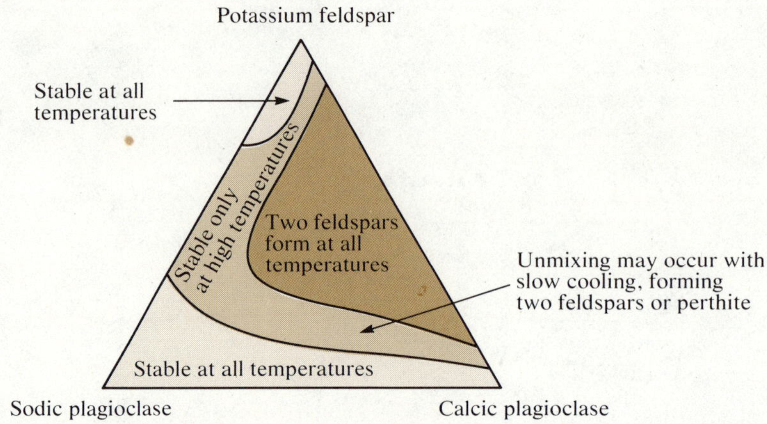

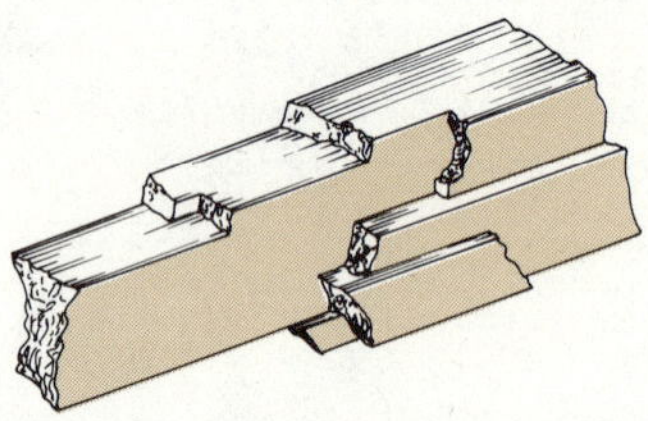

FIGURE 2–23
Plagioclase striations appear on only one of the two cleavage directions.

AMPHIBOLE

Amphibole is a family name of another very large and complex group of minerals. Here, only *hornblende,* $Ca_2Na(Mg,Fe^{II})_4(Al,Fe^{III},Ti)(Al,Si)_8O_{22}(O,OH)_2$, the main rock-former in the group, will be discussed. Its composition is even more complex than that of augite. It is black, has a hardness of 5–6, and has two cleavages that meet at oblique angles, distinguishing it from augite.

MICA

The micas are easily recognized by their colors and perfect cleavage. The common types and their formulas are:

Biotite (black mica):

$K(Mg,Fe)_3(AlSi_3O_{10})(OH)_2$

Muscovite (white mica):

$KAl_2(AlSi_3O_{10})(OH)_2$

QUARTZ

Quartz, SiO_2, is generally clear or white but may have any color. It is recognized by its hardness (7) and lack of cleavage.

As in many other minerals, the color of quartz can be caused by very minor amounts of impurities. How this can occur is shown in the diagrams of crystal defects in Fig. 2–25. In other cases, color is caused by radiation as described earlier.

FIGURE 2–24
Cleavage surface of perthite, showing intergrowth of two types of feldspar. (Photo from Smithsonian Institution)

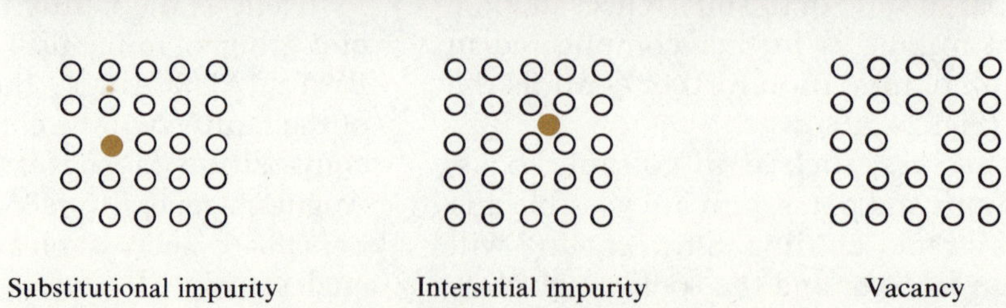

FIGURE 2–25
Some types of crystal defects shown diagrammatically.

SUMMARY

The crust of the earth is composed almost entirely of eight elements: oxygen, silicon, aluminum, iron, calcium, sodium, magnesium, and potassium.

Elements are materials that cannot be subdivided by ordinary chemical methods. A little over 100 are known. Atoms are the smallest units of an element. They are composed of electrons, neutrons, and protons.

Minerals are

1. naturally occurring, crystalline,
2. inorganic substances with
3. a definite small range in chemical and physical properties.

Crystalline substances have orderly internal structures.

A crystal is a solid form bounded by smooth planes which give an outward manifestation of the orderly internal structure.

A rock is

1. a natural aggregate of one or more minerals, or
2. any essential and appreciable part of the solid portion of the earth (or any other part of the solar system).

There are about 2000 minerals; only about 20 are common, and fewer than 10 form over 90 percent of all rocks.

Ions are atoms that have gained or lost electrons and so have an electrical charge.

Silicon and oxygen, the two most abundant elements, form the SiO_4 tetrahedron, which is the basic building block of the rock-forming minerals. The shape of the SiO_4 tetrahedron is determined by the relative sizes of oxygen and silicon ions, and consists of one small silicon ion surrounded by four large oxygen ions.

SiO_4 tetrahedra are linked by sharing oxygen ions, and the resulting arrangements determine the structures of silicate minerals. The small number of rock-forming minerals is determined by the limited number of ways that SiO_4 tetrahedra can join together and how the other abundant ions can fit into these structures.

The nucleus of an atom contains most of the mass of the atom and is composed of protons and neutrons. Protons are heavy and have a positive electrical charge. Neutrons have slightly more mass than protons and are electrically neutral. The nucleus is surrounded by a cloud of light, electrically negative electrons.

An atom is mostly empty space.

The number of electrons equals the number of protons, so an atom is electrically neutral. The number of protons determines the identity of an element and is called the atomic number of an element. The number of protons and neutrons is the atomic mass number. The number of neutrons in the atom of an element may vary, giving rise to isotopes.

Ionic bonding is due to the attraction of unlike electrically charged ions. Covalent bonding results when atoms complete their outer electron shells by sharing electrons. In metallic bonding, the atoms are closely packed and the electrons are free to move throughout the metal.

Ions of similar size can substitute for each other in certain minerals.

In radioactive decay, an atom changes into an atom of another element. In this process, alpha, beta, and gamma radiation may be emitted. Alpha particles are helium nuclei (ions) and so consist of two protons and two neutrons. Beta particles are high-velocity electrons. Gamma rays are electromagnetic waves somewhat higher in frequency than X rays.

Half-life is the time required for one-half of the remaining radioactive atoms to decay. This rate is unaffected by temperature, pressure, and so on, and therefore radioactivity can be used to determine age.

In alpha emission, an atom loses the two protons and two neutrons that make up an alpha particle. In beta decay, an electron is lost. In electron capture, an electron from an inner shell unites with a proton to form a neutron.

Mass and energy are related by the equation

$$E = mc^2$$

The main radioactive elements used in geologic dating are carbon 14, potassium 40, rubidium 87, uranium 235, uranium 238, and thorium 232. The main radioactive sources of energy in the earth are uranium 235, uranium 238, thorium 232, and potassium 40.

Primary cosmic rays are nuclei (atoms that have lost their electrons) moving at nearly the speed of light. The earth's magnetic field captures some of the less energetic cosmic rays and in this way forms the Van Allen radiation belts.

Carbon 14 is formed when a neutron collides with a nitrogen atom and knocks a proton out.

Minerals are best identified by their physical properties, such as cleavage, fracture, crystal form, hardness, color, streak, luster, specific gravity, taste, solubility, magnetism, and fluorescence. Cleavage is a smooth break caused by planes of weakness in a mineral. Conchoidal fracture is the hollow, rounded type of break that occurs in glass. Hardness is the resistance of a mineral to scratching. Streak is the color of the powdered mineral. Luster is the way a mineral reflects light. The main types are metallic and nonmetallic. Specific gravity is the ratio of the mass of a substance to the mass of an equal volume of water.

The main rock-forming minerals are:

Feldspars
- plagioclase
  - calcic plagioclase, $CaAl_2Si_2O_8$
  - sodic plagioclase, $NaAlSi_3O_8$
- potassium feldspar (orthoclase, sanidine), $KAlSi_3O_8$

All mixtures of calcic and sodic plagioclase are possible. Limited mixing of potassium feldspar with plagioclase is possible at high temperatures, but only a few of these mixtures are stable at ordinary temperatures. The unmixed feldspar produced by slow cooling of these mixes is called perthite.

Olivine, $(Mg, Fe)_2SiO_4$

Pyroxene
- augite, $(Ca,Na)(Mg,Fe^{II},Fe^{III},Al)(Si,Al)_2O_6$

Amphibole
- hornblende, $Ca_2Na(Mg,Fe^{II})_4(Al,Fe^{III},Ti)(Al,Si)_8O_{22}(O,OH)_2$

Mica
- biotite (black mica), $K(Mg,Fe)_3(AlSi_3O_{10})(OH)_2$
- muscovite (white mica), $KAl_2(Al,Si_3O_{10})(OH)_2$

Quartz, SiO_2

REVIEW EXERCISES

1 What are the four most abundant elements (weight percent), in order?

2 Define *mineral.*

3 Distinguish between *crystal* and *crystalline.*

4 Why are there only eight important rock-forming minerals?

5 Can you relate the perfect cleavage in mica to its internal structure?

6 How are minerals distinguished from rocks?

7 Distinguish between *silicon* and *silicate.*

8 Why are the physical properties of graphite and diamond so different?

9 What factors determine whether one element can substitute for another in a mineral?

10 Why are silicon-oxygen tetrahedra the building blocks of most minerals?

11 Describe the units of which atoms are made.

12 What are ions, and what is ionic bonding?

13 What is an isotope?

14 Which are the geologically more important radioactive elements?

15 Define *half-life.* Why was radioactive heating more important in the early history of the earth than it is now?

16 What happens to an atom when it decays by alpha emission?

17 What are cosmic rays, and what is their role in the formation of carbon 14?

SUGGESTED READINGS

BERRY, L. G., and MASON, B. *Mineralogy.* San Francisco: W. H. Freeman and Co., 1959, 612 pp.

DIETRICH, R. V., and SKINNER, B. J. *Rocks and Rock Minerals.* Somerset, N.J.: John Wiley and Sons, 1979, 319 pp.

ERNST, W. G. *Earth Materials.* Englewood Cliffs, N.J.: Prentice-Hall, 1969, 150 pp. (paperback).

HURLBUT, C. S., JR., and KEINI, C. *Dana's Manual of Mineralogy,* 19th ed. New York: John Wiley and Sons, 1977, 532 pp.

LOEFFLER, B. M., and BURNS, R. G. "Shedding Light on the Color of Gems and Minerals." *American Scientist* (1976), Vol. 64, No. 6, pp. 636–47.

POUGH, F. H. *A Field Guide to Rocks and Minerals.* Boston: Houghton Mifflin Co., 1955, 349 pp.

RAPP, G., JR. *Color of Minerals.* ESCP Pamphlet Series, PS-6. Boston: Houghton Mifflin Co., 1971, 30 pp.

- uniformitarianism
- magma
- pumice
- scoria
- ash grain
- lapilli
- block
- bomb
- crater
- caldera
- welded tuff
- obsidian
- lava plug (dome)
- pillow lava
- mudflow
- Bowen's reaction series

3

- tephra
- pyroclastic rock
- tuff
- pyroclastic breccia
- agglomerate
- epiclastic
- basalt
- felsite
- andesite
- dike
- shield volcano
- blocky lava (aa)
- ropy lava (pahoehoe)
- cinder cone
- composite volcano
- differentiation
- continuous series
- discontinuous series
- intrusive
- extrusive
- concordant
- discordant
- batholith
- stock
- plutonic igneous rock
- sill
- laccolith
- porphyritic
- stoping
- pegmatite

Igneous Rocks

THE STUDY OF IGNEOUS ROCKS DEVELOPED FROM THE STUDY OF mineralogy. It began with a controversy started by Werner, who, as noted earlier, contributed much to mineralogy. Werner was a proponent of neptunism, the theory that rocks were precipitated in layers from a universal ocean. Thus, he taught that all rocks are what would now be termed sedimentary rocks. He thought that granite, metamorphic rocks, and volcanic rocks were all precipitates from this universal ocean. The active volcanoes, such as those in Italy, were explained as caused by the burning of coal beds. The eventual downfall of neptunism came from the proof that basalt, which is now known to be the most common volcanic rock, is a volcanic rock and not a chemical precipitate. One reason basalt was thought to be crystallized from a solution was that it commonly forms six-sided prisms that resemble crystals such as those of quartz. The basalt prisms are the result of contraction during cooling (see Fig. 3–3).

Abraham Gottlob Werner (1749–1817) began teaching mineralogy at Freiberg in 1774. He was an extraordinary teacher, and, as a result, students came to him from all over Europe. It was his students who spread his theories far and wide because he published almost nothing. His aversion to writing was so strong that in his later years, he would not even write letters. His great success in mineralogy was because he was an organizer, and he organized what was known about minerals into a useful classification. His interest in organization was such that he was said by some to buy books so that he could arrange them in his library.

Neptunism, in a sense, was an attempt to organize rocks into a simple, easily understood system. Probably this simplicity was the great appeal of neptunism, for no further study was necessary. A rock's place in the onion-skin layers of the earth could be determined by merely identifying and naming the rock.

The neptunists had a number of problems, such as what happened to the water in the universal sea. They had no answer to this question. Another problem was that enough was known about chemistry at the time to prove that the volume of rocks assumed to have precipitated was much greater than could have been in solution in the sea. No one seems to have been troubled by this.

The main problem, however, was the origin of basalt. Unknown to the antagonists of the neptunism controversy, the volcanic origin of basalt had been demonstrated in 1774, which was before Werner had taken up neptunism. The story begins with Jean Étienne Guettard (1715–1786), who, after visiting Auvergne, an area in central France, recognized the ancient volcanoes there (see Fig. 3–1). Guettard published this observation in 1752. In 1770, he published the idea that basalt is not lava; by this, he meant that the rocks he had seen at Auvergne were different from the similar rocks with no remaining vestige of volcanoes that were called basalt. Ancient basalt is commonly found with sedimentary rocks, and erosion has removed the more obvious indications of its volcanic origin, such as volcanic cones. Thus, in a sense, Guettard was on both sides of the basalt and neptunism questions. Nicholas Desmarest (1725–1815) was another Frenchman who studied the Auvergne

A.

B.

FIGURE 3–1
The volcanoes of Auvergne, France. A. Chaine des Puys. B. Puy de Pariou in foreground and Puy de Dôme in Chaine des Puys. (Photos courtesy Maurice Krafft, Équipe Vulcain)

volcanoes. His studies, begun in 1763 and published in 1774, clearly showed that the lava of Auvergne is basalt. This should have stopped the controversy before it started, but these studies went unnoticed for years.

The opponents of the neptunists were called plutonists because they believed that igneous rocks originated deep in the earth, the home of Pluto, the god of the underground. Plutonism was just one aspect of **uniformitarianism**, the idea that present processes occurring on earth can be used to interpret ancient rocks. Uniformitarianism was the idea of James Hutton (1726–1797), a Scotsman. He first stated it in 1788 and then published a book *Theory of the Earth, with Proofs and Illustrations* in 1795. The book was written in a difficult style and so was not widely read. After Hutton's death, in 1802 his friend Playfair wrote a readable book, *Illustrations of the Huttonian Theory.*

The defeat of neptunism did not occur because of the work of the plutonists as much as it did from the studies of neptunists. Two of Werner's former students visited Auvergne. The first was Leopold von Buch (1774–1853), a German, who was one of the most influential geologists of the time and certainly the most traveled. He visited central France in 1802 and recognized the volcanic nature of the basalt there, but he did not concede that Werner's theory was completely wrong. From that point on, he gradually abandoned more and more of the neptunist ideas as his travels brought additional evidence before him. The other visitor to Auvergne was a Frenchman, Jean Françoise D'Aubuisson de Voissins (1769–1819). De Voissins abandoned neptunism in 1804 after this visit. The journey was suggested by the referees of a paper embracing neptunism that he had published the year before. This was a dramatic change. Thus, neptunism was challenged while Werner was still alive, and soon after his death plutonism was widely accepted.

One reason this controversy went on so long was the limited means available to study rocks. The hand lens, which was the only tool for close examination, worked reasonably well on rocks with large grains such as granite, but it could give little insight as to the nature of fine-grained rocks such as basalt. Chemistry was developing at this time, so some analyses were available to aid the study of fine-grained rocks.

Although the microscope was available, its adoption by geologists was slow. Microscopic study is now one of the most important ways rocks are studied. The technique used is to slice chips of rocks to the thinness of paper so that they are nearly transparent, and to examine these **thin sections**, as they are called, by passing light through them. The first thin sections were made by William Nicol, who also invented the Nicol prism, a device that polarizes light. These thin sections were of petrified wood and were described in 1831. No further use was made of thin sections for about 25 years until Henry Clifton Sorby came to the University of Edinburgh and saw Nicol's thin sections. He clearly realized their value in the microscopic study of rocks. Interestingly, Sorby's work was not quickly accepted because many geologists said in effect that mountains cannot be studied under the microscope. In 1862, Sorby chanced to meet the German Ferdinand Zirkel (1838–1912) while on an excursion in Germany. Zirkel's adoption of Sorby's microscopic study of rocks put the Germans in the forefront of this phase of geology, and lead to the general acceptance of microscopic study of rocks.

The only other common way natural rocks are studied is by chemical analysis. This method is applied mainly to rocks too fine-grained for microscopic study. Chemical studies and classification of rocks probably reached a peak near the turn of the present century. They may again become important because modern methods of instrumental analysis may make chemical analysis of rocks as cheap and easy as microscopic study.

Experimental study of igneous rocks has contributed much to our understanding. Such studies began early in the eighteenth century but did not become important until much later. Fouqué and Michel-Lévy did significant work in France, starting in the 1880s. Perhaps the most important work was that of N. L. Bowen (1887–1956), who worked mainly at the Geophysical Laboratory of the Carnegie Institution in Washington, D. C. Bowen's work on basalt melts is described in this chapter. Much of this was done at atmospheric pressure and thus may not duplicate conditions within the earth where most rocks crystallize. After World War II, furnaces were developed that could melt rocks under pressure, so the later studies of Bowen and his co-workers came much closer to duplicating natural conditions.

Igneous rocks are rocks that solidify from a melt or **magma**. A magma is a natural, hot melt composed of a mutual solution of rock-forming materials (mainly silicates) and some volatiles (mainly steam) that are held in solution by pressure. Magmas may contain some solid material, but are mobile. The relative abundance of igneous rocks is shown in Fig. 3–2.

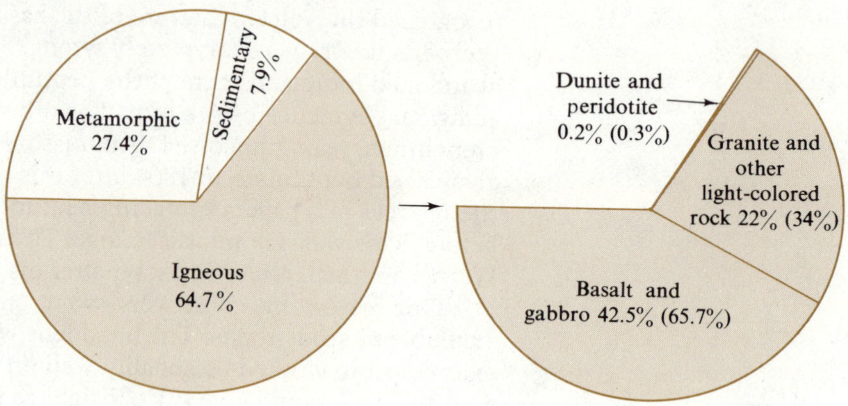

FIGURE 3–2
Relative abundance of igneous rocks in the earth's crust. For example, basalt and gabbro form 42.5 percent of all rocks and 65.7 percent of the igneous rocks. The igneous rock types are described later in this chapter. (Data from Ronov and Yaroshevsky, 1969)

VOLCANOES

A great deal can be learned about igneous processes from study of volcanoes. They are, after all, the only direct evidence we have for the existence of magma in the crust.

The active volcanoes today are of several different types, and the differences among them seem to depend on the composition of their magmas, more particularly on the behavior of the volatiles in the magmas. The most abundant volatile in magma is water, which escapes from volcanoes in the form of steam. Carbon dioxide is a common volcanic gas, but the sulfur gases (hydrogen sulfide and the oxides of sulfur), because of their strong odors, are the gases most easily noted near volcanoes. In addition, minor amounts of other gases, such as carbon monoxide, hydrochloric and hydrofluoric acid, ammonia, hydrogen, and hydrocarbons, are also released. Ordinary air also escapes from some volcanoes, especially those with porous rocks.

Volcanic emanations are believed to have played a major role in the formation of the oceans and the atmosphere (see Table 3–1). Recent eruptions at Hawaii are estimated to release 1–2 percent gas during the early stages of eruption and about 0.5 percent during later stages. Laboratory studies suggest that, at depth, magmas may contain up to 5–8 percent dissolved water.

A simple analogy will illustrate the behavior of the water and other gases in volcanic magma. Most magmas are believed to contain at least a few percent water in solution. The water is dissolved in the magma just as carbon dioxide gas is dissolved in a bottle of soda pop. Most magma is able to dissolve this water because the magma is under considerable pressure due to the weight of the overlying rocks. The gas is dissolved in the soda pop as a result of the pressure inside the bottle. When the bottle is opened, the pressure is reduced and the gas begins to bubble out of the liquid. In the same way, when magma reaches the surface, the confining pressure is reduced and the gas, in this case mainly steam, is released.

TABLE 3–1
Composition of volcanic gas at Hawaii.

Gas	Volume (%)
Water	70.75
Carbon dioxide	14.07
Hydrogen	0.33
Nitrogen	5.45
Argon	0.18
Sulfur dioxide	6.40
Sulfur trioxide	1.92
Sulfur	0.10
Chlorine	0.05
Total	99.65

SOURCE: After Shepherd, 1921, in G. A. Macdonald, "Physical Properties of Erupting Hawaiian Magmas," *Geological Society of America Bulletin* (1963), Vol. 74, No. 8, pp. 1071–78.

If the magma has a low viscosity, the gas merely escapes or forms bubbles in the resulting rock. The tops of lava flows can commonly be distinguished from the bottoms by the presence of these bubbles at the top (Fig. 3–3). In some cases, the bubble holes are filled by material deposited by the fluids, or are filled at a later time. The fillings are commonly silica materials such as opal or chalcedony, but many other minerals also occur here. Large filled holes of this kind are actively sought by rock collectors who cut and polish them.

If a fluid magma is erupted very suddenly, it may form a froth of tiny bubbles which quickly solidifies to form the light, glassy rock **pumice**. As a result of the many bubble holes, pumice is light enough to float. The formation of pumice can be likened to the opening of a bottle of warm soda pop or champagne, which results in the production of a froth. A rock with coarser bubbles is called **scoria**.

More violent events occur when more viscous magmas are erupted to the surface and the gas attempts to escape. In this case, the plastic rock is shattered by the pressure of the expanding gas. The gas may expand so rapidly that the rock explodes, blowing the fragments high in the air.

The fragments, or **pyroclasts**, produced in this way are named according to their size and shape (Table 3–2). Pyroclasts smaller than 2 mm are called **ash grains**, those between 2 and 64 mm are termed **lapilli**, and those larger than 64 mm are called **blocks** if they are angular and **bombs** if they are rounded or show other evidence of having been melted or partly melted when they were ejected. The term *ash* implies burning, but of course these rocks do not result from burning. The term was first applied when volcanoes were thought to result from subterranean burning, and so is an historical mistake. Unconsolidated deposits of these materials are called **tephra**. The rocks formed by this process are termed **pyroclastic** ("fire-broken") **rocks**. Pyroclastic rocks are named according to the size of the fragments without reference to composition. Consolidated ash is called **tuff**, consolidated lapilli is called **lapilli tuff**, a pyroclastic rock composed of blocks is called **pyroclastic breccia** (see Fig. 3–4), and one compound of bombs is called **agglomerate**. The main problem with these rocks is distinguishing them from normal sedimentary rocks formed by erosion of a volcanic area. The latter rocks are called **epiclastic tuff**, **epiclastic breccia**, or the like, depending on the size of the fragments. In general, field study is needed to distinguish between these two types of accumulation.

Volcanic bombs are formed by pyroclastic material that is still molten when thrown from the volcano. Again, the name is not really appropriate. Bombs are

TABLE 3–2
Classification of pyroclastic material.

	Fragment	Unconsolidated Deposits	Rocks
General Terms	PYROCLAST—fragment ejected by volcanic action	TEPHRA	PYROCLASTIC ROCK
	BLOCK—angular, solid when ejected	BLOCK TEPHRA	PYROCLASTIC BRECCIA
Pyroclasts <64 mm	BOMB—rounded or other evidence suggests melted or partly melted when ejected	BOMB TEPHRA	AGGLOMERATE
Pyroclasts 2–64 mm	LAPILLUS	LAPILLI TEPHRA	LAPILLI TUFF
Pyroclasts <2 mm	ASH GRAIN	ASH	TUFF

NOTE: The term *epiclastic* is used if the rock or deposit is composed largely of material eroded from volcanic terranes, that is, epiclastic tuff.

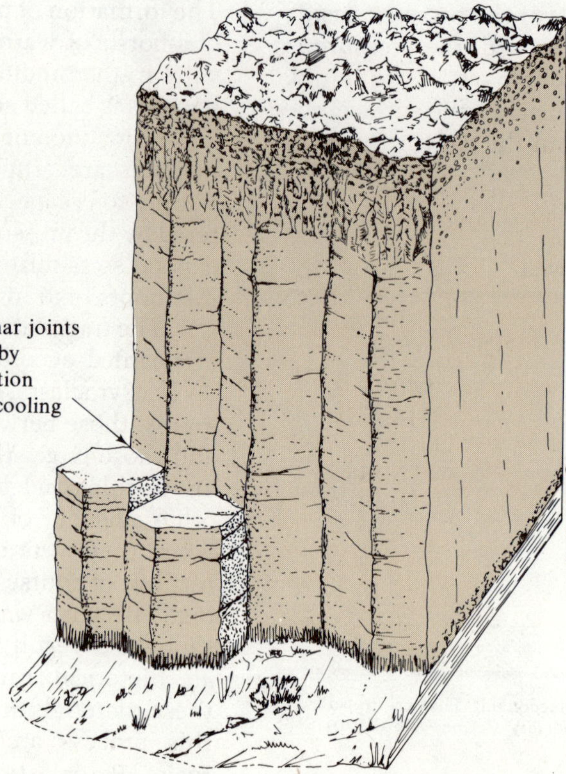

FIGURE 3–3
A section through a lava flow.

recognized by their shapes. A liquid assumes a streamlined shape while in flight. Most volcanic bombs are somewhat viscous when thrown from a volcano and solidify during flight. Most spin during flight and so have an elongate shape (Fig. 3–5). Some are still fluid when they hit the ground, and these flatten out.

Volcanic eruptions range from quiet to violent, as will be evident from the following descriptions. The examples used are currently active volcanoes and those that are recent enough that erosion has not greatly modified them. A volcano that has erupted in historical time is generally considered an active volcano; but only at one or two closely studied, well-instrumented volcanoes is it possible to predict eruptions, and even at these, warning comes just before eruption, and the place of eruption is known only generally. Thus, designating active volcanoes is difficult.

The rocks covering the surface of much of the western United States are volcanic, and many of these are millions of years old. This may seem old but is relatively young by geologic standards. Many of the original features of these rocks have been removed by erosion, but they are readily recognizable as volcanic. Less easily recognized are buried volcanic terranes. Such rocks are exposed by uplift and erosion in all parts of the earth, and the age of these rocks may be hundreds

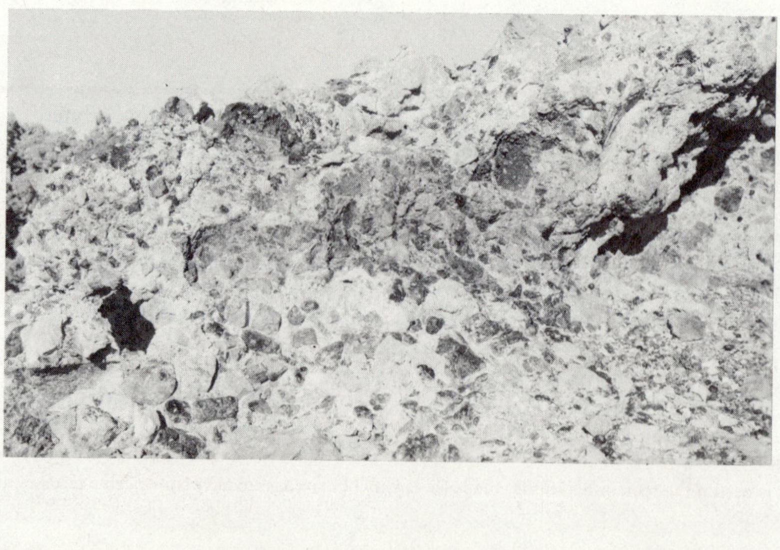

FIGURE 3–4
An outcrop of volcanic breccia in the Thomas Range, Utah. This rock is composed of coarse, angular fragments of volcanic rocks in a fine-grained matrix. The dark fragments are obsidian, and the light are pumice. (Photo by M. H. Staatz, U.S. Geological Survey)

A. B.

FIGURE 3–5
Volcanic bombs. A. An unusual volcanic bomb 4 m long. (Photo taken in 1902 in southern Idaho by I. C. Russell, U.S. Geological Survey) B. Typical volcanic bombs from Hawaii. Sizes range from a couple of centimeters to about a meter. Many other shapes occur.

of millions of years. Thus, study of present volcanoes will help in interpretation of ancient rocks—an important principle in geology.

Present-day volcanoes occur in relatively few areas, such as the rim of the Pacific Ocean, the Mediterranean Sea, and the mid-Atlantic ridge. Most of these areas are associated with sea-floor spreading (Fig. 3–6).

It will be convenient to use rock names in the following paragraphs. Dark-colored volcanic rock is called **basalt**; light-colored, **felsite**; and intermediate between the two is **andesite**. The naming of igneous rocks will be described later in the chapter.

The quietest eruptions are lava flows that come from dikes or dike swarms. The lava is basalt and apparently very fluid. These quiet flows can produce great thicknesses and cover huge areas. An example is the Columbia River Basalt of southeastern Washington and northeastern Oregon that covers an area of over 80,000 km² and is more than 1525 m thick at places. Thus, this unit constitutes at least 125,200 km³ of basalt. The source of this basalt field was four widely

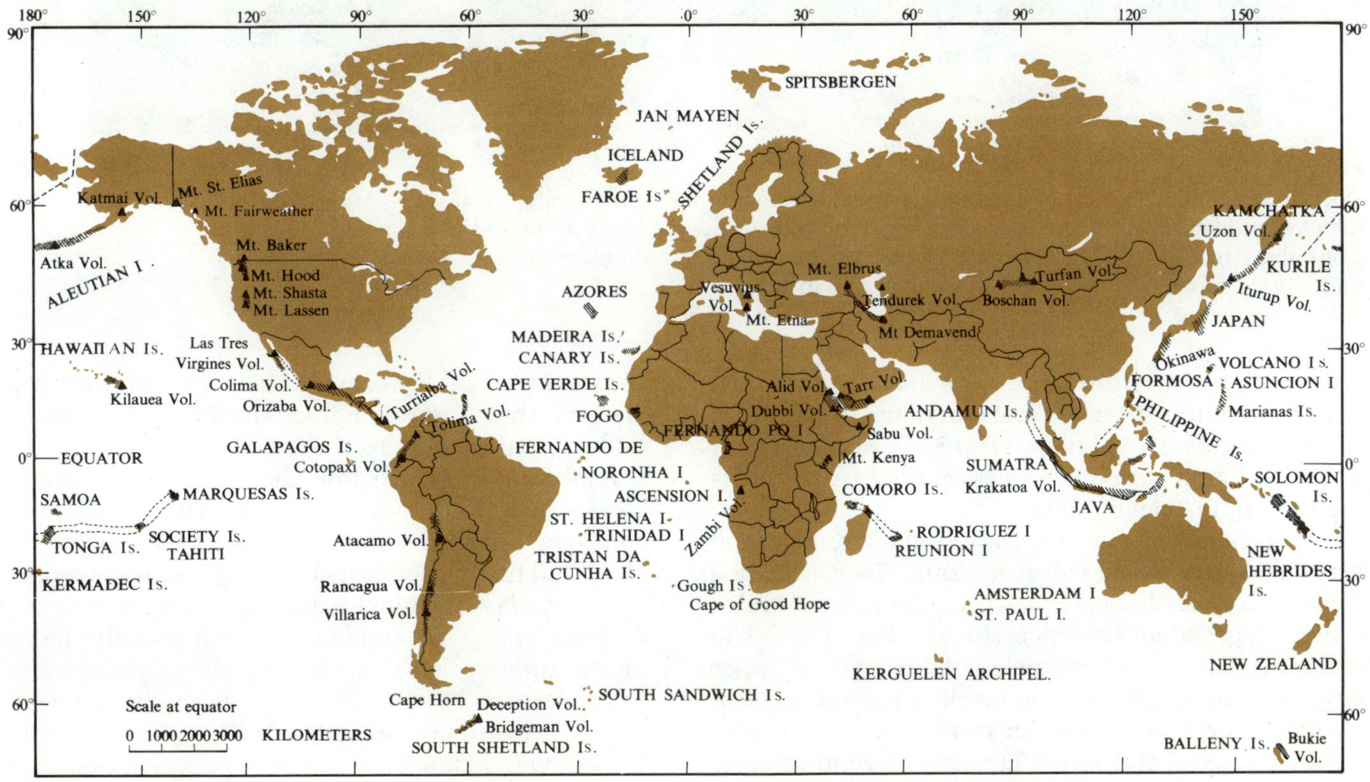

FIGURE 3–6
Volcanic areas of the world. (From G. A. Waring, U.S. Geological Survey Professional Paper 492, 1965)

FIGURE 3–7
Oblique aerial view looking east near the confluence of the Snake and Imnaha rivers on the Oregon-Idaho border. The bedded rocks on the valley sides are basalt flows of the Columbia River Basalt. The lower parts of the valleys are cut in metamorphic rocks that underlie the basalt. (Photo from U.S. Geological Survey)

separated dike swarms. The very fluid basalt must have flowed like water because individual flows can be followed for over 160 km (see Fig. 3–7). Similar basalt fields are found at several places on the earth, such as the Deccan Plateau of India.

At other places, similar basalt erupts from central vents, and large **shield volcanoes** form. They are called shield volcanoes because the very fluid magma builds gentle slopes that resemble shields (Fig. 3–8). The Hawaiian Islands are examples of this type of quiet eruption whose most violent displays are fountains of incandescent lava thrown up from lava pools in the craters (Fig. 3–9 and 3–10). Despite the gentle slopes, the Hawaiian Islands rise a total of 8500 m from the floor of the Pacific to 4200 m above sea level.

The fluid basalts discussed so far are easily recognized by their ropy surfaces (Fig. 3–11). However, even in Hawaii not all flows are of this type. Some are more viscous; their surfaces harden during flowage and are broken into blocks by the traction of the fluid lava several centimeters below the crust. These are the blocky lava flows (see Fig. 3–12). The terms *blocky* and *ropy*, although somewhat descriptive, are not too good, and many geologists use the Hawaiian words *aa* and *pahoehoe* for blocky and ropy lava, respectively. At some places, the top of a flow may solidify, and the melted interior flows on, forming a lava tube or tunnel. Examples can be seen at many places, including the Modoc lavas in northeastern California and Craters of the Moon in Idaho.

Cinder cones probably result if the lava is more viscous than that which forms blocky flows. Pyroclastic material thrown out of the vent builds up a cone around the vent. A central depression around the vent

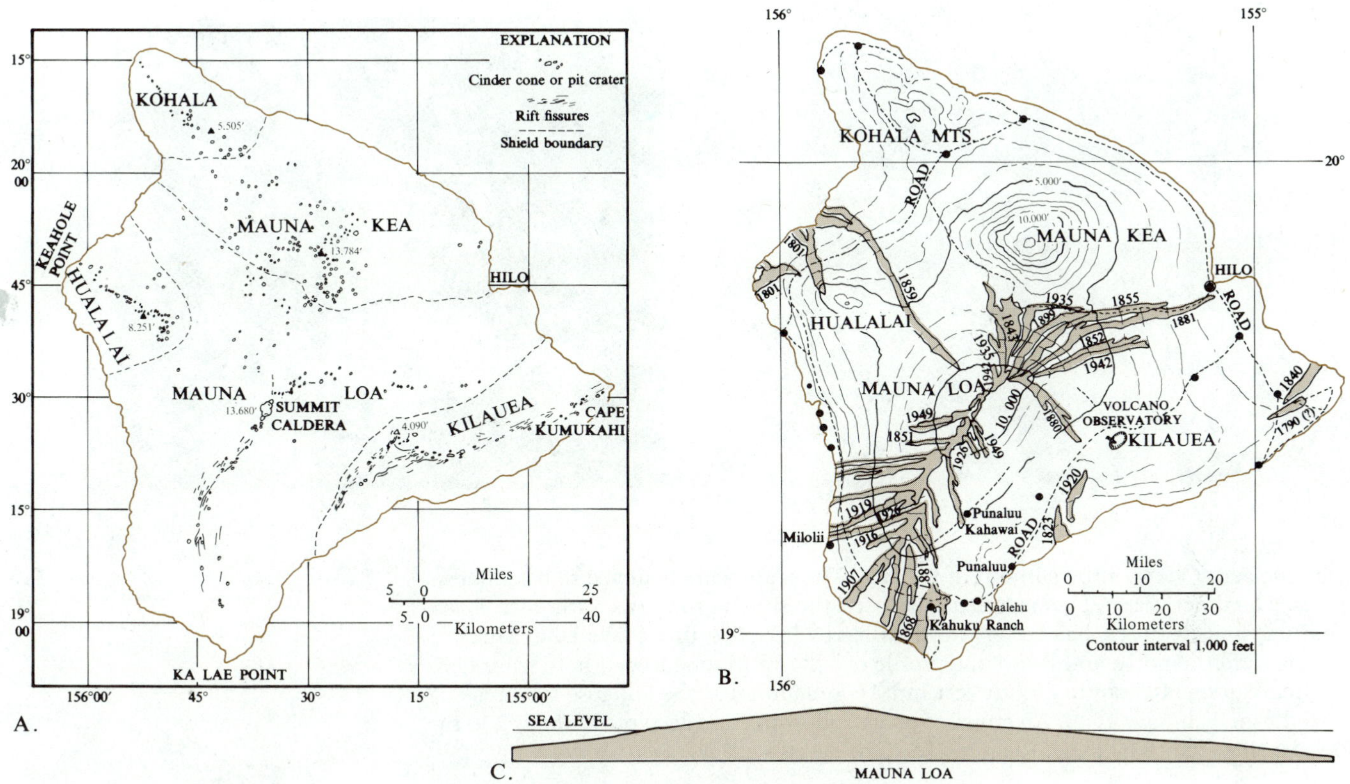

FIGURE 3–8
Maps of the island of Hawaii. A. Cinder cones, pit craters, rift fissures, and the margins of the five component volcanoes. B. Historic lava flows through 1950 (color). C. Cross section of Mauna Loa, a shield volcano, without vertical exaggeration. (Parts A and B from C. K. Wentworth and G. A. Macdonald, U.S. Geological Survey Bulletin 994, 1953; part C from K. O. Emery, and others, U.S. Geological Survey Professional Paper 260A, 1954)

remains at the summit. A cinder cone is composed of tuff, breccia, ash, cinders, blocks, and bombs (Fig. 3–13). The slope is generally near 30°, which is the steepest slope that can be made of loose material. Cinder cones are common features and, because the loose material is easily eroded, are generally young features. They are rarely very high because these slopes are so unstable; such loose slopes are commonly tiring and frustrating to climb. Cinder cones are commonly made of basalt or andesite.

The very high volcanic mountains are generally **composite volcanoes**, so called because they are made of a combination of pyroclastic material and lava. Such volcanoes are generally made of andesite (intermediate composition) and combine explosive eruptions with outpouring of lava. Mt. Shasta and Mt. Rainier (Fig. 3–14) are examples. The lavas cover the pyroclastic material and form the skeleton that enables the mountain to grow large. The lava is more viscous here, and the vents are blocked from time to time by congealing

FIGURE 3–9
Lava fountains 92 m high on Kilauea, 23 September 1961. Note the trees burning around the active margin of the flow fed by the fountains. (Photo by D. H. Richter, U.S. Geological Survey)

A.

B.

FIGURE 3–10
A. Oblique aerial view of the summit of Kilauea. The craters are believed to be collapse features. The inner crater, Halemaumau, has been the site of eruptions with lava fountaining and, at times, has been partially filled by lava, forming a lava lake. This view is toward the north and shows the profile of 4204-m Mauna Kea. (Photo from U.S. Geological Survey) B. Looking southwest into Halemaumau at the collapse features formed during February 1960. An eruptive phase followed. The deep pit is about 230 m below the rim. (Photo by D. H. Richter, U.S. Geological Survey)

lava. These plugs cause the gas pressure to build up until the pressure exceeds the strength of the plug, and a greater or lesser explosion results. Under these conditions, new vents may form on the sides of the volcano. Obviously, as the magma becomes more viscous, the explosive eruptions became more numerous and/or more destructive. Crater Lake, Oregon (Fig. 3–15), was formed by the explosive destruction of the top of a composite volcano similar to Mt. Rainier and Mt.

FIGURE 3–11
Ropy or pahoehoe lava flow near McCartys, New Mexico. The downstream end of the flow is to the right. (Photo by H. R. Cornwall, U.S. Geological Survey)

Shasta. Interestingly, the volume of material blown out of Crater Lake is much smaller than the volume of the missing top of the mountain, suggesting that after explosive discharge of the magmatic gases, the magma subsided. Most craters and *calderas,* as large craters are called, are formed by subsidence.

Krakatoa, between Java and Sumatra, has a somewhat similar history. It was a volcano 800 m above sea level that formed within a ring of islands marking the rim of an older volcano that lost its summit, probably in much the same way as Crater Lake formed. In 1883, the island of Krakatoa erupted explosively, destroying the island. The ash from this great explosion, which rose high into the atmosphere, was the cause of beautiful sunsets all over the world long after the eruption.

Lavas ranging in composition from andesite to felsite tend to be viscous, and they cause explosive eruptions. Plugs tend to block the vents, resulting in violent explosions of various kinds. One type was just described. In 1902, Mt. Pelée on Martinique Island was in eruption, and a large spine formed in the vent, plugging it. This spine was pushed up about 450 m, forming quite a spectacle. Finally, the pressure of the magmatic gases was enough to enable the gas-charged magma to break through. The resulting eruption was one of the most destructive ever witnessed, and led to the recognition of an important type of pyroclastic rock. (Recall the earlier description of the formation of pumice.) At Mt. Pelée, the magmatic gas expanded

A.

B.

FIGURE 3–12
Blocky or aa lava. A. Blocky lava flow and cinder cone. B. Steep front of a flow of viscous, blocky volcanic glass (obsidian). (Photo from U.S. Forest Service)

FIGURE 3–13
Aerial view of Parícutin, a large cinder cone that began in a cornfield in 1943 and was active for about 9 years. The lava flow spread from vents beneath the cone. (Photo from Chicago Natural History Museum)

FIGURE 3–14
Mt. Rainier, a massive composite volcano in the Cascade Mountains, Washington. (Photo from Washington State Department of Commerce and Economic Development)

FIGURE 3–15
Crater Lake, Oregon, fills a depression formed partly by explosive eruptions and mainly by subsidence of the magma. Wizard Island is a small volcanic cone with a crater at its summit. (Photo from Oregon State Highway Department)

so violently that instead of forming a frothy pumice, the expanding bubbles completely shattered the pumice, producing a mixture of hot gas and small glass fragments. Such a mixture is quite fluid because of the gas and can move very rapidly down even a gentle slope. When it comes to rest, the hot glass fragments weld together, forming **a welded tuff**. Eruptions of this type are called *glowing clouds* (**nuées ardentes**) or *glowing avalanches.* It was an eruption of this kind from Mt. Vesuvius that buried the city of Pompeii in A.D. 79.

The 1902 eruption of Mt. Pelée was witnessed by a number of people. They reported that the hot cloud traveled at about 160 km/hour. It killed all but about 4 of the perhaps 30,000 inhabitants of St. Pierre. The survivors were a prisoner in a dungeon, two people who were near the edge of the cloud, and one who was spared although all around him died. In the harbor, all but two ships were sunk. Most of the witnesses were from those ships. They reported a sudden darkness and hot dust mixed with water which caused a short, intense heat and a feeling of suffocation. Many of these people were badly burned.

Many welded tuffs closely resemble *obsidian* (volcanic glass) or felsite, and only recently has the abundance of welded tuffs been recognized. Their abundance on the surface of the continents may be as great as that of basalt. All welded tuffs are not associated with dome volcanoes. Much of Yellowstone National Park is covered by welded tuffs, and the geysers there probably get their heat from the cooling lava.

Lava plugs or domes, such as the spine at Mt. Pelée, are believed to form during the waning phase of some composite volcanoes. Mt. Lassen in California is of this type (Fig. 3–16). The plug or dome is composed of very viscous lava, and its extrusion is probably very similar to the squeezing of toothpaste from a tube. At Mt. Lassen the plug dome where exposed has steep, smooth cliffs with vertical grooves that developed when the dome was squeezed up. The eruptions of 1914–1917 were small; the most violent events were mudflows caused by melting snow.

FIGURE 3–16
Mt. Lassen in eruption. (Photo by B. F. Loomis, National Park Service)

FIGURE 3–17
Pillow lava at Depoe Bay, Oregon.

Underwater volcanic eruptions are common, and lavas sometimes flow into water bodies. This leads to rapid chilling of the lava, and causes the lava to subdivide into numerous interconnected lobes that look like a pile of pillows in cross section. The term **pillow lava** is applied to the resulting rocks (Fig. 3–17). Such rocks are abundant on the ocean bottoms.

Erosion may remove the fragmental rocks from a volcano, leaving the lava that congealed in the central conduit as shown in Fig. 3–18.

Volcanic eruptions cannot yet be predicted; however, recent studies have detected the movement of magma at a few very closely studied, well-instrumented volcanoes. The movement is shown by many thousands of shallow, very weak earthquakes. The location of these earthquakes may ultimately reveal the magma plumbing system. It also appears that the magma moves upward to shallow magma chambers just before eruption. The latter hypothesis is suggested by the swelling of a volcano just before eruption. This is detected by sensitive tilt meters. After an eruption, the magma apparently returns to the depths. This is indicated by the tilt meters and, at Hawaii, by the draining of the lava lakes that form in the craters there (see Fig. 3–19).

ENVIRONMENTAL HAZARDS OF VOLCANOES

The destruction caused by volcanoes is of two kinds: the direct destruction by material issuing from the volcano (Fig. 3–20) and the indirect damage caused by mudflows and floods. The former has already been described in the discussion of explosive volcanoes, such as Mt. Pelée and Krakatoa. There is little defense against such rapid destruction except evacuation at the first indication of activity by the volcanoes. The first eruption may be of the glowing cloud (nuée ardent) type and may occur anywhere on the volcano. In such cases, there may be no escape. In the quieter eruption of basaltic volcanoes, such as those in the Hawaiian Islands, the fluid basalt moves at a few kilometers an

FIGURE 3–18
Shiprock, New Mexico, is probably part of an ancient volcano. Dikes radiating from it can be seen on the left skyline and the right foreground. (Photo from New Mexico Department of Development)

hour, and generally there is enough time to evacuate. This may seem of little value if your home and fields are inundated by either lava or airborne ash, but at least you are alive and able to start over.

Indirect damage from volcanoes can also be swift and without warning. Many volcanoes are high mountains and, even in the tropics, are glacier-covered. An eruption under a glacier can cause extensive melting

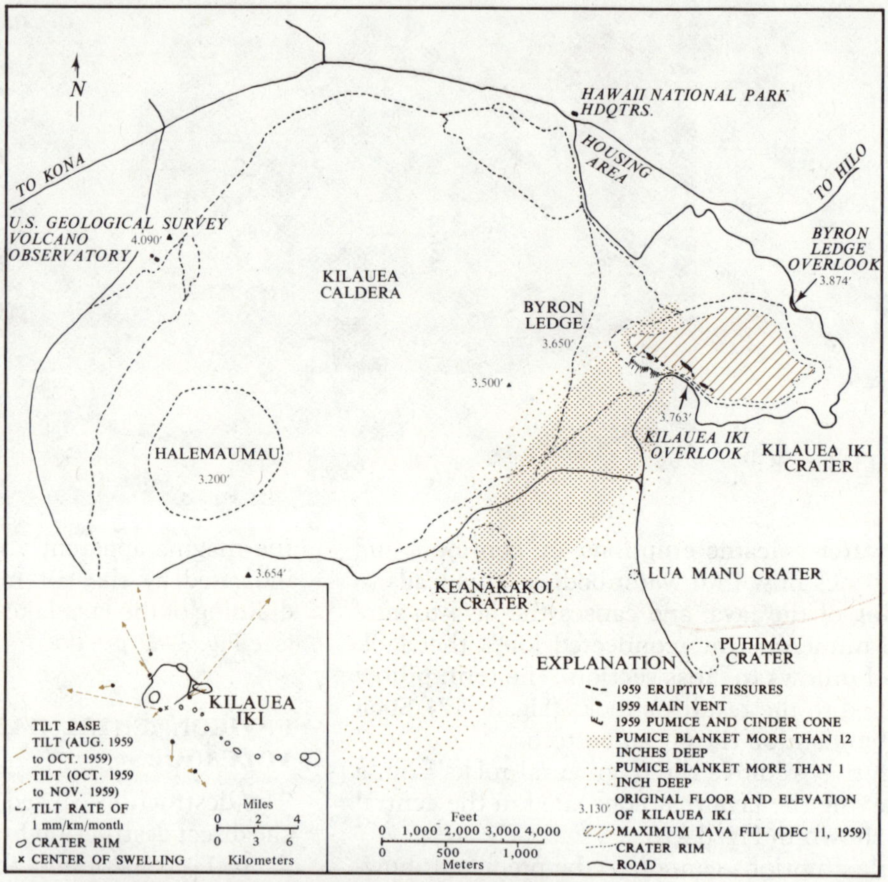

FIGURE 3–19
Map of Kilauea volcano, Hawaii, showing the principal features of the 1959 eruption of Kilauea Iki. The small-scale inset map illustrates the tilt pattern on the summit of Kilauea prior to eruption. [From J. P. Eaton and D. H. Richter, *Geotimes* (1960), Vol. 5]

FIGURE 3–20
In early 1973, an eruption began at Heimaey, an island that is part of Iceland. The cone in the left background is composed of pyroclastics which also partially cover the houses. (Photo from Consulate General of Iceland)

that results in floods. Also, many volcanoes, especially those that have been inactive for long periods, have large lakes in their depressed summit areas. Crater Lake in Oregon is an example. An eruption that breeches the lake, or an eruption into the lake, can cause the draining of the lake and so cause floods. The eruption of a submarine or shallow-island volcano such as Krakatoa can cause flooding by seismic sea waves.

Probably the largest areas inundated by volcanoes are those swept by mudflows. Although some mudflows are concurrent with the eruption, most occur at the next rainy season. Although the latter mudflows can be predicted, many lives are lost because many people refuse to leave their homes and fields. Mudflows are probably the most important way that pyroclastic volcanic rocks are distributed from volcanoes and thus are a normal geologic process. A pyroclastic eruption can deposit ash and coarser material on the slopes of the volcano and on other slopes downwind. This loose material is not stable and at the next rain becomes mud that can flow even on very gentle slopes. Rain commonly accompanies volcanic eruption, and although we do not know why, it may be related to the large amounts of steam that are part of most eruptions. Thus, mudflows can accompany eruptions. Pyroclastic eruptions commonly kill vegetation by covering low plants, and the falling debris strips the leaves from trees. Without vegetation, the runoff is increased, and the possibility of mudflows is greater.

Mount St. Helens in the Cascade Range in southern Washington State erupted in 1980, causing widespread damage. Its last previous eruption was in 1857, and at that time it had been active between about 1800 and 1857. Study of Mount St. Helens and its deposits shows that it has erupted 15 times in the last 4500 years, and that a typical eruptive period lasts about 100 years, with 100–500 quiet years between eruptive intervals.

The eruption in 1980 began with a number of shallow, mild earthquakes, and a week later, on 27 March, steam was noted issuing from the summit. The earthquakes and eruptions continued, with rock and ice carried at times to heights of over 5 km. Fine rock material was carried many kilometers, and sulfur dioxide as well as steam issued from the crater. During this time, the crater enlarged, and an unusual type of earthquake activity called harmonic tremors revealed that magma was moving in the volcano. By late April, it was noted that the north side of the crater was rising and bulging northward. Apparently, a dome was being emplaced, and the uplift continued at over a meter per day.

The big eruption, which occurred on 18 May, began with two closely spaced earthquakes that caused the bulge, now 125 m, to landslide. The landslide released the pressure on the dome or plug, and a blast of incandescent gas and ash resulted. The blast was estimated at 50 m/s, and it devastated an area of 500 km^2. It knocked down trees and seared them with 200°C temperatures. Between 2 and 3 km^3 of landslide and pyroclastic material moved up to 20 km down stream valleys, filling some to depths of 60 m. After the initial blast, the ash and steam rose vertically to heights of as much as 19 km and were carried hundreds of kilometers to the east. Fine ash several centimeters deep accumulated at many places downwind. A few hours after the blast, mudflows and debris flows fed by melting glaciers moved into river valleys, causing some flooding and warming of the river water. About 400 m was removed from the top of Mount St. Helens by the blast, and the summit crater was enlarged to a diameter of about 2 km and a depth of 1.5 km (Fig. 3–21).

CRYSTALLIZATION OF IGNEOUS ROCKS

One of the most important concepts in igneous geology is the **reaction series**, first described by Bowen, a U.S. geochemist who studied silicate melts in the laboratory. This series shows the sequence of minerals that form in the crystallization of basaltic melt, which is the most common type of volcanic rock. It also suggests how the earliest-formed crystals might be separated from the melt or magma, forming rocks with a composition different from that of the original melt. Such a process is called **differentiation**, and its role will be discussed when the origin of magmas and of the various types of igneous rocks is considered. Although Bowen originated the idea of a reaction series from studies of experimental melts, the general sequence of crystallization of most rocks was already known as a result of intensive microscopic studies. The reaction series is an oversimplification but is very useful (see Fig. 3–22).

The reaction series really consists of three series. The right-hand side is a **continuous series**; that is, all compositions of plagioclase from entirely calcic to entirely sodic and all compositions in between occur. The left-hand side is a **discontinuous series**; that is, the changes from one mineral to the next occur in discrete steps. Each of the minerals named in the left-hand series is actually a continuous series in composition of that mineral group. The lower series is only a sequence

FIGURE 3–21
1980 eruption of Mount St. Helens. A. On 10 April, before the eruption, the mountain was symmetrical and snow-covered. B. Eruption on 18 May. C. Summit area after the eruption. D. Mudflow fills the Toutle River valley after the eruption. (Photos courtesy U.S. Geological Survey; parts A, B, and D by Austin Post)

of crystallization. The arrangement shows the relative, but not exact, sequence of crystallization: calcic plagioclase and olivine tend to crystallize at the same time, and the lower series is last to crystallize. It should be emphasized that Bowen's series holds for only some basaltic magmas, but, as shall be seen, basalt magma is very important in igneous geology.

The interpretation of the crystallization of a basaltic magma in terms of Bowen's series is as follows. In general, the first mineral to crystallize from the melt is

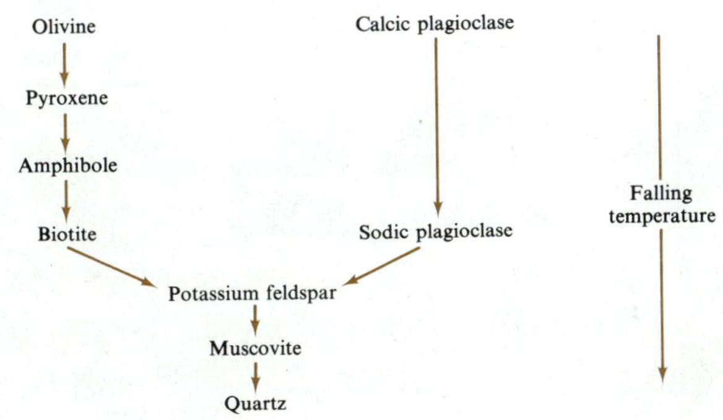

FIGURE 3–22
Bowen's reaction series, showing the general sequence of mineral formation with falling temperature in a basaltic melt.

olivine. In most cases, at the same time, or a little later, calcic plagioclase begins to crystallize. The discontinuous series continues with falling temperature; the olivine reacts with the melt to form pyroxene, which at the same time also starts to crystallize directly from the melt. Plagioclase continues to crystallize during falling temperature; the composition, however, continuously becomes more sodic. If cooling is slow enough, these two processes go on: the minerals on the left side react with the melt to form the next-lower mineral, while that mineral is crystallizing from the melt, and the plagioclase continuously reacts with the melt to form more plagioclase that is more sodic in composition. This process ends when all the ferromagnesian minerals and plagioclase are formed. Then, the potassium feldspar-muscovite-quartz series begins. This series may in part overlap the sodic part of the plagioclase series.

Another way to explain the discontinuous series is to consider what happens on heating one of the minerals. When one of the discontinuous series minerals is heated, it does not melt at a single temperature the way ice or similar substances do. Instead, when an amphibole, for example, is heated, it begins to melt when a certain temperature, depending on its composition, is reached, and melting continues over a range of temperature. If the remaining solid is examined during this first stage of melting, by quick quenching to prevent reaction during slow cooling, it would be seen that the amphibole has been transformed to a pyroxene and a liquid, just the reverse of what takes place during the slow cooling of a magma. On further heating, a temperature is reached at which further melting would occur. As before, melting would take place over a range in temperature, resulting in a melt and olivine. Finally, a temperature would be reached at which all the olivine would melt.

The evidence for the reaction series is abundant. Under the microscope, for instance, pyroxene crystals with jackets of amphibole are seen, recording a discontinuous reaction that was interrupted, probably by cooling too rapidly for the reaction to be completed (see Fig. 3–23). In differentiated rock bodies, slow cooling and low viscosity have combined to preserve some of the main points of the reaction principle. In such rock bodies, the early-crystallized olivine sank to the bottom and so was removed from contact with the main mass of the magma, allowing progressively lighter-colored rocks to form in the upper parts of the body (see Fig. 3–24).

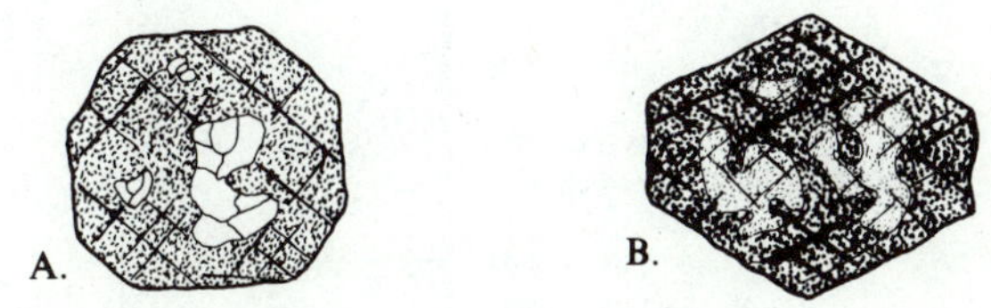

FIGURE 3–23
Evidence for the Bowen reaction series seen in microscopic study of igneous rocks. A. Augite crystal with remnants of olivine inside it. B. Hornblende crystal with augite remnants. In both cases, the reaction was incomplete, thus preserving the remnants of the earlier mineral. Note the cleavage angles in augite and hornblende. The minerals shown here are typically about 3 mm in diameter but may be much larger or smaller.

Some geologists believe that such differentiation explains the great variety of igneous rocks, but this is probably an oversimplification. The reaction series applies to basaltic magmas, and the vast areas of basalt lava certainly indicate that basalt magmas are common in the earth's crust. But, as shall be seen later, rocks of granitic composition are also abundant. The vast amount of granitic rocks and their volcanic equivalents suggests that these rocks also represent important magma types in the crust, because they are far too abundant to be derived from basalt by differentiation. Complete differentiation of basalt can produce, at the most, 15 percent granitic-composition rocks. This points out a major problem of geology: the origin of magmas. Understanding the crystallization of basalt gives some insight into the genesis of rock types, as shown in the next section.

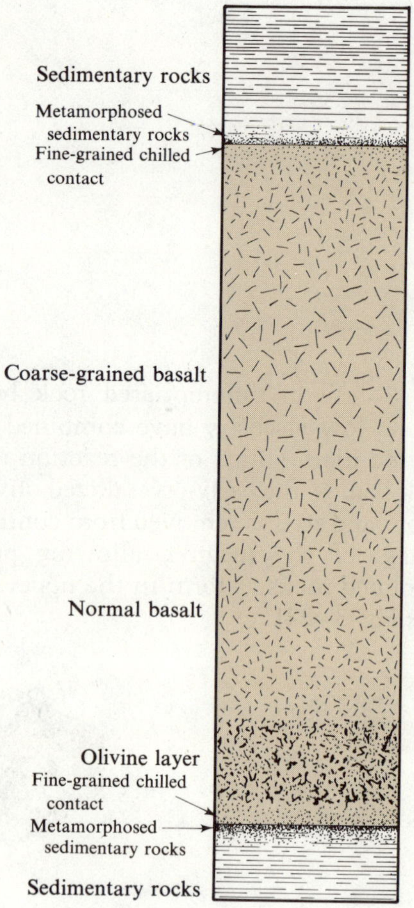

FIGURE 3–24
Differentiation in a basalt sill. A narrow column through a sill is shown here. The olivine layer develops by the sinking of the early-formed, heavy olivine crystals.

CLASSIFICATION

Igneous rocks are classified on the bases of composition and texture. The classification is an attempt to show the relationships among the various rock types; hence, it is an attempt at natural or genetic classification.

Composition, in the classification of igneous rocks, follows the reaction series. Most rock names came into use long before Bowen described the reaction series. These names had been applied to the more common rocks. It is of more than passing interest that these common groups of rocks fall into the various steps in the differentiation of a basaltic magma according to Bowen's reaction series. This correlation suggests that the compositional part of igneous rock terminology may be genetic. This is shown in Fig. 3–25.

Because almost all coarse-grained rocks contain feldspar, the most common mineral family, the type of

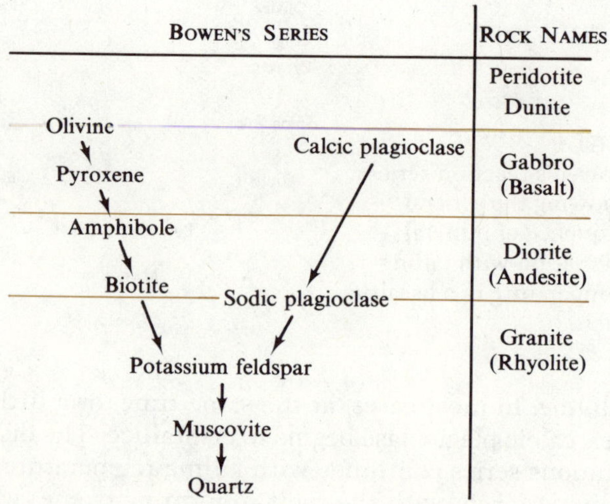

FIGURE 3–25
This diagram shows in a very general way how differentiation by separation of successively formed crystals could result in several rock types originating from a single basalt magma. (Volcanic rock types are in parentheses.) It also illustrates why minerals widely separated in the reaction series, such as quartz and olivine, rarely appear in the same rock.

feldspar is the most important factor in the composition. Of secondary importance is the type of dark mineral which generally accompanies each type of feldspar. See classification diagram Fig. 3–42.

In a similar way, the texture tells much about the cooling history of an igneous rock. In igneous rocks, the texture refers mainly to the grain size. Rocks that cool slowly are able to grow large crystals; but quickly chilled rocks, such as volcanic rocks, are fine-grained. To see the significance, the mode of occurrence of igneous rocks must be considered.

TEXTURE AND OCCURRENCE

Igneous rocks occur in two ways, either as **intrusive** (below-the-surface) bodies or as **extrusive** (surface) rocks. The ultimate source of igneous magma is probably deep in the crust or in the upper part of the mantle, and the terms *intrusive* and *extrusive* refer to the place where the rock solidified. Intrusive igneous rocks can be seen only when erosion has uncovered them. They are described as **concordant** if the contacts of the intrusive body are more or less parallel to the bedding of the intruded rocks, and **discordant** if the intrusive body cuts across the older rocks (Fig. 3–26).

The largest discordant bodies are called **batholiths**. These are very large features, the size of mountain ranges such as the Sierra Nevada in California, which

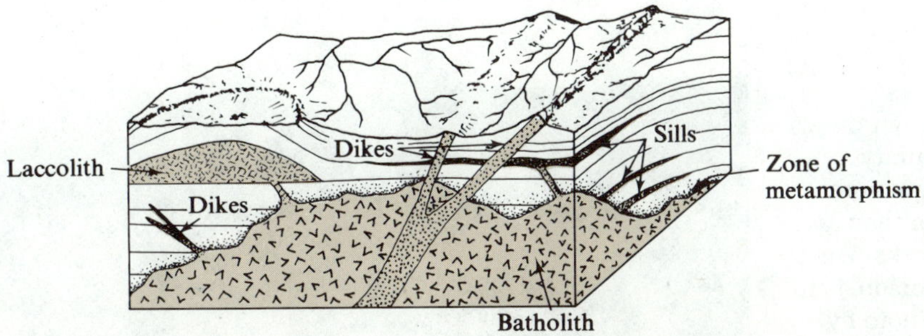

FIGURE 3–26
Intrusive igneous rock bodies. The laccolith and sills are concordant with the enclosing sedimentary beds, and the batholith and dikes are discordant. The heat from the batholith metamorphoses the surrounding rocks.

is nearly 650 km long by about 80 km wide. Many other ranges, both larger and smaller, also contain batholiths (Fig. 3–27). Because batholiths are large, and also because they probably were emplaced at least a thousand meters below the surface, they cooled very slowly. This slow cooling permitted large mineral grains to form; therefore, it is not surprising that batholiths are composed mainly of granitic rocks with crystals large enough to be seen easily (Fig. 3–28). Smaller bodies of coarse-grained igneous rocks are termed **stocks** or **plutons**. Coarse-grained igneous rocks are called **plutonic igneous rocks** because they were once believed to have crystallized at great depth.

Dikes are tabular, discordant intrusive bodies. They range in thickness from several centimeters to about a thousand meters, but generally are of the order of several meters to tens of meters thick (see Figs. 3–29 and 3–30). They are generally much longer than they are wide, and many have been traced for kilometers. Most dikes occupy cracks and have straight, parallel walls. Because the intrusive rock is commonly resistant, dikes generally form ridges when exposed by erosion.

FIGURE 3–28
Granite has a coarse, interlocking texture formed by crystallization from a melt. (Photo from Ward's Natural Science Establishment, Inc., Rochester, N.Y.)

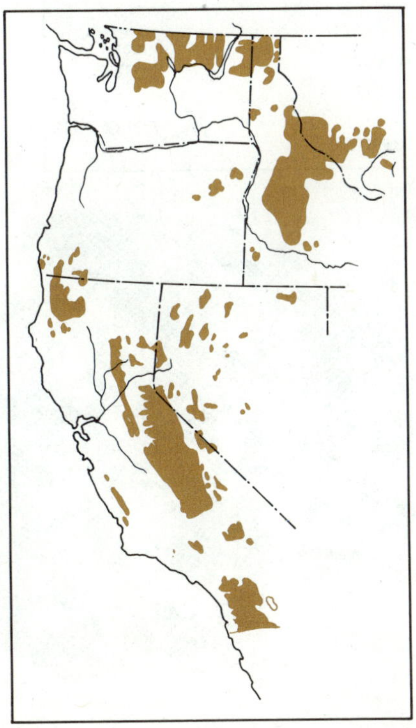

FIGURE 3–27
Batholiths of the western United States.

The corcordant intrusive bodies are **sills** and **laccoliths**. These are very similar and are intruded between sedimentary beds. Sills are thin and do not noticeably deform sedimentary beds (see Fig. 3–31). Laccoliths are thicker bodies and up-arch the overlying sediments, in some cases forming mountains.

Dikes and sills are small bodies compared to batholiths and have much more surface area for their volume; thus, they cool much more rapidly and are commonly fine-grained. Laccoliths are generally fine-grained but, depending on their size, may be coarse-grained.

The extrusive rocks result from lava flows and other types of volcanic activity. These rocks commonly cool rapidly and are fine-grained, or glassy (Fig. 3–32) if cooled so rapidly that no crystallization occurs.

Many igneous rocks are a mixture of coarse and fine crystals, and this texture is called **porphyritic** (Fig.

FIGURE 3–29
Dikes cutting through horizontal sedimentary rocks. The dikes are more resistant to erosion than are the sedimentary rocks. West Spanish Peak, Colorado, is in the background. (Photo by G. W. Stose, U.S. Geological Survey)

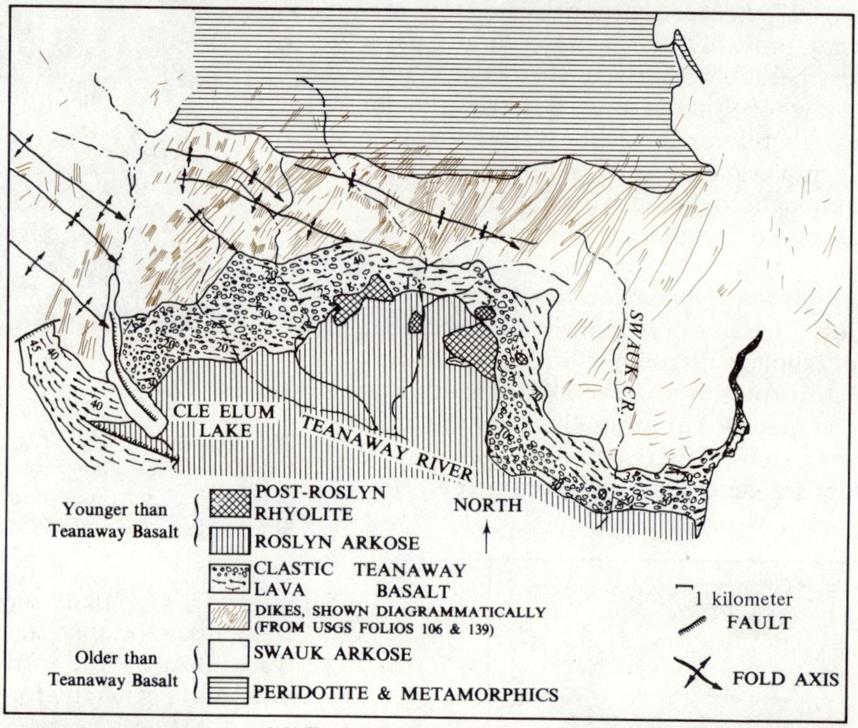

FIGURE 3–30
Teanaway dike swarm, Cascade Mountains, Washington. At places, the dikes, shown diagrammatically here, comprise over 90 percent of the surface, recording much dilation. The dikes were the sources of the Teanaway Basalt. [From R. J. Foster, *American Journal of Science* (1958), Vol. 256]

FIGURE 3–31
The black horizontal band is a sill. The inclined band is a crosscutting dike. The dark stripes in the sandstone are carbonaceous sediments. Taylor Glacier region, Victoria Land, Antarctica. (Photo by W. B. Hamilton, U.S. Geological Survey)

FIGURE 3–32
Obsidian, or volcanic glass. The round, shell-like fractures are conchoidal fractures. (Photo from Ward's Natural Science Establishment, Inc., Rochester, N.Y.)

3–33). Such a texture, in general, records a two-step history of the rock. The large crystals form as a result of slow cooling, perhaps in a deep magma chamber. Then, the magma is moved higher in the crust to form a dike, sill, or flow. The remainder of the rock cools rapidly, resulting in a fine-grained or even glassy matrix surrounding the early-formed larger crystals. A rock with a coarse matrix and some much larger crystals also may be described as porphyritic. It is doubtful whether all such rocks have a two-step history as outlined here.

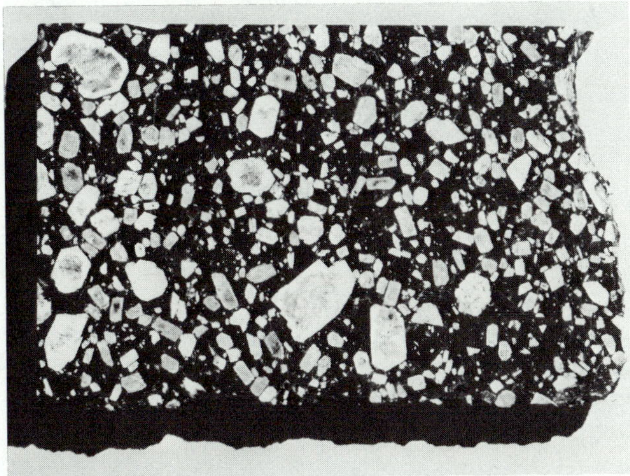

FIGURE 3–33
Porphyritic texture. The light-colored feldspar crystals grew during slow cooling, and subsequent chilling probably produced the fine-grained dark matrix. (Photo from Ward's Natural Science Establishment, Inc., Rochester, N.Y.)

The temperature of basalt lava is in the range from 1000°C to 1200°C, felsite lava is about 900°C, and granitic rocks are believed to be emplaced at about 820°C. Advancing basalt flows at Hawaii have been measured at temperatures as low as 745°C. The cooling time of igneous rocks is not well known. Batholiths are believed to solidify in 1 to a few million years. Dikes, depending on size, probably solidify in several tens of thousands of years. Basalt lava is known to have hardened to a depth of over 3 m in 2 years. Rocks are poor conductors of heat; hence, intrusive rocks take a very long time to cool because the heat must be conducted away by the surrounding rocks. Lava flows lose their heat to the atmosphere more rapidly.

A summary of textural terms will be of help in learning to recognize igneous rock types:

Glassy—no crystals.
Fine-grained—grains can be seen in sunlight with a good hand lens.
Coarse-grained—crystals easily seen.
Pegmatitic—crystals over 1 cm.
Porphyritic—generally are named for the texture of the groundmass, giving such terms as *porphyritic basalt* for a dark, fine-grained rock with some larger crystals, and *porphyritic granite* for a granite with some much larger crystals.

ORIGIN OF MAGMA AND PLATE TECTONICS

Igneous rocks must originate deep in the earth. To find out how and where magmas form, volcanic areas should be considered. The volcanic regions associated with plate tectonics are active geologically and are a good place to study.

At the mid-ocean ridges, new basaltic ocean floor is created. Submarine volcanism there has been observed and photographed by geologists working in small research submarines. At a few places, the mid-ocean ridge rises above sea level, forming islands, such as Iceland, where basalt volcanism can be observed. The mid-ocean ridges are also areas of high heat flow. Heat-flow measurements taken all over the earth reveal that at the mid-ocean ridges, the flow of heat energy from the interior is about twice the average of the rest of the ocean. This localized heat flow apparently is the cause of the mid-ocean ridges and their volcanic activity.

The ocean-floor basalt is about 5 km thick, and this layer and the overlying sedimentary and volcanic rocks are called the **oceanic crust**. The layer below the crust is called the **mantle**. It is from this layer that the oceanic basalt must originate. The mantle is believed to be composed largely of the minerals olivine and pyroxene, which, depending on the amounts of each, form the rock called *peridotite*, or *dunite*. Seismic studies show that in parts of the upper mantle, the rocks are soft because seismic waves are attenuated. The temperature and pressure at that level are such that the mantle rocks are partially melted. At the places where the heat flow from the deep interior is high—the mid-ocean ridges—more of the rocks may be melted. Rocks expand when they are heated and when they are melted, and this expansion may be the reason for the mid-ocean ridges. The hot rocks take up more volume, so the upper mantle pushes the sea floor upward, forming the mid-ocean ridges. The mantle rocks below the ridges, although partially melted, are still quite rigid (Fig. 3–34).

The origin of the basaltic magma that rises to the surface at the mid-ocean ridges is probably the partially melted material from the upper mantle. Rocks do not have a single melting point as simpler materials do. Ice and copper, for instance, at any given pressure, will each melt at a certain temperature. Many minerals and

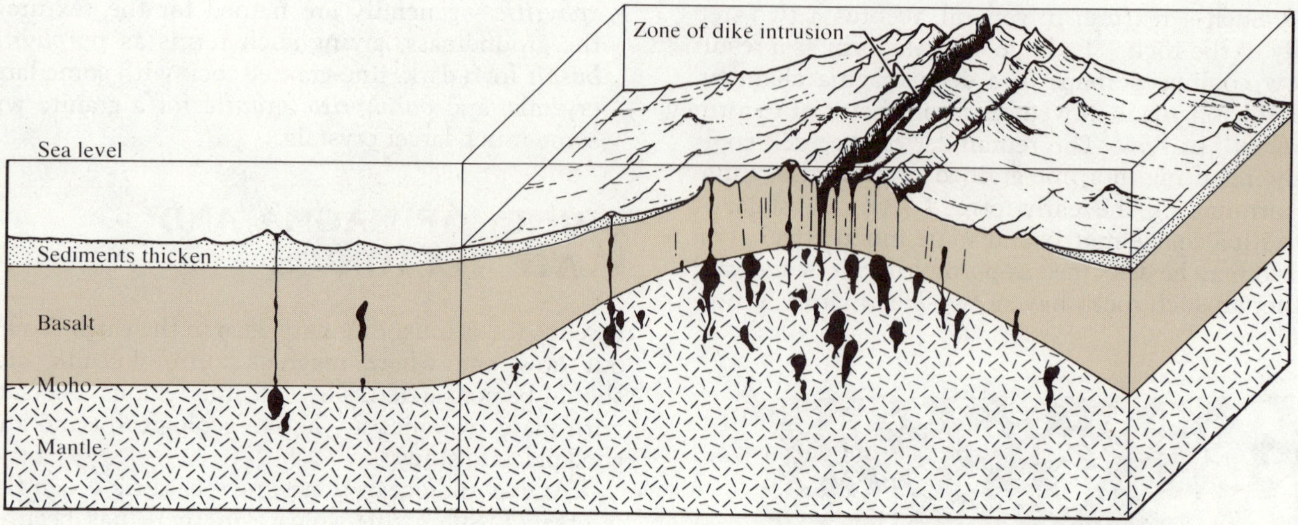

FIGURE 3–34
Ocean-floor basalt originates at mid-ocean ridges by partial melting of mantle rocks. (Drawing by Robert W. Tope)

most rocks behave differently, and melting occurs over a wide temperature range. Partial melting of the mantle peridotite can form a magma of basaltic composition. A basalt magma formed in this way would be less dense than the rest of the mantle and so would rise to cause submarine volcanism on the ocean floor.

The volcanic island arc volcanoes erupt lava of andesite composition. This magma, too, may be caused by partial melting. At the volcanic island arc-trench areas, seismic studies show that the basaltic ocean crust and its overlying sediments are carried deep into the upper mantle (Fig. 3–35). In this case, it is the basalt and the sediments that are partially melted at the higher temperature and pressure into which they are pushed in the upper mantle. The magma generated in this way may not always reach the surface and cause volcanic activity. These less fluid magmas may form intrusive bodies such as batholiths, as suggested in Fig. 3–35.

MOVEMENT OF MAGMA IN THE CRUST

Once the magma is generated, the next question is, Why does it rise in the crust? Several suggestions have been made. Rocks, like most substances, expand when

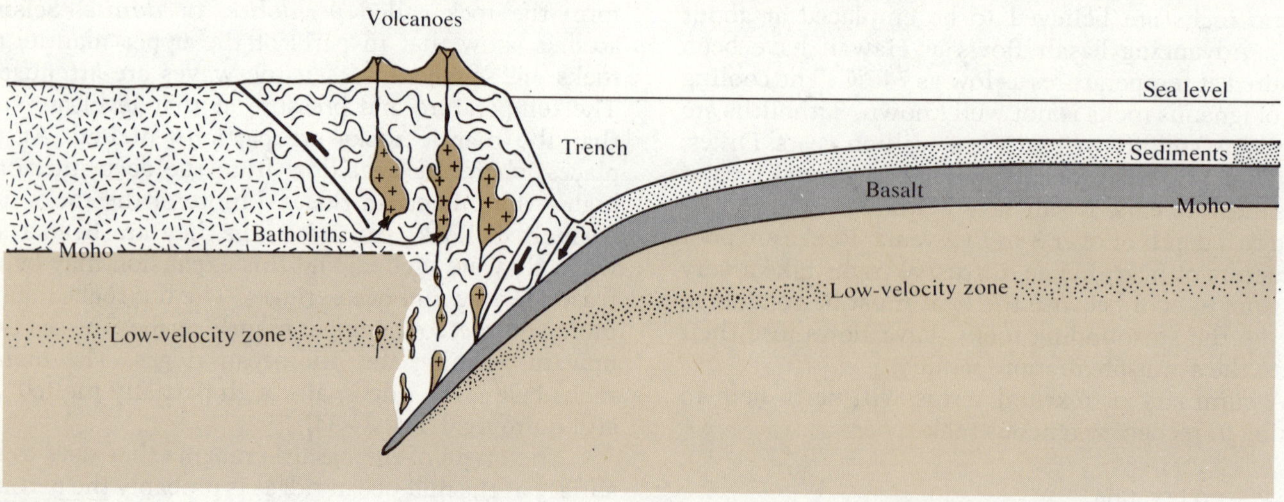

FIGURE 3–35
Batholiths and volcanoes form at a convergent plate boundary from partial melting of ocean-floor basalt and sedimentary rocks. (Drawing by Robert W. Tope)

they are melted. Thus, magma is less dense than the surrounding rocks and tends to rise, much as oil mixed with water rises and floats on the top.

Salt domes may illustrate intrusion caused by gravitational rise of lighter material. Salt domes occur in many parts of the world, and because some are associated with oil fields, they have been closely studied. Salt domes occur in sedimentary rocks; the salt originates from salt beds formed by the evaporation of seawater. Salt, or halite as the mineral is called, has a specific gravity of 2.2. The specific gravity of sedimentary rocks increases with depth of burial because the weight of the overlying rocks compresses the deeper rocks. At the Gulf Coast of the United States, the near-surface sedimentary rocks have a specific gravity of 1.9, at 600 m the specific gravity is 2.2, and at 3000 m it is 2.4. In this area, the salt beds are very deep and thus are much lighter than the surrounding rocks. The salt of lesser density tends to rise and float to the top. This seems unlikely because we are used to considering rocks as strong, rigid materials. It can be easily shown, however, that rocks are not strong enough to support themselves. A moment's thought will show that rocks have finite strength and that the weight of overlying rocks will exceed this strength at some depth. Below this depth, rocks behave as viscous fluids. In the present case, salt behaves as a viscous fluid and the overlying sedimentary rocks are very weak. To start the formation of a salt dome, some initial irregularity is probably necessary, such as a high point on the salt bed, or a weakness or break in the overlying sediments. The rising salt may remain connected to the salt bed or might, in some cases, become completely detached from the salt bed. The tops of many salt domes are mushroom-shaped (Fig. 3–36). Some salt domes may have risen up to 13 km in the Gulf Coast. The mechanics of salt dome formation are complex, and some downward movement of the surrounding sedimentary rocks may have occurred.

Magmas are less viscous than salt and, like salt, are less dense than the rocks at depth. Thus, the same mechanism that may account for the rise of salt domes may also account for the rise of some magmas. The internal structure of salt domes is somewhat similar to the internal structures of some intrusive igneous rocks, further suggesting a similarity. The heat of igneous rocks could weaken the surrounding rocks and thus assist the rise.

A possible source of pressure within the magma is the expansion of the volatiles. This was discussed earlier in connection with the surface manifestations of volcanism. In the present context, the volatiles in the newly formed magma would tend to expand as the magma rises because the confining pressure (weight of overlying rocks) is reduced. This might crack the enclosing rocks, allowing the magma to rise further.

Expanding gas bubbles might also be able to erode the enclosing rocks. This principle is used in a type of drilling and may be the mechanism by which some volcanic rocks reach the surface. Possible examples are the cylindrical, pyroclastic-filled conduits found in some volcanic areas.

Another possible way magma can work its way through the crust is to react with the enclosing rocks. The magma could, by chemical reaction, dissolve or partially dissolve the rocks above the magma. Some melting or partial melting is also possible. For melting to take place, a similar amount of crystallization would have to occur elsewhere in the magma, because magmas do not have enough excess heat energy to do much melting. Excess heat energy would have to be in the form of high temperature, but lava at the surface is generally very close to its melting point, indicating that, in general, magmas do not have excessively high temperatures. This process would tend to change the composition of the magma by assimilation of parts of the enclosing rocks. Differentiation would probably also be going on in the magma.

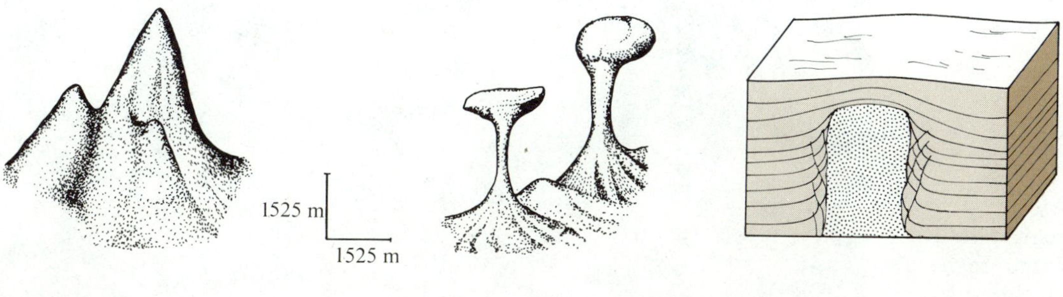

FIGURE 3–36
Salt domes showing some common shapes and the fractures caused in the surrounding rocks by their rise.

EMPLACEMENT OF BATHOLITHS

The classic explanation of the emplacement of batholiths in the crust is **stoping**. The name comes from the mining term for the process of removing ore overhead. The hot magma may cause blocks of the overlying enclosing rock to be broken off and fall through the melt. Such a process requires a low-viscosity magma and would allow much contamination of the magma by reaction with the stoped blocks. Most batholiths contain many inclusions, probably of the enclosing rocks, giving some credence to this hypothesis (see Figs. 3–37 and 3–38).

The emplacement of batholiths is a major problem of geology. The great size of a batholith like the Sierra Nevada immediately raises the question, What happened to the large volume of rocks the batholith displaced? Traditionally this question has been answered by two opposing points of view. One school of thought holds that the batholith made room for itself by a combination of stoping and pushing aside the preexisting rocks. The many included blocks, presumably of prebatholith rocks, found in some batholiths, and the zone of the deformed older rocks that surrounds most batholiths are the main evidence for this viewpoint. The opposing view is that the batholith was created more or less in place by the metamorphism of the preexisting rocks that accompanied the mountain-building event during which the batholith was emplaced. This possibility is included in the rock cycle (Fig. 1–2). The evidence for this theory is the metamorphic rocks that surround most batholiths. In some cases these zones of high-grade metamorphic rocks are very wide, and in other cases the heat from the crystallizing magma is enough to account for the smaller halo of metamorphism.

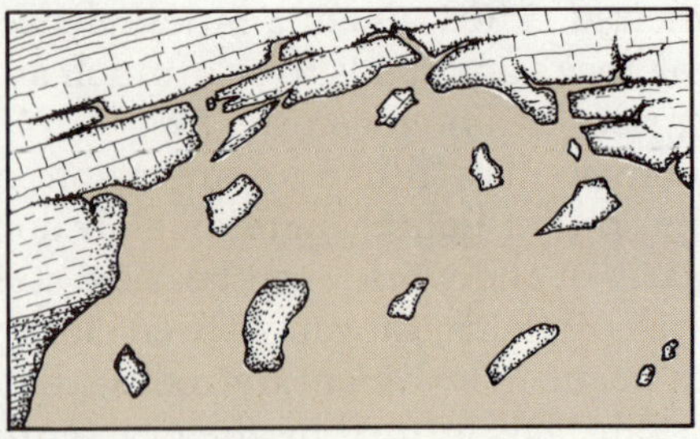

FIGURE 3–38
Stoping. Forceful injection of the magma breaks fragments from the roof, which sink through the magma, reacting with it.

It seems unreasonable to think that a huge batholith the size of the Sierra Nevada was all emplaced at the same time. Recent studies have shown that this batholith is composed of a number of much smaller plutons (see Fig. 3–39). The finer-grained margins of these plutons that resulted from more rapid cooling indicate that at no time was the entire batholith molten. Radioactive dates indicate that the emplacement of the batholith took place over a few tens of millions of years.

Because of their large surface area, batholiths traditionally have been considered to be thick bodies extending deep into the crust. Some recent seismic and gravity studies suggest that batholiths may be relatively thin features. These two interpretations of batholiths are shown in Figs. 3–26 and 3–40.

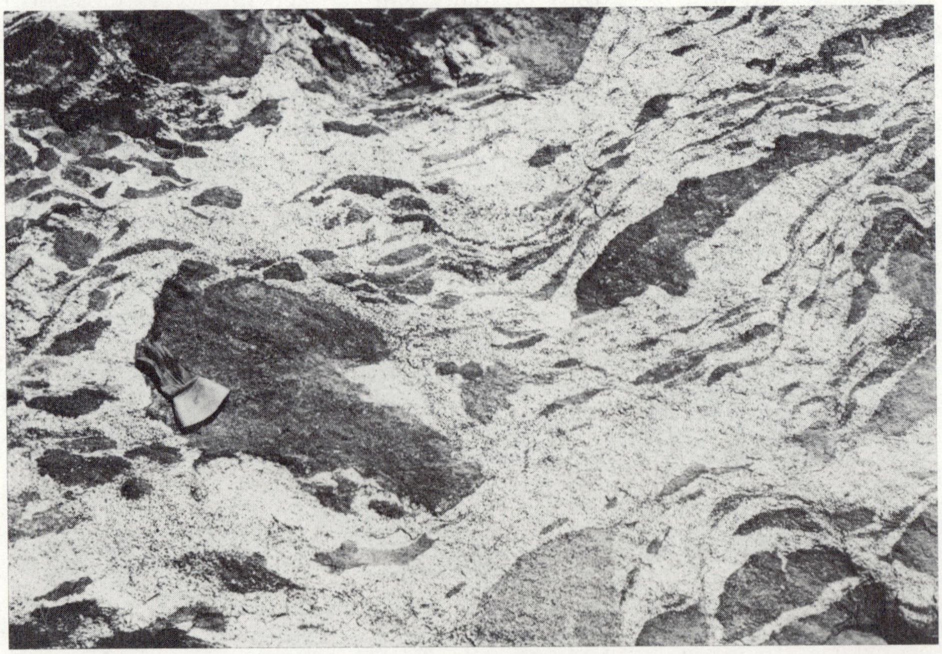

FIGURE 3–37
Granite with many dark inclusions in Yosemite National Park, California. The glove shows the scale. (Photo by F. C. Calkins, U.S. Geological Survey)

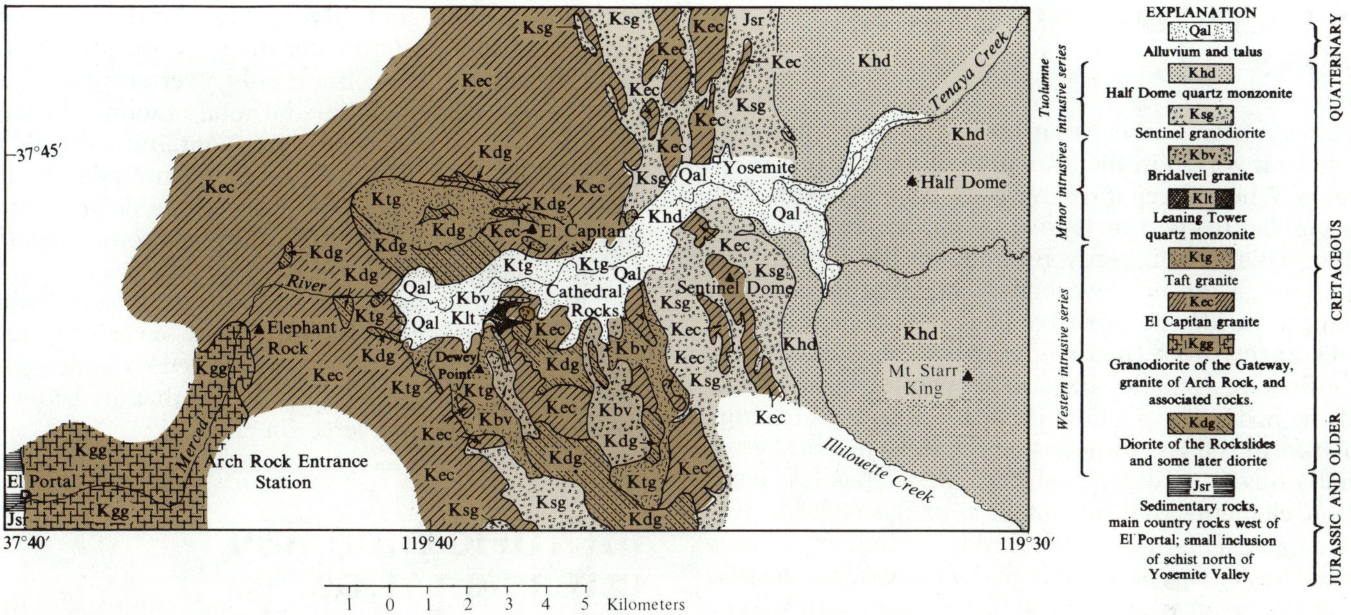

FIGURE 3–39
Generalized geologic map of Yosemite Valley in the Sierra Nevada, California. The Sierra Nevada batholith is made of many smaller intrusive bodies as shown on this map. Granodiorite and quartz monzonite are types of granitic rocks included under the term *granite* in this text. (From F. C. Calkins and D. L. Peck, California Division of Mines and Geology Bulletin 182, 1962)

All of these data have been incorporated into one recent theory. This theory holds that the batholith is formed by the bubblelike rise of the detached pods of magma generated in the upper mantle or deep crust. These pods of magma form a thin batholith near the surface. At some places, the magma reaches the surface and forms extensive volcanic rocks, and some batholiths intrude volcanic rocks that are similar in composition to the main batholith. These magma pods tend to spread out near the surface and so might account for the deformation that surrounds many batholiths. In passing through the crustal rocks, the magma would react with those rocks and could assimilate some of them by reaction and partial melting. In any case, the rocks through which the magma passes would be metamorphosed. Erosion of the thin layer of granitic rocks would expose an extensive metamorphic terrane similar to that found in many mountain ranges. Thus, this theory would explain most of the data of batholiths, and also the origin of extensive areas of metamorphic rocks. Like batholiths, these metamorphic terranes are difficult to explain because the known rise in temperature with depth is not enough to account for the amount of metamorphism.

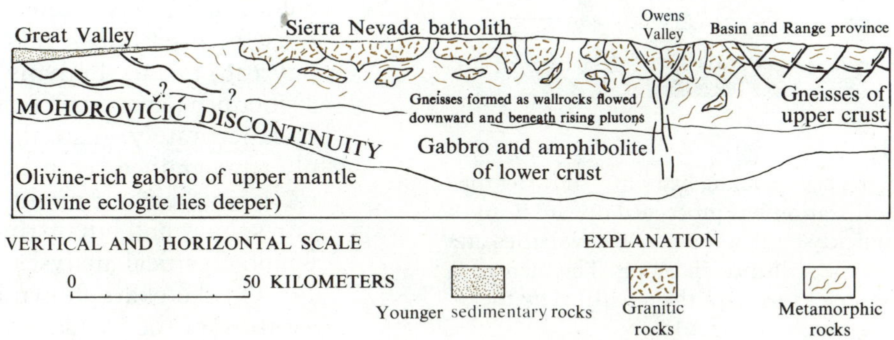

FIGURE 3–40
Hypothetical cross section through the Sierra Nevada, based on the idea that the granitic rocks of the batholith moved through the metamorphic rocks to their present positions. This cross section is based in part on seismic data. Compare with Fig. 3–26, which shows a thicker batholith. (From W. Hamilton and W. B. Myers, U.S. Geological Survey Professional Paper 554C, 1967)

PEGMATITES AND QUARTZ VEINS

The water and other volatiles (gases) in magma may also be involved in the formation of pegmatite and ore veins. When a deep intrusive body, such as a batholith, is emplaced, the rim is quickly chilled and solidifies first. This is commonly recorded by the fine-grained margins of an intrusive body. The volatiles are thus trapped inside the intrusive body; and as it solidifies, presumably from the margin inward, the volatiles are confined to a smaller and smaller volume in the center of the body. The residual fluid is rich in the elements that form the last minerals on the Bowen series, especially quartz, feldspar, and mica, as well as less common elements that do not fit into the rock-forming minerals. This water-rich residual fluid eventually escapes, perhaps as a result of the increase in pressure that can result when the volatiles are concentrated in a small volume. The increase in pressure may cause cracks that are filled by this material (Fig. 3–41). Contraction due to cooling may also help to cause the cracks. The fluid has a low viscosity because of the volatiles, and this increases the mobility of the ions. As a result, very large crystals form. The resulting very-coarse-grained rocks are called **pegmatites**, and they may have the same composition as the enclosing magma or may contain unusual minerals. In the latter case, the fluids are believed to concentrate the minor elements to form the unusual minerals. Pegmatites commonly occur as dikelike masses in or near intrusive bodies. They are actively sought because they contain large and often beautiful minerals, such as beryl, tourmaline, topaz, fluorite, all of the rock-forming minerals, and many more.

The residual fluids are probably also responsible for the formation of quartz-metal veins, but the process is not yet understood. The last fluid in a batholith would be rich in quartz and in the elements such as copper, lead, zinc, gold, silver, and sulfur that do not fit into the crystal structures of the rock-forming minerals. A batholith may contain only a very tiny percentage of these elements, but the total amount of metal may be quite large. Quartz veins containing metals in the form of sulfide minerals are found around some intrusive bodies (Fig. 3–41). These veins do not grade into pegmatites, so they must form by a somewhat different, although probably related, process.

In still other cases, these fluids may cause widespread alteration of the intrusive body or the surrounding rocks. Much more study is needed to understand the actions of the magmatic volatiles that are believed to cause all of these diverse effects.

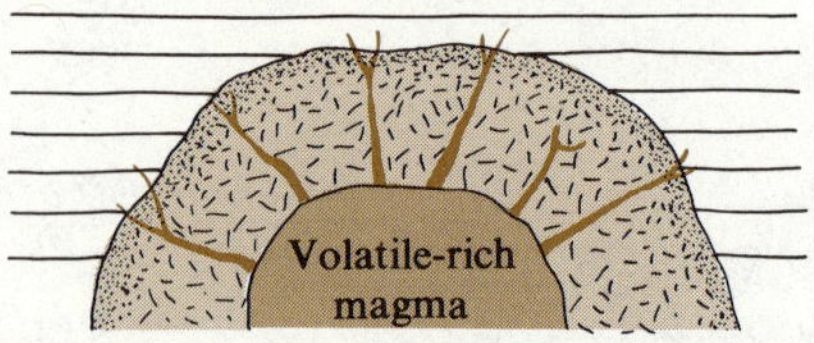

FIGURE 3–41
Cross section of a crystallizing batholith showing the origin of veins. Crystallization has proceeded inward from the fine-grained, quickly chilled margin. The volatiles are concentrated in the still-molten center. The increased pressure causes cracks to form that are filled by the metal-bearing, quartz-rich volatiles.

IDENTIFICATION AND INTERPRETATION

Identification of the coarse-grained igneous rocks will present no serious problems if one is able to identify the minerals and estimate their relative amounts. The first clue comes from the color of the rock, which, following the charts (Figs. 3–42 and 3–43), will suggest which minerals should be looked for. In the use of the term *light color,* it should be understood that any shade of red or pink, no matter how dark, must be considered a light color. In the same way, any shade of green should be considered a dark color. Overall color should not be used in place of good mineral determination, because exceptions are common enough to be troublesome (especially on examinations).

In a sense, the fine-grained, or volcanic, rocks are more difficult to identify than are the coarse-grained rocks because less data, in the form of identifiable minerals, is available; but because of this, the classification is less rigid. The procedure is similar in that color is used as the first step and, if no porphyritic crystals are present, the only step. If the porphyritic crystals can be identified, they are used in the same way as are the identifiable minerals in coarse-grained rocks; but because all potential minerals may not be present, color must be considered in deciding on a name. Care must be used not to mistake filled gas holes for porphyritic crystals. In many cases, it is necessary to have a chemical analysis in order to name a fine-grained rock accurately. Also, the volcanic rocks are not simply fine-grained equivalents of the coarse-grained rocks, as the chart might lead one to believe, but contain some small but distinct variations in composition, which chemical analyses will detect.

On the chart shown as Fig. 3–42, granite is used for most of the coarse-grained, quartz-bearing rocks. A

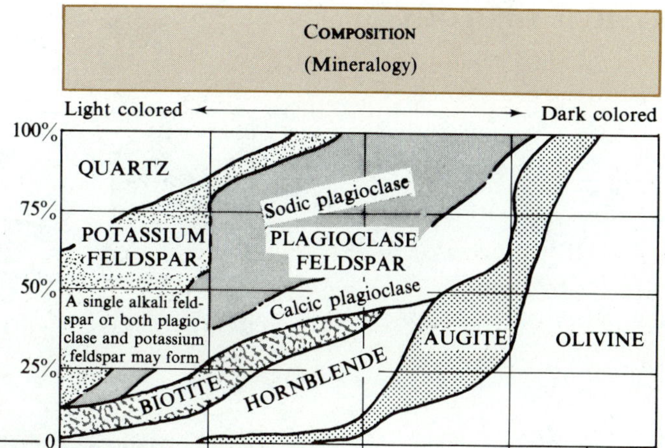

TEXTURE (Crystal size)					
Very coarse (Pegmatitic)	Pegmatitic Granite	Pegmatitic Diorite	Pegmatitic Gabbro	—	—
1 cm — Coarse°	Granite*	Diorite	Gabbro	Peridotite	Dunite
0.1 mm — Mixed sizes: Coarse matrix with some larger crystals (Porphyritic)	Porphyritic† Granite	Porphyritic† Diorite	Porphyritic† Gabbro	—	—
Mixed sizes: Fine matrix with some larger crystals	Porphyritic† Rhyolite	Porphyritic† Andesite	Porphyritic† Basalt	—	—
0.1 mm — Fine°	Felsite if light in color	Andesite if intermediate	Basalt	—	—
0 — Glassy	Obsidian (Pumice if frothy)		Tachylite	—	—

Minor accessory minerals are omitted.

*Granite is used in the widest sense. Several names are in common use, depending on the type of feldspar.

†Modifier porphyritic is generally omitted unless crystals are very prominent.

°Coarse and fine are used in the sense that phaneritic and aphanitic are used in petrography.

FIGURE 3–42
Chart showing the classification of the igneous rocks. Composition is indicated vertically, and texture is indicated horizontally. The upper part of the figure shows the range in mineral composition of each rock type.

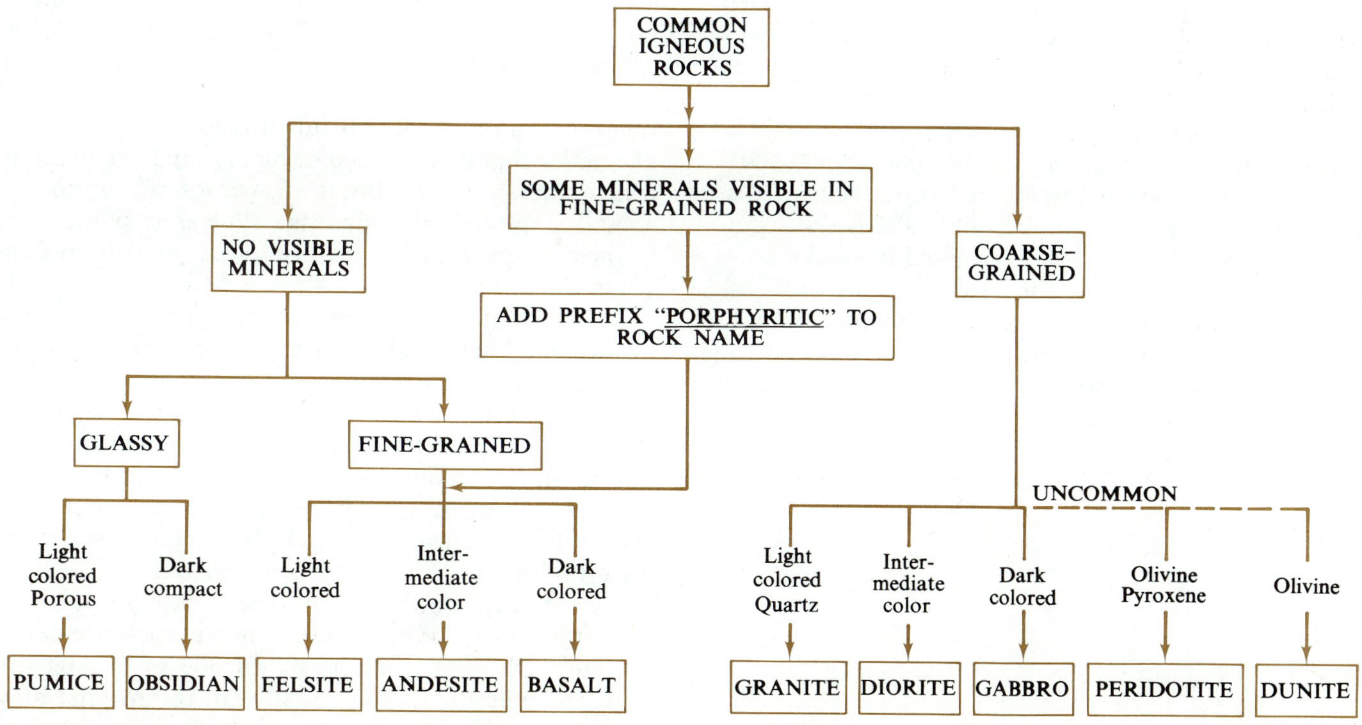

FIGURE 3–43
Flow diagram showing the identification of igneous rocks.

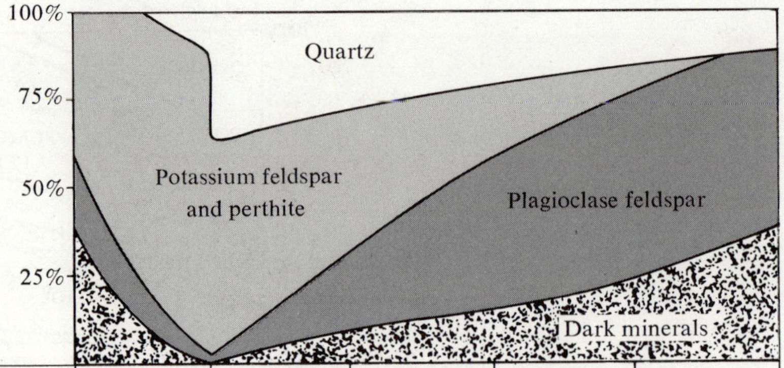

Coarse-grained	Syenite	Granite	Quartz monzonite	Granodiorite	Quartz diorite or Tonalite
Fine-grained	Trachyte	Rhyolite	Quartz latite	Quartz latite or Dacite or Rhyodacite	Dacite

FIGURE 3–44
Subdivision showing the names in use for the granitic rocks. This chart is an amplification of the left end of Fig. 3–42.

number of other names are also in use for these rocks, depending on whether they contain one feldspar, two feldspars, or mixed feldspar (perthite) (see Fig. 3–44). Diorite generally has more dark minerals than does granite, and the feldspar is plagioclase. Quartz diorite is distinguished from granite in that it contains only plagioclase feldspar. The latter distinction, however, is not always easy to make.

Understanding the formation of a rock is much more important than simply naming it. It is necessary to discuss classification so that the meaning of rock names and the relationships among the various rock types are understood. Therefore, when studying a rock, try to interpret its origin, following mainly the discussion in the section on texture and occurrence. Thus, the coarse texture of a granite implies that the specimen came from a large, slowly cooled intrusive body that crystallized deep in the earth and has been exposed by uplift and deep erosion. A pumice sample must have come from a volcano, perhaps from an explosive eruption, and study of other nearby exposures may enable reconstruction of the volcano. When actual outcrops are studied, all of the information discussed here can be used in deciphering the origin and history of the rocks.

SUMMARY

Igneous rocks were once melted and have solidified from magma. Magma is a natural, mobile, hot melt composed of rock-forming materials (mainly silicates) and volatiles (mainly steam) that are held in solution by pressure. It may or may not contain suspended solids.

In the sudden eruption of low-viscosity magma, the expanding gases form a froth, resulting in the light, glassy rock pumice. Coarser bubble holes result in the rock scoria. In eruption of more viscous magmas, the gas pressure shatters the rock into pyroclastic fragments: volcanic ash if sand-sized or smaller, lapilli if larger but less than 64 mm. Those larger than 64 mm are called blocks if angular and bombs if rounded. The pyroclastic rocks formed from these materials are tuff if composed of ash, lapilli tuff if made of lapilli, pyroclastic breccia if composed of blocks, and agglomerate if made of bombs. Tephra is a general term for unconsolidated pyroclastic material. Volcanic bombs are formed by pyroclastic material that is still molten when thrown out.

The quietest eruptions are very fluid basalt lava flows from dikes or dike swarms that can produce great thicknesses and cover huge areas. A shield volcano is formed by eruption of very fluid basalt lava from a central vent. Fluid basalt forms ropy or pahoehoe lava flows. More viscous basalt lava breaks into blocks, forming blocky or aa lava flows. Cinder cones are formed by pyroclastic material thrown out of a vent by the eruption of still more viscous magma.

Composite volcanoes are the very high volcanic mountains made of a combination of pyroclastic material and lava. If the lava is so viscous as to plug the vent, the pressure of the volatiles in the magma may build up until an explosion results. The more viscous the magma, the more numerous or destructive the explosions.

Very explosive eruptions of hot gas and shattered glass fragments are called nuées ardentes (also glowing clouds or glowing avalanches); this fluid pyroclastic material forms the rock named welded tuff when the hot glass fragments come to rest and weld together. Welded tuffs are relatively abundant and may or may not be associated with dome volcanoes.

Pillow lava forms when lava flowing into water bodies is chilled relatively rapidly.

Eruptions cannot yet be predicted, but magma movements are detected by many shallow, weak earthquakes, by tilt-meter measurements, and so forth.

Volcanic destruction is of two kinds: direct damage by the material erupted; and indirect damage by mudflows of pyroclastic material, and by floods from melted glaciers and breached lakes.

Bowen's reaction series shows the sequence in the crystallization of a basalt magma.

Separation of early-formed crystals may form rocks with a composition different from the original melt; this process is called differentiation.

In a continuous mineral series, the full range of compositions of that mineral group occurs in sequence from one end member to another. In a discontinuous mineral series, the change is from one mineral group to another in separate, discrete steps.

With falling temperature, Bowen's reaction series has the following sequence of crystallization: a discontinuous series of olivine to pyroxene to amphibole to biotite, concurrently with a continuous series of calcic to sodic plagioclase, followed by the sequence potassium feldspar, muscovite, and, lastly, quartz.

Igneous rocks are classified by composition and texture. The composition classification follows the reaction series. Texture refers to the grain size and indicates the mode of occurrence and cooling history of the rock.

Intrusive igneous rocks solidify below the surface of the earth; extrusive igneous rocks solidify on the surface. Plutonic igneous rocks are coarse-grained rocks that crystallize deep in the earth.

Intrusive igneous rocks are discordant if they cut across the bedding of the rocks they intrude, and concordant if they are more or less parallel to the bedding. Discordant intrusives are batholiths (very large, coarse-grained bodies), stocks or plutons (smaller coarse-grained bodies), and dikes (tabular, usually parallel-walled, linear, fine-grained bodies). Concordant intrusives are sills (thin, usually fine-grained bodies intruded between sedimentary layers without noticeable deformation) and laccoliths (thicker, fine- to coarse-grained bodies which up-arch the sedimentary layers above them).

Extrusive rocks result from lava flows or other volcanic activity. They are fine-grained or, with very rapid cooling, glassy.

The textural terms are:

Glassy—no crystals.
Fine-grained—grains can be seen in sunlight with a good hand lens.
Coarse-grained—crystals easily seen.
Pegmatitic—crystals over 1 cm.
Porphyritic—coarse crystals in a fine-grained or glassy matrix, or some very coarse crystals in a coarse-grained rock. Porphyritic texture may record a change in conditions during the cooling history of the rock.

The mid-ocean ridges are areas of basalt volcanism. The basalt apparently originates from partial melting of upper-mantle rocks. The andesite magma of volcanic island arcs probably originates from partial melting of the descending slab of ocean-floor basalt and its overlying sediments.

Magma most likely rises because the melted rock expands and is less dense than the surrounding rocks. Salt domes from buried salt beds are also less dense than the surrounding rocks, and may illustrate another case of gravitational rise of lighter material. Expansion of the volatiles in magma or erosion by expanding gas bubbles may help the magma to rise.

Most large batholiths are composed of a number of smaller plutons. Traditionally, batholiths have been thought to extend deep into the crust and to have been emplaced by stoping and pushing aside the preexisting rocks. Stoping is a process by which blocks of preexisting rocks above the magma fall through the melted magma, allowing the magma to rise. Another theory is that batholiths may be relatively thin features. They may be emplaced by the bubblelike rise of detached pods of magma, which metamorphose the rocks through which they pass.

Quartz veins and very-coarse-grained pegmatite masses may result from the residual fluid in magma, which is composed of trapped volatiles, elements to form the last minerals of Bowen's series, and less common elements which do not fit into the common rock-forming minerals and therefore may form unusual minerals.

In general, coarse-grained igneous rocks are identified by identifying minerals and estimating their relative amounts. The type of feldspar is important. Fine-grained igneous rocks are identified primarily by color or chemical analysis. Study of the texture and composition of the rock enables interpretation of its formation.

REVIEW EXERCISES

1. Define *igneous rock.*
2. Describe the formation of pumice.
3. Describe the types of volcanoes. Include how they are formed and which is steepest and why.
4. What hazards do volcanoes pose?
5. Describe the crystallization of a basaltic melt.
6. Describe the crystallization of plagioclase in a magma.
7. What happens when a crystal of calcic plagioclase is melted?
8. Is the classification of most igneous rocks genetic? Discuss.
9. How can a sill be distinguished from a flow?
10. What cooling history does a porphyritic texture imply?
11. What is a plutonic igneous rock?
12. What is a batholith; what is the usual composition and what problems do batholiths present?
13. Name a concordant igneous body.
14. In what type of occurrence (sill, flow, batholith, etc.) would you expect to find glassy, fine-grained, porphyritic, and coarse-grained igneous rocks?
15. Describe the formation of ore veins.
16. On the basis of what two characteristics can igneous rocks be classified?
17. What minerals are most likely to be visible in a felsite porphyry?
18. How does pegmatite form?
19. What are the main volcanic gases?
20. What is the probable origin of the mid-ocean ridge basalt?
21. What is the probable origin of andesite at volcanic island arcs?

SUGGESTED READINGS

GENERAL

DIETRICH, R. V., and SKINNER, B. J. *Rocks and Rock Minerals.* Somerset, N.J.: John Wiley and Sons, 1979, 319 pp.

ERNST, W. G. *Earth Materials.* Englewood Cliffs, N.J.: Prentice-Hall, 1969, 150 pp. (paperback).

IGNEOUS PRINCIPLES

MASON, B. *Principles of Geochemistry,* 3rd ed. Chapter 5, "Magmatism and Igneous Rocks," pp. 96–148. New York: John Wiley and Sons, 1966.

TUTTLE, O. F. "The Origin of Granite." *Scientific American* (1955), Vol. 192, No. 4, pp. 77–82. Reprint 819, W. H. Freeman and Co., San Francisco.

VOLCANOES

ASPINALL, W. P.; SIGURDSSON, H.; and SHEPHERD, J. B. "Eruption of Soufrière Volcano on St. Vincent Island, 1971–1972." *Science* (1973), Vol. 181, No. 4095, pp. 117–24.

BULLARD, F. M. *Volcanoes of the Earth.* Austin, Tex.: University of Texas Press, 1976, 579 pp.

CRANDELL, D. R., and MULLINEAUX, D. R. *Volcanic Hazards at Mount Rainier, Washington.* U.S. Geological Survey Bulletin 1238, 1967, 26 pp.

EATON, G. P., AND OTHERS. "Magma beneath Yellowstone National Park." *Science* (1975), Vol. 188, No. 4190, pp. 787–96.

EATON, J. P., and MURATA, K. J. "How Volcanoes Grow." *Science* (1960), Vol. 132, No. 3432, pp. 925–38.

HEIKEN, G. "Pyroclastic Flow Deposits." *American Scientist* (1979), Vol. 67, No. 5, pp. 564–71.

KITTLEMAN, L. R. "Tephra." *Scientific American* (1979), Vol. 241, No. 6, pp. 160–77.

MACDONALD, G. A. *Volcanoes.* Englewood Cliffs, N.J.: Prentice-Hall, 1972, 510 pp.

———. "Volcanology." *Science* (1961), Vol. 133, No. 3454, pp. 673–79.

MOORE, J. G. "Mechanism of Formation of Pillow Lava." *American Scientist* (1975), Vol. 63, No. 3, pp. 269–77.

PECK, D. L.; WRIGHT, T. L.; and DECKER, R. W. "The Lava Lakes of Kilauea." *Scientific American* (1979), Vol. 241, No. 4, pp. 114–28.

SHEETS, P. D., and GRAYSON, D. K., eds. *Volcanic Activity and Human Ecology.* New York: Academic Press, 1979, 644 pp.

WILLIAMS, R. S., Jr., and MOORE, J. G. "Iceland Chills a Lava Flow." *Geotimes* (1973), Vol. 18, No. 8, pp. 14–17. Describes the use of seawater in an attempt to save a fishing port from the hazards of a lava flow.

- weathering
- physical weathering
- chemical weathering
- frost wedging
- frost heaving
- needle ice
- exfoliation
- spall

4

- oxidation
- hydrolysis
- pH
- acid rain
- hydration
- dehydration
- base exchange
- spheroidal weathering
- core stone
- chelation
- microclimatic effects
- cation
- vesicle
- respiration
- soil
- residual soil
- transported soil
- soil profile
- A horizon (zone of leaching)
- duff
- B horizon (zone of accumulation)
- parent material
- hardpan
- laterite
- soil creep
- sediment pollution
- dose dependency
- trace element

Weathering and Soils

THROUGHOUT THE 4.6 BILLION YEARS OF EARTH HISTORY, geologic processes have been continually modifying the surface of the earth. For sure, the intensity of processes, and even the processes themselves, have varied considerably during geologic time. Nevertheless, there has been a continuous and relentless battle between those processes that tend to build continents and those that tend to wear them down. In this chapter, emphasis is placed on the first stages of the processes that tend to break rocks down. The group of physical, chemical, and biological processes that cause changes to take place in rocks and minerals at or near the surface of the earth are collectively known as **weathering**. They are particularly important to us because as residents on the land surface of the earth, we are every day in intimate contact with the weathering products of rocks. Along with air and water, soil, which is essentially the product of weathering, is one of our most necessary resources. Agricultural success in the last 7500 years has been a major factor in the success or failure of several cultures.

Throughout earth history, weathering has had a significant part in forming new rocks. The sand of beaches; the gravel, clay, and silt in rivers; the sand moved by the wind to form great fields of dunes; and the rock debris carried by glaciers are all sediment, mostly derived from weathering processes prior to being transported by wind, water, or ice. Unconsolidated deposits of sediment are often important as resources. For example, the great thickness of unconsolidated sediment in the San Joaquin Valley of California holds significant groundwater resources, and unconsolidated sediment of the right composition and size is important for construction material. Another aspect of the significance of weathering becomes apparent when it is realized that 75 percent of the rocks exposed on the surface of the earth are sedimentary rocks composed of weathered products of previously existing rocks. These sedimentary rocks contain virtually all of our energy resources, as well as many other resources.

TYPES OF WEATHERING

Weathering is a group of physical, chemical, and biochemical processes that operate on or near the surface of the earth to break rocks down. Weathering processes may be grouped according to whether the process tends to break the rock down mechanically (**physical weathering**) or attack it chemically (**chemical weathering**). In many instances, both groups of weathering processes are quite likely to be operating simultaneously.

PHYSICAL WEATHERING

The most common types of physical weathering include **frost wedging**, **frost heaving**, **expansion by plant roots**, and **exfoliation**. In addition, there are a variety of other physical processes such as daily variation in temperature, fire, and chemical expansion that facilitate or cause weathering.

Frost Wedging All rocks are fractured to some degree. In climates with sufficient precipitation or other sources of water and frequent freeze-thaw cycles, the process of water freezing and expanding in fractures is a significant weathering process. What happens is that water, upon freezing, expands approximately 9 percent. This expansion may exert considerable pressure on the walls of fractures, causing the rock to break (Fig. 4–1A). The size of the weathered blocks of rock depends on the spacing of fractures in the rocks. With intensive frost action and many cycles of freezing and thawing, this process of frost wedging may produce slopes covered with blocks of broken rock (Fig. 4–1B). The number of times a freeze-thaw cycle occurs is significant because each refreezing exerts a new set of forces on the rocks, thus furthering weathering. In North America, areas with the highest frequency of freeze-thaw cycles generally tend to be the mid-latitudes, as southern-latitude regions are too warm, and a freeze may last for months in the northern latitudes. However, there are numerous exceptions to this general picture in southern mountain and northern coastal areas, where there may be many days that both freeze and thaw.

Frost Heaving Frost heaving, like frost wedging, is associated with water freezing in fractures in rocks or pore spaces of unconsolidated surficial materials, especially frost-susceptible silts or soil. Frost heaving is a ground-ice phenomenon and involes the development of a lens, or mound, of ice that expands and forces the rock and soil upward (Fig. 4–2). This process is important in areas that freeze for extended periods of time and thaw only in the spring.

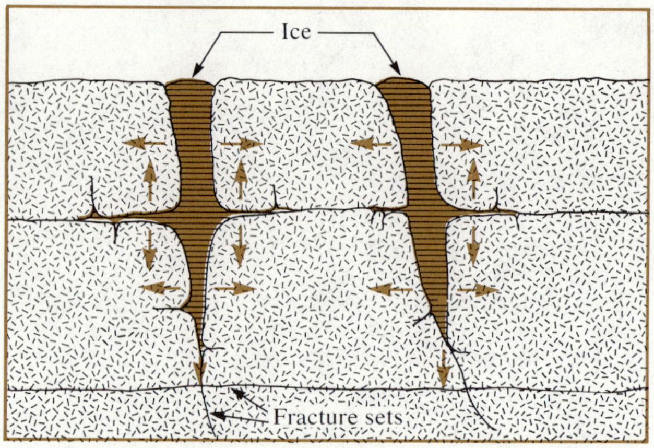

A.

B.

FIGURE 4–1
A. Idealized diagram of frost wedging. B. Slope in the Crazy Mountains of Montana composed of large blocks produced by frost wedging.

Frost heaving may also be an important physical weathering process where the temperature moves back and forth across the freezing point numerous times per year. On the piedmont of the southeastern United States as well as other areas, a phenomenon known as **needle ice** occurs during the winter. The phenomenon involves the formation of ground ice, which eventually breaks through the surface and may extend upward several centimeters (Fig. 4–3). This process tends to heave up the soil and cause considerable disturbance to the soil particles each time the soil freezes.

Expansion by Plant Roots As roots of trees and other plants begin to grow in rock fractures, the root system forces its way into the fractures and forces them further apart. Many sidewalks and streets as well as houses have been damaged by this process (Fig. 4–4).

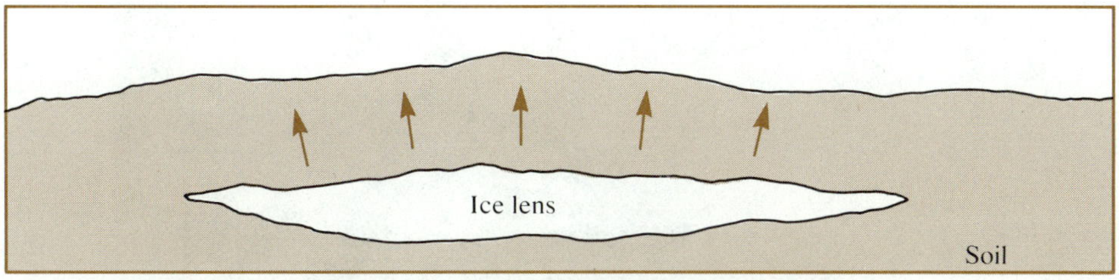

A.

B.

FIGURE 4–2
A. Idealized drawing illustrating frost heaving. B. Fence posts uplifted by frost heaving. (Photo courtesy Don Johnson)

Exfoliation Exfoliation is a physical weathering phenomenon associated with homogeneous rocks that have been buried deeply and are under pressure from the overlying rocks. These rocks, when exposed at the surface of the earth by rapid erosion, tend to expand. As the rock expands, fractures tend to form parallel to the surface of the earth. These fractures are known as **exfoliation fractures**, and the rock tends to break, or **spall**, along the zones of weakness. The process of exfoliation often leads to the development of beautiful domes (Fig. 4–5) or dangerous slopes, with the fractures inclined downslope providing slip planes along which landslides might occur (Fig. 4–6).

FIGURE 4–3
Needle ice at Turkey Run, Indiana. The growth of such ice often lifts soil particles. (Photo courtesy Don Johnson)

Exfoliation is most commonly found on intrusive igneous rocks such as granite, which are relatively inelastic. The Sierra Nevada of California contains numerous examples of exfoliation, and other good examples may be found in the southern Appalachian Mountains and many other areas.

Other Types of Physical Weathering Numerous physical processes may in one way or another facilitate physical weathering. For example, forest or brush fires cause physical weathering as the heated surface layer of rocks expands more rapidly than deeper parts of the rock, facilitating further weathering and development of rounded corners on blocks of rocks due to heat-induced spalling.

Weathering may also be due to the daily variation in temperature. In areas that experience a large daily variation in temperature, it might be expected that the

A.

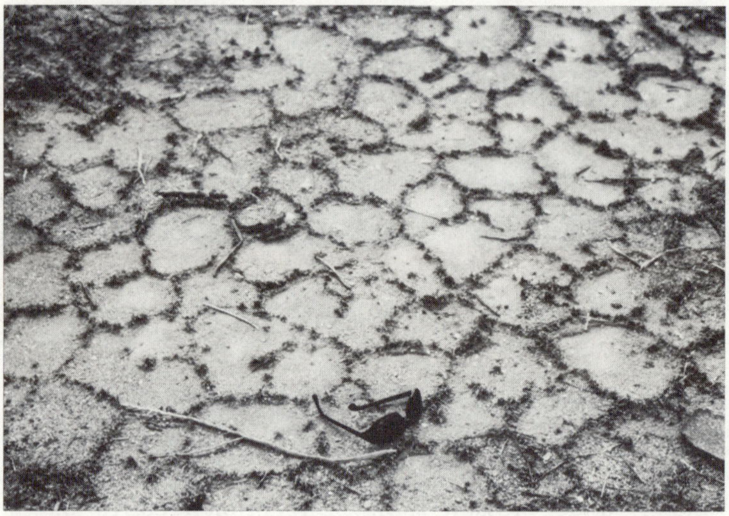

B.

FIGURE 4–4
A. Tree roots growing in fractures and along other planes of weakness in bedrock. B. Vegetation growing along mud cracks in the Mojave Desert near Palmdale, California. Both of these photographs illustrate the ability of plants to become established along planes of weakness either in rocks or in a soil, thus facilitating mechanical weathering.

FIGURE 4–5
Exfoliated dome. Half Dome, Yosemite National Park, California.

FIGURE 4–6
Exfoliation fractures inclined toward a highway in Yosemite National Park, providing convenient slip planes along which landslides may occur.

outermost layer of rocks will expand differentially during the hot temperatures and contract during the times of cooler temperatures, slowly weakening the rock and eventually causing small failures, granular disintegration, or spalling. Research to demonstrate this has not been successful, even though the argument seems reasonable. Therefore, the importance of daily variations in temperature in physical weathering remains debatable.

Other interesting physical weathering processes include the expansion of minerals as they change phases to form new materials that take up a larger volume. For example, feldspar and biotite, when they chemically weather to form clays, may expand, and this expansion may cause a physical weathering in the rocks around them. This process may be very significant since feldspars are the most common group of rock-forming minerals. Notice, however, that this phenomenon involves a chemical reaction necessary to form the clay. This will be discussed in greater detail when chemical weathering is considered.

In favorable circumstances, earth materials such as halite (NaCl, or common table salt), calcite ($CaCO_3$), and gypsum ($CaSO_4 \cdot 3\,H_2O$) are highly soluble in water and therefore may be found in solution. As solutions migrate through rocks the minerals may be precipitated, and as the crystals grow in rock fractures they may exert a force on the surrounding rocks, thus causing a physical weathering. Again, this process is associated with chemical reactions.

CHEMICAL WEATHERING

In many parts of the world, chemical weathering is more rapid than physical weathering. Furthermore, there is a definite interaction between the two, as physical weathering often prepares or leads the way for chemical weathering. This results because chemical weathering is dependent upon the surface area of exposure, and rocks fractured or broken by physical weathering or other processes will have a much larger surface area of exposure for chemical weathering (Fig. 4–7). The most important processes of chemical weathering are **oxidation**, **hydrolysis**, **hydration** and **dehydration**, **base exchange**, **spheroidal weathering**, and **chelation**.

Oxidation Oxidation is defined as the addition of oxygen or the loss of an electron to an element or compound. Oxidation products from iron-bearing minerals produce the characteristic brown, yellow, or red color of many soils. Oxidation is an important weathering process because it facilitates weakening, which generally breaks rocks down. The weathering process produces changes as the iron oxidizes, causing a partial collapse of the crystal lattice of the mineral. Thus, the oxidation process makes the mineral more susceptible to other weathering processes. See Fig. 4–8 for an illustration of the weathering by oxidation of an iron silicate mineral. A specific example is the oxidation of pyrite (FeS_2):

$$\boxed{\text{Pyrite}} + \boxed{\text{water}} + \boxed{\text{oxygen}} \rightleftharpoons \boxed{\text{iron oxide}} + \boxed{\text{sulfuric acid}}$$

$$FeS_2 + H_2O + 15\,O \rightleftharpoons 2\,Fe(OH)_3 + 4\,H_2SO_4$$

FIGURE 4–7
The rate of chemical weathering is dependent upon the surface area available. The more a rock is fractured or otherwise broken up by physical weathering, the more points of entry there are along which chemical weathering may occur. This idealized diagram dramatically shows the increase in surface area as the number of fractures in a square block increases.

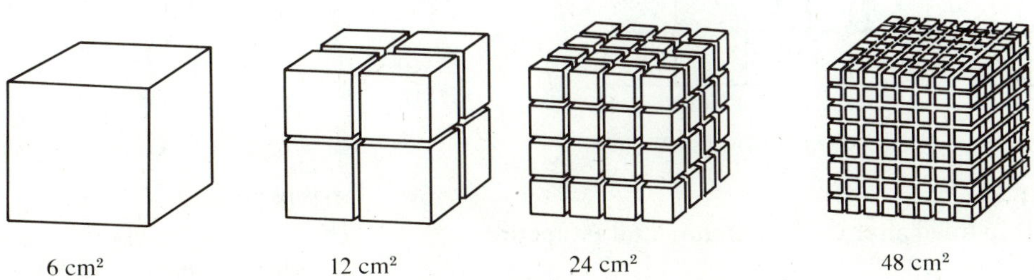

This reaction is of particular importance in the mining of coal because pyrite is often associated as waste and its oxidation results in the production of acid water, which pollutes streams and groundwater resources.

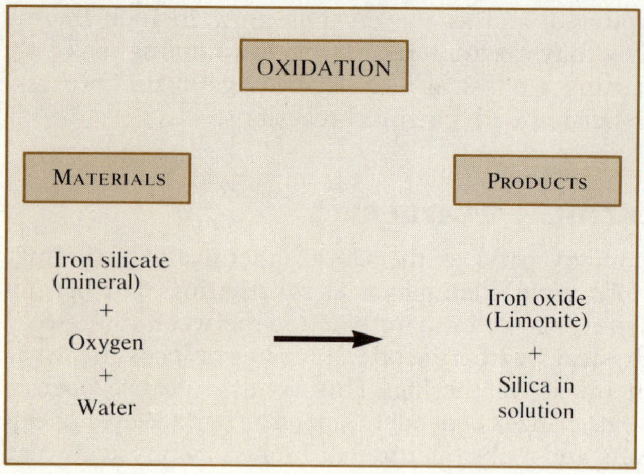

FIGURE 4–8
Oxidation.

Hydrolysis Hydrolysis is the most important of all of the chemical weathering processes because it is the major chemical weathering process of the silicate and carbonate minerals, which comprise over 90 percent of all of the rock-forming minerals. Hydrolysis involves the universal solvent water plus carbon dioxide (CO_2), which combine at surface conditions (because the carbon dioxide is soluble in water) to form a weak acid:

Carbon dioxide in atmosphere + water ⇌ carbonic acid

$$CO_2 + H_2O \rightleftharpoons H_2CO_3$$

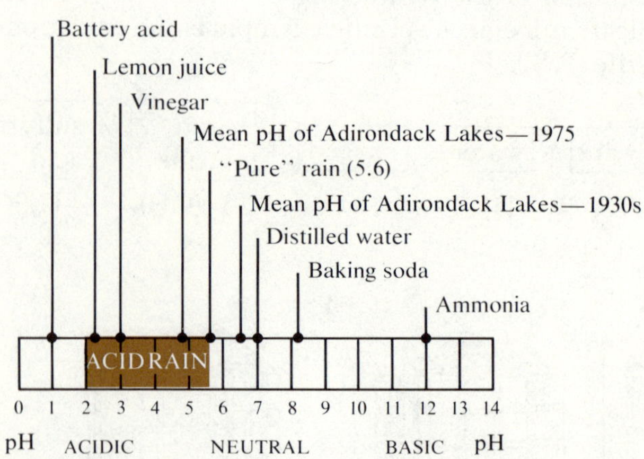

FIGURE 4–9
pH scale. (Modified after U.S. Environmental Protection Agency, 1979)

Thus, unpolluted rainwater is slightly acidic (Fig. 4–9) and can react with various substances to produce distinct products as follows:

Carbonic acid + mineral → major product

Carbonic acid + calcite → solution

Carbonic acid + feldspars → clays

The general equation to explain the chemical weathering of calcium carbonate ($CaCO_3$) is

Calcium carbonate + carbonic acid ⇌ calcium ion in solution + bicarbonate ion in solution

$$CaCO_3 + H_2CO_3 \rightleftharpoons Ca^{+2} + 2\,HCO_3^-$$

$$\upharpoonleft\downharpoonright$$

$$CO_2 + H_2O$$

carbon dioxide gas + water

and the solubility of calcium carbonate varies with the amount (partial pressure) of dissolved carbon dioxide and the acidity (**pH**) of the solution. Increasing the amount of dissolved carbon dioxide and thus the acidity will drive the equation to the right, dissolving more calcium carbonate. The generalized relationship between acidity and amount of dissolved calcium carbonate is shown in Fig. 4–10. The partial pressure or

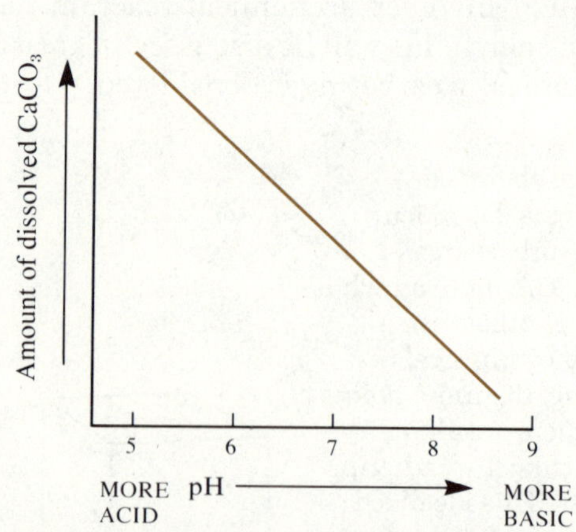

FIGURE 4–10
The relationship between amount of dissolved calcium carbonate ($CaCO_3$) and pH of the solution (water).

amount of carbon dioxide gas that can dissolve in water is temperature-dependent: warm water contains less carbon dioxide than cold water. Thus, calcium carbonate should dissolve more easily in cold than in warm water. To some extent this is true, but the total amount of water available is also important. For example, limestone (calcium carbonate) carvings in Egypt weather very slowly, and after centuries little damage can be seen. On the other hand, bring the same limestone carvings to New York City, where there is an abundance of precipitation and where rainwater combines with higher than normal concentrations of carbon dioxide, nitric oxide, and sulfur dioxide to form acid precipitation. In just a few years, the damage from weathering will be readily discernible.

Pure calcium carbonate will dissolve completely, leaving no residuals. Because limestone is seldom pure, residuals of clay and iron oxides are often left, explaining the origin of the red soils found in areas dominated by limestone bedrock.

Hydrolysis of feldspar minerals produces weathering products including clay minerals; silica in solution; and ions of potassium, sodium, or calcium in solution. As one example, consider the hydrolysis of the potassium-rich feldspar orthoclase ($KAlSi_3O_8$). The general reaction, *not balanced,* is

$$\boxed{\text{Orthoclase}} + \boxed{\text{carbonic acid}} = \boxed{\text{clay kaolinite}}$$
$$KAlSi_3O_8 + H_2CO_3 = Al_2Si_2O_5(OH)_4$$
$$+ \boxed{\text{silica in solution}} + \boxed{\text{potassium ions in solution}}$$
$$+ H_4SiO_4 + K^+$$

Considering the silicates as a whole, the amount and type of product (especially the clays) formed by hydrolysis are quite variable. Clays are finely crystalline, hydrous alumino silicates, and the particular type of clay that forms depends primarily on climate, parent material, and drainage. Furthermore, clays are environmentally important minerals, as they may greatly affect the engineering properties of soils.

In recent years, the acidity (pH) of precipitation has decreased to the point that it is less than 4.5 over most of the eastern United States. Burning of fossil fuels (coal, natural gas, and oil) annually releases approximately 50 million metric tons of sulfur and nitrogen oxides into the atmosphere. These pollutants, through a series of complex reactions, may be converted to acids that return to the earth with snow or ice as acid precipitation commonly known as **acid rain** (Fig. 4–11).

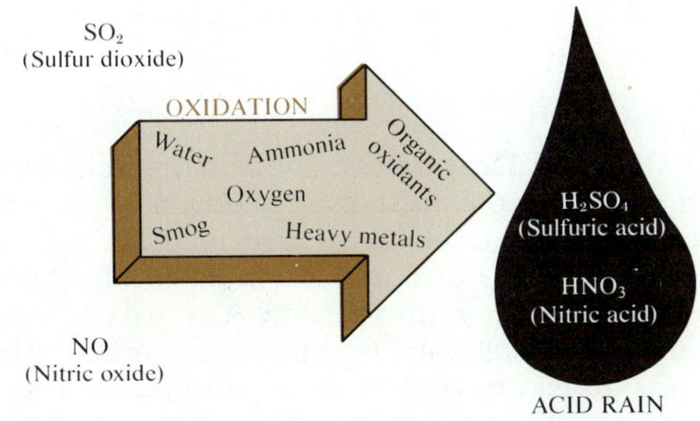

FIGURE 4–11
Idealized diagram showing how sulphur dioxide and nitric oxide may oxidize and eventually form acid rain. (From U.S. Environmental Protection Agency, 1979)

Precipitation is acid if the pH is below the natural expected value of 5.6. The pH scale (Fig. 4–9) ranges from 0 to 14. Battery acid, with a pH of 1, is very acid, whereas ammonia, with a pH of 12, is very basic. The scale is logarithmic; therefore, a pH of 3 is 10 times as acidic as a pH of 4 and 100 times as acidic as a pH of 5.

The acid rain problem is potentially very serious, as acid water accelerates adverse chemical weathering including leaching of nutrients from soils; leaching of heavy metals from pipes; and destruction of stone statues, monuments, and building stone in urban centers. Biological consequences of acid rain are even more severe: many lakes in the northeastern United States have become so acid that fish life cannot be supported, and agricultural yields are reduced as acid attacks foliage. In the United States, some areas—depending on the chemical composition of rocks and soil, climate, and vegetation—are particularly sensitive to acid rain (Fig. 4–12). The most sensitive regions

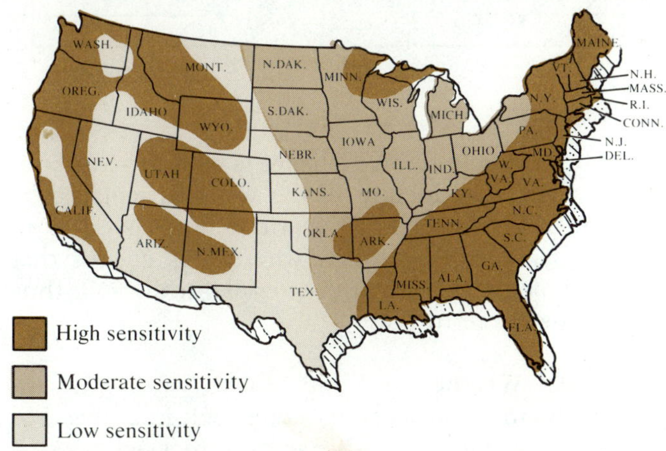

FIGURE 4–12
Areas in the United States that now are or will shortly become sensitive to acid rain. (From U.S. Environmental Protection Agency, 1979)

are those characterized by granitic rocks or soils lacking a capacity to buffer or chemically neutralize the acid.

Hydration and Dehydration Hydration and dehydration involve the addition or subtraction of entire water molecules to or from a mineral. This may include an increase in volume as a mineral is hydrated. Hydration reactions involve only a relatively small number of minerals, and the process is most effective on the surfaces and edges of the mineral grains. Therefore, hydration is not nearly so important as some other types of chemical weathering.

Base Exchange Base exchange occurs when an ion in solution is exchanged with an ion in a mineral; for example, sodium and calcium are readily exchangeable (Fig. 4–13). This is a weathering process because ions are not the same size, and exchange of larger ones for smaller ones and vice versa can lead to weaknesses in the crystal lattice of the mineral, which further facilitates other weathering processes.

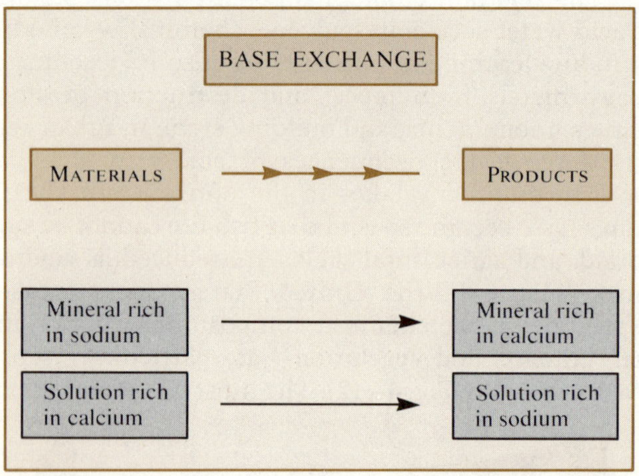

FIGURE 4–13
Base exchange.

The principle of ion exchange is used in water softeners where the calcium in the water is replaced by sodium in the softener. This principle is also used in the application of fertilizers where it is desired that certain ions exchange with materials in the soil, thus improving the fertility.

Spheroidal Weathering Spheroidal weathering involves chemical weathering along rock fractures. For example, consider a massive granitic rock with a systematic fracture system roughly dividing the rock into cubic blocks (Fig. 4–14). As chemical weathering proceeds inward along the fractures, the rate of attack on corners is eventually twice that on the sides, and thus the corners will become rounded. As the process continues, rings or shells form such that the outer shells are thinner and more completely weathered (Figs. 4–15 and 4–16). Eventually, the blocks (which may be quite large) may weather out of the cover soil material, and the center parts of the spheroidal weathered boulders, known as **core stones**, may be left on the surface of the ground.

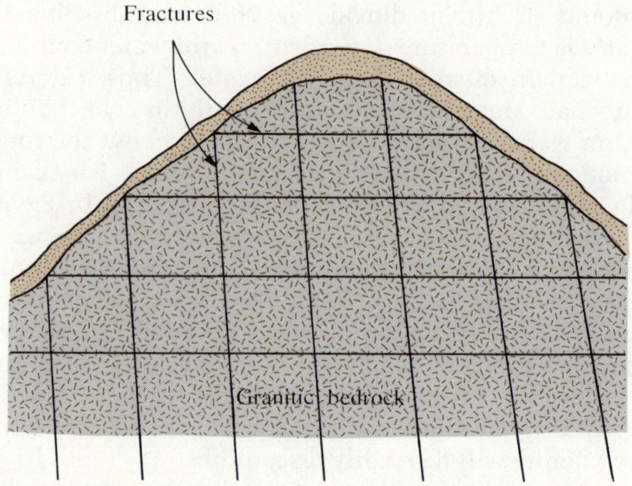

FIGURE 4–14
Idealized diagram showing a fractured granitic rock that could be quite susceptible to spheroidal weathering. The fracture system divides the rock roughly into cubic blocks, and as chemical weathering proceeds inward along the fractures the rate of attack on corners is twice that on the side, and thus corners will become rounded, and eventually shells or rings will form around centers of unweathered rock.

Chelation Certain biological processes such as excretion by lichens from chelating agents which produce a particular type of weathering. The importance of chelating agents in chemical weathering is just beginning to be understood. However, for a long time it has been known that lichens growing on rock surfaces or on soils can cause a good deal of weathering. One way that chelating agents may facilitate weathering is by rendering certain substances more soluble than they would be were these agents not present. Thus, the increased solubility would facilitate a more rapid chemical weathering.

RATES OF WEATHERING

Rates of weathering are variable depending upon such factors as climate, mineral assemblage, rock structure, and biological activity. The listing in Table 4–1 of the total dissolved solids for several rivers in North

FIGURE 4–15
Core stones of unweathered rock surrounded by concentric rings of rock that is weathered to a greater extent near the outside. The location of this spheroidal weathering is the San Gabriel Mountains of southern California. Note the rock hammer for scale.

FIGURE 4–16
An example of a large spheroidal weathered boulder. Indio Hills, California.

TABLE 4–1
Total dissolved solids for selected rivers in North America.

River	Solids (ppm)
Sacramento River, California	152
Columbia River, Pacific Northwest	125
Colorado River, Yuma, Arizona	711
Rio Grande River, Texas	881
Mississippi River, New Orleans	223
Tombigbee River, South Atlantic	155
Hudson River, North Atlantic	116

SOURCE: Data from U.S. Geological Survey Professional Paper 440–6, 1963.
NOTE: 100 ppm is approximately 0.01 weight percent.

America can be used to illustrate this variability for chemical weathering. Because of the many possible variables involved, the rate of weathering of rocks in a particular area may change quickly over a short distance.

Climates characterized by warm temperatures and abundant precipitation facilitate chemical weathering, whereas areas that are dry and cold are more likely to experience physical weathering (Fig. 4–17). Overall, the rates of weathering are considerably greater in humid climates than in dry climates. This is readily apparent in arid and semiarid regions, where both physical and chemical weathering may be relatively slow. Microclimatic effects are also important. For example, in the Northern Hemisphere south-facing slopes receive more direct sunlight, experience a higher rate of evaporation, are drier, have more freeze-thaw cycles, retain snow cover for a shorter time, and have less soil development and less vegetation than north-facing slopes. As an example, consider Frijoles Canyon in Bandelier National Monument, New Mexico. The canyon is oriented approximately east-west, and the rock type is a consolidated volcanic ash erupted about 1 million years ago. The canyon is strongly asymmetrical (Fig. 4–18) with a gentle north-facing slope and a steep south-facing slope. The north-facing slope has more water, more soil, more vegetation, and less variability in extremes than the steep south-facing slope, which the Pueblo Indians extensively excavated into cave dwellings. The harsher

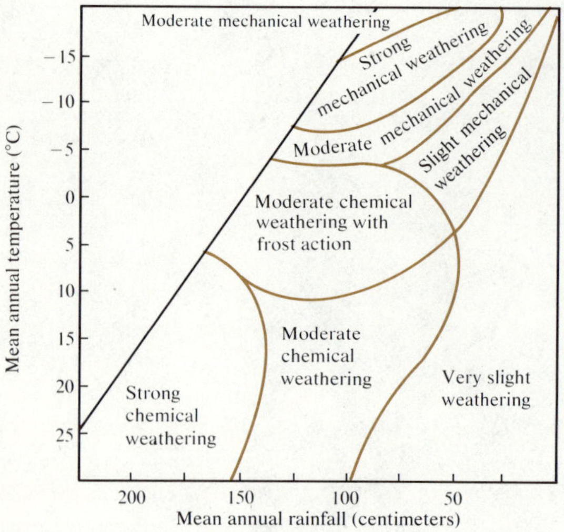

FIGURE 4–17
Intensity of weathering as roughly dependent upon the mean annual temperature and mean annual rainfall. (After L. C. Peltier; reproduced by permission from *Annals of the Association of American Geographers,* Vol. 40, 1950)

microenvironment of the south-facing slope favors mechanical weathering and development of a cliff, whereas the soil-covered north-facing slope with the greater retention of water favors chemical weathering and slope processes that lead to a more gentle slope. It is apparent that the observed asymmetry of the canyon results in part from the different weathering and slope processes reflecting microclimatic factors.

Silicate minerals are the main building blocks of rocks, and the reason why some silicates weather faster than others may be examined by comparing the strength of the chemical bonds between various positive ions (*cations*) with oxygen (i.e., their chemistry), and the environment in which the minerals form. Table 4–2 contains a list of the relative strengths of

TABLE 4–2
Relative strengths of bonds between selected elements (cations) and oxygen in common rock-forming minerals.

Bond	Relative Strength
Silicon-oxygen	2.4
Aluminum-oxygen	1.7
Iron(Fe^{+3})-oxygen	1.4
Magnesium-oxygen	0.9
Iron(Fe^{+2})-oxygen	0.9
Calcium-oxygen	0.7
Sodium-oxygen	0.4
Potassium-oxygen	0.3

SOURCE: Modified after G. D. Nicholls, Science Progress, U.S1, 1963.

bonds between selected cations (common in rocks) and oxygen. The silicon-oxygen bond is by far the strongest, partially explaining why quartz (SiO_2), composed entirely of silicon-oxygen bonds, is very resistant to weathering. As aluminum is substituted for silica and potassium added to the structure such as in the feldspars (as for example orthoclase, $KAlSi_3O_8$), there are fewer strong bonds and the mineral is less stable and therefore less resistant to weathering. One of the silicate minerals least resistant to weathering is olivine [$(Fe,Mg)_2SiO_4$], which has fewer silicon-oxygen bonds and more iron-oxygen and magnesium-oxygen bonds.

Environments where minerals form also reflect the stability and resistance to weathering of a mineral. For example, clay minerals (hydrated alumino silicates) that form at near-surface conditions as weathering products from feldspars and other minerals are relatively stable compared to minerals such as augite or olivine that form under conditions of relatively high temperature and pressure. Quartz also forms under relatively low temperature and pressure compared to olivine and therefore is more stable at near-surface conditions.

FIGURE 4–18
Frijoles Canyon in New Mexico. The canyon is strongly asymmetrical with a gentle north-facing slope (right) and steep south-facing slope (left).

Another important factor in mineral stability is the solubility of minerals in the weak acids commonly found at near-surface conditions. One way to evaluate this is to compare the percentage of an element in the parent rock of a drainage basin to the percentage of the same element leaving the basin in chemical solution, either in groundwater or streamflow. This type of study generally shows the following relative mobility sequence (from most mobile to least mobile): calcium, sodium, magnesium, potassium, silica, iron (Fe^{+3}), aluminum. Silica (as quartz) is only slightly soluble under the normal-range soil acidity; mobility of iron and aluminum are extremely low, explaining why intensive chemical weathering produces soils rich in oxides of iron and aluminum and devoid of almost everything else.

The effect of mineral chemistry and climate is also quite pronounced. For example, limestone (composed almost entirely of the mineral calcite, $CaCO_3$) weathers quite rapidly in the humid environment. Thus, limestones in the eastern United States often form valleys. In the arid environment, limestone is quite resistant to weathering and is often a ridge former (Fig. 4–19). In this case, the rate of weathering is controlled primarily by the available precipitation.

Rock structure, defined very generally to include the texture and fractures in rocks, is important when considering the rates of weathering. For example, some rocks such as basalts have **vesicles**, or holes through which water can migrate, facilitating weathering. Rocks that are highly fractured weather quickly because every fracture is a plane of weakness along which water may migrate, facilitating chemical weathering.

BIOLOGICAL ACTIVITY

Biological activity, especially vegetation, significantly influences rates of weathering. This results because plants produce organic acids and influence the amount of carbon dioxide in a soil. Thus, more calcium carbonate dissolves in soils beneath vegetation than where vegetation is lacking because there is more carbon dioxide (from plant respiration), which favors the dissolution. Another example comes from Hawaii, where it has been shown that basalt rock chemically weathers faster where lichens are growing on the rock. The role of the lichens is to increase locally the carbon dioxide (by respiration), aiding in the development of carbonic acid.

Despite the numerous and complex variables involved with weathering, some attempts at generalization can be made. As an example, Table 4–3 contains a list of the relative stability of selected earth materials assuming a humid climate.

SIGNIFICANCE OF WEATHERING

Weathering is geologically important because it is one of the processes that cause a general lowering of the landscape by modification of landforms. Although very few landforms are produced directly by weathering, weathering does contribute to the modification of landforms in several ways. For example, interesting forms consisting of hollows or pockets produced by chemical weathering are visible in many rock types (Fig. 4–20).

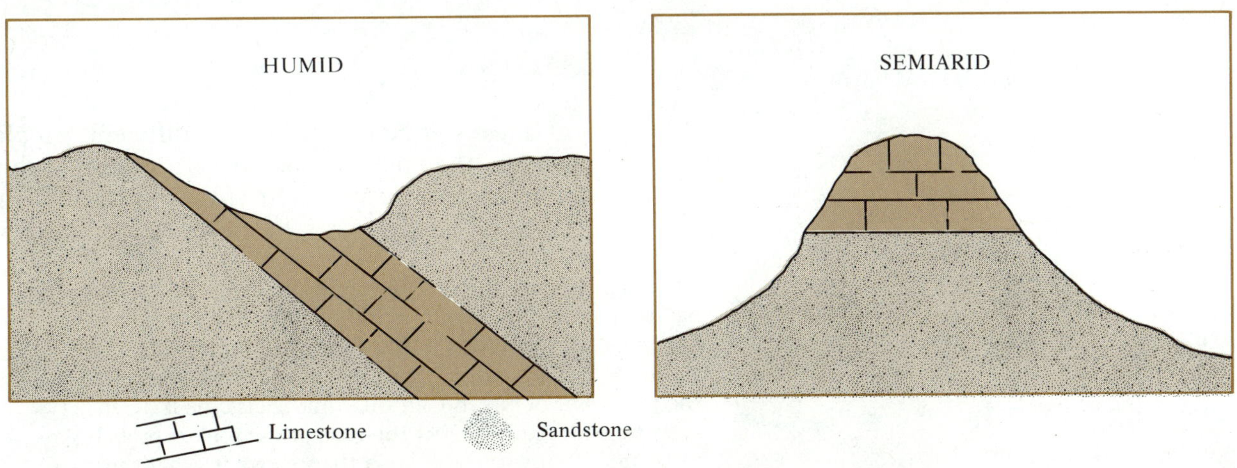

FIGURE 4–19
Idealized diagram showing that limestone may form valleys in a humid climate, but ridges in a semiarid region. The difference has to do with the availability of water to dissolve the limestone.

TABLE 4–3
Highly generalized relative stabilities and rates of weathering for selected earth materials (assuming a humid climate).

Material	Stability	Weathering rate
*Oxides of iron and aluminum	Stable at Surface Conditions ↑	Weathers Slowly ↓
Quartz		
Clays		
Muscovite		
Orthoclase		
Biotite		
Sodium-rich plagioclase		
Hornblende		
Augite		
Calcium-rich plagioclase		
Olivine		
Calcite	Unstable at Surface Conditions	Weathers Quickly

*Are weathering products and therefore are quite stable at surface conditions.

Weathering is important in preparing material for transport by wind, water, and ice. The sand on beaches of southern California is weathered from granitic rocks in mountain areas. The first process is the actual weathering of granite, which releases the quartz and feldspar. The small grains of mineral materials (sand) are then transported by running water to the coastal environment, where they are further distributed along the beach by coastal processes. The sand found in sand dunes of the arid southwestern United States was also originally prepared for transportation by weathering processes. Glaciers move tremendous amounts of material, most of which is probably picked up as loose, weathered products. Thus, weathering produces the raw materials that eventually (after transport and deposition) recombine to produce sedimentary rocks.

Another significant aspect of weathering is its importance to the formation of soil. Certainly weathering has to be classified as the most important of the soil-forming processes. Humid environments, where weathering processes are rapid, are usually characterized by thick soils, whereas arid and semiarid regions, characterized by slow weathering, typically have very thin soils.

A.

B.

SOILS

Soil means different things to different people. For example, the engineer may define soil as earth materials that may be removed without blasting. And, to

FIGURE 4–20
Chemical weathering processes may produce a variety of bizarre-looking forms of rocks. A. Solutional pits formed in a granitic rock (syenite) in the Crazy Mountains of Montana. B. This rather strange-shaped rock is located in the piedmont of North Carolina. The rock type is again syenite, and weathering at the base of the rock is probably more rapid below the soil layer than above it. That the rock is slightly notched near the base suggests that the level of soil has been higher in the past. Perhaps what is shown here is the effect of soil erosion following removal of indigenous vegetation approximately 200 years ago.

the soil scientist and geologist, soil is earth material that has been modified by physical, chemical, and biological processes such that it will support rooted plant life. This latter definition recognizes that soils consist of several types of materials, including weathered rock, organic material, water, and various gases (mostly air). The precise percentage of these various components in any one soil depends upon such factors as climate, topography, geology, and nature and extent of vegetation.

TYPES OF SOILS

There are many ways of classifying soils. Engineers use a classification based upon particle size and the presence or absence of organic material. Soil scientists use a rather elaborate classification based upon 10 broad orders of soils, with complex terminology (Table 4–4). Geologists have not generally agreed upon any one classification, and no classification exists that is satisfactory for all purposes.

From a geologic standpoint, soils may be grouped in terms of the origin of the materials upon which the soils form. Thus, it is recognized that weathering is the first step in soil-forming processes. If the more insoluble weathering material over bedrock remains in place and is further modified by organic processes to form a soil, such a soil is termed **residual** (Fig. 4–21). For example, the red soils of the piedmonts of the southeastern United States have formed in place upon igneous and metamorphic bedrock. On the other hand, if the parent material of the soil is transported by wind, water, or ice (glaciers) and then deposited and further modified by weathering and organic processes, then the soil is referred to as **transported** (Fig. 4–21). The fertile soils from the glacial deposits in the Midwest as well as soils that form on floodplains are examples.

TABLE 4–4
General properties of soil orders used with the 7th Approximation.

Order	General Properties
Entisols	No horizon development (azonal); many are recent alluvium; synthetic soils are included; are often young soils.
Vertisols	Include swelling clays (greater than 35 percent) that expand and contract with changing moisture content; generally form in regions with pronounced wet and dry seasons.
Inceptisols	One or more of horizons have developed quickly; horizons are often difficult to differentiate; most often found in young but not recent land surfaces; have appreciable accumulation of organic material; most common in humid climates but range from the arctic to the tropics; native vegetation is most often forest.
Aridisols	Desert soils; soils of dry places; low organic accumulation; have subsoil horizon where gypsum, caliche (calcium carbonate), salt, or other materials may accumulate.
Mollisols	Soils characterized by black, organic-rich A horizon (prairie soils); surface horizons are also rich in bases; commonly found in semiarid or subhumid regions.
Spodosols	Soils characterized by ash-colored sands over subsoil, accumulations of amorphous iron-aluminum sesquioxides and humus; are acid soils that commonly form in sandy parent materials; found principally under forests in humid regions.
Alfisols	Soils characterized by a brown or gray brown surface horizon, an argillic (clay-rich) subsoil accumulation with an intermediate to high base saturation (greater than 35 percent as measured by the sum of cations, such as calcium, sodium, magnesium, etc.); commonly form under forests in humid regions of mid-latitudes.
Ulfisols	Soils characterized by an argillic horizon with low base saturation (less than 35 percent as measured by the sum of cations); often have a red yellow or reddish brown color; restricted to humid climates and generally form on older landforms or younger, highly weathered parent materials.
Oxisols	Relatively featureless, often deep soils, leached of bases, hydrated, containing oxides of iron and aluminum (laterite) as well as kaolinite clay; primarily restricted to tropical and subtropical regions.
Histosols	Organic soils (peat, muck, bog).

SOURCE: After Soil Survey Staff, 1960, *Soil Classification: A Comprehensive System.*

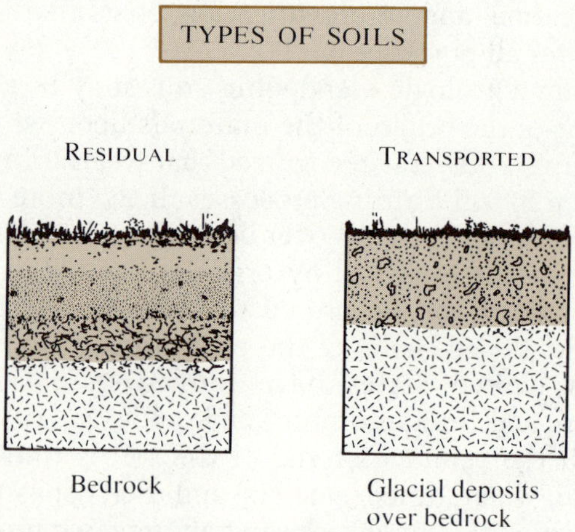

FIGURE 4–21
Residual and transported soils.

SOIL PROFILES

Vertical and horizontal movement of material in the soil system produces a distinct soil layering, or **soil profile**, which is divided into zones, or **horizons** (Fig. 4–22). The **A horizon**, or **zone of leaching**, is a dynamic zone characterized by interactions between migrating water, organic material, and weathered rock. The uppermost part of the A horizon may contain a considerable amount of organic debris. In forested areas, the upper layer of organic material (known as **duff**) is especially thick, containing both undecomposed and partially decomposed debris. Directly below the organic debris is usually a dark-colored horizon that contains a considerable amount of mineral material as well as much organic material. The A horizon is known as the zone of leaching because as water moves through the horizon, it picks up nutrients and other materials and carries them in chemical solution or fluid suspension to the B horizon. Carbonate and sodium salts and other soluble material are carried in chemical solution, and fine solid materials such as clay are also moved downward as water slowly percolates through the upper horizon.

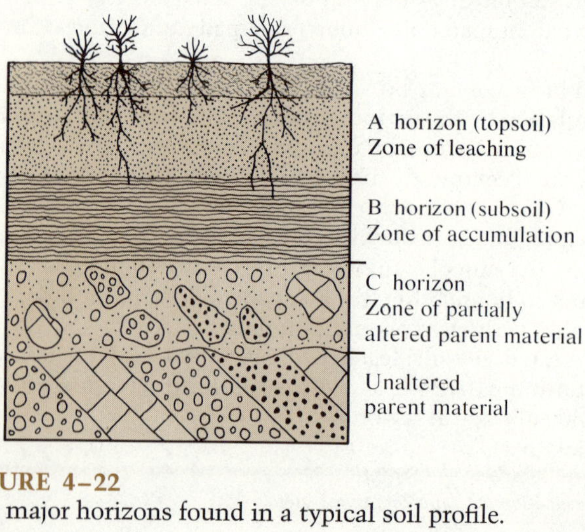

FIGURE 4–22
The major horizons found in a typical soil profile.

The **B horizon** is called the **zone of accumulation** because much of the material that moves downward from the leaching of the A horizon is deposited there. The material from which the soil develops is termed the **parent material**, and in the B horizon only those minerals resistant in the parent material are still recognizable. The minerals that weather readily generally are converted to new minerals such as clay, or to soluble salts.

It is worthwhile to mention here that the word *clay* has several meanings. We often speak of clay-sized particles when we refer to solid particles less than 1/256 mm in diameter. However, we also refer to clay minerals which are generally the weathering products of silicates such as the feldspars. There are many types of clay minerals, some of which tend to expand and contract readily with gain or loss of water. Clay minerals may accumulate in the B horizon as thin clay skins surrounding mineral grains or as a definite layer that is difficult to cultivate and is therefore known as **hardpan**.

The B horizon generally grades into the **C horizon**, which is a **zone of partially altered parent material** that grades into unaltered, fresh parent material.

The A, B, and C horizons of a soil form the soil profile. In a humid climate, the development of the soil profile takes place in an open system. That is, there is a constant inflow and outflow of material in the various horizons until an equilibrium of action is reached, at which time the thicknesses of the horizons remain constant. In other words, as the soil profiles develop there is considerable movement of materials from the upper to lower horizon and vice versa, but eventually there is an equilibrium reached when the amount of material leaching down, the amount of material accumulating in the B horizon, and the amount of material moving through the soil reach a rough balance. At this time, the profile development is basically time-independent and will not change unless some of the major independent variables responsible for soil formation change.

FACTORS IN SOIL FORMATION

Five major factors are primarily responsible for the development of soils. These are climate, topography, parent material, plant and animal activity, and time. Of these five factors, climate and topography are by far the most important.

CLIMATE

The effect of climate on the development of soils is pronounced. For example, consider the development of soil profiles in semiarid climates, humid climates, and under tropical rainforest conditions. Soils in semiarid regions tend to be thin. These soils contain abundant nutrients and usually need only water to be very productive. Soils in dry, hot areas may contain deposits of caliche (calcium carbonate, $CaCO_3$) in the lowest part of B horizon and also may have a relatively high amount of alkali. Soils in humid areas are primarily produced by chemical weathering and usually are thick. However, these soils may be nutrient-poor, as the nutrients are leached down to the B horizon or out of the soil in solution. A layer of clay (hardpan) may develop in the B horizon. Soils in the humid climate usually are acid and silica-rich because the abundant vegetation favors the formation of acids in soils, and silica is relatively insoluble in an acid environment. In remarkable contrast to the arid and humid conditions, soils in very hot, wet climates typical of rainforests are usually very thick. These soils are generally extremely nutrient-poor and quite rich in silica oxides of iron and aluminum, and acid. It is interesting to note that soils in these areas usually cannot support vegetation other than that which has specifically evolved to adapt to that soil.

A type of soil known as **laterite** is particularly dependent upon climatic factors for its formation. Laterites are soils rich in hydrated iron and aluminum oxides. These soils may be important resources because if the aluminum oxides become very concentrated, then they actually may be mined for their aluminum content. The concentrated aluminum oxide is known as **bauxite**.

There has been considerable speculation over how lateritic soils develop. One school argues that they form through intensive leaching, such as might occur in tropical rainforest environments. A problem with lateritic soils produced by intensive leaching is that with the abundant vegetation of a rainforest, the soils are acidic and it is difficult to remove the silica, which is not very soluble in an acid environment. One attribute of lateritic soils is that they are silica-deficient. This may be reconciled by looking at the plants. Some particular types of plants such as bamboo do remove a considerable amount of silica from the soil, and this might be the answer.

A second school holds that the development of lateritic soils is best in areas characterized by pronounced wet and dry seasons (*savannas*). The development would be as follows. During the wet season, the upper part of the soil horizons would be acidic due to leaching of alkaline materials and accumulation of organic matter. During the dry season, however, evaporation becomes important and there is a net movement upward of alkaline salts. Thus, at the end of the dry season the upper part of the A horizon is quite alkaline. During the first rains of the next wet season, a little silica, which is more soluble in an alkaline environment, goes into solution and is removed from the system. In this way, silica is selectively removed, until the remaining soil is silica-poor. During this process, the iron and aluminum oxides remain in the soil and eventually a lateritic soil develops.

TOPOGRAPHY

Soil profiles tend to vary considerably with topography. Consider Fig. 4–23, which is an illustration of a typical slope for a humid environment. At the crest of the slope where the profile is relatively flat, the soil is well drained and the soil profile will be well developed into A, B, and C horizons. Where the slope steepens, the soil cover becomes thinner and the profile less developed. This results because the slope materials are moving downhill in response to hillslope processes such as *soil creep* (slow flowage of soil). At the base of the slope, there may be a poorly drained soil, and as a result there may be an increase in organic material in the upper part of the profile as well as high water-table conditions.

PARENT MATERIAL

The parent material can affect the development of soil. This is particularly true when comparing transported material such as floodplain deposits with in-place bedrock. However, everything else considered, parent

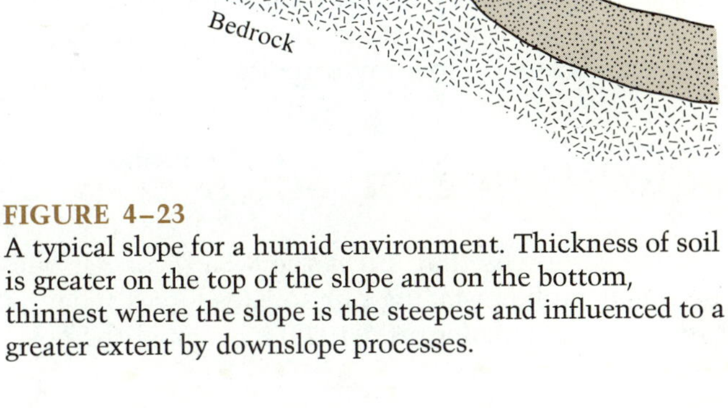

FIGURE 4–23
A typical slope for a humid environment. Thickness of soil is greater on the top of the slope and on the bottom, thinnest where the slope is the steepest and influenced to a greater extent by downslope processes.

material is not as important as either climate or topography in producing the differences seen in soils.

If the parent material contains iron, then the soil may take on an interesting variety of colors depending upon local drainage conditions. For example, with good drainage and aeration the soil is well oxidized, producing a red soil. This is the type of soil typically found on hilltops on the piedmont of the southeastern United States, as well-drained soils in mountainous areas, and as soils developed on limestone terrane. With poor drainage conditions, oxygen may be deficient, and as a result the iron is reduced and the soil is likely to be a dull gray. A yellow soil suggests that the soil is probably moist much of the time and that the iron is hydrated as well as oxidized.

PLANT AND ANIMAL ACTIVITY

Biological activity is an important factor in soil formation. This is particularly true for forest soils where there is a considerable thickness of organic debris. The layer of debris greatly increases the amount of precipitation that infiltrates. Therefore, there is a longer period of time in which organic matter is in contact with water, which leads to the development of organic acids, which in turn facilitate weathering processes.

TIME

Time is another important variable in the development of soils. In general, the thickness of the B horizon increases with time as clay is translocated down from the A horizon, and the color becomes more pronounced (a deeper red). After several hundred thousand years, some soil profiles may reach a dynamic equilibrium state, after which they change little with time. In a humid environment, soil-forming processes may be rapid, and therefore development of soil profiles may not take nearly as much time as it would in a more arid region. Factors such as soil fertility, however, do change with time. The fertile (nutrient-rich) transported soils of the midwestern United States are primarily associated with young glacial deposits. On the other hand, older residual soils found on the piedmont of the southeastern United States are highly leached and thus are low in nutrients.

SIGNIFICANCE OF SOILS

Soils are resources that affect the wealth of nations. The major problem facing our soil resources today is erosion. W. C. Lowdermilk (1943) has stated the position well: "One generation of people replaces another, but productive soils destroyed by erosion are seldom restorable and never replaceable."

Erosion has a twofold environmental effect. First, it depletes soils resources (Fig. 4–24). Second, the water resources (lakes, streams, rivers) that the eroded soil particles move into are degraded. In Fig. 4–25, the relationship between sediment yield and drainage area for various land uses is shown. Notice that the smallest sediment yields are associated with wooded and posturbanization conditions, and the highest yields are from areas severely exposed during construction. Sediment is becoming such a serious problem that some authorities now believe that sediment pollution may be one of our greatest pollutants. In many areas, sediment is choking the streams; filling in lakes, reservoirs, ponds, canals, drainage ditches, and harbors; burying vegetation; and generally creating a nuisance that is extremely difficult to remove (Fig. 4–26).

The solution to sediment pollution includes sound conservation practices. Conservation is particularly important in urbanizing areas where tremendous quantities of sediment are produced during construction. A main objective of sediment control during construction is to keep the sediment on-site by utilizing diversions to collect runoff, and sediment basins to collect the sediment.

EXPANSIVE SOILS

In the United States, soils rich in expansive clay minerals are causing $2 billion of damage yearly. Expansive clays swell with water and contract with drying. Expansive soils may be recognized following rainstorms by the swelling, or by contraction cracks during the dry months. This process can lead to cracks in house foundations, and even walls (Fig. 4–27). Minimizing hazards associated with expansive soils is usually centered around identifying soils with a high shrink-swell potential and avoiding them whenever possible. Special engineering techniques may be utilized to minimize the damage.

SOILS AND HEALTH

The soil we cultivate to provide nutrients for subsistence can and does influence our chances of developing a serious health problem such as heart disease or cancer. This results because soils contain a variety of **trace elements**, released by weathering or human-induced pollution, that affect in a positive or negative way our health and well-being. However, disease, defined as an imbalance resulting from a poor adjust-

FIGURE 4–24
Severe soil erosion has resulted in the loss of a valuable resource. (Photo by F. M. Roadman, U.S. Department of Agriculture)

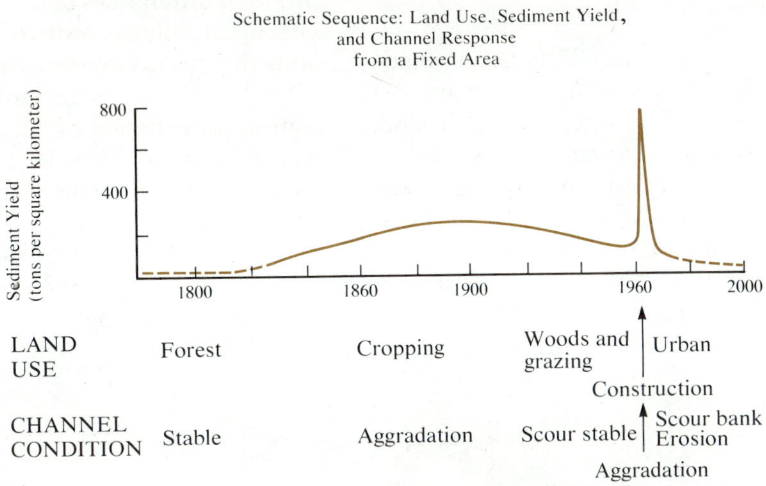

FIGURE 4–25
The effect of land use change on sediment yield in the piedmont region of the United States, from the beginning of extensive farming through urbanization. (After M. G. Wolman, *Geografiska Annaler,* Vol. 49A, 1966)

ment between an individual and the environment, seldom is a one-cause–one-effect relationship, and therefore possible effects of trace elements on health reflect only one part of a multidimensional problem.

All living tissue is composed primarily of a few bulk elements such as oxygen, carbon, hydrogen, nitrogen, and calcium. In addition, there are other elements living tissue requires to function properly. These trace elements, present in minute quantities but important in regulating dynamic processes of life, include copper, zinc, iodine, iron, and selenium, among others. Each trace element has a whole spectrum of possible effects

FIGURE 4–26
Sediment pollution. (Photo by W. B. King, U.S. Department of Agriculture)

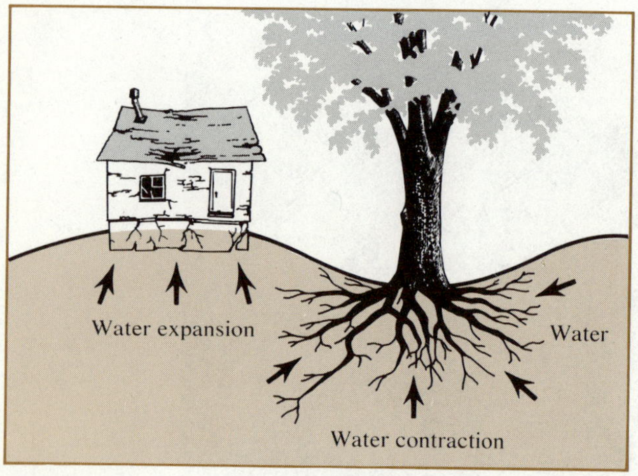

FIGURE 4–27
Results of expansion and contraction of soil on a house. Expansion occurs as water collects underneath the foundation, and contraction may take place during the dry season as trees planted too near the house draw up most of the available water. (Drawing by Robert W. Tope)

on a particular plant or animal. This is a **dose-dependency relationship** (Fig. 4–28). For example, selenium (when concentrated) is toxic to cattle and sheep, has no observable effect in many cases, and is definitely beneficial to animal production in some areas when it is added to feed. The apparent contradictions are resolved by recognizing that the first case represents oversupply of selenium, the second signifies a balance, and the third suggests a deficiency of selenium in the natural environment. Fluoride is another good example. Concentrations of about 1–5 ppm fluoride are beneficial to tooth and bone formation in humans, but at concentrations of 8–20 ppm problems of excessive bone development may occur. At very high concentrations, fluoride is the main ingredient of some rodent poisons. Unfortunately, the critical threshold concentrations for beneficial or harmful effects associated with many trace elements are not established.

Studies in Wales, England, and Iran have shown relationships between cancer and soil characteristics. In Wales and England, it has been found that rates of stomach cancer are related to the amount of organic material in soil as measured by the weight loss of a dry soil after burning (ignition loss) and the concentration of zinc, cobalt, and chromium in the soil. Zinc may be significant, as it is active in the process of gastric digestion, and cobalt is important because it is known to have carcinogenic properties.

High rates of esophageal cancer in northern Iran near the Caspian Sea are extremely variable, with marked differences occurring over short distances. The strongest relationship between the cancer incidence and the environment centers on the general soil type. The highest rates of cancer are associated with the saline (salty) soils of the eastern end of the so-called cancer belt. These soils are in an area of little rainfall. To the west, the rainfall increases dramatically and the soils become more and more leached. Perhaps the leaching removes carcinogenic materials, thus leading to the reduction in the incidence of esophageal cancer.

An interesting pattern of heart disease mortality and possible relationship to soils have been reported

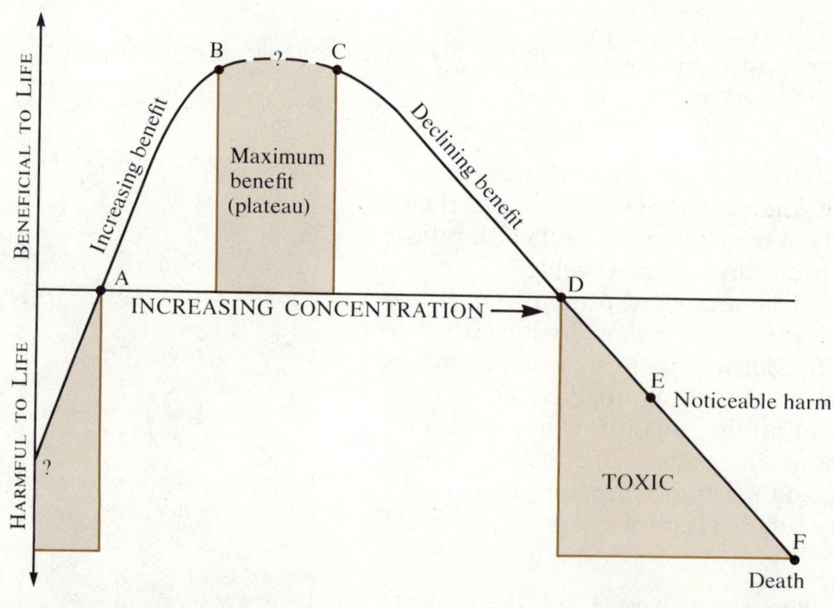

FIGURE 4–28
A generalized dose-response curve.

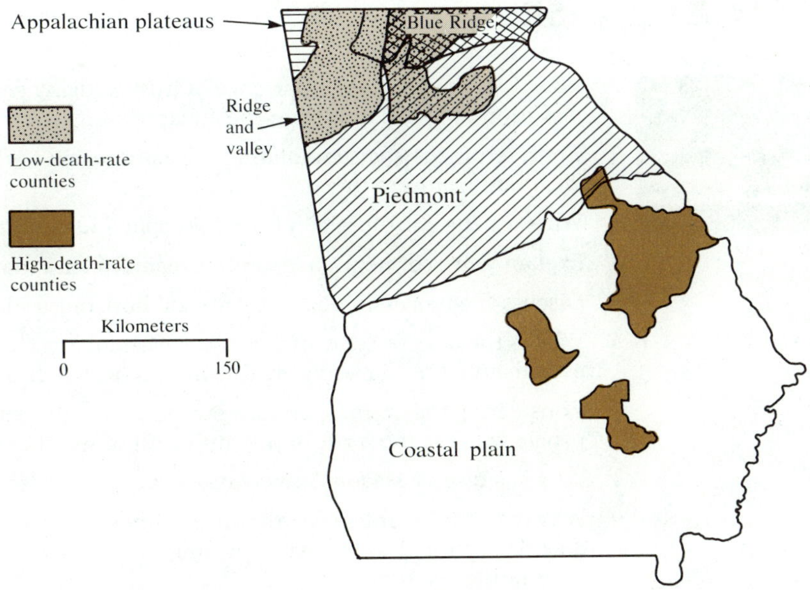

FIGURE 4–29
Map showing groups of counties in Georgia having high and low death rates due to cardiovascular disease. (After Shacklette, Sauer, and Miesch, U.S. Geological Survey Professional Paper 574C, 1970)

from Georgia. The range in the death rates from the 159 counties in Georgia is as great as can be found anywhere in the United States. Nine counties in the mountains of northern Georgia with low death rates of approximately 600 per 100,000 population of males 35–74 years of age were compared, geochemically, with 9 high-death-rate (approximately 1200 deaths per 100,000) counties in central and southern Georgia (Fig. 4–29). Thirty trace elements were analyzed, and the results did not suggest that the counties with high death rates are associated with dangerous concentrations of harmful trace elements. Rather, several trace elements thought to have beneficial effects on heart disease are concentrated in the low-death-rate areas of Georgia. Thus, the tendency for higher mortality due to heart disease in Georgia is apparently due to subtle relationships associated with a deficiency rather than an excess of certain trace elements.

SUMMARY

Weathering is a group of physical, chemical, and biochemical processes that break rocks down at or near the surface of the earth. Common types of physical (mechanical) weathering are frost wedging, frost heaving, expansion by plant roots, and exfoliation. The most important chemical weathering process is hydrolysis, which involves partial or complete solution of minerals upon attack by weak acids. Other types of chemical weathering include oxidation, hydration and dehydration, base exchange, and chelation. Rates of weathering are variable depending on such factors as climate, mineral type, rock structure, and biological activity. They are greater in humid than in arid regions.

Weathering is important for several reasons: (1) it is one of the processes that causes a general lowering of the landscape; (2) it prepares material for transport by wind, water, or ice; and (3) it is the major process involved in the formation of soil.

There are several ways to define soil and several ways to classify various types of soils. Unfortunately, no one classification is satisfactory for all purposes.

Two major factors in the development of soils are climate and topography. Others are parent material, biological activity, and time.

Vertical and horizontal movement of material in a soil produce a distinct profile, or layering of the soil into distinct zones characterized by specific processes.

Soils are resources that affect the wealth of nations. A major global problem is erosion, which depletes the soil resource while degrading water quality.

Expansive soils, which swell and contract with the addition or loss of water, are a major natural hazard, causing $2 billion in property damage each year in the United States.

The soil we cultivate for food can and does influence rates and patterns of chronic disease such as cancer and heart disease. This results because soils contain a variety of trace elements released or removed by weathering and human-induced pollution that in a positive or negative way affect the health and well-being of people.

REVIEW EXERCISES

1 Why does the relative importance of frost wedging generally increase with higher elevations and distance from the equator?

2 Considering the sparse rainfall in semiarid regions, why is hydrolysis an effective weathering process there?

3 What are the optimum conditions for spheroidal weathering to occur?

4 Explain why sediment yields from urbanized land may be less than for farmland.

5 Discuss the processes necessary for soil horizons to develop.

6 Why is the dose or concentration of a particular trace element so important in establishing the element's significance in human health?

7 Considering the entire spectrum of weathering processes and their importance to people, what is the most important result of weathering?

8 Why is a disease seldom a one-cause–one-effect relationship?

9 Why does the chemical weathering of limestone depend upon how much carbon dioxide is present in the water passing through the limestone and thus facilitating the weathering reactions?

10 Considering a chronic disease such as cancer and the geochemical environment, what do the case histories discussed in the text have in common?

11 Compare and contrast the relative importance of chemical and physical weathering with different climates.

12 What factors control the rate of weathering?

13 What is exfoliation, and how does the process work?

14 Why does quartz weather slowly compared to biotite and feldspar?

15 What factors control the color of soils?

16 How does laterite form?

17 Why is soil erosion such a serious environmental problem?

18 How might trace elements affect the rates of cancer and health problems?

19 What differences might be expected between transported and residual soils?

SUGGESTED READINGS

BASILE, R. M. *A Geography of Soils.* Dubuque, Iowa: William C. Brown Publishers, 1971, 152 pp.

BIRKLAND, P. W. *Pedology, Weathering, and Geomorphic Research.* London: Oxford University Press, 1974, 285 pp.

BRADY, N. C. *The Nature and Properties of Soils,* 8th ed. New York: Macmillan Co., 1974.

BRIDGES, E. M. *World Soils.* Cambridge: At the University Press, 1970, 89 pp.

BUOL, S. W.; HOLE, F. D.; and McCRACKEN, R. J. *Soil Genesis and Classification.* Ames, Iowa: Iowa State University Press, 1973, 360 pp.

HUNT, C. B. *Geology of Soils.* San Francisco: W. H. Freeman and Co., 1972, 344 pp.

KELLER, W. D. *Principles of Chemical Weathering.* Columbia, Mo.: Lucas Brothers, 1957.

SAUER, H. I., and BRAND, F. R. "Geographic Patterns and Risk of Dying." In H. L. Cannon and H. C. Hopps, eds. *Environmental Geochemistry and Health.* Geological Society of America Memoir 123, 1971, pp. 131–50.

STEILA, D. *The Geography of Soils.* Englewood Cliffs, N.J.: Prentice-Hall, 1976, 222 pp.

UNDERWOOD, E. J. "Geographical and Geochemical Relationships of Trace Elements to Health and Disease." In D. D. Demphill, ed. *Trace Substances in the Environment,* Vol. 4. Columbia, Mo.: University of Missouri Press, 1971, pp. 3–11.

- clastic rock
- mechanical weathering
- chemical weathering
- conglomerate
- breccia
- sandstone
- quartzose sandstone
- arkosic sandstone (arkose)
- coal
- lithification
- cementation
- compaction
- recrystallization
- interlocking texture
- bedding
- mud cracks

5

- graywacke
- siltstone (mudstone)
- shale
- fissility
- claystone
- nonclastic rock
- limestone
- coquina
- chalk
- dolomite
- chert
- gypsum
- evaporite
- rock salt
- diatomite
- ripple mark
- cut and fill
- cross-bedding
- graded bed
- sorting
- porosity
- permeability
- slump features
- continental shelf
- continental slope
- submarine canyon
- turbidity current
- geosyncline (geocline)
- ooze

Sedimentary Rocks

THE STUDY OF SEDIMENTARY ROCKS BEGAN VERY EARLY. Leonardo da Vinci around 1500 and Nicholas Steno in 1668 noted that sedimentary rocks had originally been deposited in horizontal layers, that the fossils they contain were once living animals, and that seas had once covered much of what is now land. These were major observations, and had they been generally recognized, geology would have developed much earlier than it did. During the eighteenth century, a number of attempts were made to classify and organize the sedimentary layers. Study of these layers and their contained fossils ultimately led, in the early nineteenth century, to a geologic time scale and to deciphering the geologic history of the earth.

The study of sedimentary rocks themselves, their composition and formation, came much later. Study of sedimentary rocks may have lagged behind study of igneous rocks because sedimentary rocks seemed to present fewer problems than controversial rocks such as basalt. Study of how sedimentary rocks form probably started in earnest when uniformitarianism was generally accepted. Uniformitarianism, as was described in the last chapter, in this case means that to understand ancient rocks, geologists must study how similar rocks are forming today. Uniformitarianism is central to all geology. It began with James Hutton, late in the eighteenth century, but did not become widely accepted until it was popularized by Charles Lyell (1797–1875). Lyell wrote a textbook that was first published in 1830 and went on through many editions. He extended and illustrated uniformitarianism and was responsible for its acceptance.

Even though the scene was set for study of the sedimentary rocks with the acceptance of uniformitarianism and the development of microscope techniques, study of sedimentary rocks developed slowly. The microscope was used first on igneous and metamorphic rocks. Study of these rocks gave insight into internal processes of the earth and so was an important research area. The practical geologists of the nineteenth century were engaged in mining, and thus they, too, were mainly interested in igneous and metamorphic rocks. A few geologists did do important work on sedimentary rocks, but it took a new economic incentive to give the study its rightful place in geology. Petroleum is found in sedimentary rocks, and its importance in the twentieth century encouraged detailed study of all aspects of sedimentary rocks.

Sedimentary rocks are formed at or near the surface of the earth. They are composed of rock fragments, weathering products, organic material, or precipitates (Fig. 5–1). Most are deposited in beds, or layers. They comprise a very small volume of the earth, only about 8 percent of the crust (Fig. 5–2). Despite their small volume, they cover about 75 percent of the surface.

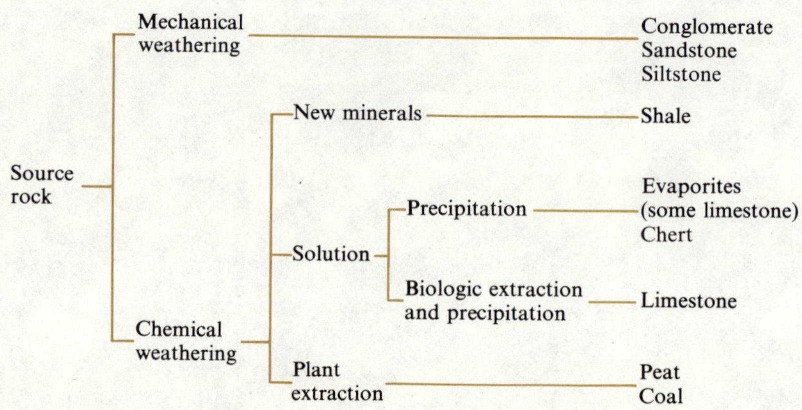

FIGURE 5–1
Simplified chart showing origins of sedimentary rocks.

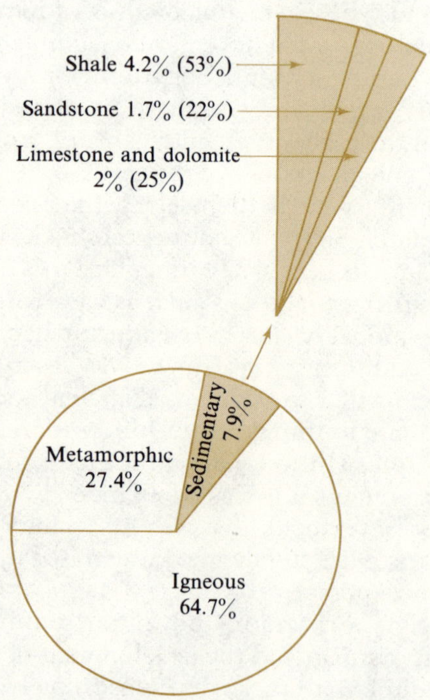

FIGURE 5–2
Relative abundance of sedimentary rocks in the earth's crust. For example, shale forms 4.2 percent of all rocks, and 53 percent of the sedimentary rocks. (Data from Ronov and Yaroshevsky, 1969)

CLASSIFICATION AND FORMATION

As with the igneous rocks, it is more important to interpret the formation of sedimentary rocks than merely to name them. Knowing the origin of the materials that compose a sedimentary rock and understanding the origin of its sedimentary features will permit such interpretation.

According to the way they were formed, sedimentary rocks can be classified as marine, lacustrine (lake-deposited), glacial, eolian (wind-deposited), fluvial (river-deposited), and so on. They also can be classified by type as limestone, chert, quartzose sandstone, and so forth, or by their mode of origin as clastic, chemical precipitate, or organic. In practice, these are blended to produce the practical classification that follows. The origin of common sedimentary rocks is shown in Fig. 5–1.

CLASTIC ROCKS

Clastic rocks are composed of rock fragments or mineral grains broken from any type of preexisting rock (Fig. 5–3). They are subdivided according to fragment size. Commonly, sizes are mixed, requiring intermediate names such as sandy siltstone. They are recognized by their clastic texture. The fragments originate from mechanical weathering.

Composition is not used in the general classification of clastic sedimentary rocks because the composition may be affected by several factors. The examples of this are almost endless, but a few simple ones will illustrate. A quartz sandstone may be formed if the source area is mainly quartz (perhaps an older quartz sandstone), or a quartz sandstone may be the result of prolonged weathering, leaving only quartz fragments. Large amounts of feldspar in a sandstone may imply rapid deposition and burial before chemical weathering could decompose the feldspar, or it might imply a cold climate in which chemical weathering is very slow. In many studies, composition is used to subdivide sandstones as indicated in Table 5–1.

The actual formation of shale is somewhat more complex than indicated in Table 5–1. Much of the clay that gives a shale its fissility apparently develops or is mechanically reoriented after deposition. This is suggested by the lack of geologically young shale. Young fine-grained clastic rocks are mainly mudstones.

Clastic texture
Note open spaces.

Interlocking texture. Developed in chemical precipitates such as some limestones and in many igneous and metamorphic rocks.

FIGURE 5–3
Recognition of textures is very important in the identification and interpretation of rocks.

TABLE 5–1
Clastic sedimentary rocks.

Size	Sediment	Rock
> 2 mm	Gravel	*Conglomerate*—generally has a sandy matrix. May be subdivided into roundstone and sharpstone conglomerates (*breccia* is a synonym) depending on the fragment shape (see Fig. 5–4).
1/16–2 mm	Sand	*Sandstone*—recognized by gritty feel. (See Figs. 5–5 and 5–6.) Generally designated as coarse, medium, or fine if well sorted. Have also been subdivided on the basis of composition; the more important types are: *Quartzose sandstone* (mainly quartz) *Arkosic sandstone* (*arkose*)—over 20 percent feldspar *Graywacke*—poorly sorted, with clay or chloritic matrix
1/256–1/16 mm	Silt	*Siltstone* (*mudstone*)—may be necessary to rub on teeth to detect grittiness, thus distinguishing it from shale.
< 1/256 mm	Clay	*Shale*—distinguished from siltstone by its lack of grittiness and its *fissility* (ability to split easily on bedding planes). Rocks composed of clay that lack fissility are called *claystone*.

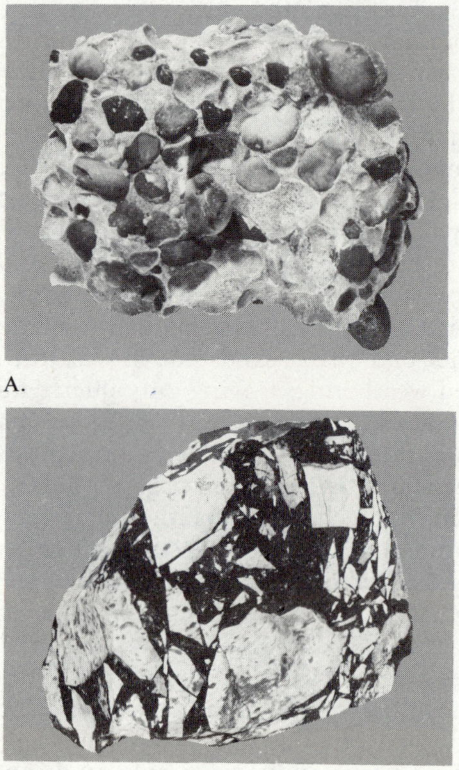

A.

B.

FIGURE 5–4
Conglomerate. A. Rounded pebbles in a sandstone matrix. B. Sharpstone conglomerate (breccia). The angular pebbles in this rock were probably not transported far compared with the pebbles in A. (Photos from Ward's Natural Science Establishment, Inc., Rochester, N.Y.)

FIGURE 5–5
Sandstone. The bedding is too thick to show in this specimen. (Photo from Ward's Natural Science Establishment, Inc., Rochester, N.Y.)

NONCLASTIC ROCKS

Nonclastic rocks are formed by chemical precipitation, by biologic precipitation, and by accumulation of organic material. These processes extract specific materials from their surroundings, generally seawater, and precipitate these substances, forming rocks. The rocks are classified mainly by composition (see Table 5–2). As with the clastic rocks, these rocks are commonly mixed, both among themselves and with the clastic rocks.

Limestone is composed of calcite. It is recognized by effervescence with dilute hydrochloric acid. It is generally of biologic origin and may contain fossils (Fig. 5–7A). A rock composed mainly of fossils or fossil fragments is called *coquina* (Fig. 5–7B).

Limestone can form in many ways, as shown in Table 5–3. Most limestone probably originates from organisms that remove calcium carbonate from seawater. The remains of these animals may accumulate to form the limestone directly, or they may be broken and redeposited. Calcite can also be precipitated chemically. The amount of carbon dioxide in seawater controls the amount of calcite that can remain in solution. If the amount of carbon dioxide is reduced by warming the water, as would occur in shallow tropical water, calcite may be precipitated. Marine plants also remove carbon dioxide from seawater and aid in calcite precipitation. In addition, bacteria can aid chemical precipitation of calcite by making the water more alkaline. Decaying organic material can increase the amount of carbon dioxide and so retard precipitation of calcite. Limestone is presently being formed by chemical precipitation on the shallow Bahama Banks where the factors discussed are favorable.

Chalk is soft, white limestone formed by the accumulation of the shells of microscopic animals. The shells are composed of calcium carbonate, and chalk is recognized by effervescence with acid.

Dolomite is composed mainly of the mineral dolomite. It is recognized by effervescence with dilute hydrochloric acid after scratching (to produce powder); it will also react (without scratching) with concentrated or warm, dilute hydrochloric acid. Dolomite is generally formed by replacement of calcite, presumably soon after burial. Some of the calcium atoms in calcite are replaced by magnesium atoms. The reduction in volume in this replacement may produce irregular voids and generally obliterates fossils.

Chert is miscroscopically fine-grained silica (SiO_2). It is equivalent to chalcedony (see the mineral tables in Appendix A). It is unfortunate that many names have been applied to the fine-grained varieties of silica. These names were based on minor differences such as color and luster that cannot be substantiated by modern methods, and there is little agreement among mineralogists on their naming. Both chert and chalcedony may contain opal and fine-grained quartz. Dozens of names have been applied to chalcedonic silica;

FIGURE 5–6
Sandstone outcrop. (Photo from New Mexico State Tourist Bureau)

of these, a few may be useful, for instance agate for banded types, and flint for dark gray or black chert.

Chert originates in several ways. Some may precipitate directly from seawater in areas where volcanism releases abundant silica. Most comes from the accumulation of silica shells of organisms. These organisms remove dissolved silica from seawater to form their shells or skeletons. These silica remains come from diatoms, radiolaria, and sponge spicules, and are composed of opal. Opal is easily recrystallized to form chert. Thus, much chert is recrystallized, making the origin difficult to discern. In recrystallization, the silica may replace other materials, and fossils replaced by silica are common. If this occurs in limestone, beautifully preserved fossils with delicate features intact can be recovered by dissolving the limestone with acid. Chert may form either beds or nodules.

Gypsum forms from the evaporation of seawater. Rocks formed in this way are termed *evaporites.* The identification of gypsum is described in Appendix A.

TABLE 5–2
Nonclastic sedimentary rocks.

Rock	Principal Texture	Principal Composition	How Recognized
Limestone	Fine	Calcite, $CaCO_3$	Effervesces with dilute hydrochloric acid. Generally of biologic origin and may contain fossils. *Coquina* is limestone composed mainly of fossils or fossil fragments.
Chalk	Fine	Calcite, $CaCO_3$	Soft, white limestone formed by the accumulation of microscopic shells. Effervesces with dilute hydrochloric acid. Compare with diatomite.
Dolomite	Fine	Dolomite, $CaMg(CO_3)_2$	Effervesces with dilute hydrochloric acid after scratching (to produce powder). May have irregular holes.
Chert	Fine	Silica, SiO_2	Hardness 6 or 7. Similar to chalcedony. *Flint* is black or dark gray.
Gypsum	Fine to coarse	Gypsum, $CaSO_4 \cdot 2\,H_2O$	Hardness 2. One good cleavage and two poorer cleavages.
Rock salt	Fine to coarse	Halite, NaCl	Cubic crystals and cleavage may be visible. Salty taste.
Diatomite	Fine	Silica, SiO_2	Soft, white rock formed by the accumulation of microscopic shells composed of silica. Distinguished from chalk by lack of effervescence.
Coal	Fine to coarse	Carbon, C	Soft, black, burns.

A. B.

FIGURE 5–7
Limestone. A. Fossiliferous limestone, a fine-grained, white limestone with abundant fossil fragments. B. Coquina, a variety of limestone composed almost entirely of fossil fragments. (Photos from Ward's Natural Science Establishment, Inc., Rochester, N.Y.)

TABLE 5–3
Origins of the more important nonclastic sedimentary rocks.

Rock	Biologic	Chemical		Remarks
		Evaporite	Other	
Limestone	×	× Rare	×	
Dolomite		× Rare		Forms mainly by replacement
Chert	×		× Precipitation from volcanic silica	Most biologic chert recrystallizes from opal
Gypsum		×		
Rock salt		×		
Chalk	×			
Diatomite	×			
Coal	×			

Rock salt is composed of halite. It is recognized by its cubic cleavage and taste (see Appendix A). Rock salt is deposited when restricted parts of the sea are evaporated.

Diatomite is soft, white rock composed of the remains of microscopic plants. Because the remains are composed of silica, it is distinguished from chalk by lack of effervescence.

Coal is formed by the accumulation of plant material.

Evaporite deposits are formed by the evaporation of seawater (see Table 5–4). Gypsum and rock salt are the main rocks formed in this way. When seawater is evaporated at surface temperatures, such as in a restricted basin, the first mineral precipitated is calcite. Dolomite is the next mineral precipitated, but only very small amounts of limestone and dolomite can be formed in this way. Evaporation of a kilometer column of seawater would only produce a few centimeters of limestone and dolomite. After about two-thirds of the water is evaporated, gypsum is precipitated; and when nine-tenths of the water is removed, halite forms. During the last stages of precipitation, potassium and magnesium salts form. Thick beds of rock salt imply evaporation of large amounts of seawater.

The rocks just described are the most important, but by no means the only, sedimentary rocks. Any mixture of types is possible, and because any weathering product may form a sedimentary rock, endless variety is possible. Some less abundant types include economically important deposits, such as iron oxides, phosphorous rocks, bauxite (aluminum ore), and potash.

Shale, limestone, and sandstone make up over 99 percent of all sedimentary rocks. Because feldspars are the most abundant minerals, their main weathering product, shale, is the most abundant sedimentary rock. Limestone is next in abundance, and sandstone is last (see Fig. 5–2). The abundance of shale is somewhat less than is predicted from the abundance of clay-forming silicate minerals, suggesting that some clay is deposited in the deep-sea basins. Clay particles are very small and sink slowly; they can be carried thousands of kilometers by gentle currents.

LITHIFICATION

The transformation of a sediment into a rock is called **lithification**. Several processes are involved. **Cementa-**

TABLE 5–4
Composition of seawater.

Element	Percentage	Origin	Rock Formed
Chlorine	0.019	Volcanoes	Rock salt (halite)
Sodium	0.011	Chemical weathering	
Magnesium	0.001	Chemical weathering	Dolomite
Sulfur	0.0009	Volcanoes	Gypsum
Calcium	0.0004	Chemical weathering	Limestone (calcite)
Potassium	0.0004	Chemical weathering	Potassium salts

tion is the deposition by groundwater of soluble material between the grains and, as might be expected, is most effective in coarse-grained, well-sorted permeable rocks. The main cementing agents are:

Calcite—recognized by acid test. (Do not mistake the effervescence of cement for the overall effervescence of limestone.)
Silica—generally produces the toughest rocks.
Iron oxides—color the rock red or yellow.

Other cements, such as dolomite, are possible. In certain poorly sorted rocks, clay, which generally colors the rock gray or green gray, may be thought of as forming the cement.

Compaction is an effective lithifier of fine-grained rocks such as shale and siltstone. Compaction generally comes about from the weight of overburden during burial, and the reduction in volume due to compaction and squeezing out of water may amount to over 50 percent in some shales.

Recrystallization probably is important in producing chert and in transforming calcite-bearing mud into limestone. Recrystallization produces an **interlocking texture**.

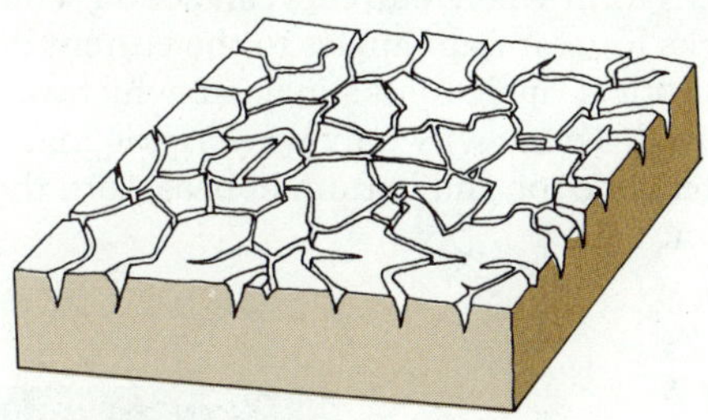

A.

B.

FIGURE 5–8
A. Mud cracks, which form as a result of drying mud flats, can be used to indicate top and bottom. B. Mud cracks in Death Valley, California. (Photo by J. R. Stacy, U.S. Geological Survey)

FEATURES OF SEDIMENTARY ROCKS

The most noticeable feature of an outcrop of sedimentary rocks is the **bedding**, which records the layers in the order of deposition, with the oldest at the bottom. In some instances, the beds are too thick to show in a small outcrop, and, for the same reason, many hand specimens do not show obvious bedding.

Study of the type of fossils and of sedimentary features can tell much about the environment of deposition and, in some cases, the postdepositional history. **Mud cracks** on bedding planes record periodic drying and suggest shallow water and, perhaps, seasonal drying (Fig. 5–8). **Ripple marks** also suggest shallow water with some current action, but are also known to form in deep water. Detailed study of ripple marks can show type and direction of current (see Figs. 5–9 and 5–10). Current action can also cause localized scouring or cutting to produce **cut-and-fill features** (Fig. 5–11). Another type of current feature is **cross-bedding** (Figs. 5–12 and 5–13). These features also can be used to recognize beds that have been overturned by folding or faulting (Fig. 5–14). Currents can also align mineral grains and fossils and so impart linear internal structures to rocks that can be used to determine the current direction.

If a mixture of sediments is suddenly deposited into a sedimentary basin, then the large fragments sink faster than the small ones. This does not violate Galileo's famous experiment, because here we are concerned with falling bodies in a viscous medium in which velocity of descent is controlled mainly by the size. The type of bedding produced by this kind of sedimentation is called **graded bedding** (Fig. 5–15); this can be demonstrated experimentally by putting sediments of various sizes in a jar of water, shaking, and then allowing the contents to settle. Graded beds can form in many ways, such as the sudden influx of sediment by a storm or seasonal flow of rivers. Their formation at the base of the continental slope is described in the next section.

The **sorting** of a sediment is a measure of the range in size of its fragments. If the fragments that compose a sediment are all similar in size, the sediment is well

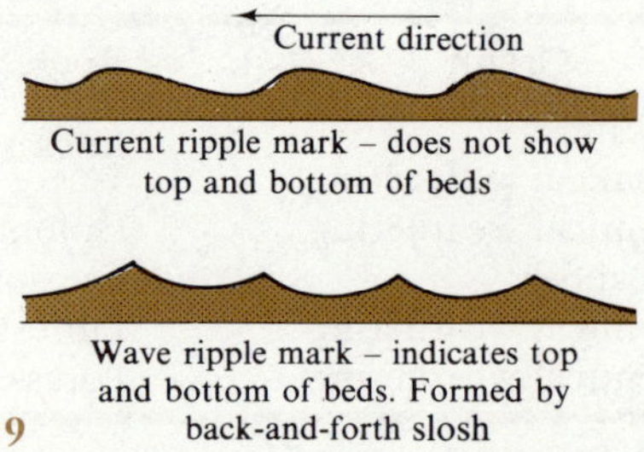

FIGURE 5–9
Ripple marks form where currents can act on sediments. Ripple marks form at right angles to the current direction, and asymmetrical ripple marks indicate which way the current flowed. To see why only wave ripple marks can be used to determine top and bottom of beds, turn the page upside down.

FIGURE 5–10
Oscillation ripple marks, Erie County, Ohio. Note the presence of ripple marks in successive layers. (Photo by D. F. Demarest, U.S. Geological Survey)

FIGURE 5–11
Scouring produced by localized currents. A. Initial deposition. B. Channel cut by current. C. Deposition is resumed, filling the channel. Such a feature records an interruption in deposition and can be used to distinguish top and bottom. D. Channel filled with conglomerate. Note that the conglomerate is more resistant to weathering than the sandstone. Channel is about 3 m deep. Daggett County, Utah. (Photo by W. R. Hansen, U.S. Geological Survey)

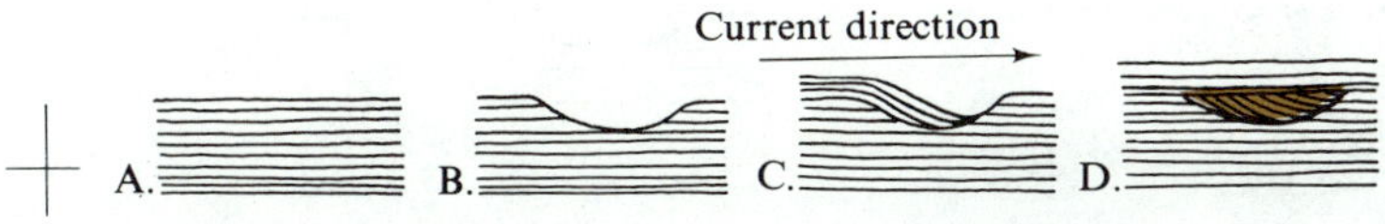

FIGURE 5–12
Development of cross-beds. A. Initial deposition. B. Channel cut by current. C. Channel filled by deposition from one side. D. Normal deposition after channel is filled. Cross-beds also can be used to distinguish top and bottom.

FIGURE 5–13
Cross-bedding. A. In river-deposited coarse sandstone. (Photo by H. E. Malde, U.S. Geological Survey) B. In wind-deposited sandstone in Zion National Park, Utah. (Photo by H. E. Gregory, U.S. Geological Survey)

sorted. The sorting of a sediment determines other textural features. The **porosity** of a sedimentary rock is a measure of its empty or void space, and its **permeability** is a measure of the interconnections of the pore spaces. These properties become quite important when one is concerned with recovering oil or groundwater from an underground reservoir. The amount of pore space in sedimentary rocks is quite varied but surprisingly large. If packing of uniform spheres is considered (and most sand grains are rounded), the closest possible packing is found to contain 26 percent void space, and

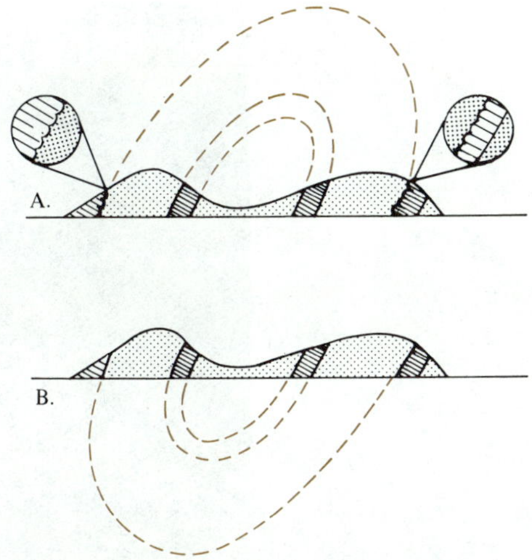

FIGURE 5–14
Alternate interpretations of the structure of outcrops of folded sedimentary rocks. The ripple marks in part A show that it is the correct interpretation. In the absence of ripple marks, or other sedimentary features that indicate top and bottom, it would not be possible to choose between the two interpretations.

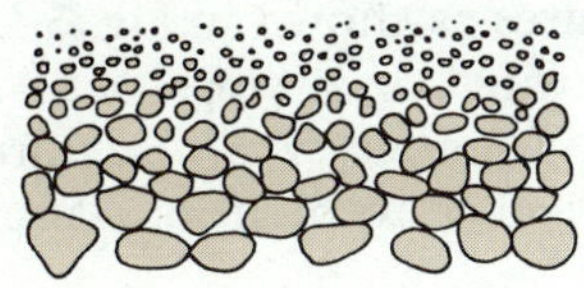

FIGURE 5–15
Graded bedding, produced by the more rapid settling of coarse material than fine material.

the loosest, 48 percent (see Fig. 5–16). The permeability will vary depending on the amount of pore space, the sorting, the cementation, and the size of the particles. A shale may have over 30 percent porosity, but the surface tension of water will prevent flowage through the tiny openings.

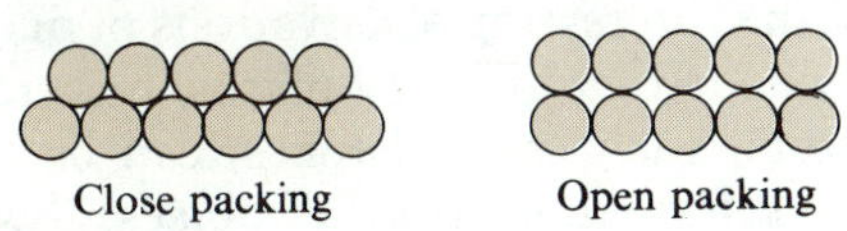

FIGURE 5–16
The packing controls the porosity, as illustrated here in the idealized case of spheres.

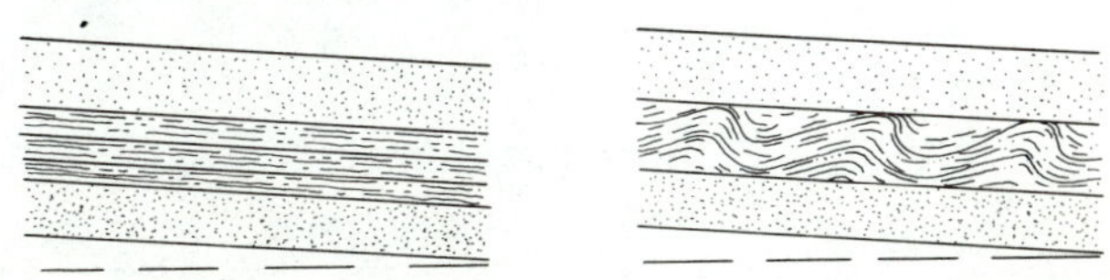

FIGURE 5–17
Soft sediments can slide down even very gentle slopes, developing slump features.

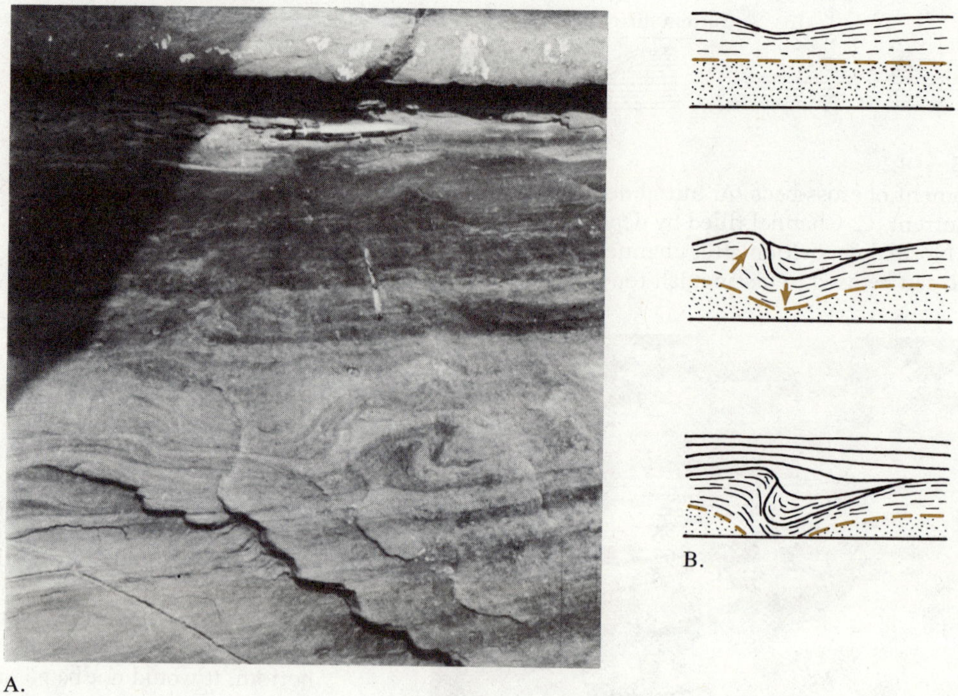

FIGURE 5–18
A. Photo of deformed bedding. B. Diagram showing how deformation of soft rock forms the structure shown in the photo.

Slump features as shown in Fig. 5–17 form if sediments are deposited on a slope or if they are tilted while still soft. Another type of deformation of soft rock is the vertical movement of sediment shown in Fig. 5–18.

DEPOSITION

Most sedimentary rocks are deposited in oceans, and these marine rocks, easily dated by their contained fossils, record much of the earth's history. In the present section, a few of the more important marine environments will be considered.

One of the interesting observations of marine sedimentary rocks is that at many places, individual sedimentary units, such as sandstones, can be traced for hundreds of kilometers. At other places, the beds are irregular and have limited horizontal extent. To see how this comes about, we will consider how marine sedimentary rocks are formed. For simplicity, the discussion will be limited to clastic rocks—sandstones and shales. As will be seen, the origin of the fragments that compose these rocks is weathering and erosion. These fragments are delivered to the ocean by rivers.

Because most marine sediments consist of material delivered by rivers, most sedimentary rocks form near the continents. Sampling and drilling in the oceans show this to be true. If one wishes to interpret the thick accumulations of old sedimentary rocks exposed in today's mountain ranges, the **continental shelves** and the **continental slopes** should be studied (Fig. 5–19).

The continental shelf is part of the continent, and its extent is determined by seismic studies of the crustal structure. However, the 100-fathom (183-m) depth line is generally close to the edge of the continental

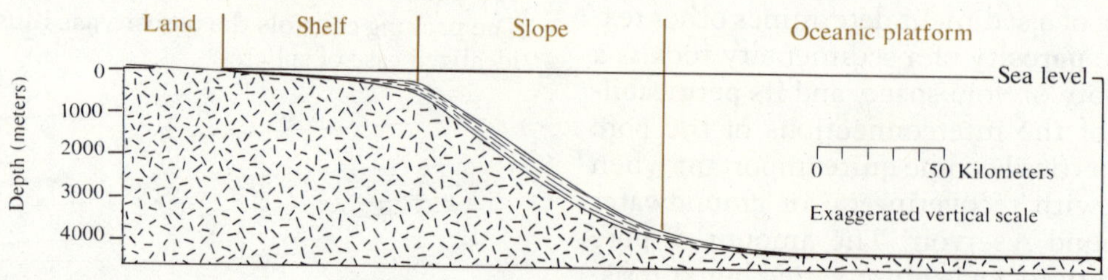

FIGURE 5–19
Continental margin showing typical continental shelf, continental slope, and oceanic platform. The vertical scale is exaggerated.

shelf. The continental shelf is almost flat; it has an average slope of about 2 m/km. The break in slope between the shelf and the continental slope generally begins at a depth of about 120–150 m, and the range is from near sea level to 460 m. The average depth of the shelves is near 60 m. The continental slope extends on the average to a depth of 3.2 km or more, with a slope of 2°–6°. This is a very steep slope compared to the mountain ranges of the earth.

The continental shelves for the most part are covered with clastic sediments. One would expect that, in general, the coarsest material should be found near shore or near river mouths and the finer silt and clay should be farther from the source (see Fig. 5–20). This

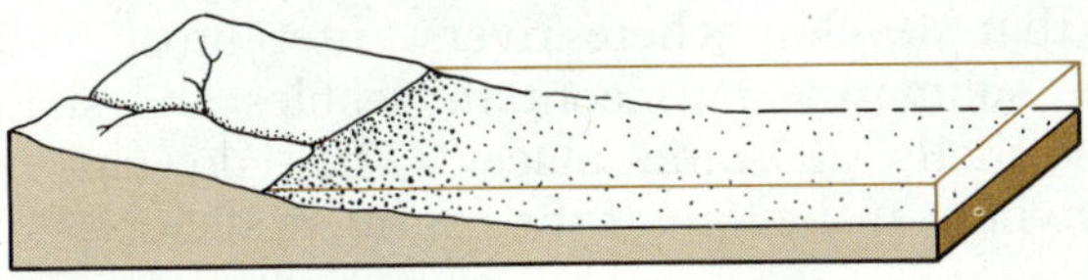

FIGURE 5–20
Idealized diagram showing coarse material being deposited near the shore and finer material further out.

is probably true in a very general way, but the many departures suggest the situation is not simple (see Fig. 5–21). One very obvious reason is that during the recent glacial time, the oceans were lower because much water was held in the glacial ice. Apparently, too, tidal and other currents are fairly effective in moving and sorting clastic material on the relatively shallow shelves. This type of reworking of the sediments may account for the uniform beds extending over large areas that are found on the present continents. In any event, pebbles and conglomerate are not uncommon far out on the shelf.

The present coastal plains on the Atlantic and Gulf coasts are underlain by what are apparently continental shelf deposits. Uplift has exposed the landward parts of these sedimentary accumulations, and seismic studies and deep drilling indicate that they extend onto and thicken on the present continental shelf. On the Gulf Coast, these deposits are several kilometers thick, and their features and fossils indicate shallow-water deposition. The latter suggests slow subsidence.

Much less is known about the structure and formation of the continental slopes. Limited sampling shows the surface, at least, to be covered mainly with fine sediments, presumably swept off the shelves. A more important process in the building of the slopes may be submarine landslides or mudslides. Following earthquakes, submarine telephone cables may be broken. The cables do not all break at the time of the earthquake but at later times, farther down the slope. In some cases, many cables have been broken and the time of breaking is accurately known. This, plus the burial of some of the cables reported by the repair ships, strongly suggests that submarine landslides move down the continental slopes.

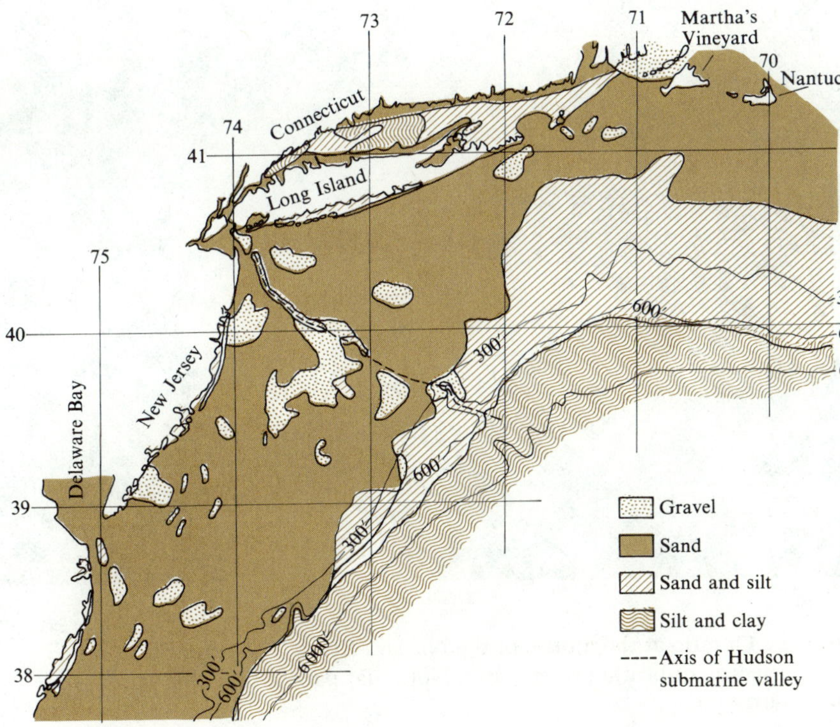

FIGURE 5–21
Bottom sediments on part of the continental shelf in northeastern United States. The many departures from the idealized case shown in Fig. 5–20 are probably caused by bottom currents and by deposition when sea level was lowered during glacial times. (Adapted from F. P. Shepard and G. V. Cohee, *Geological Society of America Bulletin*, Vol. 47, 1936)

Another feature of continental shelves and slopes is the **submarine canyons**. Submarine canyons have many of the features of river canyons, and some are larger than such river canyons as the Grand Canyon. The origin of such features has been the subject of lively speculation, and not everyone is yet satisfied with the explanations offered. At first, submarine canyons were thought to have formed by ordinary river erosion at a time when sea level was lower than it is at present. Many of the submarine canyons have heads near the mouths of large rivers, making this a reasonable hypothesis. However, many of them extend to depths of over 3 km, and sea-level changes of this magnitude in the recent past seem impossible. The submarine landslides mentioned previously offer the most likely explanation. Such landslides consist of sediment-laden water. Because they are heavier than the surrounding water, these landslides or mudflows move along the bottom and are capable of eroding the soft bottom material. These muddy waters are called **turbidity currents**. They can move down very gentle slopes and can transport fairly large pebbles. Landsliding of unconsolidated material on the continental slopes can account for the canyons there; and turbidity currents initiated by stream discharge, storms, or other currents can explain the submarine canyons on the continental shelves. As well as explaining the canyons, these processes can also account for the building of the continental slopes, for seismic studies show the slopes to be composed of sedimentary rocks.

Turbidity currents and submarine landslides can also account for the smooth ocean bottoms at the foot of the continental slope at many places (see Fig. 5–22). These smooth surfaces are apparently similar to the fans that develop where rivers enter broad valleys. These submarine fans contain pebbles and shallow-water shells at some places. They do not occur everywhere at the foot of the continental slope.

FIGURE 5–22
Turbidity current in Lake Ontario at the mouth of the Niagara River. The turbidity current is caused by the sediment-bearing river water. Note that part of the river water is carried by longshore currents.

A.

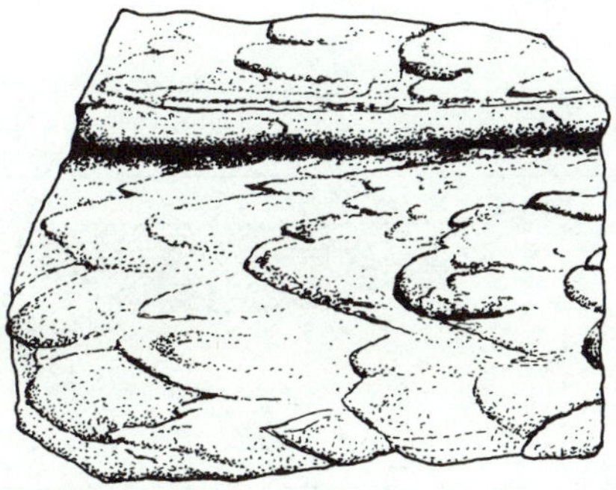

B.

FIGURE 5–23
Depositional features of turbidity currents. A. Photo. The bottom of a bed. B. Diagram. The lobelike casts point in the opposite direction to current movement. Crossing the specimen is the cast of a groove, probably caused by a boulder carried by the turbidity current. The specimen is about half a meter in diameter.

To further evaluate the possible existence and importance of turbidity currents in the geologic past, we will look at their depositional features. Turbidity currents, because of their density and speed, can transport large fragments. These larger fragments and the main mass of fine material are deposited fairly rapidly. The larger, heavier fragments settle first, forming graded beds. The eroding ability of turbidity currents suggests that they may erode or otherwise deform the top of the underlying bed. This is shown in Fig. 5–23.

It was noted long ago by geologists that many mountain ranges are formed of bent and broken layers of sedimentary rocks. These sedimentary rocks had been deposited as flat layers in shallow seas. It is known that the seas were shallow because the rocks have features that commonly form in shallow water, such as mud cracks, and the fossils they contain are organisms that live in shallow waters. The mountain ranges, then, began as elongate basins of deposition, and these areas of sedimentation were named **geosynclines**, now commonly shortened to **geoclines**. The sedimentary rocks in geosynclines are much thicker than nearby rocks of similar age. Geosynclinal mountain ranges are commonly near the margins of continents, and so some of them may have formed in the continental shelf-continental slope area just described.

The sediments of continental shelves tend to be sandstones near the coast, with shale or limestone accumulating farther from shore, although many exceptions are found (Fig. 5–24). The rocks that underlie the continental slope are generally clastics, shale and sandstone, commonly intermixed. Few or no volcanic rocks are found in the continental shelf-continental slope area.

In most geosynclinal mountain ranges, volcanic rocks are found interbedded with sedimentary rocks in some areas. This suggests that in these cases, convergent plate boundaries may be involved. Volcanic island arcs form where oceanic plates collide with continental plates and so are also at the margins of continents (Fig. 5–25). The volcanic borderlands can be the source of both volcanic rocks and clastic sediments. Geosynclinal clastic sediments containing volcanic rocks generally are very poorly sorted and therefore have a wide variety of sizes and types of fragments. Such rocks contrast with the sediments on continental shelves that tend to be well sorted. It is later plate movements that are believed to crumple the geosynclinal sediments.

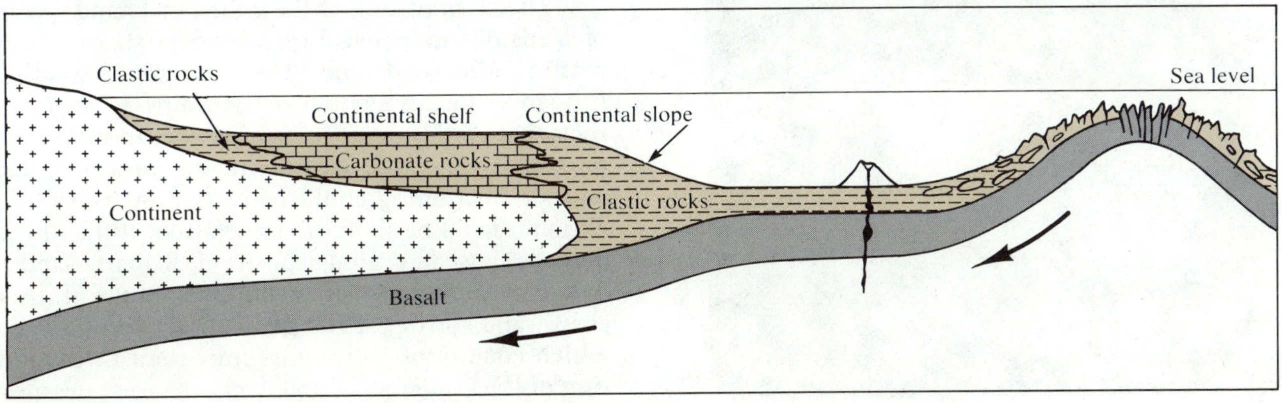

FIGURE 5–24
Development of thick sections of sedimentary rocks at a continental shelf and slope, and much thinner layers of sediments and volcanic rocks on the ocean floors and at mid-ocean ridges. (Drawing by Robert W. Tope)

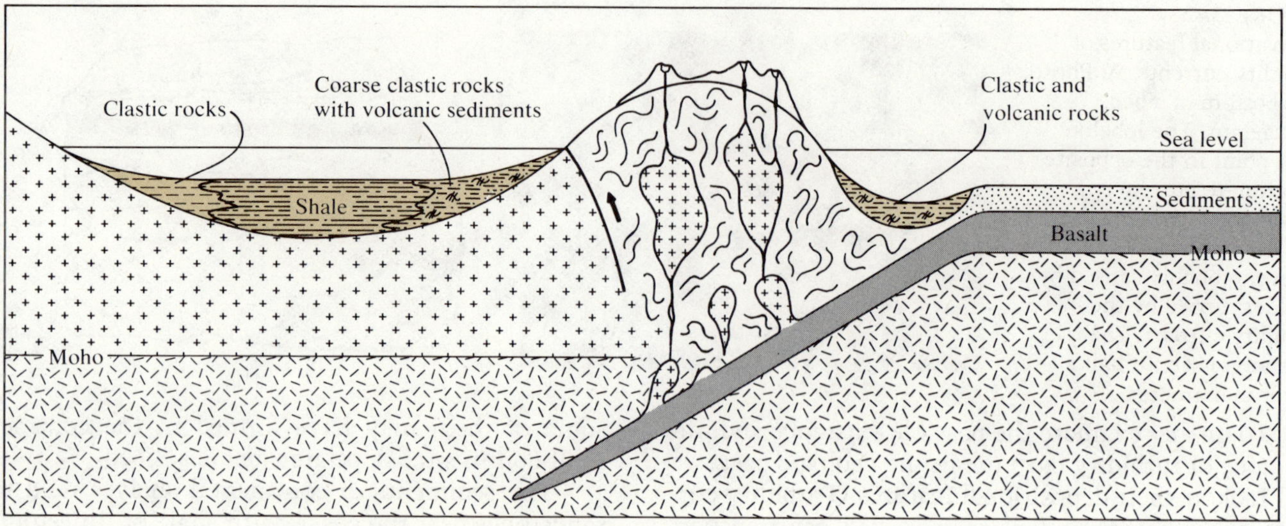

FIGURE 5–25
Development of thick sections of sedimentary and volcanic sedimentary rocks at a convergent plate boundary. (Drawing by Robert W. Tope)

In the deep ocean, far from land, the sediments consist mainly of wind-carried dust, largely of volcanic origin, and the fallen shells of tiny animals that live near the surface. Sediments of this type that are composed of more than 30 percent biologic material are called **oozes** (Fig. 5–26). The rate of accumulation of oozes is very slow compared with other sedimentary rocks. It is estimated that it requires between 1000 and 10,000 years to deposit 1 mm. The volcanic dust settles so slowly that most of it is oxidized while sinking to the bottom, and so it is termed *red clay.* Some features of deep-ocean sediments are shown in Fig. 5–27.

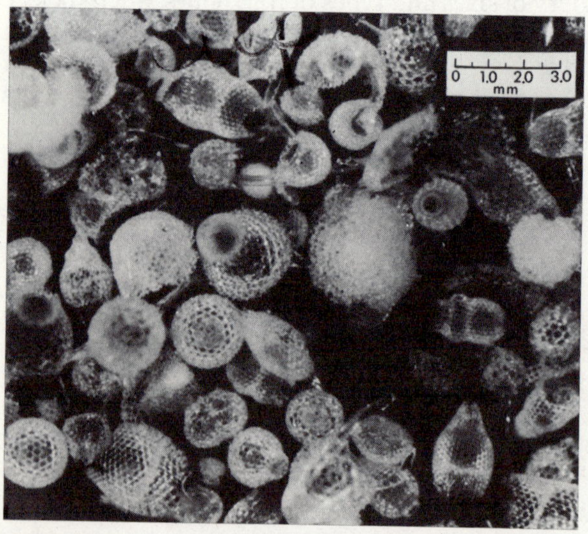

FIGURE 5–26
Radiolarian ooze from a depth of 5205 m, about 950 km south of Hawaii. (Photo from D. R. Horn, Lamont-Doherty Geological Observatory)

INTERPRETING SEDIMENTARY ROCKS—SOME SIMPLE EXAMPLES

Interpreting geologic history from the study of sedimentary rocks is an important activity carried on by many geologists. These studies are conducted at many levels, both in the field and in the laboratory. The few examples given here may help you to interpret the rocks in your area. These examples are simple, but a little imagination will reveal the many other possibilities and the many complexities possible. They will be limited to the clastic sedimentary rocks, especially sandstones.

The composition of the grains in a hand specimen of sandstone may reveal the rock type that was eroded to form the sandstone. The amount of weathering influences the mineralogy of the grains, so the grains of rock that could reveal the source are, in general, only present if very little weathering has occurred. Such a lack of weathering could be caused by a dry or cold climate, rapid or short transportation, or rapid burial, perhaps caused by rapid deposition. A quartz sandstone could reveal the opposite conditions. Study of the size, shape, and sorting of the grains may help in deciding which conditions were most important in the formation of the sandstone. Large grains suggest nearness to the source or short distance of transportation. Angular fragments, unless very resistant to abrasion, have not traveled far. The sorting can reveal, among other things, the mode of transportation. Wind-transported

A. B. C.

FIGURE 5–27
Ocean-bottom photographs. A. Hydrographer Canyon off the east coast of the United States. Soft coral and other bottom organisms at a depth of 725 m. B. Hydrographer Canyon showing brittlestars and other organisms. C. Deep seabottom in Romanche Trench. (Photos courtesy R. M. Pratt)

sediments are well sorted, and glacially deposited rocks have an extreme size range. All of these features can be seen with a hand lens, but the complete study of a hand specimen in the laboratory requires the use of the microscope and other instruments.

Much more can be learned at the outcrop of the rock layer. Here, the bedding can be seen. The thickness and uniformity of the beds can reveal much about their origin. The irregular bedding of river deposits contrasts with the regularity of many shallow-water marine sedimentary rocks. The many features of sedimentary rocks, such as ripple marks, cross-beds, and graded bedding, can also be studied at the outcrop. These features and their interpretation were described earlier. Fossils can also be collected at the outcrop. The environment of deposition can, in many cases, be determined from study of fossils, because the types of animals that live near shore are different from those in deep water; they also differ between warm and cold water. Perhaps the most important use made of fossils is to determine the age of the sedimentary layers. The deformation, if any, of the beds can also be seen in many outcrops.

The next level of study of a sedimentary layer is to trace its areal extent by mapping its outcrops. By combining the interpretations made at each outcrop, it is possible to reconstruct the basin in which the beds were deposited. An example of such an interpretation is shown in Fig. 5–28. In this example, the source area for the Morrison Formation is revealed from the current directions shown by the cross-bedding, the increase in thickness near the source, and the coarseness of the grains near the source. Note also that the Morrison Formation is exposed in only a small part of the basin and everywhere else it has either been removed by erosion or is covered by younger rocks. Most geologic interpretations must be made on the basis of such fragmentary evidence.

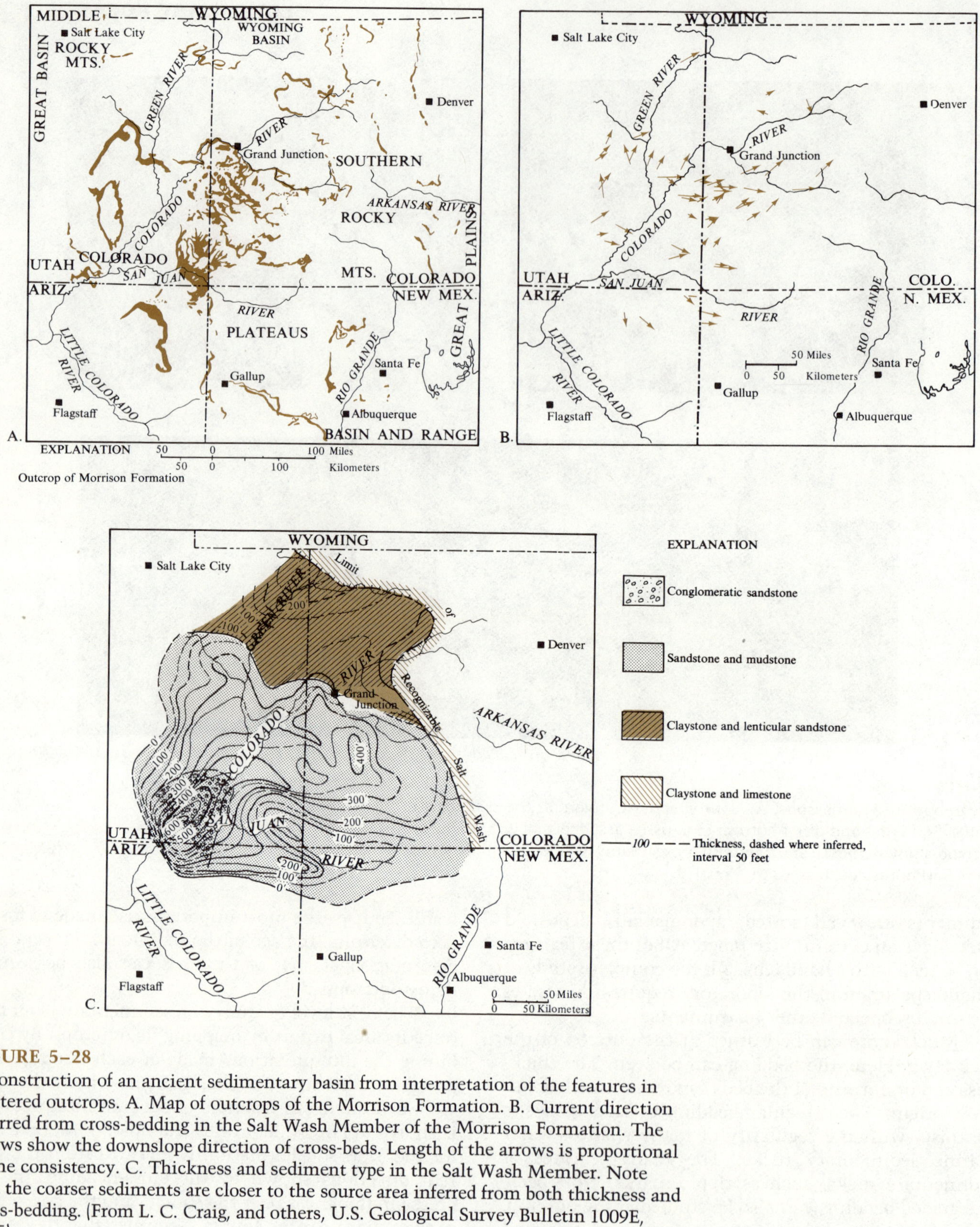

FIGURE 5–28

Reconstruction of an ancient sedimentary basin from interpretation of the features in scattered outcrops. A. Map of outcrops of the Morrison Formation. B. Current direction inferred from cross-bedding in the Salt Wash Member of the Morrison Formation. The arrows show the downslope direction of cross-beds. Length of the arrows is proportional to the consistency. C. Thickness and sediment type in the Salt Wash Member. Note that the coarser sediments are closer to the source area inferred from both thickness and cross-bedding. (From L. C. Craig, and others, U.S. Geological Survey Bulletin 1009E, 1955)

SUMMARY

Sedimentary rocks are composed of rock fragments, weathering products, organic material, or precipitates; and most are deposited in beds, or layers. They compose only about 8 percent of the earth's crust, although they cover about 75 percent of the surface.

Lithification is the transformation of sediment into a rock. Cementation is deposition of soluble material, such as calcite, silica, and iron oxides, between the grains by groundwater. Compaction lithifies fine sediments by weight of overburden during burial. Recrystallization produces the interlocking texture of chert and limestone.

Clastic rocks are composed of fragments or minerals broken from any type of preexisting rock; clastic rocks are generally subdivided by fragment size, not by composition, except for some sandstone types:

Conglomerate—fragments over 2 mm. May be subdivided into roundstone or sharpstone conglomerate (breccia) if fragments rounded or angular, respectively.

Sandstone—fragments between 1/16 and 2 mm. Some compositional types are quartzose sandstone (mainly quartz), arkosic sandstone, or arkose (over 20 percent feldspar), and graywacke (poorly sorted, clay or chloritic matrix).

Siltstone (mudstone)—fragments 1/256–1/16 mm. Gritty.

Shale—fragments less than 1/256 mm (clay). Not gritty; has fissility (ability to split on bedding planes). Nonfissile clay rocks are claystone.

Nonclastic rocks are formed by chemical or biologic precipitation, and by accumulation of organic material:

Limestone—composed of calcite, in some cases from chemical precipitation but usually of biologic origin. If composed mainly of fossils, called coquina.

Chalk—soft, white limestone formed by accumulation of the calcium carbonate shells of microscopic animals.

Dolomite—formed by replacement of calcite with dolomite.

Chert—fine-grained silica. Some precipitated directly from seawater, most from accumulation of silica shells of organisms.

Evaporite—a deposit formed by evaporation of seawater. The precipitation sequence is calcite, dolomite, gypsum, rock salt, and lastly potassium and magnesium salts. Gypsum and rock salt (halite) are the main rocks formed this way.

Diatomite—soft, white rock composed of silica remains of microscopic plants.

Coal—formed by accumulation of plant material.

Shale, limestone, and sandstone make up over 99 percent of all sedimentary rocks. Shale is most abundant because of the abundance of feldspars.

Bedding of sedimentary rocks records the layers in order of deposition, with the oldest at the bottom. Mud cracks record periodic drying and can show top and bottom of deformed beds. Ripple marks can show type and direction of currents, and wave ripple marks can show top and bottom. Cut-and-fill features are caused by localized scouring and can show top and bottom. Cross-bedding, another current feature, also shows top and bottom. Graded bedding is formed by larger fragments settling faster than smaller fragments; bottom is indicated by the larger fragments.

Porosity is a measure of the amount of void space in a sediment; permeability is a measure of the interconnections of the void spaces.

Continental shelves, the almost-flat portions of the continent below sea level, are generally covered with clastic sediments. Continental slopes, the steep slopes between the continental shelves and the deep-ocean platforms, are mainly covered with fine sediments. Slopes and shelves have some large submarine canyons. These may be caused by turbidity currents, underwater landslides consisting of heavy, sediment-laden water which can erode the soft bottom material.

Many mountain ranges are formed of bent and broken layers of sedimentary rocks. These sedimentary rocks are much thicker than nearby rocks of the same age. Geosyncline, or geocline, is the name given to the elongate basin in which the rocks accumulated. Geosynclines generally form at continent-ocean borders and may form at continental shelf-continental slope areas or at convergent plate boundaries.

In the deep ocean, the sediments are composed of slowly settled, wind-carried dust, largely of volcanic origin, and fallen shells of tiny animals. Sediments that have more than 30 percent biologic material are called oozes.

Sedimentary rocks are important in the interpretation of geologic history through study of grain size and shape, depositional features, size and shape of depositional area, fossil study including dating, and deformation of the area.

REVIEW EXERCISES

1 Why is quartz a common mineral in sandstones?

2 What is the origin of chert?

3 What are the common cements in sandstones?

4 On what basis are clastic sedimentary rocks classified?

5 How can top and bottom of the beds be determined in a group of vertical sedimentary beds?

6 How might sediments deposited in shallow water differ from those deposited in deep water?

7 What features could be used to distinguish turbidity-current-deposited sediments?

8 What geologic history might be inferred from each of the following rocks?
- (a) A feldspar-rich sandstone
- (b) A well-rounded conglomerate
- (c) A sandstone composed almost entirely of quartz
- (d) A limestone with abundant fossils

9 What kinds of fossils would you expect to find in rocks deposited in the following environments?
- (a) Tropical shallow water
- (b) Deep water
- (c) Temperate near shore
- (d) Marsh

SUGGESTED READINGS

GENERAL

DIETRICH, R. V., and SKINNER, B. J. *Rocks and Rock Minerals.* Somerset, N.J.: John Wiley and Sons, 1979, 319 pp.

ERNST, W. G. *Earth Materials.* Englewood Cliffs, N.J.: Prentice-Hall, 1969, 150 pp. (paperback).

SEDIMENTARY ROCKS

FREEMAN, T. *Field Guide to Layered Rocks.* ESCP Pamphlet Series, PS-3. Boston: Houghton Mifflin Co., 1971, 43 pp.

HEEZEN, B. C. "The Origin of Submarine Canyons." *Scientific American* (1956), Vol. 195, No. 2, pp. 36–41. Reprint 807, W. H. Freeman and Co., San Francisco.

HSÜ, K. J. "When the Mediterranean Dried Up." *Scientific American* (1972), Vol. 227, No. 6, pp. 26–36. An excellent article on the interpretation of climate and environment from sedimentary rocks and other geologic features.

KUENEN, P. H. "Sand." *Scientific American* (1960), Vol. 202, No. 4, pp. 94–106. Reprint 803, W. H. Freeman and Co., San Francisco.

LAPORTE, L. F. *Ancient Environments.* Englewood Cliffs, N.J.: Prentice-Hall, 1968, 116 pp. (paperback).

PETTIJOHN, F. J., and POTTER, P. E. *Atlas and Glossary of Primary Sedimentary Structures.* New York: Springer-Verlag, 1964, 370 pp.

POTTER, P. E., and PETTIJOHN, F. J. *Paleocurrents and Basin Analysis.* Berlin: Springer-Verlag; published in United States by Academic Press, New York, 1963, 296 pp.

SHROCK, R. R. *Sequence in Layered Rocks.* New York: McGraw-Hill Book Co., 1948, 507 pp.

ZENGER, D. H. "Dolomitization and Uniformitarianism." *Journal of Geological Education* (1972), Vol. 20, No. 3, pp. 107–24.

- pressure
- stress
- shear
- chemically active solution
- contact (thermal) metamorphism
- regional metamorphism
- foliation
- dynamic metamorphism

6

- hornfels
- marble
- dolomite marble
- quartzite
- progressive metamorphism
- retrogressive metamorphism
- serpentine
- serpentinite
- soapstone
- talc schist
- slate
- schist
- gneiss
- greenschist
- amphibolite
- fault gouge
- fault breccia
- mylonite
- metamorphic grade
- index mineral
- isograd
- metamorphic facies
- equilibrium assemblage
- blueschist

Metamorphic Rocks

METAMORPHIC ROCKS ARE ROCKS OF ANY TYPE OR COMPOSITION that have been changed in either mineral composition or texture.

The study of metamorphic rocks developed later than the study of igneous rocks, although the same methods of study are used for both rock types. Metamorphic rocks are the more difficult of the two to study because of their complex structures and their variations in composition and mineralogy. Hutton was the first to recognize clearly that rocks could undergo changes, although his concept of metamorphism was crude in the light of later knowledge. By the middle of the nineteenth century, French geologists had established the two main types of metamorphism, **contact metamorphism** and **regional metamorphism**. These terms are described later.

Even before Werner, the idea developed that metamorphic rocks are the oldest rocks. Their complex structures and their locations in the cores of many mountain ranges and under the oldest sedimentary rocks suggest their antiquity. It is now known that although the oldest rocks are metamorphic rocks, not all metamorphic rocks are old. The discovery of deformed fossils in metamorphic rocks showed this as early as 1814 when De Charpentier found fossils in mica schist in the Alps. This discovery helped to convince some neptunists that Werner's age assignment was wrong. In 1881, Reusch found fossils in metamorphic rocks in Norway, and this discovery accelerated the study and understanding of metamorphic rocks. At the present time, the mapping of metamorphic areas is aided by the use of radioactive dating. Mapping complex structures without some means of recognizing the ages of the rocks is a difficult undertaking.

Around the turn of the century, studies of the intensity or grade of metamorphism were made from field mapping and microscopic study. Studies were made by Barrow in Scotland in 1893, Becke and Grubenmann in the Alps in 1904, Goldschmidt at Oslo in 1911, and Escola in Finland in 1915. These men established the principles involved in the mineralogical changes that occur in metamorphism. Refinements of this work, especially using the results of more recent laboratory studies, is an ongoing type of research.

AGENTS OF METAMORPHISM

Metamorphic rocks are rocks that have been changed while in the solid state either in texture or in mineral composition by any of the following: heat, pressure, directed pressure (stress), shear, or chemically active solutions. Because any rock of any type or composition can be subjected to any or all of these agents, the variety of metamorphic rocks is endless. Note that the changes that take place in the lithification of a sedimentary rock are, according to the definition, metamorphic. Such changes, however, are not considered metamorphic, but are arbitrarily excluded.

The changes that occur during metamorphism are the result of an attempt to reestablish equilibrium with the new conditions to which the rock is now subjected. Again note the similarity to weathering. Weathering occurs when rocks that formed deep in the earth are subjected to surface conditions. However, when surface-formed rocks, such as shales, are metamorphosed, the changes that occur proceed in the opposite direction and higher-temperature minerals are formed.

Many factors promote or retard metamorphic reactions. Large surface area promotes chemical reaction, so fine-grained rocks react faster than coarse-grained rocks. Glass is less stable than crystalline material and so also reacts faster.

Heat is an important agent in causing metamorphism. It is well known that as one goes deeper into the earth in mines or drill holes, the temperature increases. Thus, deep burial alone can cause the temperature to rise, but most metamorphic rocks probably form at places where the temperature rise with depth is greater than normal.

Pressure also is the result of burial. The pressure is due to the weight of the overlying rocks. At the increased temperature and pressure deep in the earth, rocks lose much of their strength and so behave plastically. This causes the pressure due to the weight of overlying rocks to act uniformly in all directions, just as the pressure due to the weight of the water in a swimming pool acts in all directions.

Directed pressure (stress) is also commonly present during metamorphism. The effect of this is to fold the more or less plastic rocks. Stress may deform rocks, and flattened or stretched pebbles or fossils record some of this deformation.

Shear results when the rocks are broken and moved by directed pressure, as in faults. In metamorphic rocks, the movement is commonly distributed on many closely spaced shear surfaces (Fig. 6–1).

Chemically active solutions are very important in metamorphic reactions. Even if temperature and pressure are high enough to cause a metamorphic reaction to occur, some water is generally necessary to allow the reaction to take place. Without water to act as a catalyst, the reaction may actually be too slow to occur. Water is generally present in most rocks, and only a thin film on the grain boundaries is necessary. The water may contain other materials that cause or promote the reactions, and it can also bring in or remove elements from the rock. Small amounts of water may also greatly reduce the strength of rocks undergoing metamorphism.

The hot waters that cause low-grade metamorphism have apparently been found in some areas in the quest for geothermal power. Hot springs are relatively common and are spectacular at places such as Yellowstone National Park. Recent study of the geothermal area near the Salton Sea in southern California shows that the hot waters there have caused metamorphism in the surrounding sediments.

TYPES OF METAMORPHISM

Metamorphic rocks can be subdivided into three genetic groups, although there is gradation among the groups:

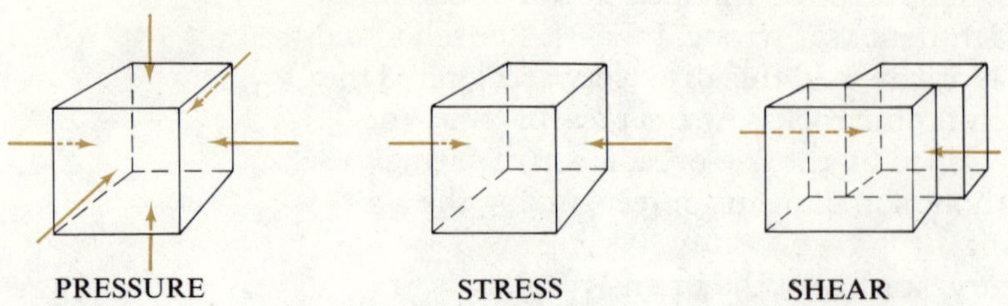

FIGURE 6–1 Pressure, stress, and shear.

1. **Thermal** or **contact metamorphic rocks**. These rocks generally are found at the margins of intrusive igneous bodies such as batholiths. The rocks are termed **hornfels**.
2. **Regional metamorphic rocks** are so called because they generally occupy large areas. These rocks form deep in the crust, and their presence at the surface reveals much uplift and erosion. The rocks generally have recrystallized under stress, so the new minerals grow in preferred orientation. These rocks are said to be **foliated**, and the most common are *gneiss* and *schist*.
3. **Dynamic metamorphic rocks** result from breaking and grinding without much recrystallization.

FORMATION AND IDENTIFICATION OF METAMORPHIC ROCKS

For classification and identification, metamorphic rocks can be subdivided into two textural groups:

1. **Foliated**—having a directional or layered aspect
2. **Nonfoliated**—homogeneous or massive rocks

NONFOLIATED METAMORPHIC ROCKS

Because the nonfoliated rocks are the simpler, we will consider them first. There are two types of nonfoliated metamorphic rocks. The first type consists of thermal or contact metamorphic rocks called **hornfels** (singular). Hornfels generally occurs in narrow belts around intrusive bodies and may originate from any type of parent rock. Sometimes these are called *baked rocks;* the name is appropriate, because their formation is quite similar to the baking of clay pottery in a kiln. They are generally fine-grained, tough rocks that are difficult to identify without microscopic study unless the field relations are clear. They range from being completely recrystallized rocks with none of their original features preserved to being slightly modified with most of the original features preserved (Fig. 6–2). In some of these rocks, new minerals grow. At the early stages of growth, the irregular growths produce what are called spotted rocks. Calcite-bearing contact rocks are generally quite spectacular, containing large crystals of garnet and other minerals.

The second type of nonfoliated metamorphic rock develops if the newly formed metamorphic minerals are equidimensional and so do not grow in any preferred orientation. The best examples of this are the

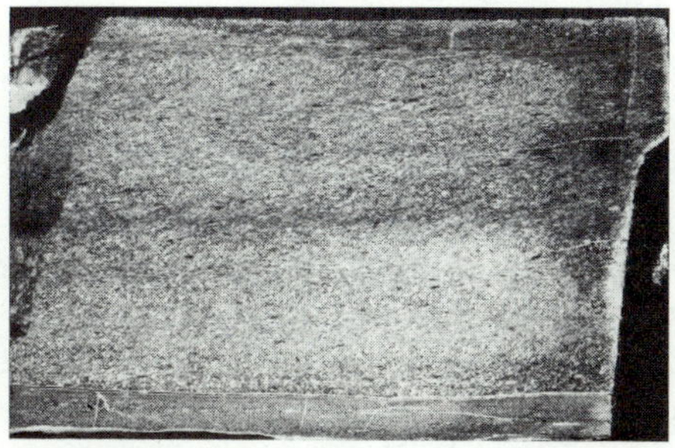

FIGURE 6–2
Hornfels, showing bedding and graded bedding. This polished specimen is just over 5 cm in diameter. (Photo by F. C. Armstrong, U.S. Geological Survey)

rocks that result when monomineralic rocks such as limestone, quartz sandstone, and dunite are metamorphosed. In the case of limestone, which is composed of calcite, no new mineral can form; thus, the calcite crystals, which are small in limestone, grow bigger, develop an interlocking texture, and become **marble** (see Fig. 6–3). Marble can be distinguished from limestone only by its larger crystals and lack of fossils; being composed of calcite, both effervesce in acid. Marble may have any color; pure white calcite rocks are more apt to be marble than limestone. Impure limestones, as noted previously, develop new minerals, particularly garnet. Dolomite marble also occurs.

FIGURE 6–3
Marble showing coarse interlocking texture. (Photo from Ward's Natural Science Establishment, Inc., Rochester, N.Y.)

Because quartz sandstone similarly can form no new minerals, the quartz crystals enlarge and intergrow to form **quartzite**. Quartzite is distinguished from chalcedony, chert, and opal by its peculiar sugary luster. The term *quartzite* is best reserved for the metamorphic rock, although quartz sandstone is called quartzite by some and the metamorphic rock is called *metaquartzite.*

Both marble and quartzite may form under stress, and so the crystals may be aligned. However, because the crystals are equidimensional, this preferred orientation can be detected only by microscopic examination. Thus, although these rocks are termed nonfoliated, some may have a hidden foliation.

So far, we have been concerned with **progressive metamorphism**, which is metamorphism caused by increased temperature, pressure, and other agents of metamorphism. The next example is one of **retrogressive metamorphism**, which is metamorphism at a temperature lower than that of the rock's original formation. Dunite is composed mainly of olivine, which is the highest-temperature mineral in Bowen's series. When dunite is subjected to the same metamorphic conditions that produce the rocks discussed earlier, it too is metamorphosed, but at a temperature lower than that of its formation. The olivine of dunite is changed to the mineral *serpentine,* and the metamorphic rock produced is called *serpentinite.* In other instances, instead of serpentine, *talc* is the metamorphic mineral that forms. If nonfoliated, the talc rock is called *soapstone;* if foliated, it is called *talc schist.* Slight differences in the composition of the parent rock or of the water solutions that cause this type of metamorphism probably determine which mineral is formed.

FOLIATED METAMORPHIC ROCKS

Foliated rocks are the more common of the metamorphic rocks (see Fig. 6–4) and are the product of regional metamorphism. Regional metamorphism generally takes place in areas undergoing deformation, and these very active areas are generally exposed in the cores of mountain ranges. Regional metamorphic rocks are produced by the same type of forces that make mountain structures. The forces that fold and fault the shallow rocks exposed on the flanks of mountain ranges can provide the stress fields under which the deep-seated metamorphic rocks, now exposed in the uplifted cores of mountain ranges, were recrystallized.

One example of the rocks produced by increasing metamorphic grade will show most of the foliated metamorphic rock types. These same rock types will form when many other parent rocks are metamorphosed; the differences in original chemical composition will change the relative amounts of the metamorphic minerals only somewhat.

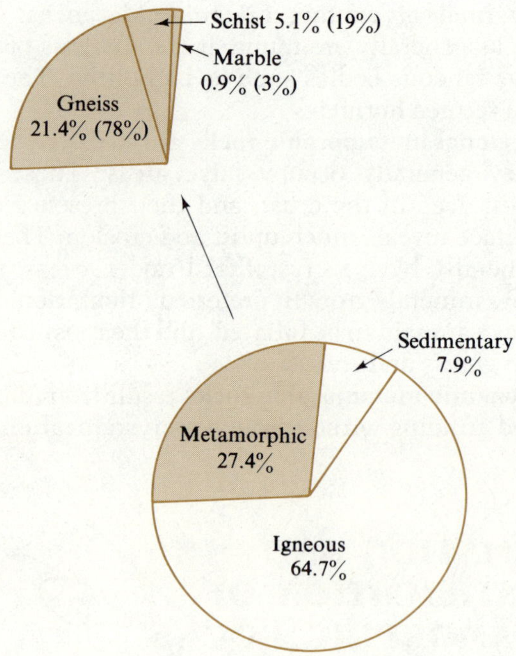

FIGURE 6–4
Relative abundance of metamorphic rocks in the earth's crust. For example, gneiss forms 21.4 percent of all rocks and 78 percent of the metamorphic rocks. (Data from Ronov and Yaroshevsky, 1969)

Shale is composed mainly of clay. During lower-grade metamorphism the clay is transformed to mica, and at higher grades of metamorphism to feldspar. Which micas and which feldspars form depends on the bulk chemical composition. Compare this with chemical weathering and note that weathering is, in a sense, retrogressive metamorphism. Because mica is a flat, platy mineral, it tends to grow with its leaves perpendicular to the maximum stress, forming a preferred orientation (see Figs. 6–5 through 6–7). Microscopic examination of some foliated rocks shows that the bedding planes are slightly displaced in the plane of the foliation. This suggests that foliation, at least at some places, forms in shear directions. The shear direction is in general at about 45° to the maximum stress in uniform materials. Thus, foliation may form perpendicular to the direction of maximum stress or at about 45° to that direction.

The sequence in the metamorphism of a shale is given in Table 6–1. The rock names are applied for the overall texture of the rock and not strictly for the mineral transformations which occur over a range in temperature and pressure (see Figs. 6–8 through 6–10). Most *schists* contain feldspar, but the name *gneiss* is reserved for rocks with much more feldspar than mica. Other rocks that have the same general bulk chemical

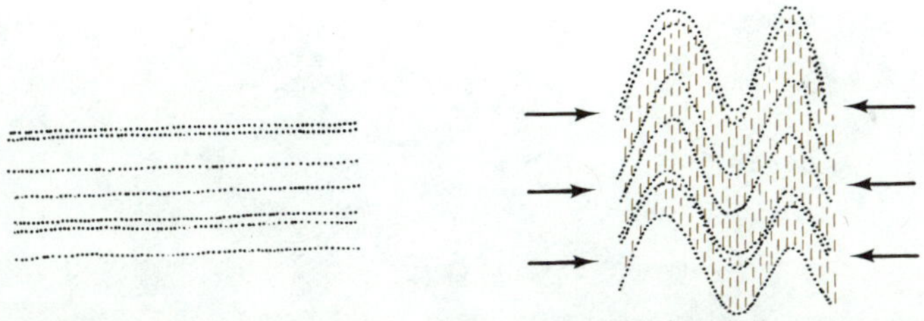

Original flat-lying sedimentary rocks.

Mica, shown diagrammatically by short dashes, grows perpendicular to the forces that may fold the rocks during metamorphism. Note that the foliation may form at an angle to the bedding.

FIGURE 6–5
Development of foliation.

FIGURE 6–6
Foliation and bedding in folded rocks near Walland, Tennessee. The rock type is slate. Compare with Fig. 6–5. (Photo by A. Keith, U.S. Geological Survey)

composition, thus producing the same metamorphic rocks, are certain pyroclastic and volcanic sedimentary rocks, many sandstones, arkose, granite, and rhyolite. The coarse-grained rocks on this list are rarely the parents of slate or fine-grained schist but, in general, remain more or less unaffected until medium-grade metamorphism is reached. Occurrence of slate in areas where the nonshale rocks are not metamorphosed suggests that slate may be formed by a mechanical reorientation of clay particles during folding, together with limited recrystallization.

Only one other group of regional metamorphic rocks must be considered among the most common metamorphic rocks. These rocks are formed by the metamorphism of basalt, certain pyroclastic and volcanic sedimentary rocks, gabbro, graywacke, and some calcite-bearing or dolomitic sedimentary rocks. The sequence is given in Table 6–2.

DYNAMIC METAMORPHIC ROCKS

Dynamic metamorphism results when rocks are broken, sheared, and ground near the surface where the temperature and pressure are too low to cause any significant recrystallization. Thus, these rocks are commonly associated with fault zones, but there are all gradations between these rocks and ordinary schist, especially lower-grade schist formed from coarse-grained rocks. The fine-grained, ground rock in a fault

TABLE 6–1
Metamorphism of shale.

Sedimentary Rock	Low-Grade Metamorphism	Medium-Grade Metamorphism	High-Grade Metamorphism
Shale →	**Slate** →	**Schist** →	**Gneiss**
Clay	Clay begins to be transformed into mica. Mica crystals are too small to see but impart a foliation to the rock. May also form by mechanical rearrangement of clay alone.	Mica Grains are larger, so the rock has a conspicuous foliation.	Mica has transformed largely to feldspar, giving the rock a banded or layered aspect.

FIGURE 6-7
Typical exposure of slate on Prudence Island, Rhode Island. Bedding is nearly vertical and is shown by color differences. Foliation is perpendicular to the bedding. (Photo by J. B. Woodworth, U.S. Geological Survey)

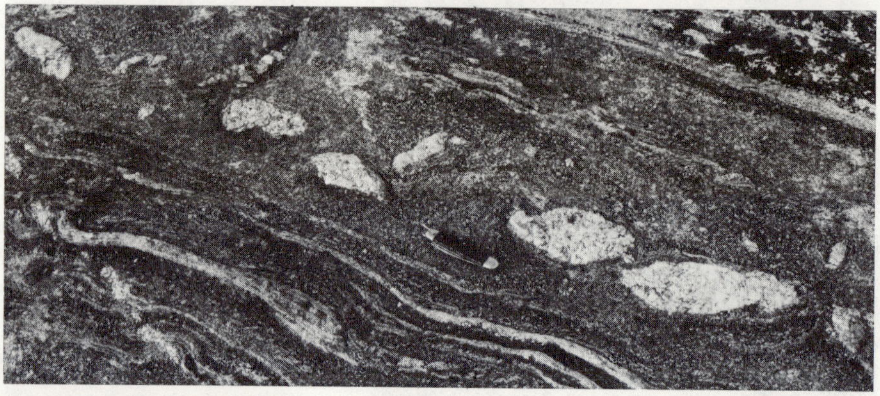

FIGURE 6-8
Outcrop of foliated rock. The rock is an impure marble, and the foliation is parallel to the bedding. The light-colored pods above the knife are quartzite, and the pods formed when a sandstone bed broke brittlely when the rocks were folded. (Photo by J. A. Donaldson, Geological Survey of Canada, Ottawa)

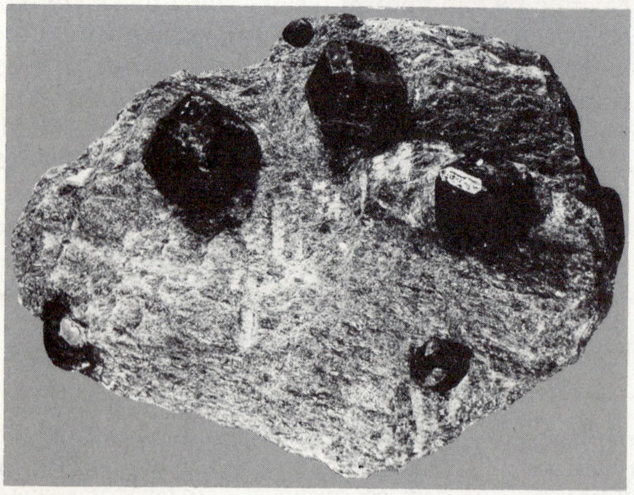

FIGURE 6-9
Garnet-mica schist. The large garnet crystals grew during metamorphism. Foliation does not show clearly in the orientation pictured. (Photo from Smithsonian Institution)

zone is called **fault gouge** or **fault breccia** if it has little cohesion, and similar cohesive rocks are generally called **mylonite**. As the amount of recrystallization of such rocks increases, they become gradational with schist and gneiss. The origin of these rocks is clear when they are encountered in the field, but they are very difficult to identify in a hand specimen.

An unusual type of dynamic metamorphism can occur at impact craters caused by meteorites.

The origin, texture, and mineralogy of the common metamorphic rocks are summarized in Figs. 6-11 through 6-13.

INTERPRETATION OF METAMORPHIC ROCKS

Study of a metamorphic rock should reveal what the parent or original rock was, and under what conditions or in what environment it was metamorphosed. At

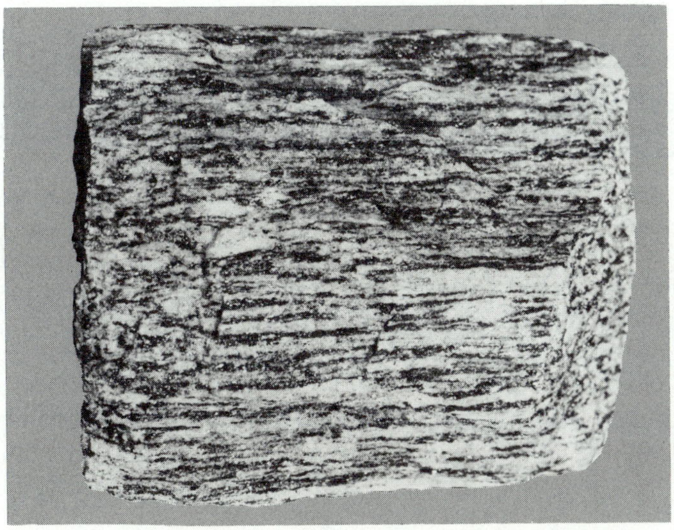

A.

B.

FIGURE 6–10
Gneiss. A. The foliation is shown by discontinuous layers of dark- and light-colored minerals. B. The layers in this gneiss are more irregular. (Photos from Ward's Natural Science Establishment, Inc., Rochester, N.Y.)

TABLE 6–2
Metamorphism of basalt.

Parent Rock	Low-Grade Metamorphism	Medium- and High-Grade Metamorphism
Basalt →	**Greenschist** →	**Amphibolite** (Amphibole schist)
	Fine-grained, foliated Chlorite (green "mica") Green amphibole Quartz	Coarse-grained, foliated Plagioclase Dark amphibole

some places, a layer of metamorphic rock can be followed, or **mapped**, to areas where it is not metamorphosed. In this way, the parent rock can be determined, as well as the changes that take place as the degree of metamorphism changes. Such studies, coupled with laboratory experiments, can yield much information about conditions and processes deep in the crust. That is, if we know the conditions of formation of a mineral or, better, an assemblage of minerals, when we find rocks composed of such minerals, we know the conditions under which the rocks formed. The problem is complicated by the diversity of natural rocks in that they contain many more elements, which are the variables in experiments, than can be studied easily in experiments. It is also not possible to reproduce the effects of geologic time in the laboratory, nor is it generally possible to detect the effects of solutions or volatiles that may have catalyzed the reactions in natural rocks. Thus, although we are very rapidly gaining an understanding of metamorphic processes, many problems remain.

To interpret metamorphic rocks fully, one must know whether they contain equilibrium mineral assemblages or whether conditions changed before the metamorphic reactions were complete; whether material was added or subtracted during metamorphism; and whether the rocks were subjected to one or more periods of metamorphism. One can think of many more possibilities that must be considered: for example, if the reactions take place during deep burial, will the rocks revert to something like their original state during slow uplift? No doubt some such effect may operate in some instances, but most of the chemical reactions involved go very slowly in the reverse direction. If this were not so, nothing but weathering products would be visible at the surface.

METAMORPHIC GRADE

The concept of **metamorphic grade** was introduced in the discussion of the formation of metamorphic rocks. Metamorphic grade is an interpretation based on field and laboratory evidence and should be further examined. The first attempts to map and study grade of metamorphic rocks consisted of mapping mineral zones. This was done by restricting the study to a single parent rock type and recording the first occurrence of certain **index minerals** as the metamorphic grade increased (Fig. 6–14 and Table 6–3). This simple method has been improved by considering assemblages of several minerals and by comparing such equilibrium assemblages in rocks of many compositions.

METAMORPHIC FACIES

A **metamorphic facies** can be defined as a group or assemblage of minerals that formed in response to an environment. Using mineral assemblages is an obvious improvement over index minerals in determining that the rock reached equilibrium. **Equilibrium mineral assemblages** are determined from field studies. At each place studied, rocks of a given composition are traced from areas of little or no metamorphism to areas of intense metamorphism. The minerals in the rock change as the metamorphic conditions or environment changes. Mineral assemblages found at many places, especially if in the same sequence, are assumed to be equilibrium assemblages.

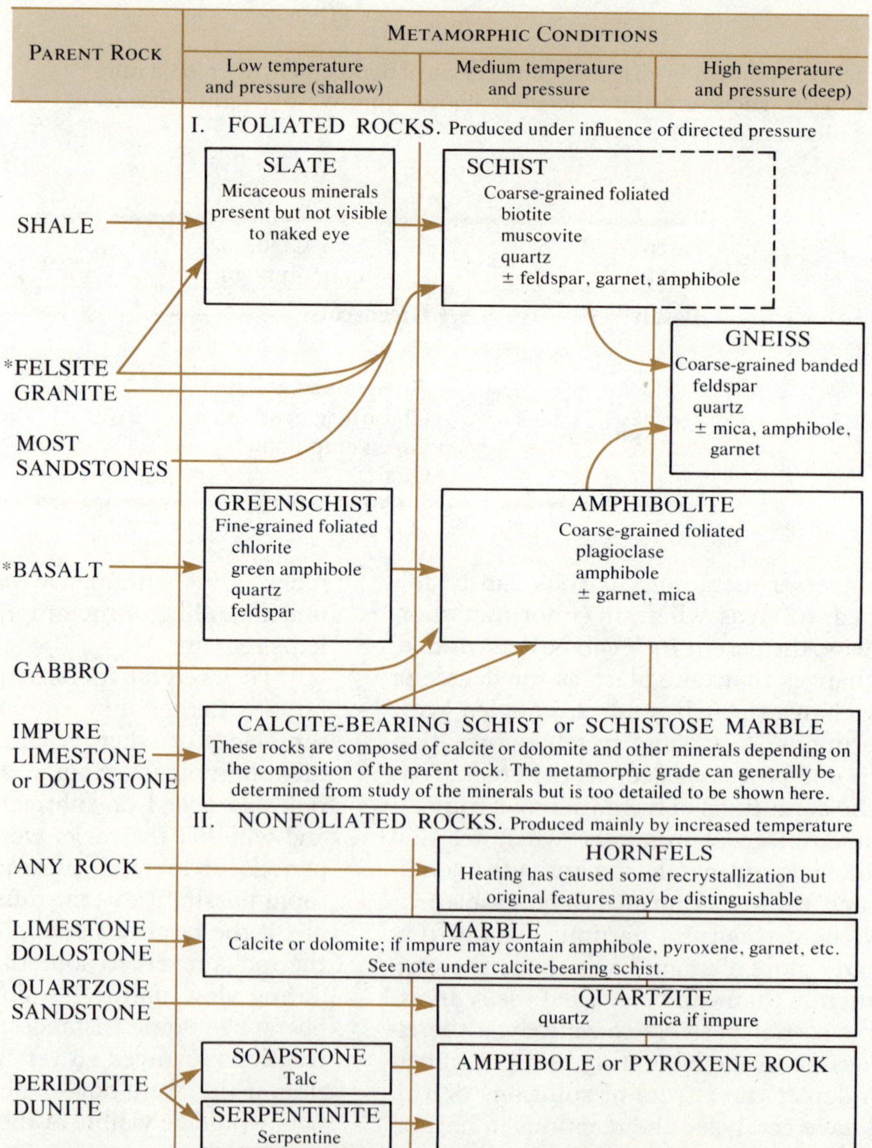

*Including fragmental volcanic rocks of this composition.

FIGURE 6–11
A generalized chart showing the origin of common metamorphic rocks.

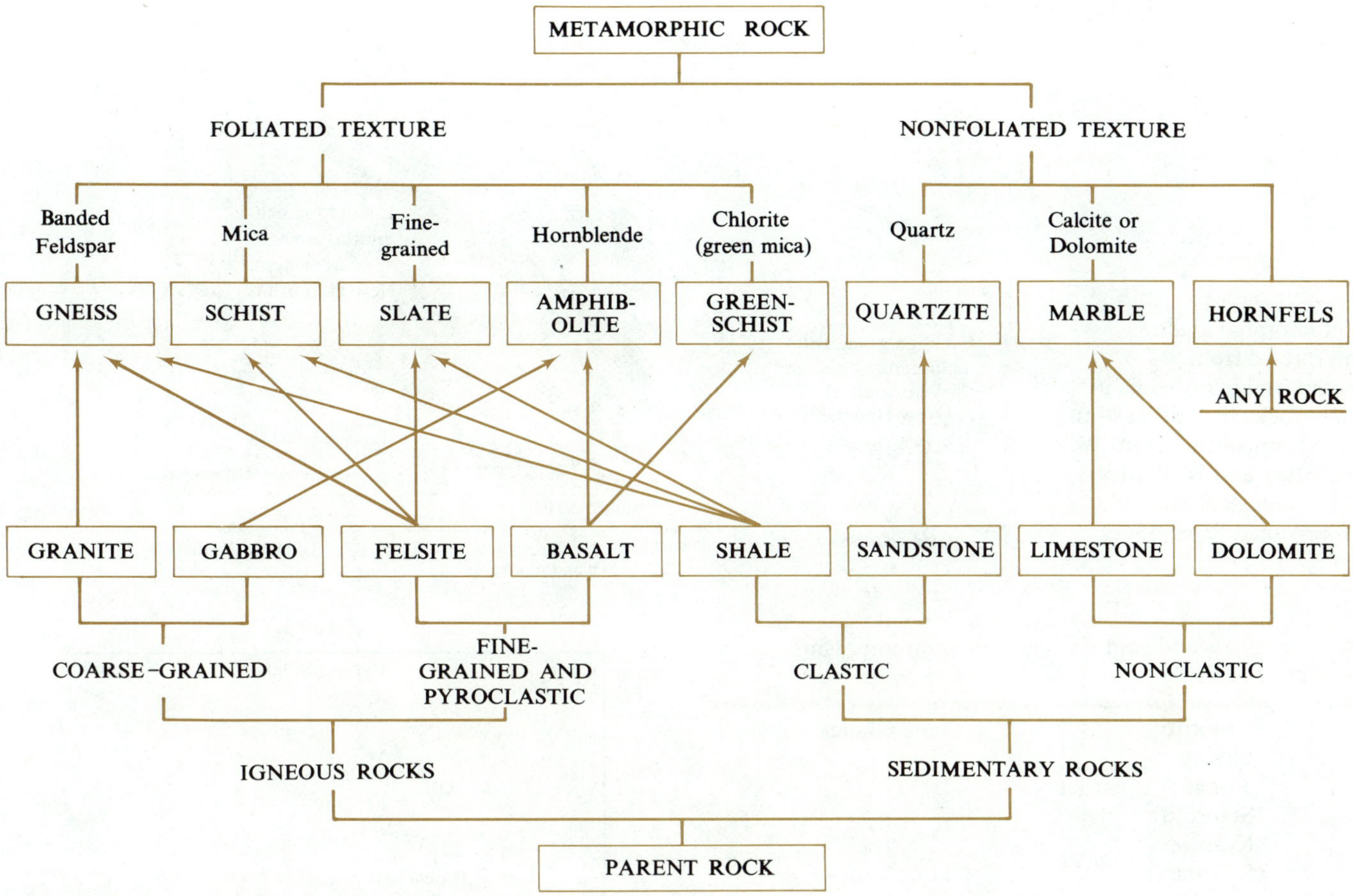

FIGURE 6–12
Origin of common metamorphic rocks. Not all possible parent rocks are shown in this simplified diagram.

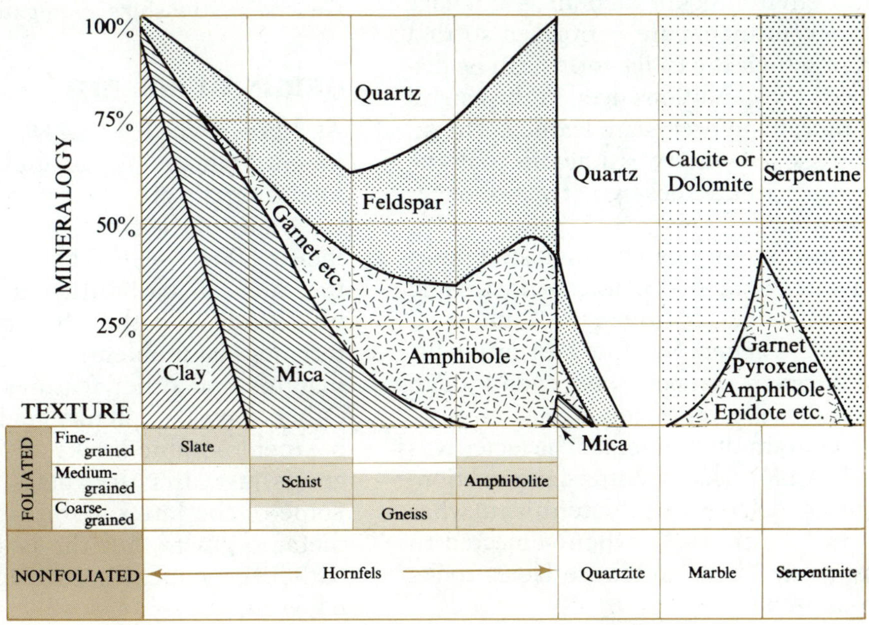

FIGURE 6–13
Texture and mineralogy of common metamorphic rocks.

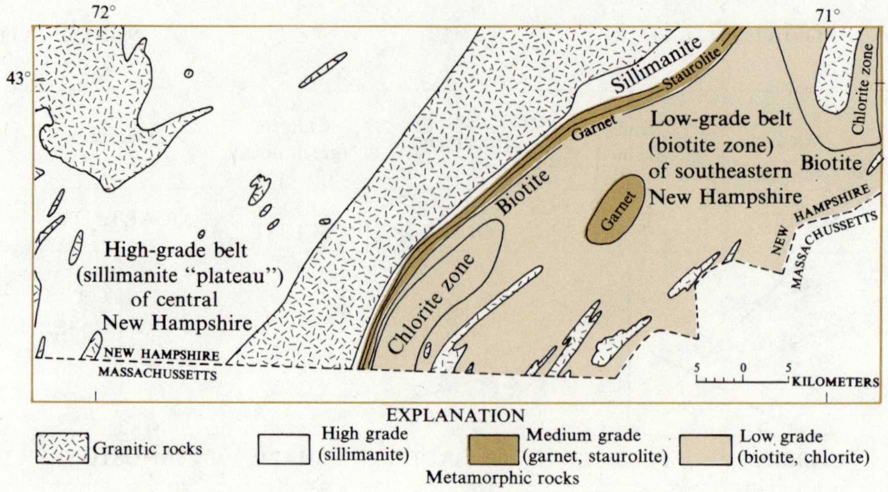

FIGURE 6–14
Metamorphic grade as interpreted from metamorphic minerals in shaly rocks in southeastern New Hampshire. (From W. Hamilton and W. B. Myers, U.S. Geological Survey Professional Paper 554C, 1967)

TABLE 6–3
Index minerals and grade of regional metamorphism developed in shale.

Chlorite	Low Grade
Biotite	↑
Almandine garnet	
Staurolite	
Kyanite	↓
Sillimanite	High Grade

A line on a map showing the first occurrence of each of the index minerals is an **isograd** and is the boundary between mineral zones. That is, the first occurrence of biotite is the boundary between the chlorite zone and the biotite zone.

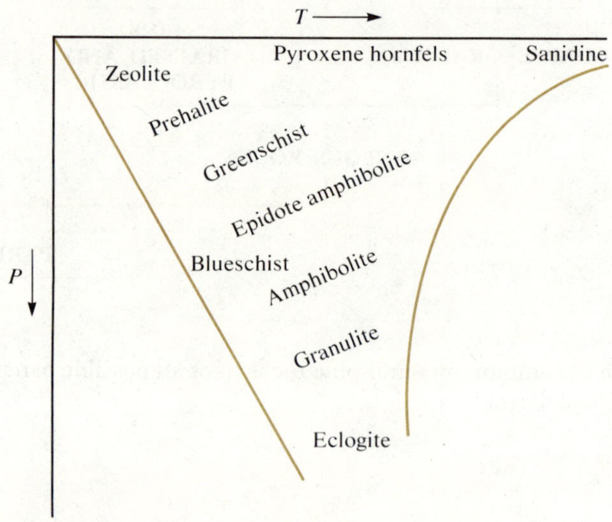

FIGURE 6–15
General relationships among metamorphic facies.

Once equilibrium assemblages or metamorphic facies are recognized, the next step is to learn under what conditions or environments such an assemblage forms. Laboratory experiments are performed so that the environment in which the facies forms can be discovered. In this way, the conditions deep in the earth's crust where metamorphism occurs are learned. The result of many of these laboratory studies is that the amount of water vapor in the rocks greatly influences which mineral assemblage forms.

The relationships among the common metamorphic facies are shown in Fig. 6–15, and the facies and the minerals in each facies are given in Table 6–4. Note that the facies are generally named for minerals, but that rocks of different compositions, even though metamorphosed under the same conditions or environment, may not contain that mineral. The facies was named from studies of rocks of some composition, shale, for example, and later work determined what minerals formed in basaltic rocks when subjected to the same environment. Thus, sanidinite facies rocks need not contain sanidine.

ORIGIN OF GRANITE

As shown in Fig. 6–16, at the highest grades of metamorphism some melting may occur, and so metamorphic processes may merge with igneous activity. Many of the small lenses of granitic rocks found in melting (Fig. 6–17). The origin of larger bodies of granite, such as batholiths, in high-grade metamorphic nite, such as batholiths, in high-grade metamorphic terranes is a problem. Some show chilled, fine-grained margins and other features that are clearly igneous. Others have completely gradational contacts and could be metamorphic rocks or could have intruded the gneiss just after the gneiss formed and was still hot. Some of the latter granites are probably the result of metamorphism, but the isotope composition of most batholiths indicates an origin in the lower crust or upper mantle.

TABLE 6–4
Typical mineral assemblages formed in metamorphic facies for rocks of shale and basalt composition.

Metamorphic Facies	Rocks of Shale Composition												Rocks of Basaltic Composition													
	Clay	Chlorite	Stilpnomelane	Muscovite	Biotite	Garnet	Sillimanite	Cordierite	Andalusite	Sanidine	Orthopyroxene	Quartz	Zeolites	Smectite	Prehnite	Pumpellyite	Chlorite	Epidote	Lawsonite	Amphibole	Garnet	Orthopyroxene	Clinopyroxene	Omphacite	Quartz	Plagioclase
Zeolite	×	×										×	×	×												
Prehnite-pumpellyte	×	×	×									×			×	×										
Greenschist		×		×								×					×	×		(Actinolite) ×					×	(Albite) ×
Blueschist (glaucophane schist)		×		×		(Spessartine) ×						×							×	(Glaucophane) ×					×	
Epidote amphibolite		×		×	×	(Almandine) ×						×						×		(Hornblende) ×	(Almandine) ×				×	(Albite) ×
Amphibolite				×	×	×	×					×								(Hornblende) ×	×				×	(Andesine) ×
Pyroxene hornfels					×			×	×			×											×		×	(Labradorite) ×
Granulite					×	×	×	×				×										×	×		×	(Labradorite) ×
Sanidinite							×	×		×	×	×											×		×	(Labradorite) ×
Eclogite	None known																				(Pyrope) ×			×		

NOTE: This table shows how elements are rearranged into different minerals in different metamorphic environments.

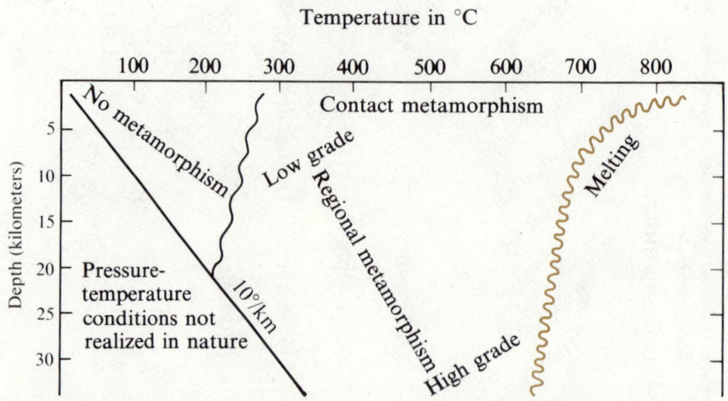

FIGURE 6–16
Diagram showing the conditions under which metamorphism occurs. Pressure is proportional to depth. Biotite is believed to crystallize between 200°C and 300°C, and sillimanite between 600°C and 700°C, or very near the melting point of granite.

FIGURE 6–17
The small, irregular occurrence of massive granitic rock that cuts across the foliation may have formed by partial melting. The granitic rock is localized in a minor fault zone where the foliation is disturbed. Dannemora Mountain, New York. (Photo by A. W. Postel, U.S. Geological Survey)

METAMORPHIC ROCKS AND PLATE TECTONICS

The active areas of plate tectonics—mid-ocean ridges and convergent plate boundaries—are also places where metamorphism occurs. At mid-ocean ridges, the rocks near the crest are metamorphosed, as shown in Fig. 6–18. The crest is an area of active basalt volcanism, so it is not surprising that the nearby rocks are subjected to metamorphic conditions. Volatiles, especially water, from the volcanic activity and the heat that generates the magma are probably important causes of this metamorphism.

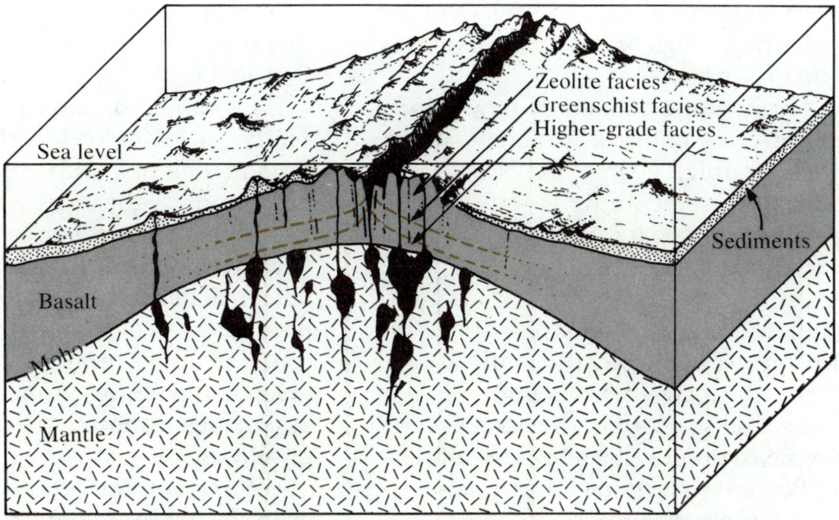

FIGURE 6–18
Metamorphism at a mid-ocean ridge. (Drawing by Robert W. Tope)

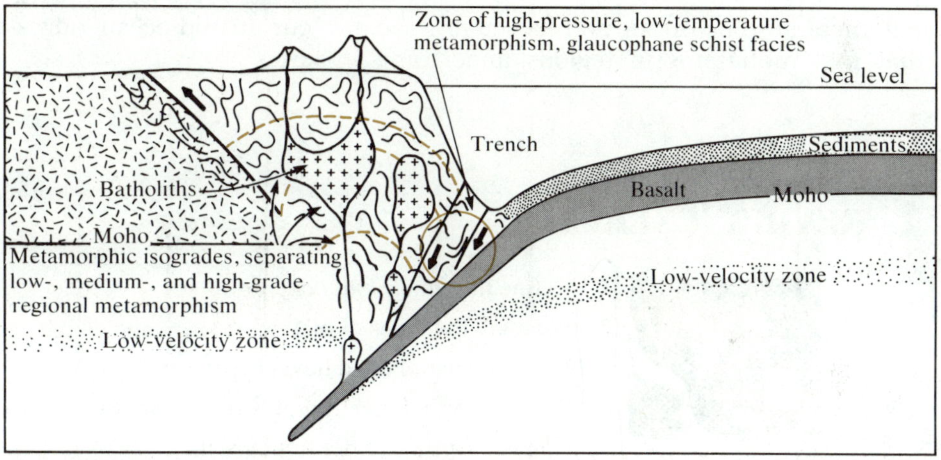

FIGURE 6–19
Metamorphism at a convergent plate boundary. (Drawing by Robert W. Tope)

Where an oceanic plate collides and moves downward under a continental plate, the resulting igneous and metamorphic rocks are easier to study because they are above sea level. The most common metamorphic rocks are the regional metamorphic rocks, and they are formed at such places. The heat and volatiles that cause the metamorphism are apparently associated with the rising magma that forms the volcanoes and the batholiths, as shown in Fig. 6–19. Near the slab of descending ocean floor is a zone of high-pressure–low-temperature metamorphism, shown in Fig. 6–19. This limited region is where the glaucophane schist facies rocks form. These rocks commonly contain blue-colored minerals, such as glaucophane, and are called *blueschist* by some geologists.

SUMMARY

Metamorphic rocks are rocks that have been changed, either in texture or mineral composition, by heat and pressure, usually as a result of burial, directed pressure (stress), shear, or chemically active solutions, especially water.

There are three genetic groups with gradation among them:

1. Thermal or contact metamorphic rocks, called hornfels, generally are found at margins of intrusive igneous bodies.
2. Regional metamorphic rocks occupy large areas and reveal much uplift and erosion after their formation. They are usually foliated (the new minerals have preferred orientation because of recrystallization under stress). The most common are gneiss and schist.
3. Dynamic metamorphic rocks result from breaking and grinding without much recrystallization.

Textural groups are: foliated (directional or layered aspect) and nonfoliated (homogeneous or massive).

There are two types of nonfoliated rocks:

1. Thermal or contact metamorphic rocks, called hornfels, fine-grained, tough baked rocks in narrow belts around intrusive bodies
2. The development of equidimensional new minerals, usually from monomineralic parent rocks. Limestone into marble; quartz sandstone into quartzite; dunite into serpentinite if serpentine is the new mineral, soapstone if talc is the new mineral (foliated talc rock is talc schist)

Progressive metamorphism is caused by increased pressure and temperature, and so forth, as in cases of marble and quartzite. Retrogressive metamorphism is at lower temperature, and so on, than the parent rock's formation; the olivine in dunite is the highest-temperature mineral in Bowen's reaction series, and it is metamorphosed at a temperature lower than that of its formation.

Foliated metamorphic rocks are caused by the stresses that fold mountains in regions undergoing deformation. Rocks are named largely for texture and not strictly for minerals. Examples are given in Tables 6–1 and 6–2.

Dynamic metamorphism occurs in fault zones or, rarely, at meteorite impact craters. Noncohesive ground rock in a fault zone is fault gouge or fault breccia. Similar cohesive rocks are mylonite. These grade into schist and gneiss.

Interpretation of metamorphic rocks gives valuable insight into earth processes and history. Metamorphic grade is determined by mapping the first occurrence of index minerals in metamorphic rocks of a given composition. A metamorphic facies is a group or assemblage of minerals that formed in response to an environment.

Metamorphic processes may merge with igneous, as in the origin of some granite bodies. Much metamorphism is associated with the igneous processes that occur at mid-ocean ridges and convergent plate boundaries.

REVIEW EXERCISES

1. Define *metamorphic rock.*
2. What are the types of metamorphism, and which agents are most important in each?
3. What is foliation, and how does it form?
4. What is contact metamorphism, and what types of rocks are formed?
5. How does a shale differ from a slate?
6. What is the low-grade metamorphic equivalent of a basalt?
7. Serpentinite forms from metamorphism of what rock?
8. Soapstone forms from metamorphism of what rock?
9. How is the metamorphic grade established?
10. How might the field relationships differ between an igneous and a metamorphic granite?

SUGGESTED READINGS

GENERAL

DIETRICH, R. V., and SKINNER, B. J. *Rocks and Rock Minerals.* Somerset, N.J.: John Wiley and Sons, 1979, 319 pp.

ERNST, W. G. *Earth Materials.* Englewood Cliffs, N.J.: Prentice-Hall, 1969, 150 pp. (paperback).

METAMORPHIC ROCKS

DIETRICH, R. V. "Migmatites: A Résumé." *Journal of Geological Education* (1974), Vol. 22, No. 4, pp. 144–56.

FYFE, W. S.; TURNER, F. J.; and VERHOOGEN, J. *Metamorphic Reactions and Metamorphic Facies.* Chapter 1, "Concept of Metamorphic Facies," pp. 3–20; Chapter 7, "Mineral Assemblages of Individual Metamorphic Facies," pp. 199–239. Geological Society of America Memoir 73, 1958.

HIGGINS, M. W. *Cataclastic Rocks.* U.S. Geological Survey Professional Paper 687, 1971, 97 pp.

MASON, B. *Principles of Geochemistry,* 3rd ed. Chapter 10, "Metamorphism and Metamorphic Rocks," pp. 248–82. New York: John Wiley and Sons, 1966.

- seismic waves
- earthquake focus
- fault
- compressional (P) wave
- shear (S) wave
- core
- mantle
- seismograph

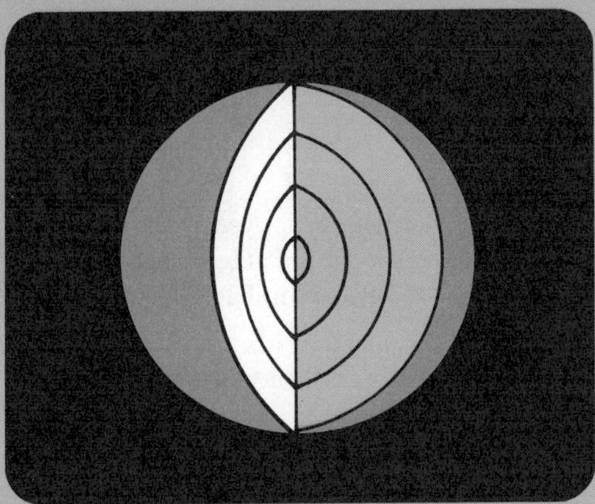

7

- shadow zone
- Mohorovičić discontinuity (Moho)
- vibration frequency
- low-velocity zone
- phase change
- continental crust
- oceanic crust
- Bouguer anomaly
- free-air anomaly
- isostasy
- declination
- inclination
- Curie temperature
- polar wandering
- self-exciting dynamo
- thermal gradient
- heat flow

The Earth's Interior

OUR CAPABILITY FOR DIRECT OBSERVATION OF ROCKS OF THE earth's interior is extremely limited. From excavations we can sample rocks of the earth's crust to a depth of only a few hundred meters, from deep mines to a depth of approximately 2000 m, and in deep-oil-well drilling to a depth of nearly 10 km. Because of the shallow nature of these probes and the fact that it is 6350 km to the center of the earth, we have very limited first-hand information on the earth's interior. In 1798, Henry Cavendish determined that the average density of the earth is 5.52 g/cm^3. Study of surficial rocks showed their density to range from 2 to 3.3 g/cm^3; thus, it became clear that the earth must have an inhomogeneous distribution of density, which probably represents an inhomogeneous distribution of rock composition. Construction of a model for the earth's interior requires integration into a consistent picture of various measurements, including the velocity of seismic waves, gravity, magnetic field, and heat flow, and estimates of compositional variations throughout the earth. Although none of these data yield unique solutions, they all place limits on the range of models that can be proposed.

EARTHQUAKES AND SEISMIC WAVES

The earth is a dynamic body in which rock masses shift relative to one another, producing earthquakes within the crust and below the region of the oceanic trenches in the upper mantle to a depth of 700 km. Earthquakes are generated as the slow shifting of rock masses causes increasing elastic deformation until the rocks fracture, releasing much of the stored elastic energy as seismic waves of various types which propagate away from the area of the initial fracture, called the **focus** (or **hypocenter**). An analogy is the slow pulling of a rubber band, which stores elastic energy, until it breaks, rapidly releasing elastic energy. A fracture in the earth propagates away from the focus for tens, sometimes hundreds, of kilometers. F. Reid proposed a useful model for the generation of earthquakes. He visualized shifting rock masses causing elastic deformation along a narrow zone. When the fracture strength of the rock is exceeded, the rock ruptures, releasing elastic energy and sending waves of compression and rarefaction (elastic seismic waves) outward from the focus (Fig. 7–1). After the elastic energy has been released, there is a displacement of geologic features on each side of the fracture, or **fault**.

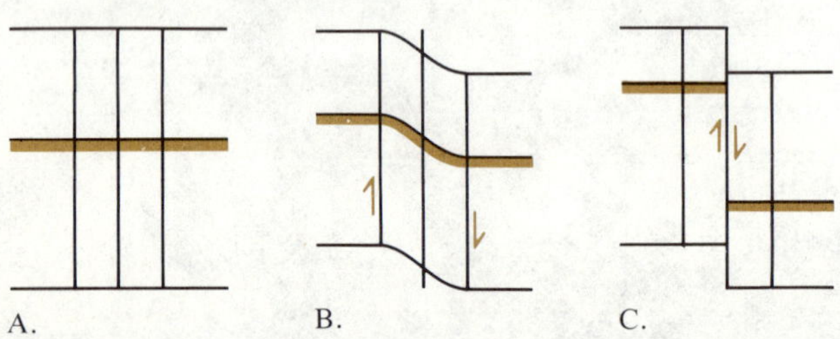

FIGURE 7–1
A. Rock mass prior to initial movement. B. The movement of rock masses preceding an earthquake can be thought of as a slow shifting of rock masses concentrated along a narrow zone. C. When fracture strength of the rock is exceeded, a fracture is initiated and the elastic energy stored in the narrow zone is released, producing rapid displacement of rock masses near the fracture.

In the study of the internal structure of the earth, two types of seismic waves, the **compressional wave** and the **shear wave**, are of particular interest. The compressional wave, called a **P wave**, travels by a small elastic compression of the rock mass that propagates in the direction of motion (Fig. 7–2A). The shear wave, called an **S wave**, travels by elastic shear deformation at right angles to the direction of propagation (Fig. 7–2B). The two seismic waves travel with unequal

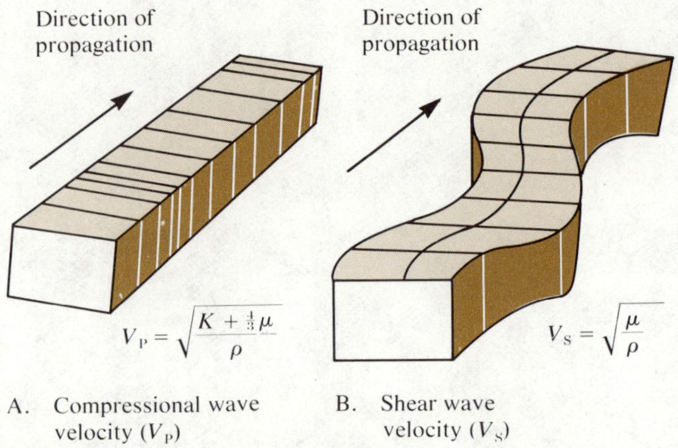

FIGURE 7–2
Two types of waves generated by an earthquake, the compressional wave and the shear wave. A. The compressional wave travels by small elastic compression of the rock mass in the direction of wave propagation. B. The shear wave travels by a small elastic deformation of the rock mass at right angles to the direction of propagation. The two types of waves are related to density and the elastic properties of the rocks in different ways, as shown by the formulas, thus providing information about the distribution of density and elastic properties within the earth. K = bulk modulus, μ = modulus of rigidity, ρ = density.

velocity; the compressional wave is approximately 40 percent faster than the shear wave.

A **seismograph** is a geophysical instrument that records small earth motions associated with seismic waves. Ideally, earth motions would be recorded by a pen fixed external to the earth and recorded on a chart attached to the earth (or vice versa), thus recording earth motion relative to a fixed object. Because a pen cannot be fixed external to the earth, the next best arrangement is for the pen to be on a weighted arm loosely attached to the earth, very much as a pendulum is attached to a building or frame. The inertia of the weighted arm causes the pen to remain relatively fixed as the earth moves. Weighted arms of a seismograph can be positioned so that they record motions in the vertical, north-south, and east-west directions (Fig. 7–3). By analyzing the amplitude of the earth motion in each of these three directions, a complete picture of the ground motion can be constructed. At any seismograph station, the first wave to arrive is the compressional wave, followed at some time by the shear wave because the two waves travel with different velocity. If the velocities of the two waves are known, the time between the arrival of the two waves is a measure of the distance to the earthquake focus. If data from several seismograph stations are analyzed, the position of the earthquake can be determined (Fig. 7–4). The po-

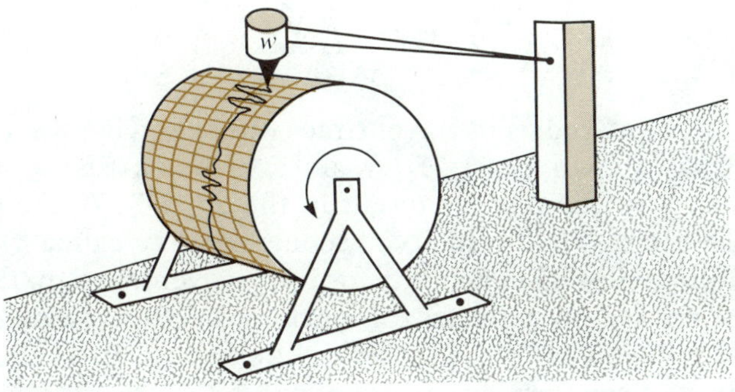

FIGURE 7–3
The recording of seismic waves at an observation point on the earth is done by a seismograph. Elastic waves produce small ground motions which are recorded continuously as the recording drum moves with the earth while the weighted recording pen (w) remains loosely attached to the earth and stationary.

sition of the **epicenter**, which is the point at the surface directly above the focus, can also be calculated.

Seismic velocity is not constant within the earth. Seismic velocities of compressional and shear waves

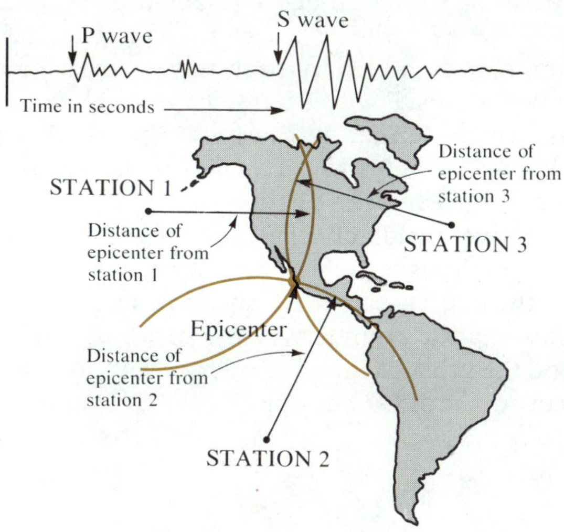

FIGURE 7–4
Because the compressional wave, commonly called the P wave, travels faster than the shear wave, commonly called the S wave, the P wave arrives at the seismographic station first and is recorded by a deflection on the seismograph chart recorder as shown in the figure. The S wave is recorded by a similar deflection but at a later time. The two types of waves travel at different velocities; thus, the time difference of their arrivals is a measure of the distance to the earthquake focus and epicenter. The location of an earthquake can be determined when the distance of an earthquake from several seismographic stations is calculated. The point of intersection of circles whose radii are equal to the distances from the seismographic station to the point of origin of the earthquake locates the focus and the epicenter.

are related to the elastic properties of rocks and to their density. Because elastic properties, density, and rock composition within the earth are not uniform, seismic velocity changes with depth and rock type. A simple example will show how variation in seismic velocity can be determined.

When seismic waves travel through the earth, they encounter layers of different rock type, density, and elastic properties, and thus have different seismic velocities. When a wave encounters a boundary between two layers of different velocity, the wave may be refracted (bent) as it passes across the boundary just as light is refracted passing across an air-water boundary. A wave may also be reflected from the boundary. In the general case, a wave is both reflected and refracted.

Earthquake waves that pass through the earth encounter many important discontinuities. One such discontinuity is shown by the variation in the paths of compressional (P) waves in the earth (Fig. 7–5). P waves are recorded from an earthquake focus out to stations located 102° from the focus. No direct P waves are recorded from 102° to 143°, called the **shadow zone**. Waves recorded from 143° to 180° arrive with velocities indicating that they have passed through a slower-velocity medium. These data have been used to demonstrate that the earth contains a central core region of lower P-wave velocity which controls the paths of P waves as shown in Fig. 7–5. S waves do not pass through the core, indicating that it is fluid, because fluids do not transmit S waves.

The natural source for the generation of elastic waves within the earth is earthquakes, but geophysicists can also use artificial sources such as explosions. By use of artificial sources, geophysicists can study the velocity structure of the outer parts of the earth anywhere without waiting for an earthquake to supply the needed energy. If elastic waves are generated in a simple two-layered medium where the upper layer has a

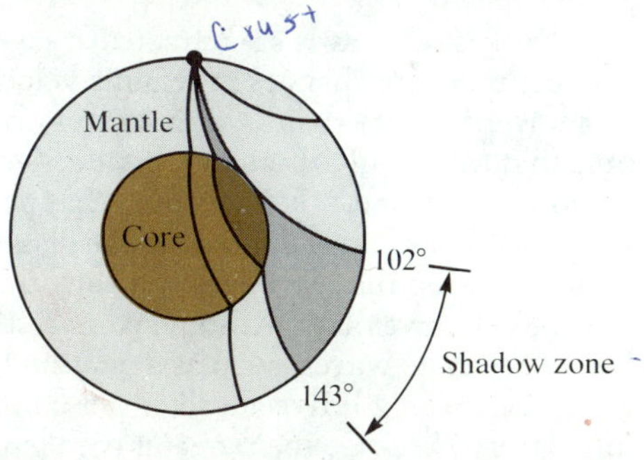

FIGURE 7–5
P waves that travel through the earth are not recorded between 102° and 143° from the earthquake focus, because P waves are refracted by more dense material in the inner part of the earth. This dense material makes up the earth's core, which begins 2900 km below the surface.

slower velocity than the lower layer, waves traveling in the upper layer will arrive first at the stations closest to the source, but at some distance waves that travel through the lower layer back to the surface will arrive first at the recording station because the velocity in the lower layer is greater (Fig. 7–6). By plotting travel time of P waves versus distance to recording stations, the velocity of the waves traveling through both layers can be determined. Also the thickness of the upper layer can be found (Fig. 7–6).

The velocity of the upper layer (layer 1 in Fig. 7–6) is 1/slope of the line passing through the origin and connecting the time-distance plots for the closest stations. The velocity of the lower layer (layer 2 in Fig. 7–6) is 1/slope of the line that connects travel-distance plots for more distant stations that does not pass through the origin. This line intersects the travel-time axis at intercept T. From the values of T, V_1, and V_2, the thickness of the upper layer, Z, can be calculated from the formula

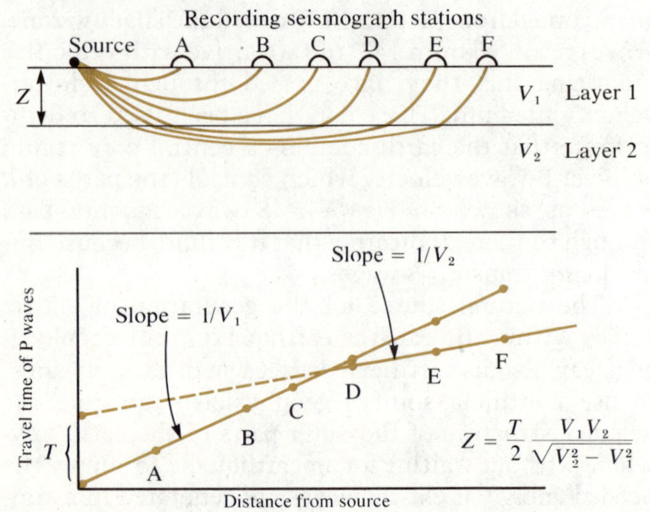

FIGURE 7–6
In layered rocks, seismic waves are refracted at layer interfaces where there are changes in seismic velocity. Here, layer 1 has a lower velocity than layer 2. Waves originating at the source that travel only through layer 1 arrive at recording stations A, B, and C before waves that pass through the upper part of layer 2, because the distance traveled is shorter, even though waves traveling through layer 2 travel at higher velocity. At some point farther from the origin of the seismic waves, such as at station D, waves that travel through layer 2 arrive ahead of those that travel only through layer 1, because their greater path length is offset by the fact that they travel with higher velocity. From data taken at many recording stations at varying distances from the origin of the seismic waves, a scientist can determine the velocity of layers 1 and 2 and the thickness (Z) of layer 1.

$$Z = \frac{T}{2} \frac{V_1 V_2}{\sqrt{V_2^2 - V_1^2}}$$

From studies of travel times of earthquake waves, it was shown nearly 70 years ago that a first-order change in velocity is present in the outer 5–70 km of the earth. This velocity discontinuity is called the **Mohorovičić discontinuity** for its discoverer and marks the separation between the earth's **crust** and the **mantle** below. The Mohorovičić discontinuity, also called the **M-discontinuity** or **Moho**, separates crustal rocks that commonly have P-wave velocities of about 6–7 km/s from upper-mantle rocks that have velocities of about 8–8.2 km/s.

Study of seismic waves from seismographic stations around the world has shown that earthquake-generated seismic waves have complex paths when they pass through the earth. P and S waves break up into many different types of waves as they encounter the important discontinuities within the earth. They may travel through the earth back to the surface, where they can be reflected back into the earth to arrive again at the surface. These types of reflected waves have paths as shown in Fig. 7–7 and are designated as **PP** or **SS waves**. Other P waves may be reflected from the core-mantle boundary back to the surface and are called **PcP waves**. P waves refracted through the core to the surface are termed **PKP waves**. The study of these complex waves helps to better define the velocity discontinuities within the earth.

The two major changes in seismic velocity described previously define the first-order divisions within the earth—the crust, the mantle, and the core. The core-mantle boundary is at a depth of about 2900 km, and the crust-mantle boundary varies from a depth of 5 km to about 70 km (Fig. 7–8). The variability of

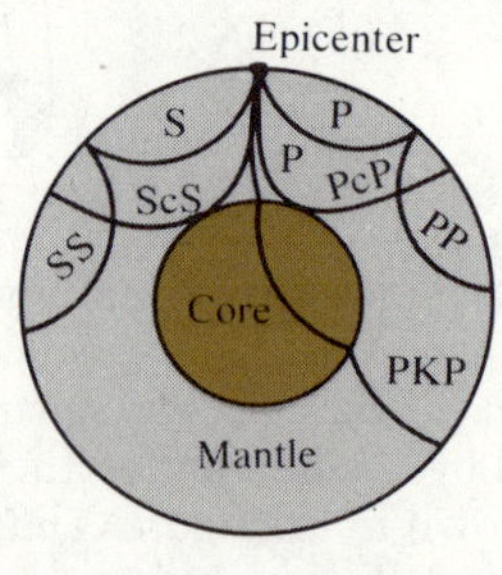

FIGURE 7–7
P waves and S waves take many different paths through the earth. Both types of waves may be refracted and reflected from different boundaries in the earth. Some P waves are refracted at the core-mantle boundary (PKP waves). Other P and S waves are reflected at the same boundary (PcP and ScS waves), whereas some are reflected at the surface of the earth (PP and SS waves).

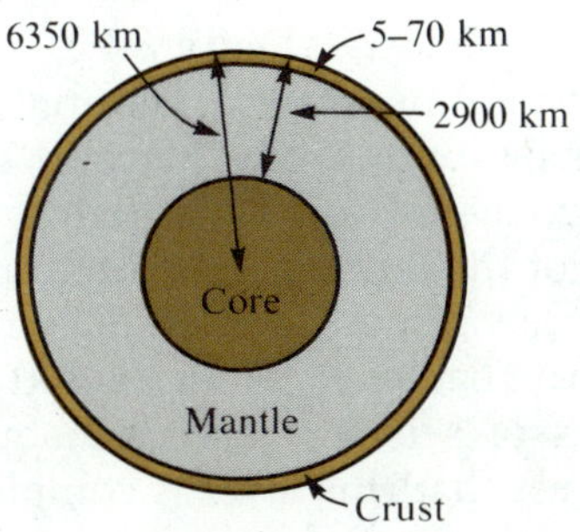

FIGURE 7–8
From the study of the velocity of P and S waves that travel through the earth, two first-order discontinuities in velocity have been shown—one at a depth of about 5–70 km, the other at a depth of approximately 2900 km. These discontinuities divide the earth into an outer layer called the crust, a thick middle layer called the mantle, and a central core.

crustal thickness, that is, the depth to the Moho, is commonly related to surface elevation. Where mountains are highest the Moho is deepest, and where surface elevation is lowest, as in the oceans, the Moho is shallowest. This variation reflects the fact that the crust is less dense than the mantle and therefore "floats" on the underlying mantle, very much like an iceberg floats in seawater. This relationship is an expression of Archimedes' principle of buoyancy applied to the earth and is called the **principle of isostasy**.

DENSITY AND COMPOSITION OF THE EARTH

The study of seismic waves indicates a first-order layering of the earth into crust, mantle, and core, but such study has not yielded a unique solution as to what materials constitute these layers. The velocity equations for P and S waves have three variables—the density and two elastic properties, the shear modulus and compressibility. Thus, even though the velocities of the seismic waves can be determined, the density and elastic properties of the material cannot be determined uniquely. However, other characteristics of the earth are also related to the density distribution, for example, the earth's moment of inertia and vibration frequency. The earth's moment of inertia, which relates mass and distance from the axis of rotation of the earth, can be calculated from astronomical data. This calculation indicates that the most dense material in the earth must be concentrated toward the center of the earth. Also, the entire earth vibrates from the energy released during an earthquake. One way in which it vibrates is to expand and contract radially with displacements of less than a millimeter. The frequency of these vibrations is dependent upon the velocity of the waves as well as the density of the different layers within the earth. By trying to match various density distributions against known P- and S-wave velocities, vibration frequencies, and moment of inertia, a density distribution for the interior of the earth can be constructed. Although there is some uncertainty about exact densities, the limits are reasonably small, perhaps only a few percent except for the inner part of the core. The density distribution for the earth is shown in Fig. 7–9.

From Fig. 7–9, it is clear that density changes at the boundaries between the crust, mantle, and core are significant, as might be expected because these boundaries also mark significant changes in seismic velocity.

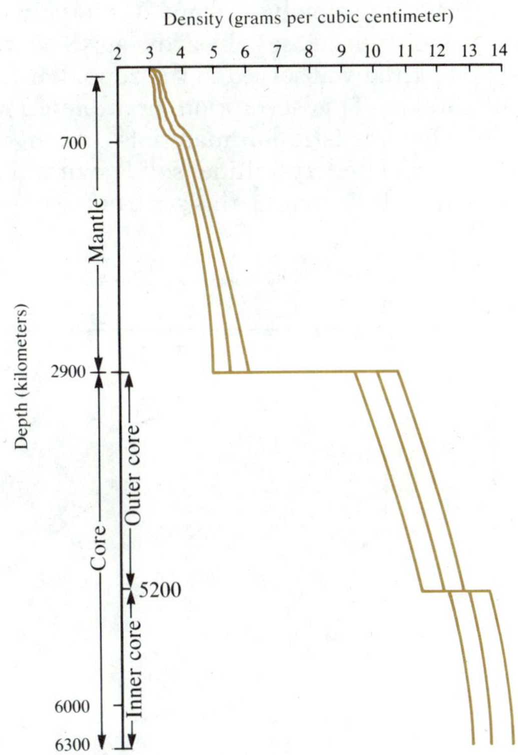

FIGURE 7–9
The density distribution within the earth. This distribution can be determined within a few percent by matching the density distribution obtained from vibration frequencies and seismic wave velocities. The major changes in density occur at the boundaries determined from seismic wave velocities—the Moho and the core-mantle boundary. Significant changes in density also occur within the upper 700 km of the earth.

In Fig. 7–9, notice that there are other places within the earth where both seismic velocity and density change rapidly. These changes occur within the core and within the upper 700 km of the earth. The change of seismic velocity within the core was first suggested by analysis of P-wave travel times with very weak amplitude that arrived in the shadow zone. The data suggested the presence of an inner core of increased P-wave velocity, and modeling of this inner core to match vibration frequencies and moments of inertia suggests an increase in density. This increase can be explained if the inner core is solid, rather than liquid like the outer core.

Changes in P- and S-wave seismic velocity and density in the outer 700 km of the earth are particularly important because they reflect changes within the mantle that affect dynamic processes near the earth's surface (see Fig. 7–10). Several important changes take place at about 70–100, 250, 400, and 700 km. Seismic velocity and density increase sharply at the Moho, then increase more slowly to about 70–100 km, where seismic velocity decreases. In addition, S waves are slowed and partially absorbed in the zone that extends to about 250 km. This absorption, or **attenuation**, of S waves is characteristic of materials tested in the laboratory that are crystalline solids with a small amount of liquid. Therefore, this zone of lower seismic velocity, called the **low-velocity zone**, is thought to be formed by crystalline mantle material with a small amount (perhaps only a few percent) of liquid. The low-velocity zone probably has a slightly lower density than the rest of the mantle, a feature important in the mantle dynamics discussed later. Seismic velocity increases slightly between 250 and 400 km, and the characteristics of seismic wave transmission of this material suggest that this layer is completely solid. At 400 km, there is a marked increase in seismic velocity over a very narrow zone. From 400 to 650 km, there is a gradual increase in seismic velocity and density until both increase markedly at about 650–700 km. Below 700 km, seismic velocity and density increase gradually to the core. At a depth of 700 km, the mantle can be conveniently divided into an apparently more homogeneous lower mantle and an upper mantle with considerable variation in seismic velocity and density.

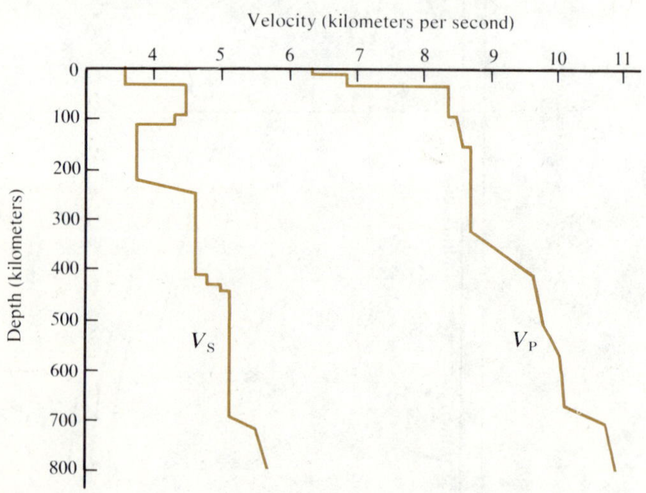

FIGURE 7–10
Changes in P- and S-wave velocity in the outer 700 km of the earth indicate that several important changes take place at about 70–100, 250, 400, and 700 km. In general, seismic velocity increases with depth, with rapid increases at 10–40, 350–400, and about 700 km. However, between about 100 and 250 km beneath the oceans, S-wave velocity decreases. This interval is called the low-velocity zone.

This zonation of the upper mantle is not uniform laterally. The transitional zones at 400 and 700 km appear to be present throughout the mantle; however, the lateral continuity of the low-velocity zone is not clear. The low-velocity zone appears to be well developed below oceanic areas, but whether it exists beneath continental areas is uncertain. Some workers have suggested it is present under continents but at greater depth, perhaps at 150–250 km, whereas other workers doubt its existence. The presence or absence of a low-velocity zone beneath continental areas is still an unsolved problem.

As has been shown, variations of seismic velocity and depth can be determined with some accuracy; however, the composition of the crust, mantle, and core is not so easily established. Evidence for the composition of the different parts of the earth comes from a variety of sources. First, data from the composition of meteorites, thought to be parts of disrupted planets, and, second, calculation of cosmic abundances of the elements yield a probable composition of the earth as a whole. Third, the mineral assemblages that form the rocks of the interior of the earth must be distributed in such a way to explain the seismic velocity and density distributions calculated for the earth.

From these pieces of information, the closest fit to the data for the mantle is an ultramafic rock (i.e., a rock composed of iron and magnesium silicates with a low silica content). In the upper part of the mantle, above the low-velocity zone it is probably peridotite, consisting dominantly of olivine and pyroxene, with some dunite, consisting mostly of olivine. Laboratory experiments indicate that at the temperatures and pressures in the uppermost mantle, peridotite has the appropriate seismic velocity and density. Within the low-velocity zone, the temperature distribution in the earth (to be discussed later) just reaches the melting point of some of the minerals in peridotite, and a small amount of melt is present in this part of the mantle. The sharp increase in seismic velocity and density at 400 km is the result of a phase change of the mineral olivine in peridotite to spinel, a mineral with a more densely packed structure. A phase change takes place when the internal structure of a mineral is altered, without a change in chemical composition, in response to changes in temperature and pressure. Again, laboratory experiments have confirmed earlier theoretical studies that indicated such a phase change was expected and would produce the densities and seismic velocities observed. The change in seismic velocity and density between 650 and 700 km is not as well known as that at 400 km, but a phase change of spinel to a more dense structure is suspected. The gradual increase in density and seismic velocity below 700 km suggests that no more important phase changes occur in the mantle and that the increase in density and seismic velocity is due simply to an increase in pressure.

The density for the ultramafic rocks in the mantle ranges from 3.2 g/cm^3 at the top to about 5 g/cm^3 at the base. Because the average density of the earth is 5.5 g/cm^3, greater than average densities occur in the core. Some meteorites consist of iron and nickel, and this observation has led to the suggestion that the core of the earth consists of iron and nickel. Pressures are so great in the core that only special laboratory shock wave experiments can reach the appropriate pressures, and then for only a fraction of a second. Studies of this type have indicated that nickel is too dense and that iron, while less dense, still has a density that is too high. The best suggestion is that the core consists of iron and nickel mixed with some lighter, but abundant element, such as silicon or sulfur, to produce the correct density (Fig. 7–11).

The crust of the earth is more varied compositionally than either the core or the mantle because many processes operate on crustal rocks to change their compositions locally. Seismic velocity data indicate that the crust is divided into two different types, an oceanic and a continental crust. Oceanic crust generally can be divided into three layers of different seismic velocity. These seismic velocities are thought to correspond to approximately 0.5 km of sedimentary rocks, 1–2 km of basalt, and 3 km of gabbro. These thicknesses vary from place to place and must be considered only as averages. Oceanic crust is thin, usually

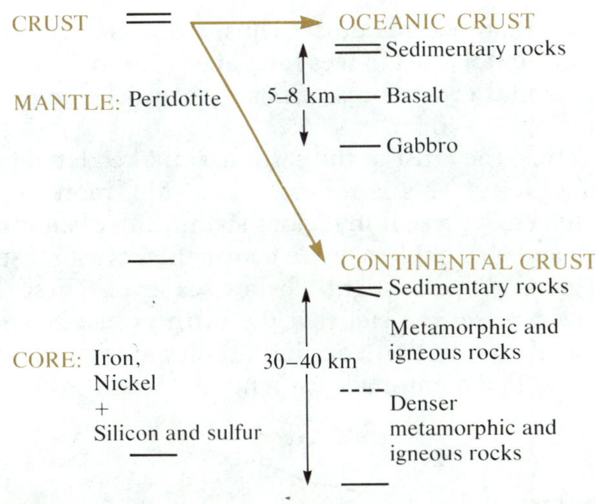

FIGURE 7–11
Evidence from the distribution of seismic velocity and density with depth, coupled with bulk chemical composition of the earth derived from models of planetary chemistry, sets limits on the variation in composition of the earth's interior. These data suggest that the mantle consists of ultramafic rocks such as peridotite and dunite, and that the core is composed of iron and nickel mixed with lighter elements such as silicon and sulfur. The crust is more compositionally varied than the mantle and consists of two basic crustal types, oceanic and continental crust. Oceanic crust is generally 5–8 km thick and consists of a thin sediment cover overlying layers of basalt and gabbro. Continental crust averages 30–40 km thick and consists of igneous and metamorphic rocks that are lighter and richer in silicon than oceanic crust.

about 5 km thick, and is overlain by 5 km of water; thus, the Moho is about 10 km below sea level.

Continental crust is very complex and is composed of metamorphic and igneous rocks overlain at some places by varying thicknesses of sedimentary rocks. Generally, seismic velocity is low near the surface (P-wave velocities 4–5 km/s) to high near the base (P-wave velocities 6–7 km/s), but there is considerable lateral variation, and in some places low-velocity zones are present between zones of higher velocity. Rocks present within continents are highly varied and cannot be typified by a single rock type or even a few simple rock types. In general, rocks of the continental crust are more silica-rich than those of the oceanic crust; thus, the igneous rocks range from diorite to granite in composition, with rare gabbro and less-silica-rich rock types present. The metamorphic rocks consist of metamorphosed igneous and sedimentary rocks whose compositions are generally richer in the lighter elements such as silicon, potassium, and sodium than oceanic crust. Thus, the average density of continental crust ranges from about 2.6 to 2.8 g/cm³. Continental crust is also thicker than oceanic crust, averaging 35–40 km.

Thus, the crust of the earth has marked lateral inhomogeneity. This is reflected in a thin, more dense oceanic crust present in regions significantly below sea level; and a thick, less dense continental crust present in regions above or slightly below sea level. These differences reflect the fact that the earth's crust behaves as if it is in flotation equilibrium on a denser mantle (Fig. 7–12), a point made earlier.

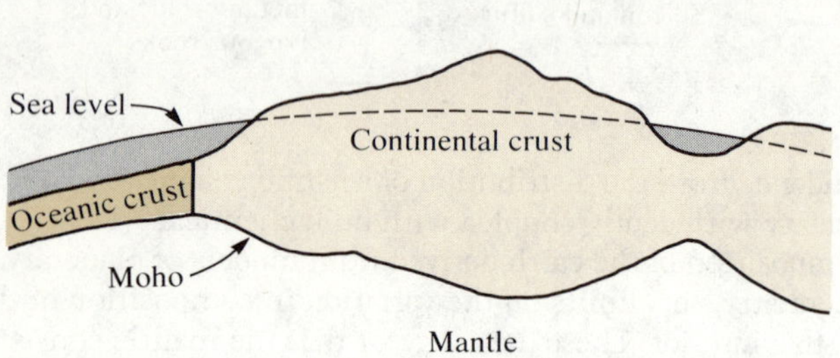

FIGURE 7–12
Continental and oceanic crust behave as if they are in flotation equilibrium on the underlying mantle. Continental crust is less dense but thicker than oceanic crust and stands higher, generally averaging 0.5 km above sea level, whereas the average elevation of oceanic crust is 5 km below sea level. The thicker the crust, the deeper it extends into the mantle, analogous to an iceberg.

Thus far, only a generalized and incomplete view of the interior of the earth has been given. The earth is a highly differentiated body with a dense iron-rich core, an ultramafic mantle, and a lighter, more silicon-rich, inhomogeneous crust. Variations in seismic velocity and density are most numerous in the outer 700 km of the earth. Lateral rather than radial inhomogeneity becomes more common in the outer 400 km, and the most pronounced lateral inhomogeneities are in the crust and uppermost mantle. These inhomogeneities are a reflection of the dynamic processes that manifest themselves in sea-floor spreading and plate tectonic processes.

GRAVITY AND THE STRUCTURE OF THE CRUST AND UPPER MANTLE

If the earth were a nonrotating, homogeneous, spherical body, the acceleration of gravity (measured, for example, in centimeters per second per second) at its surface would be everywhere the same. The acceleration of gravity (g) for such a body is given by the formula

$$g = G\frac{m}{R^2}$$

where G is the universal gravitational constant, m is the mass of the earth, and R is the radius of the earth. The earth, however, is not homogeneous, but varies in composition and density from its surface to its center. If these variations formed spherical zones within the earth, the acceleration of gravity measured at the surface would still be constant, because the mass below any point on the sphere would be the same.

The earth also rotates, and, because of the centrifugal force due to rotation, it has a shorter radius at the poles than at the equator, giving it an ellipsoidal shape. This shape causes lateral changes in the measured value of gravity with latitude. The accepted value of gravity at the equator is 978.049 cm/s², and the value of gravity varies with latitude by the formula

$$g = 978.049(140.0052884 \sin^2\phi - 0.0000059 \sin^2 2\phi)$$

reaching more than 983 cm/s² at the poles.* Thus, the value of gravity can be calculated for any point on the earth's surface, assuming the earth is a simple ellipsoidal body.

The earth, however, is not a simple rotating ellipsoid where all gravity readings can be made on an ideal ellipsoidal surface; it has local topographic relief. Mountainous areas rise above the theoretical ellipsoidal surface just as the surface of the oceanic crust falls below it. By correcting measured values of gravity for topography and the associated rock mass, all values of gravity can be referred to a theoretical ellipsoidal surface, and any difference that remains must be related to differences in mass distribution below the recording stations. A difference between the corrected value of gravity and that calculated at the same point for an ideal ellipsoidal earth is called a **Bouguer anomaly** (Fig. 7–13). A negative Bouguer anomaly indicates a mass deficiency, that is, that there is a body of rock of lower density than expected below the surface. By mapping out the shape and magnitude of the anomaly, the size and density of the anomalous mass can be calculated.

REGIONAL GRAVITY ANOMALIES AND ISOSTASY

Gravity studies over continents and oceans have indicated that Bouguer anomalies over mountainous areas are negative, and that those over most oceanic areas are positive. If a theoretical earth model with a laterally

*ϕ is a symbol for latitude.

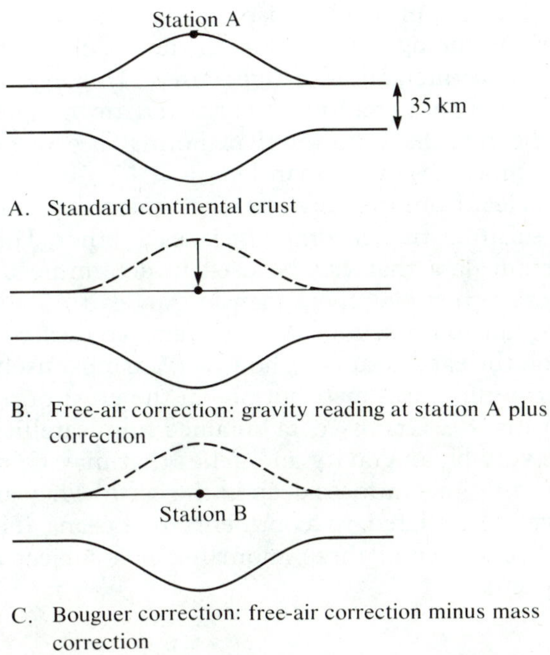

FIGURE 7–13
Measurement of gravity over areas of topographic relief varies from the value of gravity calculated assuming the earth to be a perfect ellipsoid. A. Over mountains, the measured value of gravity is too low because the observation is made further from the center of the earth and because of the mass attraction below station A. B. Correction for the elevation alone must be added to the reading at station A, and is called the *free-air correction.* C. Correction for the mass below station A must be subtracted from the free-air correction, and is called the *Bouguer correction.* After making these two corrections, the value of gravity is as if it had been measured at station B beneath the mountains. The free-air and Bouguer corrections yield a value of gravity that is too low because the mountain is compensated for at depth by a low-density root which displaces more dense mantle. Such observations yield data for density distributions within the earth.

uniform crust is assumed, these Bouguer anomalies reflect a mass deficiency below the mountains and a mass excess below the oceans. The explanation for these mass differences is a lateral variation in density and thickness of the earth's crust, a feature previously described from the variations in seismic velocity. Because oceanic crust is thin, the more dense mantle is closer to the surface than expected from a uniformly thick crust (Fig. 7–14). That part of the denser mantle below the oceanic crust and above the base of the assumed uniform theoretical crust gives rise to the positive Bouguer gravity anomaly. Below the mountainous area, there is thicker crust than the crust of the theoretical model that displaces some denser mantle, yielding a negative Bouguer anomaly.

Studies such as these have demonstrated that as a first approximation, continental and oceanic crust are in flotation equilibrium on a dense mantle. Furthermore, gravity studies over mountainous areas have indicated that flotation equilibrium is probably achieved for mountains whose width is of the order of a few tens of kilometers. Thus, as a general rule the configuration of the Moho at depth is the inverse of the topographic relief at the surface. The Bouguer anomaly profile also has the general configuration of the Moho (Fig. 7–14). As indicated previously, this characteristic of crustal rocks to be in general flotation equilibrium on the dense mantle is the principle of isostasy.

There are some exceptions to the concept that all topographic relief is balanced by inverse relief on the Moho. In some cases, the balancing mass deficiency or excess is not related to crustal thickness, but rather to changes in density of the upper mantle. For example, in some parts of the Basin and Range Province of the western United States the average elevation is more than 2 km (Fig. 7–15). A crustal thickening would be expected to balance this relief. Gravity surveys indicate a negative Bouguer anomaly, as would be expected. Seismic studies in the areas, however, indicate

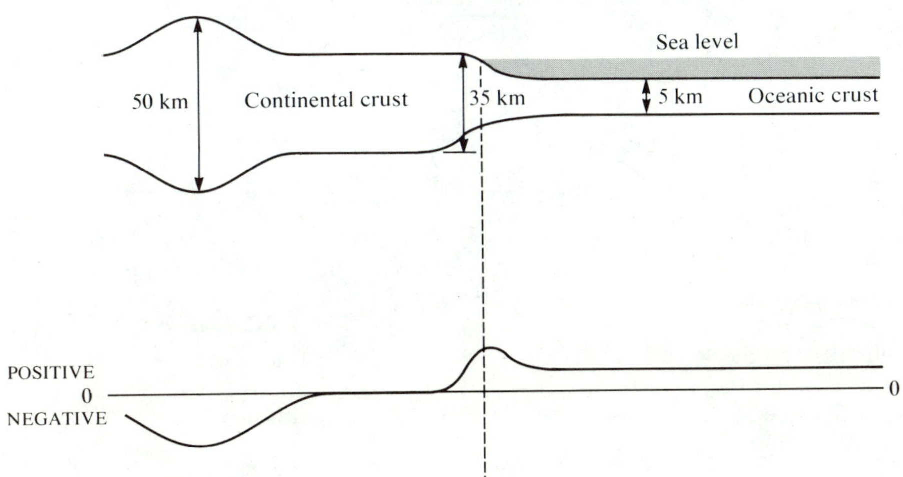

FIGURE 7–14
Bouguer gravity anomalies are commonly negative over topographically high mountains and positive over areas below sea level, such as the ocean basins.

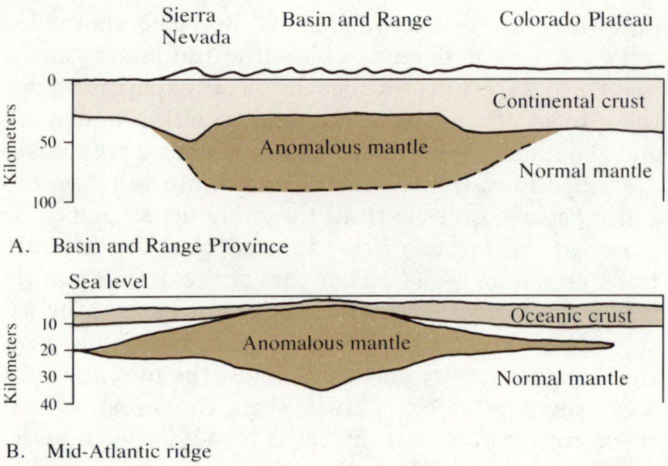

FIGURE 7–15
Not all topographic relief is compensated for by an inverse change in the elevation of the Moho. Compensation can also take place by changes in the density of the upper mantle as a result of increase or decrease in temperature. A. The Basin and Range Province of the western United States has an average elevation of about 2 km, but seismic studies indicate that the Moho is unexpectedly shallow. B. The mid-Atlantic ridge beneath the Atlantic Ocean likewise has thin oceanic crust but has elevations above the deeper part of the oceans underlain by thicker oceanic crust. Gravity data indicate that both areas are isostatically balanced, but compensation is supplied by an upper mantle that is hotter and less dense than normal.

that the crust is thinner than normal. These studies also show that the upper mantle has abnormally low seismic wave velocities where P-wave velocities are about 7.8 cm/s rather than 8 cm/s. The low velocity is correlated with lower than normal density for the upper mantle to a depth of more than 60 km (Fig. 7–15). Thus, the principle of isostasy is still valid, but the mass balancing deficiency is in the upper mantle and not in a thickening of light crust. This type of low-density upper mantle is characteristic of areas where the mantle is hotter than normal because of dynamic processes in the mantle.

It is clear from the previous discussions that gravity and seismic studies complement each other. They both obtain data that can be used to determine the character of crust and upper mantle. Any earth model must explain both sets of data. Gravity studies conducted on the earth's surface, however, are most useful for determining mass distributions in the crust or uppermost mantle. Gravity data obtained from satellites indicate very broad gravity anomalies that may be related to variations in mass deep in the earth. They are not obviously related to continents or oceans (Fig. 7–16). The source of these anomalies is a subject of much speculation.

THE EARTH'S MAGNETIC FIELD

The earth's magnetic field has been recognized for hundreds of years, and its shape and strength are well known. Despite these facts, the origin of the earth's magnetic field remains a matter of speculation.

Given a magnetic needle that is free to swing, such as in a compass, the needle will rotate and align parallel to the direction of the magnetic field. The magnetic field of a bar magnet is visually displayed when iron filings, each filing acting like a compass needle, surround a bar magnet (Fig. 7–17A). The earth's magnetic field has a form very similar to that of a bar magnet

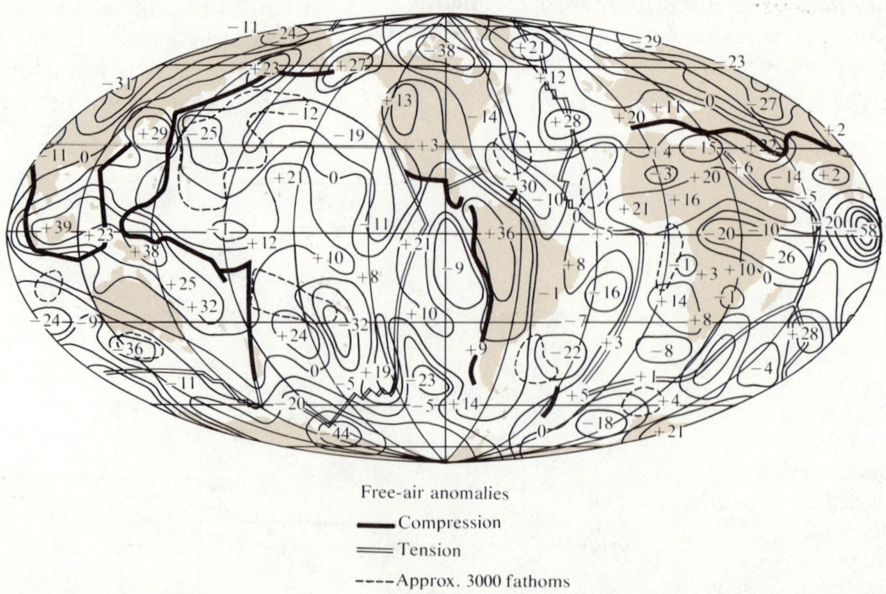

FIGURE 7–16
Gravity anomalies over the earth as determined from satellite observations. (After W. M. Kaula, *Science,* vol. 169, pp. 982–985, Sept. 4, 1970. Copyright 1970 by the American Association for the Advancement of Science. Map by Robert W. Tope)

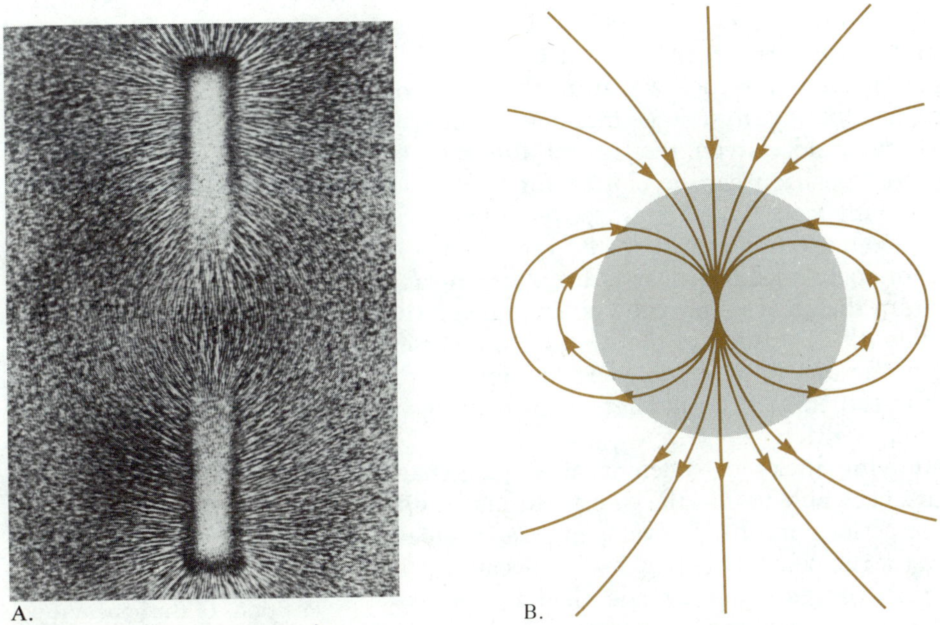

FIGURE 7–17
A. Magnetic field of a bar magnet shown by the alignment of iron filings on paper. (Photo from Kodansha Ltd.) B. The earth's magnetic field has a configuration as if a bar magnet were placed in the center of the earth.

positioned within the core (Fig. 7–17B). The north-seeking end of the compass defines a north pole that is not coincident with the rotational axis of the earth but lies approximately 11½° from it. Thus, the north end of a compass needle does not point to the geographic north pole. The difference between these two directions is called the **declination** (Fig. 7–18). The declination is different for different positions on the earth. For example, in parts of California the compass needle points far to the east because the geographic pole lies west of the magnetic pole. The declination is thus about 16° west. In this way, the horizontal component of the magnetic field can be mapped out.

If a magnetic needle is free to swing in the vertical plane, the **inclination** of the magnetic field can be determined. From Fig. 7–17, it is clear that the inclination of the earth's field varies from zero at the magnetic equator to 90° at its pole.

PALEOMAGNETISM AND POLAR WANDERING

Certain rocks rich in iron-bearing minerals can become magnetized in the direction of the earth's magnetic field when the rocks are formed. Iron-bearing minerals crystallizing from a lava become magnetized in a direction parallel to the earth's field once they pass below a certain temperature, called the **Curie temperature**. Different minerals have different Curie temperatures. For example, the Curie temperature of magnetite is 578°C, and for hematite it is 680°C. Iron-bearing minerals precipitated from solutions below the Curie temperature also can acquire a magnetization of the earth's field at the time of their formation. Sedimentary rocks may acquire a magnetization at the time of formation, by the alignment of magnetic minerals during deposition. Once these minerals and rocks are formed, they lock in the direction of the earth's magnetic field at the time of their formation. Studies of magnetism locked into ancient rocks (paleomagnetic studies) have shown magnetic fields different from the present field. By studying the declination and inclination of these ancient fields, paleomagnetists can determine that large bodies of rocks such as continental fragments have moved relative to the present magnetic pole or moved relative to one another. By studying these rocks in the

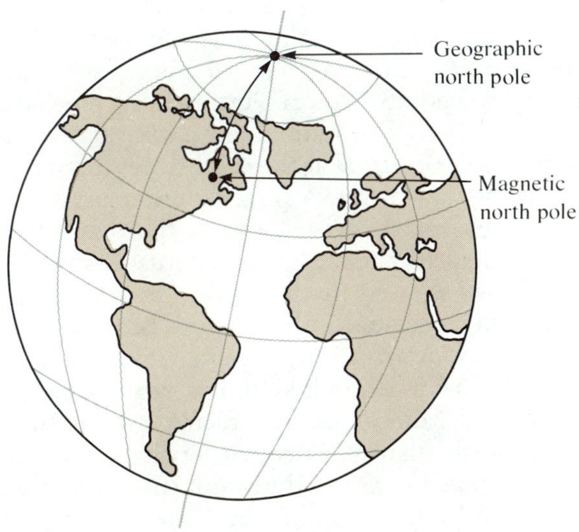

FIGURE 7–18
The positions of the geographic north pole and the magnetic north pole are not the same. The difference between them at present is 11½°. The difference between the direction to these two poles anywhere on the earth is called declination.

laboratory, the direction of the magnetic field that was present at the time the rocks formed can be determined. The distance the rock was from the magnetic pole can be found from the inclination of the paleomagnetic field. Given the age of the rock, a magnetic pole position can be plotted for that time. Three important facts have been discovered from the study of paleomagnetic data. First, the magnetic pole positions for rocks of different ages are different and different from the present magnetic poles. Second, the north and south directions of the field have reversed positions numerous times in geologic history. Third, rocks of the same age from different continents give different pole positions.

By studying rocks of different ages, paleomagnetists have been able to plot the pole positions for different time periods and have developed **polar wandering** curves for the continents (Fig. 7–19). Because it is believed that the earth's magnetic field is approximately fixed with respect to the earth's rotational axis, it is not the pole that wanders; rather, the rock masses making up the outer part of the earth move with respect to the pole. In addition, paleomagnetic studies from the different continental areas and commonly from different parts of the same continent show different polar wandering paths. These data indicate that the continents, and at times parts of continents, have moved not only with respect to the magnetic pole but also with respect to one another.

A surprising discovery from paleomagnetic studies has been that the earth's magnetic field has reversed its polarity many times during geologic history. Studies of paleomagnetism in a thick sequence of volcanic rocks showed that some of the volcanic rocks had a paleomagnetic field that pointed north, whereas in other rocks it pointed south. Furthermore, precise dating techniques such as radiometric dating showed that these reversals occurred at the same times in many different sequences of rocks from different parts of the world. These data demonstrated that the earth's magnetic field had reversed its polarity many times. The data also showed that reversals occur about every half million to several million years. Because present dating techniques can only resolve time periods this short in the last few hundred million years, a magnetic reversal time scale can be constructed only for the last part of geologic history. Paleomagnetists can construct a very accurate and detailed reversal time scale for the past few million years, but at increasingly greater times, the detail for short reversal periods becomes lost. Reversal data have been important in the development of the concepts of sea-floor spreading and plate tectonics.

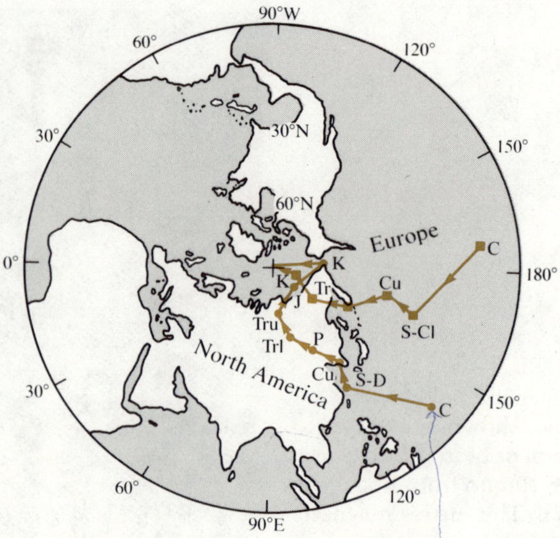

FIGURE 7–19
Positions of the apparent polar wandering paths for North America (circles) and Europe (squares). The positions of the poles are determined from areas on the continents that have been stable relative to one another within each continent. The letter symbols designate the various time periods: K = Cretaceous, Tr = Triassic, Tru = Upper Triassic, Trl = Lower Triassic, P = Permian, Cu = Upper Carboniferous, S-D = Silurian-Devonian, S-Cl = Silurian-Lower Carboniferous, C = Cambrian. [After N. W. McElhinny, *Paleomagnetism and Plate Tectonics* (Cambridge: At the University Press, 1973)]

ORIGIN OF THE EARTH'S MAGNETIC FIELD

The origin of the earth's magnetic field is poorly understood. Even though it can be described as a field similar to a bar magnet, it cannot be generated by such a simple mechanism. The Curie temperature for magnetic minerals is reached at depths of only 20 or 30 km below the surface; thus, minerals below the crust or upper mantle cannot be permanently magnetized.

The best explanation offered to date is that of W. M. Elsasser. In 1958, Elsasser proposed that the fluid core of the earth acts as a **self-exciting dynamo**. Electrical current is generated in a dynamo in power plants by rotating an electrical conductor in a magnetic field. Elsasser argued that if a weak stray magnetic field were present, movement in the fluid iron core would generate an electrical current, which would then produce its own magnetic field. Thus, as long as there was movement in the fluid core, the core would generate electrical currents from its own magnetic field; hence the term *self-exciting dynamo* (Fig. 7–20). Data from

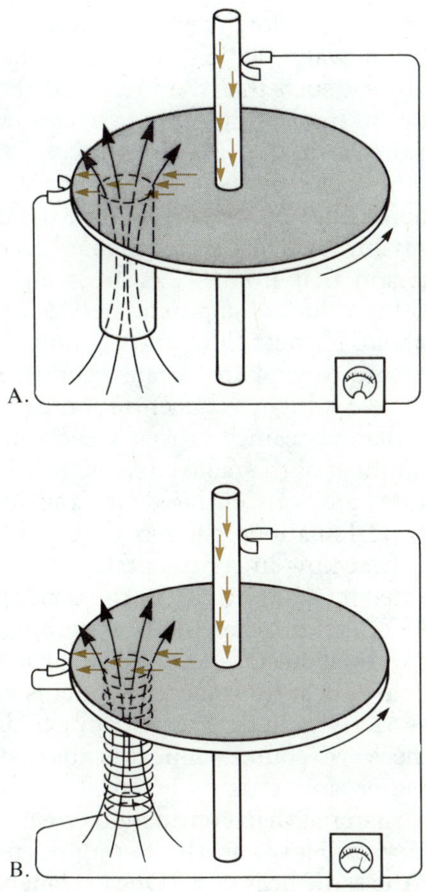

FIGURE 7–20
A. A simple dynamo generates an electric current (colored arrows) when a copper disk is rotated through the magnetic lines of force of a bar magnet. B. If the bar magnet is replaced by a coil, this will induce a current in the disk in exactly the same way provided there is a current in the coil. If the current induced in the disk is fed back into the coil, a self-contained system for generating current by simply spinning the disk (a self-exciting dynamo) is developed. The liquid core is thought to flow in such a way as to develop a self-exciting dynamo and generate the earth's magnetic field. (From "The Earth as a Dynamo" by W. M. Elsasser. Copyright © 1958 by Scientific American, Inc. All rights reserved)

seismology indicate that part of the core is fluid. Geochemical data suggest that the core is probably composed largely of iron. Thus, the mechanism seems feasible even though the details are poorly understood.

The present magnetic field of the earth is close to the earth's axis of rotation. Motion within the fluid core, then, might be expected to be closely related to the axis of rotation. Thus, the assumption that the magnetic poles have always been close to the geographic poles is assumed to be valid.

TEMPERATURES WITHIN THE EARTH

The temperature distribution within the earth is constrained by certain interpretations about the nature of the earth, such as a partially melted low-velocity zone, certain phase changes in the upper mantle, a solid mantle below the low-velocity zone, and a fluid outer core. The geothermal gradient (variation of temperature with depth) is, however, poorly known.

THERMAL GRADIENT

Measurements made from drill holes and deep mines give a general idea of the change of temperature with depth (**thermal gradient**) in the outer part of the earth. These data suggest that temperature increases 20°C–30°C per kilometer depth. Gradients of this order cannot exist for very great depths in the earth, because temperatures would be far above the melting temperature of the rocks in the lower crust and upper mantle. Temperatures of 800°C–1200°C would be predicted by these gradients at a depth of 40 km, which would be at the base of the crust or within the upper mantle. At 100 km, these gradients would yield temperatures of 2000°C–3000°C, which are far above the melting temperature of mantle rocks at those depths. This means that the internal temperatures cannot be determined by a simple extrapolation of near-surface gradients.

Several pieces of information imply specific temperatures at various depths within the earth. Crustal and upper-mantle rocks to a depth of about 100 km under the oceans are solid, whereas from about 100 to 250 km they may be partially melted. Laboratory experiments determining the pressure-temperature relations of melting of peridotite suggest temperatures ranging from 1100°C at the top to 1250°C at the bottom of the low-velocity zone. The presence of a low-velocity zone beneath continents is less well established. The absence of a low-velocity zone would suggest lower temperatures in the upper mantle beneath continents than at the same depth beneath the oceans. Some data suggest a low-velocity zone beneath continents, but with its top at a deeper level, perhaps 150–200 km, indicating that melting temperatures are reached at a deeper level beneath continents. In either case, the data suggest a different geothermal gradient within the upper mantle beneath continents and oceans (Fig. 7–21).

At 400 and 700 km, two phase changes take place in the mantle. Both have been investigated experimen-

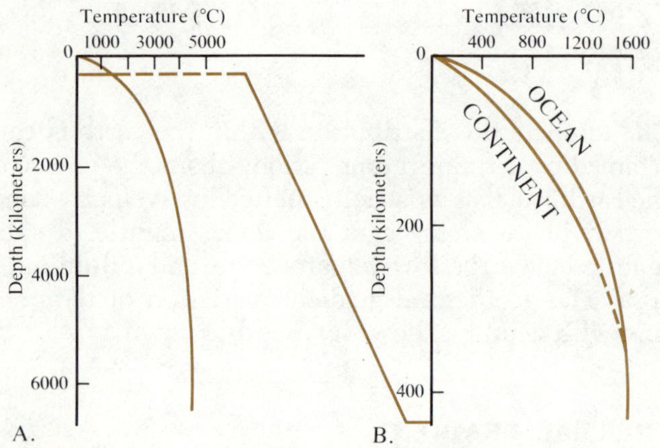

FIGURE 7–21
A. The temperature increases downward in the earth, rapidly at first and then more slowly. B. The temperature profile beneath continents and oceans is different to a depth of about 400 km.

tally. At 400 km, the phase change of olivine to spinel takes place at about 1500°C. At 700 km, the phase change of spinel to a more dense phase occurs at about 1900°C. The remainder of the mantle is solid, and no phase changes are suggested from the seismic velocity data.

At the core-mantle boundary, there is a sharp change from solid mantle to fluid core. Pressures at these depths are beyond pressures that can be achieved experimentally; thus, the melting temperatures for rocks of the core cannot be determined directly. Theoretical considerations indicate that temperatures at the core-mantle boundary should be about 3700°C. Temperatures in the inner core are uncertain, but probably not above 4500°C.

HEAT FLOW

Because temperature increases at depth in the earth, heat must flow from the interior to the surface. The heat flow can be calculated at the surface by measuring the thermal gradient (ΔT) and the thermal conductivity (K) of the rocks. The heat flow (q) is calculated from the expression

$$q = K\Delta T$$

Measurements of heat flow are made in oceanic areas by dropping a probe into the sediment on the ocean floor. The temperature difference between two points a known distance apart is measured. The probe also collects some of the sediment so that a conductivity measurement may be made. From these data, the heat flow can be calculated. On land, drill holes are commonly used. Probably the major problem in obtaining correct heat-flow measurements is the movement of groundwater, which can change the temperature of the rock such that the measured thermal gradient is not the gradient due only to heat flow from the interior of the earth.

Over the past 10 years, a great many heat-flow measurements have been made all over the earth. One of the surprising facts discovered is that heat flow from continents and that from ocean areas are about the same, yielding values of approximately 1.2×10^{-6} cal/cm²/s, or about 1.2 heat-flow units (h.f.u.). This is not an easy feature to explain, because continental crust appears to have a higher concentration of radioactive elements than oceanic crust. Geochemical data suggest that most of the radioactive elements in continental crust are concentrated in the upper few kilometers. Calculations suggest that nearly 30 percent of the heat flow in continental areas comes from heat generated by radioactivity in the upper part of the continental crust, and that the heat coming from the interior is only about 0.7 h.f.u. Surficial heat flow is thus the total of heat from the interior plus heat generated by radioactivity in the crust. Trying to distinguish between these two sources of heat is not easy.

Because oceanic crust is believed to contain less radioactive material than continental crust, the important question is, How can the two different types of crust have the same heat flow? One explanation is that the thermal gradient is higher beneath the oceanic than the continental areas (Fig. 7–21). A higher thermal gradient under the oceans would increase the amount of heat from the interior reaching the surface. The heat flow in oceanic areas would then have a smaller component of heat generated by crustal radioactivity than in continental areas. Present data seem to suggest that the similar heat flow in the oceans and on the continents is coincidence.

Other characteristics of the outer part of the earth indicate that the different geothermal gradients beneath continents and oceans extend into the upper mantle at least a few hundred kilometers. Seismic data indicate the presence of a low-velocity layer beneath oceanic areas at a depth of from 100 to 250 km. The low-velocity layer is interpreted as a zone where the geothermal gradient has reached the melting point of some constituents of peridotite. Below continents, the low-velocity zone is poorly defined and may either be deeper and thinner or nonexistent; the data are not clear. This suggests that the geothermal gradient below continents may not reach the melting curve for peridotite, or may reach it at a deeper level. In either case, the gradient below continental areas would be less than that beneath oceans. The difference in geothermal gradients does not appear to extend to the 400-km discontinuity.

SUMMARY

The earth is a layered and chemically differentiated planet. From the study of seismic waves generated principally by earthquakes, it can be subdivided into an outer crust 5–70 km thick, a mantle extending to a depth of about 2900 km, and a core extending from 2900 km depth to the center of the earth, at about 6350 km.

The crust of the earth is made of both oceanic and continental crust, bounded at the base by a marked increase in seismic velocity called the Mohorovičić discontinuity, or Moho. Oceanic crust is about 5 km thick and has a density between approximately 3 and 3.2 g/cm^3. It is made up of three layers: layer 1 at the top consists of thin sedimentary rocks, layer 2 consists of basaltic lava, and layer 3 consists of gabbro. Continental crust is thicker (20–70 km) and less dense (2.6–2.8 g/cm^3) than oceanic crust. Structurally and compositionally, it is much more complex. In general, rocks that make up the continental crust consist of igneous rocks ranging in composition from diorite to granite; metamorphic rocks; and sedimentary rocks rich in silicon, potassium, and sodium.

The earth's mantle consists of ultramafic rocks whose density ranges from about 3.2 g/cm^3 at the top to about 5 g/cm^3 at the base. Chemically, the mantle consists of minerals rich in iron and magnesium and poor in silicon, such as olivine and pyroxene. In the upper part of the mantle they combine to form peridotite, but at greater depth the minerals undergo phase changes that result in denser mineral species. Seismic velocity generally increases with depth in the mantle. However, from about 70 to 250 km beneath the oceanic crust seismic velocity decreases in a low-velocity zone where temperature-pressure conditions probably cause a small amount of melting.

The core of the earth does not transmit S waves and has a marked decrease in P-wave velocity from the mantle, and therefore is thought to be fluid. Some evidence suggests the presence of a solid inner core. The density of the core suggests that the core is composed of iron and nickel mixed with some lighter and abundant element such as silicon or sulfur.

The measured acceleration of gravity varies over the earth because the earth is a rotating ellipsoid, and is greater at the poles than at the equator. Gravitational acceleration also varies because of unequal thickness of the lighter rocks that make up the crust. Topographically high mountains have negative Bouguer gravity anomalies, which indicate that they are underlain by an abnormally thick crust. Also, seismic data suggest that the Moho is deeper under high mountains. These data indicate that topographic relief of the crust is compensated by inverse relief of the Moho, so the crust of the earth behaves as if it "floats" on the denser mantle, a concept called the principle of isostasy.

The earth's magnetic field is a dipole field similar to that which would be created by a bar magnet placed within the core. The present north magnetic pole lies $11\frac{1}{2}^\circ$ away from the geographic north pole. Although the angle between the geographic and magnetic poles has probably changed in the past, current theories for the generation of the earth's magnetic field suggest that they have always remained fairly close. The earth's magnetic field is thought to be related to convective flow of an electrically conducting fluid in the earth's core, and thus should retain some close relationship to the earth's rotational axis.

Minerals in rocks can become magnetized in the direction of the earth's magnetic field during their formation. By studying the magnetism of ancient rocks, scientists have been able to show that the earth's magnetic field has reversed its polarity many times during geologic history. They also have shown that the continents have moved relative to the magnetic poles and also relative to one another.

The temperature distribution in the earth is not well known. Measurements near the surface suggest that the thermal gradient (change of temperature with depth) is about 20°C–30°C/km. This gradient cannot persist at depth because it predicts melting of mantle rocks at depths of only 40 km; thus, the thermal gradient must decrease with depth. Temperatures are probably about 1100°C–1250°C within the low-velocity zone at 70–250 km depth, where partial melting should be present. Temperatures increase slowly to about 3700°C at the core-mantle boundary, and to no more than 4500°C in the inner core.

REVIEW EXERCISES

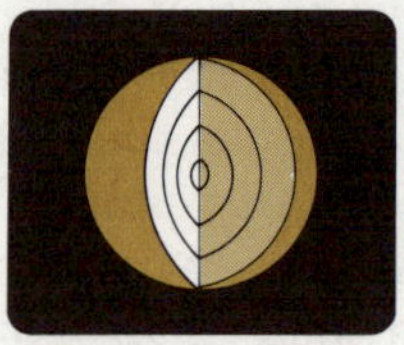

1 What is the difference between a P wave and an S wave?
2 What is one way that seismic waves are generated?
3 What are the three major divisions of the interior of the earth?
4 What is the low-velocity zone?
5 How do we know that density of rocks in the earth generally increases downward?
6 What are the two types of crust in the earth?
7. Why is the acceleration of gravity greater at the poles than at the equator?
8 Why do topographically high mountains have a negative Bouguer anomaly?
9 What is the principle of isostasy?
10 How do we know that the earth's magnetic field has reversed many times during geologic history?
11 How does evidence from the paleomagnetic studies of rocks suggest that the continents have moved relative to one another?
12 What is heat flow?
13 How is the heat flow in the earth related to the thermal gradient?

SUGGESTED READINGS

BOLT, B. "The Fine Structure of the Earth's Interior." *Scientific American* (1973), Vol. 228, No. 3, pp. 24–33.

BOLT, B., compiler. *Earthquakes and Volcanoes.* San Francisco: W. H. Freeman and Co., 1980, 154 pp.

GARLAND, G. D. *Introduction to Geophysics.* Toronto: W. B. Saunders Co., 1979, 494 pp.

———. *The Earth's Shape and Gravity.* Oxford: Pergamon Press, 1965, 183 pp.

GUBBINS, D. "Theories of the Geomagnetic and Solar Dynamos." *Reviews of Geophysics and Space Physics* (1974), Vol. 12, No. 2, pp. 137–54.

RICHTER, C. F. *Elementary Seismology.* San Francisco: W. H. Freeman and Co., 1958, 768 pp.

RINGWOOD, A. E. *Composition and Origin of the Earth.* Publication No. 1299, Research School of Earth Sciences, Australian National University, 1977, 65 pp.

———. *Composition and Petrology of the Earth's Mantle.* New York: McGraw-Hill Book Co., 1975, 618 pp.

STACEY, F. D. *Physics of the Earth,* 2nd ed. New York: John Wiley and Sons, 1977, 414 pp.

STRANGWAY, D. W. *History of the Earth's Magnetic Field.* New York: McGraw-Hill Book Co., 1970, 168 pp.

- continental drift
- mid-ocean ridge
- oceanic trench
- sea-floor spreading
- abyssal plain
- continental rise
- continental slope
- continental shelf

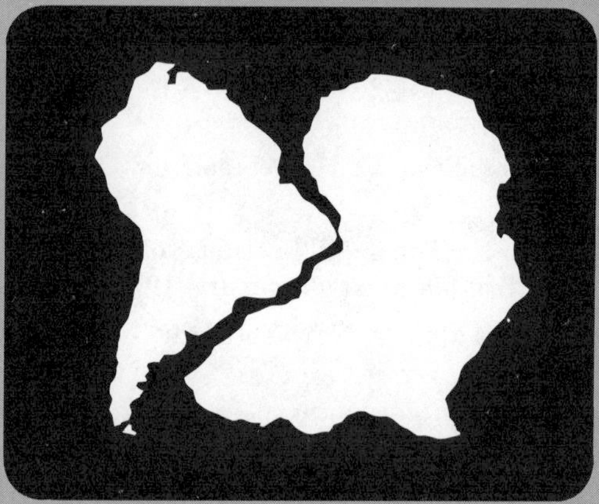

8

- island arc
- marginal sea
- magnetic anomaly
- magnetic reversal
- median valley
- transform fault
- Benioff zone
- subduction
- lithosphere
- asthenosphere
- subduction zone
- plate
- plate tectonics
- hot spot
- plume
- plume track
- volcanic (magmatic) arc
- Andean-type arc
- marginal sea
- strike-slip fault

Plate Tectonics

ALTHOUGH SOME EARLIER WORKERS PROPOSED CONCEPTS THAT might be regarded as the birth of **continental drift**, Alfred Wegener's work in a series of papers from 1912 to 1929 is generally considered the beginning of the concept as it was developed to about 1960. Evidence accumulated, particularly in the Southern Hemisphere, to suggest that the present continents of India, Australia, South America, Africa, and Antarctica had been one large continent; early workers called it Gondwanaland. This continent was centered near the South Pole about 250 million years ago. From that time to the present, it fragmented, and most of the pieces drifted slowly northward to their present positions. Many workers envisioned the continents as moving through the oceanic terranes as ships moving through the water, a concept that appeared impossible mechanically and was correctly criticized by many people. Because of mechanical problems and the fact that much of the data was from the Southern Hemisphere, North American and many European geologists were opposed to the concept or, at least, skeptical.

Arthur Holmes in his textbook on physical geology published in 1944 suggested that not only did the continents drift, but so did the oceanic crust (Fig. 8–1). Similar ideas were proposed by Harry Hess in 1962 in a work that hypothesized that oceanic crust was created at **mid-ocean ridges**, moved laterally carrying continents along with the oceanic crust, and that at **oceanic trenches** the oceanic crust plunged back in the earth's mantle. Hess viewed the process as a giant convective system, and his work along with that of Holmes is generally considered as the birth of the ideas of **sea-floor spreading** and continental drift as we know them today. Confirmation of Hess's early ideas came quickly, and in a series of articles published from 1963 to 1968 the concepts of sea-floor spreading and horizontal movement of oceanic and continental crust were developed and well documented.

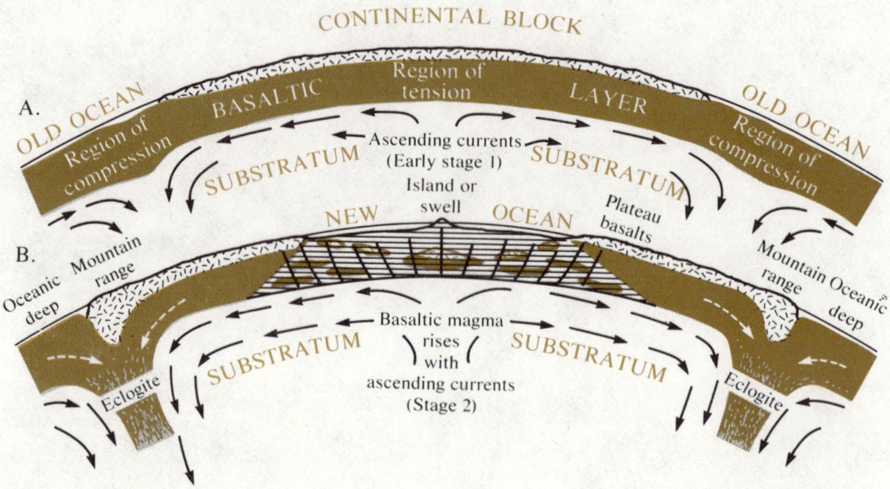

FIGURE 8–1
Hypothetical scheme for continental drift, published by Arthur Holmes in 1944. These diagrams illustrate Holmes's concept of subcrustal currents and how they create new oceanic regions and zones of compression.

OCEAN-BASIN MORPHOLOGY

The concepts of sea-floor spreading and the motion of the earth's crust were developed largely from data gathered from the oceans; thus, it is important to discuss briefly the morphology of the ocean basins.

The Atlantic ocean basin is perhaps the simplest to describe (Fig. 8–2A). A topographic cross section of the Atlantic ocean bottom (Fig. 8–2B) contains a broad ridge in its central part flanked by two broad, flat plains. The ridge varies in width from 2000 to 4000 km, and its crest averages about 2 km below sea level. In a few places, the ridge crest rises above sea level, as in Iceland. Although the local topography of the ridge is rugged, the average elevation gradually decreases toward the flanking basins. Because of its median position within the ocean basin, the ridge is commonly called a mid-ocean ridge. The flat plains flanking the ridge lie at an elevation of about 5 km below sea level and are called **abyssal plains**. The plains may be interrupted by high, steep-sided hills or mountains which may reach above sea level. Many of these topographic features are volcanic cones.

Nearer the continents is a gentle slope, perhaps 1° or less, that rises toward the continent, followed by a steeper slope, about 5°, where the ocean floor rises to about 200 m below sea level. The gentle slope is called the **continental rise**, and the steeper slope is called the **continental slope**. Above the continental slope to the shoreline is again a gentle slope, termed the **continental shelf**.

The mid-ocean ridge and abyssal plains are underlain by oceanic crust whereas the continental shelf is underlain by continental crust. Somewhere beneath the deeper continental slope and rise, the transition between continental and oceanic crust takes place, but the details of this transition are poorly documented.

Most ocean basins contain a similar mid-ocean ridge that is locally offset along transverse valleys and scarps trending approximately perpendicular to the trend of the ridge. There are two prominent exceptions: (1) the Pacific Ocean (Fig. 8–3) where the broad ridge is in the eastern part of the basin off Central America and disappears at the Gulf of California, with a short segment reappearing off the coast of northwestern United States and southwestern Canada; and (2) the Arctic Ocean where there are two prominent parallel ridges (Fig. 8–4).

Abyssal plains are present in all oceans, and they have different shapes and sizes depending upon the shape and proximity of the adjacent continents. The abyssal plains of the Pacific Ocean (Fig. 8–3) are broken by numerous isolated topographic hills and mountains that cluster locally into a broad group called the mid-Pacific mountains. In some places, the abyssal plains are traversed by long, linear ridges. Some of these ridges are dotted by higher hills and mountains which are volcanic cones, such as the Hawaiian-Emperor chain. Other ridges are not volcanic, such as the Norfolk ridge of the southwest Pacific.

In the western Pacific, and parts of the Indian and Atlantic oceans, there are arcuate alignments of active volcanoes that rise above sea level and crown a broader continuous ridge. Because of their arcuate shape, they are called **island arcs**, and are most typically developed in the Kuril, Japan, and Mariana island chains of the western Pacific (Fig. 8–3). Between these island arcs and the adjacent continents, the ocean bottom is deep, varying between 2 and 5 km. These small areas of deep

A.

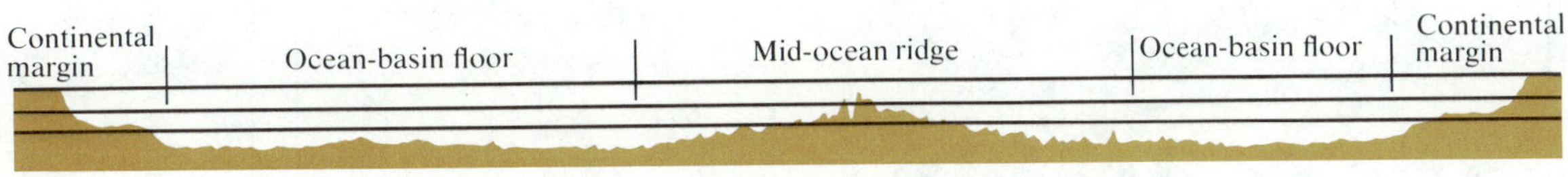

B.

FIGURE 8–2
A. Artist's conception of the North Atlantic Ocean. (From a painting by Heinrich C. Berann; courtesy Aluminum Company of America) B. Schematic cross section of the North Atlantic showing the three major subdivisions of the ocean region: the mid-ocean ridge, the ocean-basin floor, and the continental margin. (After B. C. Heezen and others, 1959)

water on the concave side of island arcs are called **marginal seas**. Locally, they may be subdivided by one or more linear ridges at shallower depths, such as in the Philippine Sea. Most marginal seas are underlain by oceanic crust, whereas the ridges within them are underlain by rocks similar to those of the island arc or by thin continental crust.

On the convex side of most island arcs are deep, narrow trenches that parallel the arcs. These trenches contain the deepest water in the oceans, locally deeper than 11 km below sea level. Similar trenches are present along continental areas that contain active volcanoes, such as Peru and northern Chile. In some areas, such as south of Alaska, where arcuate vol-

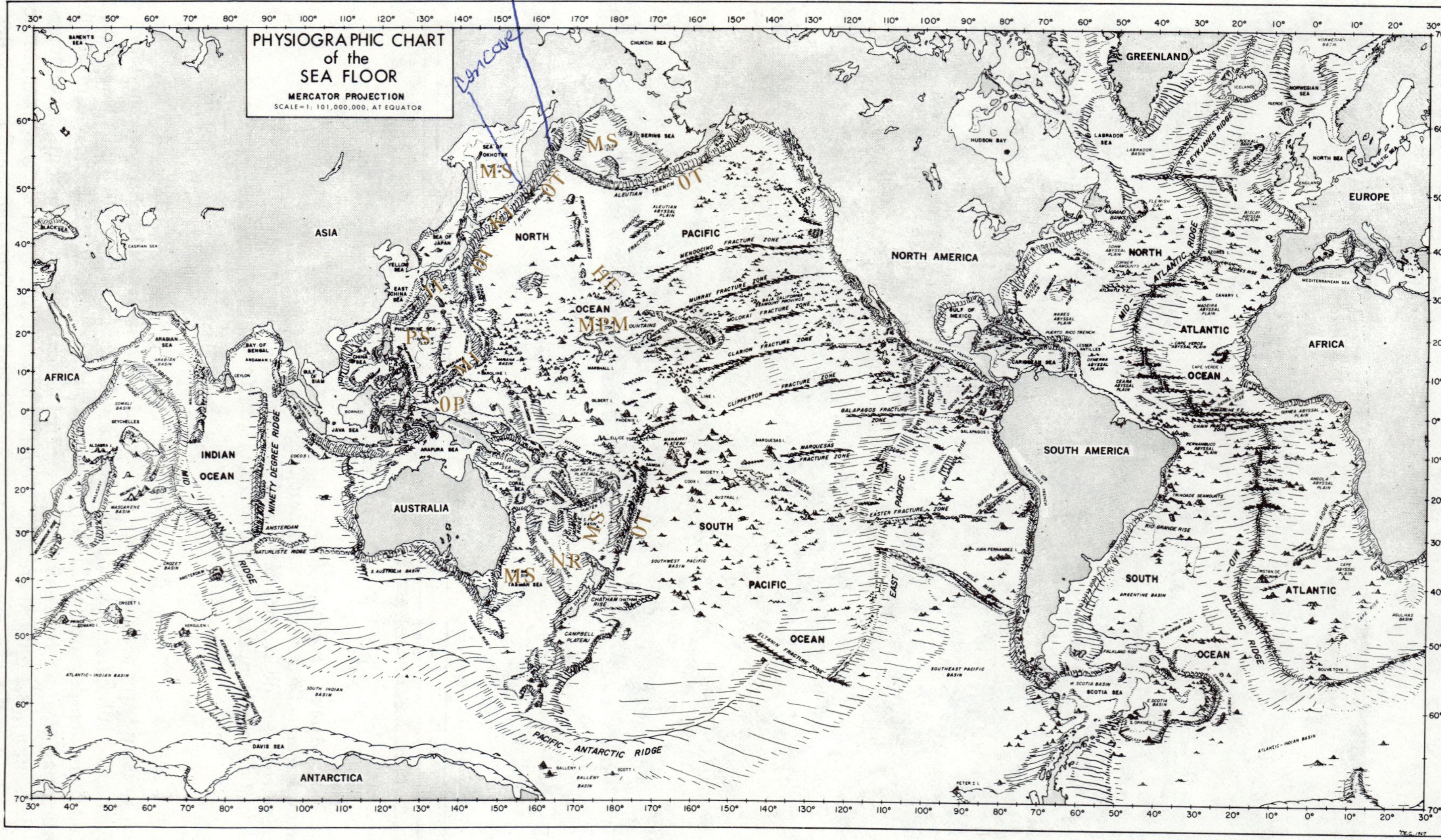

FIGURE 8–3
Artist's conception of the major morphological features of the earth's ocean regions. The following features are lettered in the figure: MPM = mid-Pacific mountains, HE = Hawaiian-Emperor chain, NR = Norfolk ridge, KI = Kuril island arc, JI = Japan island arc, MI = Mariana island arc, MS = marginal sea, PS = Philippine Sea, OT = oceanic trench, OP = Ontong Java Plateau. (Adapted from Hubbard Scientific Co.)

FIGURE 8–4
Artist's conception of the morphological features of the Arctic Ocean region. (Drawing by Robert W. Tope)

canoes are close to the edge of continents, there appears to be no trench, but geophysical studies have shown a trench is present but has been filled by young sediments. In these regions where trenches are present, there is no continental rise, and the continental slope is much steeper and often irregular.

Within some oceans, large islands are flanked by even broader areas of shallow water. Some of these areas, such as Madagascar and adjacent regions, are underlain by continental crust. Other areas, such as the Ontong Java Plateau of the west Pacific, may be underlain by a thicker, elevated oceanic crust. Morphology alone is not enough to identify the type of crust that underlies a particular region of shallower water.

Most of the morphological features of the ocean basins are a direct reflection of the processes that created and modified the ocean basins. Most of these features can be explained in terms of the dynamics of sea-floor spreading and horizontal movement of the earth's crust.

SEA-FLOOR SPREADING

Two important and seemingly unrelated pieces of geologic information were discovered in the late 1950s and early 1960s: (1) the earth's magnetic field has reversed its polarity several times in the last few million years, and (2) surveys of the northeast Pacific Ocean showed that the oceanic crust has a peculiar pattern of parallel striped **magnetic anomalies** (deviations in the earth's geomagnetic field due to local disturbances caused by magnetized rocks). Careful dating and paleomagnetic studies have now developed a magnetic polarity sequence that is well established for the last 5 million years, moderately well established back to about 100 million years, and is currently being extended back into the older geologic record (Fig. 8–5). In 1963, F. Vine and D. Mathews put together a hypothesis relating magnetic anomalies of the oceanic crust and **magnetic reversals** (Fig. 8–6). Studying the magnetic anomalies from a part of the mid-Atlantic ridge south of Iceland, they recognized that the anomaly patterns are symmetrical about the central part of the ridge, which is marked by a **median valley**. They reasoned that the oceanic crust was moving apart along this median valley and that molten rock was intruding along the zone between these diverging pieces of crust. As the molten rock crystallized, it formed new oceanic crust and, once it cooled below the Curie temperature, became magnetized in the direction of the earth's field at that time. During a time period when the earth's field was reversed, new ocean crust would acquire a magnetization of reversed polarity. Thus, the oceanic crust would be formed at the center of the mid-ocean ridge and, during formation, would be split apart with half of the newly formed crust on each side of the spreading center, each half of the ocean crust carrying the record of magnetic polarity reversals. Thus, the ocean crust was like a conveyor belt carrying the record of polarity reversals on both sides of the spreading center (Fig. 8–6). Recent studies have confirmed these hypotheses.

Studies of magnetic anomalies spread rapidly to other oceans, where it was discovered that paired anomaly patterns, symmetrically disposed about oceanic ridge axes, are present in all the oceans. Furthermore, because of the irregular time interval of polarity reversals in the geologic past, magnetic anomalies representing certain time intervals could be recognized in each ocean and were assigned numbers. Once the age of a particular anomaly was dated in one ocean, the dating could be extended to all oceans where that anomaly was recognized. Anomalies were dated by

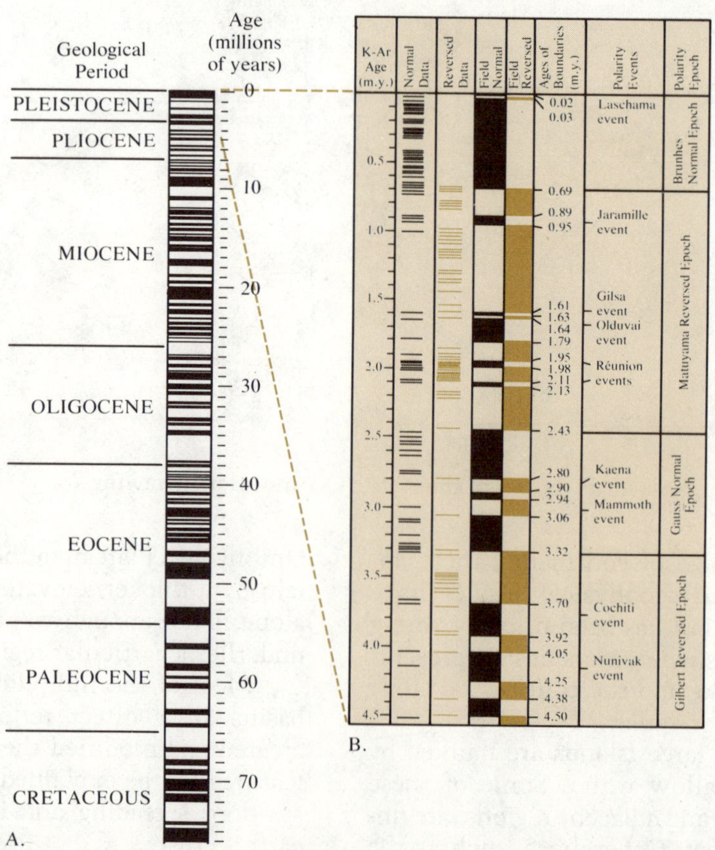

FIGURE 8–5

A. Times of normal (colored bars) and reversed (white bars) magnetic polarity established for the last 75 million years. (After J. R. Heirtzler and others, *Journal of Geophysical Research,* Vol. 73, p. 2119, 1968, copyrighted by the American Geophysical Union) B. For the past 4.5 million years, dating of magnetic reversals can be done on a much finer scale. [Data from A. Cox (1969), *Science,* 163, 237–45. © 1969 by the American Association for the Advancement of Science]

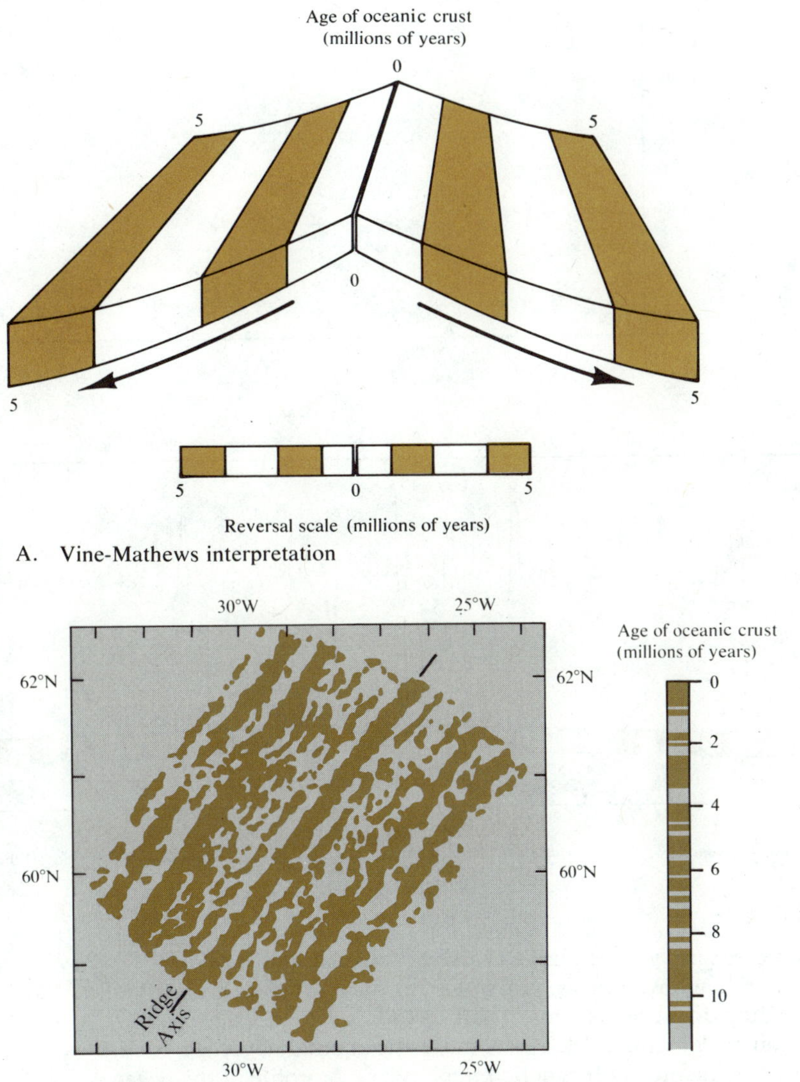

FIGURE 8–6
A. The hypothesis of Vine and Mathews explained the magnetic anomaly patterns present in the ocean basins by sea-floor spreading. B. The magnetic anomaly pattern of the Reykjanes ridge south of Iceland shows an example of the type pattern present near spreading ridges. [Part B from F. J. Vine, "Magnetic Anomalies Associated with Mid-Ocean Ridges," in *The History of the Earth's Crust*, by Robert A. Phinney (ed.) (copyright © 1968 by Princeton University Press), fig. 6, facing p. 82. Reprinted by permission of Princeton University Press. Drawings by Robert W. Tope]

drilling through the overlying sedimentary layer and dating the fossils from the oldest sediments that overlay the top of the oceanic crust. When the ages determined from fossils were correlated with the absolute time scale, which gives geologic time in millions of years, a spreading rate could be calculated by dividing the distance from the spreading axis by the absolute age of the anomaly (Fig. 8–7). It was quickly discovered that spreading rates for different parts of a single ocean are different, and that rates vary from ocean to ocean. Total spreading rates vary from 1 cm/year or less to nearly 20 cm/year in some places.

Topographic studies of the oceanic ridges associated with spreading axes have shown that ridges vary from 2000 to nearly 4000 km in width and have ridge crests approximately 2 km below sea level. For slow-spreading ridges, such as the mid-Atlantic ridge, which spread at only a few centimeters per year, the spreading axis is marked by a median valley, but for fast-spreading ridges no median valley is present, such as along parts of the east Pacific ridge system. The ridge is also marked by offsets along its axis that follow fracture zones. Studies of magnetic anomalies show that the spreading centers are also offset by these fracture zones, most of which trend at right angles to the ridge. Early interpretations suggested that these fracture zones offset the ridge and its associated magnetic anomaly patterns. However, J. T. Wilson in 1965 proposed that the spreading axis, magnetic anomalies, and the topography of the ridges were inherently discontinuous and that the offsets along the fracture zones were a product of sea-floor spreading. He reasoned that spreading and creation of oceanic crust took place along a discontinuous axis and that each segment of the spreading axis was connected to the adjacent segment by a fracture zone, which he termed a **transform**

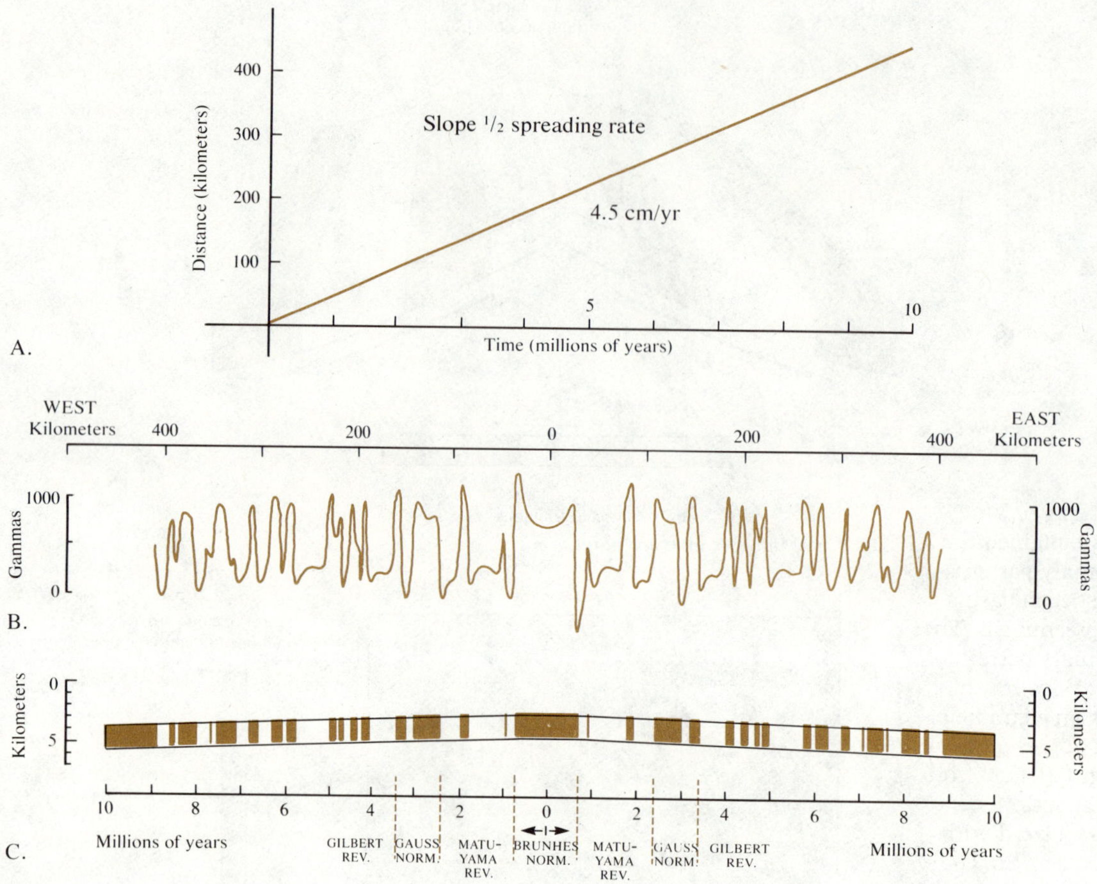

FIGURE 8–7

Many magnetic profiles present in the ocean basins can be explained by sea-floor spreading using the Vine-Mathews hypothesis. A. By plotting distance of an anomaly of known age from the spreading ridge against time, a half spreading rate, usually referred to as the spreading rate, can be determined. B. Model magnetic anomaly over the Pacific Antarctic ridge (gammas are a measure of magnetic intensity). C. A profile of the ocean crust with normal (Norm.) and reversedly polarized (Rev.) crustal segments that could be the causative bodies for the observed magnetic anomalies. [Parts B and C from J. R. Heirtzler, "Evidence for Ocean Floor Spreading across the Ocean Basins," in *The History of the Earth's Crust,* by Robert A. Phinney (ed.) (copyright © 1968 by Princeton University Press), fig. 4, p. 94. Reprinted by permission of Princeton University Press]

fault (Fig. 8–8). He reasoned that spreading along the axis would cause displacements on the transform faults which would be opposite to the displacement inferred by assuming that the axis and magnetic anomalies were once continuous and later offset by the faults. Furthermore, he reasoned that because oceanic crust is formed at the ridge axis, only the part of the transform fault that connected the two ridge axes would be active. Parts of the transform faults that passed beyond the zone between the two active spreading axes would be inactive (Fig. 8–8). Examination of earthquake activity along oceanic ridges shows that most earthquakes occur along the spreading axis and along the segment of transform faults that connects the spreading axes. Seismological studies of individual earthquakes have shown that those earthquakes associated with the spreading axis can be explained by faults generated by extending the oceanic crust approximately perpendicular to the spreading axis. These studies have also shown that earthquakes along the active segments of transform faults can be explained by faults with horizontal displacement and with a sense of displacement that confirms Wilson's hypothesis of the motion on transform faults.

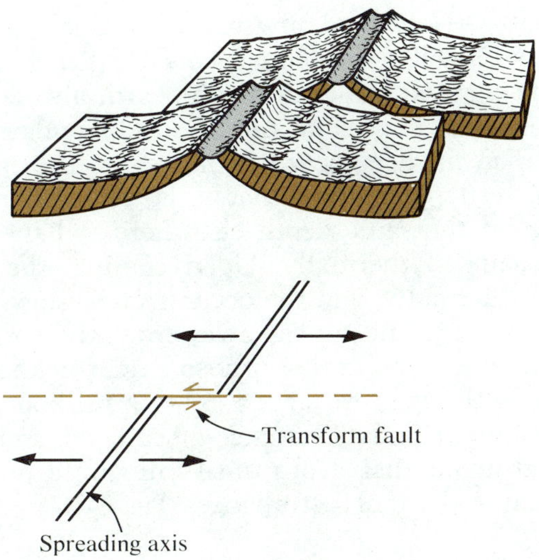

FIGURE 8–8
Transform faults are approximately at right angles to spreading ridges and connect spreading axes.

SUBDUCTION

In some parts of the world, particularly near deep-sea trenches, earthquake activity extends to depths of more than 700 km. This contrasts greatly with earthquake activity associated with spreading axes and transform faults, which is generally 10–15 km or less. Studies of these deep earthquakes showed that their hypocenters defined a zone that began at the site of the oceanic trenches and dipped at varying angles away from the ocean basins. This dipping zone of earthquake activity, called the **Benioff zone** in honor of Hugo Benioff, who is given credit for its discovery, has been carefully examined in a few places, such as along the Tonga-Kermadec region in the southwest Pacific (Fig. 8–9). Studies of earthquakes in this region showed that the velocity of seismic waves generated by deep earthquakes was highly variable, depending upon the path they took to reach the surface. This velocity distribution could be satisfied by assuming that a thick slab, approximately 100 km thick, with seismic properties similar to those of the oceanic crust and its upper mantle, was present below the trenches. These studies indicated that this slab consisted of the oceanic crust and its upper mantle that had moved downward into the earth's mantle. Further data from seismic waves suggested that the integrity of the slab may be lost as it is resorbed into the mantle at depths of about 700 km. The process of oceanic crust and upper mantle being taken into the mantle and resorbed is called **subduction**.

Studies on the deformational properties of rocks have indicated that at increased temperatures, crystalline rocks may behave in a fluidlike or ductile manner. This type of ductile flow is similar to the movement of glacial ice, which remains a crystalline material while it flows. Experiments on rocks thought to represent upper mantle material suggest that they will flow at rates appropriate for those of sea-floor spreading at temperatures of about 1100°C–1200°C. Thus, rocks below the depth where these temperatures are reached in the earth can be regarded as behaving somewhat like a very viscous, ductile material. Rocks at shallower depths will not flow significantly over the same length of time. This concept of mantle behavior is used to subdivide the upper mantle into a more rigid upper part, called the **lithosphere**, and a more ductile, fluidlike, but very viscous lower part, called the **asthenosphere**. Beneath the oceans, the base of the lithosphere is probably close to the top of the low-velocity zone. The fluidlike behavior of the asthenosphere extends

FIGURE 8–9
The distribution of deep-focus earthquakes (dots) within the mantle near oceanic trenches defines a dipping zone called the Benioff zone. P-wave velocities (V_P) recorded at different stations from deep-focus earthquakes indicated that a slab about 100 km thick which contains the Benioff zone in its upper part is present in the upper mantle.

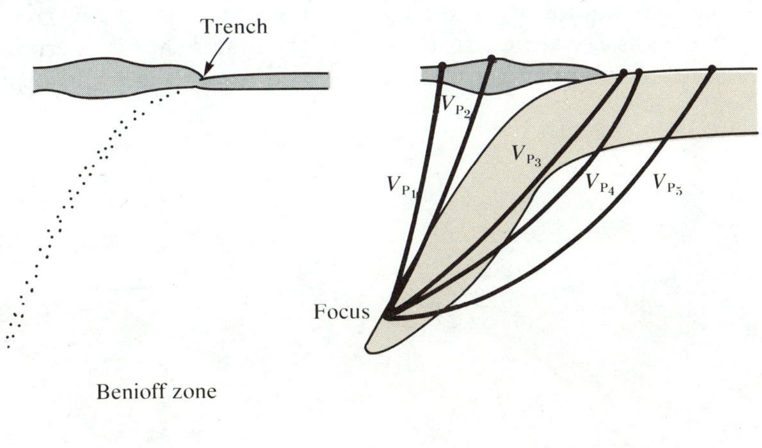

throughout and below the low-velocity zone, and may extend through the entire mantle. Beneath the continents, the critical temperatures for defining the base of the lithosphere are probably reached at greater depth than beneath the oceans, although seismic evidence for a low-velocity zone may be lacking. Thus, the lithosphere consists of both oceanic and continental crust and a variable thickness cf upper mantle.

Thickness of the lithosphere is variable and is largely the result of the temperature distribution in the earth. At the spreading ridges in the oceans, the lithosphere is only a few kilometers thick because hot, molten—and thus very ductile—material directly underlies the spreading axis (Fig. 8–10). As the oceanic crust is formed and moves away from the spreading axis, it and the underlying mantle material cool, and the level of ductile behavior becomes deeper. Thus, the thickness of the lithosphere increases with time and distance from the ridge. At approximately 50–60 million years after its formation, the oceanic lithosphere is roughly 80–100 km thick and consists of 5 km of oceanic crust and 75–95 km of upper mantle. Lithosphere older than 50–60 million years remains at about constant thickness.

Thickness of lithosphere beneath continents is less well established, and in some cases is difficult to identify from seismic evidence. Early ideas suggested that the lithosphere in continental areas was thicker than beneath oceans and could reach depths of much more than 150 km. More recent work also suggests that the lithosphere beneath continents is probably thicker than beneath oceans, extending to a depth of 150 km or more. The uncertainty is caused by the difficulty in establishing the depth to the low-velocity zone and the imprecise knowledge of temperatures beneath continents. In fact, the seismic data in some areas do not demand the presence of a low-velocity zone. This uncertainty remains and has important implications not only for the thickness of lithosphere, but also for the mechanism and configuration of mantle motion. Whatever its thickness, continental lithosphere consists of a variable thickness of continental crust (10–70 km) and an unknown thickness (but probably 150–200 km) of upper mantle (Fig. 8–10).

The thermal structure in the earth also controls the topography of oceanic areas. Spreading ridges stand higher than the deep-ocean floor because they are underlain by upper mantle that is hotter than mantle material at the same depth elsewhere and the rocks have expanded thermally. Upon cooling, the rocks contract thermally, and the oceanic crust slowly subsides as it ages and moves away from the spreading axis. At the same time, lithospheric thickness increases with age. At about 50–60 million years, oceanic crust stands 5 km below sea level, and it remains at about that depth until influenced by other structural or thermal disturbances (Fig. 8–11).

PLATE TECTONICS

From examination of the seismicity map of the world (Fig. 8–12), which shows the epicenters (map location) of earthquakes during 1961–1967, it is clear that most of the earthquakes lie along narrow, continuous, and interconnecting zones. Study of these earthquake zones demonstrates that nearly all earthquakes are associated with either a spreading ridge, transform fault, or **subduction zone**. Spreading ridges and transform faults are marked by less seismic activity than the subduction zones. All earthquake activity deeper than a few tens of kilometers is confined to subduction zones. These three types of zones separate large areas that are seismically quiet and are regions of little or no internal deformation. Combining these observations, the earth's lithosphere can be divided into six large and several smaller **plates** which are bounded by three fundamental types of zones that form the plate boundaries (Fig. 8–13). The basic concept of **plate tectonics** is that lithosphere is created at spreading axes, slides horizontally past itself along transform faults, and plunges into the mantle at subduction zones. The rate of subduction and resorption of lithosphere must approximately bal-

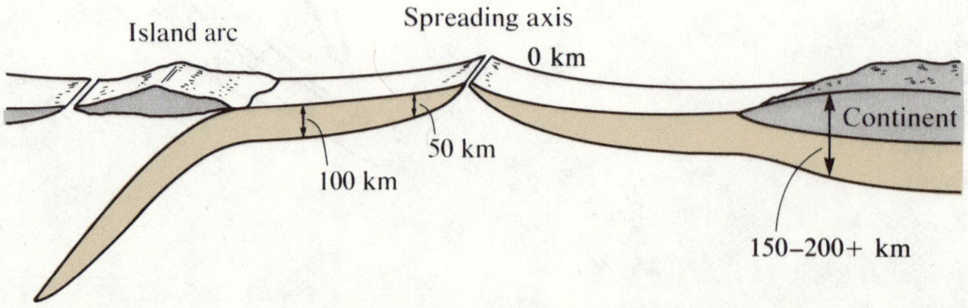

FIGURE 8–10
The variable thickness of the lithosphere. Thickness of lithosphere is related largely to temperature within the earth. Generally, the lithosphere thickens away from the spreading axis to about 100 km, and may be considerably thicker beneath continents.

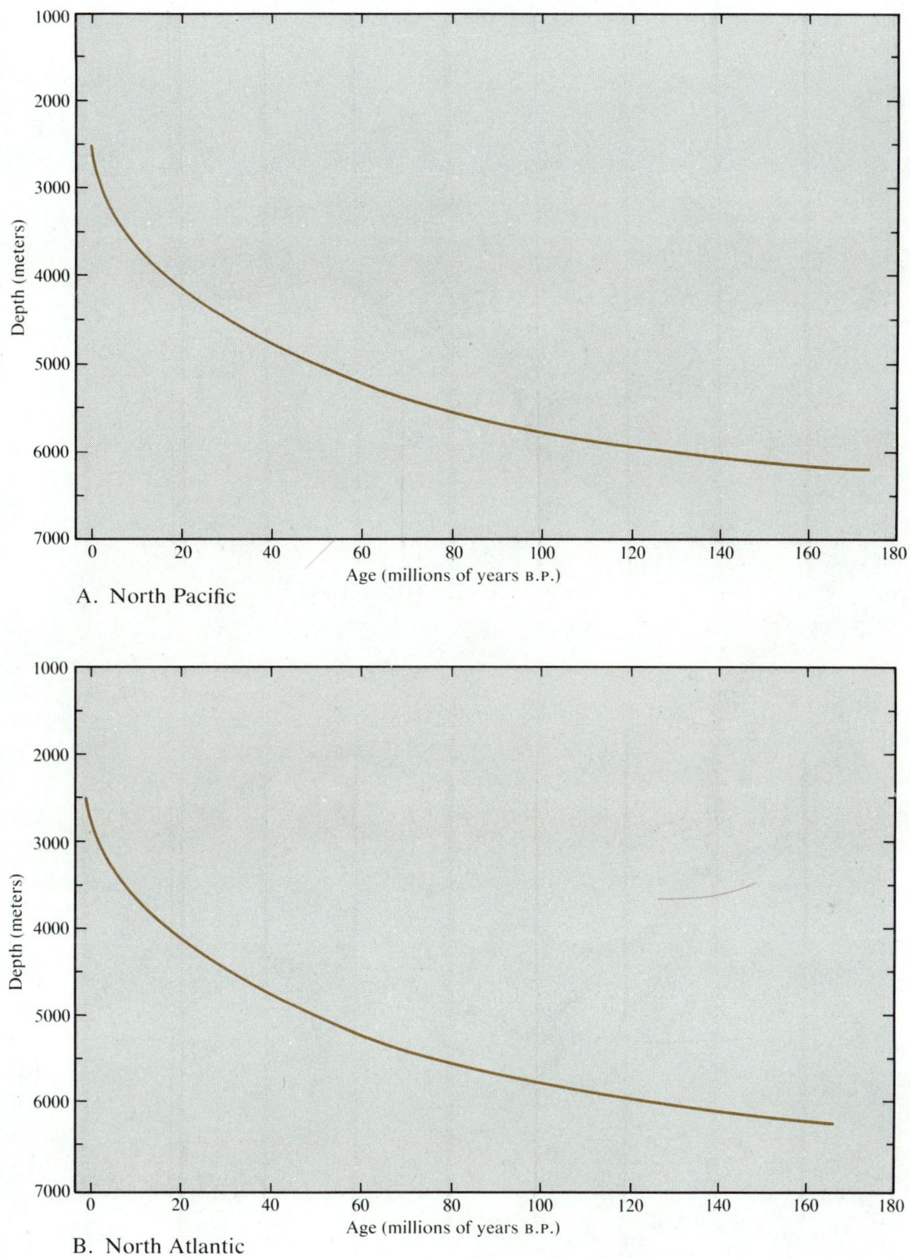

FIGURE 8–11
Generalized curves showing mean depth of sea floor versus age for the North Pacific and North Atlantic. (After B. Parsons and J. Sclater, *Journal of Geophysical Research,* Vol. 82, p. 822, 1977, copyrighted by the American Geophysical Union)

ance the rate of formation of lithosphere, as there is little evidence to support significant expansion or contraction of the earth during the present episode of seafloor spreading, which extends back at least 180 million years.

The plates consist of lithosphere generally 150–200 km thick beneath continents and 100 km thick beneath oceans, except where the thermal gradient is high. All lithosphere, except near a spreading axis, consists of both upper mantle and crust, but the crust in some plates is entirely oceanic, such as the Pacific plate; whereas in other plates the crust is both continental and oceanic, such as the African plate. Only a few of the small plates contain a crust that is dominantly continental, such as the small plates of the Middle East.

All motion between the plates is relative because there is no fixed frame of reference on the earth from which absolute motion can be determined. Hot spots may permit the determination of absolute motion (hot spots will be considered later in the chapter). The direction of relative motion between two plates at a spreading axis is determined by the direction of the transform faults that offset the spreading axis. Relative motion must be parallel to the transform faults and in most cases is at right angles to the spreading axis. The relative motion between two plates across a subduction boundary is not fixed by the geometry of the

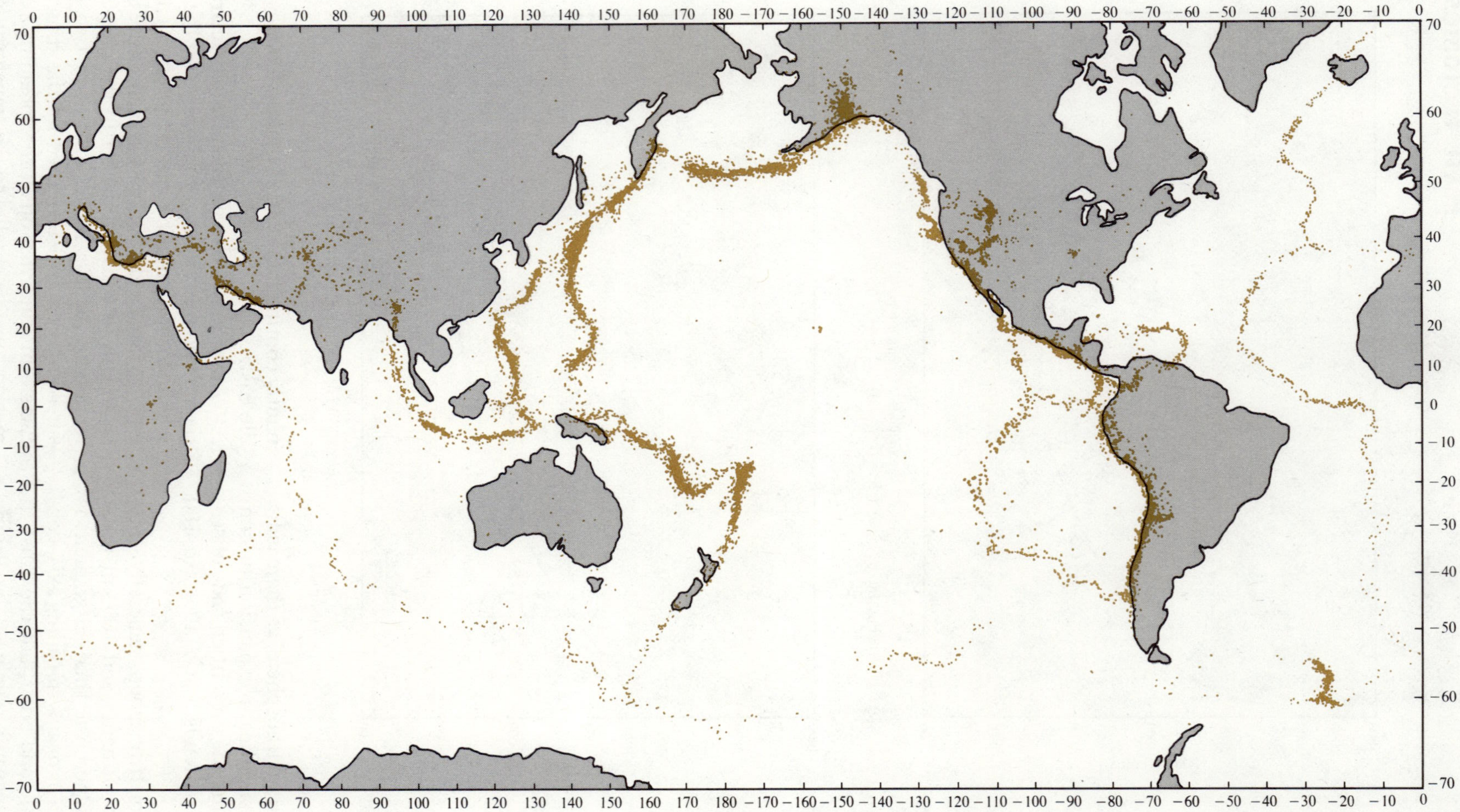

FIGURE 8–12
Worldwide distribution of all earthquake epicenters for the period 1961–1967 as reported by the U.S. Coast and Geodetic Survey. (After M. Barazangi and J. Dorman, 1968)

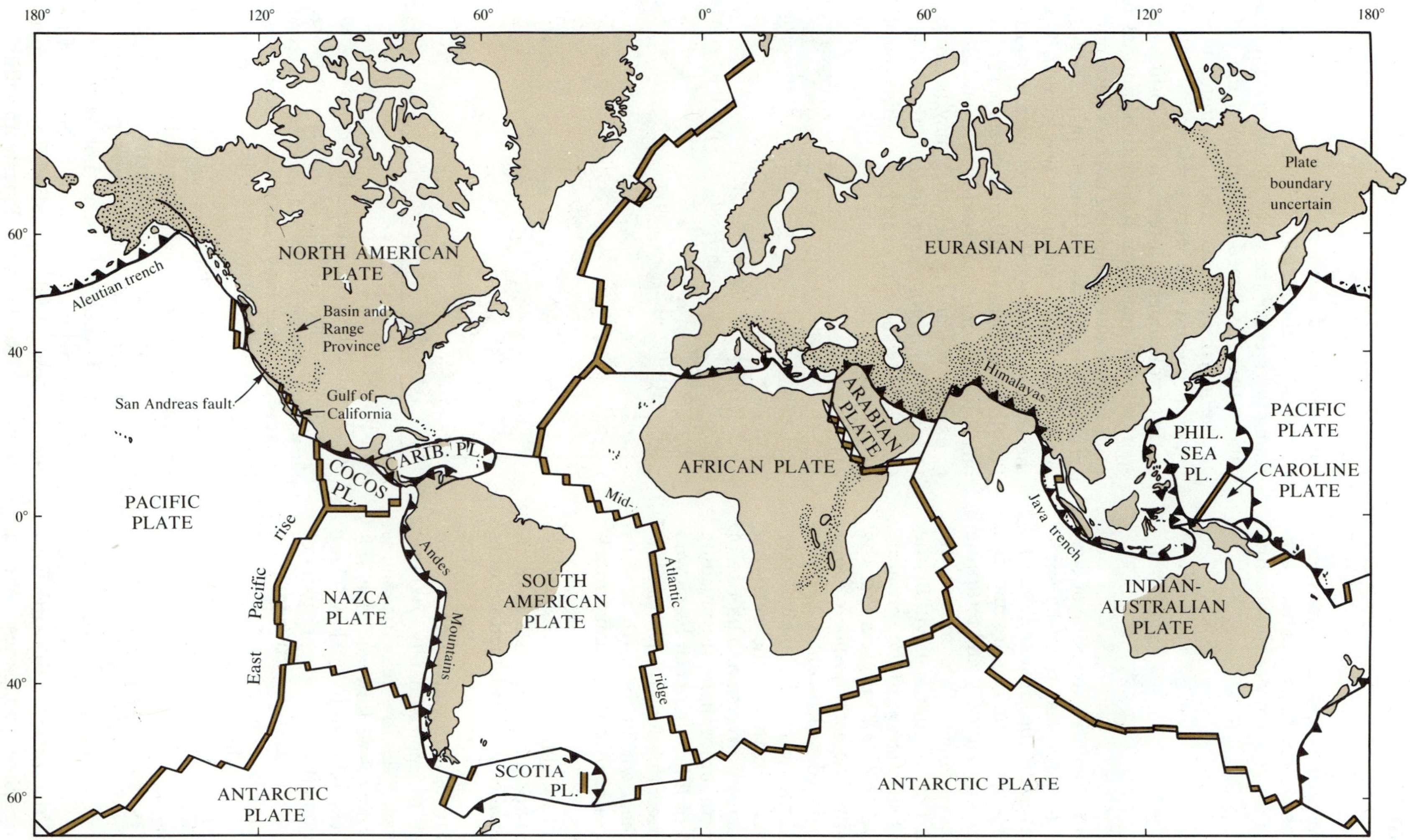

FIGURE 8–13
Simplified map of the lithospheric plates of the world. Plate boundaries are divided into spreading axes (double lines), transform faults (single lines), and convergent or subduction zones (barbed lines on overriding plate) Stippled areas within continents are regions of active deformation away from the plate boundary. (Modified slightly from W. Hamilton, 1977)

boundary as it is for spreading axes and transform faults. The relative motion at a subduction zone can be determined, but not from the geometry of the subduction boundary alone.

CONVECTION

Provided the earth is neither rapidly expanding or contracting, plate tectonics requires that there be some convective circulation involving the crust and mantle. Lithosphere is formed at a spreading axis by material rising from the mantle, moving laterally, and returning to the mantle in subduction zones. To complete the circuit, there must be a flow of material within the mantle from the zones of subduction to the spreading axes. Information from seismic data suggests that the lithosphere behaves rigidly. The asthenosphere, however, is thought to behave in a ductile manner. Because of the presence of partial melting in the low-velocity zone, the density is lower than that of the overlying lithosphere, creating a density inversion. The rise of the asthenosphere to form lithosphere at spreading axes and the subduction of lithosphere can be viewed as nature's attempt to correct this density inversion. As long as there is enough heat within the earth to form partial melting in the upper mantle, this density imbalance will survive.

The depth of the return convective flow within the mantle is poorly known. Some scientists have argued that because earthquakes along subduction zones extend to a depth of nearly 700 km, indicating that the subducted lithosphere extends to that depth, the return flow must also extend to approximately 700 km. Others have argued that only the low-velocity zone is involved in significant convective flow and that significant return flow can extend to only 250 km. There have been some suggestions that even deeper parts of the mantle may be involved in the convective process. The fact is that not enough is known about the state and processes of the upper mantle to be sure.

HOT SPOTS AND MANTLE PLUMES

Volcanic centers occur at the surface of the earth both within plates and along plate boundaries. Within oceanic areas, particularly the Pacific Ocean, some volcanoes lie at the ends of oceanic ridges that are not related to transform faults or spreading axes. The best-studied example is the Hawaiian Islands, which lie at the eastern end of a long ridge with two straight segments called the Hawaiian-Emperor ridge (Fig. 8–14). Along the crest of the ridge are submerged extinct volcanoes that developed sequentially, the oldest being at the northwest end and becoming progressively younger toward the active Hawaiian Islands. This chain of volcanoes is interpreted as forming over a **hot spot**, a small region of higher than normal temperatures that produced melting in the mantle and volcanic activity above it. As the Pacific plate moved over the hot spot, a chain of volcanoes was formed. The kink in the volcanic chain reflects a change in motion of the Pacific plate. Other volcanic centers of similar type are present along or near spreading axes and are connected to extinct volcanic ridges that may be similar to the Hawaiian-Emperor chain. Iceland, the Azores, and the Galápagos Islands are examples of this type of volcanic center (Fig. 8–15).

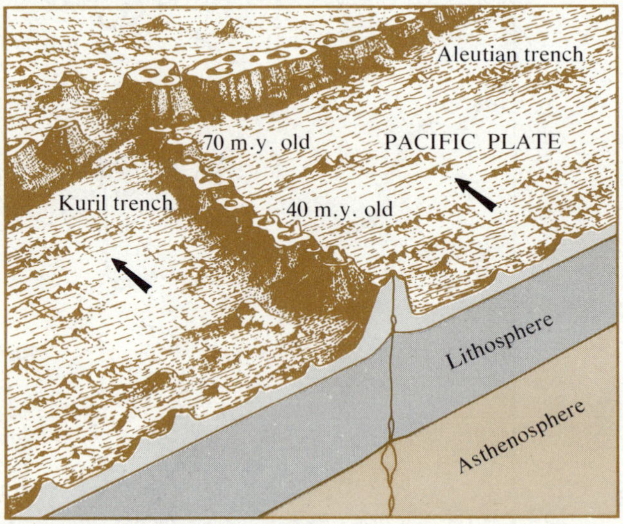

FIGURE 8–14
The Hawaiian-Emperor ridge is thought to have been formed by movement of the Pacific plate over a hot spot in the mantle. It is formed by a chain of volcanic cones that get progressively older away from the active volcanoes of the Hawaiian Islands. (Drawing by Robert W. Tope)

Because the plates move relative to these hot spots, the hot spots are interpreted as having their origin deep in the mantle, perhaps beneath the convective flow associated with plate movement. It is hypothesized that they represent hot spots localized in the deeper mantle which cause the rise of less dense mantle upward along cylindrical channels, called **plumes**, unlike the more elongate upwelling associated with spreading axes. At shallower depths, melting takes place, and the magma continues upward to form volcanoes at the surface. Many plumes and their associated linear volcanic chains, called **plume tracks**,

FIGURE 8–15
World map showing the locations of some probable hot spots. The modern plate boundaries are also shown. (Map by Robert W. Tope)

have been recognized. Whether all the volcanic centers proposed as surface expressions of plumes really lie above plumes remains to be established. The importance of plumes and their plume tracks is that they can be used as a possible frame of reference for describing plate motions. If these plumes are fixed in time, they could be considered as forming a reference frame for absolute rather than relative plate motion. Using the best examples of hot spots, a hot spot frame of reference can be used to plot plate motions. Present data suggest that hot spots do move relative to one another but that they move much more slowly than the overlying plates. Some scientists have suggested that hot spots and plumes may be the mechanism that initiates spreading and begins the development of new plate tectonic boundaries. The nature of hot spots and plumes is poorly understood, and considerably more study of these features is necessary before their full significance is known.

RELATIVE MOTION OF THE PLATE MOSAIC

The plate mosaic of the earth is an integrated system in which the relative motions of plates are interrelated. The African and Antarctic plates are nearly surrounded by spreading axes, yet there are no subduction zones within these plates. Because these two plates are enlarging by the formation of oceanic lithosphere around them, it must be concluded, first, that the spreading axes are moving away from one another and, second, that the oceanic lithosphere created at these ridges must be subducted elsewhere (provided the earth is not expanding).

The rate of subduction must be related to the rate of spreading, integrated around the earth. From a study of plate motions on the earth, it is clear that both plates and their boundaries move relative to one another (Fig. 8–16). As plate boundaries become inactive or are newly formed in one part of the globe, they may affect the relative motion at all other plate boundaries. Plates move relative to hot spots, and over long periods of time hot spots may move relative to one another. Thus, before absolute plate motions can be resolved, scientists must establish a frame of reference external to the earth. With advances in satellite technology and astronomical measurement techniques, it will not be too long before such an external frame of reference can be established and absolute plate motions determined.

SUBDUCTION-RELATED PHENOMENA

Subduction boundaries are associated with volcanic activity, the formation of small areas of oceanic lithosphere, and deformation leading to the formation of continental material. The last item will be treated in detail in Chapter 9. Volcanism is not randomly distributed on the earth's surface but occurs mostly on or near plate boundaries. Volcanic rocks associated with spreading axes are dominantly basalt and in general can be distinguished chemically from basaltic rocks present in oceanic islands related to hot spots. Trans-

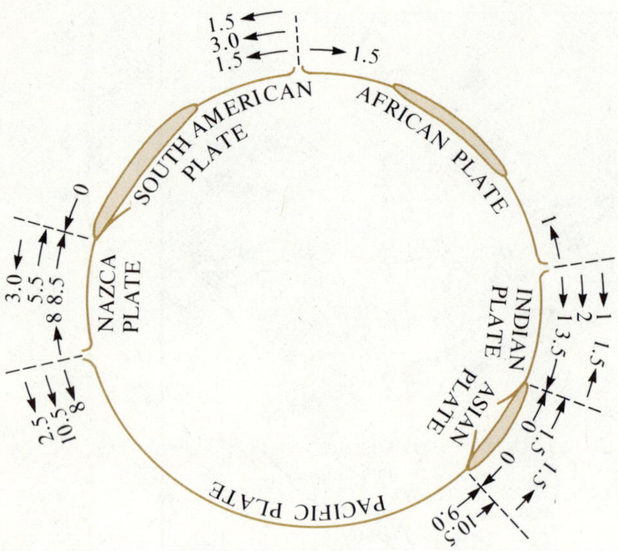

FIGURE 8–16
A great circle path around the earth approximately along the equator showing the relative motions between the plates and their boundaries. Velocities are given in centimeters per year. Velocities shown that are closest to the earth's surface are approximate spreading and convergent rates given by X. Le Pichon (1973) and recomputed into the great circle path. Middle and upper velocities are the relative motions of the plates and plate boundaries, respectively, relative to a fixed African plate.

form boundaries are not commonly associated with volcanoes. Subduction zones, however, are usually associated with linear volcanic belts that occur 100–200 km away from the subduction boundary and in the overriding plate (Fig. 8–17). Because the subduction zones commonly form an arcuate trace at the earth's surface, the belt of volcanoes is also curved and is called a **volcanic or magmatic arc**. Arc volcanic rocks are characterized by andesite but may contain a rather wide range of rock types.

Adjacent to subduction zones where oceanic lithosphere is present in both plates, such as in the southwest Pacific, the volcanic arc is built upon oceanic lithosphere and forms an **island arc** (Fig. 8–18). Island chains such as the Marianas and Tonga islands of the western Pacific region are examples of island arcs. Adjacent to subduction zones where oceanic lithosphere is subducted beneath continental lithosphere, such as along the west coast of South America, the volcanoes are built upon continental lithosphere and are called **Andean-type arcs** (Fig. 8–18). In some cases, Andean-type arcs may be largely covered by water and superficially resemble island arcs. Sumatra and the eastern end of the Aleutian Islands are examples of the latter, and geologic and geophysical studies are required to distinguish the two.

Volcanic activity is also associated with subduction-related activity where two plates carrying continental lithosphere are converging. In the Middle East, India, and southern Asia, such a zone is present. This and other such zones are considered in more detail in Chapter 9.

The origin of the igneous rocks associated with subduction zones is complex, and there is no general agreement on the process or processes involved. The volcanoes are positioned above that part of the subducted plate that is at a depth of approximately 100 km (Fig. 8–19). This is the depth at which the subducted lithosphere enters the asthenosphere, and it seems reasonable to assume that some melting of the lithosphere would occur at this point. There is considerable debate as to whether this magma reaches the surface, or whether the volcanoes are actually fed from higher sources only secondarily related to the subduction zone.

Formation of small regions of oceanic lithosphere is related to subduction, resulting in **marginal seas**. The Japan Sea, Lau Basin, and parts of the Philippine Sea are examples (Fig. 8–20). Under certain conditions, which are poorly understood, the lithosphere above a subduction zone and within the volcanic arc is stretched. As the lithosphere stretches, a process similar to sea-floor spreading forms new oceanic lithosphere. Sometimes magnetic anomaly patterns like those formed at oceanic spreading axes are created. Where an island arc has extended to form a marginal sea, all lithosphere is oceanic. Where an Andean-type volcanic arc has extended and oceanic lithosphere developed, an arc may form that superficially resembles an island arc, but the volcanic arc is built on continental lithosphere. Japan is an example of a volcanic arc formed in this way.

Not all marginal seas are formed by spreading related to subduction. Some are thought to have developed by the initiation of subduction within oceanic lithosphere, trapping older oceanic crust behind a younger island arc. The Bering Sea north of the Aleutian arc contains magnetic anomalies older than the island arc. The magnetic anomalies are oriented such that they are not compatible with spreading behind the arc.

The spreading process in marginal seas is associated with rising hot upper mantle, a process similar if not identical to that at oceanic spreading axes. Spreading in marginal seas is associated with higher than normal thermal gradients and shallow oceanic sea floor, and as the oceanic lithosphere in marginal seas ages, it cools and subsides, as does oceanic lithosphere at true oceanic spreading axes. Because of the high thermal gradient associated with spreading, young marginal seas are areas of thin lithosphere.

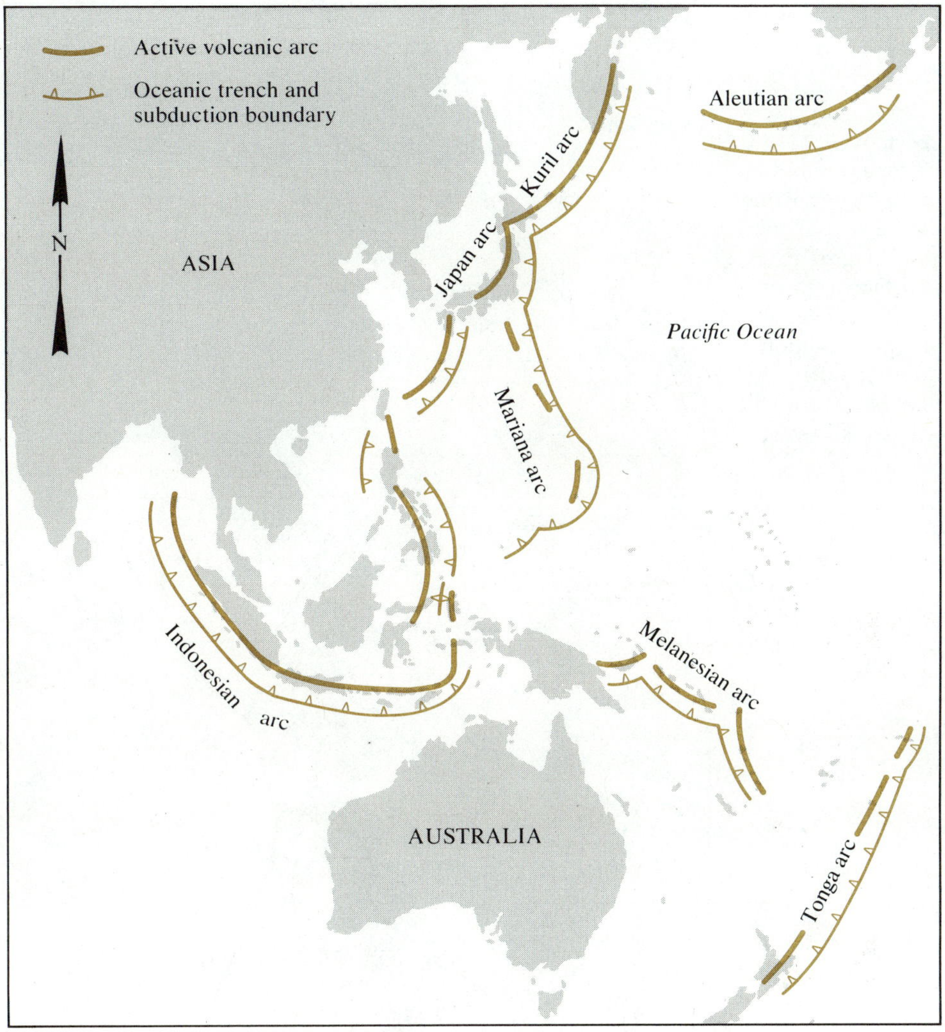

FIGURE 8–17
Generalized map of the western Pacific Ocean region showing active island arc volcanic areas and oceanic trenches that mark the surface expression of active subduction zones. (Map by Robert W. Tope)

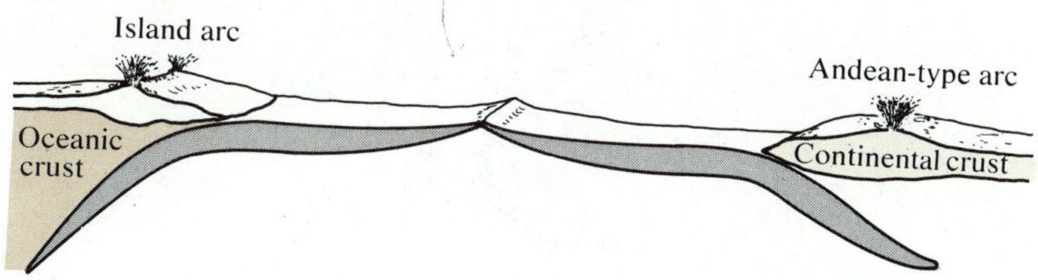

FIGURE 8–18
Volcanic arcs may be constructs on either oceanic or continental crust. Volcanic arcs built on oceanic crust are commonly called island arcs, and those built on continental crust are commonly called Andean-type arcs.

Not all regions above subducted lithosphere are under extension. Where the direction of subduction is oblique to the plate boundary, zones of faults where the motion of the rocks on either side is horizontal and parallel to the fault (called a **strike-slip fault**) may be present. In Sumatra, a large fault zone cuts through the island and is parallel to the volcanic arc. Subduction in the trench to the southwest has a more northerly direction than the northwest-trending plate boundary. The fault has a relative sense of displacement consistent with north-directed subduction, so the rocks on the west side of the fault have moved northwest (Fig. 8–21).

Some areas above subduction zones are in compression. In parts of the Peruvian Andes above the subducted Nazca plate, numerous earthquakes occur within the South American plate. They extend 700–800 km east of the Peru-Chile subduction zone. Analysis of these earthquakes has indicated that the western edge of the South American plate is under

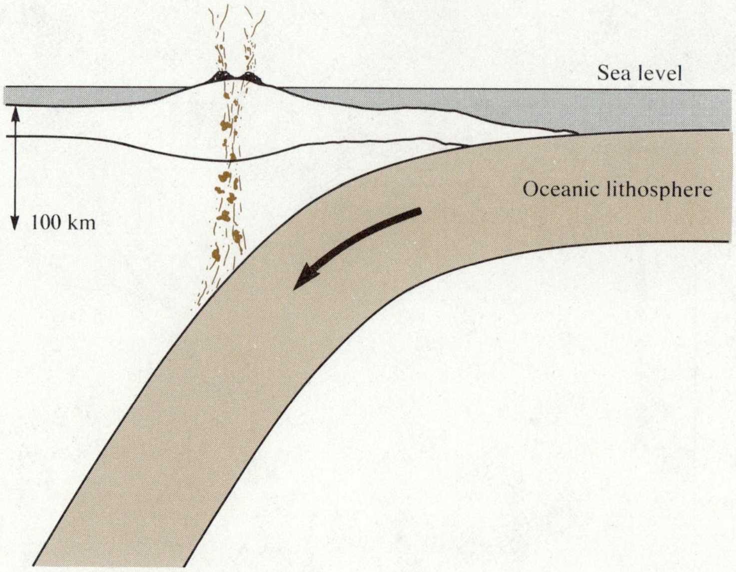

FIGURE 8–19
Volcanic arcs above subduction zones generally occur above the region where the Benioff zone is at about 100 km depth. Magma that rises to the surface to form the volcanic arc may be generated along the Benioff zone and modified during its ascent, or may be generated and modified at shallower levels.

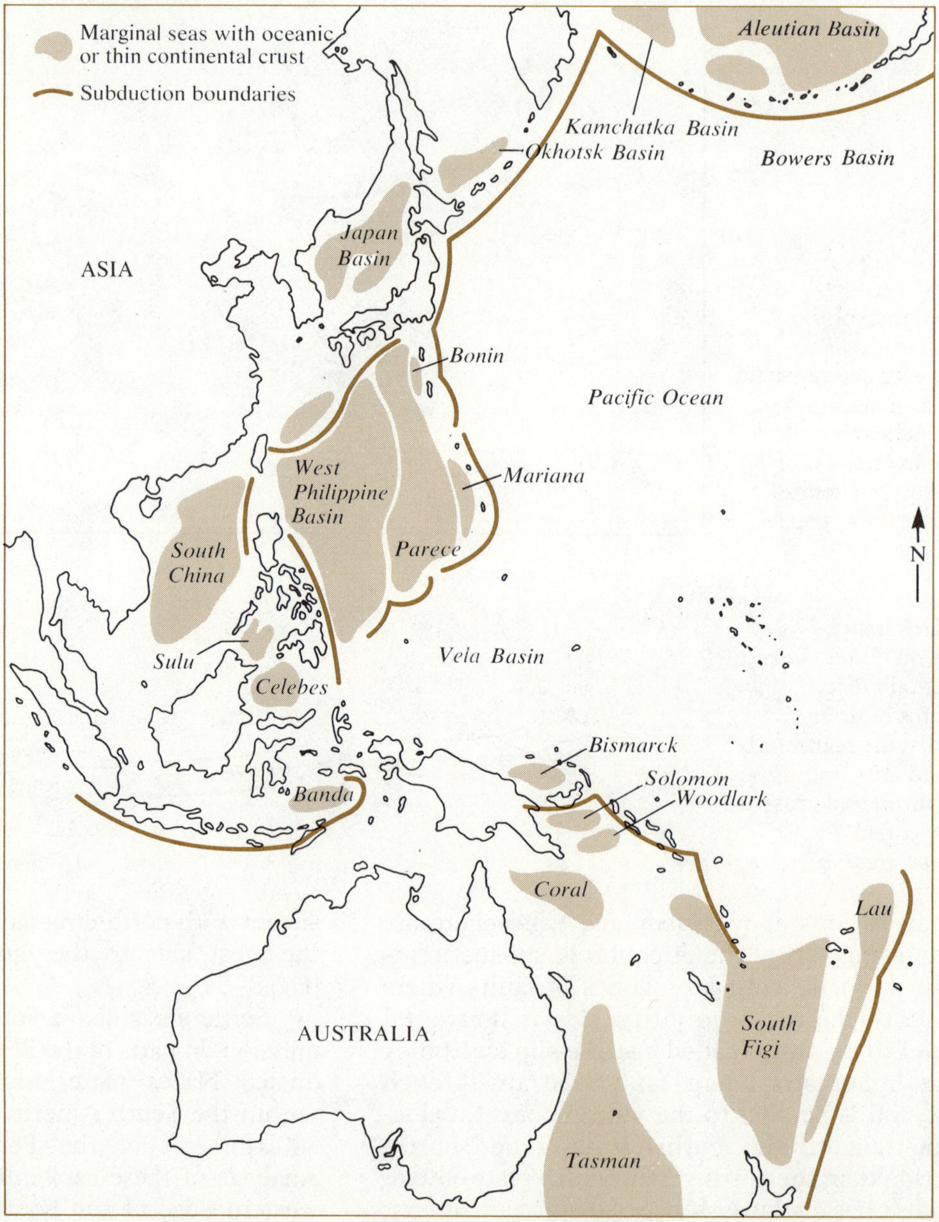

FIGURE 8–20
Map showing the location of the major marginal seas of the western Pacific region. (Map by Robert W. Tope)

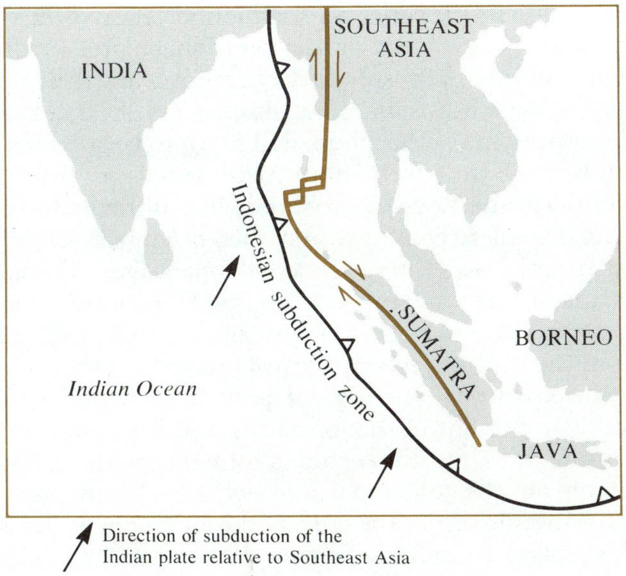

FIGURE 8–21
Oblique subduction of the Indian plate at the Java trench is associated with large strike-slip faults in Sumatra and Burma within the overriding plate. (Map by Robert W. Tope)

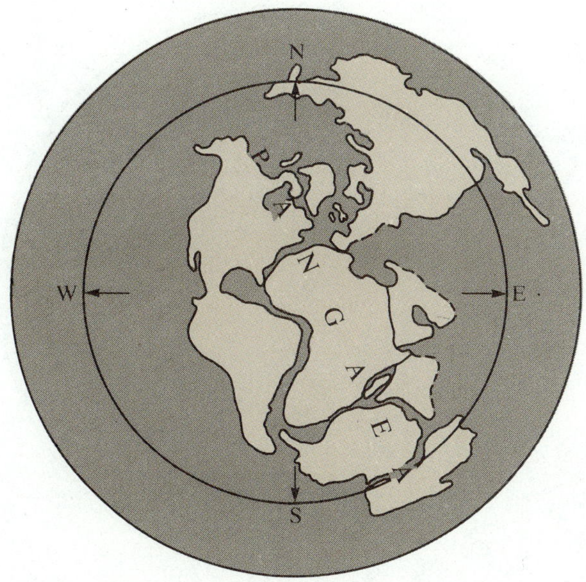

FIGURE 8–22
Continental reconstruction for about 200 million years ago showing one major continent, called Pangaea.

horizontal compression and that the rocks are being shortened. As yet, it is not clear why some subduction zones are overlain by regions of extension whereas others are overlain by regions of compression.

THE MODERN EPISODE OF PLATE TECTONICS

The oldest oceanic lithosphere presently observed in the oceans is approximately 180 million years old. Because all oceanic lithosphere is formed at spreading axes, or in marginal sea spreading, we can reverse this process and move the oceanic "conveyor belt" in reverse. Anomalies of the same age on opposite sides of the spreading axes can be brought together along transform faults. As different-aged anomalies are brought together, a picture of the earth develops at different times. By reversing spreading activity back to about 200 million years, a view of the earth at the beginning of the modern episode of plate tectonics is obtained.

Tectonic reconstruction shows that 200 million years ago, one major continental area—**Pangaea**—existed, and was surrounded by one large ocean (Fig. 8–22). Using the anomaly patterns to reconstruct the breakup of Pangaea into the present system of continents and oceans, the following major events can be determined. The Atlantic Ocean began to open in several stages, the oldest being the central segment east of the United States, which began about 180 million years ago, followed by the South Atlantic at 120 million years ago (Fig. 8–23), then the North Atlantic in a complex series of events about 60 million years ago (Fig. 8–24). Northward movement of India and Australia began by continental breakup and sea-floor spreading about 120 million years and 50 million years ago, respectively, as the present Indian Ocean began to form. At about 50 million years ago, India and Australia became parts of a single plate, the Indian plate. It is clear that as the Indian plate moved northward and away from the spreading axis in the south, older ocean floor must have been subducted north of this plate. In fact, comparing continental positions at 180 million years ago with those now, a major oceanic terrane must have wholly disappeared in the region between the eastern Mediterranean and eastern India (Fig. 8–25). A consequence of subduction of this ocean has been the collision of continental fragments such as India, resulting in mountain building in the Himalayan mountain chain, a process that will be examined in Chapter 9.

As the Atlantic, Indian, and southern Pacific oceans opened, subduction closed the ocean south of Asia and reduced the area of the region now occupied by the Pacific. This is a direct consequence of sea-floor spreading if no significant expansion of the earth has occurred. It requires that thousands of kilometers of oceanic lithosphere have been subducted in the Pacific. An expression of this subduction is the widespread active and ancient volcanic belts that rim the Pacific in what has been termed the **ring of fire**.

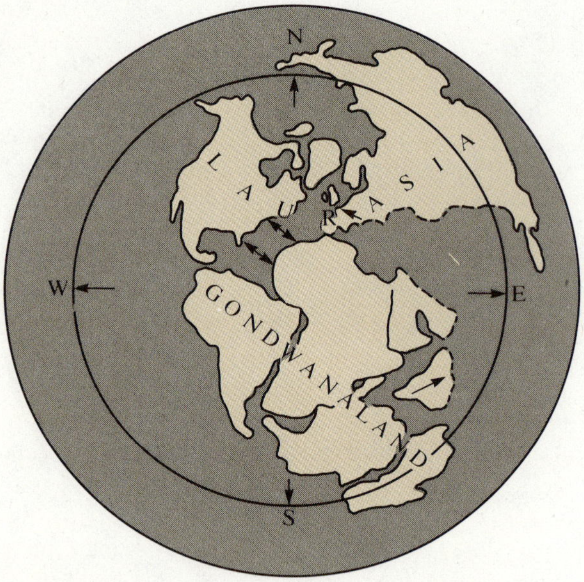

FIGURE 8–23
Continental reconstruction for about 135 million years ago. Continental separation has led to the formation of two large landmasses called Laurasia and Gondwanaland. Reconstruction is just prior to the opening of the South Atlantic at about 120 million years ago.

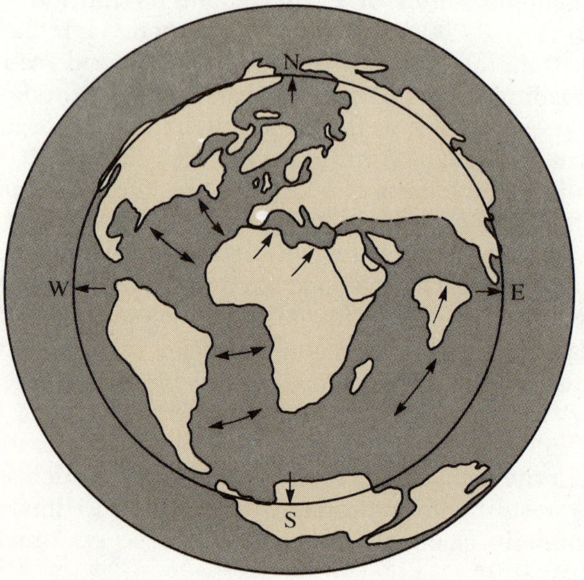

FIGURE 8–24
Continental configurations at about 65 million years ago. Reconstruction is just prior to opening of the North Atlantic at about 60 million years ago.

Another consequence of plate tectonics is that collisions between continents, or between continents and island arcs, are inevitable. Most lithospheric plates contain both oceanic and continental crust. The subduction of oceanic lithosphere is a stable process because oceanic lithosphere is denser than the mantle in the low-velocity zone. As subduction proceeds, a volcanic arc is developed on the overriding plate. Subduction of oceanic lithosphere may continue until all adjacent oceanic lithosphere has disappeared. At that time, the continental lithosphere will begin to be subducted, but because the continental crust is less dense than the mantle in the low-velocity zone, it is difficult to subduct. The plate boundary may then become inactive or other processes may become operative. Through events of this type, two continents, or a continent and an island arc, may be brought into contact. Evidence from the geologic record suggests that this process has occurred throughout a great part of geologic history and is important in the formation and development of continents. Thus, the record of former oceanic terranes remains in the collisional zones within continents.

Most deformation within the lithosphere occurs along plate boundaries which are the loci of modern deformational activity. Modern plate boundaries as well as geologic studies into ancient plate boundary activity indicate that deformation may extend more than 1000 km from a boundary. In collisional settings, deformation may extend into both plates, and in a noncollisional setting, under certain conditions, deformation can take place, but only in the overriding plate, such as in the modern Andes. Transform and spreading boundaries can also affect the lithosphere for great distances from the boundary, principally where one or both plates contain continental crust.

By studying recent plate tectonic activity, we can learn about tectonic processes that are presently operative and about the geologic consequences of plate tectonics. These studies can be extended back in time

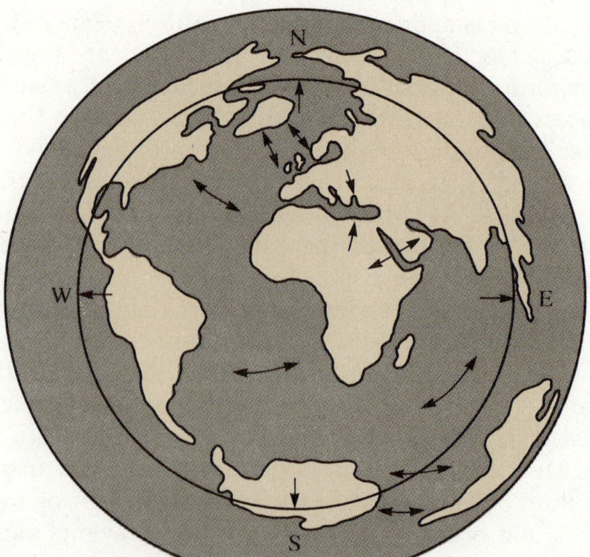

FIGURE 8–25
Present configuration of continents and lithospheric plates.

180 million years, which is as far as the magnetic anomalies present in the modern oceans permit. From these studies, we learn the rules whereby we can make plate tectonic reconstructions and recognize how the geologic evolution of an area can be related to plate tectonics prior to 180 million years ago.

SUMMARY

Ocean-basin morphology is generally simple, consisting of a broad mid-ocean ridge flanked by flat, deep abyssal plains, except in the northern Pacific Ocean, which lacks a mid-ocean ridge. The ocean basins are dotted with numerous volcanic peaks and locally contain linear volcanic and nonvolcanic ridges and plateau regions. Parts of some oceans contain arcuate linear volcanic chains called island arcs, behind which are small semienclosed ocean basins called marginal seas. Adjacent to island arcs and some continents are narrow, deep oceanic trenches more than 11 km deep. The continental margin is the transition zone between the oceans and the continents and consists of a continental rise, slope, and shelf.

Concepts that led to the formulation of sea-floor spreading and plate tectonics were developed largely from studies of the oceans. Geophysical studies have shown that magnetic anomalies present in the oceans can be explained by sea-floor spreading. Oceanic crust diverges at oceanic ridges, and molten rock is intruded at the spreading axis. As the molten rock crystallizes, it forms new oceanic crust. Once the minerals cool below their Curie temperature, they become magnetized in the direction of the earth field at that time. The ocean crust continues to spread, splitting the bodies of magnetized rocks and causing them to be symmetrically disposed about the spreading axis. Reversals of the earth's magnetic field are thus recorded on the sea floor as magnetic anomalies.

In other areas, the earth's crust slides past itself horizontally along strike-slip faults called transform faults. At deep-ocean trenches, earthquake activity along the Benioff zone reaches to depths of nearly 700 km. Benioff zones mark the location of subduction of oceanic crust and part of its upper mantle. Most of the world's earthquakes occur at spreading axes, transform faults, and subduction zones, and these three features form an interconnecting network outlining a first-order system of plates. From these types of data, the theory of plate tectonics was developed: oceanic crust is created at spreading axes, slides horizontally past itself along transform faults, and plunges back into the mantle at subduction zones. The base of the moving plates is within the upper mantle at a depth of about 100 km beneath the oceans, perhaps deeper beneath the continents. Rocks above this depth are thought to be more rigid and are called the lithosphere. Rocks below this depth are more ductile, flow in slow convective currents, and comprise the asthenosphere.

Plates that lie above subduction zones commonly contain long linear belts of volcanic rocks called volcanic or magmatic arcs. They lie 100–200 km from the surface trace of the subduction zone and are thought to be related to the melting of the downgoing lithosphere or lithosphere that overlies the subducted slab. The region behind the volcanic arc may be subjected to spreading and form marginal seas, whereas in other regions the rocks may be under compression and form linear deformed belts.

Some persistent volcanism is present at local centers of magmatic activity called hot spots, both within plates and at the boundaries of plates. The hot spots are thought to lie above localized centers of upwelling called plumes. Where plumes lie below moving lithospheric plates, they cause linear volcanic belts, or plume tracks, which become progressively younger in one direction.

The present oceanic crust is no older than about 180 million years. A record of the plate motions over the past 180 million years is contained in this ocean floor, and from this record plate motions can be reconstructed for the past 5 percent of earth history. These reconstructions confirm the early concepts of Wegener, Holmes, and Hess that the continents have moved relative to one another.

REVIEW EXERCISES

1 What is the difference between the continental shelf, slope, and rise?

2 What are the abyssal plains?

3 How did the discovery of reversals of the earth's magnetic field and the magnetic anomaly patterns on the ocean floor lead to the discovery of sea-floor spreading?

4 How can the rate of sea-floor spreading be calculated?

5 What is a transform fault?

6 What are the three types of plate boundaries?

7 Describe the process of subduction.

8 What is the difference between the lithosphere and the asthenosphere?

9 How are plates defined?

10 Describe the fundamental concept of plate tectonics.

11 Describe mantle convection.

12 What is a hot spot?

13 What are plume tracks?

14 What major geologic features are related to subduction?

SUGGESTED READINGS

BURKE, C. A., and DRAKE, C. L., eds. *The Geology of Continental Margins.* New York: Springer-Verlag, 1974, 1009 pp.

BURKE, K. C., and WILSON, J. T. "Hot Spots on the Earth's Surface." *Scientific American* (1976), Vol. 235, No. 2, pp. 46–57.

DEWEY, J. "Plate Tectonics." *Scientific American* (1972), Vol. 226, No. 5, pp. 56–68.

HAYES, D. E., ed. *The Tectonic and Geologic Evolution of Southeast Asian Seas and Islands.* Washington, D.C.: American Geophysical Union, 1980, 326 pp.

LE PICHON, X.; FRANCHETEAU, J.; and BONNEN, J. *Plate Tectonics.* Amsterdam: Elsevier Scientific Publishing Co., 1973, 300 pp.

MATHEWS, S. W. "This Changing Earth." *National Geographic* (1973), Vol. 143, No. 1, pp. 1–37.

MCELHINNY, M. W. *Paleomagnetism and Plate Tectonics.* Cambridge: At the University Press, 1973, 358 pp.

SCLATER, J., and TAPSCOTT, C. "The History of the Atlantic Ocean." *Scientific American* (1979), Vol. 240, No. 6, pp. 156–65.

TOKSÖZ, M. N.; VYEDA, S.; and FRANCHETEAU, J., eds. *Oceanic Ridges and Arcs: Geodynamic Processes.* Amsterdam: Elsevier Scientific Publishing Co., 1980, 538 pp.

WILSON, J. T., compiler. *Continents Adrift and Continents Aground.* San Francisco: W. H. Freeman and Co., 1976, 230 pp.

9

- oceanic crust
- continental crust
- sheeted dike complex
- ophiolite
- body force
- transmitted force
- stress
- normal stress
- shear stress
- principal stresses
- strain
- shear strain
- brittle mode
- normal fault
- thrust fault
- strike-slip fault
- oblique-slip fault
- ductile mode
- confining pressure
- fold
- hinge
- hinge surface
- anticline
- syncline
- overturned fold
- recumbent fold
- flexural-slip fold
- décollement zone
- passive fold
- deformed belt
- mountain belt
- noncollisional deformed belt
- foreland fold and thrust belt
- collisional deformed belt

The Earth's Crust and Mountain Building

THE EARTH'S CRUST IS DIVIDED INTO OCEANIC AND CONTINENTAL crust and forms the outer part of the lithosphere. Oceanic and continental crust are compositionally and structurally different. These differences are manifest in the topography of the earth's surface, where oceanic crust stands at an average of 5 km below sea level and continental crust stands at an average elevation of 0.5 km above sea level. Oceanic crust beneath the oceans is not known to be older than Jurassic (approximately 180 million years old), whereas continental crust contains rocks as old as 3.9 billion years, and most of its rocks are older than 180 million years. Oceanic crust is formed at a spreading axis and has a relatively simple history until it is subducted or involved in complex deformation at a subduction boundary. In contrast, continental crust cannot be subducted easily (or, at best, only a small amount can be subducted); thus, continental crust may contain records of many episodes of deformation. The total amount of continental crust has probably increased with time. In this chapter, the differences between oceanic and continental crust are examined, along with the way in which the crust deforms and forms mountains, and the complex history recorded within continental crust.

OCEANIC CRUST

Changes in seismic velocity with depth suggest that oceanic crust consists of three layers. Drilling in the deep ocean has penetrated the upper layer, layer 1, which consists of approximately 0.5 km of sedimentary rocks, and a few hundred meters into the second layer, layer 2. The sedimentary rocks consist of very-fine-grained material composed of the calcareous and siliceous shells of tiny floating organisms and very fine nonorganic clay-sized material. The upper part of layer 2 consists of basaltic volcanic rocks. Deeper parts of the oceanic crust have not been penetrated by drilling. Dredging on topographic scarps along ocean ridges and fracture zones has yielded a great variety of rocks besides those found in drilling. Serpentine, serpentinized peridotite, gabbro, diabase, and amphibolite have been recovered. The parts of the oceanic crust represented by these rocks cannot be determined from dredging alone.

Within deformed belts in many continental areas are thick sequences of rocks which may represent fragments of oceanic crust caught in mountain-building processes. These rocks have appropriate seismic velocities and thicknesses to match the seismic layering of oceanic crust. They consist of 0.5 km of fine-grained sediments overlying basaltic volcanic rocks similar to those found in deep-sea drilling. The basalts pass downward into volcanic rocks cut by numerous basaltic dikes that were subvertical when emplaced. In some areas, the dikes form the bulk of this part of the sequence and are referred to as a **sheeted dike complex**. The volcanic rocks and sheeted dikes are approximately 1–2 km thick and are thought to be correlative with oceanic layer 2. Below the sheeted dikes are 2–4 km of gabbro, commonly layered, and a basal part consisting of layered ultramafic rocks such as peridotite and dunite. These rocks are thought to correlate with oceanic layer 3. Rocks correlated with layers 2 and 3 have textures that indicate they crystallized from a magma, whereas ultramafic rocks that lie farther below in the sequence have textures and structures that indicate they were never molten, but deformed as ductile crystalline solids. The deformed ultramafic rocks are correlated with the upper mantle, and the Moho would be represented by the basal part of the gabbro layer.

The sequence of rocks just described is termed an **ophiolite**, and when the full sequence is present, the ophiolite is interpreted as a fragment of oceanic crust. Studies of ophiolites from many different mountain ranges around the world have shown that although the sequence of rock types is almost always present, the thicknesses of the various layers is variable. This variation in layer thickness can be interpreted to mean that oceanic crustal layers have more variable thickness than commonly assumed from seismic data, or that not all ophiolites are fragments of deep oceanic crust. Some ophiolites may represent oceanic crust from marginal seas or modified oceanic crust beneath island arcs. Considerable study of each ophiolite and its associated rocks and structural setting may be necessary before a correct paleogeographic interpretation can be made.

Most oceanic crust is thought to be generated at a spreading axis. From drilling, geophysical studies along spreading axes, deep-sea observations in scientifically equipped submarines, and onland studies of spreading axes and ophiolites, the following genesis of oceanic crust can be inferred. Upwelling hot mantle beneath a spreading axis is partially melted as the confining pressure decreases. This partially melted mantle has a low velocity and thus forms a shallow low-velocity zone and asthenosphere at a depth of only a few kilometers below the spreading axis. Basaltic magma rises through fractures in the overlying thin oceanic crust to be extruded onto the sea floor as lavas. The lava flows may build up to a thickness of a kilometer or more. Magma that cools within fractures before it reaches the surface forms dikes cutting the lower parts of the lava flows. Because the spreading axis is an area of divergence between two plates, the space occupied by the magma within the dikes balances the amount of divergent movement of the oceanic crust. Whether the divergent movement of oceanic crust is caused by the forceful shouldering aside by the intruding magma, or the magma passively intrudes into the zone of separation caused by divergent motion of the oceanic crust, is not clear. As spreading continues, oceanic crust is accreted to both plates as they move away from the spreading axis. Chambers of basaltic magma below the spreading axis crystallize minerals which settle to the bottom of the chamber, first as layered ultramafic rocks and then as layered gabbro. Some gabbro is formed at the top of the chamber as the oceanic crust above cools. As cooling continues, the magma chamber becomes solidified by both crystallization at its top and crystal settling at its base. As further cooling takes place, the ductile, but nonmolten mantle of the asthenosphere cools, and as its ductility is reduced, it becomes added as a gradually thickening layer to the oceanic crust and forms the base of the lithosphere (Fig. 9–1).

Sedimentation on the oceanic crust begins as soon as the crust is formed. Thus, oceanic crust can be dated by the age of the oldest sedimentary rocks that rest on the basalt. The age of these sedimentary rocks also dates the magnetic anomalies that formed at that time.

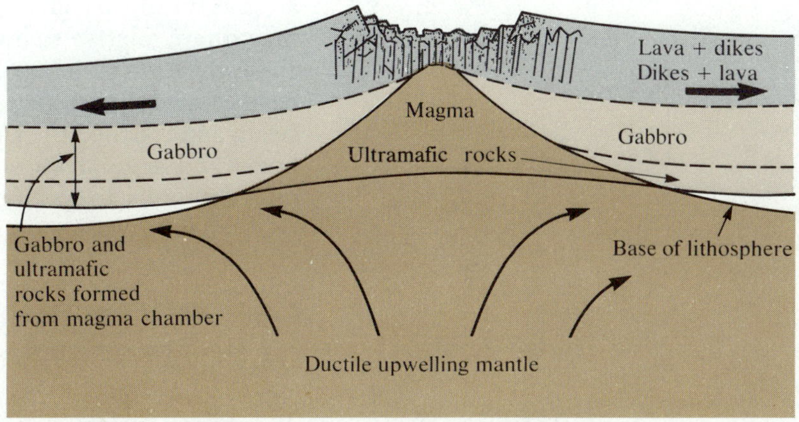

FIGURE 9–1
Oceanic crust is formed at spreading ridges above upwelling mantle. Magma partially melted from the mantle extrudes as mainly basaltic lava at the surface. The lava is intruded by dikes at depth. Crystallization of magma beneath these rocks forms gabbro and ultramafic rocks which rest on deformed mantle rocks that were never molten. Oceanic crust is formed progressively as plates separate at a spreading axis.

CONTINENTAL CRUST

The formation, composition, and evolution of continental crust is much more complicated than that of oceanic crust. Whereas oceanic crust is largely a transitory material on the face of the earth, often wholly subducted not long after its formation, continental crust is not readily subducted, and from the time of its formation its history is recorded in the material added to it and modified by subsequent events. Geologic mapping has demonstrated that continental crust at the earth's surface consists of deformed rocks that lie in belts generally 1000 km or more long and 100–1000 km wide. In many places, these deformed rocks have been eroded and are overlain by the flat-lying sedimentary rocks, indicating continental stability since their deposition. In some places, these flat-lying rocks are more than 1.5 billion years old.

To understand some of the structural and compositional features of continental crust, a discussion of mountain belts and deformed belts is necessary. At the end of this chapter, we will return to a general discussion of the nature of continental crust.

ROCK DEFORMATION

Forces are generated within the lithosphere largely because of plate motion and plate interaction. Where these forces become large enough, rocks of the lithosphere deform by one of two modes, **brittle** or **ductile**. The physical environment, which includes the temperature, pressure, strain rate (rate of deformation), and other factors, governs whether a rock will deform in a brittle or ductile mode. Rocks exposed at the earth's surface have been deformed in both brittle and ductile modes. It is necessary to understand how rocks deform in order to develop concepts of how to interpret geologic structures, mountain building, and their relations to plate tectonics.

FORCE AND STRESS

Forces within rock masses are either **body forces** or **transmitted forces**. Body forces act on rock particles throughout the rock uniformly and do not require the force to be transmitted from one point to another. They are present throughout a rock even if no external forces are applied to it. Forces generated by inertia and gravity are body forces. Transmitted forces are forces generated in one part of the rock and transmitted by interaction between the rock particles to some more distant point.

Stress is force per unit area acting across a surface within a rock mass. For example, a 2-kg weight placed on a 1-cm^2 area would exert a stress of 2 kg/cm^2 (Fig. 9–2). Within a rock, a force acting across a planar surface can be resolved into two components, one perpendicular to the surface and one parallel to the surface. When these forces are divided by the area of the surface upon which they act, they define the stress. The stress acting perpendicular to the surface is the **normal stress**, and the stress acting parallel to the surface is the **tangential** or **shear stress**.

Within any stressed body, there are three mutually perpendicular surfaces across which there are only normal stresses. The shear stresses on these surfaces are zero. These three planes are called the **principal planes of stress**, and the three normal stresses are the

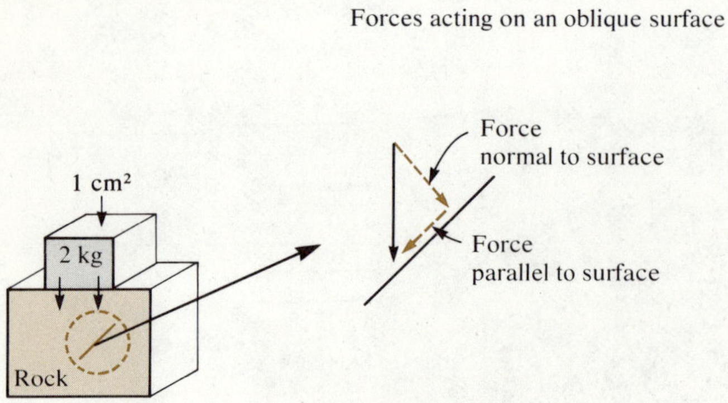

$$\frac{\text{Normal force}}{\text{area}} = \text{normal stress} = \sigma$$

$$\frac{\text{Parallel force}}{\text{area}} = \text{shear stress} = \tau$$

FIGURE 9–2
Within a rock, body forces across any oblique surface can be resolved into a normal and a parallel component. Normal and shear stresses are determined by dividing the normal and parallel components of the forces by the area of the surface on which the forces act.

principal stresses. These three principal stresses are the **maximum**, **minimum**, and **intermediate** principal stresses (Fig. 9–3). The normal stress across any plane within a stressed body cannot exceed the maximum principal stress, or be less than the minimum principal stress (Fig. 9–3). Compressive stresses cause particles within the rocks to move closer together, whereas tensile stresses cause particles within the rocks to move farther apart.

Strain is a measure of the amount of deformation a material has undergone. Strain can be measured by the ratio of an initial length of a line to its final length, or the change in the length of a line to its initial length (Fig. 9–4). This can also be expressed as a percentage change in the length of a line. The ratios or percentages are the linear measure of the strain. In addition, lines may rotate relative to one another. The tangent of the angle through which two lines originally at 90° apart have rotated is a measure of the angular component of strain and is called the **shear strain** (Fig. 9–4). Both the linear and angular measures of strain must be determined in order to specify the total strain.

EXPERIMENTAL ROCK DEFORMATION

An understanding of how rocks deform must come from experimental laboratory studies, because rocks deform so slowly that, in general, they cannot be studied as they deform naturally. For example, our normal experience with rocks is that they are hard, brittle materials; however, folds can be seen within rocks that indicate that these rocks deformed as ductile materials. Hot tar deforms in a ductile mode in a period of time short enough that the deformation can be seen taking place. Ice in glaciers also deforms in a ductile mode over a period of time that can be observed, but over a much longer period of time than the hot tar. Rocks deform as ductile materials over time periods so large—of the order of 100,000–10 million years—that the deformation cannot be observed. Laboratory studies attempt to get around this problem by changing the variables of the deformation environment, such as temperature, so that rocks can deform as ductile materials in shorter periods of time.

Rocks tested experimentally are usually cut into cylinders and deformed by driving a piston down on

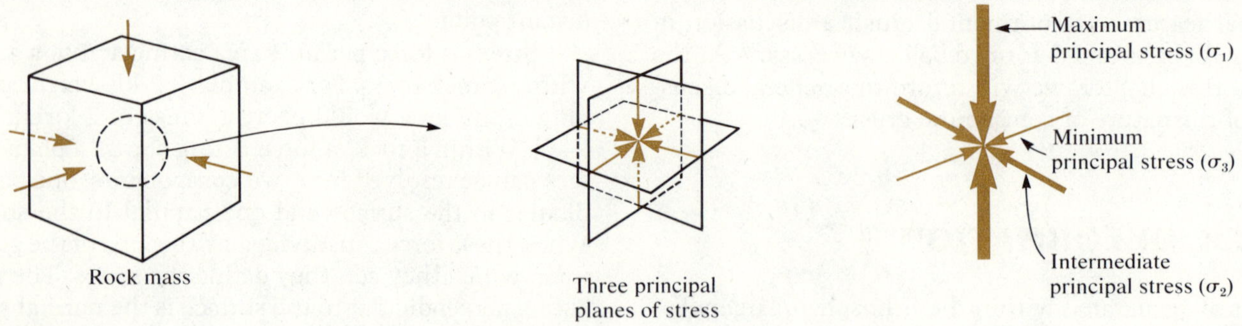

FIGURE 9–3
Within any stressed body of rock, there are three mutually perpendicular planes that have only normal stresses. These three normal stresses are called the principal stresses. One of them reaches the maximum value for any normal stress in the rock (σ_1, the maximum principal stress), and another reaches the minimum value for any normal stress in the rock (σ_3, the minimum principal stress). The third has an intermediate value (σ_2, the intermediate principal stress).

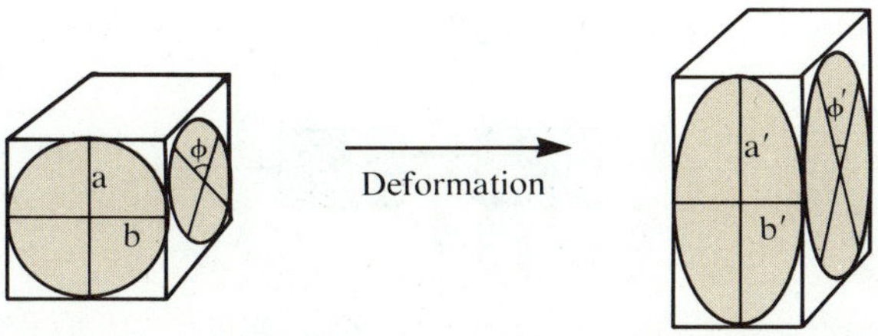

$$\text{Linear strain for line a} = \frac{a' - a}{a}$$

$$\text{Linear strain for line b} = \frac{b' - b}{b}$$

$$\text{Shear strain} = \tan(\phi - \phi')$$

FIGURE 9–4
Strain can be measured by the changes in the lengths of lines and the angular change between lines. One measure of the linear change in a line length is to compare the change in length to the original length. The angular change between lines is commonly measured as the tangent of the angle through which two lines originally at 90° have rotated.

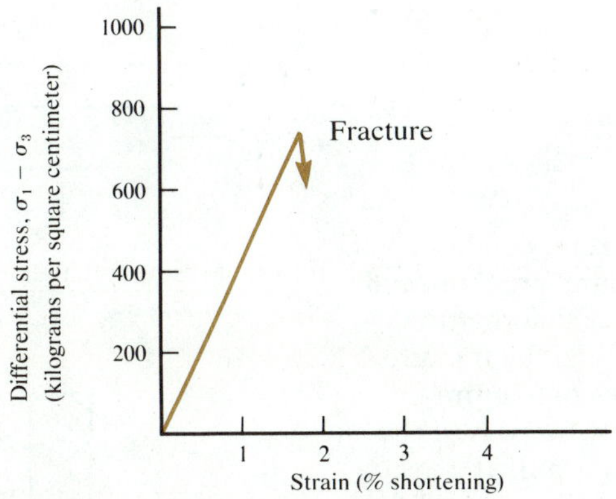

FIGURE 9–5
As rocks are subjected to stress, the first 1–2 percent strain shows a linear increase with increasing stress. If the stress is removed before fracture occurs, all the strain is recoverable. This type of behavior is called elastic. If the rock deforms only elastically then fractures with increasing stress, the rock behaves in a brittle mode.

one end of the cylinder (a **compression test**) or by pulling on the ends of the cylinder (an **extension test**). Both types of tests generate stresses within the material, and changes in relative distance between particles within the rock that can be measured as strain. By placing the cylinder of rock within a pressure chamber and surrounding it with a small furnace, the pressure surrounding the cylinder can be increased. This simulates increased depth in the earth. Increased temperatures are also used to simulate conditions deeper in the earth. The one variable in the physical environment that cannot be reproduced in the laboratory is time, because experiments lasting 100,000 years or longer are not feasible to conduct. Experimentalists attempt to evaluate this problem by varying the strain rate, that is, the speed at which the piston compresses or extends the sample. In this way, the effect of varying the time of a deformation from a few seconds to several days is evaluated, and the results are extrapolated to geologic time of 100,000–10 million years. The extrapolation is indeed enormous, but it is the only way the effects of strain rate can be studied.

BRITTLE MODE

In their initial stages of deformation, where the strain is only 1–2 percent, all rocks exhibit an elastic behavior. When the stress is removed, all the strain is recovered (Fig. 9–5). If a rock fractures after some elastic behavior, the rock is termed *brittle.* Earthquakes are thought to result from a slow buildup of elastic strain which is released in part by fracture and movement along faults, and by the generation of elastic waves.

Most faults are the geologic expression of brittle deformation. Experimental studies of brittle fracture have shown that two types of fractures are produced in stressed rocks and occur when the difference between the maximum and minimum principal stresses exceeds the fracture strength of the rock. **Tension fractures** form perpendicular to the minimum principal stress and lie in a principal plane of stress (Fig. 9–6A). Motion of material on both sides of the fracture is perpendicular to the fracture surface, so the distance between the two sides increases. **Shear fractures** commonly make angles of approximately 30° to the maximum principal stress, and the intermediate principal stress lies within the fracture plane. Motion of material on both sides of the fracture is parallel to the fracture surface (Fig. 9–6B). Because of the symmetry of the stress field, two fracture surfaces are possible. Applying these data to faults observed in the earth, the direction of motion along the fault surface can be used to determine the orientation of principal stresses.

The three simplest orientations of the principal stresses occur where one of the three stresses is vertical (Fig. 9–7). Where the maximum principal stress is vertical, faults are formed that dip at about 60° from the horizontal. Rocks above the fault, termed the **hanging wall**, move down relative to the rocks below the fault, termed the **footwall**. Faults of this type are classified as **normal faults**. Rocks cut by normal faults are thinned vertically and lengthened horizontally. Where the minimum principal stress is vertical, faults that form dip approximately 30° from the horizontal, and the hanging wall moves upward relative to the foot-

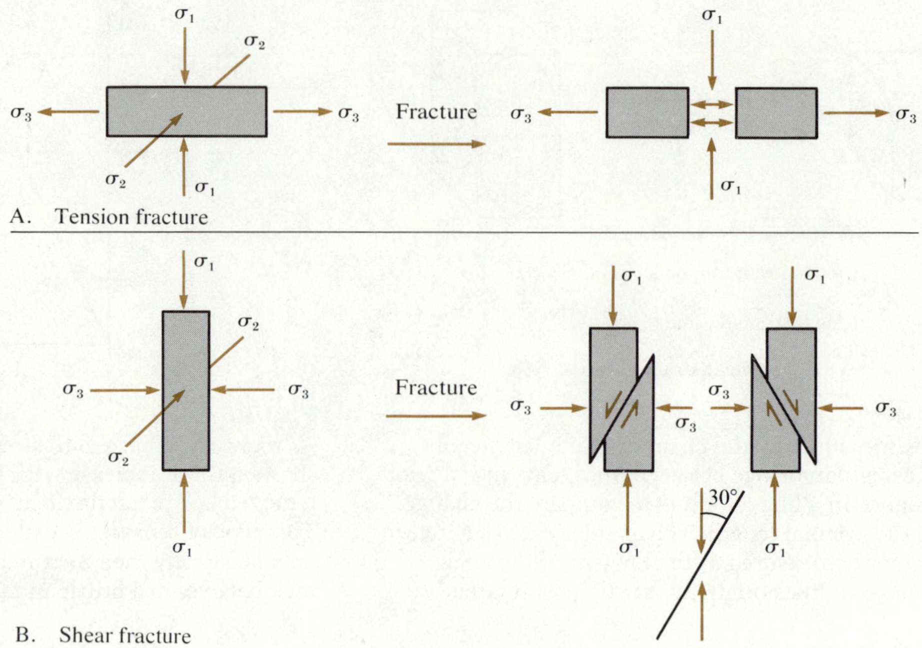

FIGURE 9–6
Under certain conditions of temperature, pressure, and duration of deformation, rocks will fail by fracture. Two types of fractures commonly form. A. If the specimen is pulled, tension fractures form perpendicular to the minimum principal stress. B. If the specimen is pushed or compressed, a shear fracture forms that makes an angle of about 30° to the maximum principal stress and contains the intermediate stress axis.

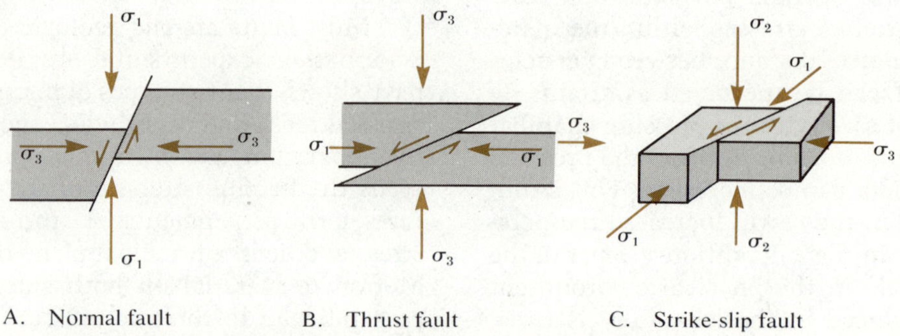

FIGURE 9–7
Fracturing of rocks in the earth produces faults that are classified according to the orientation of the fault and the type of relative motion that occurs between the blocks on either side of the fault. Where the fault surface dips, the block above the fault is called the hanging wall, the block below, the footwall. Where the hanging wall has moved down, the fault is called a normal fault (A). Where the hanging wall moves up, it is called a thrust fault (B). Faults with horizontal motion are called strike-slip faults (C).

wall. These faults are classified as **thrust faults**. Rocks cut by thrust faults are thickened vertically and shortened horizontally. Where the intermediate principal stress is vertical, faults dip vertically, and the relative motion of the two sides of the fault is horizontal. These faults are called **strike-slip faults**. Strike-slip faults neither thicken nor thin the rocks vertically, but cause only horizontal movement of the rock bodies. Orientations where none of the principal stress axes is vertical can produce faults of any possible orientation and direction of relative motion. These faults are termed **oblique-slip faults** and may cause varying degrees of thickening, thinning, or horizontal motion.

DUCTILE MODE

Under certain conditions, rocks undergo large magnitudes of permanent strain without fracture, and behave in a ductile mode. Experimental rock deformation studies have defined the conditions necessary for ductile deformation. This work gives insight into how rocks deform as ductile materials in the earth.

Conditions of increased pressure and temperature, and decreased strain rate, favor ductile deformation in rocks. In the experimental apparatus, the rock sample is surrounded by a fluid or solid whose pressure can be controlled (Fig. 9–8). Increased pressure on the mate-

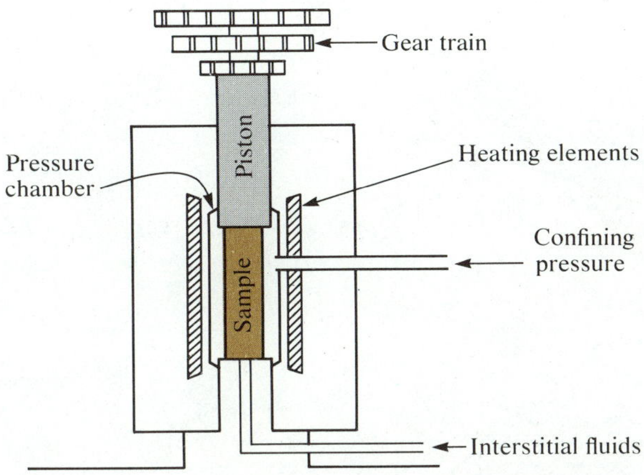

FIGURE 9–8
To examine how rocks deform under the variable conditions in the earth, laboratory equipment is designed to try and reproduce these conditions. The sample is placed in a pressure chamber which can simulate confining pressure. It is surrounded by a furnace to test the effect of temperature. A gear train drives the piston at different speeds to evaluate the effects of time. Interstitial fluids may be pumped into the sample independently from the pressure vessel to test their effects on rock behavior.

rial surrounding the sample simulates the pressure a piece of rock in the earth's crust experiences from the surrounding rocks, and is called the **confining pressure**. In general, confining pressure and temperature increase with depth in the earth. Increase in temperature is simulated in the apparatus by a small furnace surrounding the sample. By changing the speed with which a rock sample is shortened or lengthened, the sensitivity of deformation to strain rate, that is, changes in the length of time over which the deformation takes place, can be evaluated. Because the length of geologic time cannot be reproduced in the laboratory, the data from experiments run over time periods of from seconds to weeks must be extrapolated to geologic lengths of time.

Experimental studies have shown that increased confining pressure permits rocks to deform in a ductile mode. Increase in temperature increases the ductility of rocks and decreases the stresses necessary to produce ductile strain. Decreasing the strain rate, that is, increasing the length of time over which the deformation occurs, produces the same result as increase in temperature. Fluids in a rock produce chemical effects that also increase ductility and decrease stresses necessary to cause nonrecoverable strain. Thus, conditions of increased confining pressure and temperature with the presence of small amounts of fluids and very slow geologic strain rates make rocks susceptible to ductile deformation within the crust and mantle.

When rocks deform in a ductile mode, they behave as viscous fluids with very high viscosities, at least to a first approximation. The viscosity, that is, the ability of a material to flow, is measured in poises. Water has a viscosity of about 1 poise, and its fluid behavior is obvious. Honey might have a viscosity of 10^5 poises; that is, it is 100,000 times more viscous than water. Cold tar might have a viscosity as high as 10^{12} poises, and its fluid character is not obvious. Rocks have effective rather than true viscosities (because they are not true viscous fluids) of the order of between 10^{19} and 10^{28} poises. This would mean that, in order to observe their fluid character, they would have to be observed over a period of 100,000–1 million years. Rocks are thus very sticky fluids. Within the realm of our time frame they are only brittle materials, but given the geologic time scale they may behave in a ductile mode and should be regarded as fluids. Silly Putty is a substance whose mode of deformation changes from brittle to ductile with a change of strain rate of only seconds. The same effect would occur in rocks, but change in the strain rate would have to be about 1 million years.

FOLDING

Whereas brittle deformation in rocks produces faults, ductile deformation produces **folds**. During a ductile deformation where strain within the rock is inhomogeneous, that is, where the magnitude or orientation of strain varies from place to place, preexisting surfaces become distorted into folds.

The line that connects all the points of maximum curvature on a folded surface is called the **hinge** (Fig. 9–9). The surface containing the hinges for successive

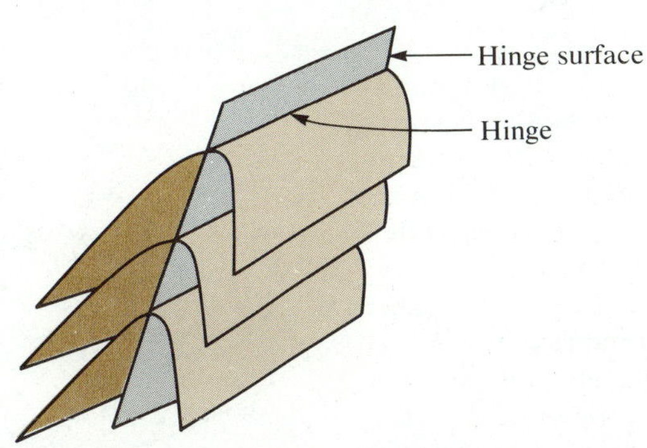

FIGURE 9–9
The line that connects the points of maximum curvature on folded layers is called the hinge of a fold. The surface that connects all the hinges of successive folded layers is called the hinge surface.

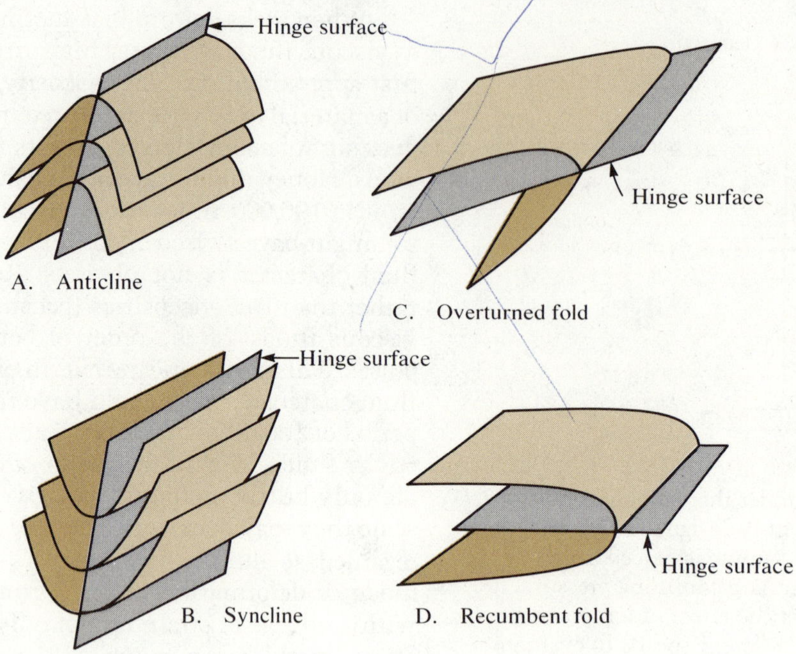

FIGURE 9–10
Some types of folds.

surfaces is called the **hinge surface**. A fold in which the folded surface opens downward is an **anticline** (Fig. 9–10A), and where the folded surface opens upward, the fold is called a **syncline** (Fig. 9–10B). Where the hinge surface is inclined and both sides of the folded surface dip in the same direction, the fold is an **overturned fold** (Fig. 9–10C). If the hinge surface is horizontal, the fold is a **recumbent fold** (Fig. 9–10D).

Folds can be classified in many different ways, but a very simple classification will be adopted here. Two types of folds will be discussed: (1) where the strain is concentrated along the surfaces being folded, called a **flexural-slip fold**; and (2) where the strain has no relation to the surfaces being folded, called a **passive fold**.

FLEXURAL-SLIP FOLDS

In rocks containing preexisting parallel surfaces, such as bedding in sedimentary rocks, there are commonly very marked changes in viscosity between layers of different composition. When subjected to sufficient stresses, rock sequences of this type will form a series of anticlines and synclines. The layer thickness of the more viscous layers, that is, the stiffer layers, often remains constant throughout the fold. Folding of this type where layer thickness remains constant is accompanied by slip or viscous flow between the layers (Fig. 9–11). An analogy is the folding of a telephone book. For the book to be folded, the pages must slip over one another. There is little if any internal defor-

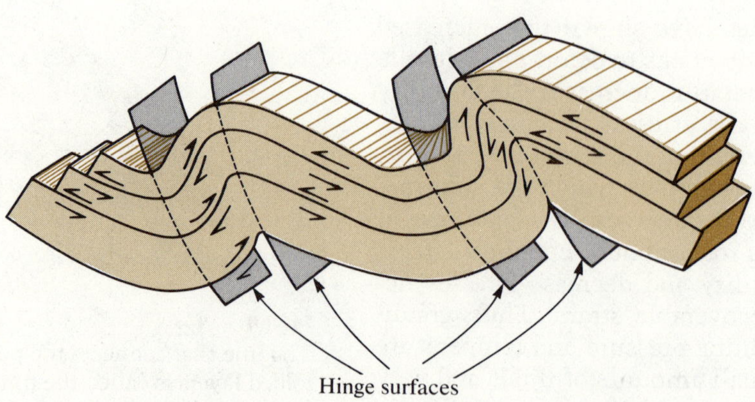

FIGURE 9–11
Flexural-slip folds contain slip or viscous flow between the surfaces of the layers being folded. Notice that the sense of slip reverses itself across the hinge surfaces of each fold. The sense of slip is fixed geometrically, just as it is when a telephone book is flexed. Also notice that for the layer thickness to remain constant, the form of the fold must change along the hinge surface.

mation of the pages, because most of the deformation is taken up by slip between the pages. Because the layers slip over one another as they are flexed, this type of folding is called *flexural-slip folding.*

There are two important consequences of flexural-slip folding:

1. Flexural slip folds require that the layers being folded are shortened in a direction perpendicular to their hinge surfaces.
2. The geometry of flexural-slip folds is that they change form rapidly along the hinge surface.

The first point results from the fact that the layers being folded have little internal strain, and thus the length of the folded layer measured along the arc of the fold is the same before and after folding. To estimate how much shortening has taken place, the folded length of a layer is compared to a planar layer of the same length (Fig. 9–11). The second point results from the fact that to keep layer thickness constant in a folded sequence, layers must be shortened by different amounts. Because all layers begin with approximately equal lengths, for flexural-slip folds to form there are changes in geometry in order to accommodate changes in the amount of layer shortening. Flexural-slip folds commonly become detached from adjacent layers along some surface, called a **décollement zone**, and the folds disappear below the décollement zone (Fig. 9–12).

PASSIVE FOLDS

In some rock sequences, the individual layers have about the same viscosity. Under such conditions, preexisting planar surfaces such as bedding have no mechanical significance, and the lengthening and shortening (i.e., strain) within the rocks crosses the layering as if it were not there. In fact, in folds of this type the only significance of the layering is that it marks the fold form. The folds can be identified because the shape of each layer is identical to the one above and below. This geometry produces layers whose thickness is greatest in the area of maximum curvature and least between regions of opposite curvature (Fig. 9–13). These folds are the result of ductile deformation where the layering is mechanically passive; hence the name *passive folds.*

Passive folds are also the result of shortening in the rock. Strain measurements made on passive folds have indicated that the rocks have been shortened in a direction approximately perpendicular to the hinge surface. The magnitude of the shortening, however, cannot be determined from the fold geometry as is the case for the flexural-slip folds. Shortening must be determined independently by measurements on deformed objects whose original shape is known, such as fossils or pebbles in a conglomerate.

SUMMARY OF ROCK DEFORMATION

The physical environment of rocks during deformation controls the mode of deformation. As the stresses increase, all rocks deform elastically for the first 1–2 percent strain. Under conditions of low confining pressure, low temperature, and high strain rate, rocks tend to fracture and relieve the stresses in a brittle mode. Under conditions of high confining pressure, high temperature, and low strain rate, they deform in a ductile mode by folding. Thus, both faulting and folding are responses to release of stress, but under different physical conditions. There are exceptions to the gen-

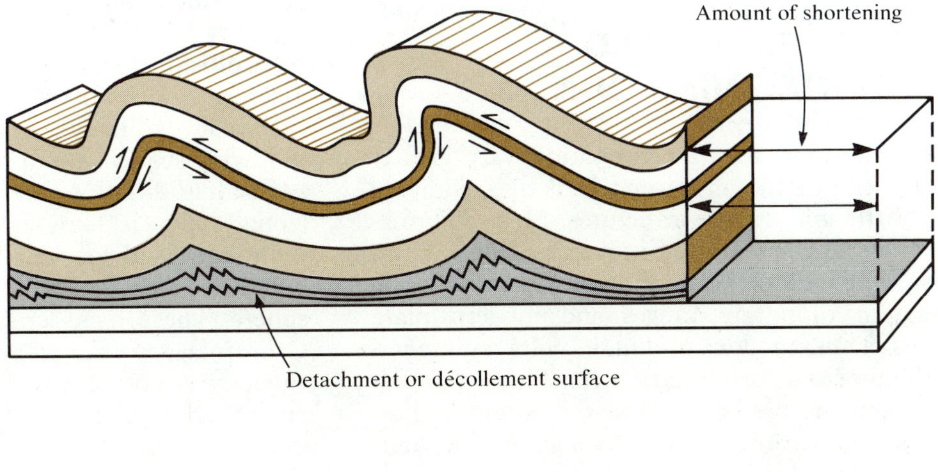

FIGURE 9–12
Flexural-slip folds commonly become detached along some surface at depth, called the décollement zone. The folds are not present below the décollement zone. The amount of shortening in the folds can be calculated by comparing the length of a folded layer to the same length of a planar (unfolded) layer.

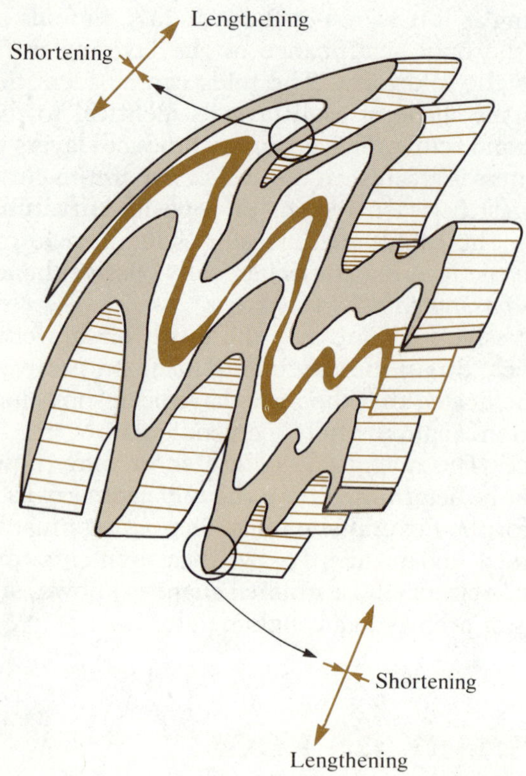

FIGURE 9–13
In passive folds, the viscosity of each layer is the same or nearly the same, and each layer has a similar folded form. Strain within these folds generally has the maximum shortening nearly perpendicular to the hinge surface, and the maximum lengthening within the hinge surface. Fold form is relatively constant for long distances along the hinge surface, and layer thickness does not remain constant (compare to flexural-slip folds in Figs. 9–11 and 9–12).

eral statements just made, but a full treatment of rock deformation and its relations to the physical environment is beyond the scope of this book.

MOUNTAIN BELTS AND CRUSTAL DEFORMATION

Physical conditions during deformation vary greatly throughout the earth's crust and lead to a complex array of brittle and ductile structures. Most deformation tends to be concentrated in narrow belts within the crust, and present belts of deformation coincide with modern mountain ranges and modern plate boundaries. Thus, modern mountain belts are coincident with modern deformational belts. Although modern mountains are present both in the oceans and on continents, the oceanic mountains such as the mid-ocean ridges and linear volcanic plume tracks are not considered here, as they have been covered (Chapter 8). Mountain belts on continents are transitory features, eventually worn away by erosion. However, even after the mountainous topography has disappeared, a record of their existence remains as a long, narrow zone of deformed rocks within the crust. On continents, evidence for ancient mountain belts more than 3 billion years old is present, whereas no oceanic mountain belts older than 180 million years can be found. Therefore, the study of ancient mountain belts is confined to the continents. By comparing modern mountain belts on continents and their relation to plate tectonic activity with older, more deeply eroded, inactive deformed belts, geologists can begin to reconstruct a three-dimensional picture of deformed belts and how they were formed.

Deformed belts have a great variety of internal structures, and no single assemblage of structures is characteristic of all mountain belts. There is no way that the entire spectrum of deformed belts can be presented here; in fact, there is still so much geologic data yet to be gathered that the full range of deformed belt structures and their evolution is not known. Deformed belts can be simply classified as related to **convergent**, **divergent**, and **transform** regions.

DEFORMED BELTS RELATED TO CONVERGENT REGIONS

Probably most of the major deformed belts within continents are the result of convergence at plate boundaries. Two types of deformed belts within this group will be discussed: **noncollisional** and **collisional**. The noncollisional type is a deformed belt related to subduction of oceanic lithosphere alone. The collisional type is a deformed belt related to the convergence of : (1) two island arcs, (2) an arc and a continent, or (3) two pieces of continent after they have come in contact at a subduction boundary.

Noncollisional Deformed Belts Geologic studies of a deformed belt that formed between 200 and 70 million years ago in the western United States show the following sequence of rocks and structures, from west to east (Fig. 9–14). Along the west coastal area is a zone (zone 1) up to 100 km wide of thrust-faulted and folded sedimentary rocks, locally mixed with fragments of ophiolites representing pieces of ancient oceanic lithosphere. These rocks crop out in the Coast Range of California, for example. This zone is interpreted as a subduction complex similar to rocks and structures present above modern subduction boundaries. Subduction of oceanic lithosphere was eastward, and the brit-

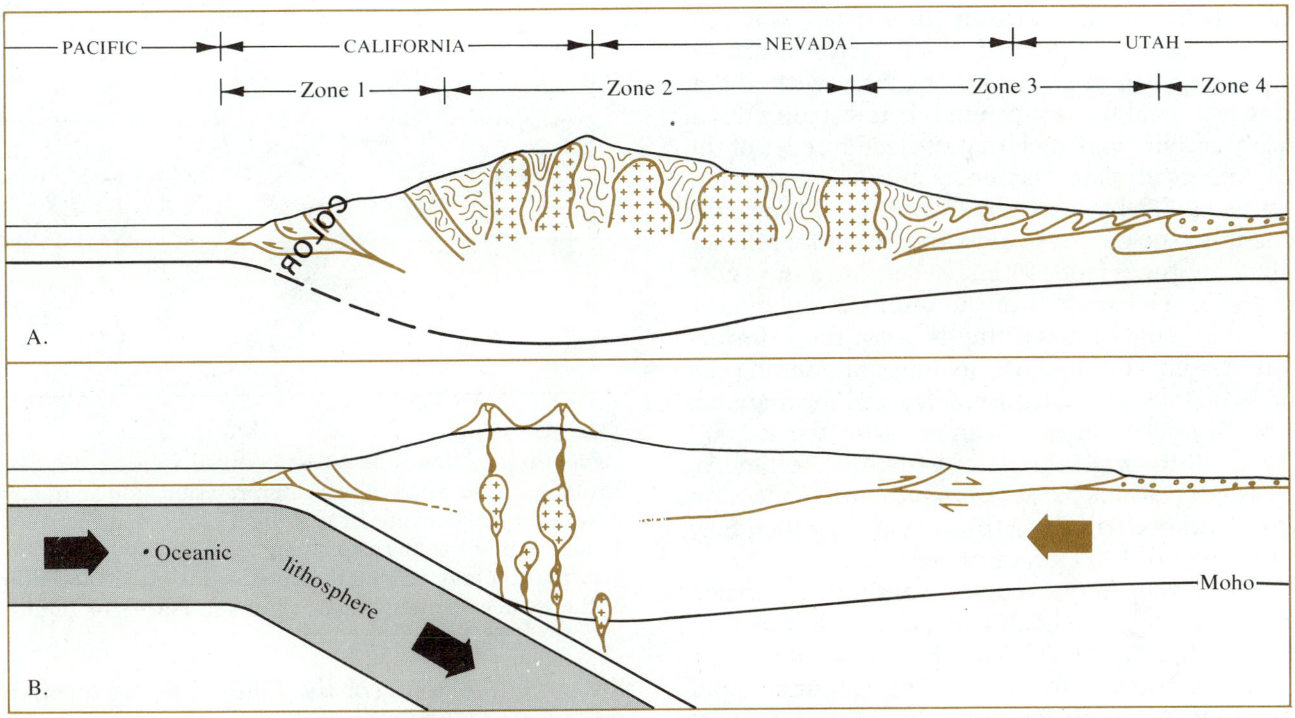

FIGURE 9–14
A. An east-west cross section through the late Mesozoic structures of the western United States with younger deformation removed. B. A plate tectonic cross-sectional interpretation of how the geologic features may have formed along a convergent plate boundary. (Drawing by Robert W. Tope)

tle thrust-faulted structure was facilitated by low temperature and high strain rate. Folds are of both the flexural-slip and passive type, indicating changes of physical environment during the deformation.

The second zone to the east (zone 2) consists of sedimentary and volcanic rocks intruded by large bodies of granitic igneous rocks and deformed in a ductile mode. The sedimentary, volcanic, and granitic rocks of the time interval considered (200–70 million years ago) are interpreted as the eroded core of an Andean-type volcanic arc. They crop out in the Sierra Nevada of California. The ductile deformation of rocks in this zone is the result of high temperature and probably of moderate strain rate, as well as of moderate confining pressure. The physical environment is conducive to ductile deformation, and zone 2 has a dominance of folding over faulting. Where temperatures were high and sequences had uniform rock types, passive folds tend to predominate over flexural-slip folds.

Zone 3, farther east, consists of older continental crust, dated at 1.7 billion years and containing metamorphic rocks overlain by a thick sequence of sedimentary rocks. Structures in this zone are dominated by west-dipping thrust faults. These rocks are now exposed from central Nevada to central Utah. This zone is interpreted as a thick sequence of continental shelf rocks deposited before the continental margin became the site of an active convergent plate margin. The thrust faults indicate shortening of the sedimentary rocks. Folds associated with the thrust faults tend to be flexural-slip types because these rocks were deformed under conditions of low temperature, and low to moderate confining pressure where layering in the sedimentary rocks retained its mechanical importance. This zone is called a **foreland fold and thrust belt**.

Zone 4 consists of undeformed sedimentary rocks overlying the older metamorphic rocks of the North American continent. These rocks were far enough away from the active convergent margin that they were never subjected to stresses large enough to fracture or fold them. Thick sedimentary rocks present in this zone adjacent to the deformed rocks of zone 3 were deposited during the deformation. Material in these sedimentary rocks was eroded from the deformed zone to the west, thus indicating that the deformed zone did form topographically high mountains. Thick sequences of sedimentary rocks formed in this position are commonly called **foredeep deposits** because they were deposited in a subsiding area adjacent to the foreland fold and thrust belt.

The overall interpretation of this deformed belt is that it represents the leading edge of the North American plate beneath which oceanic lithosphere was subducted. Igneous rocks related to subduction intruded the overlying lithosphere, carrying heat with them. Heating of crustal rocks increased their ductility. Plate boundary activity was such that the leading edge of the North American plate was under increased compressive stresses. The hotter zone (zone 2) beneath the magmatic arc deformed in a ductile fashion while the two flanking zones (zones 1 and 3) deformed in a more brittle mode. The amount of shortening in the ductile region is difficult to determine because the deformation is largely by passive folding. Shortening by flexural-slip folding and thrust faulting in the marginal fold and thrust belt (zone 3) can be estimated at 100–200 km. Shortening in the thrust-faulted terrane of zone 1 is not possible because it represents subduction of oceanic lithosphere along that boundary, which may be of the order of 1000 km or more.

The Peruvian Andes appear to be a modern analog to the ancient deformed belt along the western part of North America (Fig. 9–15). The Nazca oceanic plate is being subducted eastward beneath South America, and the associated volcanic arc caps the present Andean mountain range. It can only be inferred that ductile shortening is presently active at depth beneath the volcanic arc because deformation occurs far below the present land surface. Earthquake studies beneath the Andes suggest that the rocks at depth are presently being shortened horizontally and thickened vertically. This part of the Andes would correlate with zone 2 in the western United States. Along the eastern margin of the Andes, folds of the flexural-slip type and west-dipping thrust faults are present and may represent a developing marginal fold and thrust belt. This would correlate with zone 3 in the western United States. Thus, the Andes may provide a picture of what the deformed belt of the western United States may have looked like 200–70 million years ago. The structures exposed at the surface in this older deformed belt may be similar to structures now forming at depth beneath the modern Andes. Structures analogous to zone 1 of the western United States are poorly developed in the continental slope of Peru.

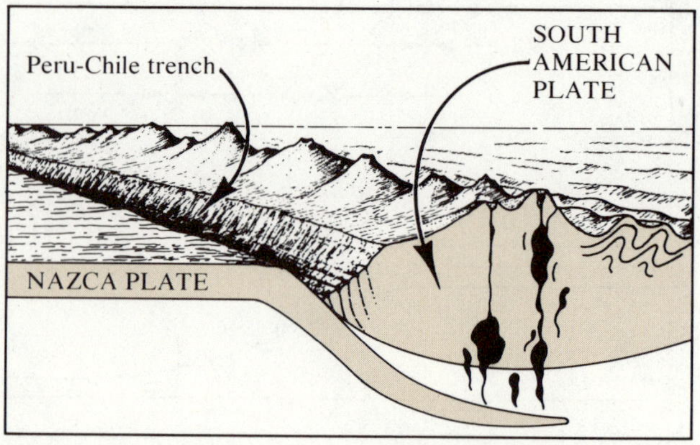

FIGURE 9–15
Generalized oblique and cross-sectional view of the Peruvian Andes showing an interpretation along a modern noncollisional convergent margin. The relations between subduction of the Nazca plate, igneous activity, and deformation may be a modern analog for Mesozoic features of the western United States, shown in Fig. 9–14. (Drawing by Robert W. Tope)

Collisional Deformed Belts Three basic types of collisions may occur: (1) collision of two island arcs, (2) collision of an island arc and a continent, and (3) collision of two continents. A variety of collisional geometries are possible depending upon which boundary or boundaries of the collisional units are active and which way the subduction is inclined.

1. Arc-arc collisions—Collision between two island arcs may be an important process in the geologic record, and at least one collision of this type is developing currently. South of the Philippines, two active island arcs, Halmahera and Sangihe, are approaching each other as the intervening oceanic area is being subducted beneath both arcs (Fig. 9–16). Their thrusted and folded subduction complexes are now in contact, and the arcs are being deformed beneath the Molucca Sea. Northward, the collision is in a more advanced stage. The two arcs have collided in the mountainous southern island of the Philippines, and subduction has ceased between them.

Even though arc-arc collisional processes produce deformation and a thickened crust, the combined arcs are still flanked on both sides by oceanic areas. Because oceanic crust is susceptible to subduction (no oceanic crust older than 180 million years is present in modern oceans), the ultimate fate of any island arc terrane is to collide with a continental margin and become a part of the continent. For this reason, arc-arc types of collisions are viewed as a stage of island arc development. A more fundamental process is arc-continent collision.

2. Arc-continent collisions—Three examples of arc-continent collisions are currently active, and many examples have been described from the geologic record. In the western United States, a well-documented example of an arc-continent collision took place about 250 million years ago, prior to the Andean-type deformation described previously. From central Nevada to the west, three zones of rocks can be identified (Fig. 9–17). In central Nevada, the rocks are shallow-water, marine sedimentary rocks and were deposited upon North American continental crust. In west-central Nevada, there is a narrow zone of sedimentary rocks of the same age, but these rocks were deposited on

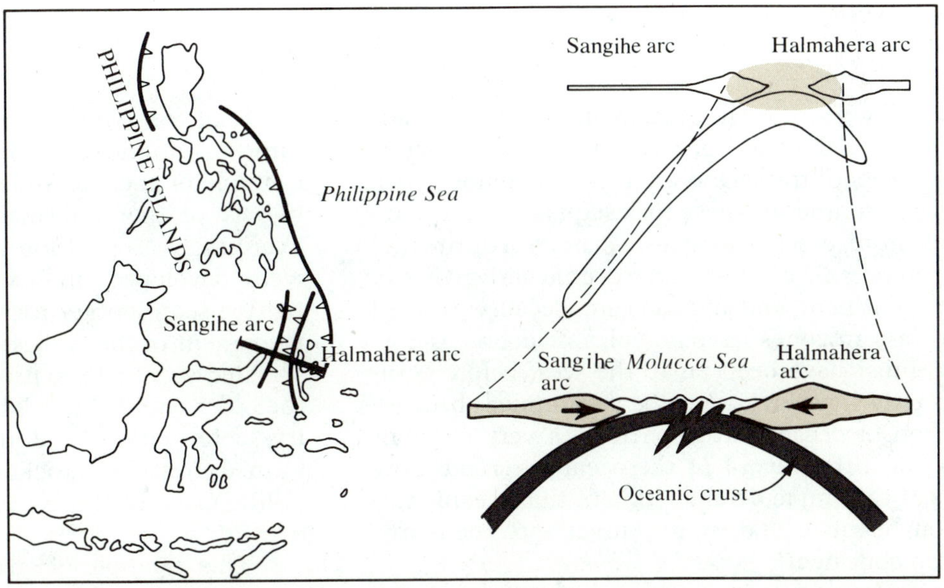

FIGURE 9–16
A modern example of an arc-arc collision. South of the Philippine islands, the Sangihe and Halmahera island arcs are approaching each other as the oceanic lithosphere between them is being subducted. Beneath the Molucca Sea between the arcs, the thrust-faulted and folded sedimentary rocks in the two overriding plates have already come into contact, but the main bodies of the arcs have not yet collided. (After E. A. Silver and J. C. Moore, *Journal of Geophysical Research,* Vol. 83, pp. 1681–91, 1978, © by AGU. Drawing by Robert W. Tope)

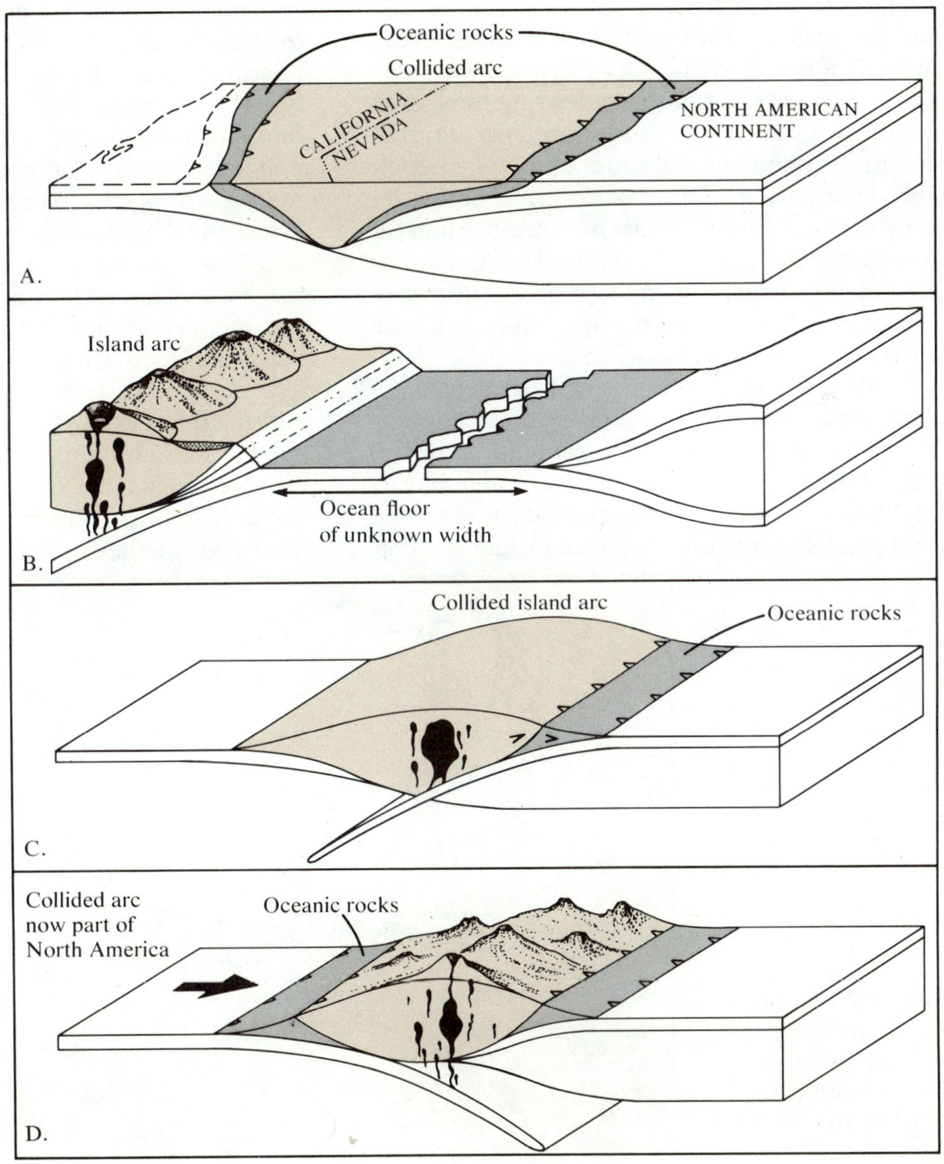

FIGURE 9–17
Schematic interpretation of part of the western United States at the end of the Paleozoic and beginning of the Mesozoic.
A. Distribution of rock sequences in early Mesozoic time. B. Plate tectonic scheme near the end of the Paleozoic. C. Collision of an island arc in earliest Mesozoic time and oceanic rocks, with the island arc becoming part of the North American plate.
D. Continued convergence by subduction of oceanic lithosphere beneath a newly accreted island arc. (Drawing by Robert W. Tope)

oceanic crust. Farther west, rocks in western Nevada and northern California consist of a thick sequence of volcanic rocks of the island arc type. By analogy with modern environments, these rock sequences can be interpreted from east to west as a passive continental margin, an oceanic terrane with material derived partly from the continent, and an island arc. Because rocks in the two eastern zones have no volcanic debris, the island arc must have been far to the west before collision. Exactly how far cannot be determined, because subduction of oceanic lithosphere is a very efficient process and little record of the oceanic terrane now remains at the surface. Subduction of this oceanic terrane brought this island arc in contact with the North American continent.

Deformation was concentrated along the continental margin and the east side of the former arc where thrust faults with a large magnitude of eastward displacement are present. However, within parts of the arc no deformation is evident. The collision took place on a west-dipping subduction zone, leading to closing of the oceanic terrane between the continental margin of North America and the volcanic arc. Even though significant deformation took place during collision, geologic evidence suggests that no topographically high mountains were formed. Following collision, marine sedimentary rocks were deposited across the collision boundary and within the arc, indicating that much of the terrane was below sea level.

Shortly after collision, subduction-related igneous activity developed within the collided arc. Subduction was east-directed at this time, as oceanic crust to the west of the collided arc was subducted under North America. The magmatic arc was of Andean type because during collision the older arc terrane had become a part of the North American continent. This is the same magmatic arc that was discussed earlier as an example of ancient Andean-type arcs. During the long history of the Andean arc (extending intermittently over nearly 150 million years), topographic mountains were developed, indicated by the disappearance of marine sedimentary rocks in the area early in the development of the Andean arc. Igneous activity was associated with deformation and metamorphism within the older collided arc. The processes of igneous activity, deformation, and metamorphism slowly transformed these older rocks into a thicker and more crystalline crust of the type characteristic of most continents.

This example of an arc-continent collision illustrates four important points. First, arc-continent collisions enlarge continents. Second, deformation during collision may produce deformed belts of rocks, but not topographically high mountains. Third, plate boundary relations may change rapidly. In this example, an island arc/marginal sea/continental margin paleogeography changed to an Andean paleogeography. Fourth, younger igneous activity, deformation, and metamorphism transform collided arcs and oceanic terranes into typical continental crust.

Three examples of active arc-continent collisions are present in west Pacific and Indian Ocean areas. In the Indonesian arc, lithosphere of the Indian plate is being subducted northward beneath the Asian plate. The Australian continent is part of the Indian plate, and in the eastern part of Indonesia it is entering the subduction zone (Fig. 9–18). Continental margin and oceanic rocks are being thrust-faulted and folded along the collision front as the Indonesian arc terrane rides onto the northern edge of the Australian continent.

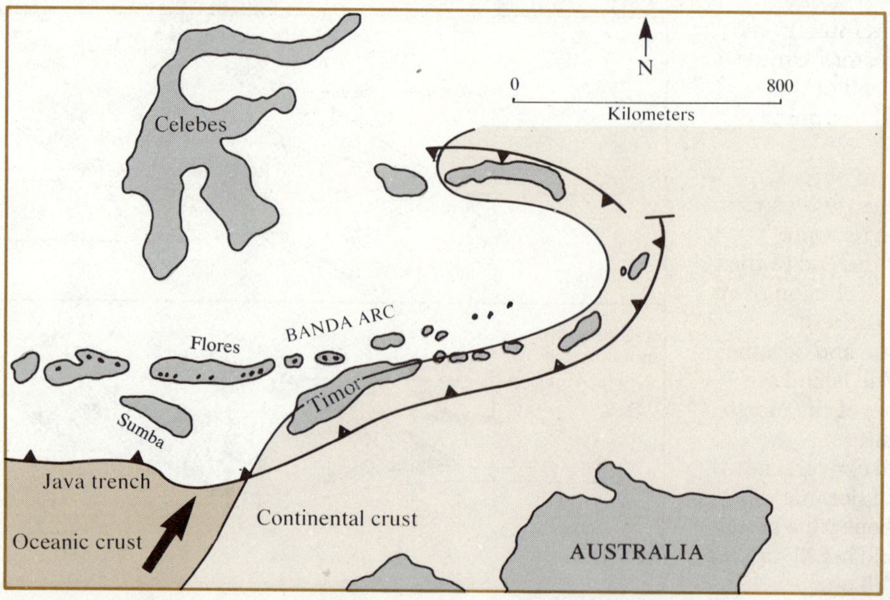

FIGURE 9–18
Sketch map of the eastern Indonesian region showing the modern active subduction of continental lithosphere and arc-continent collision.

Another example occurs in eastern New Guinea, also a part of the Australian continent, where the New Britain island arc, formed above a north-dipping subduction zone, has collided along its western end with northern New Guinea (Fig. 9–19). Finally, the continental crust of western Taiwan has collided with the northern part of the Philippine island arc along the east-dipping Manila subduction zone. In Taiwan, there has been considerable crustal thickening by thrust faulting and folding, leading to the formation of both geologically and topographically high mountains (Fig. 9–20). The deformed belt formed in Taiwan consists of east-dipping thrust faults and flexural-slip folds constructing a typical marginal fold and thrust belt. Shortening for this deformed belt may be more than 100 km.

3. Continent-continent collisions—Numerous continent-continent collisions are present in the geologic record, and a major collision is currently underway in Asia along the Alpine-Himalayan mountain belt. The Indian continent has collided with the continental crust of China, leading to the formation of the Himalayan mountain chain and the world's highest mountains (Fig. 9–21).

By studying the magnetic anomalies and history of the Indian Ocean, the movement of the Indian continent, which is a part of the Indian plate, can be reconstructed. After its breakaway from the Antarctic continent, India moved northward until it made contact with the continental crust of China about 40 million years ago. The oceanic crust that lay between India and China was subducted northward under China prior to collision. The north margin of the Indian continent was a passive continental margin. Convergence between the Indian and Asian plates continued following collision, but because only continental crust was present, simple subduction could no longer take place. Instead, convergence has been taken up during the last 40 million years by shortening of the northern margin of the Indian plate by thrust faulting and by both ductile and brittle deformation of the Asian continental crust.

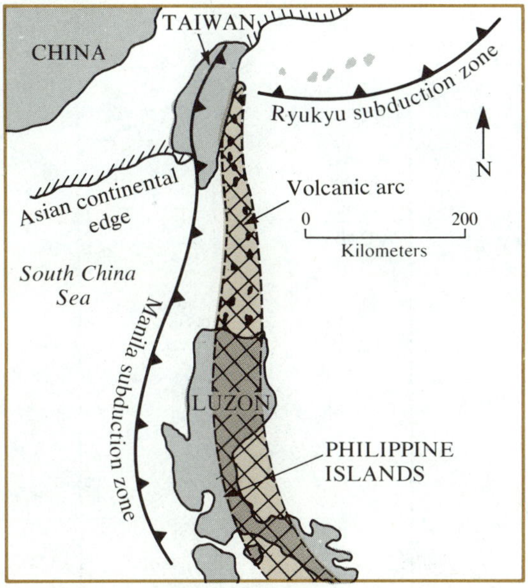

FIGURE 9–20
Simplified sketch map of the island arc collision with the Asian continent along the eastern coast of Taiwan. The eastern half of the island of Taiwan consists of a west-directed sequence of thrust faults and folds caused by the subduction of the easternmost edge of the Asian continent.

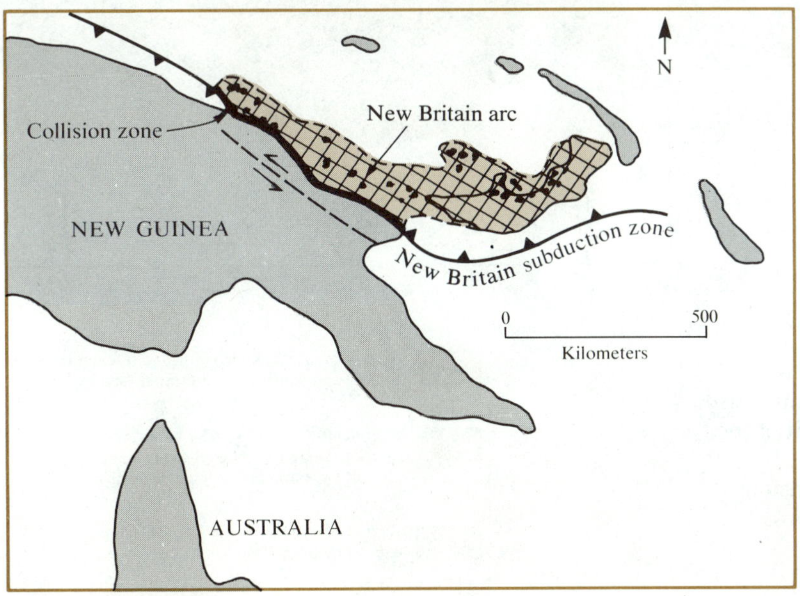

FIGURE 9–19
Simplified sketch showing features related to the recent collision of the New Britain island arc with New Guinea, which forms the northern edge of the Australian continent. Only the tectonic features related to the collision are shown. Dots are volcanoes related to the youngest part of the New Britain island arc.

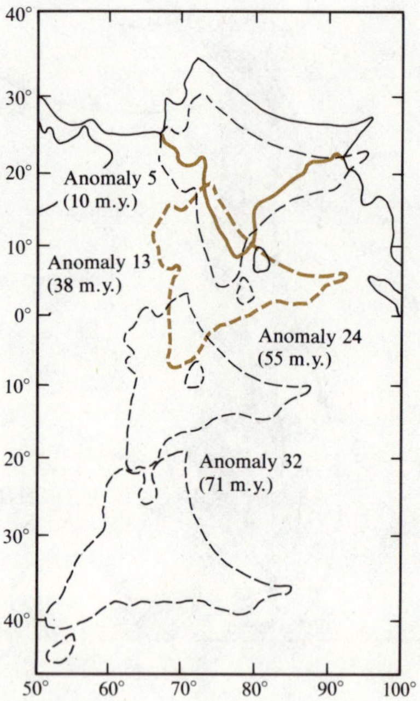

FIGURE 9–21
Several positions of India relative to Eurasia during the last 71 million years. Collision occurred at about 40 million years ago when India was in the position shown by the color outline. (From P. Molnar and P. Tapponnier, 1975)

As the northern margin of the Indian continent was shortened by thrust faulting, the thrust faults carried thick plates of continental crust and their overlying sediments relatively southward as the rest of the Indian continental crust moved northward beneath them. The stacking of these thrust plates and the related shortening and thickening of the rocks make the present high mountains of the Himalaya. Studies of earthquakes in the Himalaya indicate that the process is still active.

In Asia, the deformation caused by the continued convergence is a complex pattern of strike-slip, thrust, and normal faults (Fig. 9–22). In addition, ductile deformation has probably taken place beneath Tibet, thickening the continental crust by folding and forming the highest plateau region in the world. Fault movements indicate that the Indian continental crust is pushing into the Asian continental crust, forcing it to move aside along strike-slip faults. Normal faulting occurs where movement of the Asian crust is marked by tension. The fault patterns and movement within the Asian plate have been modeled as if the Indian continental crust is a rigid block indenting a plastic and ductile Asian plate. This model suggests that the effects of continent-continent convergence following collision can be responsible for producing mountains

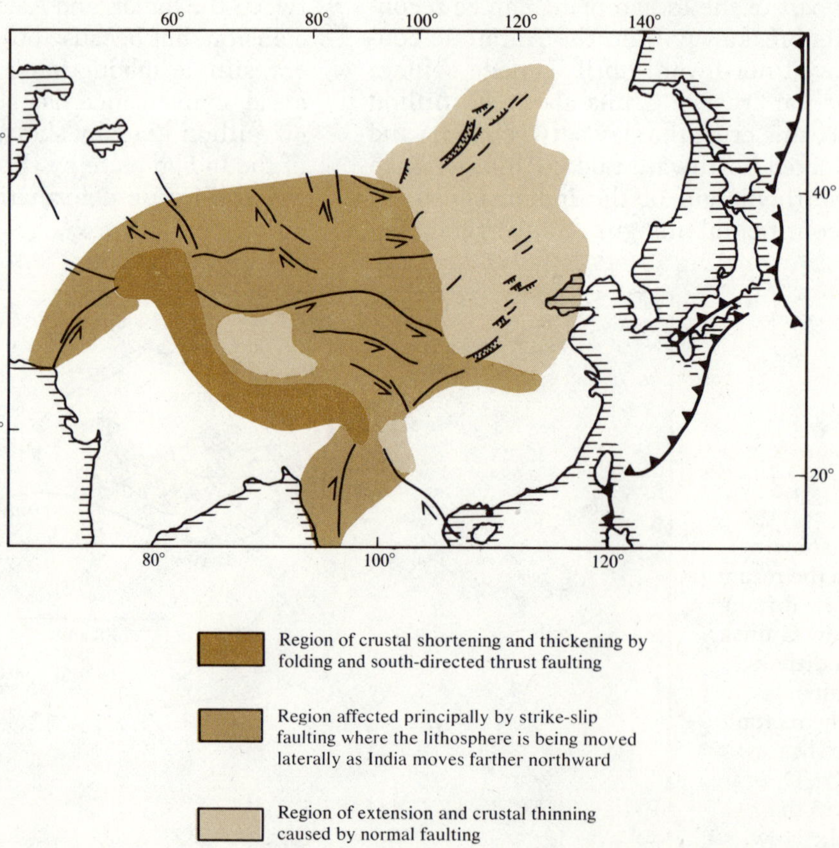

FIGURE 9–22
Distribution of deformational styles within Asia caused by the continued northward movement of the Indian continental lithosphere after collision with Asia about 40 million years ago.

of diverse types for more than 2000 km from the collision boundary. Thus, not only are the Himalaya Mountains a product of the collision, but so are many of the mountain ranges in Asia.

DEFORMED BELTS RELATED TO TRANSFORM REGIONS

Transform faults by themselves do not produce great topographic relief because the motion between the two rock masses on either side is horizontal, with no vertical component. In some cases, however, transform faults are associated with either compression or tension. This adds a vertical component to their displacements and produces some topographic relief although the major motion along the fault has been horizontal.

The Pyrenees of southern France and northern Spain may be an example of a mountain range produced along a transform boundary associated with convergence (Fig. 9–23). The Pyrenees mark the boundary between the small Spanish plate and the larger European plate, and the major motion between these two plates along the Pyrenees has been by strike-slip faulting. Convergence along the boundary is responsible for the wide belt of deformation that produced a central zone of rocks, locally deformed in a ductile mode, which are flanked by zones of thrust faults directed away from the central core area. The component of convergence at the boundary thickened the crust and produced a deformed belt and topographically high mountains.

DEFORMED BELTS RELATED TO DIVERGENT REGIONS

Divergent motion or extension at plate boundaries or within plates can also produce topographically high mountains. The internal structure of these mountains is very different from that of mountains produced by convergence, and extension of the crust is largely reflected at the surface by brittle faulting of the type that produces normal faults. Because normal faults have a large component of vertical motion, they can produce topographic relief at the earth's surface.

Tilting or rotation of the faulted blocks about a horizontal axis is commonly associated with normal faults. As the crust extends, rocks near the surface deform by brittle failure, but at depth, where temperature is higher, deformation would be by ductile flow. A connection between these two modes of deformation would cause the brittle faults to gradually die out into zones of horizontal ductile flow at depth.

In the western part of the United States, the Basin and Range Province is an area presently undergoing extension; it is also an area of mountains separated by valleys—hence the name Basin and Range. The westernmost mountain range, the Sierra Nevada of California, is essentially one large, west-tilted fault block bounded by normal faults along its east side (Fig. 9–24).

Mountains produced by extension are usually bounded on one or both sides by normal faults of large displacement and contain numerous smaller normal faults within the mountains. Folding is rare, and ductile deformation of wide areas nonexistent. Folding

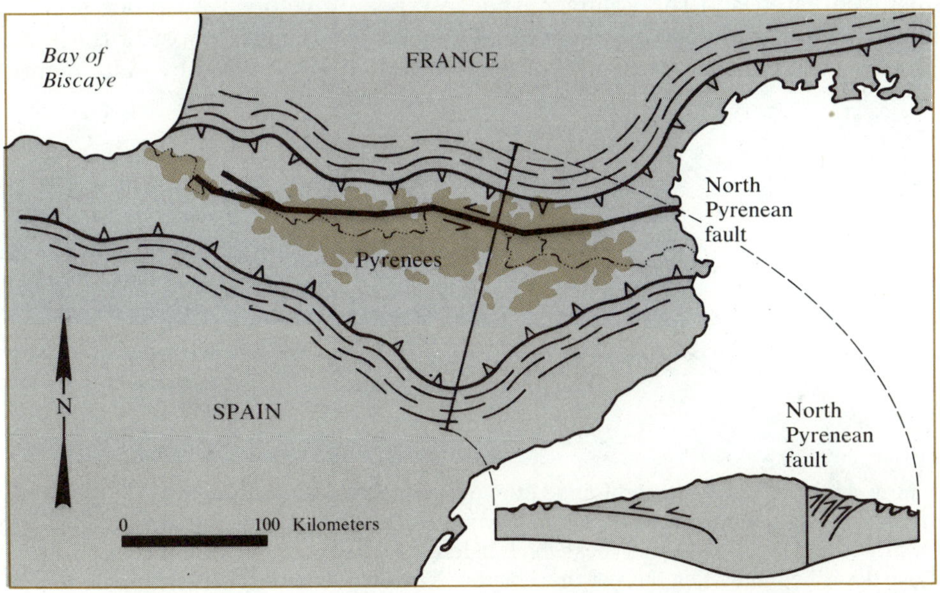

FIGURE 9–23
The main structural elements of the Pyrenees, which formed by convergence along a transform boundary marked by the north Pyrenean fault. The Pyrenees chain consists of folded zones along its north and south margins (short line segments.). Elevation of the mountains is caused by crust thickening by thrust faulting both to the south and to the north, and by more ductile deformation and folding in the axial region (color area). (Drawing by Robert W. Tope)

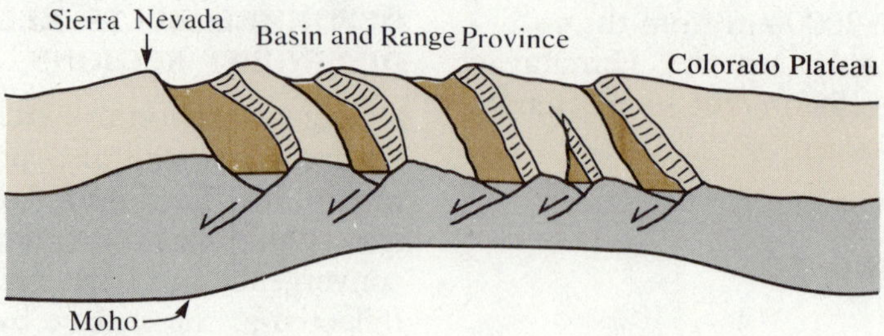

FIGURE 9–24
The Basin and Range Province of the western United States consists of generally north-south-trending ranges and valleys. Locally, the ranges reach more than 4000 m in elevation. These mountain ranges are the result of crustal extension and thinning caused principally by normal faulting.

within mountains may be present but is related to earlier periods of deformation and unrelated to the extension that produced the topographic relief. This is clearly seen where flat-lying sedimentary or volcanic rocks have been faulted and tilted in the mountain ranges. Under these conditions, the originally horizontal layers of these rocks have been faulted and rotated but not internally deformed by folding (Fig. 9–25).

Crustal extension produces crustal thinning and thus should be associated with general subsidence, not significant elevation above sea level. Many extensional areas do subside below sea level, where they may be buried by sedimentary rocks and never produce topographically high mountains. Other areas, such as the Basin and Range Province, have been elevated during extension to form high mountains. The Basin and Range Province appears to be underlain by upwelling hot upper mantle, which causes crustal elevation, even though the crust is being thinned. The Basin and Range structure has been superimposed on the site of the first arc-continent collision and the younger Andean-type deformed belt, which were discussed earlier. This history emphasizes the complex evolution of continental areas, which may be reworked by many different deformational events and subjected to repeated plate boundary activity.

ANCIENT DEFORMED BELTS

The examples presented previously for the formation of geologic and topographic mountain belts have all been taken from the last 250 million years of earth history, and many are related to plate tectonic interactions for which the sea-floor record still exists. Geologic studies of the continents indicate that the mountain belts are formed from deformed belts of rocks of various ages, some as old as 3.9 billion years. Most of the deformed belts developed in the last 1 billion years have rock associations, structures, and various geophysical properties that suggest they were developed by processes not unlike those active today. That is, most of these belts can be interpreted to have developed by plate tectonic processes.

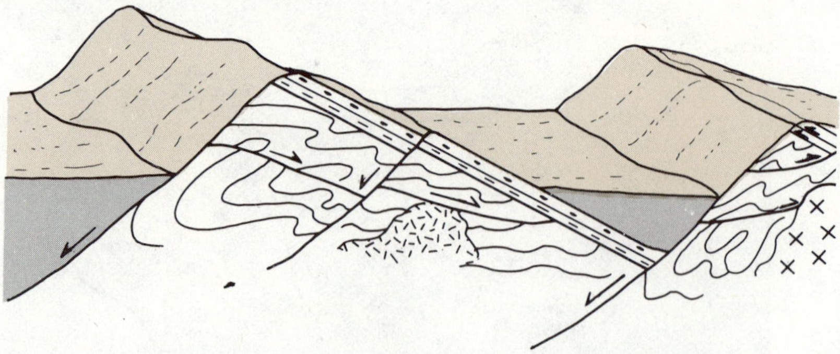

FIGURE 9–25
Mountain ranges within the Basin and Range Province are the result of differential movement and rotation of crustal blocks along normal faults. Within the faulted blocks are complex folds and faults of other types which belong to older periods of deformation, such as those described in earlier parts of this chapter. The older complex structures are unrelated to the normal-faulted mountain ranges.

Deformed belts of rocks that developed prior to 1 billion years ago are generally deeply eroded. Rocks and structures in many of these deformed belts have elements similar to those of the younger belts; however, they are missing some of the important rock associations used to infer a plate tectonic origin. For example, the ophiolite assemblage of rocks used to infer the presence of oceanic lithosphere is very rare. High-pressure metamorphic mineral assemblages used to infer subduction are not recognized either. Igneous rock associations similar to modern island or Andean-type arcs are present but are often difficult to relate to plate tectonic boundaries. These problems in interpreting the evolution of many ancient deformed belts has led to several schools of thought. Some workers have argued that plate tectonics in its present form did not exist prior to 2 or 2.5 billion years ago. Others have argued that plate tectonics did exist, but that the deformed belts are so poorly studied and understood that the correct plate tectonic interpretation cannot be developed. Between these two extremes are those who hold that early plate tectonic processes were operating but at different rates and on a thinner lithosphere than exists today, and thus the resultant deformed belts had a different form. Some of these processes may have been peculiar to the early part of earth history and are not operative today. Clearly, there is no agreement on the interpretation of this important part of earth history, which lasted more than 3 billion years.

FIGURE 9–26
The continental crust of North America is composed of belts of rocks that were deformed, metamorphosed, and intruded by igneous rocks. These deformational belts span long time periods (shown by the numbers in billions of years ago) and are all formed by multiple periods of deformation. They contain the record of lithosphere generation and evolution of more than 3.9 billion years. Other continental regions consist of similar belts of deformed rocks.

CONTINENTAL CRUST

Continental crust has been formed and has evolved over the past 3.9 billion years. Study of continental crust indicates that, at the surface, it consists of deformed belts of rocks that have formed at different times (Fig. 9–26). For the past 2 or perhaps 3.9 billion years, the continents have grown by the collision of island arcs underlain by oceanic crust at old plate boundaries as new material generated in the mantle has been added to the continents. These rocks may be intruded during later igneous events, metamorphosed, and redeformed by ductile deformation to form the metamorphic and igneous rocks that make up the bulk of continental crust at the surface.

Most deformed belts are composed of redeformed continental crustal rocks generated during previous episodes of deformation. The addition of new material to the continents is usually less than half, and sometimes only a very small percentage, of the rocks in the deformed belt. Deformed belts of different ages clearly crosscut one another in continental crust. In general, continental crust consists of rocks whose densities are less than those of oceanic crust, such as igneous rocks richer in silica than oceanic basalt, sedimentary rocks, and their metamorphosed equivalents. Because these light rocks cannot be subducted easily, they remain in the overriding plates and are swept together at subduction boundaries and accreted to the continents. At times, the continents may be fragmented by sea-floor spreading or transform faults and later reassembled at convergent boundaries. They contain the only direct record of the earth's history prior to 180 million years ago.

SUMMARY

The earth's crust is composed of oceanic and continental crust. Oceanic crust is relatively simple, consisting of three layers: (1) sedimentary rocks at the top underlain by (2) basaltic lavas and sheeted dikes, and (3) gabbro at the base. Present oceanic crust is no older than about 180 million years, approximately 5 percent of the earth's history. Continental crust is much more complex, containing belts of deformed sedimentary, metamorphic, and igneous rocks that mark the sites of former and present mountain belts. It contains rocks more than 3.9 billion years old, and thus the record of most of earth history not recorded in oceanic crust.

Continental crust consists of deformed belts of rocks of different ages. These belts indicate that the rocks, when subjected to stress, have behaved in both a brittle and a ductile mode. Experimental laboratory studies of rock deformation have provided insight into how rocks behave in the earth's crust. Variations in confining pressure, temperature, and strain rate make rocks behave either as brittle materials, producing faults, or as ductile materials, producing folds. Rocks usually deform as ductile materials under conditions of increased temperature and confining pressure, and decreased strain rate. Under such conditions, rocks can flow as viscouslike fluids, but in the crystalline state. Their viscosities are so high that their ductile behavior cannot be observed in a lifetime.

Examination has shown that modern areas of deformation form along wide linear belts associated with modern plate boundaries. Thus, deformed belts and mountain-building processes are related to plate tectonics. Probably the widest and most common belts of deformation present within continents are related to subduction boundaries. The volcanic arcs related to subduction bring heat to shallow levels in the crust and cause the conditions necessary for the ductile behavior of crustal rocks. These belts of ductile deformation may be flanked on both sides by regions of more brittle behavior marked by both folds and faults.

Deformed belts may form in noncollisional and collisional subduction settings. The modern Andes of South America are an example of a noncollisional, subduction-related mountain belt. Collisional belts may form during arc-arc, arc-continent, or continent-continent collisions. Examples of all of these types of collisional settings can be found both at modern plate boundaries and within the ancient parts of continents. By studying both the modern active plate boundaries and older, deeply eroded, inactive deformed belts within continents, scientists can begin to build up a three-dimensional picture of the processes involved in mountain building and in the evolution of continental crust.

Some mountain ranges are related to convergence at transform boundaries, whereas others are related to spreading or divergent motion at plate boundaries or within plates. Most mountain belts and deformed belts within continents are the result of subduction, transform faulting, or divergent plate boundary activity, which may be occurring all at the same time or superimposed at different times. This is why continental crust has such a complex geologic history and evolution.

Some very ancient deformed belts, particularly those older than 2 or 2.5 billion years, do not resemble closely the younger deformed belts. It is still uncertain whether this means that the early history of the earth's crust is related to plate tectonic processes that occurred at different rates or under different conditions of temperature and pressure, or that processes other than plate tectonics are required to explain the early evolution of continental crust.

REVIEW EXERCISES

1 What is thought to be the composition of oceanic crust?

2 What is an ophiolite?

3 How can the age of formation of oceanic crust be determined?

4 What is stress?

5 Why is it necessary to conduct rock deformation experiments?

6 What is the difference between tension and shear fractures?

7 Define *normal, thrust,* and *strike-slip faults.*

8 Why do rocks appear to us as only brittle materials, when we can see that they have behaved as ductile materials within the earth's crust?

9 Is a fold a brittle or ductile response to deformation?

10 What is the difference between a flexural-slip fold and a passive fold?

11 Why do continents contain evidence of folded mountain belts more than 3 billion years old, whereas oceanic crust contains evidence of mountain belts no older than about 180 million years?

12 Mountain belts with complex internal structure are formed along plate boundaries. Cite one example and give its possible plate tectonic origin.

13 To understand the development and internal structure of mountain belts, why is it important to study both modern active plate boundaries and ancient inactive deformed belts within continents?

SUGGESTED READINGS

BIRD, J. M., and ISACKS, B., eds. *Plate Tectonics,* Washington, D.C.: American Geophysical Union, 1972, 563 pp.

CARTER, N. L.; FRIEDMAN, M. O.; LOGAM, J. M.; and STEARNS, D. W., eds. *Mechanical Behavior of Crustal Rocks.* Geophysical Monograph 24. Washington, D.C.: American Geophysical Union, 1981, 326 pp.

COX, A., ed. *Plate Tectonics and Geomagnetic Reversals.* San Francisco: W. H. Freeman and Co., 1973, 702 pp.

DEWEY, J., and BIRD, J. "Mountain Belts and the New Global Tectonics." *Journal of Geophysical Research* (1970), Vol. 77, No. 14, pp. 2625–47.

HOBBS, B. E.; MEANS, W. D.; and WILLIAMS, P. F. *An Outline of Structural Geology.* New York: John Wiley and Sons, 1976, 571 pp.

STRANGWAY, D. W., ed. *The Continental Crust and Its Mineral Deposits.* Special Paper 20. Ottawa: Geological Association of Canada, 1980, 804 pp.

TARLING, D. H., ed. *Evolution of the Earth's Crust.* New York: Academic Press, 1978, 443 pp.

WILSON, J. T., compiler. *Continents Adrift.* Readings from *Scientific American.* San Francisco: W. H. Freeman and Co., 1980, 172 pp.

WINDLEY, B. F. *The Evolving Continents.* New York: John Wiley and Sons, 1977, 385 pp.

WYLLIE, P. J. *The Dynamic Earth.* New York: John Wiley and Sons, 1971, 416 pp.

- relative dating
- superposition
- crosscutting relationships
- inclusions
- faunal succession
- era
- period
- epoch

10

- time correlation
- index fossil
- concurrent range zone
- key bed
- varve
- event correlation
- absolute dating
- radiometric dating
- isotope
- Precambrian era
- craton
- shield
- Paleozoic era
- Mesozoic era
- Cenozoic era

Geologic Time and Earth History

GEOLOGY HAS OFTEN BEEN DESCRIBED AS PHYSICS AND CHEMISTRY applied to the real world. As such, it is called a "synthetic" science, that is, one whose principles derive from other, more basic sciences. There is, however, one significant concept developed within the science of geology that is unique: the immensity of geologic time. In physics and chemistry, one deals with natural "laws" which are presumed to be independent of time. In geology, on the other hand, constantly changing interactions of these laws over great spans of time are the focus. Thus, the geologist finds a unique set of circumstances for each moment of the past.

The discovery of the great length of geologic time and of the evolution of the earth and its life through that time is one of the greatest scientific triumphs of the nineteenth and twentieth centuries. Although people in other areas of the world have allowed great lengths of time for earth history, the tradition of Western civilization has held that the earth was specially created a short time ago for the domination of humans and would soon perish. The Copernican revolution in astronomy moved the earth out of its central position in the universe, the Darwinian revolution put people in their place in the biological world, and geology has demonstrated that the span of human existence is but an infinitesimal bit of earth history. In attacking the concept of the central importance of humanity, geology has come into conflict with time-honored traditions, as did astronomy and biology at earlier dates. As with those sciences, there have been great conflicts with religion, and many in the nonscientific community still do not accept an evolutionary view of a universe of which we are but a tiny, insignificant part.

The notion of the great length of geologic time must have occurred to many classical and Renaissance scholars who saw evidence in the earth's rocks of numerous past changes in lands and seas and in living things. However, the short time period deduced from the Bible discouraged public statements about great antiquity for the earth. Indeed, in the mid-1700s, the French scientist the Comte de Buffon reasoned that the earth must be at least 75,000 years old, and was forced to publicly recant by the Church. Buffon believed that the earth was once molten because volcanoes and the high temperatures in deep mines indicated there was still much internal heat. If the earth were once molten, then by estimating its present heat and the annual heat loss, one could calculate the planet's age. Although much of Buffon's reasoning was based on incorrect premises, he did open the door to scientific consideration of a great age for the earth. In this chapter, the development of modern ideas of geologic time will be traced, followed by a brief review of the physical and biologic history of the earth.

EARLY ESTIMATES OF GEOLOGIC TIME

Before the advent of modern dating methods in the twentieth century, geologists and others interested in earth history used four major methods of estimating the earth's age (Table 10–1). In the mid-1600s, two different church officials added up all the biblical genealogies and arrived at a date of approximately 6000 years ago for the creation of Adam and Eve. Early geologists of the eighteenth and nineteenth centuries saw much evidence in rocks that implied the earth was much older, but just how much, no one yet knew.

Three classical scientific approaches were developed in the late nineteenth century (Fig. 10–1). All were based upon measuring the rate of a modern process and its cumulative effects throughout geologic history. In 1899, John Joly calculated the total amount of dissolved salts in the oceans and the annual addition of salts by the rivers of the world, and then divided the latter into the former to obtain an estimate of the earth's age. His calculation gave an age of 90 million years but failed to take into account the recycling of salt into sediments and the atmosphere. It now appears that the salinity of the oceans attained an equilibrium early in the earth's history and has since remained stable.

TABLE 10–1
Early estimates of geologic time.

Method	Age (years)
Biblical genealogies	6000
Saltiness of oceans	
$\frac{\text{Total salinity}}{\text{yearly addition}} = \text{age}$	90 million
Rate of sedimentation	
$\frac{\text{Total thickness}}{\text{yearly accumulation}} = \text{age}$	3 million–1.6 billion
Cooling of earth	
$\frac{\text{Initial heat} - \text{present heat}}{\text{annual heat loss}} = \text{age}$	25–100 million

Between 1860 and 1909, nearly a score of geologists attempted to calculate the age of the earth by determining the maximum thickness of the sedimentary rock record and dividing this number by the average annual accumulation. The results varied greatly between 3 million years and over 1.5 billion years. There are so many uncertainties in this method, such as the gaps (discontinuities) in the rock record, the amount of compaction in the conversion of sediment to rock, and the highly variable rates of sedimen-

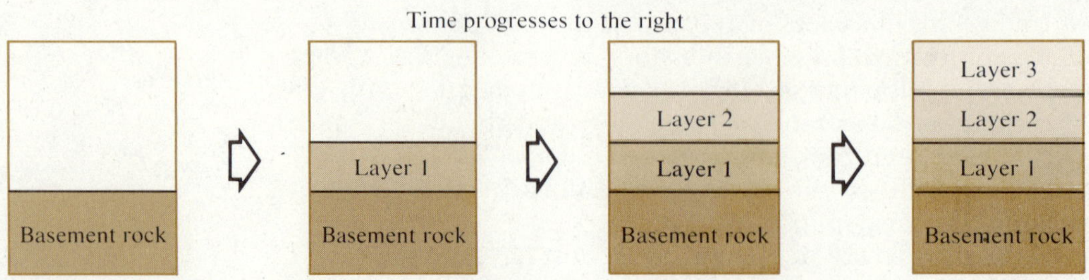

FIGURE 10–1
The law of superposition, which states that in an undeformed rock sequence, each layer is older than the one above but younger than the one below. Each layer must have a base upon which to be deposited.

tation, that it is essentially useless for accurate age determinations.

Finally, because of the ample evidence of internal heat from volcanoes and deep mines, the physicist Lord Kelvin reasoned that the earth had once been molten, and therefore if one could determine the difference between the molten and present temperatures and divide this figure by the annual heat loss, the age of the earth could be calculated. Kelvin made these calculations and arrived at ages between 25 and 100 million years. His estimates were later proved erroneous by the discovery of radioactivity, which could produce the earth's own internal heat, thus invalidating his basic premise.

RELATIVE DATING BY PHYSICAL METHODS

Before they were able to determine the ages in years of particular geologic features, geologists devised several methods of determining their relative ages, that is, their ages in relation to one another. The major techniques of **relative dating** are examined in the following paragraphs.

The first of these basic geologic principles is called **superposition** (Fig. 10–1). Originally noted in 1669 by Nicholas Steno in northern Italy, the law of superposition simply states that in any sequence of rocks undeformed by folding or faulting, each layer is older than the one above and younger than the one below. This "law" is based on the simple observation that a rock layer must be deposited on a preexisting rock layer.

Two other basic principles of relative dating were enunciated by the famous Scottish geologist James Hutton in the late 1700s. The **law of crosscutting relationships** (Fig. 10–2) states that a body of rock cutting across another rock body must be younger than the rock it cuts. The other rock must be there in the first place for it to be crosscut. The **law of inclusions** indicates that if a rock body contains fragments of another rock body, it must be younger than that other rock body. The other rock body must have been there first to provide rock fragments. Hutton noted that inclusions enable one to determine if a sedimentary rock overlying a granite was deposited on top of an older eroded granite or had been intruded from below by a younger granite (Fig. 10–3).

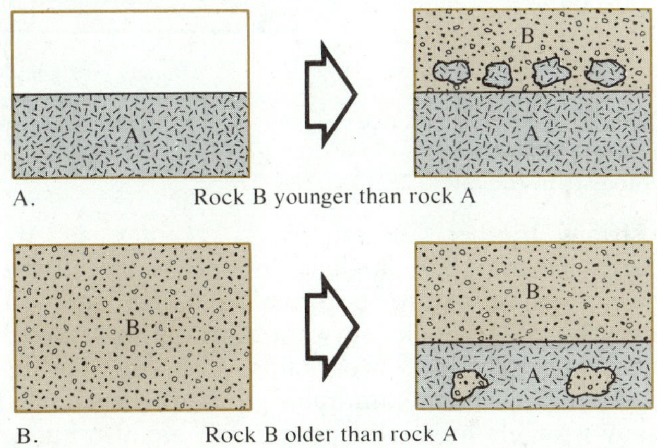

FIGURE 10–3
Two instances of a granite-sandstone contact illustrating the law of inclusions. A. The sandstone was deposited on top of the eroded granite, pebbles of which are present at its base. B. The sandstone was intruded from below by a granite that contains some unmelted pieces of the engulfed sandstone.

EARLY TIME SCALES

Utilizing the relative dating principles just discussed, early geologists attempted to work out a chronology of the earth (Fig. 10–4). In the late 1700s, Arduino in Italy and Lehmann and Füchsel in Germany observed a basic threefold subdivision of rocks based on super-

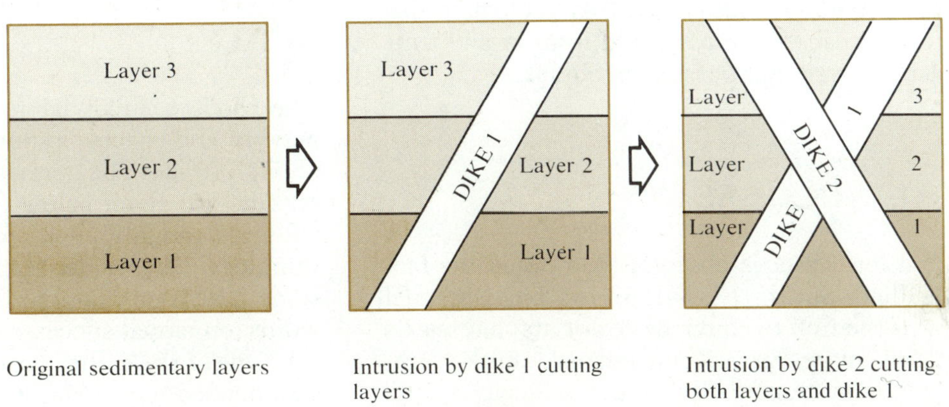

FIGURE 10–2
The law of crosscutting relationships, which states that a body of rock must be younger than any rock bodies it cuts; that is, the other rock bodies must have been present first for the rock body to cut them.

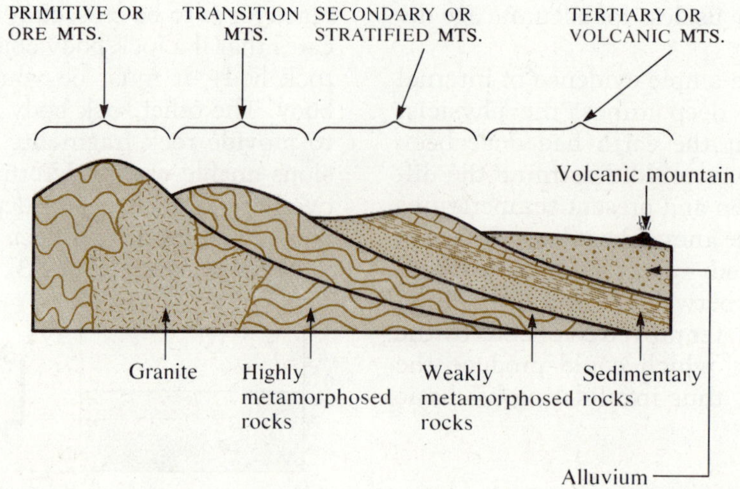

FIGURE 10–4
The basic fourfold sequence of rocks recognized by geologists in the late 1700s.

position; another German, A. G. Werner, added a fourth. Forming the highest mountains because of their resistant nature, but passing superpositionally beneath younger rocks outward from the core, are the crystalline Primitive Mountains or Rocks. Because these igneous and metamorphic rocks are the source of many valuable mineral deposits, they are also known as Ore Mountains. They lack fossils. Overlying the Primitive Rocks on the mountain flanks are less strongly metamorphosed rocks named the Transition Mountains by Werner. A few fossils occur in Transition Rocks. Still younger are the overlying Secondary or Stratified Mountains consisting of the normal sedimentary rocks sandstone, shale, and limestone. These rocks are generally quite fossiliferous. On top of the sequence are poorly consolidated sediments, the Tertiary or Alluvial Mountains. These rocks are also fossiliferous. Lava flows of modern volcanoes are included in the uppermost Tertiary sequence.

Werner assumed that this sequence was invariable and essentially worldwide because all rocks were the precipitates of a universal ocean. Hence, he believed that the type of rock—its **lithology**—indicated the age of the rock. For example, according to Werner, all limestones formed in the Secondary time span. In the early 1800s, Werner's ideas faded as geologists discovered that most rock types could form at any time in the past and were not oceanic precipitates.

FAUNAL SUCCESSION

Although a few obscure observers had noted the fact earlier, William Smith (1769–1839), an English canal builder, was the first to convince geologists that fossils (the remains of prehistoric organisms), unlike lithologies, occur systematically in the geologic record. Rather than occurring in various sequences in various places as do the kinds of rock, fossils occur in a definite, invariable sequence in the geologic column. Smith could tell the correct position of any rock body in the age sequence simply by examining the rock's fossil content. This simple, pragmatic person had found, as a by-product of careful observations made during canal construction, the key to determining the relative ages of sedimentary rocks from different areas. Two representative fossil samples from the upper end of the column, the Chalk and the London Clay, are shown in Fig. 10–5. Using the new **law of faunal succession**, Smith was able to prepare the first detailed geologic map. This map of Great Britain is a classic in the field of geology.

After the discovery of faunal succession, the present subdivisions of the geologic column were rapidly established in the middle 1800s. Smith never did learn the reason for the systematic occurrence of fossils in strata, but Charles Darwin provided that explanation in 1859 with the discovery of evolution by natural selection.

MODERN RELATIVE TIME SCALE

The modern subdivision of geologic time into **eras**, **periods**, and **epochs** is shown in the right four columns of Fig. 10–6. The names and dates of the original authors are given below each term, and the left four columns roughly show equivalency with earlier terminology. The modern time scale consists of subdivisions based on characteristic fossil content, in other words, on faunal succession.

One of the earlier terms, *Tertiary*, survives, but in an amended sense. Many names have a geographic ori-

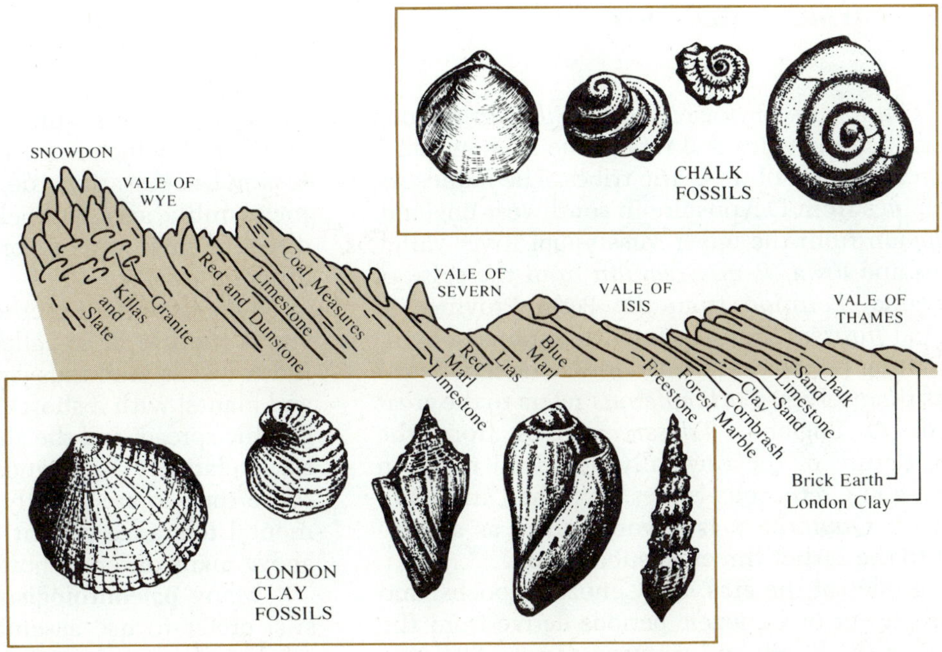

FIGURE 10–5
In the late 1700s, an English canal builder, William Smith, discovered the most important tool for relative age dating. Smith noted that fossils do not occur in a haphazard pattern in the rocks but rather in a definite and determinable sequence; hence, the age of a rock can be determined by its characteristic faunal content. (Drawing by Robert W. Tope) The cross section of England and Wales originally drawn by Smith is shown in the figure, along with some characteristic fossils from two of the units.

EARLY SUBDIVISIONS				MODERN USAGE			
Arduino 1760	Lehmann 1756 Füchsel 1760–1773	Werner ca. 1800	English Equivalents	Eras	Periods	Epochs	Alternate Periods
Volcanic alluvium	Aufgeschwemmt-gebirge	Angeschwemmt-gebirge Neues Flötzge-birge	Alluvium	CENOZOIC Phillips 1841	NEOGENE Hoernes 1853	Holocene Pleistocene	QUATERNARY Desnoyers 1829
						Pliocene Miocene	TERTIARY Arduino 1760
Tertiary			Tertiary		PALEOGENE Naumann 1866	Oligocene Eocene Paleocene	
Secondary	Flotzgeburge	Flötzgebirge	Secondary	MESOZOIC Phillips 1841	CRETACEOUS d'Halloy 1822		
					JURASSIC von Humboldt 1799		
					TRIASSIC von Alberti 1834		
				PALEOZOIC Sedgwick 1838	PERMIAN Murchison 1841		
					PENNSYLVANIAN Williams 1891		CARBONIFEROUS Conybeare & Philips 1822
					MISSISSIPPIAN Winchell 1870		
					DEVONIAN Murchison, Sedgwick 1837		
Primitive	Ganggeburge	Übergangs-gebirge	Transition		SILURIAN Murchison 1835		
					ORDOVICIAN Lapworth 1879		
					CAMBRIAN Sedgwick 1835		
		Urgebirge	Primary	PRECAMBRIAN			

FIGURE 10–6
The development of the relative geologic time scale. The four right-hand columns show the modern units: eras, which are divided into periods (in some cases with alternate sets), which in turn are divided into epochs. The four left-hand columns show the subdivisions proposed by the earlier workers and their rough correspondence to modern terms.

gin. The *Cambrian, Ordovician,* and *Silurian* periods were described in Wales and reflect the Roman name for the area and two of its Celtic tribes. The name *Devonian* comes from Devonshire in southwest England, *Mississippian* from the upper Mississippi River valley of Illinois and Iowa, *Pennsylvanian* from the state of Pennsylvania, *Permian* from the Perm Province of Russia, and *Jurassic* from the Jura Mountains of Switzerland. Other periods are named for characteristics of the rocks themselves. *Carboniferous* refers to the great coal beds of England, *Triassic* derives from the threefold nature of the originally described rocks in Germany, and *Cretaceous* comes from the Latin word for "chalk." *Quaternary* is a word coined as a complement to the earlier threefold subdivision.

The names of the eras, the Cenozoic epochs, and the alternate set of Cenozoic periods derive from the similarity of the fossils to living organisms. All rocks older than the first appearance of abundant fossils in the Cambrian are called *Precambrian.*

TIME CORRELATION BY FOSSILS

Faunal succession enabled geologists to establish the relative geologic time scale. Fossils can also be used to determine if two bodies of rock formed at the same time. This process is called **time correlation**. (There are other kinds of geologic correlation, which should not be confused with this process; for example, the determination of lithologic continuity or rock correlation.) In simplest terms, time correlation by fossils is based on the following principle: like fossils mean like ages. It is not necessarily true, however, that unlike fossils mean unlike ages, for rocks formed at the same time in different environments or geographic regions may have different fossils.

One important method of time correlation by fossils is the use of so-called **index fossils** (Fig. 10–7A). Index fossils are kinds (**species**) of organisms (animals and plants) with a short span of existence, a wide geographic spread, and the abundance necessary to be useful. Relatively few kinds of organisms qualify, and those that do are usually forms with a wide environmental tolerance or that live in the open water or air above many different bottom environments.

Many **paleontologists** (geologists who study fossils) prefer to use **assemblages of fossils** rather than single index species in making time correlations. An obvious advantage is that, to make a time correlation, it is not necessary to find one individual kind of fossil, but only a representative sampling. For example, two rocks with entirely different fossils can be correlated with each other because all their species belong to a single assemblage (Fig. 10–7B).

Finally, paleontologists have noted that each species has a span of existence, and by plotting these overlapping spans a series of unique assemblages, not occurring earlier or later in the column, can be identified (Fig. 10–8). The strata containing such unique assemblages are called **concurrent range zones**, and their use overcomes the potential problem in the preceding method that the two partial assemblages may have existed earlier or later in some areas than in others.

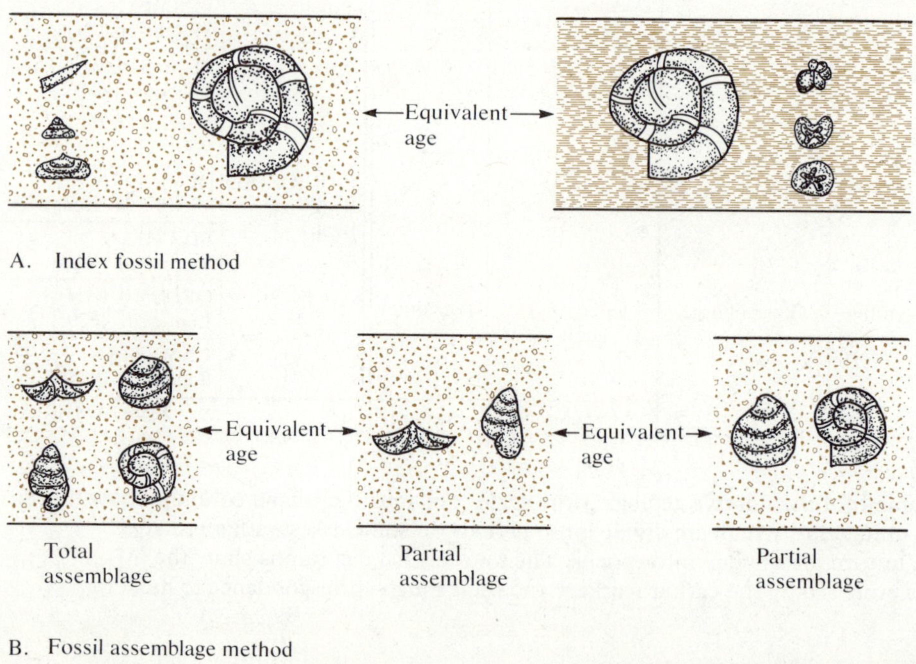

FIGURE 10–7
A. The index fossil method utilizes wide-ranging, short-lived, abundant species to make time correlations.
B. Many geologists prefer to use assemblages of species rather than index fossils. An advantage is that it is not necessary to find a single variety of fossil, but only a representative sampling, to make a time correlation. Even assemblages with no species in common can be correlated, as shown here.

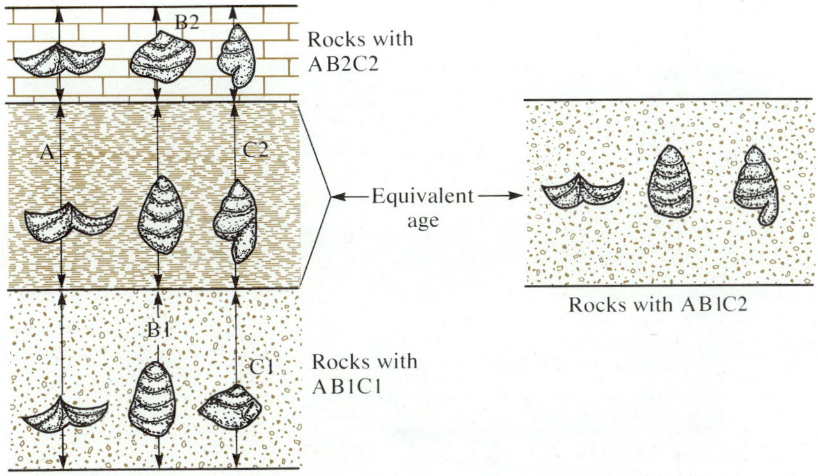

FIGURE 10–8
Concurrent range zone method. In this method, the precise vertical ranges of species are determined. As a result, there will be consecutive segments with unique fossil assemblages, as shown in the left-hand column. There are three successive assemblages, which are, from oldest to youngest: AB1C1, AB1C2, and AB2C2. A rock body with fossils AB1C2 would therefore be time-correlated with the middle zone.

PHYSICAL METHODS OF TIME CORRELATION

As noted previously, lithology is ordinarily not an indication of the age of a rock body. Particular lithologies occur whenever and wherever the proper environment occurs, as they do not undergo the progressive evolution of living things. However, some physical attributes of rocks are time-dependent and hence can be used for time correlation.

KEY BEDS

Certain physical events occur essentially instantaneously in geologic terms—in a few hours to a few years. If they are reasonably distinct from other similar occurrences, these so-called **key beds** can be used for time correlation. (Some key beds, such as an accumulation of dead shells by waves, are not time-dependent.) One of the most useful key beds for this purpose is the ash bed that results from an ash fall of a volcanic eruption. The key ash bed resulting from the eruption of Mount Mazama, a peak that collapsed to form the basin containing Crater Lake in Oregon, is an easily recognizable ash layer occurring across a large portion of the northwestern United States and the adjacent part of Canada (Fig. 10–9). A time-dependent key bed, such as a volcanic ash layer, provides a time plane that cuts across other lithologies, each of which formed in a different place at a different time.

VARVES

Another physical method of time correlation utilizes **varves** (Fig. 10–10). Varves are genetically related pairs, or couplets, of strata that result from seasonal changes in sedimentation. Common examples are glacial varves, which form in lakes at the fronts of glaciers. These varves contain a thicker, coarser summer layer formed when the ice melts and drops its heterogeneous debris, and a thinner, finer winter layer formed when the lake is frozen and the finest particles have an opportunity to settle out. Other varves occur in water bodies in regions of pronounced dry and wet seasons. In the dry season chemical precipitates form, whereas in the wet season rock particles (clastic debris) are washed into the sea from surrounding lands. Because each year is somewhat different from preceding and succeeding ones, succeeding couplets differ. Geologists can match identical couplets in different sections and hence time-correlate them.

EVENT CORRELATION

Event correlation, a relatively new physical method of time correlation, utilizes the principle that a certain physical event can have repercussions in a variety of environments. If two events can be recognized as resulting from the same cause, they can be time-correlated, as illustrated in Fig. 10–11. During the deposition of the strata shown in this figure, in two relatively short intervals sea level rose, thus drowning

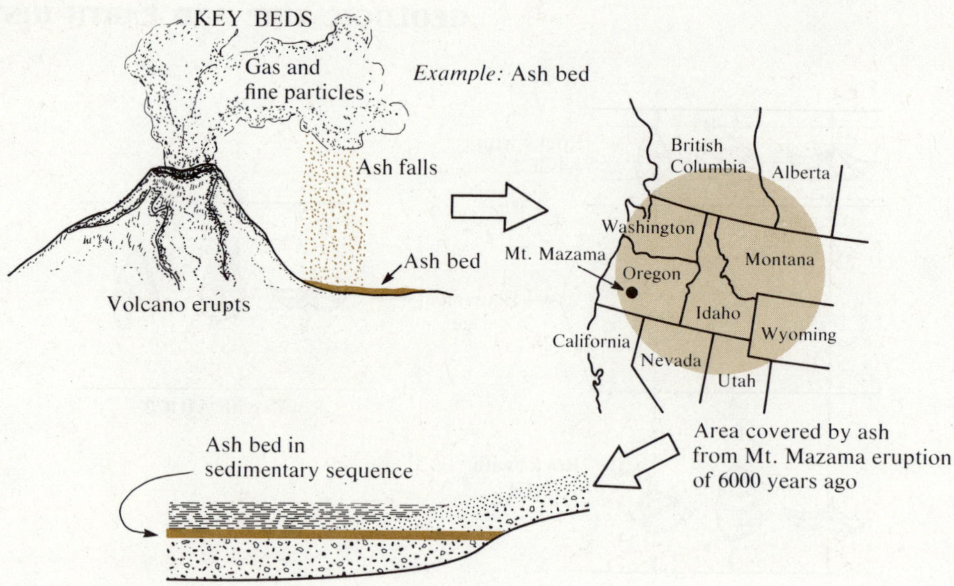

FIGURE 10–9
The area covered by the ash erupted when Mt. Mazama in Oregon underwent its final eruption and collapsed to form the caldera that today holds Crater Lake. The illustration also shows how a time-dependent ash bed typically cuts across other rock units of different ages. (After Stokes, 1966)

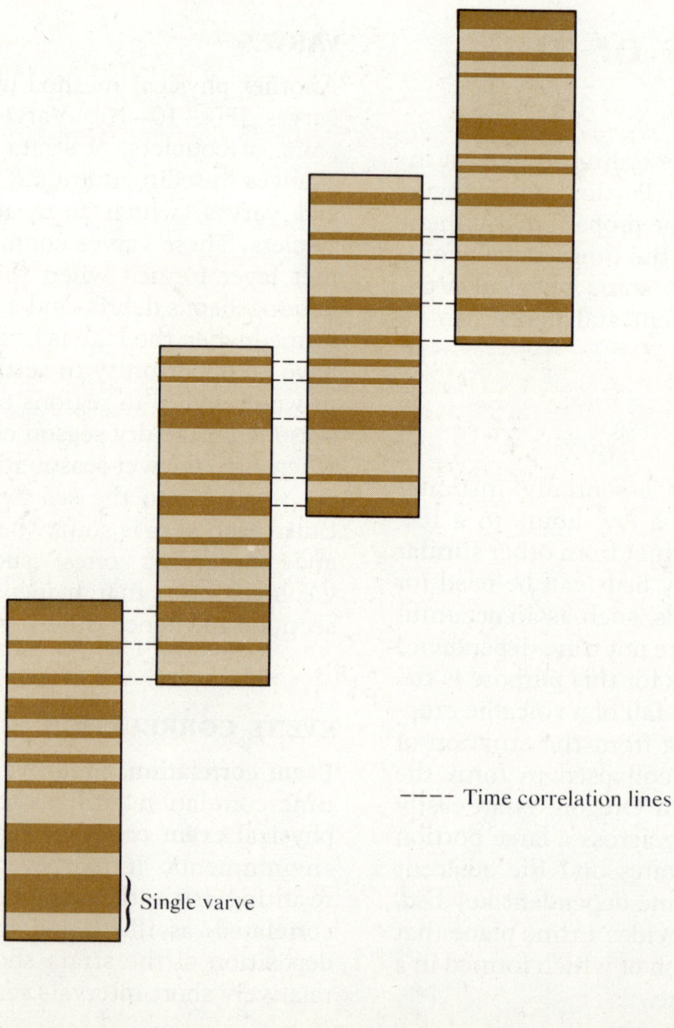

FIGURE 10–10
Varves, couplets of strata resulting from seasonal differences, such as cold versus warm or dry versus wet. Because two years are seldom identical and the sequence of different couplets changes through time, geologists can time-correlate by matching pairs of couplets.

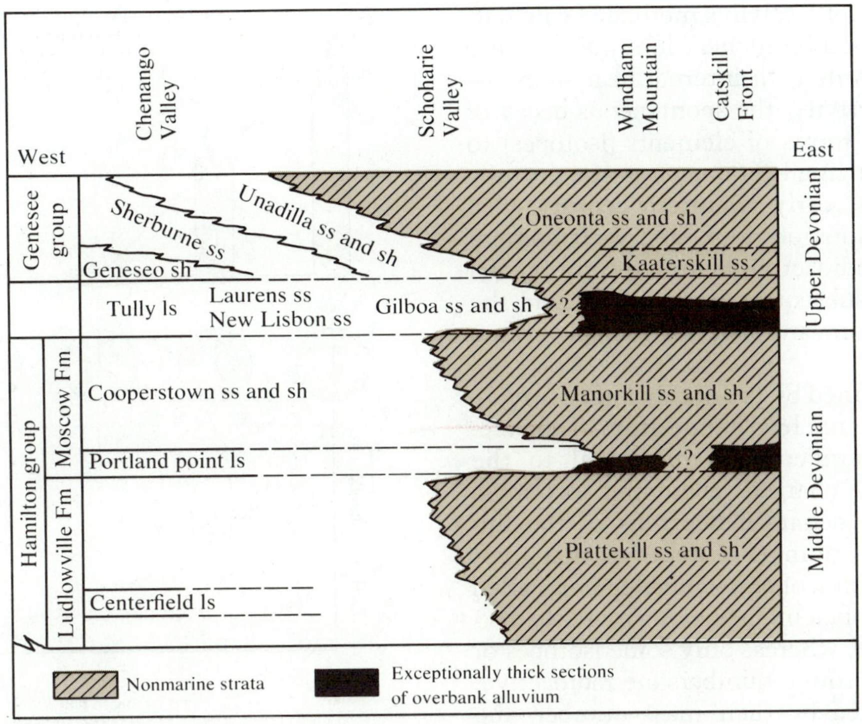

FIGURE 10–11
Event correlation is based on the principle that a certain physical event can produce repercussions in a variety of environments. This illustration shows a cross section of Devonian strata in New York. In two brief intervals, sea level rose, drowning the mouths of rivers coming from the eastern land area. In these intervals, streams dropped their loads of sediment at the heads of their drowned mouths, producing the thick alluvium shown in black. As a result, little sediment reached the western seas and limestone was deposited in the clear offshore waters. Thus, the limestones and alluvial deposits are time-correlative. (After I. N. McCave, *American Association of Physical Geologists Bulletin,* Vol. 53, 1969; from Robley K. Matthews, *Dynamic Stratigraphy.* © 1974. Reprinted by permission of Prentice-Hall, Inc., Englewood Cliffs, N.J.)

the mouths of rivers coming from the eastern land area. In these intervals, the streams dropped their loads of sediment at the heads of their drowned mouths (estuaries). Thus, clastic debris that ordinarily would have been carried into the western sea to form sandstones and shales was impounded in the estuaries. As a result, little debris was able to reach the ocean and muddy its water. Hence, the chemical precipitate limestone could form offshore. Geologists can time-correlate the thin limestones to the west with the thick sections of floodplain alluvium to the east.

PALEOMAGNETISM AND TIME CORRELATION

As noted in Chapter 8, strips of the ocean floor can be time-correlated by their normal or reversed magnetic polarity and their relative positions on either side of a spreading ridge. Discovery of this relationship has made it possible to time-correlate most areas of the ocean floor simply by towing a magnetometer behind a ship and recording the magnetic anomalies.

RADIOMETRIC DATING

All of the techniques described thus far involve relative dating, the determination of the time relationships of events with one another and with a standard of reference. Relative dating enables geologists to tell that a certain event occurred before another and after yet another one, and when in the relative time scale all three events occurred. But the question of how many years ago a particular event occurred, such as the origin of the earth, cannot be directly answered by relative dating. A means of **absolute dating**, that is, the age in years of events, is required.

As noted earlier, Lord Kelvin's method of calculating the earth's age based on its heat loss was a failure because the earth has its own internal heat source—radioactivity. Radioactivity, the spontaneous decay of unstable elements or forms of elements (isotopes) to stable elements, producing heat energy in the process, was discovered in the early twentieth century. (See Chapter 2 for definitions of atoms and their components.) Not only did radioactivity refute the notion of a cooling earth, but it subsequently proved to hold the key to the long-sought means of dating geologic events in terms of years.

An element is defined by the number of positively charged protons in the nucleus of one atom of the element (the atomic number), a figure equal to the number of negatively charged, encircling electrons. Atoms also contain uncharged neutrons in the nucleus. It is the variable number of neutrons that gives rise to the different forms of the same element, called **isotopes**. All isotopes of elements having more than 83 protons are radioactive, whereas only some isotopes of elements with lesser atomic numbers are radioactive. (Isotopes are designated by their mass number, the sum of the protons and neutrons in the nucleus.)

The decay of most substances involves two **half-lives** (Fig. 10–12). For example, a burning candle loses half of its mass in one interval (an hour, for example) and the other half in another equal interval (another hour). Radioactive decay proceeds in a different manner: one-half of the substance decays in the first half-life (an hour, for example), but only one-half of the remaining substance decays in the next half-life (another hour), and one-half of the remainder in the next half-life (yet another hour), and so on almost indefinitely until the last unstable radioactive atom decays. Each radioactive isotope has a fixed half-life that is unaffected by normal physical and chemical activities. Hence, knowing the half-life of a radioactive isotope and the amounts of the original isotope (**parent**) and the decay product (**daughter**), one can calculate the amount of time since the decay began and hence the age of the enclosing rock. This process is called **radiometric dating**.

A hypothetical example of radiometric dating will be useful in illustrating the methodology. The radioactive parent isotope in Fig. 10–13 has a half-life of 1 million years. If the ratio of parent to daughter isotope is 1/16 to 15/16, the enclosing rock has passed through four half-lives and therefore is 4 million years old.

In radiometric dating, it is very important that the enclosing rock has been a closed system, because the gain or loss of any parent or daughter isotope can cause the determination of erroneous dates. Thus, the radiometric date of a rock really reflects the last time it

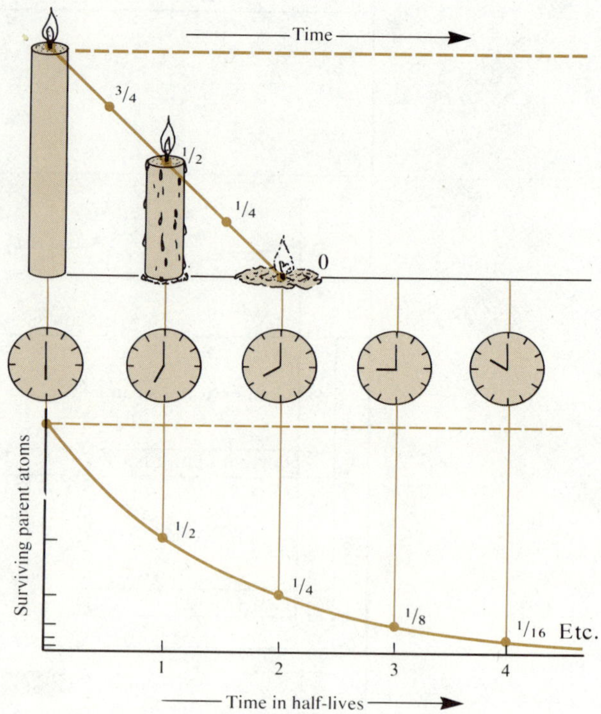

FIGURE 10–12
Radioactive substances spontaneously decay to nonradioactive ones. However, their breakdown differs from most ordinary decay processes, in which the first and second halves of a substance decay in equal intervals, as shown by the burning candle in the upper portion of this illustration. Radioactive substances have a different rate of decay, in which one-half of the *remaining amount* decays in each half-life. This exponential decay rate is depicted in the bottom part of the drawing. (After D. L. Eicher, 1968)

was a closed system—the time at which the radiometric "clock" was last set. Also, the presence of any daughter isotope in the rock when it formed, and hence daughter material not produced by decay of the parent isotope in that rock, can cause errors. Fortunately, the amount of nonradioactively produced daughter elements usually occurs in a constant proportion which can be subtracted to eliminate errors.

Major Radiometric Dating Methods There are four radioactive isotopes abundant enough in rocks of the earth's crust to serve as significant radiometric dating tools. These isotopes are given in Table 10–2, along with their half-lives, the major daughter isotopes, and the common minerals and rocks in which the isotopes occur. The uranium and rubidium isotopes have such long half-lives that they are useful only for dating very ancient rocks; too little of the daughter isotopes has accumulated in younger rocks to be accurately measured. The decays of the two radioactive isotopes of uranium to stable isotopes of lead are both long, step-

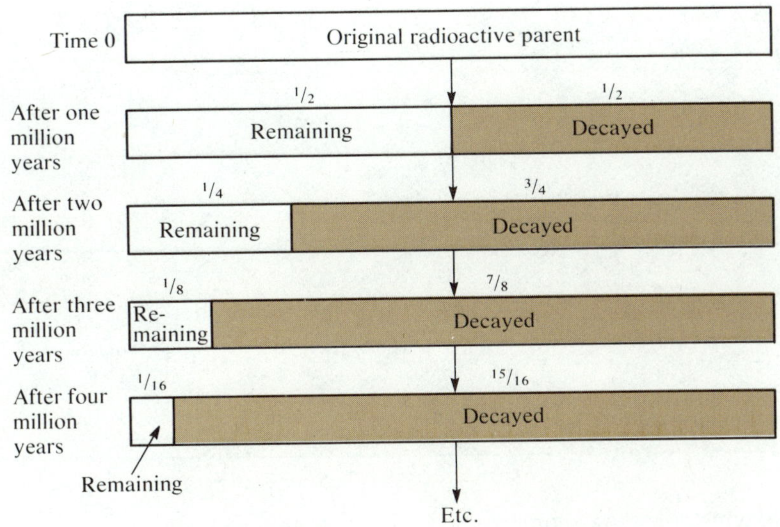

FIGURE 10–13
Diagram showing the decay of a radioactive substance with a half-life of 1 million years. In each half-life, one-half of the remaining sample decays. The decayed portion is shown in color.

TABLE 10–2
Radiometric dating methods.

Parent Isotope	Half-Life (millions of years)	Daughter Isotope	Minerals and Rocks Commonly Dated
Uranium 238	4,510	Lead 206	Zircon, uraninite, pitchblende
Uranium 235	713	Lead 207	Zircon, uraninite, pitchblende
Potassium 40	1,300	Argon 40	Muscovite, biotite, hornblende, glauconite, sanidine, whole volcanic rock
Rubidium 87	47,000	Strontium 87	Muscovite, biotite, lepidolite, microcline, glauconite, whole metamorphic rock

wise processes involving the loss from the nucleus of alpha particles (each consisting of two protons and two neutrons) and beta particles (each consisting of an electron formed by the breakup of a neutron into a proton and an electron). The decay of rubidium 87 to strontium 87 occurs in a single step of beta decay. The potassium-argon method is probably the most useful means of radiometric dating because of the moderate half-life of radiometric potassium 40, which allows its application throughout virtually the entire geologic record. The decay of potassium 40 to argon 40 occurs by electron capture; that is, an electron and a proton join to produce a neutron, thus reducing the number of protons in the nucleus by 1. Many of you probably have also heard of the carbon-14 method of radiometric dating, but this isotope has such a short half-life that it is useful primarily to archeologists working in Holocene and very late Pleistocene deposits.

Calibration of Radiometric and Relative Time Scales

As noted in the last column of Table 10–2, most of the minerals containing radioactive substances occur in igneous and metamorphic rocks (see Chapters 2, 3, and 6). Because the fossils upon which the relative time scale is based occur primarily in sedimentary rocks, there are problems in calibrating the absolute and relative time scales. Because of this, geologists must use crosscutting and superpositional relationships to relate radiometrically dated igneous and metamorphic rocks to sedimentary rocks (Fig. 10–14). Using this methodology, geologists have now calibrated the relative and radiometrically determined time scales. In Fig. 10–15, the relative lengths of all the eras are shown on the left. Note the extremely long duration of the Precambrian, which constitutes the bulk of geologic time. The five columns to the right focus primarily on the Paleozoic, Mesozoic, and Cenozoic eras. They show the progressive refinement of the calibration, beginning with Holmes's pioneering work in 1937 and ending with the 1964 Geological Society of London (GSL) time scale. Subsequent refinement has produced little alteration in the GSL dates.

ORIGIN OF THE EARTH

The remainder of this chapter is devoted to a brief overview of our planet's past. The earth is believed to

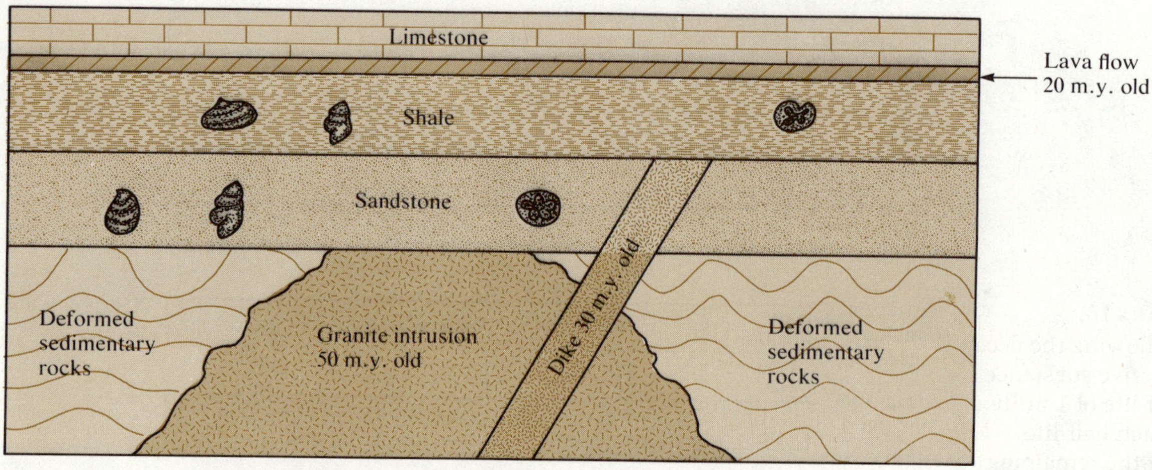

FIGURE 10–14
A hypothetical example showing how the radiometric and relative time scales can be calibrated. Here, there are three radiometrically dated igneous bodies: a 50-million-year-old granite, a 30-million-year-old dike, and a 20-million-year-old lava flow. The first two can be related to the sedimentary rocks by crosscutting relationships; the last, by superposition. The deformed sediments are cut by the granite and must be older than 50 million years, though there is no direct evidence in this section of how much older. The sandstone truncates the upper surface of the granite and is in turn cut by the dike. Thus, it is between 50 and 30 million years in age. The shale truncates the dike, and the lava flow superpositionally rests upon them both. Thus, the shale is between 30 and 20 million years old. The limestone rests upon the lava flow and must be younger than 20 million years in age, but how much is not determinable in this section.

have formed approximately 4.6 billion years ago, though no rocks of that age are known on earth. The oldest rocks are found in Greenland and date from about 3.8 billion years ago. Scientists believe the earth's age to be about 4.6 billion years because meteorites—assumed to be fragments left over from the formation of the planets—and rocks of the moon—preserved since its origin because of that body's inert nature—date from this time. Extrapolating backwards the ratios of the various isotopes of lead to a time when there was no radioactively produced lead also yields this date.

The formation of planets appears to go hand in hand with star formation, and many stages in this process can be seen in the universe today. From these observations and calculations based upon physical laws, the following discussion, illustrated in Fig. 10–16, is the probable scenario for the origin of the earth. In step 1, cold dust and gas particles collected in a region of space. Once a certain mass accumulated, the body began to rotate and flatten according to physical laws. The resultant body of matter was a **nebula**. In step 2, turbulence created eddies within the rotating nebula. Individual layers detached as the inward force of gravity and the outward centrifugal force balanced. As this process continued, planetary bodies and a large central body formed by "sweeping up" the material in their vicinities (step 3). Small-scale repetition of this process within each planetary region led to the formation of satellites. The so-called **protoplanets** of this stage were much larger than the present planets. When the central mass attained a critical size, its internal pressures and temperatures reached the point where nuclear fusion (the creation of helium from hydrogen) began, and a star—the sun—was born. The burst of radiant energy (the **solar wind**) from the nuclear reactions of the sun drove off most of the gaseous components of the protoplanets, thus reducing their sizes. Finally, the solar system assumed its modern configuration (step 4).

DENSITY DIFFERENTIATION OF THE EARTH

After formation of the primitive earth and the loss of its volatile (gaseous) components, the planet underwent a density differentiation (Fig. 10–17). Because of the earth's own internal heat, generated by gravitational pressure and radioactivity, much of the earth's interior probably became molten, and convection occurred. The very heavy, denser elements such as iron (and possibly nickel) sank to the interior to form the

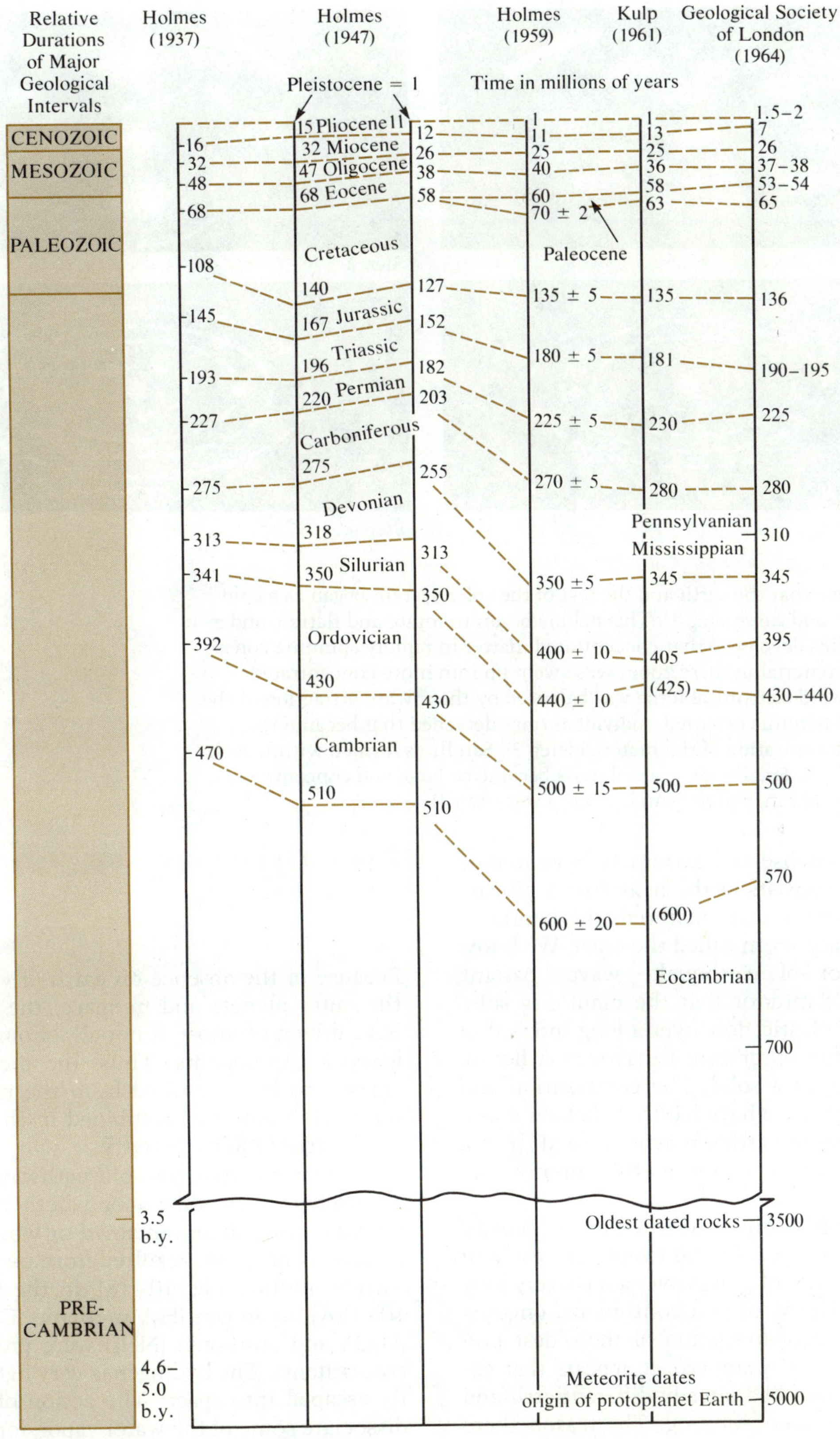

FIGURE 10–15
Calibration of relative and radiometric time scales. The Geological Society of London dates have been modified only slightly in recent years, an indication that they are essentially correct.

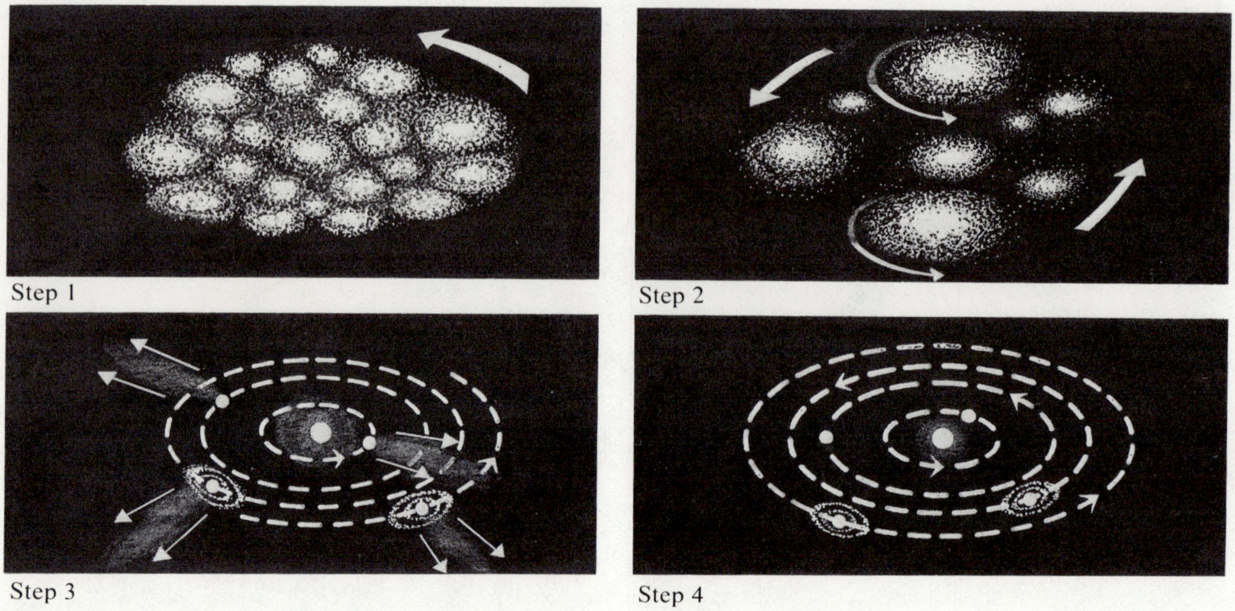

FIGURE 10–16
Most scientists believe that the earth and the rest of the solar system began as a cold, nebular cloud of dust and gases (step 1). This nebula began to rotate and flatten, and as it did so, turbulent eddies developed that concentrated matter in rapidly spinning zones (step 2). Eventually, material in these zones was swept up into more concentrated masses, and the outward centrifugal force was balanced by the inward attraction of the large central mass. When this occurred, individual rings detached that became the planets by further concentration of the material (step 3). Satellites formed within each ring by a similar process. Finally, the central mass became so large and concentrated that nuclear reactions began, and the sun became a star (step 4).

earth's **core**. A less dense and extremely large middle region, or **mantle**, consists of the heavy, dark silicate minerals. The lighter silicate materials differentiated to form a thin surface scum called the **crust**. We know from the behavior of earthquake waves passing through the earth's interior that the mantle is solid (though capable of plastic flow over a long interval of sustained stress); the outer core behaves as a liquid, and the inner core, as a solid. The composition and characteristics of the core have led to its behavior as a magnet generating the earth's magnetic field. (For a more detailed discussion of the earth's interior, see Chapter 7.)

Scientists are uncertain as to whether the density differentiation occurred as a rapid event very early in earth history, in a series of pulses over a relatively long interval of earth history, or is a continuous, ongoing process. By the time of formation of the oldest preserved rocks, 3.8 billion years ago, it appears that extensive granitic terrain had already differentiated and incipient continents were forming. The reason there are no known earlier rocks is that the crust was probably too unstable in earlier times for rocks to be preserved. Any older crust was likely melted and absorbed back into the interior.

EVOLUTION OF THE ATMOSPHERE

Because of the absence on earth of volatiles found on the outer planets and in space, the solar wind must have driven off most, if not all, of our planet's original gaseous components. Thus, the present atmosphere appears to have resulted from the release of gaseous materials trapped or combined in the rocks and from subsequent biologic activity.

There are two possible pathways to the present atmosphere, depending upon whether a remnant of the original atmosphere remained or whether most of the present atmosphere resulted from outgassing from the earth's interior (Fig. 10–18). In the former case (Fig. 10–18A), hydrogen (H_2), methane (CH_4), water vapor (H_2O), and ammonia (NH_3) were probably the major constituents. The hydrogen is very light, and thus readily escaped into space. The action of sunlight would dissociate some of the water vapor, forming free hydrogen, which would escape to space, and free oxygen (O_2), which would react with methane to form carbon dioxide (CO_2) and water, and with ammonia to form free nitrogen (N_2) and water. Nitrogen is the major

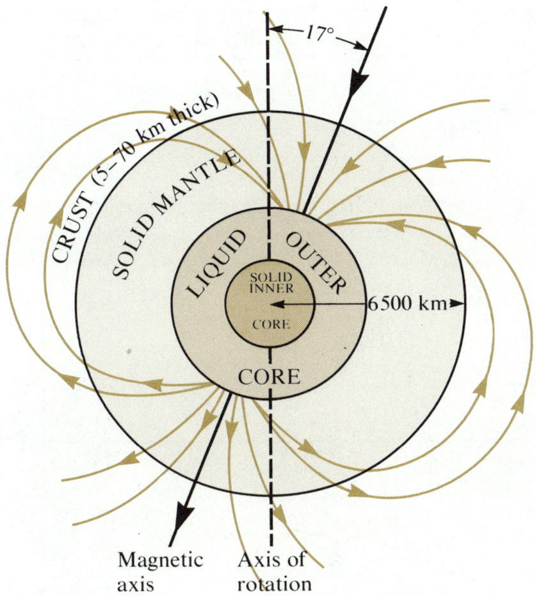

FIGURE 10–17
After its formation, the earth underwent a density differentiation, probably by thermal convection. The planet heated up under its own internal gravitational pressure and by the decay of contained radioactive substances. As a result, the heaviest elements sank to the interior to form a dense metallic core, the intermediate elements formed the moderately dense silicates of the mantle, and the lightest elements rose to the surface to form the thin crust of light silicates. The metallic core is responsible for the earth's magnetic field, which has had a profound effect on earth history, particularly in terms of the record left by plate tectonic processes.

component of the present atmosphere, and the water resulting from these reactions would add to that already present to form the oceans. Carbon dioxide is taken up by plants in their metabolism, and oxygen is released. Oxygen is the second most abundant constituent of the modern atmosphere and appears to be primarily the result of biologic activity.

In the other possible scenario (Fig. 10–18B), the gases that would escape from the structure of rocks would primarily be hydrogen, carbon monoxide (CO), carbon dioxide, water vapor, and nitrogen. As in the first possible sequence of events, hydrogen would escape to space, and sunlight would dissociate water vapor into free hydrogen and oxygen. Carbon monoxide and the other constituents would be acted upon by various energy sources to produce the first life, as shall be discussed later in the chapter. Carbon dioxide and water would be taken up by organisms once they appeared and, using the energy of sunlight, converted into organic compounds and oxygen by the process called **photosynthesis**. In **respiration**, organisms break down organic compounds, using oxygen and releasing carbon dioxide and water.

For both of the scenarios, the result would be the modern atmosphere, which is 78 percent nitrogen, 21 percent oxygen, 0.033 percent carbon dioxide, and has a variable amount (a trace to 4 percent) of water vapor and traces of several other gases including hydrogen and methane. Most of the oxygen was probably produced by photosynthesis. Also note that in both scenarios, there was virtually no oxygen in the primitive atmosphere. Hence, the earth had a reducing atmosphere (rather than the present oxidizing one) in its early history.

ORIGIN OF LIFE

Regardless of whether one starts with hydrogen, methane, water, and ammonia, or with hydrogen, carbon monoxide, carbon dioxide, water, and nitrogen, it is possible to derive the compounds that lead to the formation of living organisms. As noted in Table 10–3, life consists of six basic types of chemical compounds, each of which has a specific function and composition.

TABLE 10–3
Principal chemical constituents of life.

Compound	Functions	Composition
Water	Universal solvent	Hydrogen, oxygen
Carbohydrates	Energy source	Hydrogen, oxygen, carbon
Fats (lipids)	Energy storage	Hydrogen, oxygen, carbon
Adenosine phosphates (ADP, ATP)	Energy transfer	Hydrogen, oxygen, carbon, nitrogen, phosphorus
Proteins	Structural; facilitation of chemical reactions	Hydrogen, oxygen, carbon, nitrogen, phosphorus, sulfur (organized into 20 amino acid building blocks)
Nucleic acids (DNA, RNA)	Patterns for protein synthesis	Hydrogen, oxygen, carbon, nitrogen, phosphorus (organized into 5 nucleotide building blocks)

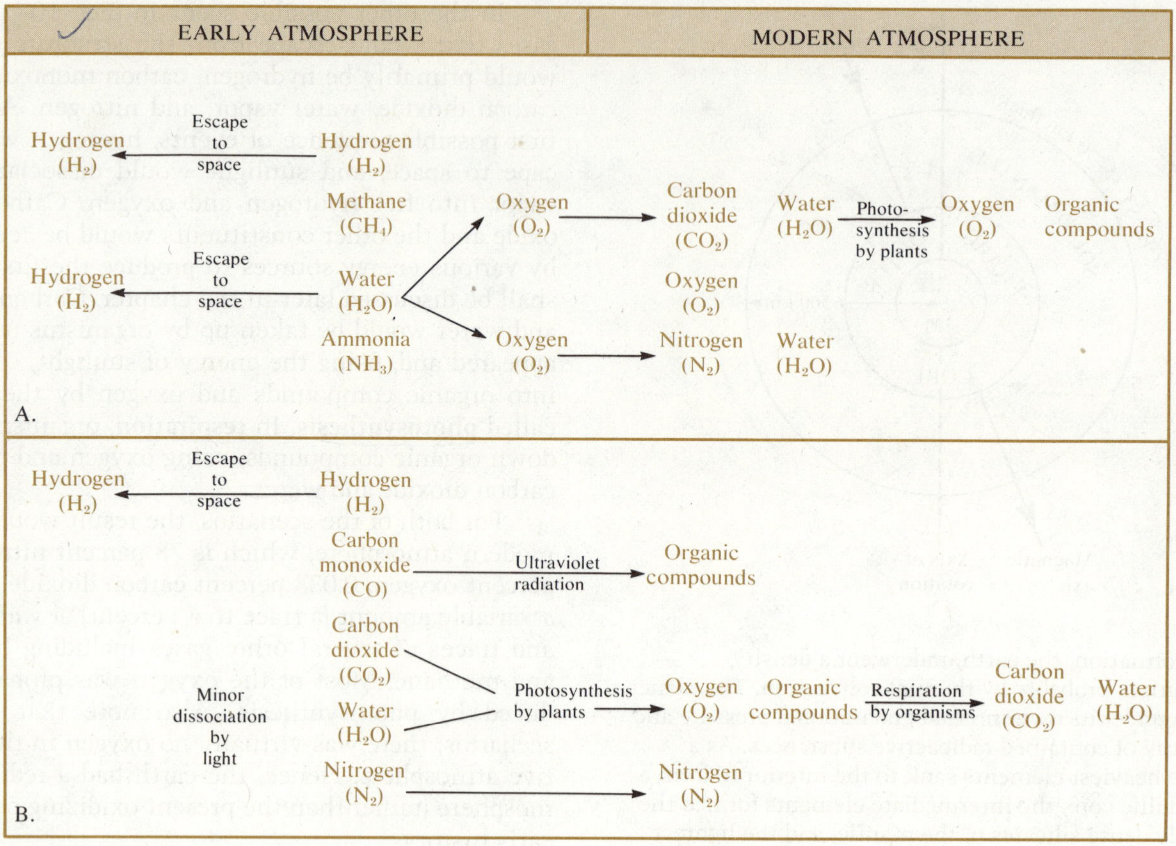

FIGURE 10–18
The two probable paths of the evolution of the earth's atmosphere.

Utilizing energy sources available on the early earth (ultraviolet light, lightning, thunderclaps, etc.), biochemists have been able to aggregate the basic compounds present in either scenario of the earth's early atmospheric-oceanic evolution and create parts or all of these building blocks of life. Although no one has yet duplicated every step in the transformation of simple inorganic compounds into a primitive organisms, many of the steps have been performed, indicating that life could certainly have arisen from the cosmically abundant substances hydrogen, methane, water vapor, ammonia, carbon monoxide, carbon dioxide, and nitrogen.

The first life was undoubtedly very simple and lived off the energy stored in the organic compounds of the "soup" in which it had evolved. The process of **fermentation**, which still exists in the primitive yeasts, probably represents survival of this ancient metabolism. In fermentation, organic compounds such as sugars are broken up to produce carbon dioxide, alcohol, and the energy necessary for an organism to function (Fig. 10–19A). Fermentation depends on an external supply of organic substances, and life would not have progressed far using it as the sole energy source. Once organisms evolved the ability to photosynthesize, they could manufacture their own organic substances from the simple organic compounds carbon dioxide and water using sunlight as a readily available energy source (Fig. 10–19B). With photosynthesis to produce organic substances and fermentation to break them apart and release energy, early organisms had a complete, but inefficient, metabolic cycle. The free oxygen released in photosynthesis was a poison to life until certain chemicals were evolved to allow its use in metabolism. Once this occurred, respiration replaced fermentation as the major process whereby organisms "burn" organic compounds to release metabolic energy (Fig. 10–19C).

THE PRECAMBRIAN

PRECAMBRIAN ROCKS

The **Precambrian** is the long interval from the origin of the earth 4.6 billion years ago to the appearance of abundant fossils of the larger marine shelled animals 600 million years ago, at the beginning of the Cam-

A. Fermentation

$$C_6H_{12}O_6 \longrightarrow 2\,CO_2 + 2\,C_2H_5OH + \text{energy}$$

Sugar — Carbon dioxide — Alcohol

B. Photosynthesis

$$6\,CO_2 + 6\,H_2O + \text{energy} \longrightarrow C_6H_{12}O_6 + 6\,O_2$$

Carbon dioxide — Water — Sugar — Oxygen

C. Respiration

$$C_6H_{12}O_6 + 6\,O_2 \longrightarrow 6\,CO_2 + 6\,H_2O + \text{energy}$$

Sugar — Oxygen — Carbon dioxide — Water

FIGURE 10–19
The major chemical equations important to the evolution of life. Fermentation is not efficient and is dependent on an external supply of organic substances. Once organisms evolved the ability to photosynthesize, they could manufacture their own organic compounds from simple, readily available inorganic compounds using sunlight as an energy source. Finally, when organisms developed the ability to utilize the free oxygen released in photosynthesis, they could use it to "burn" organic compounds and release energy. This process, respiration, is much more efficient than fermentation and is used by virtually all living things.

brian period of the Paleozoic era. Although it encompasses seven-eighths of the earth's history, the Precambrian is not as well understood as the remainder of geologic time, for which fossils allow precise, fine subdivisions. Furthermore, Precambrian rocks have been around for so long that their opportunities for deformation and destruction by plate tectonic processes have been many. Each continent, however, does contain one or more areas where extensive Precambrian rocks occur and have not been deformed since the beginning of the Cambrian. These areas are called **cratons**, and the areas within them where the Precambrian rocks are exposed and not covered by a thin veneer of younger rocks are called **shields**. The location and extent of the shields and the inferred boundaries of the cratons are shown in Fig. 10–20. As seen on this map, there appear to be no systematic relationships between shields and other geologic features such as continental boundaries. Nor do shields show any systematic relationship to plate boundaries.

Each shield records a long history of sedimentation and mountain-building events, but it has not yet proved possible to provide an accurate picture of Precambrian plate tectonics. Paleomagnetic studies and climatologic evidence indicate significant plate movements as in later eras. In the earliest Precambrian or **Archeozoic** record (from 3.8 billion to 2.5 billion years ago), the earth's crust was thinner and plate tectonic activity occurred over broader zones rather than in the present narrow marginal zones. The later Precambrian record, the **Proterozoic**, primarily records the continued operation of modern physical processes on the earth's surface. The only significant difference is the much smaller amount of atmospheric oxygen which was probably available in the Precambrian. This is reflected in a few unusual types of sediments found chiefly in the early Proterozoic, and in the lack of any larger animals and plants until just before the beginning of the Paleozoic era.

PRECAMBRIAN HISTORY

The history of North America's craton and the Canadian Shield will be examined in some detail because they are probably the most familiar to you. Radiometric dating indicates that there are at least seven separate regions of the Precambrian craton based upon the ages of final deformation (Fig. 10–21). The oldest, the **Superior Province**, is a broad east-west belt between the Great Lakes and Hudson Bay. Final deformation occurred about 2.5 billion years ago. (Isolated patches of the same age occur in the Slave Province of northwest Canada, the eastern Nain Province of Labrador,

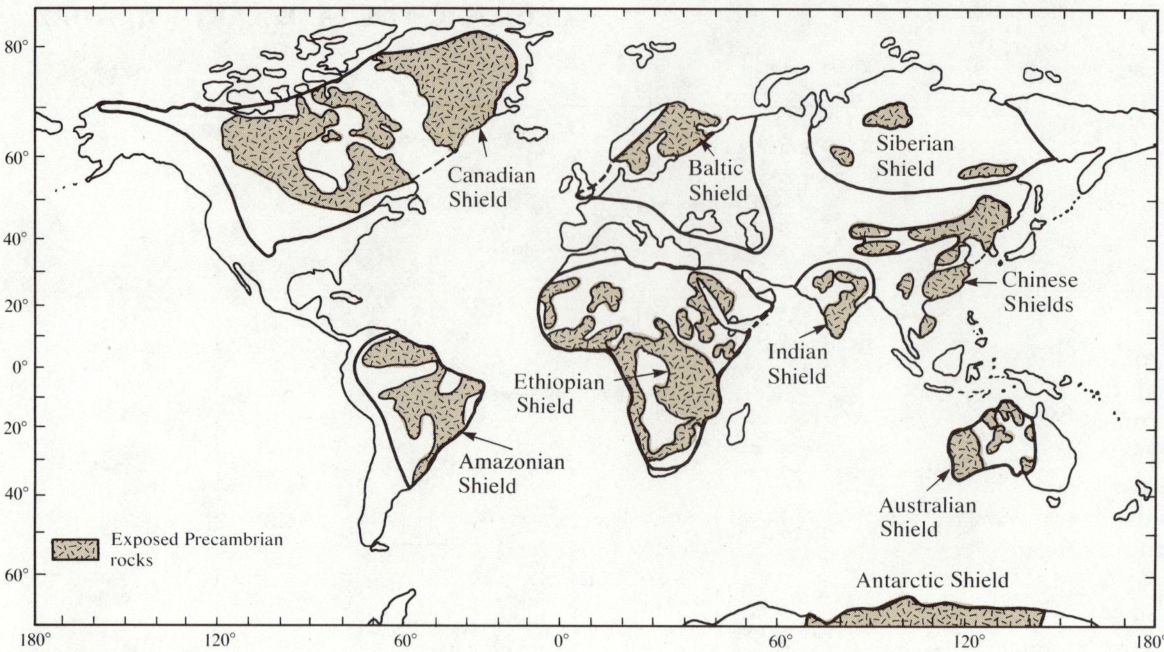

FIGURE 10–20
Cratons and shields. Many Precambrian rocks have been highly altered by subsequent plate movements, but some extensive areas have escaped intense deformation since the Precambrian. These areas are called cratons and are partially covered by a thin veneer of younger sedimentary rocks. The portions where Precambrian rocks are exposed at the surface are known as shields. (From B. Windley, *The Evolving Continents,* © 1977 by John Wiley and Sons. Used with permission)

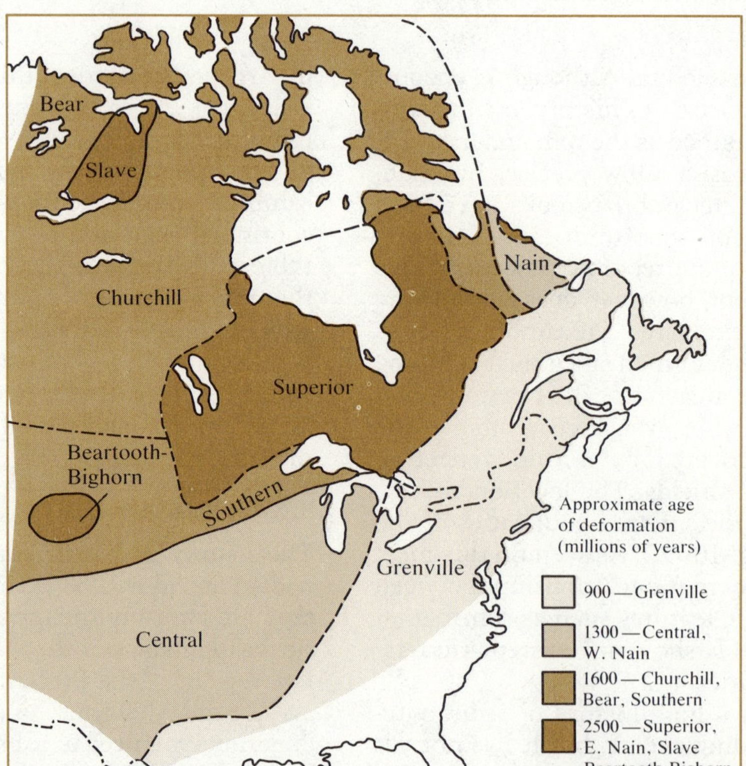

FIGURE 10–21
The North American craton, including the Canadian Shield, displays a geologic record indicating four major episodes of deformation at 2.5 billion, 1.6 billion, 1.3 billion, and 900 million years ago. Each period of deformation mainly affected a particular large region, or province, known respectively as the Superior, Churchill, Central, and Grenville provinces.

and the Beartooth-Bighorn Province of the Rocky Mountains.) Surrounding the Superior Province on all sides except the southeast is the **Churchill Province** (its southern extension in the Great Lakes area is the Southern Province, and its northwestern extension is the Bear Province). The Churchill Province was last deformed around 1.6 billion years ago. Approximately 1.3 billion years ago, the **Central Province**, which underlies much of the midwestern and southwestern United States, and the western Nain Province of Labrador were deformed. Finally, 900 million years ago the eastern margin of the craton, as well as the Appalachian area, was deformed. This area of the craton is called the **Grenville Province**. All of these periods of deformation and mountain building testify to the existence of previous plate boundaries in these regions.

In Fig. 10–22, the history of the Canadian Shield is summarized in tabular form. In the left columns, entitled "Standard Subdivisions," are shown the two traditional eras—the Archeozoic (Archean) and the Proterozoic—each broken into periods of sedimentary deposition separated by mountain building and intrusion. In reality, these are not time divisions, but rather local rock names. The Keewatin rocks are similar to oceanic ridge types, the Huronian rocks are shelf sea deposits with some glacial debris, Keweenawan rocks resemble rocks of modern rift zones, and various other types occur. In the middle columns, these are translated into modern terminology, and radiometric dates are given. The Canadian Shield and North American craton were deformed by five major periods of mountain building (orogeny). The last four—**Kenoran**, **Hudsonian**, **Elsonian**, and **Grenville**—are the ones that define the major provinces shown in Fig. 10–21. (The earliest, or **Saganagan**, affected at least part of the Superior Province, but the later Kenoran obliterated much of its effects.) The new time-unit columns at the far right recognize four eras defined on the basis of the three major orogenies. These eras, from oldest to youngest, are: Archean, Aphebian, Helikian, and Hadrynian (the last three are frequently grouped together as the Proterozoic era). The late Hadrynian partially overlaps the **Eocambrian**, a late phase of the Precambrian containing the oldest fossils of large, many-celled animals.

PRECAMBRIAN FOSSILS

Most of the Precambrian is characterized by the absence of fossils larger than microscopic size. Apparently, the larger multicellular animals and plants did not evolve until very late in the Precambrian. The only Precambrian fossils large enough to be seen with the naked eye are large, many-layered mounds called **stromatolites**. These were built by colonies of a primitive type of plant known as **blue-green algae**. In recent years, a number of assemblages of microscopic fossils have been found in Precambrian rocks (Fig. 10–23).

The oldest known fossils come from Archeozoic rocks of approximately 3.5 billion years in age in western Australia and the Fig Tree Group of South Africa of about 3 billion years in age. The fossils are of the most primitive living creatures—bacteria and blue-green algae. These tiny, one-celled plants are so primitive that they do not have a nucleus of hereditary material as do higher organisms.

The Gunflint Chert assemblage from southern Canada is from much younger rocks of Aphebian or early Proterozoic age (2 billion years in age). Bacteria and blue-green algae also occur in this unit, but there are a number of more complex one-celled creatures possibly belonging to other algal and fungal groups, though no definite nuclei have been noted. Lastly, the Bittersprings Chert of Australia contains a more diverse Hadrynian or late Proterozoic microfossil assemblage (900 million years in age). Colonial blue-green algae, flagellate protozoa (one-celled animals), and green algae with apparent nuclei preserved are elements of this assemblage.

EDIACARA FAUNA

In recent years, the latest Precambrian (Eocambrian) rocks have yielded the oldest known fossils of complex, multicellular animals. Although examples of this assemblage are known in various localities, the first discovered and most prolific locality is in the Ediacara Hills of southern Australia. No hard parts are known; the animals are preserved only as impressions in the sediment. At least six major kinds of animals can be recognized (Fig. 10–24). Two of these are assigned to the modern group of the coelenterates, which includes living jellyfish, sea anemones, and corals. Some of these Ediacara fossils closely resemble certain colonial corals called sea pens, and others represent a number of jellyfish species. Other varieties appear to be segmented worms, or annelids. One is flattened and very wormlike; the other has a head shield like the primitive arthropods called trilobites. The Ediacara form may represent a transitional species between these two great animal groups. The last two are more enigmatic fossil types. One is shieldlike and bears a faint re-

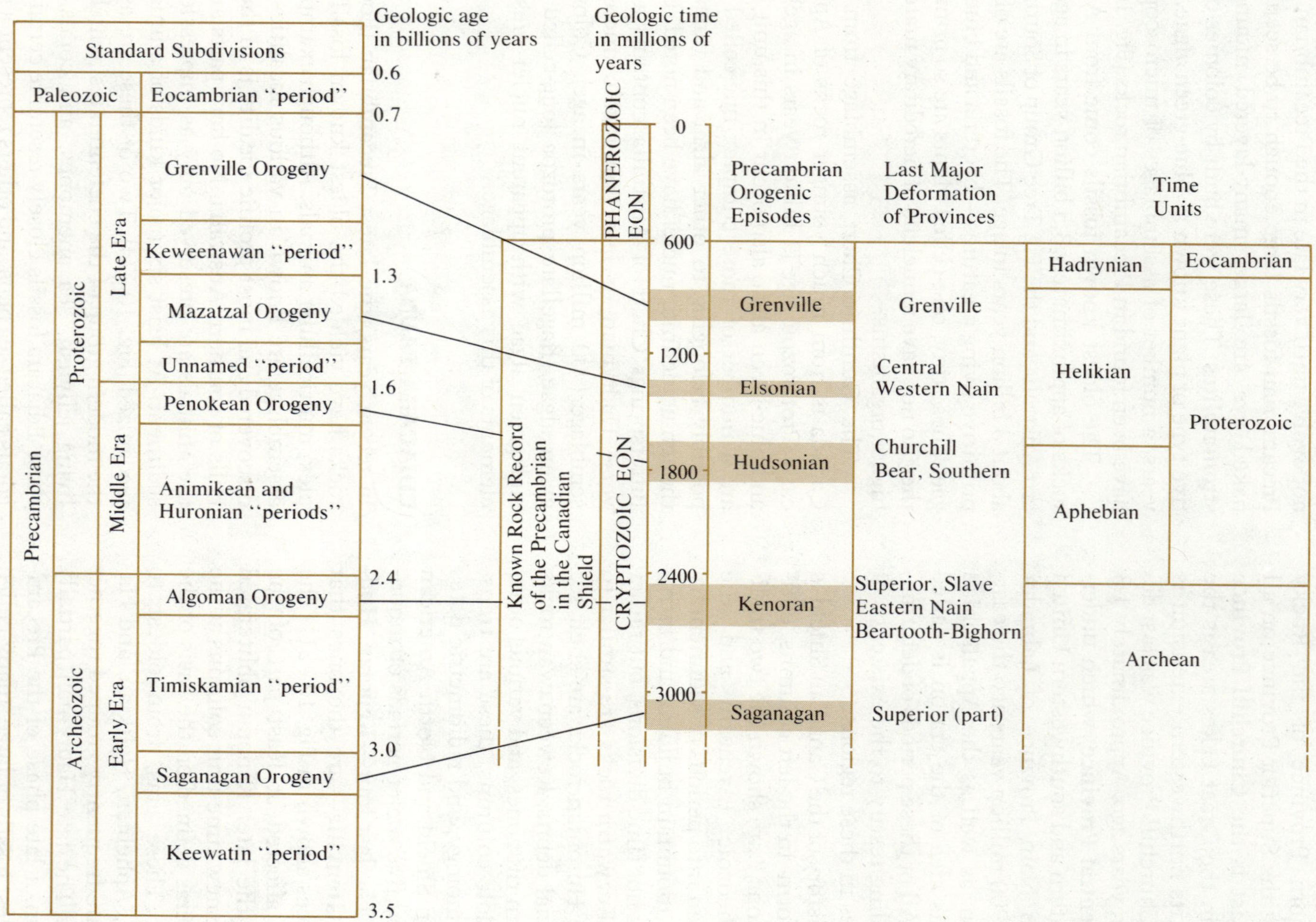

FIGURE 10–22
A chart summary of the known geologic history of the Canadian Shield. The left columns show the older standard subdivisions, which were primarily based upon local rock units. The right side shows the newer chronology, both radiometric and relative, the provinces affected, and the new time units.

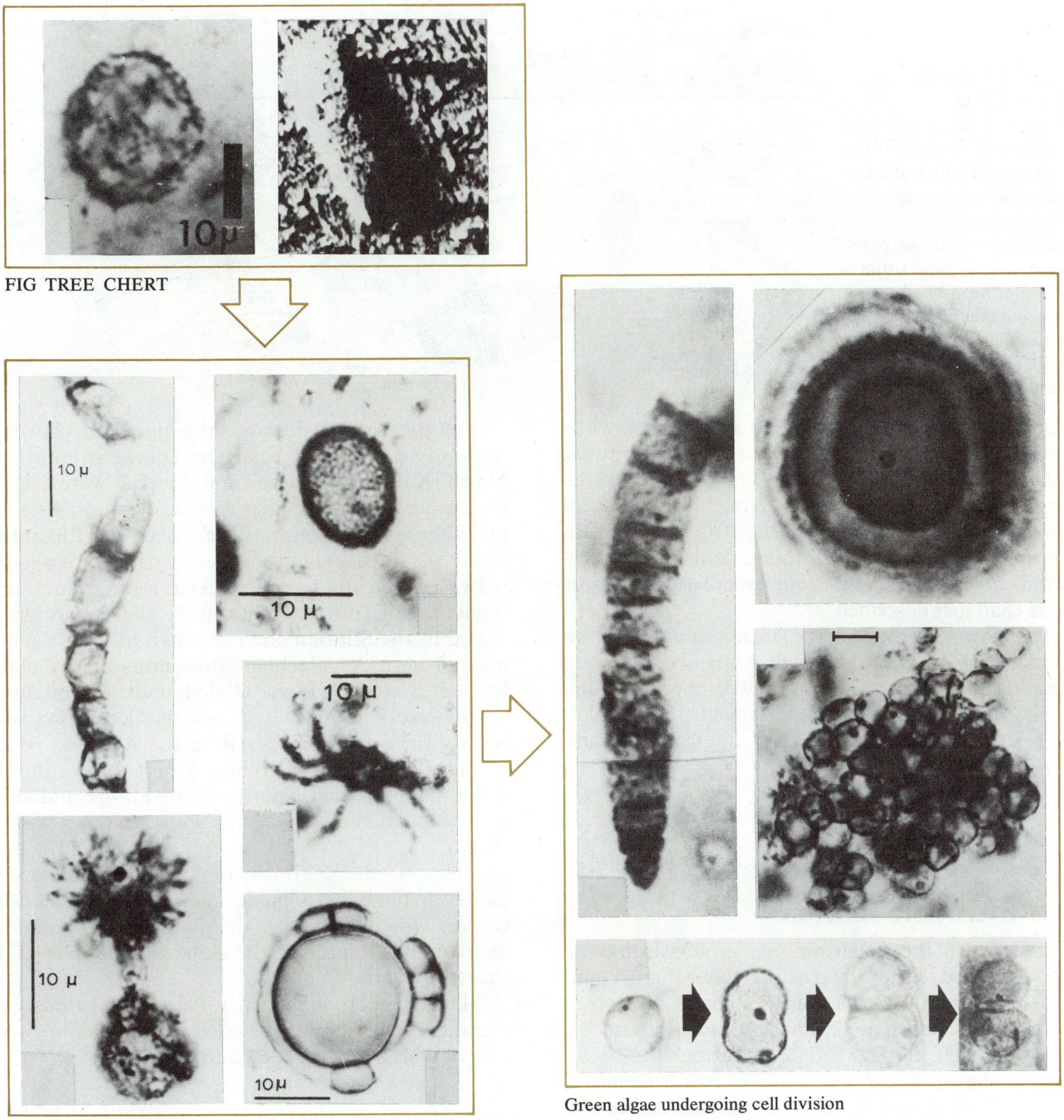

FIGURE 10–23
Three stages in the evolution of Precambrian fossils. The 3-billion-year-old Fig Tree Chert contains only one-celled bacteria and blue-green algae. The 2-billion-year-old Gunflint Chert contains a great variety of microfossils, including single and complex one-celled organisms, some of which were colonial. (Photos by E. S. Barghoorn) Finally, the 1-billion-year-old Bittersprings Chert contains, in addition to forms like those in the older rocks, more complex forms, including the first evidence of a cell nucleus. (Photo courtesy J. W. Schopf)

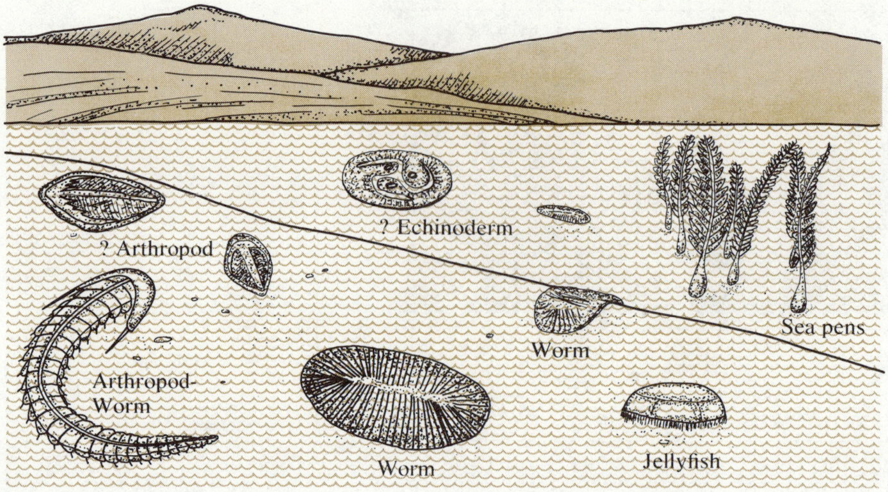

FIGURE 10–24
Best known from the Ediacara Hills of south-central Australia, this fauna consists entirely of the impressions of soft-bodied animals. Dome-shaped jellyfish, colonial corals resembling modern sea pens, flattened worms, and other worms that have head shields and resemble arthropods are present. Possible arthropods and echinoderms also occur.

semblance to early arthropods such as trilobites, while the other is flattened and has three armlike structures. An extinct type of echinoderm (the group to which modern starfish and sea urchins belong) is reminiscent of this form, but it has five armlike structures. A variety of burrows and trails of other organisms are present, indicating that the Ediacara fauna was even more diverse than just described.

The reason for the sudden, late appearance of multicellular animals is obscure, but it may be related to the atmosphere's containing sufficient oxygen to support the metabolism of complex organisms. Perhaps such a level was not reached until the latest Precambrian time.

THE PALEOZOIC

EARLY PALEOZOIC HISTORY

Beginning with the **Paleozoic era**, geologists have a reasonably good idea of the past positions of most continents. See Fig. 10–25. The early Paleozoic consists of three periods: Cambrian, Ordovician, and Silurian.

As shown in Fig. 10–25, during the early Paleozoic the southern continents of South America, Africa, Australia, and Antarctica, as well as Arabia, India, and part of southern Europe and the southern United States, were aggregated into a huge supercontinent called **Gondwanaland**. Gondwanaland came into being in the late Precambrian in a number of continental collisions, and it persisted throughout the entire Paleozoic era. Most of China is shown as part of Gondwanaland in this restoration, but this interpretation is by no means certain. Gondwanaland was situated in polar latitudes in the early Paleozoic, and extensive continental glaciation is known from west Africa.

In the early Paleozoic, the northern continents were more scattered than the southern continents. North America, Europe, and Siberia were separate landmasses. Three smaller pieces of eastern Asia were probably also separate, as shown in the illustration. North America and Europe were separated by a narrow seaway called the **Iapetus Ocean**, a forerunner of the modern Atlantic that opened in the Cambrian but closed in the Silurian and Devonian to form the original northern **Appalachian Mountains** of the eastern United States and the **Caledonian Mountains** of northwestern Europe. The Iapetus Ocean had somewhat different boundaries than the Atlantic, so fragments of both Paleozoic North America and Europe are now on opposite sides of the ocean. Europe and Siberia were separated in the early Paleozoic by the **Ural Ocean**.

The northern continents were separated from Gondwanaland by a broad seaway called the **Theic Ocean**, which closed in the late Paleozoic. The remainder of the space on the globe was taken by a huge ancestral **Pacific Ocean**.

During each of the early Paleozoic periods, most continents were widely flooded by vast shallow-water seas in which marine life flourished.

EARLY PALEOZOIC LIFE

The early Paleozoic is often called the age of marine invertebrates because the seas teemed with animals without backbones, though there was no life on land until the late Silurian. The five major varieties of early Paleozoic fossils are tabulate "corals," bryozoans, brachiopods, trilobites, and graptolites. See Fig. 10–26.

The slender tubes of the honeycomb coral, composed of lime ($CaCO_3$), are presumed to represent cavities that contained small polyplike animals similar to those of modern corals. Some paleontologists note

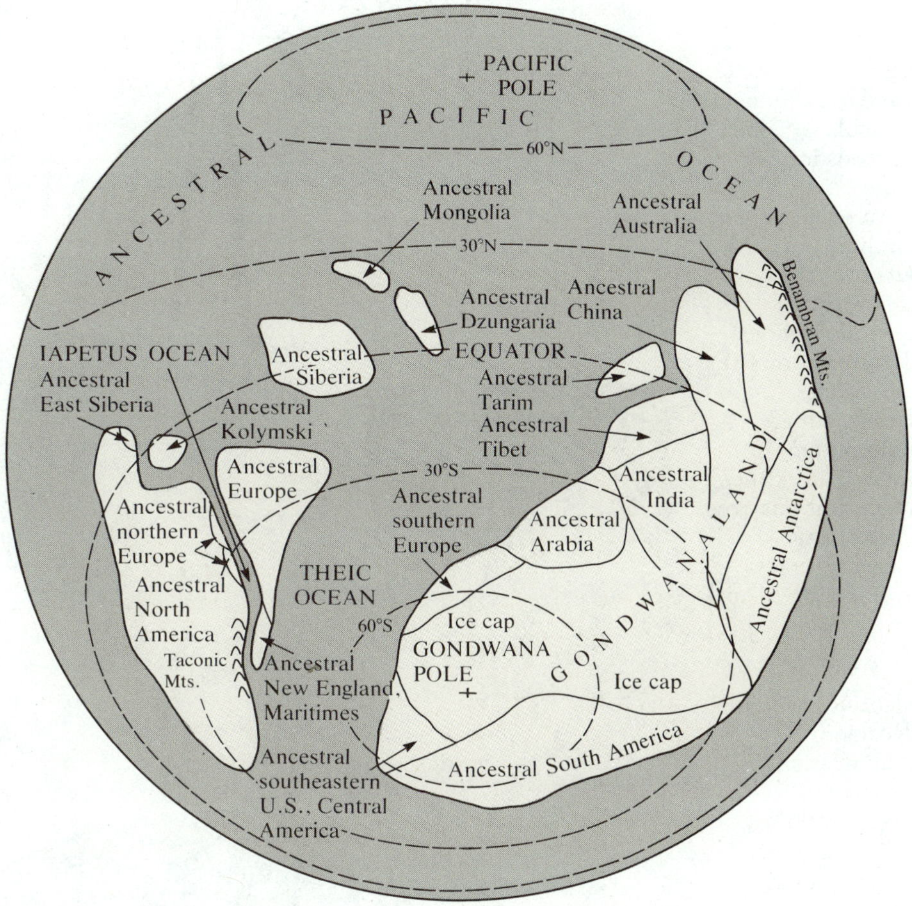

FIGURE 10–25
The early Paleozoic earth. A vast southern supercontinent, Gondwanaland, consisted of present-day South America, Africa, Arabia, India, Australia, Antarctica, and parts of southern Europe and the United States. The south or Gondwana pole was situated in present northwestern Africa, and there was continental glaciation in this region. North America, Europe, Siberia, and some other fragments were separate continents. The positions of the various parts of China are very tentative. The narrow Iapetus Ocean separated Europe and North America, and the broad Theic Ocean lay between the northern and southern continents.

the presence of pores in the tubes and believe that this group of animals, called the **tabulate "corals,"** are really sponges.

Bryozoans ("moss-animals") are tiny colonial animals, most varieties of which built little houses of lime. The remains of twiglike, stony bryozoans are so abundant in some Ordovician strata that they comprise the bulk of the rock.

Brachiopods have two valves, one top and one bottom, which house the body. The valves are hinged at the rear (top of views in Fig. 10–26) and are composed of lime. (In the superficially similar, but unrelated clams, the valves are positioned right and left and are hinged along the animal's back.) Brachiopods are among the most abundant and varied of Paleozoic fossils.

Trilobites form an extinct group of arthropods that reached its peak in the early Paleozoic. They are distant relatives of modern crustaceans and insects.

Graptolites are another extinct group; they lived mostly in the Ordovician and Silurian periods. The most common variety floated in the oceans and thus attained a wide geographic distribution, making it very useful for time correlation.

LATE PALEOZOIC HISTORY

The late Paleozoic era consists of three periods: Devonian, Carboniferous, and Permian. The late Paleozoic shows some remarkable changes from the early portion of the era. See Fig. 10–27. By its close—the Permian period—all the continents were in contact, forming a vast landmass called **Pangaea**. A broad embayment, the **Tethys Ocean**, intruded into the eastern edge of Pangaea, providing a partial separation between northern **Laurasia** and southern Gondwanaland.

The collision of the African part of Gondwanaland with southeastern North America produced the original southern Appalachian Mountains, while the collision of Europe with Siberia raised an earlier version of the **Ural Mountains**. There was also mountain building in southern Europe and along the entire western margin of the supercontinent. This area, including the present western United States and Canada, was a subduction zone-island arc complex for much of the Paleozoic (see Chapter 9).

Southern Gondwanaland was centered on the southern pole, and there was extensive continental glaciation. A smaller ice cap existed in northern latitudes in what is today eastern Siberia.

FIGURE 10–26
The early Paleozoic era was the age of marine invertebrates. Animals without backbones teemed in the seas and left abundant fossils. The tabulate "corals" and the bryozoans formed sizable colonies, though they themselves were individually small. Graptolites were also colonial, but are extinct organisms of uncertain relationships. The bivalved brachiopods, no relation to clams, are abundant Paleozoic fossils, but are relatively unimportant in modern oceans. The extinct trilobites are distant relatives of the crustaceans and so-called horseshoe crabs. (Drawing by Robert W. Tope)

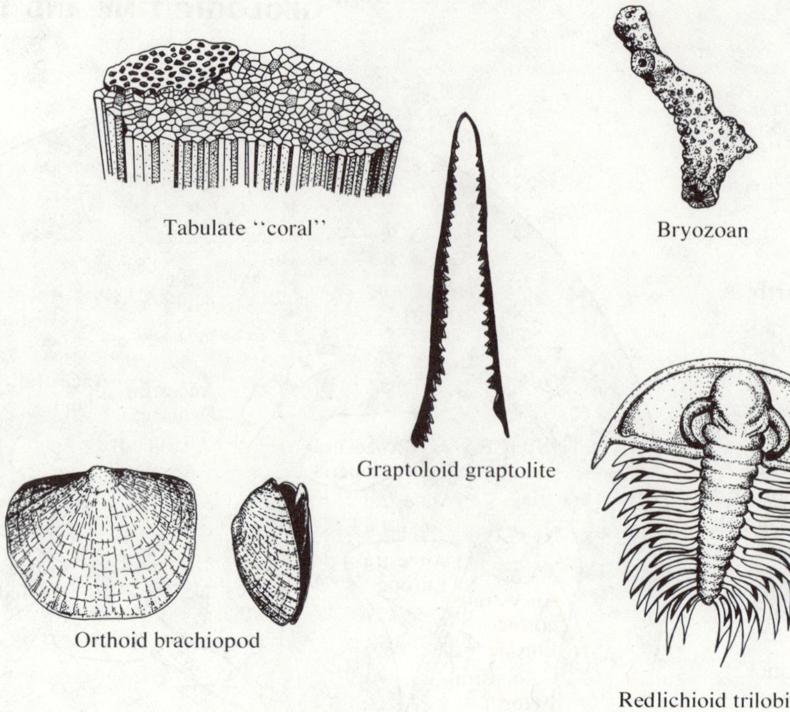

FIGURE 10–27
The late Paleozoic earth. In the late Paleozoic, continental unification reached an extreme. Gondwanaland continued from the earlier part of the era. The southern portion was in polar regions and glaciated. The northern continents of North America, Europe, and Asia also came into contact, forming the supercontinent Laurasia. As these masses came together, the original northern Appalachian, Hercynian, and Ural mountains came into existence. Laurasia and Gondwanaland also collided in the area of the Ouachita and southern Appalachian mountains, producing a single continent, Pangaea. The positions of the fragments of China are tentative. The Tethys Ocean, a great tropical embayment, extended into the eastern side of Pangaea. The western and southern margins of Pangaea also experienced mountain building induced by collision with oceanic plate margins.

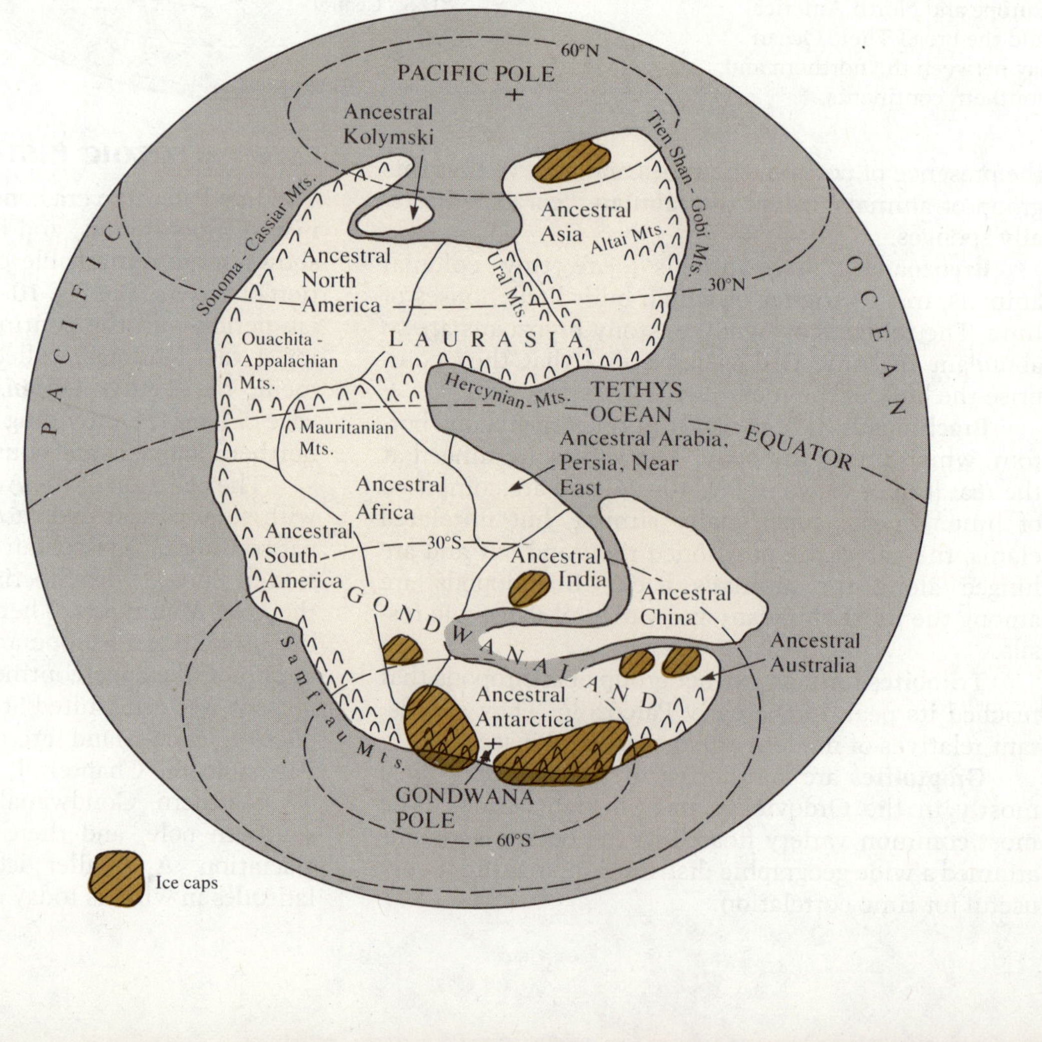

The continents were extensively flooded by shallow seas in the Devonian and Carboniferous, but in the Permian, a time of mountain building and continental unification, these seas disappeared.

LATE PALEOZOIC MARINE LIFE

Life in the seas continued to diversify through much of the late Paleozoic. There were many common types of marine organisms during this time. See Fig. 10–28. The **foraminifera**, or forams, are tiny one-celled animals that build shells of lime. One of the most abundant fossil groups of the late Paleozoic is that of the extinct fusulines. Fusulines are large for forams, their shells resembling wheat grains in size and shape.

Tetracorals are an extinct type of coral found in great abundance in late Paleozoic oceans. Their lime shells were typically cup-shaped, something like a cornucopia, or "horn of plenty." Colonial types built reefs.

Bryozoans continued in abundance, but the most important late Paleozoic types built lacy fronds of lime to support their many tiny bodies.

Brachiopods diversified into numerous common types. Two of the most characteristic are the spiny productids and the winglike spiriferids. Neither variety survives today.

Related to the living pearly nautilus are the **ammonoids** of the late Paleozoic and Mesozoic. These extinct animals probably resembled squids or octopuses with a coiled shell. With their buoyant shells, they were mostly swimmers and very widespread, hence their great importance in time correlation. The shell is divided internally by partitions into a series of chambers. In late Paleozoic types called **goniatites**, the line of junction between the partition and the shell wall is a moderately complex wavy line.

Crinoids are **echinoderms**, the group to which starfish and sea urchins belong. Their globular bodies had long, feathery arms and were attached to the bottom by a stalk. Although crinoids survive today, most are unlike the bulky Paleozoic varieties.

Finally, the **vertebrates** attained late Paleozoic success in the seas. Both bony and cartilaginous **fish** (e.g., the shark), as well as a variety of extinct types, were common.

The close of the Paleozoic was a great crisis in the history of marine life. Fusuline forams, tetracorals, lacy bryozoans, many varieties of brachiopods, goniatite ammonoids, and most kinds of crinoids became extinct, to be succeeded by their modern counterparts in the Mesozoic. The extreme continental unification of the late Paleozoic, which eliminated continental seas and brought once-separate groups into competition with one another, probably caused these extinctions.

LATE PALEOZOIC LAND LIFE

Although the first invertebrates and plants invaded the land in the late Silurian, life did not really flourish in the terrestrial environment until the late Paleozoic. See Fig. 10–29. Among the invertebrates, the arthropod insects and arachnids (spiders and their kin)

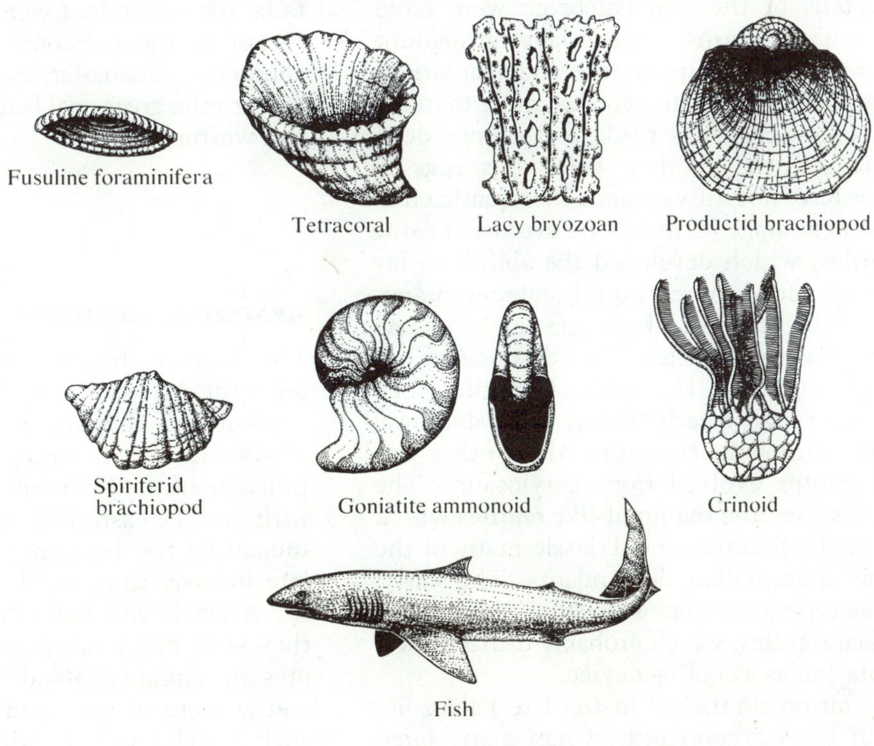

FIGURE 10–28
Life in the oceans continued to diversify through much of the late Paleozoic, but great extinctions, including most of the forms shown here, occurred at the end of the era. Among the invertebrates, the seedlike fusulines, the cup-shaped corals, the lacy bryozoans, the spiny and winglike brachiopods, the nautiluslike ammonoids, and the flowerlike crinoids dominated. Fish of many types, both bony and cartilaginous (e.g., the shark), swam in late Paleozoic seas. (Drawing by Robert W. Tope)

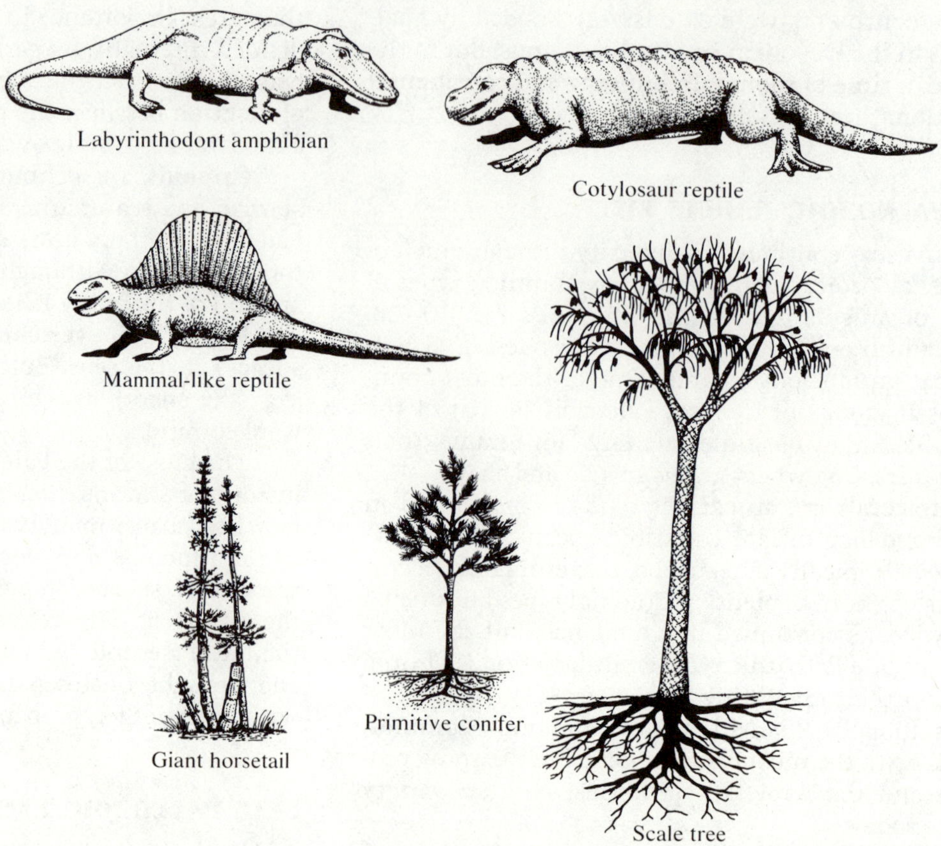

FIGURE 10–29
The first abundant land life existed in late Paleozoic times. Bulky amphibians known as labyrinthodonts evolved from lobe-finned fish. The early reptiles, cotylosaurs, resembled their amphibian ancestors. From cotylosaurs, a number of other reptiles developed, including forms ancestral to mammals. Among this group, though not directly ancestral to mammals, were the fin-backed forms. Great coal-forming swamps existed, and these were dominated by spore-bearing scale trees and giant horsetails, as well as seed-bearing conifer ancestors. (Drawing by Robert W. Tope)

abounded in the great coal swamps of the late Carboniferous. Vertebrates invaded the land at the close of the Devonian when lobe-finned fish evolved limbs to go with their lungs and became **amphibians**. The dominant amphibians of the late Paleozoic were large, bulky, semiaquatic forms called **labyrinthodonts**. These forms resembled in appearance and adaptations the living crocodiles, which are reptiles, more than the living amphibians—the frogs, toads, and salamanders.

Amphibians must lay their gelatinous eggs in water and thus have not fully escaped the aquatic environment. The next stage in vertebrate evolution came with the **reptiles**, which developed the ability to lay shelled eggs with a food source, liquid, and membranes for respiration and excretion. These eggs can be laid on land, thus freeing the reptiles from dependence on water for reproduction. The earliest reptiles, the **cotylosaurs**, were not greatly different in anatomy from their labyrinthodont ancestors. Many other varied reptilian groups evolved from cotylosaurs. The most significant were the **mammal-like reptiles** which developed over the Permian and Triassic many of the traits of their mammalian descendants. The early, more primitive types of the late Paleozoic included the large fin-backed reptiles, which probably utilized their peculiar adaptation as a cooling device.

Land vegetation flourished in the late Paleozoic, particularly in the Carboniferous, when many large trees and seed ferns were abundant. The 25–30-m-tall **scale trees** and **calamites** (giant horsetails) belong to spore-bearing, seedless plant groups represented today only by the small, herbaceous club mosses and horsetails. Also abundant were the **cordaites**, distant ancestors of the modern conifers. These trees, along with an abundance of smaller vegetation, have been compacted to form the great coal beds of the eastern United States and western Europe.

THE MESOZOIC

MESOZOIC HISTORY

The Triassic, Jurassic, and Cretaceous periods comprise the **Mesozoic era**. The Mesozoic world saw the breakup of the supercontinent Pangaea. See Fig. 10–30. In the Triassic, that great landmass existed much as in the Permian, although there is evidence of rifting in the eastern United States and western Africa, indicating the beginnings of the Atlantic Ocean. By late Jurassic time, as shown in the illustration, both the **Atlantic** and **Indian oceans** had opened, although they were much narrower and more restricted than at present. Gondwanaland fragmented into large eastern and western masses, and a number of smaller blocks, such as India, were moving northward into the Tethys

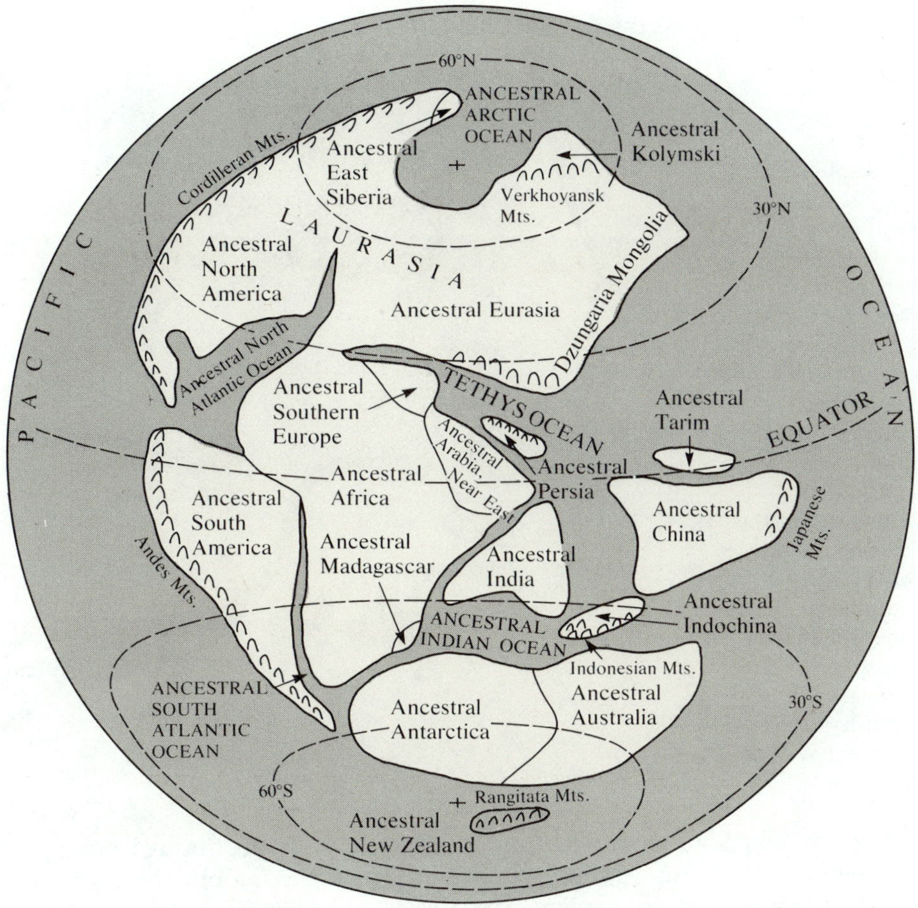

FIGURE 10–30
The Mesozoic earth. In the Mesozoic era, the fragmentation of Pangaea began. Gondwanaland ruptured into the areas known today as South America, Africa-Arabia, India, Australia-Antarctica, other smaller pieces, and probably China as well. Laurasia retained more continuity, but the ancestral North Atlantic Ocean produced a partial separation between North America and Europe. The great tropical seaway of Tethys existed in this interval, and there was extensive flooding of continents by shallow seas in the latter part of the era. Mountain building was prominent along the eastern side of the Pacific, producing the original Rocky Mountains and the Sierra Nevada—the western Cordillera.

Ocean. Laurasia was still in existence, but the growing North Atlantic Ocean was splitting off North America from Eurasia. The Cretaceous period saw the continued separation of the continental fragments and widening of the oceans. Of the Gondwanaland continents, only Antarctica and Australia remained in contact at the end of the Mesozoic. In the north, Greenland still served as a connection between the main northern landmasses.

Seas flooded the continental masses in both the late Jurassic and late Cretaceous, providing abundant homes for shallow-water marine organisms. At the end of the Mesozoic, there was much mountain building, particularly along the western margin of the Americas as the ancestors of the modern **Rocky Mountains**, **Sierra Nevada**, and **Andes** were elevated (see Chapter 9).

MESOZOIC MARINE LIFE

The marine life of the Mesozoic contrasts markedly with that of the Paleozoic. See Fig. 10–31. The mollusc group dominated, and new types of corals and echinoderms were prominent.

Sponges are very primitive animals with a body pierced by pores and supported by tiny interlocking pieces of lime or silica. They were particularly common in late Mesozoic tropical seas. A new variety of coral, the **hexacoral**, appeared in the Mesozoic and has dominated ever since. Colonial forms, such as the one illustrated, built prominent reefs in tropical Mesozoic oceans.

Among the **molluscs**, there are three main groups: the **gastropods**, or snails; the clams, **pelecypods**, or bivalves; and the **cephalopods**. There was a great variety of snails including some extinct large, high-spired forms. There were also many different types of clams including some characteristic extinct oysters. Two types of cephalopods teemed in Mesozoic seas. The fossil **ammonoids**, chiefly coiled forms with the loops in a single plane, have highly ornate **sutures**, the lines where the internal partitions meet the shell wall. In addition, the exterior of the shell is typically ornamented with ribs, nodes, keels, and so forth. **Belemnites** were squidlike forms that built an elongate, thick, cigar-shaped shell. Neither ammonoids nor belemnites survived the end of the Mesozoic.

Among the echinoderms, **echinoids** became the dominant group. Burrowing forms, such as the heart urchins, are responsible for the bulk of the fossils.

In the Mesozoic, several reptile groups secondarily returned to an aquatic life and became common in the

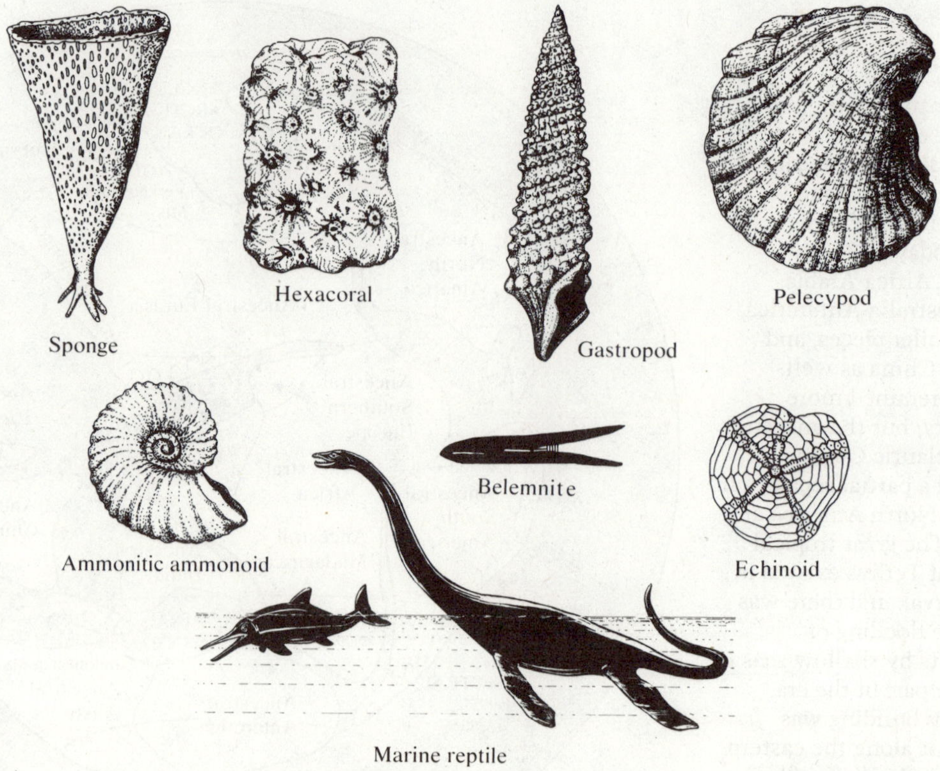

FIGURE 10–31
Marine life was abundant in the Mesozoic, but most of the major fossil groups differed from those of the Paleozoic. Particularly important were the molluscs, including the snails, clams, nautiluslike ammonoids, and squidlike belemnites. Sponges, new types of corals, and the echinoids were also common. Reptiles, the dominant vertebrates of the Mesozoic, invaded the seas in the Mesozoic and produced a number of successful types, including the sea-serpent-like plesiosaurs and the porpoiselike ichthyosaurs. (Drawing by Robert W. Tope)

world's oceans. In addition to turtles and marine lizards, there were some extinct groups such as the dolphinlike **ichthyosaurs** and sea-serpent-like **plesiosaurs.**

The close of the Mesozoic saw a number of extinctions of marine organisms, though not on the scale of those at the end of the Paleozoic. Most notable were the demises of the ammonoids, belemnites, and marine reptiles; groups within the snails, clams, and other classes and orders of animals also perished. These losses were coupled with great extinctions on the land.

MESOZOIC LAND LIFE

The Mesozoic era is often called the age of reptiles because these vertebrates dominated the land and were major marine carnivores as well. See Fig. 10–32. The age of reptiles actually consisted of two parts. In the Triassic, the mammal-like reptiles were the characteristic forms. Their hard parts show the development of the mammalian jaw and tooth structure, and it is probable that many were hairy, warm-blooded, and viviparous (non-egg-laying). In the late Triassic, after giving rise to the mammals, these reptiles were eclipsed by the **dinosaurs** and their kin.

Beginning from a primitive group of reptiles called thecodonts, the dinosaurs and their relatives, the flying pterosaurs and amphibious crocodiles, reached a Mesozoic peak. Dinosaurs were actually two groups of reptiles, many of which reached large size. Most were herbivorous, but one major group of two-legged, ferocious carnivores evolved. The herbivores fed mainly on the naked-seed or **gymnosperm** plants (cycads, conifers, ginkgoes) and the ferns, which dominated plant life of the Mesozoic until the late Cretaceous. Because the dinosaurs became extinct at the end of the era, they traditionally have been pictured as plodding, stupid beasts. This description has changed in recent years as detailed analyses of dinosaur structure and ecology indicate that at least some of them may well have been active, agile, warm-blooded animals with intelligence comparable to that of birds, one of their successful offshoots.

The number of reptiles became drastically reduced at the end of the Mesozoic, with most of the dominant types dying out. The cause of this extinction is not known, but the simultaneous demise of many marine creatures has suggested to some scientists that the cause was a global one, such as a change in the composition of the atmosphere.

In the Cretaceous, the flowering plants (**angiosperms**), with their fruit-enclosed seeds, appeared and came to dominate the land as they do today.

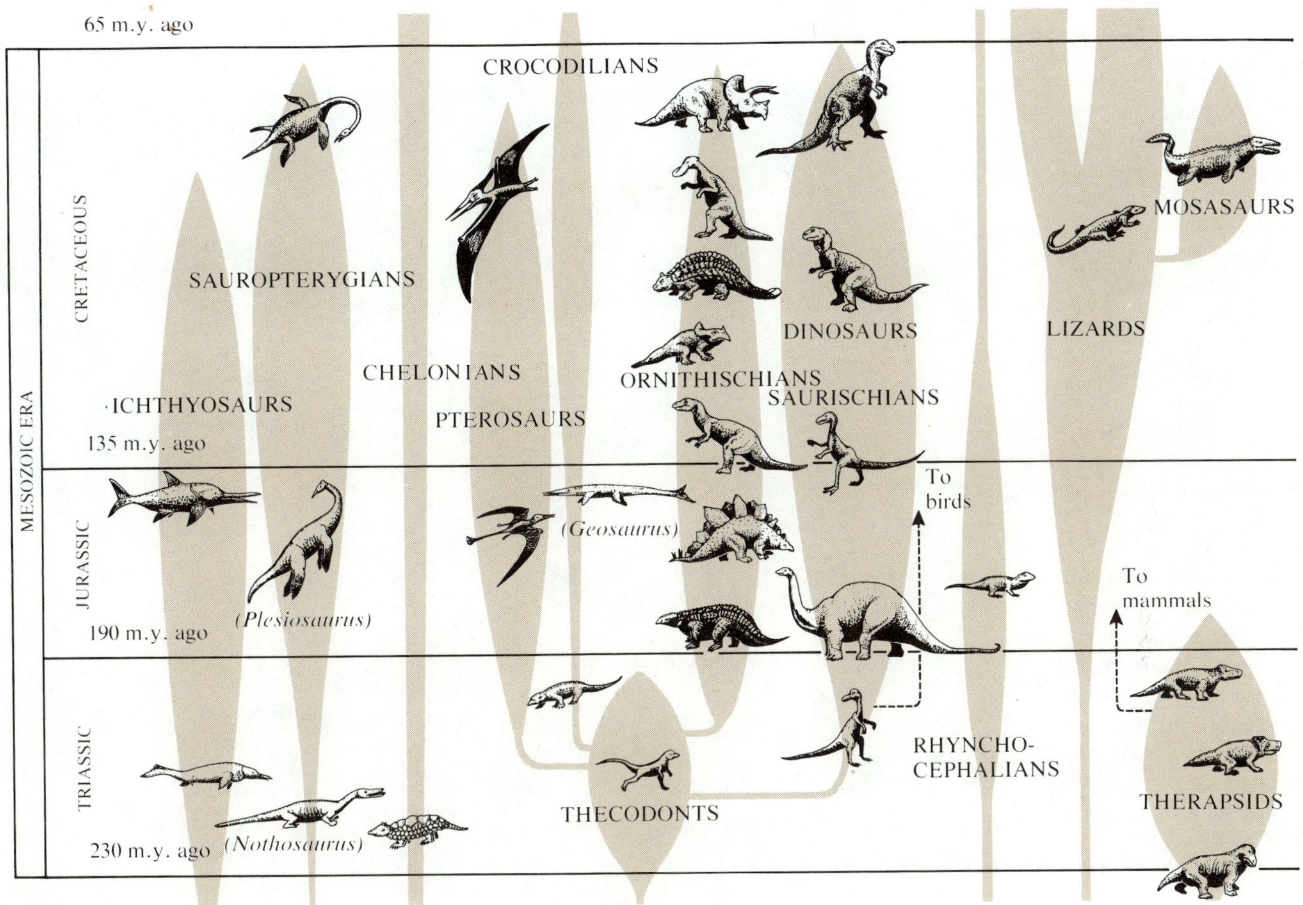

FIGURE 10–32
The Mesozoic era is often called the age of reptiles. For most of the Triassic, the advanced mammal-like reptiles, the therapsids, dominated. Near the close of the Triassic, the dinosaurs evolved from another group of reptiles, the thecodonts, and came to dominate land faunas for the rest of the era. Pterosaurs, or flying reptiles, were close dinosaur relatives. Dinosaurs, pterosaurs, and the large marine reptiles all became extinct by the end of the Mesozoic. (Drawing by Robert W. Tope)

THE CENOZOIC

CENOZOIC HISTORY

In the **Cenozoic era**, the continents gradually assumed their present positions. See Fig. 10–33. By the early part of the era, the Paleogene period, all of the Gondwanaland continents were separate entities. South America had not yet contacted North America through the Panama isthmus; Africa was approaching southern Eurasia; Turkey, Persia, India, and probably China as well were rapidly approaching their collision with Eurasia; Australia had separated from Antarctica and was overtaking southeastern Asia. In the middle and late Cenozoic, all the continental fragments south of Asia collided with it, producing the great mountain ranges of the Near East, central Asia, and Indonesia. The **Himalayas**, the earth's highest mountains today, are one result of this continent-continent collision.

North America separated from Eurasia between Greenland and Scandinavia in the early Cenozoic, but rejoined it in the late Cenozoic, the Neogene, in the Bering Sea area. Thus, the Arctic Sea became a northward extension of the Atlantic and not the Pacific. The plate motions of the Cenozoic not only produced mountains by continental collisions but also formed ranges along continent-ocean interfaces, such as the modern western Cordillera of North America and the Andes of South America (see Chapter 9). With the advent of so much mountain building, there were few continental seaways. In the late Cenozoic, particularly

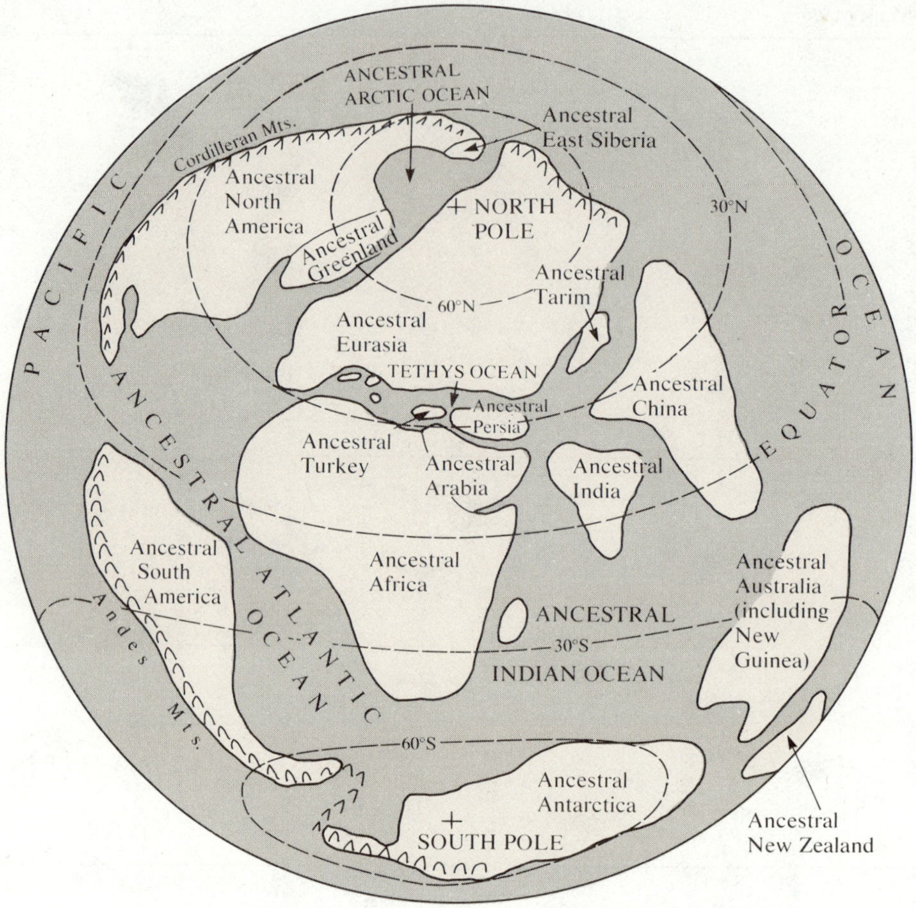

FIGURE 10–33
The Cenozoic earth. During the Cenozoic era, the continents gradually assumed their present positions. The Atlantic and Indian oceans widened by sea-floor spreading. Tethys closed as Africa, India, various continental fragments, and probably China collided with the southern margin of Eurasia, producing the Alps, Near Eastern, and Himalayan ranges. In the late Cenozoic, Australia, free from Antarctica, moved northward to collide with Asia in the Indonesian region. North America and Eurasia separated in the Greenland area in the early Cenozoic, only to reunite in the Alaska-eastern Siberia region in the late Cenozoic. Plates in the Pacific area impinged on the western Americas, producing the Cordillera and Andes.

the Pleistocene epoch, great ice sheets gripped the northern continents and the tips of southern ones.

CENOZOIC MARINE LIFE

Cenozoic marine life was dominated by microorganisms in the open oceans and by the molluscs in the shallower, nearshore waters. See Fig. 10–34. Conspicuous by their absence were the ammonoids of earlier eras.

Three major types of one-celled microorganisms were abundant in the Cenozoic. Two of these, the **diatoms**, which are plants, and the **radiolaria**, which are animals, built their tiny shells of silica (SiO_2). Diatoms and radiolaria were predominantly floating (**planktonic**) forms. The foraminifera built tiny lime shells typically divided by internal partitions. Although most varieties are bottom-dwelling (**benthonic**), a few very abundant planktonic foraminifera evolved and are important for time correlation.

Among the molluscs, the snails and clams (pelecypods) predominate. Some reached very large size, particularly among the clams. The pectens, or scallops, have a symmetry very much like that of brachiopods. Echinoids reached a peak in the Cenozoic, with the burrowing, flattened sand dollars predominating. The most successful vertebrates of Cenozoic oceans were the bony fish.

CENOZOIC LAND LIFE

The Cenozoic era is often called the age of mammals because mammals dominated land life in the youngest era. See Fig. 10–35. (It could also be called the age of flowering plants because the angiosperms enjoyed a similar status in the plant world.)

Mammals are typically characterized by their hairy covering; constant body temperature (warm-bloodedness); mammary glands for nourishing the young; and the viviparous condition, in which the shelled egg of the reptilian ancestors has been eliminated. These features ordinarily cannot be detected in fossils; what can be seen are a number of characteristic skeletal structures such as a single lower jawbone and highly differentiated teeth. Mammals evolved from reptiles in the Mesozoic, but were subsidiary to them in that era.

A great diversity of mammals developed in the Cenozoic. Mammals have taken to the air (bats) and

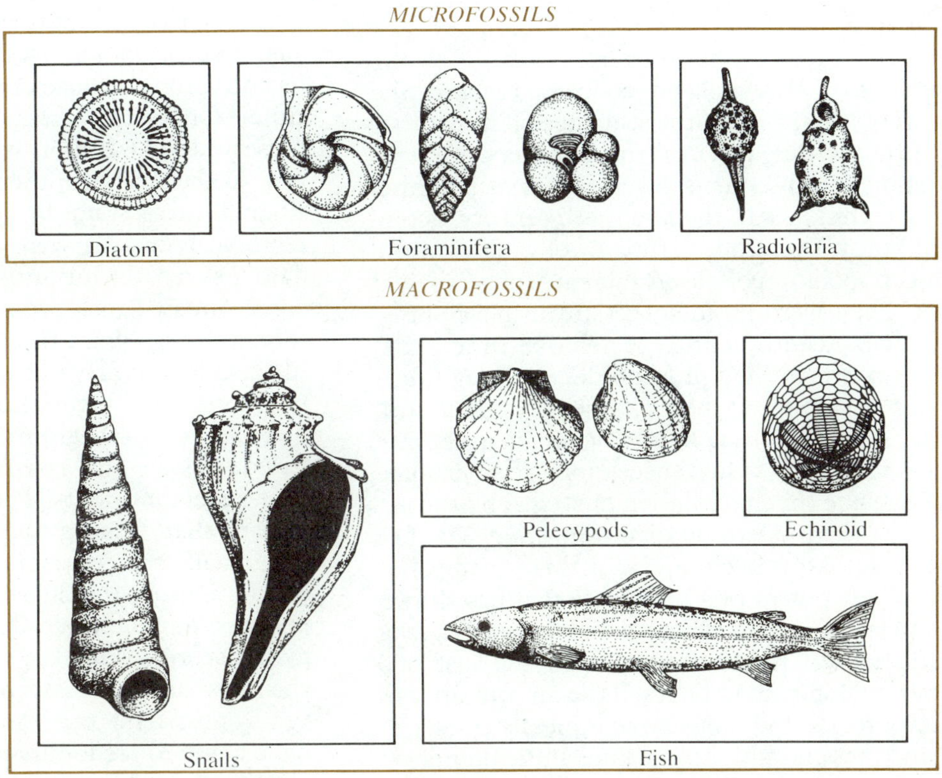

FIGURE 10–34
Microscopic marine organisms are abundant in Cenozoic oceanic sediments. The diatoms have been major photosynthesizers, whereas the foraminifera and radiolaria have been miniscule carnivores. Among the larger invertebrates, the molluscs and echinoderms predominate. A variety of snails, clams, and sand dollars occur in Cenozoic marine strata. Fish are common vertebrate fossils in such rocks, although their remains are usually fragmentary. (Drawing by Robert W. Tope)

the seas (whales, seals, sea cows) as well as developing a variety of terrestrial adaptations. Mammalian types of locomotion include running, climbing, burrowing, and hopping; their feeding types are carnivores, herbivores, and omnivores. A number of early Cenozoic groups of mammals became extinct, primarily due to the reunification of many of the continents in the Neogene, which brought about increased competition.

The **primate** group includes humans, relative late-comers in earth history, but whose activities are having a profound impact on the earth's physical features and life. The human family (**hominids**) evolved in the Neogene and spread, probably from an African homeland, throughout the world to become the dominant form of life today.

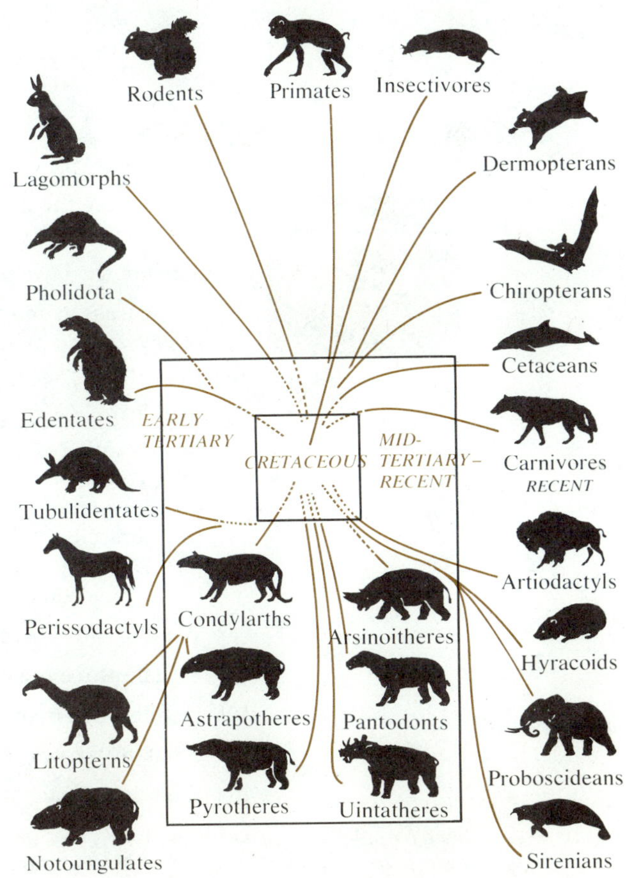

FIGURE 10–35
The age of mammals is the name commonly applied to the Cenozoic era. In the Cenozoic, the hairy, live-bearing mammals developed a great variety of adaptive types. They are runners, hoppers, crawlers, burrowers, flyers, and swimmers. They are carnivorous, herbivorous, and omnivorous. (From Edwin H. Colbert, *Evolution of the Vertebrates*, © 1969, John Wiley and Sons. Drawing by L. Darling, © 1969. Used with permission of artist and publisher)

SUMMARY

Geologists have always been concerned about the length of geologic time and the sequence of events that have occurred within it. Until the discovery of radioactivity and the subsequent development of radiometric dating techniques in the twentieth century, geologists could only give crude estimates of geologic time based on extrapolation of observable rates of geological processes. They were more successful in developing, during the nineteenth century, a relative time scale, which puts events in their proper order. Although various physical attributes of the rocks were locally useful, it was the discovery that fossils occur in a definite sequence—the law of faunal succession—that enabled the long-distance time correlation of strata and the development of the modern relative time scale. The use of radiometric dating techniques in this century has led to the development of a time scale that has shown the earth to be 4.6 billion years old and the existence of abundantly fossiliferous strata to span the last 600 million years. Unfortunately, fossils occur primarily in sedimentary rocks, and radioactive minerals chiefly in igneous and metamorphic rocks. Enough tie points are known that geologists now have a pretty good idea of the duration of the geologic eras, periods, and epochs.

Seven-eighths of earth history is encompassed in a vast Precambrian interval, the first 20 percent of which has no record because of the unstable conditions at the time. Once the crust was stable enough to be preserved—3.8 billion years ago—the rock record began. In the Archean, there are several small, thin continental masses, but by Proterozoic time modern plate tectonic processes were operating. The lack of abundant oxygen during most of the Precambrian led to some unusual rock types and the lack of others common in late geologic time. Although the fossil record of life also began early in the Precambrian, it too was limited to simple, predominantly one-celled forms until just before the beginning of the Paleozoic era. The Paleozoic saw the assembly of the continents into a vast supercontinent—Pangaea—the great diversification of shell-bearing marine invertebrates, and, in its latter half, the invasion of the land by animals and plants. During the succeeding Mesozoic era, Pangaea began to fragment, reptiles (particularly dinosaurs) and gymnosperms ruled the land, and molluscs dominated the seas. In the Cenozoic era—the last 65 million years of earth history—mammals and flowering plants have come to predominate on land as the plates gradually assumed their present positions.

REVIEW EXERCISES

1 What were the erroneous assumptions upon which were based the three major eighteenth century attempts to estimate geologic time?

2 Explain the laws of superposition, crosscutting relationships, and inclusions.

3 What was the major geologic contribution of Arduino, Lehmann, and Werner? What was Werner's major error?

4 Define the *law of faunal succession,* and give the name of its discoverer.

5 To what era of geologic time do each of the following periods belong?
  (a) Permian
  (b) Tertiary
  (c) Cambrian
  (d) Jurassic
  (e) Devonian
  (f) Cretaceous

6 What is time correlation?

7 Explain three major methods of time correlation based on fossils.

8 Define three major physical methods of time correlation.

9 What are varves, and why can they be used for time correlation?

10 What is radioactivity? How can it be used to date rocks?

11 If a radioactive parent isotope has a half-life of 5 million years and the ratio of parent to daughter isotope is 1/4 to 3/4, how old is the enclosing rock?

12 What are the four most important radiometric dating methods?

13 If a sedimentary rock group overlies a lava flow of 500 million years in age and is cut by a dike of 200 million years of age, what is the age of the sedimentary rocks?

14 When did the Precambrian-Paleozoic transition occur?

15 According to modern geologic theory, approximately when did the earth form?

16 Briefly describe the density stratification of the earth.

17 What are thought to be the main gaseous constituents of the earth's early atmosphere? How did they evolve to their present form?

18 Define *photosynthesis. Fermentation. Respiration.*

19 What are shields and cratons? List the major ones.

20 Where do the oldest rocks in North America occur?

21 What is the basic nature of the Precambrian fossil record?

22 What is the significance of the Ediacara fauna?

23 Describe the locations, motions, and disruptions of Gondwanaland through the Paleozoic, Mesozoic, and Cenozoic eras.

24 When and where in the geologic past did continental glaciation occur?

25 Name three prominent forms of marine life for each of the major geologic eras.

26 When did the original Appalachian and Rocky mountains form?

27 Name the segments of geologic time in which the amphibians, the reptiles, and the mammals dominated land life.

28 Describe a brachiopod. A pelecypod. A crinoid. An echinoid.

29 What evidence in the rocks would have indicated to early geologists that the earth is much older than the 6000-year estimate based on biblical genealogies?

30 Superposition is valid for undeformed rocks, but how does a geologist determine the relative ages of rocks in folded, overturned, or faulted strata?

31 Why is it important to distinguish time correlation from rock correlation?

32 What errors occur in radiometric dating techniques if the sample has not been a closed system?

33 Why are there no rocks known older than 3.8 billion years?

34 Give the evidence for and against each major scenario for the development of the earth's early atmosphere.

35 Why is respiration superior to fermentation as a means of energy release?

36 What is the significance of shields?

37 What are the major differences between Precambrian and later geologic history?

38 Trace the geologic history of the continental block now known as North America.

39 How does the fossil record indicate the existence of evolution?

40 When were there major extinctions in the earth's past? Do there appear to be any relationships to plate tectonics?

SUGGESTED READINGS

BERRY, W. B. N. *Growth of a Prehistoric Time Scale.* San Francisco: W. H. Freeman and Co., 1968, 158 pp.

BRASIER, M. D. *Microfossils.* London: Allen and Unwin, 1980, 193 pp.

DOTT, R. H., and BATTEN, R. L. *Evolution of the Earth.* New York: McGraw-Hill Book Co., 1976, 573 pp.

EICHER, D. L., and MCALESTER, A. L. *History of the Earth.* Englewood Cliffs, N.J.: Prentice-Hall, 1980, 413 pp.

______. *Geologic Time.* Englewood Cliffs, N.J.: Prentice-Hall, 1976, 150 pp.

GLEN, W. *Continental Drift and Plate Tectonics.* Columbus, Ohio: Charles E. Merrill Publishing Co., 1975, 188 pp.

LANE, N. G. *Life of the Past.* Columbus, Ohio: Charles E. Merrill Publishing Co., 1978, 288 pp.

LEVIN, H. L. *Life through Time.* Dubuque, Iowa: William C. Brown Publishers, 1975, 217 pp.

MCALESTER, A. L. *The History of Life.* Englewood Cliffs, N.J.: Prentice-Hall, 1977, 167 pp.

MINTZ, L. W. *Historical Geology,* 3rd ed. Columbus, Ohio: Charles E. Merrill Publishing Co., 1981, 611 pp.

SEYFERT, C. K., and SIRKIN, L. A. *Earth History and Plate Tectonics.* New York: Harper and Row Publishers, 1979, 600 pp.

WYLLIE, P. *The Way the Earth Works.* New York: John Wiley and Sons, 1976, 296 pp.

- mass wasting
- debris avalanche
- slide
- slip plane
- flow
- fall
- subsidence
- waxing slope

11

- free face (cliff)
- debris (talus) slope
- waning (wash) slope
- concave-convex slope
- oversteepening
- sea cliff
- creep
- slump
- scarp
- terracette
- mudflow
- earthflow
- debris avalanche
- solifluction
- permafrost
- landslide
- slope stability
- driving forces
- resisting forces
- factor of safety

Mass Wasting

SLOPES ARE THE MOST COMMON SUBAERIAL LANDFORM ON THE surface of the earth. Although most of the time slopes appear quite inactive, they are the result of numerous physical, chemical, and biological processes that combine to form a complex and dynamic system. Activities on slopes vary from nearly imperceptible creep of the slope materials to thundering avalanches that move at tremendous velocities. **Mass wasting** is the term applied to the group of processes that denote the movement of materials downslope. It is usually the high-magnitude, low-frequency mass-wasting processes that gain the attention of the general public. This results because large landslides have buried whole towns and villages, filled reservoirs, and dammed river valleys. In the process, thousands of lives have been lost. However, in terms of slope processes and development of slopes, the events of intermediate magnitude and frequency are likely more important. That is, processes that operate continually or frequently are probably more significant in the development of most slopes than catastrophic events that occur only rarely.

In recent years, human use and activity have been extended to the development of steep topography in urban areas. This has resulted in an increased awareness of mass-wasting processes as landslides actually turn into house slides.

Damages from landslides in the United States exceed $1 billion per year, while claiming an average of 25 lives. If failure of trenches and other excavations are included, then loss of life exceeds 100 persons per year. In one recent rainy season, the San Francisco Bay region suffered from over 400 landslides, resulting in damages of approximately $10 million. As a result of landslides and other geological problems, southern California probably employs more consultants to analyze slope stability than in any other area of geology.

Large urban centers are not the only locations with slope stability problems. Numerous areas in the Appalachian Mountains, Rocky Mountains, and Pacific coastal mountains are plagued by landslides of various types. Of particular concern has been the occurrence of debris avalanches in mountain areas. During one intense rainstorm over the Appalachian Mountains, 150 people lost their lives in debris avalanche and related flood events. In the Pacific Northwest, debris avalanching is a serious problem on steep, forested slopes, a problem made somewhat worse by logging practices that require the extensive building of roads for clear cutting operations.

The importance of studying mass-wasting processes stems from the fact that these processes are influential in the development of landscapes. Interactions between the downslope processes and running water produce river valleys, which are the most conspicuous landforms. However, if rivers alone were responsible for the valleys, the valleys would be very narrow indeed. That most river valleys are much wider than they are deep is an indication of the importance of mass-wasting processes in supplying debris to streams. Therefore, stream and mass-wasting processes work in harmony with one another to develop most of the features seen on the surface of the earth. This is, of course, a generalization, because other processes, including weathering and the work of ice and wind, are also significant in developing the landscape.

BEHAVIOR OF EARTH MATERIALS

Earth materials that make up slopes comprise a complex physical, chemical, and biological system. The materials are a mixture of rock, weathered material, and soil, with a variable amount of water and organic material. The amount of water and organic material and, to a lesser extent, the soil and rock type may vary on a slope from season to season and year to year, depending upon such factors as nature and extent of precipitation, rates of materials added to the slope by deposition or removed by erosion, and activities that affect the vegetation such as human use and activity, fire, and seasonal variation of precipitation. Complex interactions between factors involving the strength and composition of earth materials are responsible for the shape of slopes and how slopes may fail.

Slope materials may fail in several ways; they may **slide**, **flow**, **fall**, or **subside** (Fig. 11–1). Earth materials that fail by sliding generally move in a more or less continuous, cohesive unit. That is, the material in transit has a cohesion and is moving along some sort of slide or **slip plane**. Slip planes generally are zones of weakness in the soil or rock materials. They may be either fractures in the downslope direction or simply

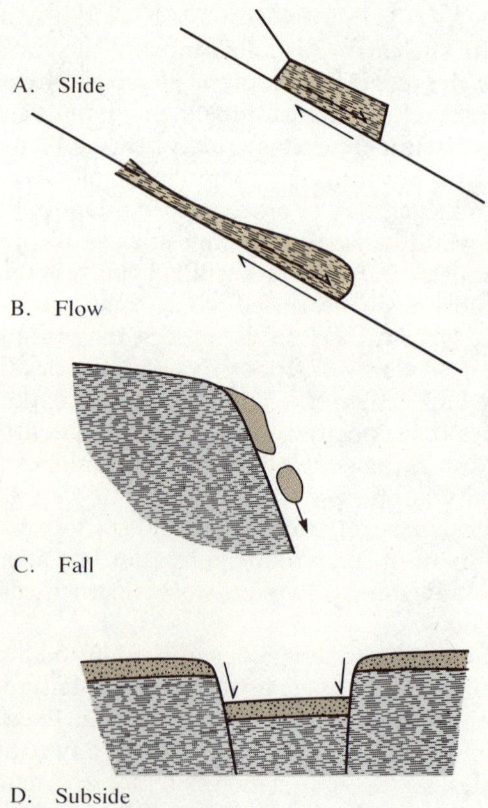

FIGURE 11–1
The four main ways earth materials may fail.

sharp breaks between the soil material and underlying rock. Figure 11–2 illustrates two of the possible slip planes that may develop.

Failure of earth materials may also take place by a flowage process. Typical flowage types of mass wasting are lobate in shape and have a characteristic hummocky topography (Fig. 11–3). Mass wasting involving flowage can take place on slopes of very low inclination with variable amounts of water in the materials. In fact, some materials appear to flow even though the amount of water in the moving materials is very low. A flowage-type mass wasting that involves very coarse material which is usually dry is illustrated in Fig. 11–4. However, these materials may also move when they are saturated or frozen.

Rockfalls occur on steep slopes when unstable blocks become dislodged. Even relatively small rockfalls, involving only several hundred cubic meters of rock, can be extremely dangerous because they move very fast. For example, in 1942 a rockfall in Beaver County, Pennsylvania, involving approximately 150 m^3 of rock crushed a bus, killing 22 persons. More recently, in 1979 several small rockfalls near Malibu, California, closed the Pacific Coast Highway for several days while the steep slopes were carefully examined (Fig. 11–5). Although rockfalls are not rare along this portion of the Pacific Coast, there was concern that these particular events might be part of a larger moving block that would eventually fail, causing a serious threat to people living downslope. This remains a serious problem, as more rockfalls and slides are expected.

Subsidence is a vertical (downward) motion of earth materials. It generally takes place when fluids are withdrawn from the earth or when there is collapse over a subterranean void.

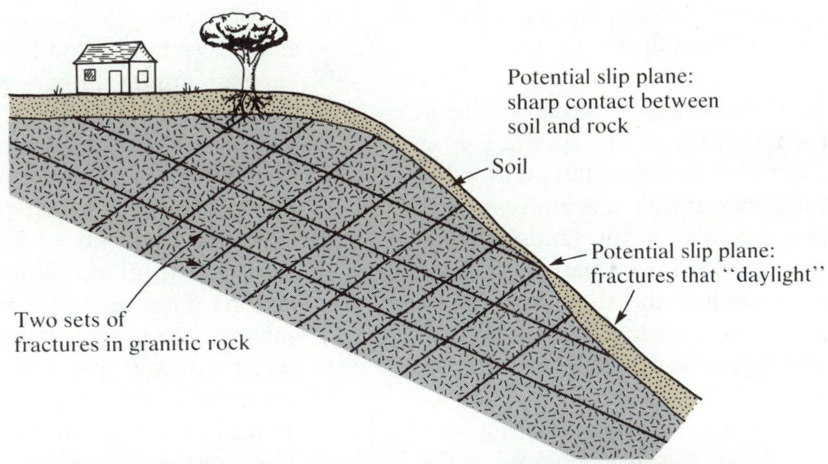

FIGURE 11–2
Two types of potential slip planes along which a slope failure may occur.

FIGURE 11–3
Hummocky topography on a large, complex earthflow along the Russian River in northern California.

FIGURE 11–4
Flowage-type mass wasting in the Crazy Mountains of Montana. The large particles, commonly the size of baseballs, form long tongues of rock debris flowing from adjacent canyon walls out onto the valley floor.

The behavior of earth materials on slopes has been classified in terms of whether the materials slide, flow, fall, or subside. However, this is somewhat artificial in that many mass-wasting events may contain several of these components. For example, some landslides involve a definite sliding in the upper part associated with a flowage in the toe area of the slide (Fig. 11–6). Also, large subsidence events are commonly associated with sliding or flowage-type events facilitated by the oversteepened slopes on the margin of the subsidence block. Therefore, when studying mass-wasting processes it must be remembered that the behavior of the slope materials may be quite variable and therefore may change within a short distance.

SLOPE MORPHOLOGY

Few slopes consist of uniform morphology from the toe to the top of the slope. Most are composed of segments that may be either concave, straight, convex, or nearly vertical. Knowledge of these slope segments is important in understanding mass-wasting processes. Figure 11–7 is a generalized profile of a slope containing four major slope elements: (1) the **waxing slope**; (2) the **free face**, or **cliff**; (3) the **debris** or **talus slope**; and (4) the **waning** or **wash slope**. Each of these slope elements will be examined in terms of the characteristic processes operating:

1. The **waxing slope** is generally a convex slope and is associated with mass-wasting processes such as slow creep of the soil or rock.
2. The **free-face** or **cliff** segment of our idealized slope is vertical or nearly vertical and thus is probably a rock outcrop. However, some earth materials such as loess (wind-blown silt) may also stand at a near-vertical angle. Processes operating on cliffs are primarily mass-wasting and weathering processes and involve rockfalls and other related processes.
3. The **debris** or **talus slope** is found at the base of the free face. It consists of materials derived by mass-wasting and weathering processes and delivered from the slope segments at higher elevations. The size of the material in the debris slope depends upon the nature of the material being delivered, and the extent of deposition depends upon the nature of the processes that remove the material from the debris slope.
4. The **waning** or **wash slope** is produced by running water, and is concave, the shape characteristic of slopes eroded by running water.

FIGURE 11–5
The steep slope in the central part of this photograph is adjacent to the Pacific Coast Highway in southern California. Lack of vegetation on the slope attests to the very recent movement of material here.

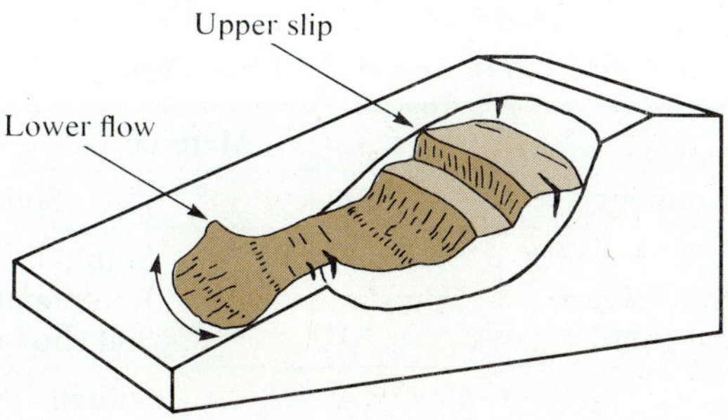

FIGURE 11–6
A landslide consisting of an upper slip motion and lower flow. (After R. Pestrong, *Slope Stability,* American Geological Institute, 1974)

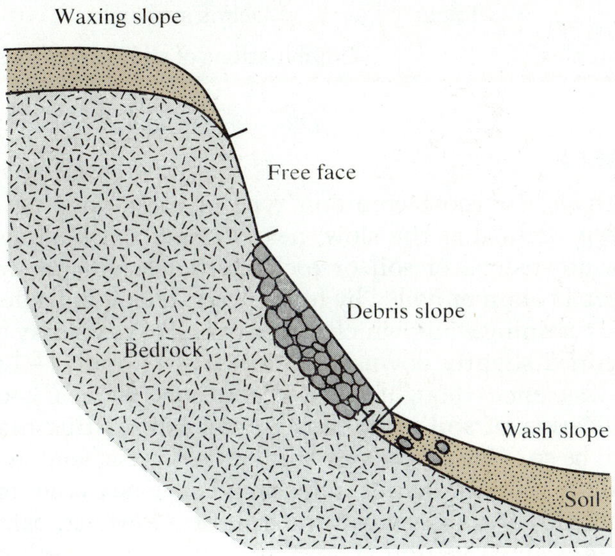

FIGURE 11–7
Four major slope elements.

The preceding discussion of idealized slope elements is highly generalized, and although a given slope may not contain all of these elements, most slopes will be constructed of two or three. A common slope in the more humid regions is the so-called **concave-convex slope**. This slope is composed of two slope elements (Fig. 11–8): the upper or convex section is a mass-wasting slope, and the lower, concave section is a wash or stream-eroded slope. In general, the convex segments are nearly always associated with mass wasting, whereas the concave are associated with processes associated with running water. This is an important point to remember, because by looking at the morphology of the slope, one can then say something about the process.

Free-face segments and debris accumulation segments on slopes are probably more common in mountainous or arid and semiarid environments. In fact, in these regions the slope morphology is generally much

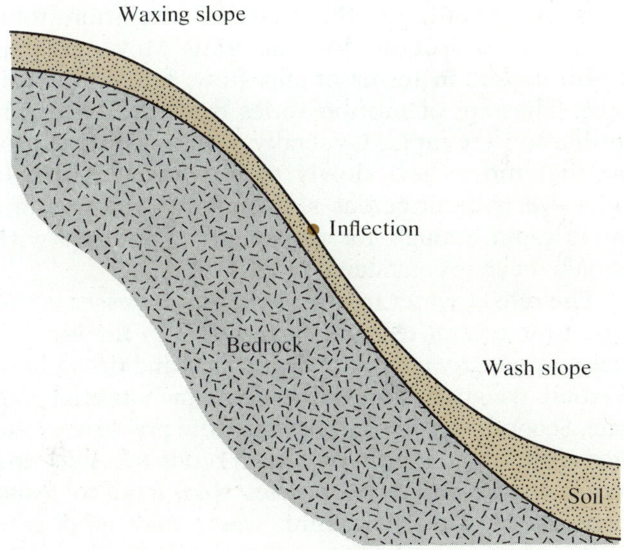

FIGURE 11–8
The concave-convex slope characteristic of humid regions. The slope is composed of two elements: the upper convex section, produced by mass wasting, and the lower concave section, associated with running water.

more pronounced and angular (Fig. 11–9). However, it is a mistake to oversimplify slope morphology in terms of one variable such as climate; for example, very angular slopes with well-developed free faces also occur in humid areas.

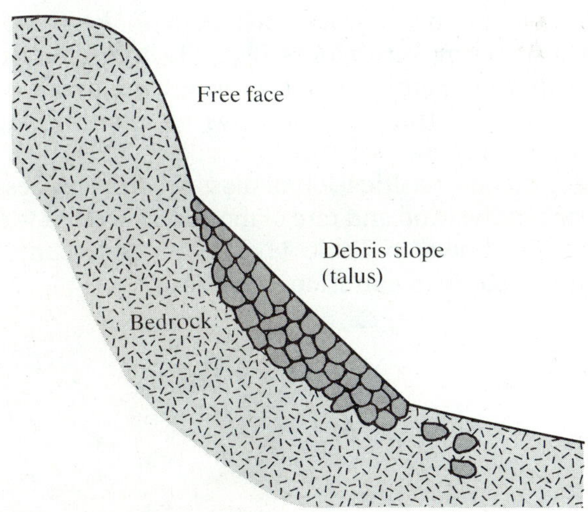

FIGURE 11–9
The general slope morphology characteristic of semiarid and arid regions.

TYPES OF MASS WASTING

The types of mass-wasting or downslope movement processes are best classified in terms of the kind and rate of movement, and the degree of saturation (total amount of water) of the slope materials. Movement has been discussed in terms of slip, flow, fall, and subsidence. The rate of motion varies from nearly imperceptible to very rapid. Generally, a mass-wasting process that moves too slowly to be viewed with the naked eye is described as slow-moving, whereas one that is rapid enough for perception of motion with the naked eye is considered rapid.

The role of water in mass-wasting processes is important for several reasons. First, water in the form of precipitation may enter slope materials and thus add to the total weight of the slope above a potential slip plane. Second, water may raise the fluid pressure in the pores of the slope materials, which tends to reduce the friction forces between particles that tend to resist downslope movement. Third, water may erode the toe of a slope, thereby increasing the angle the slope materials must maintain. Such **oversteepening** of slopes is a cause of landslides along stream channels and in the coastal environment where wave action constantly removes the toe of the slope at the junction of the sea cliff and the beach environment. Water is so important in most mass-wasting processes that some investigators have concluded that most landslides are associated with water problems. Therefore, if correction of landslides is the objective, control and drainage of surface and subsurface water become significant. To this end, rather elaborate systems of drains are used along many highway roadcuts to keep water from infiltrating into slopes (Fig. 11–10). If on your next trip you examine roadcuts in greater detail, you will learn something of the ways of retarding landslides along highways.

A general classification of mass-wasting processes in terms of the kind and rate of motion as well as water content is shown in Table 11–1. This table contains the terminology in common use today.

TABLE 11–1
General classification of mass-wasting processes.

Type of Movement	Material: Rock	Material: Soil
Slide (variable water content and rate of movement)	Slump blocks Translation slide	Slump blocks Rotational slide Soil slip (planar)
Fall	Rockfall	Soilfall
Flow — Slow	Rock creep	Soil creep
Flow	Unconsolidated materials (saturated)	
Flow — Rapid	Earthflow Mudflow Debris avalanche	
Complex	Combinations of slides and flows	

CREEP

Perhaps the most common type of mass wasting is **creep**, defined as the slow, nearly imperceptible flowage downslope of soil or rock. Surficial evidence for creep is often provided by fence posts, telephone poles, and retaining walls, which after a period of years may be inclined slightly downslope (Figs. 11–11 and 11–12). It was once thought that tree trunks were a good indication of soil creep, and in some cases this may still be so. That is, the tree trunk may have a bend in it and the roots of the tree may extend upslope as the tree itself slowly moves down the hillside. However, using trees as an indication of creep is an imprecise method, in that the slow rates of creep may be much less than the growth rate of trees, so that during its lifetime a tree will not have moved very far. Also, bends in tree trunks are more likely related to an attempt of the tree to obtain more light than an adjustment to the creep in the downhill direction. Nevertheless, trees that are

FIGURE 11–10
Control of surface and subsurface water is significant in controlling the landslide hazard on roadcuts. Shown here is a concrete drainage system designed to collect and remove water before it infiltrates into the slope or runs down the roadcut, causing erosion.

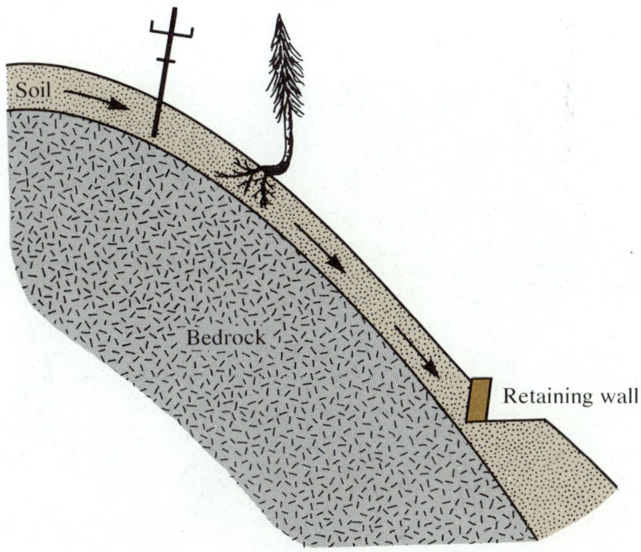

FIGURE 11–11
Some field evidence for soil creep, including tilted telephone poles, trees, and retaining walls.

FIGURE 11–13
Creep is commonly associated with slow-moving landslides. This large Douglas fir tree is slowly being split apart by slow creep taking place within an active landslide in the Cascade Mountains of Oregon.

very old and are on extremely steep slopes may tend to show the effects of creep. In fact, when creep operates along with a slow-moving landslide, some trees are slowly pulled apart (Fig. 11–13).

The common causes of creep are wetting and drying and associated swelling and shrinking, frost heaving, rainsplash, and work of burrowing animals. All of these processes, and many more, operate on slopes, tending to result in the movement of material in the downhill direction under the influence of gravity.

SLUMP

Slumping is a common type of slope failure. It is a type of landslide in which separate blocks of material slide

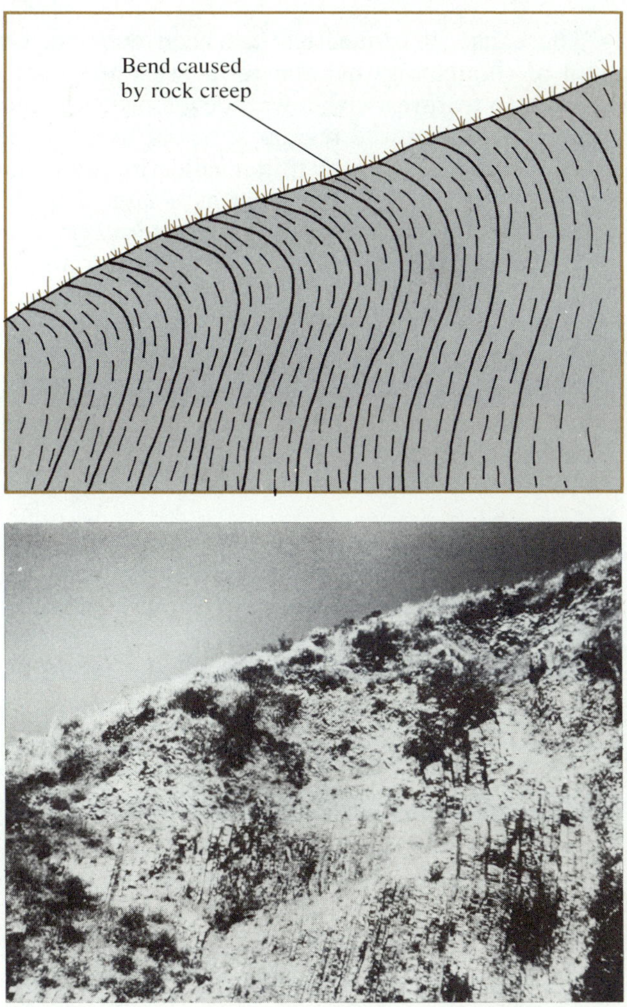

FIGURE 11–12
Rock creep is the very slow flowage of rock in the downslope direction. The process is most effective very near the surface of the earth, and as shown here, the rocks at the surface bend in a direction different from that below the surface. Thus, geologists interested in studying the orientation of sedimentary rock units might be fooled by mere surface outcrops of rocks if they have been affected by creep. (Photo courtesy James Brice)

past one another, often with some backward rotation (Fig. 11–14). Slumps generally have a well-defined slip plane at depth. The process of slumping leads to a rather chaotic topography consisting of numerous blocks separated by **scarps**. It was a series of slumps and related phenomena that caused so much trouble during the 1964 earthquake in Anchorage, Alaska (Fig. 11–15). Notice in Fig. 11–15 the well-developed slip plane between the rather stiff and the sensitive clay. The shaking of the earthquake tended to liquefy the sensitive clay, and hence the clay lost its strength, allowing the overlying stiff clay to fail in a series of slumps.

Slumps are commonly associated with such diverse phenomena as wave action or stream erosion undermining and steepening a slope, oversteepening of cut slopes in housing developments, and oversteepening of slopes for highway and other cuts. Slumping is particularly likely when there is a pronounced difference in the slope materials, such as strong, resistant rock material overlying weaker material (Fig. 11–16).

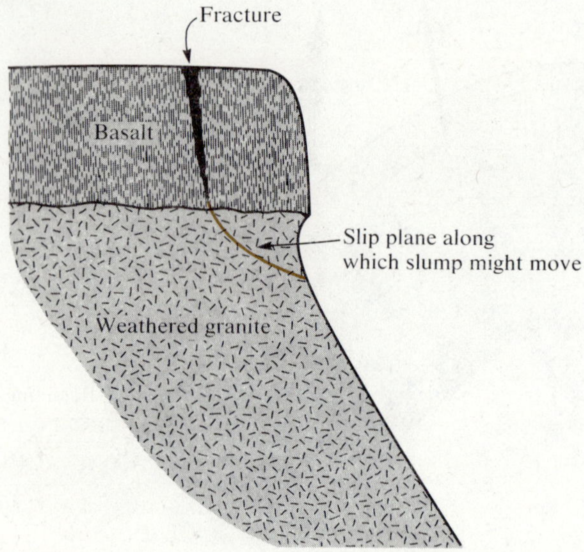

FIGURE 11–16
Pronounced differences in the slope material, such as a strong, resistant rock overlying a weaker one, are conducive to the formation of certain types of slumps.

TERRACETTES

Terracettes are minor slope features commonly found in weak rock types such as shale. They are characterized by rather regular steps in the slope (Fig. 11–17). The length of the vertical part of the step is greater than the width of the tread and is generally less than a meter.

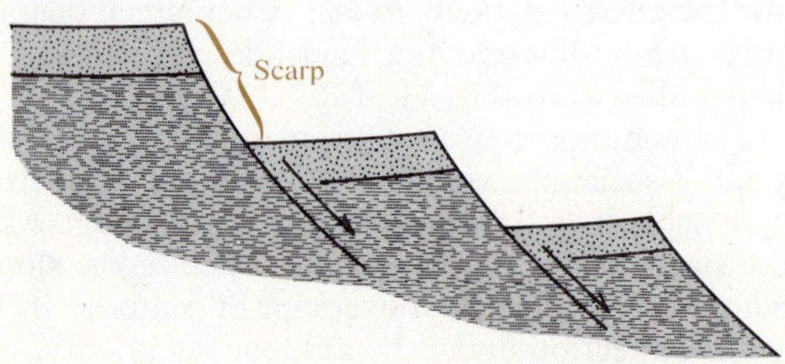

FIGURE 11–14
Slumping.

The cause of terracettes has been debated. One school of thought is that the terracettes are a set of small slump features with a well-developed slip plane. In this explanation, the regular series of terracettes is viewed as a succession of minor adjustments or failures that take place rather than as a large landslide event. A second school holds that terracettes are formed by grazing animals as they wander up and down the hillslope in search of food. This latter explanation is valid in cases in which it has been shown that the terracettes end abruptly at fence lines, where grazing ceases. Although most terracettes are probably produced by grazing animals, some are undoubtedly caused by other processes.

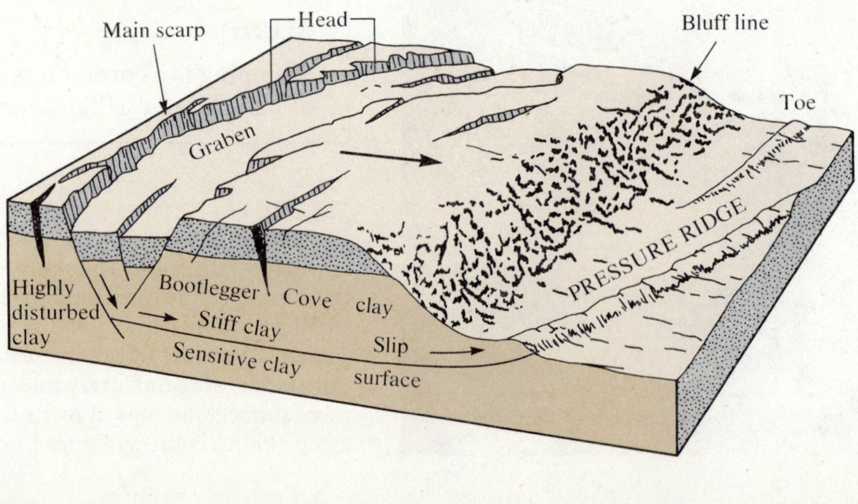

FIGURE 11–15
The way landslides developed in response to the 1964 Alaskan earthquake in Anchorage. The graben is a down-dropped block of earth material in the general chaotic topography associated with the complex landslide. (From W. R. Hansen, U.S. Geological Survey Professional Paper 542A, 1965)

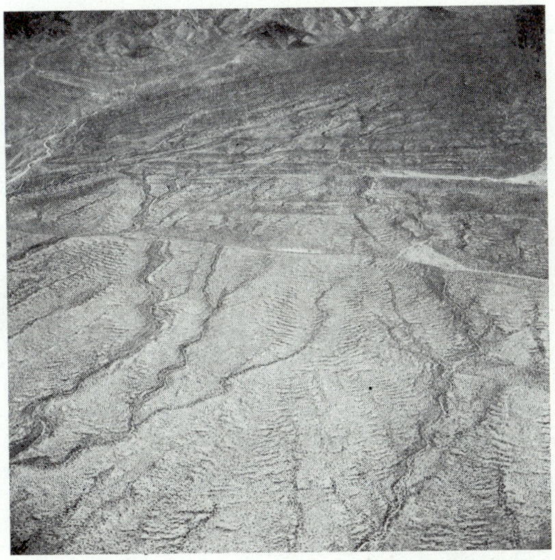

FIGURE 11–17
The numerous lines on this slope in the Indian Creek Fan are associated with small slumplike features known as terracettes. Some terracettes may form as very small slump blocks; however, most are probably formed by grazing animals and thus are perhaps more appropriately called *grazing steps.* (Photo courtesy Chester B. Beaty)

MUDFLOW

Mudflows constitute a rapid type of mass wasting characterized by a fluid motion of water-saturated debris. Mudflows are commonly associated with the semiarid environment where infrequent rainstorms of high magnitude and low frequency may mobilize debris in canyons that has been accumulating between precipitation events. The material becomes completely saturated with water, moves quickly down the canyon, and may eventually spread out over a wide area when the stream channel emerges from the mountains. Fortunately, mudflows are somewhat predictable and tend to recur from the same canyons. Thus, careful mapping may delineate locations that have experienced mudflows in the past; these areas should be avoided when development of land is considered.

Mudflows have periodically been a serious problem in southern California. Development of hillslopes and removal of indigenous vegetation by brush fires and other means have increased the frequency of mudflows (Fig. 11–18).

Gigantic mudflows have originated on the flanks of volcanoes in the Pacific Northwest and other volcanoes around the world. These flows result when highly unstable volcanic debris on the flanks becomes saturated with water and moves down the sides of the volcanoes, generally in existing river valleys. In the historic past, such flows have deposited billions of cubic meters of material in a single flow. The paths of several ancient mudflows from Mt. Rainier, Washington, are shown in Fig. 11–19. Mudflows can originate on the flanks of the volcano during an eruption and, when the mountain is obscured by smoke and other activity, may go unnoticed until they descend upon the lowlands, with possible catastrophic consequences. Over 30,000 people now live in the area covered by the old mudflow deposits of Mt. Rainier.

Effects of mudflows from a Cascade volcano were observed on 18 May 1980, when Mount St. Helens, a composite volcano in the southwestern corner of

FIGURE 11–18
Damage to property in Los Angeles caused by a mudflow. Note the enormous size of material mudflows are able to transport. (Photo from *Los Angeles Times*)

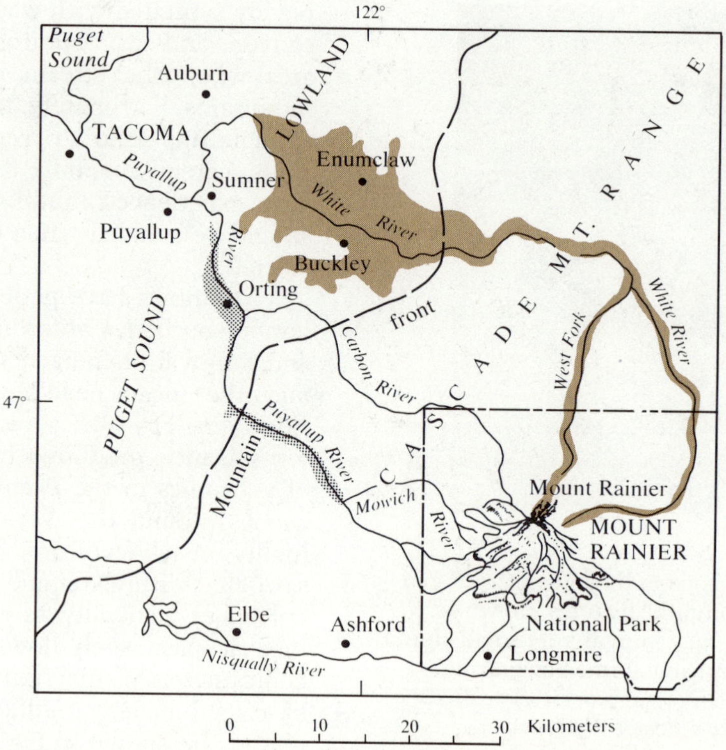

FIGURE 11–19
Map of Mt. Rainier and vicinity showing the extent of the Osceola mudflow in the White River valley (color) and the Electron mudflow (dot pattern) in the Puyallup River valley. (From Crandell and Mullineaux, U.S. Geological Survey Bulletin 1238, 1967)

Washington, erupted violently. Within minutes after the eruption started, the first of several mudflows consisting of a mixture of water, volcanic ash, rock, and organic debris (mostly logs) occurred (Fig. 11–20). The mudflows and accompanying floods raced down the valleys of the north and south forks of the Toutle River at estimated speeds of 29–55 km/hour. Water levels in the river reached at least 4 m above flood stage, and nearly all bridges along the river were destroyed. Fortunately, no large cities were in the path of the flows.

EARTHFLOW

Earthflows, another of the common types of mass-wasting processes, consist of flowage downslope of water-saturated earth materials. Earthflows differ from mudflows in that the rate of movement is generally slower, and earthflows are seldom, if ever, confined to channels as mudflows are in their source areas. A typical earthflow in the Coast Range of California is shown in Fig. 11–21. It is worthwhile to mention that many earthflows have some attributes of slumps in the upper parts of the unstable mass (Fig. 11–6). In fact, most earthflows probably start with a slumping-type motion, and then as the water content increases, the flowage takes place near the toe of the moving mass of earth material.

DEBRIS AVALANCHES

Debris avalanches generally form in steep, forested topography. They involve a rapid type of mass wasting in which all the soil and weathered debris above the bedrock surface is removed quickly, forming a rather long but narrow chute that exposes the bedrock below (Fig. 11–22). Debris avalanches generally develop following localized intense precipitation, and, in contrast to slumps and earthflows, the failure is very fast.

Debris avalanches have been common in the Appalachian Mountains as well as in the mountains of the Pacific Northwest. To some extent, the magnitude and frequency of debris avalanching may be increased by logging practices that involve the building of roads or the removal of trees. Tree loss reduces evapotranspiration and causes an increase in soil moisture, which lessens the resistance of the slope materials to sliding. Logging also accelerates runoff and erosion rates, perhaps increasing the tendency of debris avalanches to develop.

SOLIFLUCTION

The term **solifluction** is generally reserved for a slow flowage type of mass wasting associated with **permafrost** (perennially frozen ground). During the warmer part of the year, the top of the permafrost melts, form-

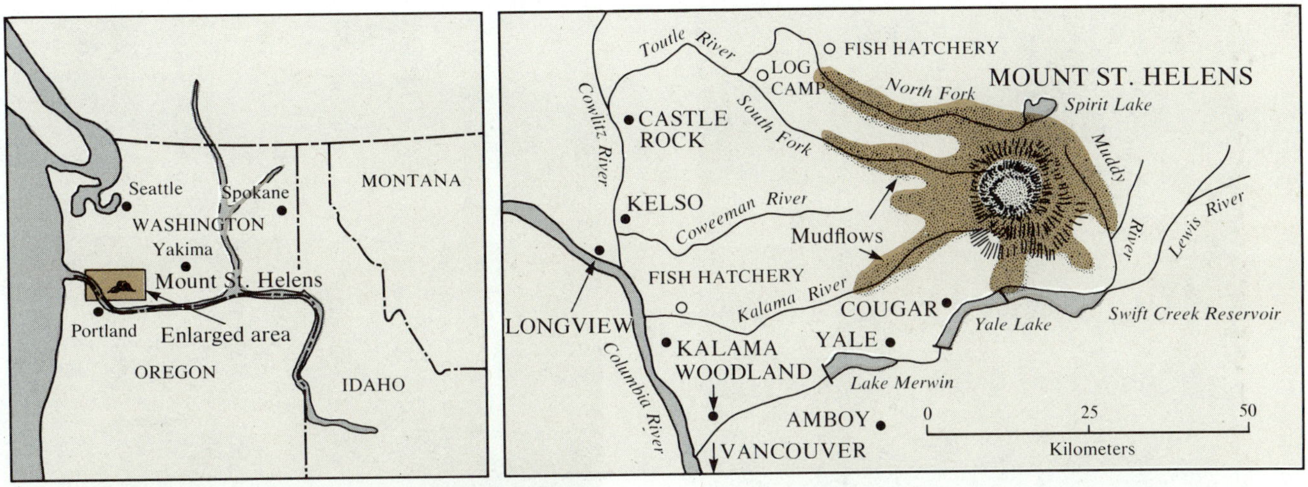

A.

B.

FIGURE 11–20
A. General area around Mount St. Helens Volcano, with enlarged area to the right. B. Mudflows in the Toutle River associated with the eruption of 18 May 1980. (Diagrams and photo from U.S. Geological Survey)

ing an active layer above the still-frozen substrata (Fig. 11–23). This active layer may on slight slopes become mobile, forming broad lobate flows. Solifluction in such an environment has caused many engineering problems, as it is difficult to anchor foundations to an area that thaws and flows seasonally.

LANDSLIDES

Landslides consist of a perceptible downslope sliding or falling of soil and rock, or a mixture of the two. Landslides vary from small features to catastrophic events. Large landslides generally have a deep-seated slip plane, and these high-magnitude events are usually preceded by days, months, or even years of low-magnitude creep, the rate of which generally increases up to the time of failure of the high-magnitude landslide. Once in motion, the material in a landslide can move rapidly downslope, and in some cases the slide mass may even become airborne and be carried on a layer of compressed air for some distance. The case histories of several large landslides will illustrate some of the interesting features of these events.

Gros Ventre Slide In 1925, the Gros Ventre River valley in Wyoming was the site of a spectacular landslide in which approximately 38 million m^3 of debris moved rapidly down a vertical distance of approximately 600 m. The slide plane was most likely a saturated clay layer interbedded with limestones and shales dipping in the down-valley direction. The generalized geology and slip plane for the Gros Ventre slide are shown in Fig. 11–24. The front of the slide moved rapidly across the valley and rose over 100 m up the opposite side before settling back down. The debris formed a dam approximately 75 m high and 1 km long, creating a lake about 8 km long. Although a considerable amount of water escaped through the dam by seepage through the landslide deposits, the lake continued to grow and there was worry about a flood from the snowmelt of the following year. However, in 1926 the snowfall was light, and there was no overtopping of the dam that year. But in 1927 the lake level rose, the dam was overtopped, and a channel approximately 15 m deep was quickly cut, causing a serious flood downstream that claimed several lives. The cause of the Gros Ventre slide was most likely the intensive rainfall

A.

B.

C.

FIGURE 11–21
An earthflow in the Coast Range of California. A. A distant view of the earthflow. B. Closer up, the flow looks like a fresh bulldozer cut in the side of the slope. The low ridges on the sides of the flow are natural levees formed by slight overbank flow during the event. C. These types of flows originate from upslope areas evacuated during the flow. Shown here are several students where the earthflow started. Notice the pronounced scarp.

FIGURE 11–22
A debris avalanche on a steep, forested slope in North Carolina. (Photo courtesy Edward Burt)

prior to the slide, which both added to the weight of the slope material and raised the fluid pressure along the potential slide plane, thereby reducing the forces that tended to hold the material together and resist the downslope sliding along the weak clay unit.

Blackhawk Landslide One of the largest prehistoric slides that has been studied is the Blackhawk landslide on the north slope of the San Bernardino Mountains in California. This particular slide is significant from the standpoint of its size and because it is one of the few slides that may have moved on a cushion of compressed air. The slide mass itself is approximately 8 km long, 3.2 km wide, and varies from 10 to 30 m in thickness. It is a tongue-shaped body of limestone derived from the Blackhawk Mountains to the south. The geologic history setting up the catastrophic event was characterized by faulting that moved the limestone over a weak, uncemented sandstone. Subsequent

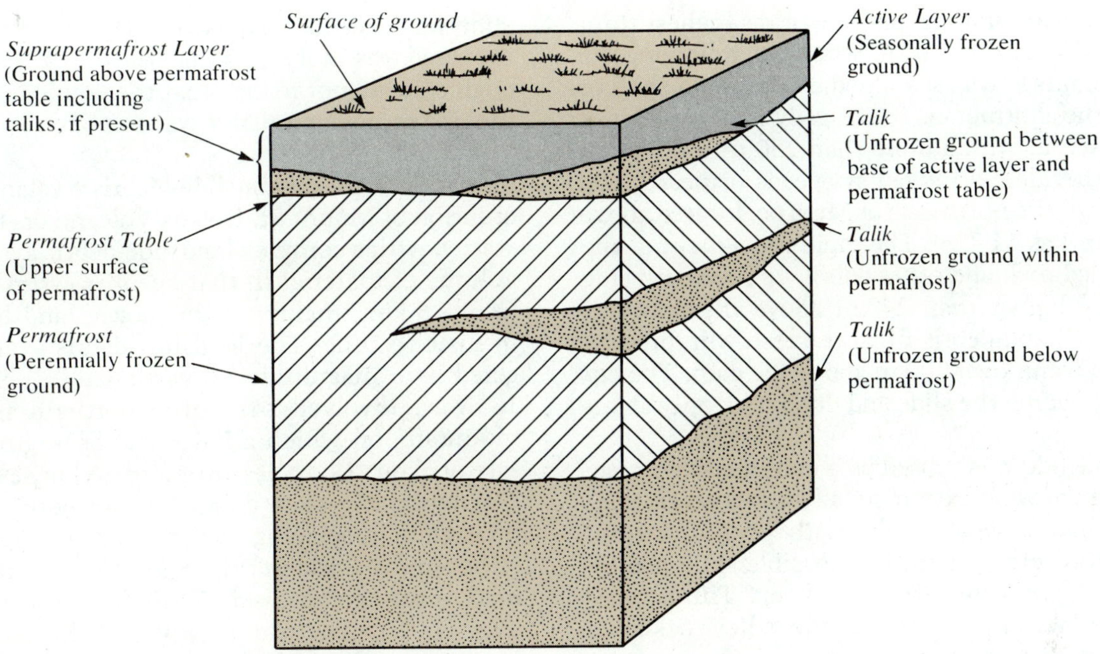

FIGURE 11–23
Terminology associated with permafrost. (From Ferrians, Kachadoorian, and Greene, U.S. Geological Survey Professional Paper 678, 1969)

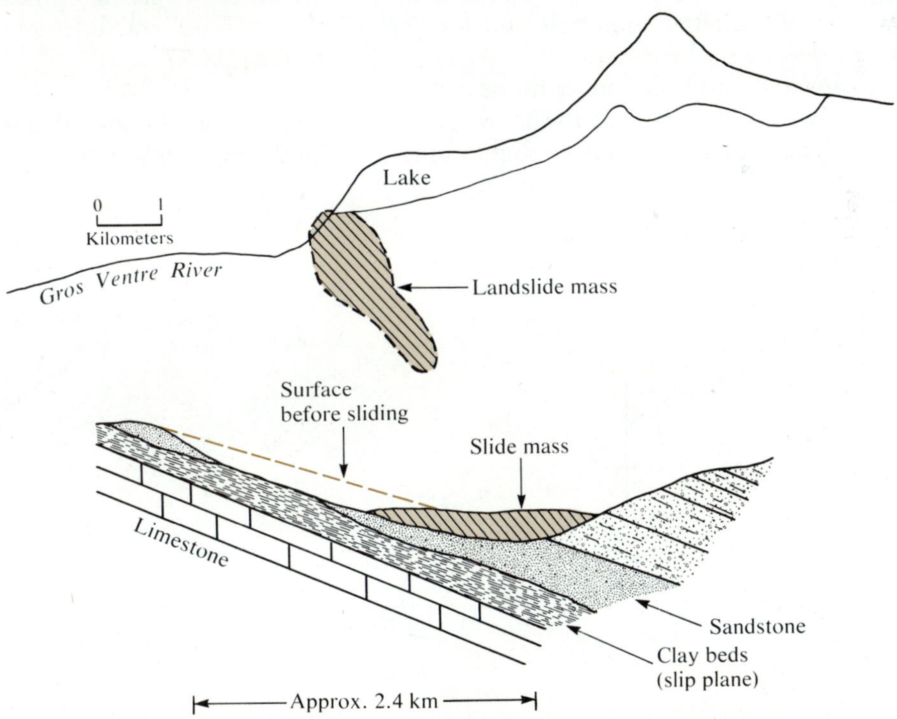

FIGURE 11–24
Idealized diagrams showing the geology of the Gros Ventre slide of 1925. (After W. C. Alden, 1928)

eroding of the weaker rock led to a steep slope, which supposedly failed catastrophically. The mass of limestone moved down the slope and became airborne, riding perhaps 100 m above the ground at 435 km/hour. The entire journey of the limestone from the Blackhawk Mountains 1200 m above the valley to the point of impact of the slide mass took 80 s.

Vaoint Dam Disaster Many landslides have been caused by complex interactions involving adverse geologic conditions, excess moisture, and human-initiated changes in the landscape and slope materials. The Vaoint Reservoir slide of 1963 is a good example of such a landslide. The event was the world's worst dam disaster; approximately 2600 lives were lost. The disaster centered around one of the world's highest thin-arch dams (266 m at the crest). Strangely, following the disaster no damage was sustained by the main shell of the dam or the abutments. The tragedy was caused by a huge landslide in which more than 230 million m³ of rock and other debris moved at speeds of nearly 100 km/hour down the north face of Mt. Toc, located above the reservoir (Fig. 11–25). The rapid movement of the slide propelled rock and other debris up the south side of the valley higher than 240 m above the reservoir level. The rock and debris filled the reservoir, producing waves of water over 100 m above the dam. The entire event including the slide and flood took only about 7 min.

The landslide was caused by a combination of factors. First, there were extremely adverse geologic conditions, including weak rocks with open fractures, with numerous clay partings as possible slide planes. Second, the topography was very steep. Third, there was increased water pressure into the valley rocks due to the impounded water in the reservoir. Water from the reservoir migrated into the bank materials, raising the pore-water pressure. Fourth, heavy rains preceded the slide up until the day of the disaster, and this increased the weight of the slope materials and further raised the water pressure in the rocks.

The Vaoint Reservoir slide did not come as a complete surprise, in that for 3 years prior to the event the rate of creep had been monitored and, in fact, a landslide was expected. However, the size of the landslide expected was much less than the disastrous event. The engineers did not realize that the whole mountain was moving in unison until it was too late.

Mount St. Helens Landslide-Debris Avalanche The eruption of Mount St. Helens Volcano on 18 May 1980 started with a complex landslide-debris avalanche. The volume of material in that event is estimated to have been 2.5 km³, making it the largest landslide on earth in historic time. The landslide-debris avalanche, triggered by a magnitude-5.0 earthquake at 8:32 A.M. on 18 May, involved the entire northern flank of the mountain, on which a bulge had been growing at the rate of about 1.5 m/day over a period of several weeks. The avalanche shot down the northern flank of the mountain, displacing water in nearby Spirit Lake; struck and overrode a ridge 8 km to the north; and then made a turn and moved 18 km down the Toutle River (Fig. 11–26). The initial failure of the bulge released internal pressure, and the volcano erupted with a lateral blast that was heard over 300 km away. At nearly the same time, a large vertical cloud rose to an altitude of about 19 km. The eruptive sequence through the landslide-debris avalanche stage is shown in Fig. 11–27.

Portuguese Bend-Abalone Cove Landslide This landslide emphasizes the impact of urbanization on

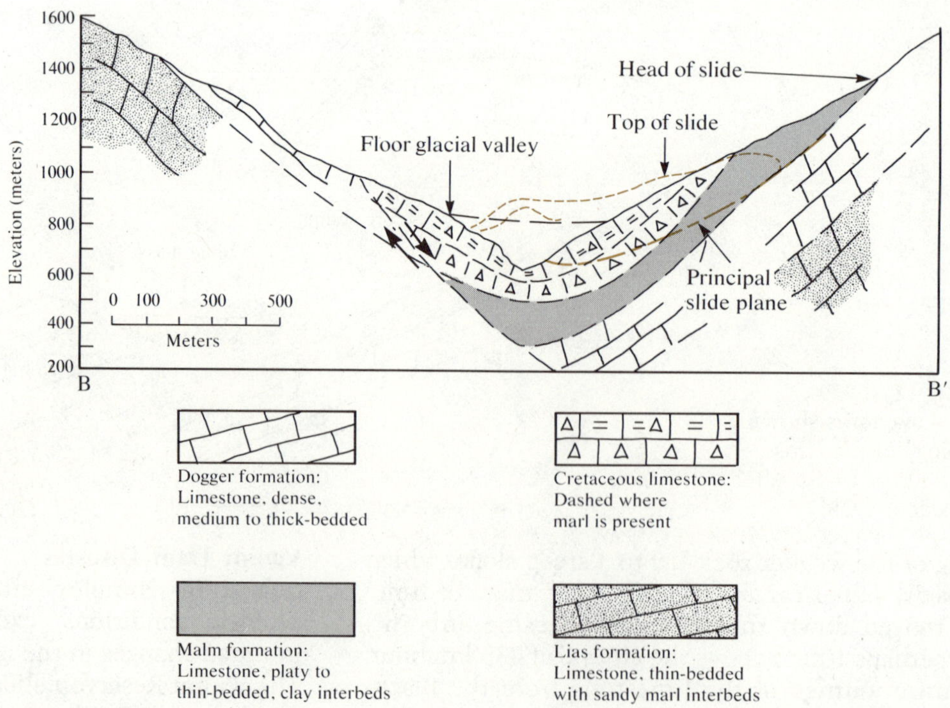

FIGURE 11–25
Generalized geologic cross section through the large landslide on the Vaoint River valley. (Modified after G. A. Kiersch, 1964)

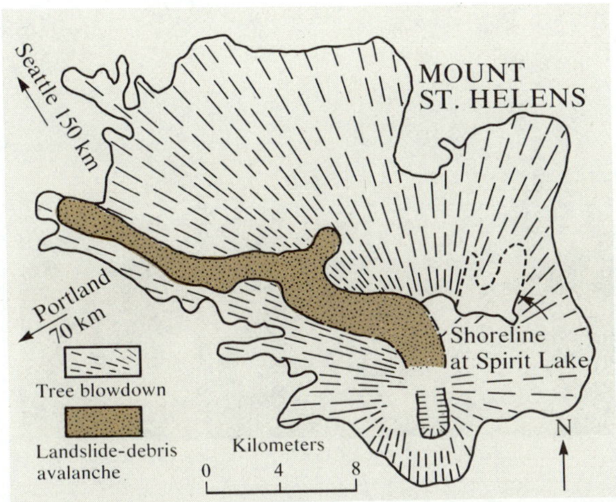

A.

B.

FIGURE 11–26
Idealized diagram (A) and photograph (B) showing the landslide-debris avalanche associated with the 18 May 1980 eruption of Mount St. Helens Volcano. The photograph is a view of the north slope of the volcano as it appeared in August 1980. The landslide-debris avalanche deposits are in the foreground. (Part A from U.S. Geological Survey; part B courtesy Harry Glicken)

slope processes. Human use of the landscape has facilitated many landslides. Slopes may be rendered unstable by removing support from the toe of a potential slide, adding fill material to a slope, or adding water to a slope. These aspects of slope stability are summarized in Fig. 11–28. The Portuguese Bend-Abalone Cove landslide is part of a larger ancient landslide that was reactivated in the late 1950s, partly in response to road building and partly by alteration of a delicate subsurface water condition by urbanization. See Fig. 11–29 for the approximate limits of the Portuguese Bend landslide, which eventually destroyed over 150 homes. The Abalone Cove landslide, which started moving in 1978, is adjacent to the Portuguese Bend slide. This recent movement has damaged homes, utility lines, roads, and a beach-front park. A vigorous program to control the landslide by pumping out excess groundwater through a series of special wells (thus reducing the pore-water pressure, which in turn increases the strength of the slope materials) is now being attempted.

SUBSIDENCE

Several interactions between geologic conditions and human use of the landscape have been involved with incidents of subsidence. Most subsidence is caused by withdrawal of fluids from subsurface reservoirs or collapse of surface and near-surface soils and rocks over subterranean voids.

Withdrawal of fluids such as oil and groundwater have caused subsidence. In all cases, the general principles are the same. Subterranean fluids in earth materials have a high fluid pressure which tends to support the earth materials above. That is, there is a buoyancy effect that tends to hold the void spaces within the rock or sediment apart. If the fluid is withdrawn, then this support is lost, and this facilitates compaction which eventually may be characterized at the surface as subsidence.

There are some differences between subsidence from withdrawal of oil and gas compared to that of groundwater. For example, extraction of oil and gas generally occurs over a small area and involves wells drilled to great depths, with considerable lowering of fluid pressure. On the other hand, with groundwater, the area of pumping may be very large (e.g., San Joaquin Valley, California). Furthermore, the wells are often much shallower than oil and gas wells and the total loss of pressure may be much less. Areas of land subsidence in California due to groundwater withdrawal are shown in Fig. 11–30. Similar maps showing subsidence could be drawn for Arizona and other states in the southwestern United States, as well as for areas in Mexico and other places around the world.

The magnitude of the subsidence is primarily dependent upon how much water pressure is lost due to pumping, the compressibility of the deposits, the thickness of the compressible beds, and the time the compressible beds have been under stress. For example, in the San Joaquin Valley of California, there has

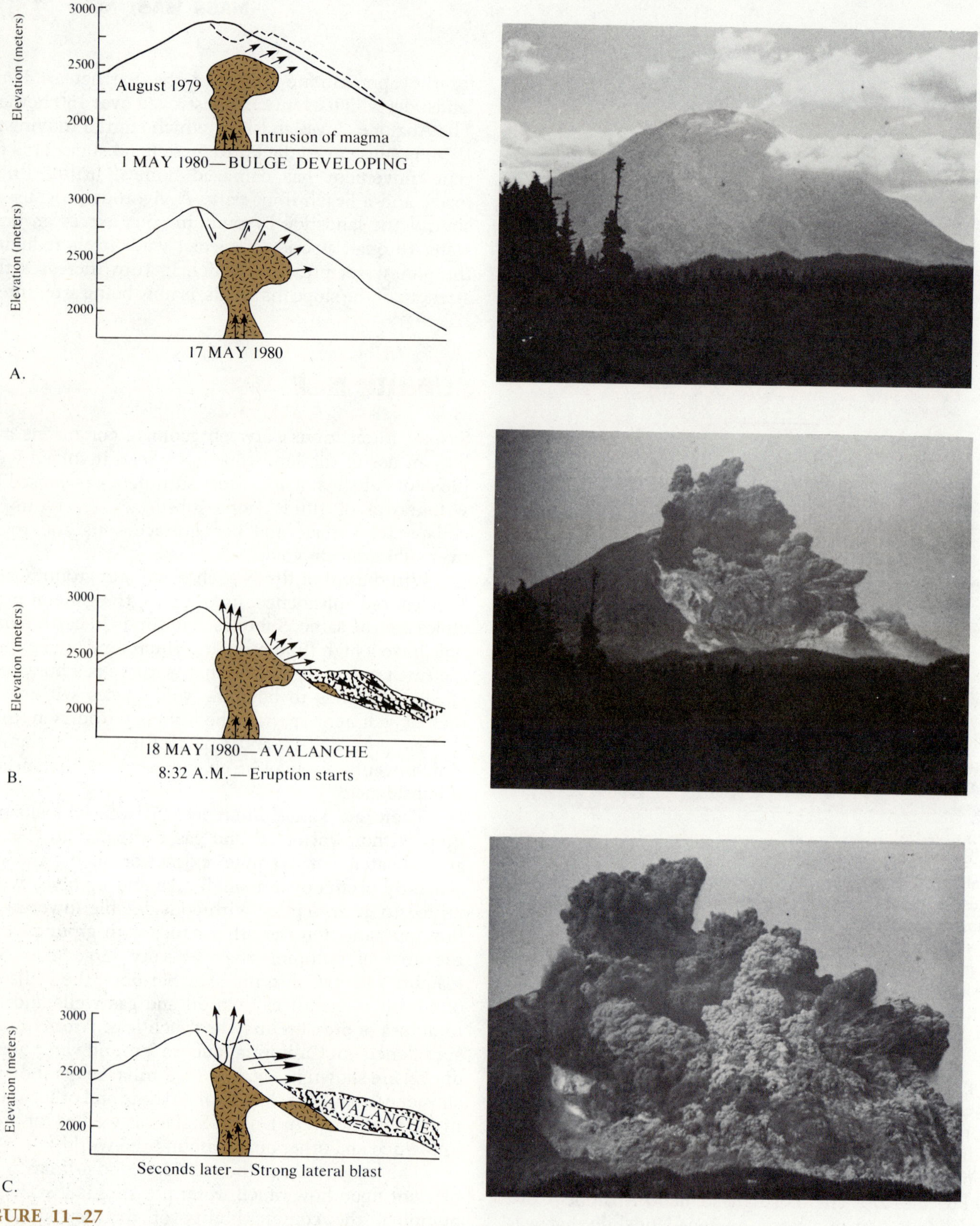

FIGURE 11–27
The eruptive sequence of Mount St. Helens on 18 May 1980. (Photos A, B, and C by Keith Ronnholm, © 1980; photo D by Robert Krimmel, U.S. Geological Survey)

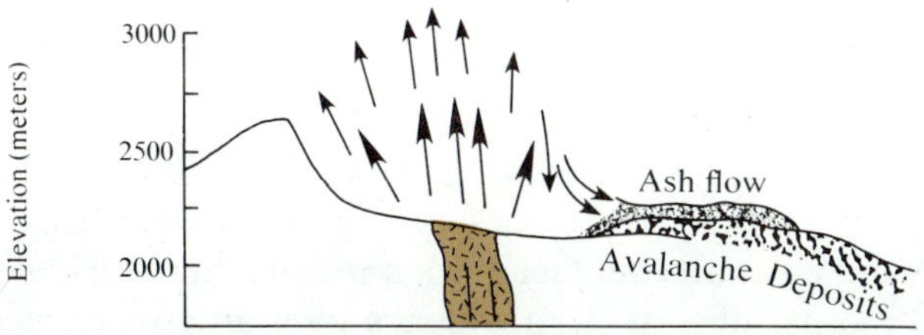

Minutes later— Full vertical eruption

D.

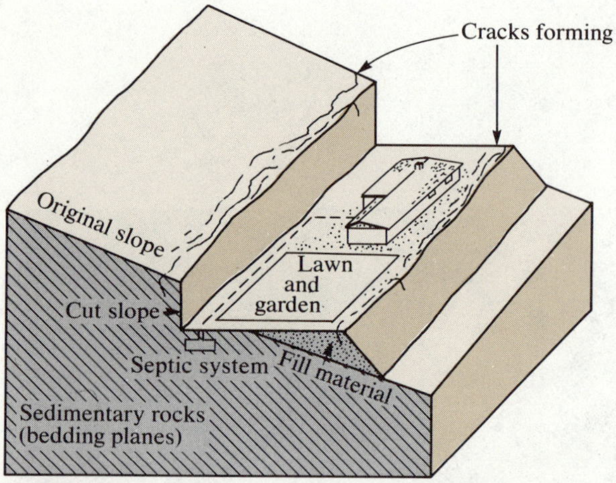

FIGURE 11–28
Several ways that slopes may be rendered unstable due to urban development. The most common ways shown here are removal of slope material producing a cut slope, placing of fill material on a slope, and adding water to a slope by watering lawns or waste disposal.

been a general mining of groundwater for a number of years. That is, the water is pumped out faster than it is being naturally replaced. As a result, there has been of the order of 8 m of subsidence over an area exceeding 9000 km^2 in the San Joaquin Valley. Pumping has been so extensive that water levels have declined over recent years by as much as 100 m or more.

Subsidence events have occurred naturally over subterranean voids. Such subsidence commonly takes place in areas underlain by limestones or bedded salt in which solutional caverns have developed. Tremendous cavities have formed when the roof of a cavern fails (Fig. 11–31).

Subsidence over mining areas has also been a problem, particularly in the coal mines of the Appalachian Mountains. Literally whole sections of towns have slowly subsided, and in some areas tilting of buildings is common. There have also been instances of subsidence over solution mines. Especially in Michigan and Virginia where salt has been mined from deep solution

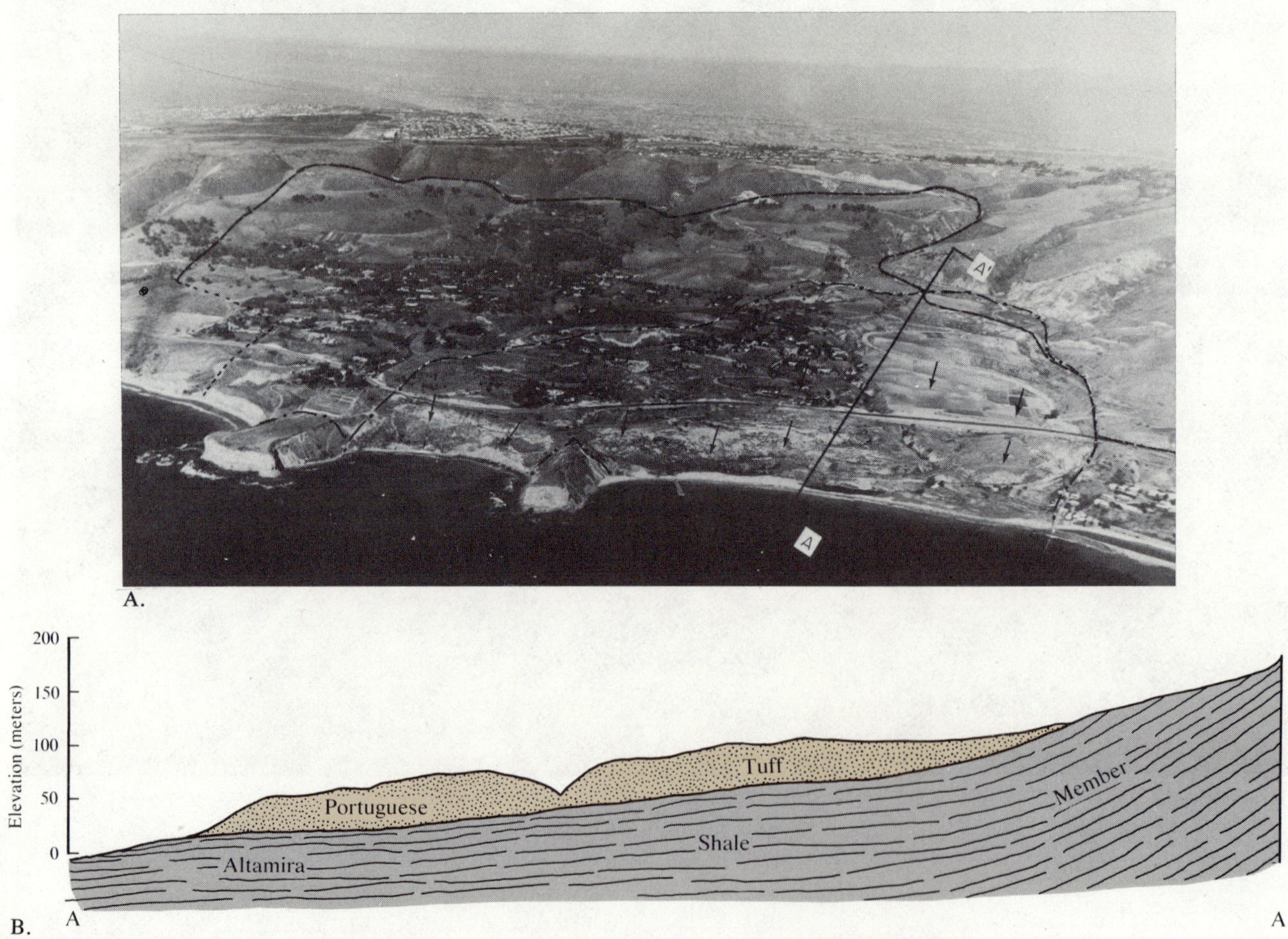

FIGURE 11–29
Photograph and diagram showing the known limits of the Portuguese Bend landslide. Arrows show the direction of movement. (Photo and cross section courtesy Los Angeles County Department of Engineering)

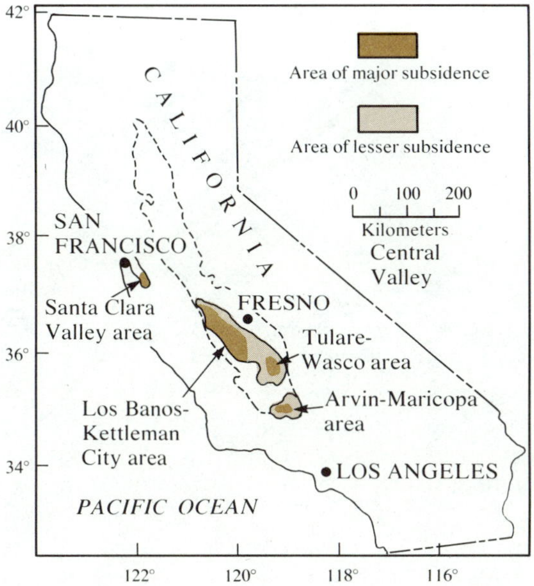

FIGURE 11–30
Principal areas of land subsidence in California resulting from groundwater withdrawal. [Reprinted by permission from W. B. Bull, "Geologic Factors Affecting Compaction of Deposits in a Land Subsidence Area," *Geological Society of America Bulletin* (1973), Vol. 84]

wells, large-scale subsidence events have occurred. The salt at depth is being mined by pumping down water relatively free of salt; the water dissolves some of the salt bed and then is pumped back to the surface, where the salt is reclaimed. The process produces large cavities below the surface, and in some cases these may work their way to the surface, causing surface subsidence failures (Fig. 11–32).

A.

B.

FIGURE 11–31
The Wink Sink. A. Depth to the water is about 10 m. Note the road and automobiles for scale. B. A view from the ground showing oil storage tanks near the sinkhole. (Photos courtesy Robert W. Baumgardner, Jr., and Texas Bureau of Economic Geology)

FIGURE 11–32
Large subsidence pit resulting from subsurface solutional mining of salt near Saltville, Virginia.

EVALUATION OF SLOPE STABILITY

The stability of a slope depends primarily upon the interaction of the forces that tend to facilitate mass wasting (**driving forces**) and those forces that tend to inhibit mass-wasting processes (**resisting forces**). A slope is stable so long as the resisting forces are greater than the driving forces. That is, the **factor of safety**, defined as the ratio of resisting to driving forces, exceeds 1 (Fig. 11–33). However, stability can change rapidly. For example, during a rainstorm precipitation may infiltrate the soil and rock, increasing the weight of the slope material and therefore causing an increase in the driving forces. At the same time, the infiltrated water may cause clay minerals to expand, as well as increase the pore pressure, both of which tend to reduce the resisting forces.

Correction of landslides once they have begun moving is a difficult process. The approach is to attack the process that caused the slope to become unstable. That is, if the slope is unstable because of an increase in pore pressure, then somehow the resisting forces must be increased. Most often this is done by draining the slope (Fig. 11–34). However, with fine-grained soils, draining is very difficult, and so the resisting forces might have to be increased mechanically by building a retaining wall (Fig. 11–35) or some other structure to increase the stability of the slope. In some instances, rather bizarre methods have been tried to stabilize slopes—freezing them, drying them out with hot air, and utilizing rock bolts to hold weak rocks to stronger rocks. Some of these methods have been successful at times, but they are expensive and their use is limited to specific instances.

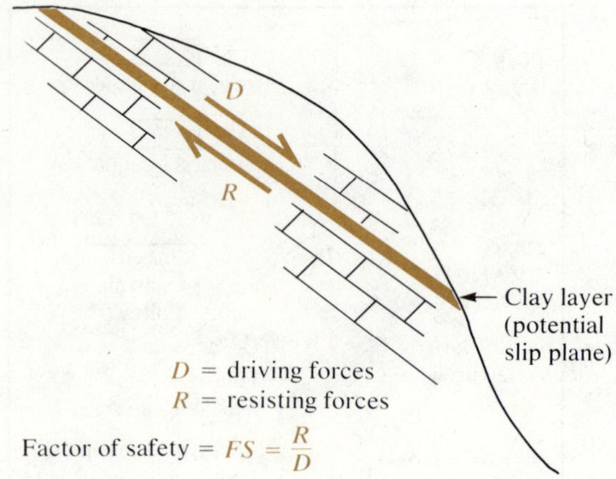

FIGURE 11–33
The factor of safety.

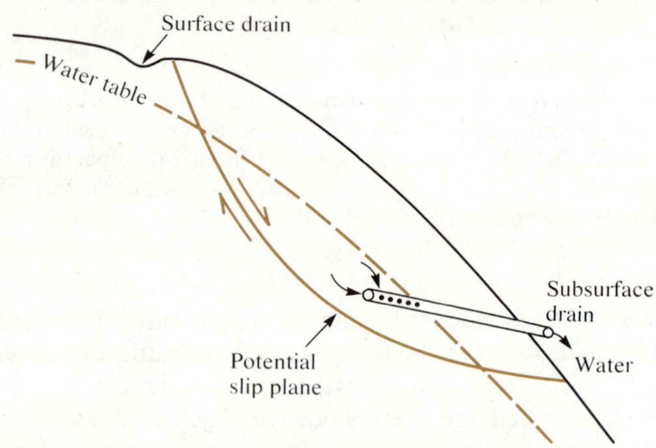

FIGURE 11–34
Idealized diagram showing how a subsurface drain that intercepts groundwater may be used to lower the pore-water pressure and thus increase the stability of a slope.

FIGURE 11–35
Retaining wall (concrete crib) used to help stabilize a roadcut.

SUMMARY

Landslides in the United States cause damage in excess of $1 billion per year and claim an average of 25 lives.

Earth materials may fail by various mechanisms, including slide, flow, fall, and subsidence. Any one mass-wasting event may include one or more of the failure mechanisms.

Natural slopes are composed of segments (concave, convex, or straight) and organized into slope elements including waxing, free-face, debris, and waning. Which slope elements are present on a given slope depends on factors such as material type, climate, and vegetative cover.

Mass-wasting or downslope movement processes are classified in terms of kind and rate of movement, and degree of saturation of slope materials.

The role of water in mass wasting is always important, and most landslides are associated with water problems. Thus, control of landslides generally involves control and drainage of surface and subsurface water.

Large landslides, involving millions or even billions of cubic meters of debris, are natural processes, but building of dams and urbanization affect their frequency, and many ancient landslides are reactivated when a delicately balanced slope stability is disrupted.

Subsidence may result from several processes, including withdrawal of fluids such as oil and groundwater, or collapse of surface or near-surface rocks over subterranean voids.

Evaluation of slope stability involves the examination of driving and resisting forces in slopes to determine the factor of safety. Landslides are easier to control prior to failure; correction of landslides once they have begun moving is a difficult process that must be matched to the mechanism that caused the failure.

REVIEW EXERCISES

1 What is the most common type of mass-wasting process?

2 Why is it believed that terracettes may be formed by grazing animals?

3 Discuss how mudflows might develop during a volcanic eruption.

4 Why do slumps commonly have some backward rotation of the blocks?

5 Why do debris avalanches develop more commonly in steep, forested land than in a more arid environment?

6 Why would an earthflow more likely occur on a shale slope than on a granite bedrock slope?

7 What is the factor of safety in slope stability, and why is the concept important?

8 Defend or criticize the statement "Mudflows are most common in humid climates."

9 The waxing slope is produced by creep, whereas the waning slope is produced by running water. What general principle concerning form and process may be learned by examining these two slope elements?

10 What is the difference between a mudflow and an earthflow?

11 Discuss the role of water in facilitating landslides.

12 List as many geologic types of slip planes as you can.

13 Compare mass-wasting processes for humid and arid conditions.

14 Defend the statement "Slope shape reflects the physical processes operating."

15 Why is creep so common?

16 What mechanisms may cause creep?

17 Why do volcanoes pose a serious mass-wasting hazard?

18 Compare and contrast the two main ideas concerning the origin of terracettes.

19 If you are contemplating purchasing a hillside home, how might you evaluate the landslide potential?

20 How does mining of groundwater cause subsidence?

SUGGESTED READINGS

CARSON, M. A., and KIRKBY, M. J. *Hillslope Form and Process.* Cambridge: At the University Press, 1972, 475 pp.

COATES, D. R., ed. *Landslides.* Reviews in Engineering Geology, Vol. 8. Boulder, Colo.: Geological Society of America, 1977, 278 pp.

FLEMING, R. W., and TAYLOR, F. A. *Estimating Costs of Landslide Damage in the United States.* U.S. Geological Survey Circular 832, 1980.

KIERSCH, G. A. "Vaoint Reservoir Disaster." *Civil Engineering* (1964), Vol. 34, pp. 32–39.

POLAND, J. F., and DAVIS, G. H. "Land Subsidence Due to Withdrawal of Fluids." In D. J. Varnes and G. Kiersch, eds. *Reviews in Engineering Geology,* 2nd ed. Boulder, Colo.: Geological Society of America, 1969, pp. 178, 269.

SCHUSTER, R. L., and KRIZEK, R. O., eds. *Landslides: Analysis and Control.* National Academy of Sciences, Transportation Research Board Special Report 176, 1978, 234 pp.

SHARP, C. F. *Landslides and Related Phenomena.* New York: Columbia University Press, 1938, 137 pp.

SHREVE, R. L. *The Blackhawk Landslide.* Geological Society of America Special Paper 108, 1968, 47 pp.

TERZAGHI, K. "Mechanics of Landslides." In S. Paige, ed. *Application of Geology to Engineering Practice: Berkey Volume.* Boulder, Colo.: Geological Society of America, 1950, pp. 82–123.

ZARUBA, Q., and MENCL, V. *Landslides and Their Control.* Amsterdam: Elsevier Scientific Publishing Co., 1969, 205 pp.

- hydrologic (water) cycle
- meteoric water
- connate water
- juvenile water
- surface water
- subsurface water
- groundwater
- direct runoff
- unconfined groundwater
- influent stream
- effluent stream
- cone of depression
- seep
- spring
- Darcy's law
- hydraulic head

12

- overland flow
- interflow
- infiltration
- hydrogeology
- zone of aeration
- zone of saturation
- water table
- soil water
- intermediate water
- capillary water
- capillary fringe
- porosity
- permeability
- aquiclude
- aquifer
- confined groundwater
- artesian water
- artesian well
- pressure surface
- perched water
- karst topography
- sinkhole
- hot springs
- geyser
- convection
- salt-water intrusion
- groundwater pollution
- sanitary landfill
- leachate

Groundwater and Related Phenomena

TO SOME PEOPLE, THE LOCATION OF GROUNDWATER IS mysterious, something to be determined by methods not far removed from witchcraft. Water witching, or dowsing, using such devices as pendulums, wires, or forked sticks to locate water is still practiced even though studies suggest that wells located by such methods are not any more successful than randomly located wells. Nevertheless, many of the witchers and dowsers have a working knowledge of the hydrogeology of their local area, which serves them in predicting areas likely to produce groundwater. Furthermore, some groundwater can be found most anywhere. For example, consider Fig. 12–1. This map of the United States shows areas of potential groundwater development. Note that they cover most of the country.

To the applied earth scientist, the location of groundwater is basically a geological problem. In areas where precipitation is limited, location of groundwater resources may be crucial to all types of development. Although vast amounts of drinkable groundwaters may be available in such regions, they usually are not uniformly distributed and are not as readily available for human use as the population concentrated in an area might desire. There are several problems that center around how people perceive the availability of groundwater, and around our understanding of this resource. First, people tend to assume that water is available when and where and in any amounts they want. They turn the faucet on and the water magically appears even though it is someone else's responsibility to make sure they have it. Second, groundwater is out of sight and therefore essentially out of mind; that is, it is mysterious. Third, compared to surface-water resources, groundwater is not easily measured, and often limited subsurface information must be used to estimate total groundwater reserves. The problem is that groundwater reserves are constantly fluctuating in response to processes that tend to input water into the groundwater system and processes that tend to remove it. Therefore, in many cases—much to the chagrin of the planner—groundwater resource pictures are broad-brushed—like a picture painted with a paint roller.

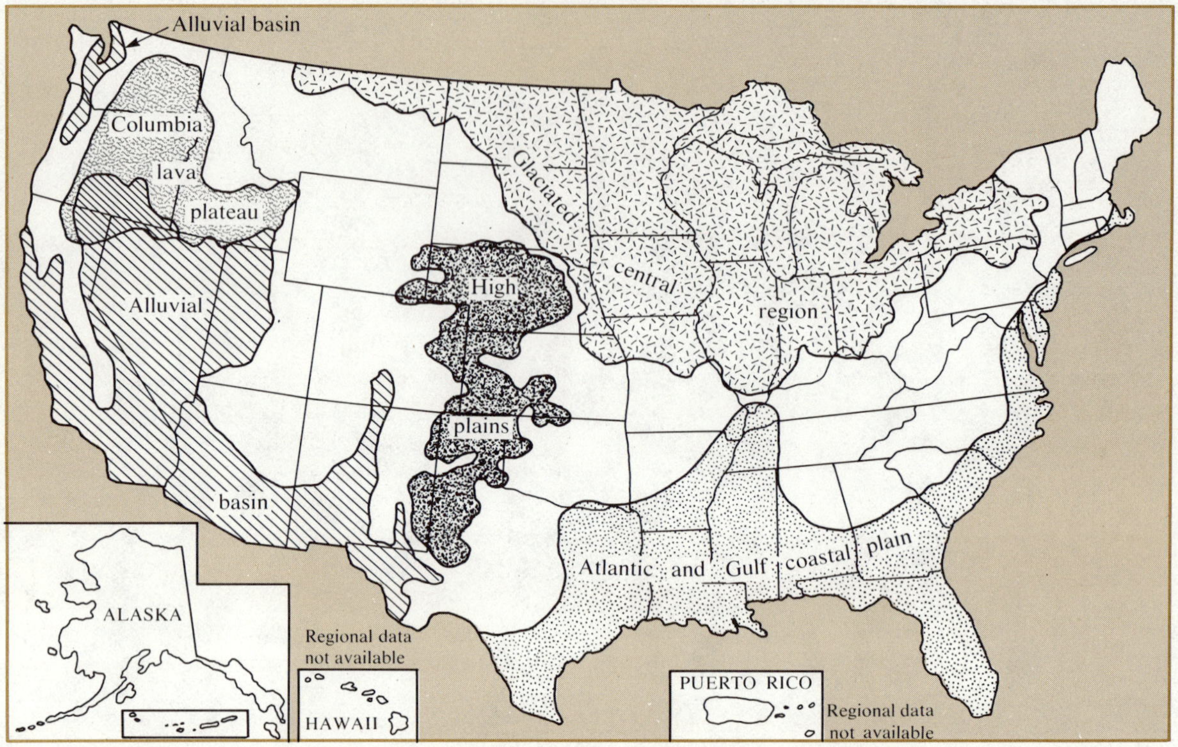

FIGURE 12–1
Areas of potential groundwater development in the United States. (From *The Nation's Water Resources,* Water Resources Council, 1968)

The drilling of wells is not a new technology. People for thousands of years have deliberately made excavations in the hope of obtaining water resources. However, only in the last several hundred years have we have gained an understanding of the origin of groundwater. We now know that one-third or less of the total precipitation that falls in a particular area will directly run off through the river system. Of course, some of the water is used by plants and animals and some evaporates, but a good deal of it infiltrates into the soil and rock to become groundwater. About 120 years ago, the French engineer Henry Darcy developed the law governing the flow of groundwater through earth materials. Since Darcy's work, many important discoveries concerning the nature of groundwater, how it moves, and how it might best be managed have been made. The fields of groundwater hydrology and hydrogeology are now well-developed disciplines.

Years ago (and today in some areas of sparse population), the only major problem associated with groundwater resources was how to locate them. Today, however, with an ever-increasing demand for water resources, coupled with an expanding population, many problems have arisen. In some areas, population densities are so great that groundwater resources are literally being mined. That is, the resource is being used faster than it can be replenished by natural processes. All sorts of waste materials are deliberately or inadvertently discharged into earth materials, and sometimes these pollute groundwater resources. As urbanization increases, the amount of land covered by impervious material (roofs, parking lots, streets, shopping centers, etc.) increases rapidly, and thus there is less area available for precipitation to infiltrate into the soil and rock and work its way down to the groundwater system. As a result, small streams in urban areas that once flowed the entire year may experience drastically reduced summer flows or even dry up. This results because groundwater seepage into stream channels is a main supply of water to streams during the dry season.

THE HYDROLOGIC CYCLE

The **hydrologic** or **water cycle** (Fig. 12–2) is a solar-driven system that constantly cycles water from the ocean to the atmosphere to the continents and back to the ocean again. Interestingly, only a small fraction of a percent of the total water on earth is in the atmosphere at any one time (Table 12–1), and yet this small amount controls the fresh-water phase of the hydrologic cycle.

TABLE 12–1
Approximate global distribution of water.

Location	Percentage of Total Water
Surface water	
Ocean	97.2
Fresh-water lakes	0.009
Saline lakes and inland seas	0.008
Rivers and streams	0.0001
Subsurface water	
Soil water	0.005
Groundwater	0.62
Other	
Ice caps and glaciers	2.15
Atmosphere	0.001

SOURCE: Data from U.S. Geological Survey.

There are many ways to classify and describe the waters in, on, and above the earth. Some of the terms that may be used are **meteoric water**, **connate water**, **juvenile water**, **surface water**, **subsurface water**, and **groundwater**. These terms are not mutually exclusive. *Meteoric water* is the water in or derived from the atmosphere. *Connate water* is water that is no longer in circulation or contact with the present water cycle. It is usually saline water trapped during the deposition of sediments and therefore is sometimes called **fossil water**. *Juvenile water* is water derived from the interior of the earth in rock-forming processes but that has not previously existed as atmospheric or surface water. It is the slow release of juvenile water by volcanic activity that is believed to be the source of most of the water in the oceans, and thus the hydrologic cycle. *Surface water* is a comprehensive term for all those waters that flow or are in storage on the surface of the lithosphere. These include of course the waters of lakes, rivers, and glaciers. *Subsurface water* is a general term for all the waters within the lithosphere. These include the water in soil, connate waters, and groundwater. The term *groundwater* is generally reserved for that part of the subsurface water within the zone of saturation. Groundwaters may be fresh or saline, and it is the fresh water that is important as the water supply for most urban and industrial and agricultural uses.

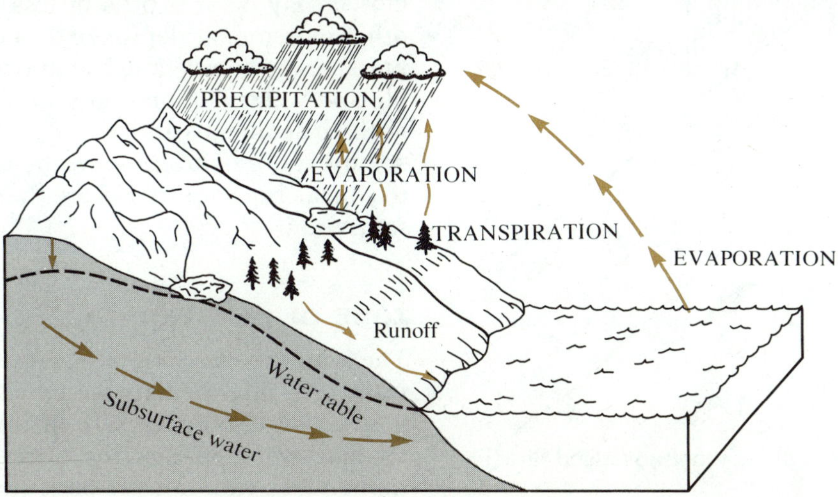

FIGURE 12–2
The hydrologic cycle.

Examination of the hydrologic cycle reveals that the movements of surface waters and groundwaters are intimately related. In the contiguous United States, approximately 30 percent of the precipitation enters into the surface-subsurface flow system. Of this, only about 1 percent reaches the ocean by way of groundwater flow; the remainder is evaporated into the atmosphere, or is utilized by plants and enters the atmosphere through transpiration. This description, however, is a gross simplification, as there are many possible interactions in which surface water may enter into the subsurface flow system, and conversely. Thus, ground- and surface-water interactions on the local scale may be quite complex, involving subtle interactions between the composition and structure of earth materials.

During and following a rainfall event, a component of the water known as the **direct runoff** moves downslope and enters stream channels. A direct runoff may be divided into two parts: the portion that flows directly on the surface of the earth (**overland flow**) and the portion that flows beneath the surface but above the groundwater table (**interflow**) (Fig. 12–3). The interflow is especially important in forested areas where surface runoff on hillslopes may be negligible (Fig. 12–3).

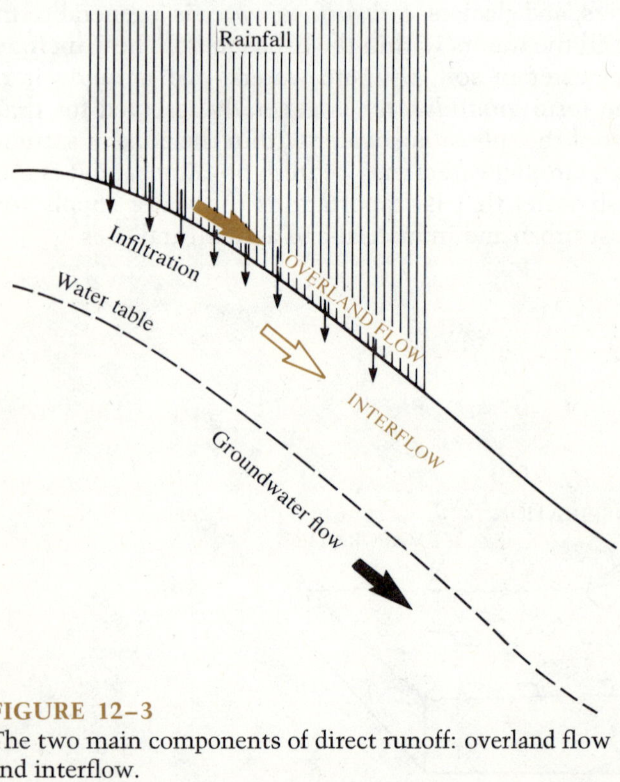

FIGURE 12–3
The two main components of direct runoff: overland flow and interflow.

The process whereby surface water enters into the soil and subsurface flow system is known as **infiltration** (Fig. 12–3). The rate at which infiltration into the groundwater system takes place depends upon such factors as the distribution and amount of precipitation; the type of soil and rocks; the topography; the amount and type of vegetation; and, lastly, the amount of rejected recharge, that is, the water that cannot enter the ground because the soil is already saturated.

HYDROGEOLOGY

Hydrogeology is a general term for the geologic aspects of groundwater occurrence, movement, and discharge. The classification of subsurface waters is shown in Fig. 12–4. The water falls into two main zones: the **zone of aeration** and the **zone of saturation**. The boundary between the zones of aeration and saturation is the **water table**, which delineates the approximate elevation of the water surface in wells that penetrate the zone of saturation. Below the water table, all the pore (void) spaces in earth materials are saturated, that is, filled completely with water. The lower boundary of the zone of saturation is not exact, because rock properties are not constant with depth. In other words, as one goes deeper in the earth, all the openings between grains in rocks or fractures eventually close because of the great weight of overlying material. At this depth, little water can exist, and motion of that water is nearly impossible. The depth to this very dense rock varies with the type of rock. For example, in areas of thick sedimentary sections it may exceed 15,000 m, while in a complex of igneous and metamorphic rocks the zone below which all fractures are essentially closed may be at depths of less than 3000 m. At the other extreme, at depths of 25 km or so the heat and pressure are great enough that open pores are impossible and water is found only in chemical combination with the minerals in the rocks (Fig. 12–5). From the standpoint of hydrogeology, the most important zones to consider in some detail are the zones of aeration and saturation.

ZONE OF AERATION

The water in the zone of aeration consists of the **soil water**, the **intermediate water**, and the **capillary water** (Fig. 12–4). Soil water is distinguished from intermediate water in that the water in soil is subject to tremendous variations in both quantity and quality. In

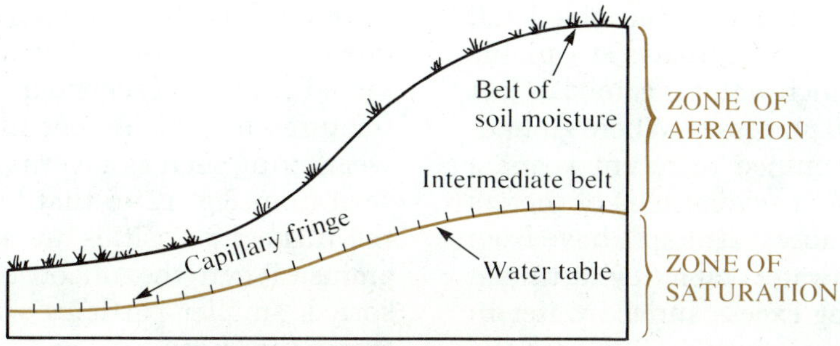

FIGURE 12–4
Classification of subsurface waters.

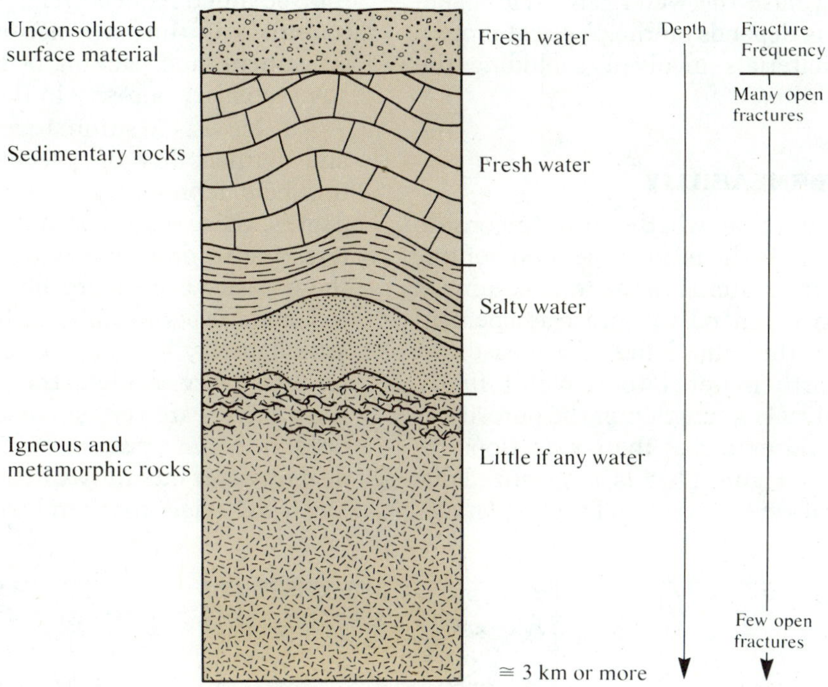

FIGURE 12–5
The occurrence of groundwater in the near-surface environment.

the upper part of the soil, direct runoff may be important through the process of interflow. Furthermore, organic material holds varying amounts of water in suspension and processes of evaporation are significant. In forested areas, soil water may go to depths of approximately 10 m, the depth to which root systems extend. Intermediate water separates the capillary water from the soil water, and its boundaries are not precise either. The thickness of the subzone of intermediate water varies from nonexistent in very wet climates to hundreds of meters in dry climates. The amount of intermediate water present at any one time depends upon the nature and extent of precipitation as well as the properties of the earth materials. If the materials are porous and if there is sufficient rainfall, then the water will percolate through the intermediate zone to the saturated zone below. The **capillary fringe** is a zone of variable thickness directly above the water table. The boundary between the intermediate zone and capillary fringe varies from quite sharp to gradual depending upon the type of earth materials involved. It tends to be gradual in fine materials such as silt and clay and may be quite sharp in porous, coarser materials such as clean gravels.

Capillary water is water that is drawn up from the water table by capillary action caused by surface tension. With porous materials, the level to which water may be drawn up is partially a function of the size of the openings. With smaller openings, water tends to be drawn up further.

ZONE OF SATURATION

The water in the zone of saturation (Fig. 12–4) is the groundwater we are interested in for water resources. Below the water table, all the pores and other openings in earth materials are saturated. The water table itself is not a stationary surface, and it fluctuates at variable rates depending upon how much water is gained or lost by natural or human-induced processes. Where groundwater has been extensively mined in recent years, a groundwater-table reduction of several tens of meters is not unusual. In other instances, attempts have been made to raise the groundwater table by artificial recharge, that is, by applying excess surface water to areas where the water is likely to infiltrate through the zone of aeration and raise the water table. How successful the recharge is depends on the physical properties of the earth materials involved, including the porosity and permeability.

POROSITY AND PERMEABILITY

Porosity of an earth material, whether it be unconsolidated sediment or rock, is the ratio of the total volume of openings in the earth material to the total volume of the material. Porosity is generally recorded as a percentage. **Permeability**, on the other hand, is a measure of the ability of an earth material to transmit fluids. Therefore, permeability is a function of the porosity as well as the nature and extent of the connections between the pore spaces. Figure 12–6 is an idealized diagram showing several types of soil and rock openings. In general, a distinction may be made between primary openings consisting of void spaces surrounding individual grains and secondary openings, which consist of fractures in rock or openings produced by chemical weathering such as caverns in limestone. It can also be seen from Fig. 12–6 that for unconsolidated material, the highest porosities are associated with well-sorted grains. When the unconsolidated material is poorly sorted, smaller particles will fill in the openings between the larger ones, and in this way the total porosity may be much reduced. The porosity in unconsolidated materials is thus a function of the size, shape, arrangement, and packing of the particles as well as of any secondary porosity in the particles themselves.

Table 12–2 summarizes some data for porosity and permeability of selected earth materials. These numbers represent near-surface conditions; at great depths, these values would be much reduced. It is interesting to note that with unconsolidated materials, the porosity is considerably greater for the finer material, and the permeability is lower. This results because although clay has a great number of openings and the total porosity is high, the interconnections between the particles are very small and these openings tend to hold the water tenaciously. Thus, clay may hold a good deal of water but allows little water to pass through it. Earth materials that tend to retard the flow of water,

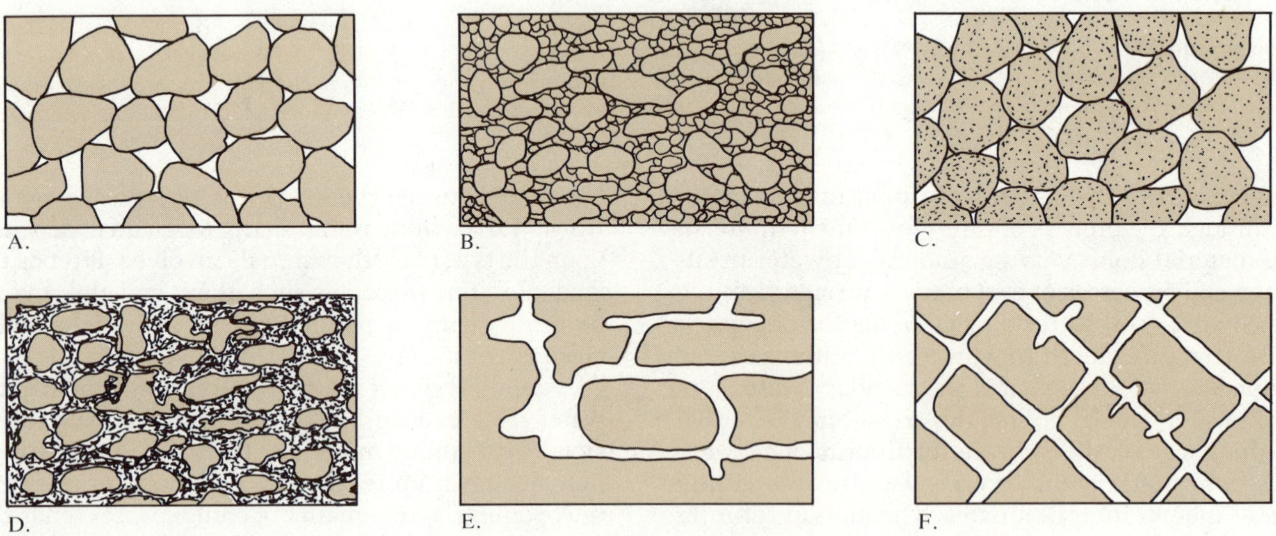

FIGURE 12–6
Several types of soil and rock openings and their relation to porosity. A. Well-sorted sediment with a high porosity. B. Poorly sorted sediment with a low porosity. C. Well-sorted sediment in which the particles themselves are also porous, resulting in a total higher porosity. D. Well-sorted sediment in which the openings between grains are partially filled with cementing material, resulting in a reduced porosity. E. Rock in which porosity is increased due to solutional openings. F. Rock that is porous due to open fractures. (From O. E. Meinzer, U.S. Geological Survey Water Supply Paper 489, 1923)

TABLE 12–2

Material	Porosity (%)	Permeability (m³/day/m²)
Unconsolidated		
Clay	45	0.041
Sand	35	32.8
Gravel	25	205.0
Gravel and sand	20	82.0
Rock		
Sandstone	15	28.7
Dense limestone or shale	5	0.041
Granite	1	0.0041

SOURCE: After Linsley, Kohler, and Paulhus, *Hydrology for Engineers* (New York: McGraw-Hill Book Co., 1958).

that is, those with very low permeabilities, are called **aquicludes**. Clays and rocks with few fractures are commonly aquicludes. These are essentially boundaries through which groundwater flow is extremely restricted (Fig. 12–7). On the other hand, a zone of earth material capable of producing groundwater at a useful rate from a well is called an **aquifer**. Aquifers, therefore, are earth materials with a relatively high porosity and permeability. Although the best aquifers are generally unconsolidated gravel, highly fractured rocks also make good aquifers.

UNCONFINED GROUNDWATER SYSTEMS

Unconfined groundwater systems are located where there are no impermeable layers restricting the upper surface of the zone of saturation. Figure 12–8 is a generalized diagram of an unconfined groundwater system. Notice that the surface of the water table has less relief than the topographic surface; otherwise, the two are quite similar in morphology. Also shown in Fig. 12–8 are two types of streams defined in terms of the relationship between the stream and water table. **Influent streams** are streams in which the flow of water is from the channel to the groundwater table and a recharge mound may exist above the groundwater table. **Effluent streams** are those streams in which the flow of water is from the groundwater system into the stream. This is the general situation necessary for perennial streams. Also shown is the recharge zone where precipitation infiltrates through the zone of aeration down to the zone of saturation. When water is pumped from a well, a **cone of depression** in the water table forms (Fig. 12–9). Overpumping of an aquifer causes the water level to lower continuously, which may eventually necessitate lowering the pump settings or drilling deeper wells. However, drilling deeper wells is not always successful, particularly when the well is in fractured rocks. Water from such wells is pumped from open fracture systems that tend to close or diminish in number and size at increasing depth.

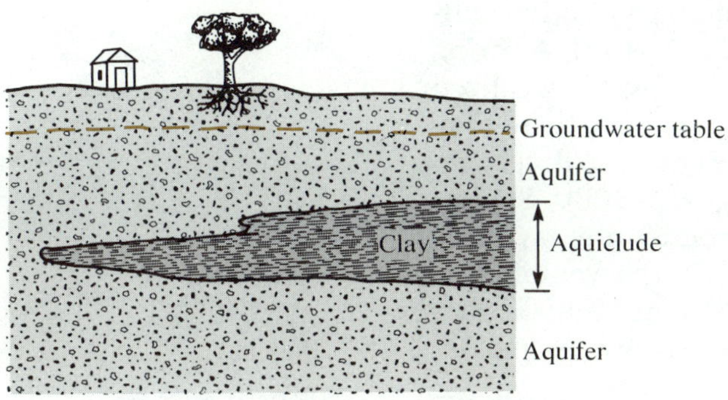

FIGURE 12–7
Aquicludes and aquifers. Aquicludes are materials with very low permeability and thus retard the flow of groundwater. Aquifers, on the other hand, are earth materials with high porosity and permeability from which groundwater may be extracted.

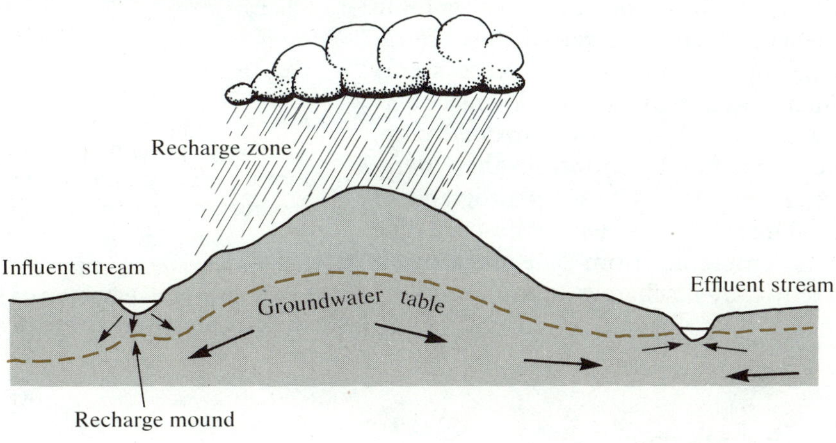

FIGURE 12–8
An unconfined groundwater system. Also illustrated are influent and effluent streams.

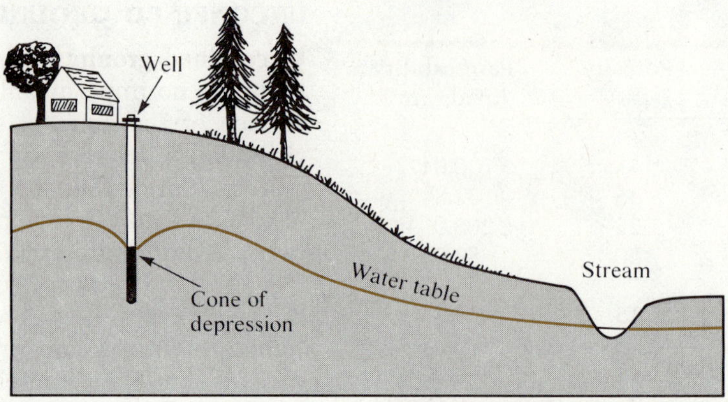

FIGURE 12–9
The cone of depression in the water table that results from pumping water from a well.

Natural discharge of groundwater by way of **seeps** or **springs** is largely controlled by geologic conditions. The difference between a seep and a spring is the magnitude of discharge. Springs produce enough discharge to form a very small stream of water, whereas seeps do not. Generally, the largest springs are found in areas underlain by limestone or volcanic rock, as both of these rock types are likely to have large subterranean voids along which large quantities of water may be transported. Basically, a seep or spring may form whenever the groundwater table intersects the earth's surface. Two ways the geologic environment may influence the development of springs are shown in Fig. 12–10. Notice that processes of erosion and the existence of some groundwater barrier such as a fault zone or zone of impermeable rock (aquiclude) greatly facilitate the occurrence of springs (Fig. 12–10A). A spring may also form if water accumulates over a localized aquiclude such as a clay layer. In this case, there may be a perched or local water table above the deeper regional groundwater system (Fig. 12–10B). Springs emerging from such perched groundwater systems may flow only part of the year, during the wet months. However, some areas have sufficient rainfall that perched water may be present the entire year and absent only during droughts.

In areas underlain by limestone and other soluble rocks, the surface water may be diverted to subterranean routes through open conduits or cavern systems (Fig. 12–11A). Such waters may emerge downslope as springs. In these areas, very large springs may develop, and in fact whole rivers may be born from the emerging groundwaters. Figure 12–11B is a photograph of the large spring at Big Spring State Park, Missouri. The flow of water is emerging from a subterranean groundwater system that discharges as much as 3.2 million m³/day.

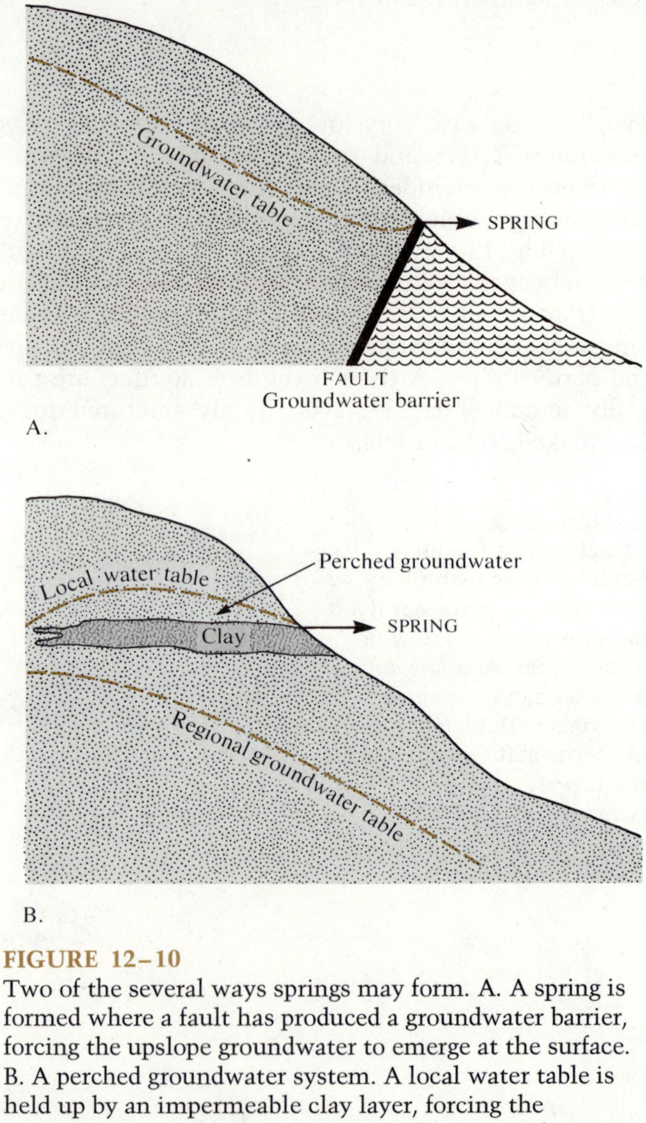

FIGURE 12–10
Two of the several ways springs may form. A. A spring is formed where a fault has produced a groundwater barrier, forcing the upslope groundwater to emerge at the surface. B. A perched groundwater system. A local water table is held up by an impermeable clay layer, forcing the groundwater to emerge as a spring at the contact between the clay and overlying materials.

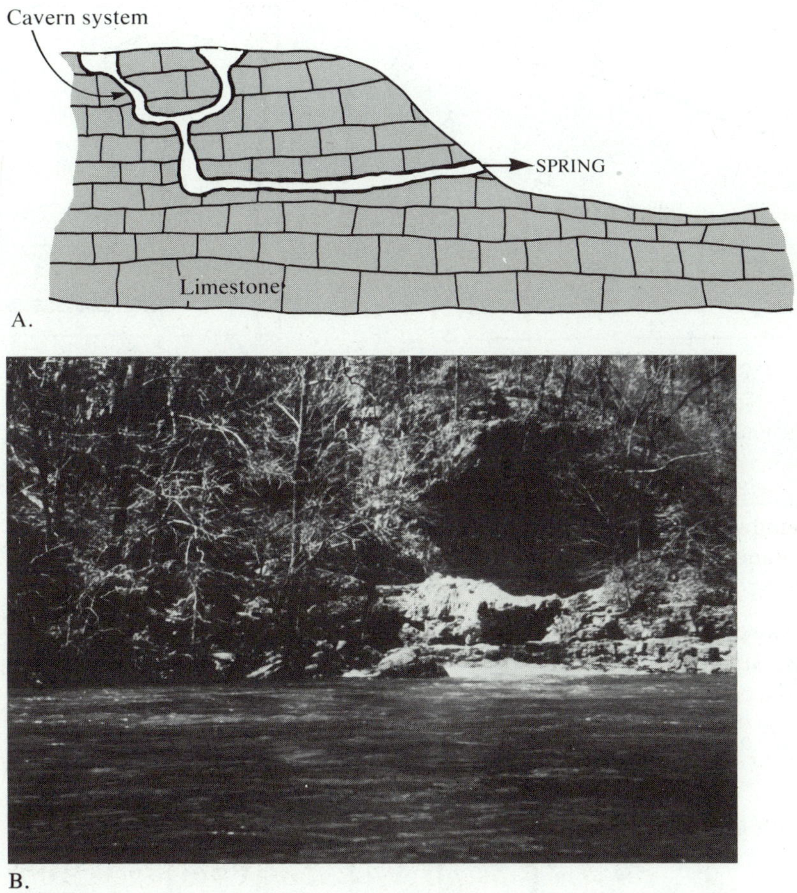

FIGURE 12–11
A. Large springs may develop in areas underlain by limestone bedrock. Cavern systems collect the water, which may be discharged by way of surface outlets. B. Big Spring at Big Spring State Park in Missouri, where approximately 3.2 million m^3 of water is discharged per day, forming a very large stream. (Photo courtesy James Brice)

MOVEMENT OF GROUNDWATER

The movement of groundwater in an aquifer is a relatively slow process when compared to other geologic processes such as flow of rivers, and rapid mass-wasting processes such as mudflows and debris avalanches. On the other hand, groundwater flow is generally faster than soil creep or the slow migration of sand dunes. In absolute terms, the rate of groundwater flow generally varies from a meter or so per year to a couple of meters per day. However, in specialized circumstances the rate of motion may be much faster, approaching that of rivers. This generally occurs in areas where the groundwater is concentrated in large subterranean voids such as those found in limestone regions.

Rates of flow of groundwater in aquifers composed of unconsolidated material or fractured rock obey Darcy's law:

$$V = K\frac{H}{L}$$

where V is the velocity of water; K is a coefficient of permeability, having the dimensions of a velocity and depending on properties of both the fluid and the material through which the groundwater flows; and H/L is the quantity known as the **hydraulic head**, or **gradient**. Some of these relations are shown in Fig. 12–12. Note that H is the difference in groundwater heights between two points (A and B in Fig. 12–12), and L is the horizontal distance between those points. Thus, the quantity H/L is the average slope of the water surface. Depending on what parameters are known, Darcy's law can be used in computing the rate of groundwater flow or the **coefficient of permeability**.

The quantity of flow per unit time Q may be calculated by

$$Q = VA$$

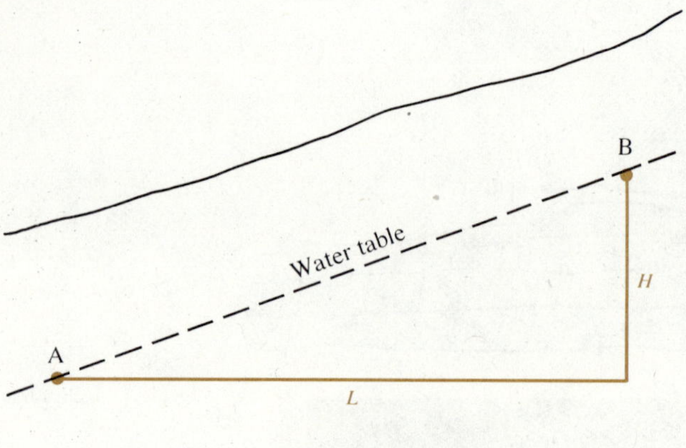

Hydraulic gradient $= \frac{H}{L}$

FIGURE 12–12
The hydraulic gradient, defined as the ratio of H (difference in elevation of the groundwater table between two points) to L (the horizontal distance between those two points).

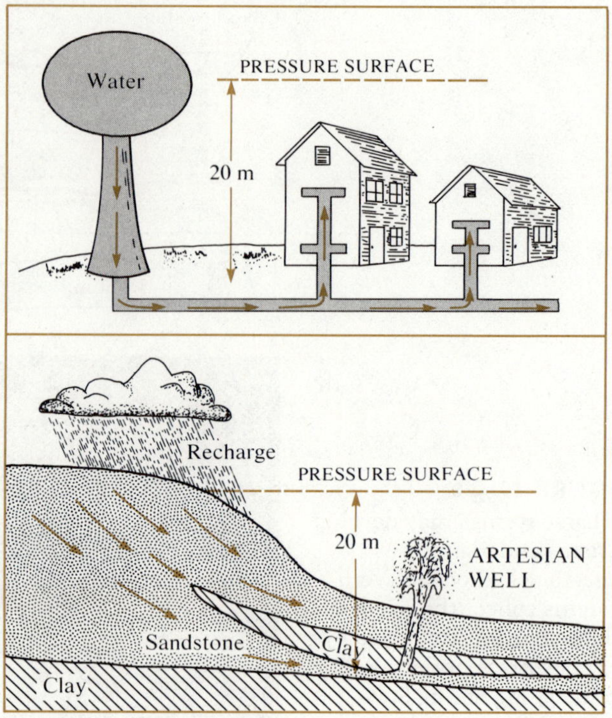

FIGURE 12–13
Idealized diagram showing how an artesian groundwater system might develop.

where A is the cross-sectional area perpendicular to the direction of flow. Substituting Darcy's law for the velocity of flow yields

$$Q = K\frac{L}{H}A$$

where Q is in cubic meters per second.

CONFINED GROUNDWATER SYSTEMS

An aquifer is said to be confined if the movement of water is restricted above and below by aquicludes. Figure 12–13 is a generalized diagram showing how a **confined groundwater system** might form. The water enters a recharge zone and moves down and laterally beneath a confining layer, in this case a clay. Thus, the water is under pressure and, if tapped, will tend to rise to the level of the recharge zone. For this reason, confined groundwater is sometimes termed **artesian water**, and wells that flow without pumping are called **artesian wells**. As shown in Fig. 12–13, this condition is analogous to a water tower's producing water pressure for homes.

Of course, both confined and unconfined aquifers may be found in the same area. For example, in the San Joaquin Valley of California, large areas are underlain by two aquifers, the upper one unconfined and the lower one confined by an aquiclude. Many of the very-high-producing wells in the valley are withdrawing water from the lower aquifer. The **pressure surface** for a confined aquifer is an imaginary surface, the level to which water would rise were it not for the confining layer. In the San Joaquin Valley, the rate of water withdrawal from the confined aquifer is greater than that of replacement by natural processes; as a result, the pressure surface has lowered by 3–4 m/year in recent times. Figure 12–14 is a generalized cross section of the San Joaquin Valley showing the upper and lower aquifers. In some areas, the total decline in water levels is about 140 m. As a result of this groundwater mining, an area of approximately 9000 km^2 has been affected by subsidence locally as much as 9 m. The mechanism by which the subsidence takes place is related to changes occurring in the packing of the particles comprising the aquifer that result from withdrawal of groundwater, as illustrated in Fig. 12–15.

An interesting phenomenon associated with confining layers is **perched water** (Fig. 12–10B). Here, the regional water table is at a relatively deep level, but part of the water filtering down through the zone of aeration is trapped by a discontinuous aquiclude such as a clay layer. Thus there is a local high water table produced. Perched water tables may also exist above other types of confining layers such as discontinuous permafrost in the higher latitudes. Perched water is commonly a seasonal phenomenon or an event that may last only a few years during times of above-normal precipitation. In other cases, perched water may be a reliable source of groundwater resources for many years.

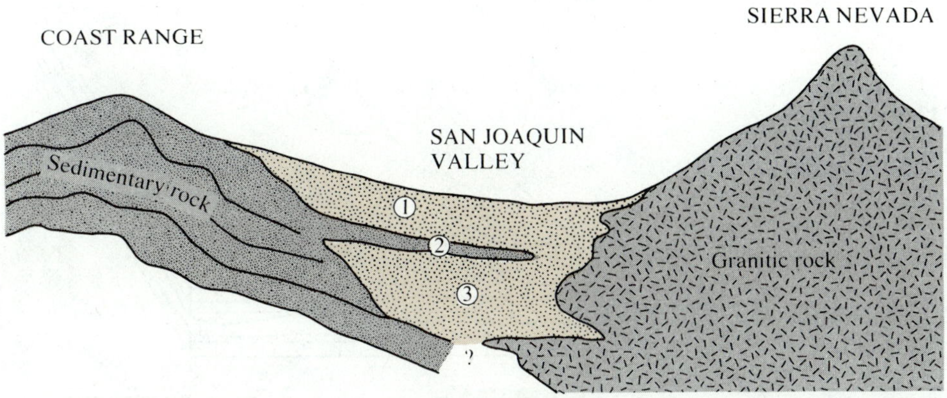

FIGURE 12–14
East-west cross section showing the San Joaquin Valley in California and the aquifers located there.

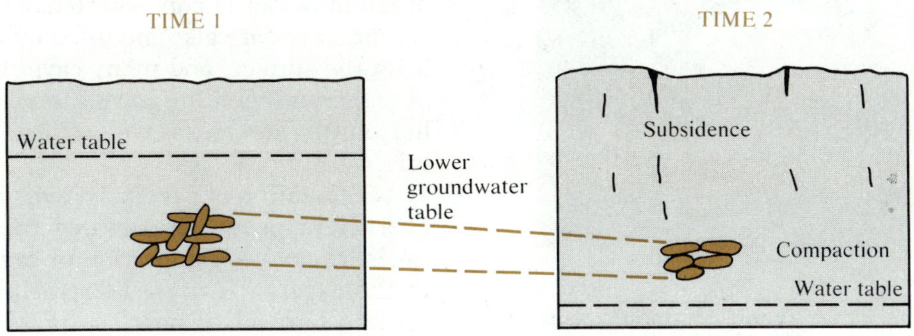

FIGURE 12–15
The mechanism by which subsidence results from withdrawal of groundwater. As the groundwater table is lowered, the effective weight of material overlying the aquifer is increased because the buoyancy effect from the pore water is lost. The flat-grained materials in the aquifer or layers between the aquifer tend to compact, resulting in a lesser thickness. As this process continues, the result at the surface is subsidence.

KARST HYDROLOGY

The term **karst topography** refers to a particular assemblage of landforms that tend to develop in soluble rocks such as limestone and marble in response to diversion of surface waters to subterranean routes (Figs. 12–16 and 12–17). Even large surface streams may be diverted, as shown in Fig. 12–18.

The major process operating in karst areas is solutional weathering of the soluble rock. Rainwater combines with carbon dioxide in the atmosphere or in soil material to form a weak acid known as carbonic acid. The acid then attacks the limestone or marble, breaking it down and carrying the weathered products off in solution (see Chapter 4). Karst topography is best developed when the limestone is dense, thin-bedded, and well fractured. The fracture surfaces and bedding planes allow the migration of water, which in turn allows for the chemical weathering that dissolves some of the limestone, producing subterranean conduits for the flow of groundwater. As a result, a series of caverns develops in many karst areas. On the surface in low areas, limestone is also dissolved, forming solutional pits known as **sinkholes**. Although most sinkholes develop by solutional processes on the surface (Fig. 12–19), some also form from collapse into subsurface voids (Figs. 12–20 and 12–21).

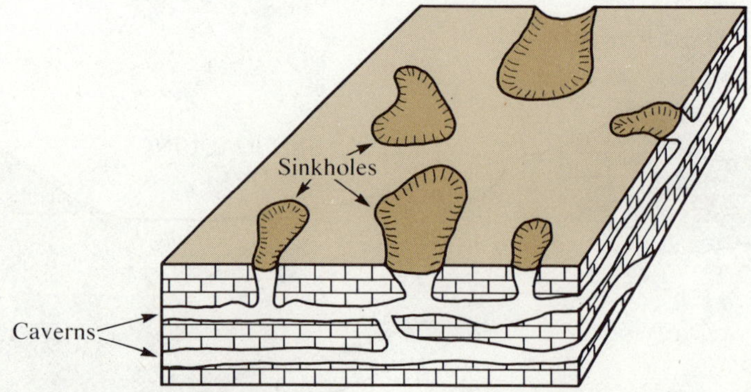

FIGURE 12–16
Some landforms associated with karst topography. The most common landforms at the surface are often sinkholes. Water from the surface is diverted to subterranean routes or caverns.

FIGURE 12–17
In karst areas, the surface of the ground commonly contains bizarre shapes of weathered limestone. Shown at the right of the hammer are linear grooves, or *striations*, formed in the rock by differential solutional weathering. The soil in limestone areas is typically quite red and is termed *terra rossa*.

The origin of caverns has been a subject of speculation for a long time. It is now believed that caverns tend to form at or near the present water table. Thus, there is a continual supply of water that is not saturated with the weathering products of the chemical solution of limestone. The fact that many caverns have several levels suggests that each level is associated with a time when the water table was at a particular level. With a general lowering of water tables, a series of relatively dry caverns are abandoned above the regional water table.

Many active cavern systems on the Mitchell Plain in southern Indiana still flood periodically. Part of Salamander Cave is shown in Fig. 12–22. Notice the fluting on the walls of the cavern. These marks are formed by running water. This cave floods periodically and thus is still in its formative stages.

Caves commonly contain interesting and beautiful formations consisting of **stalagmites**, **stalactites**, and **flowstone** (Fig. 12–23). These formations develop when dripping water, saturated in calcium carbonate, evaporates and deposits calcite. The interesting assemblage of cave formations slowly builds up by this process. The development of most stalagmites and stalactites is along fracture systems where the seepage of groundwater is concentrated. To a lesser or greater extent, caves are also modified by material washed in from the surface, and many caves contain a good deal of sediment including gravels, sand, and clay that wash in from the surface, as well as large trees and other debris.

Exploration of cavern systems is not for amateurs. Every year people are injured or lose their lives in caves; therefore, exploration of caverns should be left to the experts. It must be remembered that because caverns form at or near the present-day water level, even moderate precipitation events may completely fill them with water, trapping anyone inside.

HOT SPRINGS AND GEYSERS

Hot springs and **geysers** are related phenomena in that both are generally associated with circulation of meteoric water in the groundwater system. Hot springs are a continuous discharge of heated water, whereas geysers represent a more episodic discharge of hot groundwater under pressure. Not surprisingly, geysers and hot springs in the United States are generally restricted to the western portion of the country, where there has been recent tectonic and volcanic activity (Fig. 12–24). The heat source for hot springs and geysers is believed to be hot igneous rock bodies that may or may not still contain some partially melted rock material (magma). Water circulating in the vicinity of the hot rock then becomes heated.

A.

B.

C.

FIGURE 12–18
In karst regions, large surface streams may be diverted to subterranean routes, as illustrated by this sequence of photographs of the Lost River in southern Indiana. All three photos were taken on the same day. A. The Lost River showing a flow width of several meters on the surface. B. Approximately 2 km downstream from the view in part A, the stream is shown as it is diverted into several sinkholes, leaving behind large organic debris around the edges of the sinkholes. C. Adjacent to the large sinkholes and immediately downstream is this dry stream bed. Therefore, in just a short section of the stream the entire flow has been diverted to subterranean routes. During times of extremely high flow, the sinkholes cannot accept all of the surface runoff, and so the channel downstream from the sinkholes may contain water for a short time following precipitation events.

FIGURE 12–19
Sinkholes are commonly rather subtle features. Those shown here are identified by small groups of trees. The sinkhole in the central part of the photograph is only one of thousands formed on the karst Mitchell Plain in southern Indiana. In this region, most of the trees are in sinkholes, as slopes into these areas are often too steep to allow for plowing.

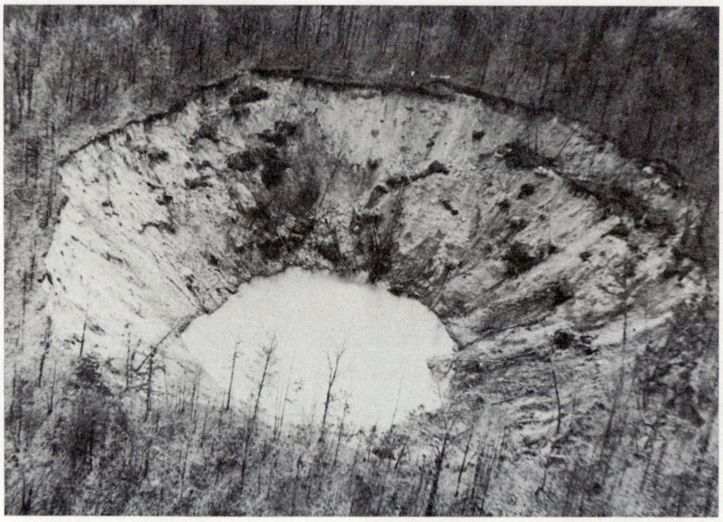

FIGURE 12–20
A very large sinkhole known as the December Giant which formed in 1972 near Montebello, Alabama. The hole is approximately 120 m wide and 45 m deep. It most likely formed by a collapse of surface material into an underlying limestone cavern. (Photo by Bill Warren, Geological Survey of Alabama)

FIGURE 12–21
Natural bridge. This structure, which now has a highway on it, may have formed by progressive collapse over a cavern system. Only a narrow arch remains.

FIGURE 12–22
Photograph taken near the entrance of Salamander Cavern in southern Indiana. The scalloped and smoothened surface of the cavern all the way to the top suggests that this cavern is often completely filled with water.

The general way geysers work is illustrated in Fig. 12–25. A geyser consists of an irregular conduit that goes down into the ground. Hot water seeps into the conduit and is heated from below. Because of the irregular shape of the conduit, the water is not able to **convect**; that is, the heated water on the bottom cannot rise and change place with the cooler water near the top. (Convection is the process whereby boiling water rolls to the surface of a heated pan, replacing the cooler water on the top.) The water at the bottom of the geyser pipe is under pressure due to the weight of the overlying water and becomes superheated (temperatures exceed the boiling point of water at atmospheric pressure). As the water in the upper parts begins to warm, it expands and overflows at the surface. As the water overflows, the pressure is reduced and the water in the bottom of the conduit flashes to steam, producing an eruption. The water then drains down, and water slowly begins to fill the conduit for the next eruption. Geysers like Old Faithful (Fig. 12–26) erupt on a very regular schedule because the flow of groundwater into the geyser conduit is at a regular rate and the heating is regular.

Hot waters of springs and geysers commonly contain dissolved material that upon evaporation or release of carbon dioxide (CO_2) deposits mineral-like compounds. The geysers of Yellowstone Park deposit **geyserite**, which is an oxide of silica. Many hot springs deposit a spongy or porous calcium carbonate material called **travertine** (Fig. 12–27).

In the western United States, there is currently considerable exploration in the hope that the use of hot rocks and hot water to generate energy may be expanded. The Geysers, located approximately 145 km north of San Francisco, is one of the largest geothermal energy fields in the world. The future of geothermal energy in the western United States is bright, as shown by the map generalizing the extent of hot crustal regions in the United States (Fig. 12–24). However, geothermal energy is not going to deliver us from our energy problems. Even in the western United States, where the prospects are the greatest, it is doubtful that geothermal energy will ever supply more than about 10 percent of our total electricity-generating needs.

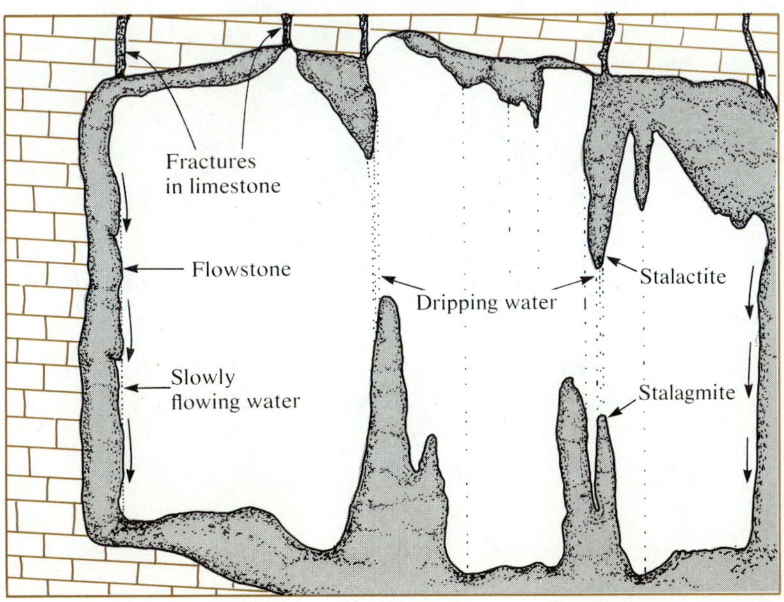

FIGURE 12–23
Cross section of a limestone cavern showing stalagmites, stalactites, and flowstone. Many of these features form as water seeps into the cavern along fractures in the limestone. As the water evaporates, calcium carbonate is deposited.

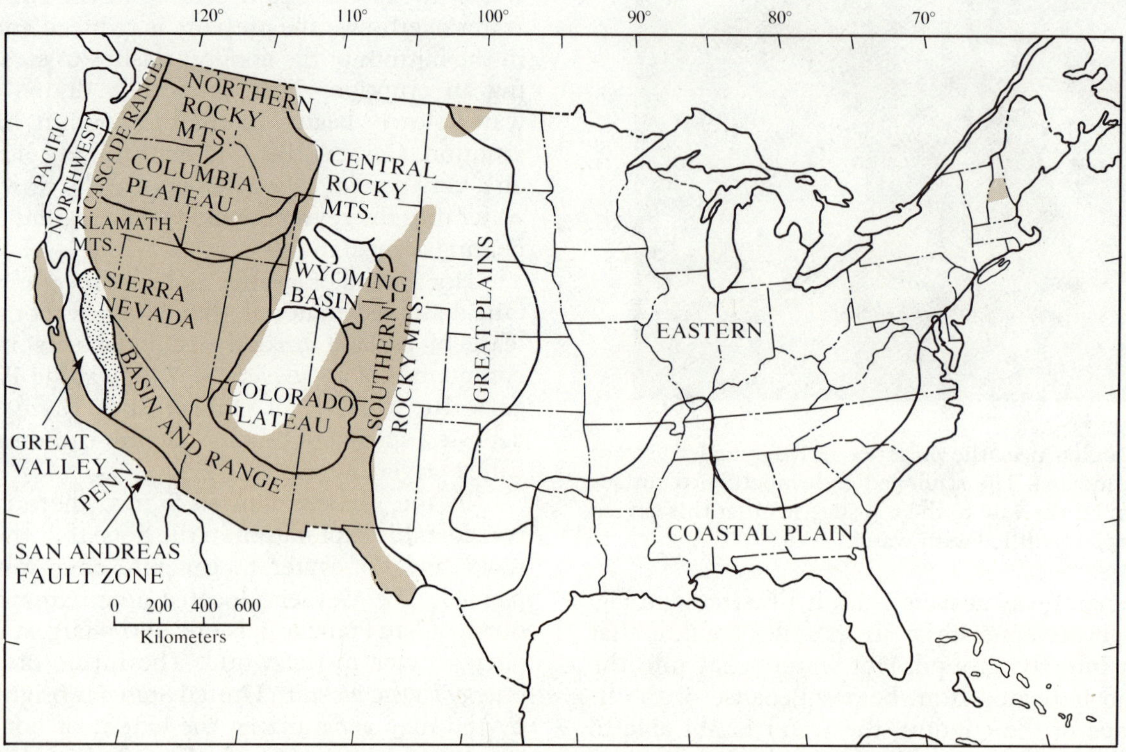

FIGURE 12–24
Map showing the generalized extent of hot (color), normal (white), and cold (stippled) crustal regions of the United States. Also shown are major physiographic provinces such as the Coastal Plain. (From D. F. White and D. L. Williams, U.S. Geological Survey Circular 726, 1975)

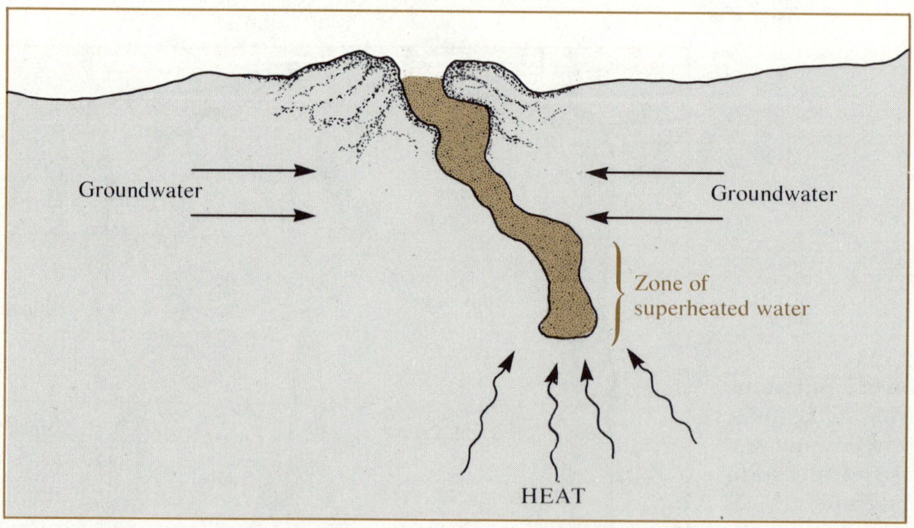

FIGURE 12–25
Idealized diagram illustrating how geysers work.

FIGURE 12–26
Old Faithful geyser in Yellowstone National Park. This geyser is very regular because the flow of groundwater into the geyser conduit is at a regular rate, and therefore the heating is regular.

GROUNDWATER PROBLEMS

Some of the problems associated with groundwater have already been touched on, including subsidence with lowering of groundwater tables. Unfortunately, the number of problems continues to increase as more demand is placed upon the groundwater resources. In coastal regions overpumping of fresh-water aquifers has led to salt-water intrusion, and with increasing human use of the land has come an increasing incidence of pollution of shallow aquifers, particularly associated with waste disposal practices.

SALT-WATER INTRUSION

Overpumping or mining of groundwater resources near the ocean has in some cases resulted in **salt-water intrusion**, which is the migration of salt water into adjacent fresh-water aquifers. Problems of this nature are best shown by examples from Florida, California, and New York, but they are not confined to these areas. Figure 12–28 is a generalized diagram of salt-water intrusion. In an unconfined or even a confined aquifer system near sea level, the surface of the water table generally is inclined toward the ocean. This helps to keep out the salt water, which is more dense and upon which the fresh water tends to float. However, with overpumping from these aquifers, there may be considerable drawdown (localized lowering of the water table), to the extent that the water-table surface may slope away from the ocean, allowing intrusion of seawater into the fresh-water aquifer. This can lead to serious degradation of the groundwater quality, requiring that the rate of pumping be reduced and perhaps that some wells be abandoned.

POLLUTION OF SHALLOW AQUIFERS

A **pollutant** may be defined as any substance, biological or chemical, an identifiable excess of which is known to be damaging to desirable living organisms. Thus, excessive amounts of heavy metals such as lead and mercury; certain radioactive isotopes; nitrogen, phosphorus, sodium, and other useful, even necessary elements; as well as certain pathogenetic bacteria and viruses are pollutants. Major differences exist between the physical, geologic, and biological environments associated with groundwater pollution and surface-water pollution. For example, consider the hydraulics of flow. In streams, there is typically a greater availability of oxygen and sunlight, which tends to degrade pollutants. Furthermore, streams flow relatively fast, facilitating rapid dilution and dispersion of pollutants. In groundwater, however, the processes are much slower and the opportunity for bacterial degradation of pollutants is generally confined to the soil or a few meters below. In addition, groundwater is commonly devoid of oxygen, a condition helpful in killing aerobic microorganisms but which may provide many happy homes for anaerobic varieties.

The ability of most soil and rock to physically filter out solids, including pollutants, is well recognized. Also known, but perhaps not so generally understood, is the capacity of clays and other selected minerals to capture and exchange some elements and compounds when they are dissociated in solutions.

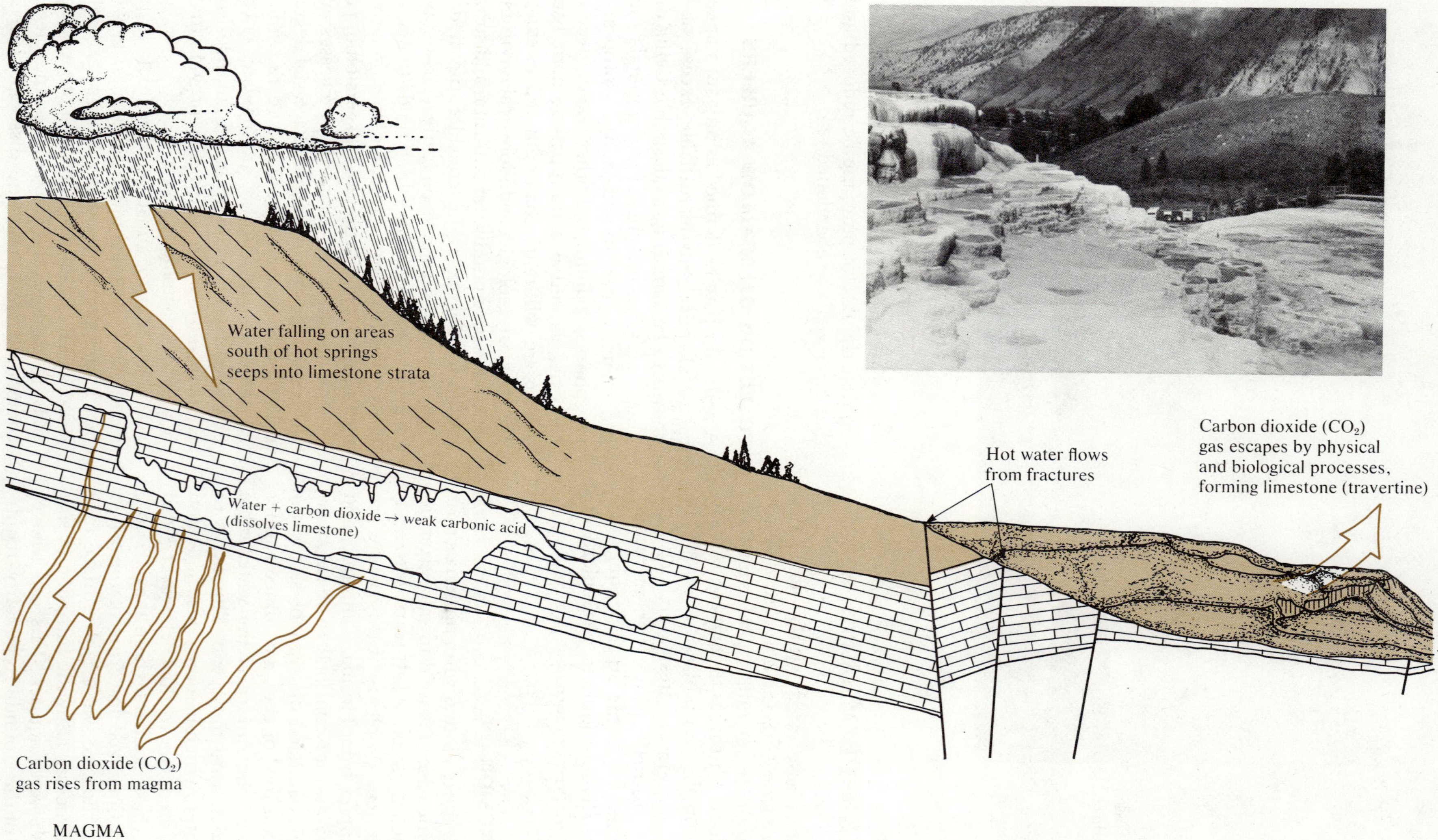

FIGURE 12–27
Photograph and idealized drawing of Mammoth Hot Springs in Yellowstone National Park. (Drawing modified slightly from a U.S. Park Service illustration)

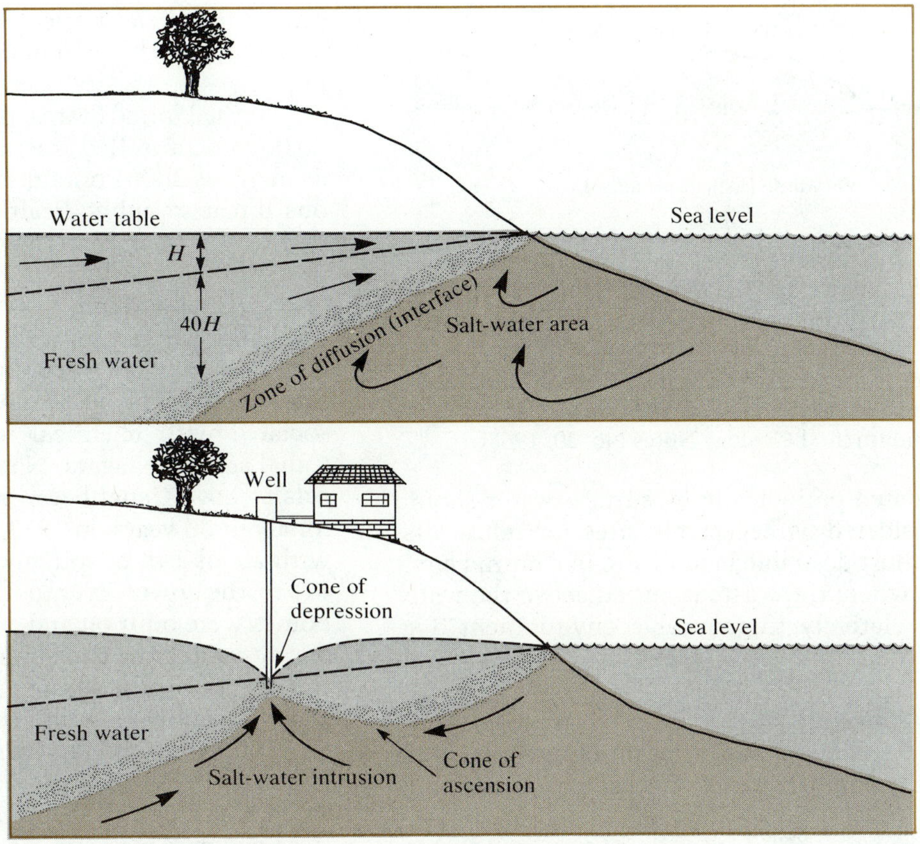

FIGURE 12–28
Salt-water intrusion into coastal aquifers. With pumping, both a cone of depression and a cone of ascension may develop, and if pumping is intensive enough, salt water is drawn up into the well.

However, like streams, the groundwaters have a limit to the amount of pollutants that may be accommodated.

The major problem of groundwater pollution is the introduction of chemical elements and compounds, and microorganisms, that do not naturally occur in aquifers that can be or are used for drinking water. Studies and case histories of groundwater pollution generally show that microorganisms seldom survive for distances of more than 30 m because they either filter out or are killed by the hostile environment. They have survived in those geologic environments where porosity and permeability conditions are high, such as in gravel and coarse sand, or bedrock where there is an abundance of open, interconnected fractures.

A complete discussion of pollution of shallow aquifers is beyond the scope of this chapter. However, the possibility of pollution of groundwater from several waste disposal systems—sanitary landfills, disposal of hazardous chemicals, and septic systems—will be considered.

Sanitary Landfill Solid waste disposal is developing into a significant environmental problem. In the United States, the **sanitary landfill** is fast becoming the most common way of disposing of solid waste. The sanitary landfill differs from open dumping in that at the landfill site, the refuse is covered each day with a layer of compacted earth material.

A major concern about sanitary landfills is pollution of shallow fresh-water aquifers. Pollution is possible because when precipitation or other water percolates through the refuse, it may pick up contaminants, forming a foul liquid known as **leachate**. Therefore, hydrologic investigation into the siting of sanitary landfills to minimize leachate production and migration is very important.

The best sites for sanitary landfills are in arid regions, where disposal conditions are relatively safe because of the dry environment. Regardless of whether the buried material is permeable or impermeable, almost no leachate is produced (Fig. 12–29). On the other hand, in humid environments some leachate will always be produced; therefore, it is advisable to try to minimize the chances of pollution of the groundwater. In general, limestone or highly fractured rock quarries and most sand and gravel pits make very poor landfill sites because these earth materials are good aquifers. Swampy areas, unless they are properly drained to prevent disposal into standing water, also make poor locations. Flat upland areas can be favorable sites provided an adequate layer of impermeable material such as clay is present above any aquifer. Finally, floodplains likely

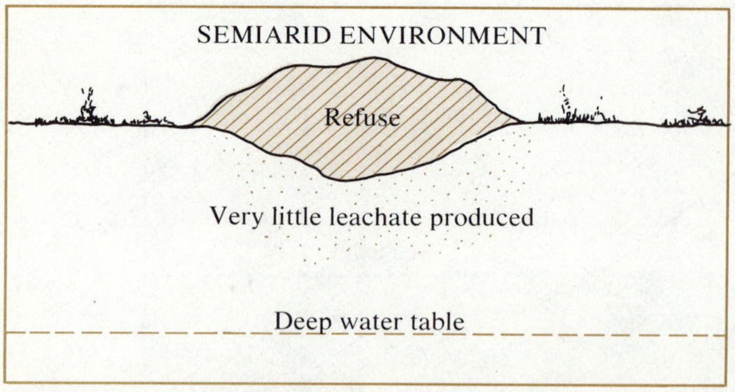

FIGURE 12–29
Sanitary landfill site in a semiarid environment. (Modified slightly from R. E. Bergstrom, Illinois State Geological Survey, Environmental Geology Notes No. 20, 1968)

to be inundated periodically by surface water should not be considered as acceptable sites for refuse disposal. The most desirable landfill site in a humid environment is where the waste is buried above the water table in a relatively impermeable environment (Fig. 12–30).

Hazardous Chemical Wastes In recent years, there has been a tremendous proliferation of new chemical compounds. Although many of these chemicals have proved beneficial to people, thousands are considered either definitely or potentially hazardous to human health.

In the United States, there are between 32,000 and 50,000 uncontrolled waste disposal sites, and of these as many as 2000 contain sufficient wastes to be a serious threat to public health and the environment. For this reason, many scientists believe that management of hazardous chemical wastes is the most serious environmental problem to ever face the United States.

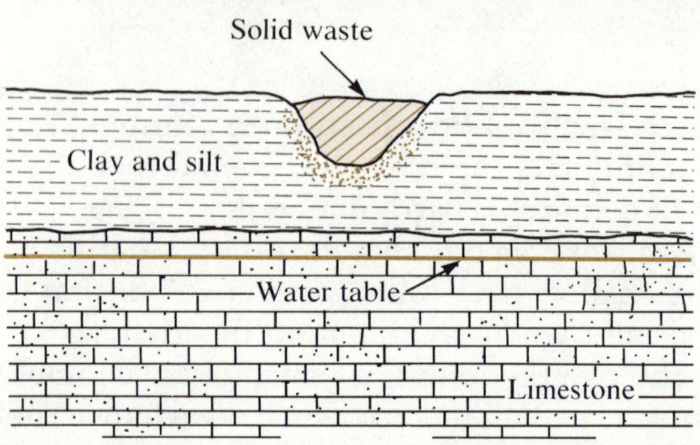

FIGURE 12–30
The most desirable landfill site in a humid environment is where the waste is buried above the water table in a relatively impermeable environment. (From W. J. Schneider, U.S. Geological Survey Circular 601F, 1970)

There are several ways that uncontrolled dumping of wastes may pollute soil and groundwater resources (Fig. 12–31). The best-known example of pollution associated with chemical waste disposal is the Love Canal case in Niagara, New York. Dumping of chemicals at Love Canal began in the 1920s and continued for about 30 years. In 1953, the dump site was covered with about 1 m of soil, and the land donated (sold for $1) to the town. Eventually, a school and about 200 homes were built on and around the site. For years, no one suspected the time bomb that was ticking beneath the site as disposal drums slowly began corroding and waste began leaking into the soil and near-surface water. Following several years of heavy snowfall, Love Canal in 1978–1980 gained national attention due to the serious pollution problem that was developing: trees and gardens were dying, dogs sniffing in the site developed sores that would not heal, rubber on children's tennis shoes disintegrated, and toxic-noxious substances oozed to the surface. It has been estimated that it may cost $100 million to clean up Love Canal and relocate residents.

Septic Systems Population movement in the United States continues to be from rural to urban or urbanizing areas. In too many cases, construction of adequate sewage systems has not kept pace with the growth. As a result, the individual septic tank (soils absorption system) which has been used for many years in both rural and urban areas continues to be an important method of sewage disposal for millions of

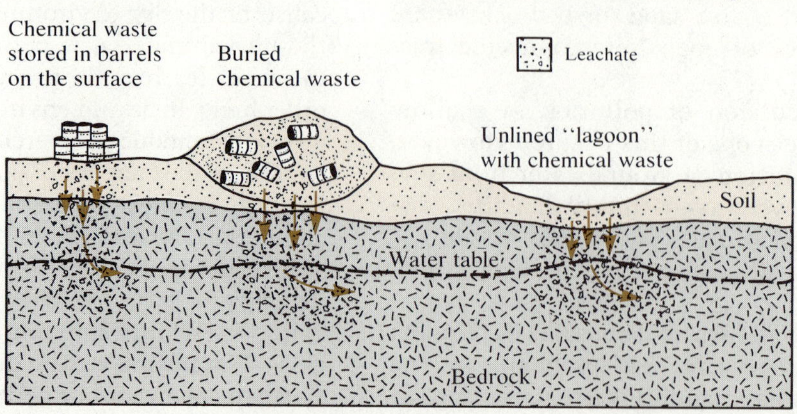

FIGURE 12–31
The ways uncontrolled dumping of chemical waste may pollute soil or groundwater. Leachate, the actual polluting material, is the result of leaking chemicals mixing with ground- or soil water.

Americans. Unfortunately, not all land is suitable for installation of septic tank disposal systems, and evaluation of individual sites is necessary and often required by law before a permit to construct such a system is issued. Figure 12–32 is a diagram of the basic parts of a septic tank disposal system.

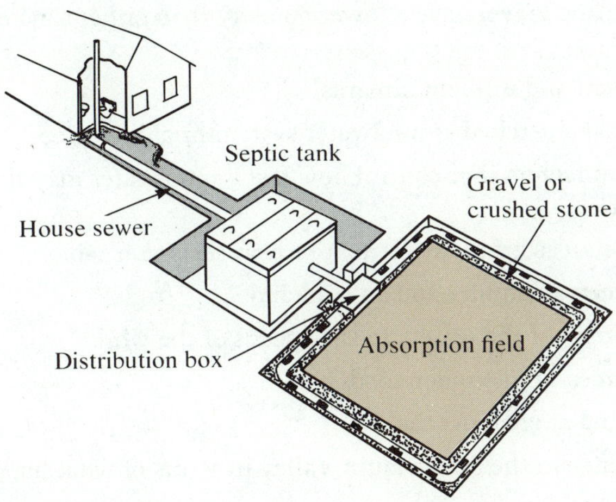

FIGURE 12–32
The septic tank sewage disposal system. (From Indiana State Board of Health)

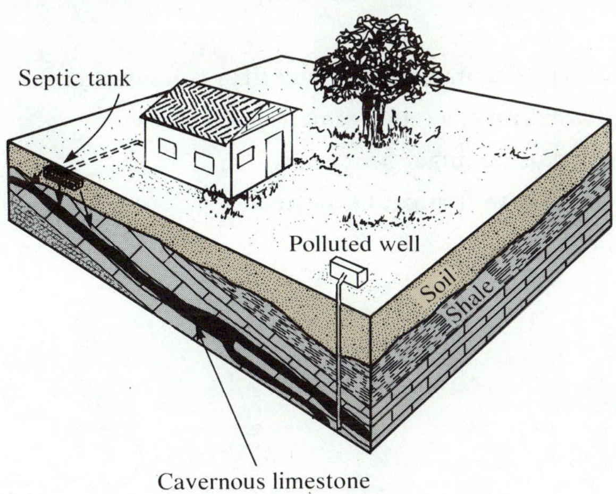

FIGURE 12–33
One way septic tank effluent may cause contamination of nearby wells. In this case, the waste directly entered the hydrologic system through small caverns in the limestones, and the polluted water is pumped up a nearby well.

Geologic factors that affect the suitability of a septic tank disposal system include type of soil, depth of the water table, depth to bedrock, and topography. Adverse conditions exist when the soil materials are relatively impermeable, the water table is high, bedrock is close to the surface, or the topography is steep. Given a combination of adverse conditions, the septic tank sewage disposal system may fail and the waste rise to the surface. Septic tank effluents may also cause contamination of wells, as shown in Fig. 12–33. A careful examination of soils, rock, and topography should be conducted before septic tank sewage systems are installed, and in fact such examination is mandatory by law in most areas.

SUMMARY

The hydrologic cycle is the solar-driven system that constantly cycles water from the ocean to the atmosphere to the continents and then back again to the ocean. Within the hydrologic cycle, the movements of surface and groundwaters are intimately related and involve subtle interactions between the composition and structure of earth materials.

Hydrogeology is the general term for geologic aspects of the occurrence, movement, and discharge of subsurface water relatively close to the surface in the zones of aeration and saturation.

Movement of groundwater is very slow compared to other geologic processes such as flow of water in rivers and rapid mass-wasting processes such as mudflows. However, it is generally much faster than soil creep and the movement of lithospheric plates. The rate of groundwater movement obeys Darcy's law, which states that the velocity is equal to the product of the coefficient of permeability and the hydraulic gradient.

Discharge of groundwater may be through natural seeps and springs or by pumping from aquifers. When groundwater is pumped out faster than it is replaced by natural processes, the particles in the aquifer may compact, resulting in surface subsidence.

Karst topography refers to a particular assemblage of landforms that tend to develop in soluble rocks such as limestone in response to diversion of surface water to subsurface routes. Of particular significance is the development of caverns at or near the water table, and solution or collapse sinkholes at the surface.

Hot springs and geysers are related phenomena generally associated with the circulation of meteoric water in the groundwater system. They are environmentally significant in that they may be associated with geothermal energy resources.

Groundwater problems include salt-water intrusion associated with overpumping near the coast, and pollution of shallow aquifers associated with waste disposal operations. In particular, the disposal of hazardous chemical waste is emerging as one of the most serious environmental problems to ever face the United States.

REVIEW EXERCISES

1 Why are ground- and surface waters so interrelated?

2 Compare and contrast the two basic types of direct runoff.

3 What is the water table, and how may its location be determined?

4 Why does a mixture of sand and gravel have a lower porosity than either sand or gravel alone?

5 Compare and contrast effluent and influent streams.

6 Discuss the various ways that a perched groundwater system might develop.

7 What are the most important factors that control how fast groundwater may flow in a particular environment?

8 Why should caverns in karst areas generally form at or near the water table?

9 What is the difference between an aquifer and an aquiclude?

10 What is thought to be the origin of the waters in the oceans of the world?

11 Why are groundwater resources so misunderstood?

12 Why is groundwater not found at great depths?

13 Explain the cause of subsidence in the San Joaquin Valley in terms of what happens to the subsurface materials.

14 Discuss the relationships between porosity and permeability for clay, sand, and gravel-sized particles.

15 How can the several levels of caverns in karst areas be explained?

16 Describe how geysers work.

17 Why may streamflow degrade some contaminants faster than groundwater flow?

18 What factors determine the feasibility of a septic tank disposal system?

19 How do you think salt-water intrusion might be reversed?

20 Explain how the site for an artesian well might be located.

SUGGESTED READINGS

BALDWIN, H. L., and McGUINNESS, C. L. *A Primer on Groundwater.* U.S. Geological Survey, 1963, 26 pp.

DAVIS, S. N., and DeWIEST, R. J. M. *Hydrogeology.* New York: John Wiley and Sons, 1966, 463 pp.

FETTER, C. W., Jr. *Applied Hydrology.* Columbus, Ohio: Charles E. Merrill Publishing Co., 1980, 488 pp.

FREEZE, R. A., and CHERRY, J. A. *Groundwater.* Englewood Cliffs, N.J.: Prentice-Hall, 1979, 604 pp.

HEATH, R. C., and TRAINER, F. W. *Introduction to Ground-Water Hydrology.* New York: John Wiley and Sons, 1968, 284 pp.

LINSLEY, R. K., Jr.; KOHLER, M. A.; and PAULHUS, J. L. H. *Hydrology for Engineers.* New York: McGraw-Hill Book Co., 1958, 340 pp.

MEINZER, O. E. *Hydrology.* New York: McGraw-Hill Book Co., 1942, 712 pp. ———.

SCHNEIDER, W. J. *Hydraulic Implications of Solid-Waste Disposal.* U.S. Geological Survey Circular 601F, 1970.

WHITE, D. F., and WILLIAMS, D. L. *Assessment of Geothermal Resources of the United States.* U.S. Geological Survey Circular 726, 1975.

- floodplain
- geomorphology
- perennial stream
- ephemeral stream
- intermittent stream
- system
- fluvial hydrology
- discharge
- centripetal drainage
- flood stage
- upstream flood
- downstream flood
- base level
- graded stream
- magnitude-frequency relationship
- extrinsic threshold

13

- turbulent flow
- convergent flow
- scour
- divergent flow
- dissolved load
- bed load
- hydrograph
- stream competence
- stream capacity
- meandering river
- braided river
- drainage pattern
- dendritic drainage
- rectangular drainage
- radial drainage
- geomorphic threshold
- geomorphic surface
- oxbow
- meander scroll
- oxbow lake
- alluvial fan
- delta
- river-dominated delta
- wave-dominated delta
- landscape evolution
- cycle of erosion
- peneplain
- rejuvenation
- dynamic equilibrium
- allometry

Rivers and Landforms

RIVERS AND LANDFORMS ARE DISCUSSED TOGETHER IN THIS chapter because the work of running water is responsible for producing most of the features of the landscape. If you have ever flown in an airplane or viewed aerial photographs, you are probably well aware of the importance of drainage features to the landscape. Whether the landscape is in the high Rocky Mountains of the western United States or the rolling hills of the piedmont of the southeastern United States, the systems of streams that drain the land are important topographic elements. The basic unit of the landscape is the drainage system known as the **drainage basin**, defined as the total area contributing water to a network of stream channels (Fig. 13–1).

The houses in which we live, the buildings where we work, and the land from which we obtain food are found on various landforms mostly produced by erosion and deposition from running water. The role of the river in our lives is emphasized every few weeks when some community in one part of the world or another is hit by a disastrous flood. In terms of lives lost and property damaged, flooding is one of our most serious so-called natural hazards. Flooding is a "so-called" natural hazard because, in fact, it is a natural process that becomes a hazard only when human use and activity crowd the river system. The river and the adjacent flatlands, known as the **floodplain**, together comprise an erosion-deposition system (Fig. 13–2) in which there is a rough balance between the load imposed by the external physical environment and the work done by the river. The floodplain is in part constructed by overbank flows (floods) that recur every year or two. We will have come a long way in alleviating damage by flood events when we recognize the floodplain and channel as a system and regulate development on floodplains to insure that human use of the land is compatible with river processes.

The work of running water is responsible for producing most of the features of our landscape, and yet many landforms are produced by other surficial processes such as glaciation, aeolian (wind) processes, and coastal processes, as well as by internal processes such as tectonics (including

A.

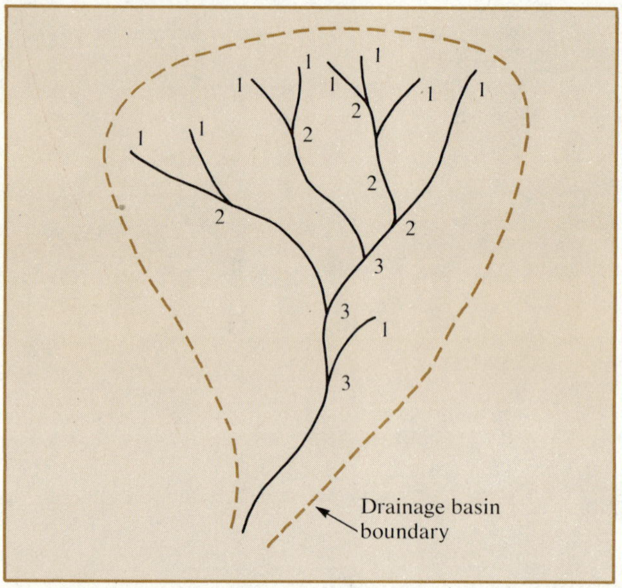

B.

FIGURE 13–1
A. A tributary canyon to Big Tujunga Canyon in southern California. The system of streams feeding the channel that emerges at the mountain front is known as the drainage basin and is the fundamental unit of the landscape. B. Idealized drawing of the drainage basin for the canyon in part A. Also shown is the system by which drainage basins are ordered. The smallest streams are designated as *first-order streams* (indicated as "1"), and whenever two first-order streams join, then downstream from that point the channel is a *second-order stream.* Whenever two equal orders meet, the order of the basin may increase. Thus, the small drainage basin shown here is third-order.

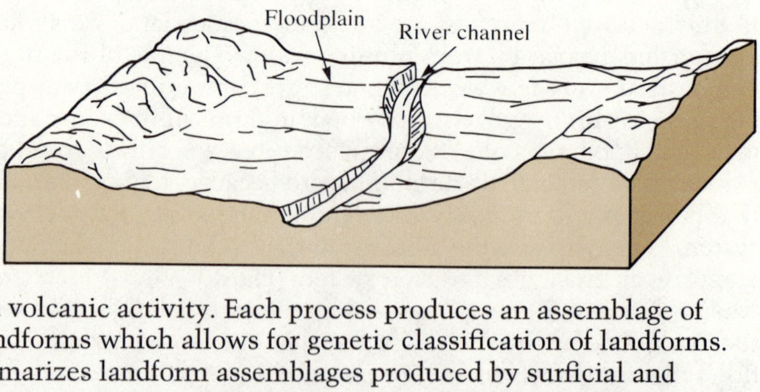

FIGURE 13–2
The floodplain-river channel system.

earthquakes) and volcanic activity. Each process produces an assemblage of characteristic landforms which allows for genetic classification of landforms. Table 13–1 summarizes landform assemblages produced by surficial and internal earth processes. It is important to recognize that most landforms, regardless of how they originate, are modified by erosion and deposition associated with running water. Examples of landforms that sometimes escape erosion by surface runoff include sand dunes and other very porous earth material, such as some cinder cones produced by volcanic processes that have such a high infiltration capacity that little, if any, water runs off to erode them. Landforms produced by glaciation, earthquakes, and most volcanic activity are soon modified by running water.

TABLE 13–1
Selected landform assemblages.

Process	Landform Assemblage
Fluvial	Channel, floodplain, valley, erosion surface, alluvial fan, delta
Glacial	Alpine—U-shaped valley, moraine, outwash plain Continental—till plain, moraine, outwash plain, esker
Eolian (wind)	Sand dune, sand sheet, deflation hollow
Coastal	Sea cliff, beach, sandspit, barrier island
Tectonic	Continent, ocean basin, mountain, fault scarp
Volcanic	Mountain chain, lava plain, crater

Geomorphology is the study of the geologic processes that produce and maintain landforms during the evolution of the landscape. The particular path that landform evolution takes depends on the interaction of many variables, including earth material type and structure, climate, vegetative cover, geologic process, and time. However, geologists are only beginning to understand the various possible interrelations of these variables and how these relationships influence the surface form of the earth.

RIVER (FLUVIAL) PROCESSES

Why are there streams and rivers? This is not as easy a question as might first be supposed. A logical first response is that streams exist because rainfall in an area exceeds the amount of water that infiltrates into the soil and rock, and therefore surface runoff occurs. This explains why there is overland flow, but it does not explain why the overland flow should be confined to a system of streams and rivers that form well-organized networks (Fig. 13–1). In fact, in some areas flowing water is not confined to channels, and much of the work of running water flows off the land as sheet flow. In very young topography, there may be little, if any, integration of drainage, and the establishment of a drainage network may take some time. For example, the gigantic (2.5-km^3) landslide-debris avalanche associated with the 18 May 1980 eruption of Mount St. Helens in Washington produced a new land surface in the Toutle River valley. There, hummocky ground with numerous depressions will retard the development of surficial drainage into a network for an undetermined period of time. Eventually, however, the isolated ponds and small streams will become organized, and the new land surface will be drained by streams. During the winter of 1980–81, there was a lot of activity on the Mount St. Helens landslide surface, suggesting that the drainage is developing fast. Therefore, the reason stream channels form networks is that geologic processes operate at differential rates on materials of varying resistance and thus cause relief. Using the Toutle River valley example again, the hummocky topography (Fig. 13–3) resulted from the landslide-debris avalanche that brought down much of the north side of Mount St. Helens, including blocks of glacial ice, soil, weathered bedrock, and more resistant rock. As the blocks of glacial ice melt, depressions may form, and as precipitation falls on the landslide mass, the weaker material will be differentially eroded by surface runoff. Small channels develop quickly, and as these grow they drain ponds, and eventually the entire slide mass will be integrated into one or more drainage systems. Similarly, but over a longer time period, following the retreat of the Pleistocene glaciers exposed glacial deposits were subjected to precipitation and runoff. Major river channels developed in response to runoff from melting glaciers, and tributaries to these channels formed as the channels eroded headward through various glacial deposits. As the headward erosion continued, ponds and lakes were drained, and generally the landscape became more and more modified by stream processes. It therefore may be concluded that relief combined with precipitation in amounts exceeding infiltration into the earth will produce overland flow. At first, small sets of channels known as **rills** will form. These are the types of channels commonly seen on the slopes of new roadcuts (Fig. 13–4). Finally, the rills develop into small stream channels which combine with others to eventually form mighty rivers such as the Mississippi, which drains about 40 percent of the conterminous United States.

In terms of the nature and extent of flow in a stream channel, there are three basic types of streams: **perennial**, **ephemeral**, and **intermittent** (Fig. 13–5). A

FIGURE 13–3
Hummocky topography resulting from the landslide-debris avalanche that brought down much of the north side of Mount St. Helens during the 18 May 1980 eruption. (Photo by Harry Glicken, U.S. Geological Survey)

FIGURE 13–4
A system of small rills developed on a roadcut. Given enough area and proper conditions, rills may eventually develop into small stream channels that, if they combine with other channels, may eventually form a drainage system.

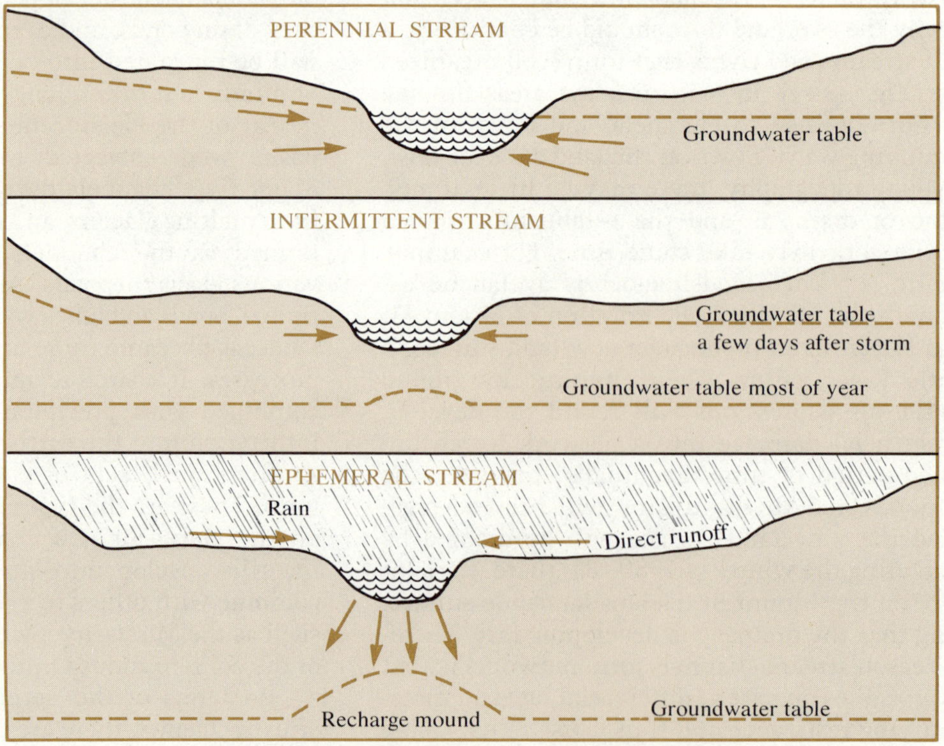

FIGURE 13–5
Perennial, intermittent, and ephemeral streams.

perennial stream is a stream that flows the entire year from the headwaters to the mouth, where it discharges into a larger stream, a lake, or the ocean. Flow of perennial streams is maintained by the seepage of groundwater into the stream channel along its entire length. That is, the perennial stream is incised to the level of the groundwater table year-round (Fig. 13–5). Ephemeral streams are streams that flow only in direct response to precipitation. An ephemeral stream is at all times above the groundwater table, and when it is flowing, water seeps out of the channel to form a recharge mound above the groundwater table (Figs. 13–5 and 13–6). Intermittent streams flow anywhere from a few days to a few months per year, in response to precipitation. An intermittent stream receives some of its water in the channel from groundwater seepage following the precipitation event. However, during most of the year the groundwater table is below the intermittent stream channel. Thus, the hydrologic behavior of intermittent streams is intermediate to that of perennial and ephemeral streams.

Perennial, intermittent, and ephemeral streams all may be found within the same drainage basin. Even in humid climates, small tributaries may be intermittent or ephemeral while the main streams are generally perennial. Streams in the semiarid and arid regions are usually of the intermittent or ephemeral variety and may be subject to flash flooding due to intense but infrequent precipitation.

STREAMS AND RIVERS AS OPEN SYSTEMS

A **system** may be defined as any part of the universe that may be isolated either in thought or in deed for the purpose of study. Open systems lack rigid boundaries to energy and materials and tend toward an "equilibrium of action" state. Streams and rivers are basically open systems in which there is a rough balance between the load imposed and the work done. The **load** refers to water and sediment delivered to the stream channel by adjacent hillslopes, and the **work** refers to the stream processes that erode, transport, or deposit the load.

The natural riverine environment is composed of three interrelated parts or phases: the fluid or water, the channel and floodplain, and the network of channels comprising the drainage basin. As the system evolves and changes, all three parts mutually adjust, and each exerts a partial control on the others. Communication between the various parts of the riverine system includes a multitude of interactions that tend to maintain a delicate balance, or equilibrium, of action within the system. Thus, to adequately cover fluvial processes, the fluvial hydrology of streams, sediment in streams, and channel and drainage patterns must be considered.

FLUVIAL HYDROLOGY

Fluvial hydrology differs from conventional hydrology in that the flow is neither uniform nor steady and the channel bed and banks are composed of material the stream is capable of eroding and transporting. The volume of water per unit time flowing by a particular point of a stream or river is termed **discharge** and is generally measured in cubic meters per second. The discharge (Q) is equal to the product of the cross-sectional area of the stream and the mean velocity and thus may be computed as follows if the width (w), depth (d), and mean velocity (v) of the flow are known:

$$Q = wdv$$

In stream systems where the discharge increases in the downstream direction, width, depth and velocity also increase downstream (Fig. 13–7). The downstream increases in width and depth are easier to understand because they are consistent with simple observation. However, it takes careful measurement to

FIGURE 13–6
Flat-floored ephemeral stream channel in the Mecca Hills of the Imperial Valley, southern California. This stream channel flows only in direct response to precipitation.

verify that the mean velocity tends to increase or remain the same in the downstream direction. At first glance, it might be expected that mean velocity should decrease as the slope of the river decreases in the downstream direction. However, the velocity of a stream is roughly proportional to the product of the depth of flow and the slope of the water surface, and in the downstream direction the water depth often increases faster than the slope decreases, allowing the mean velocity to increase. It must be emphasized, however, that velocity will not always increase in the downstream direction. For example, streams emerging from a mountain front and flowing over alluvial deposits may experience a decrease in discharge, resulting in a decrease in the velocity of the stream as well. In other instances, a stream or river may flow through a channel constriction produced by resistant rocks along the banks, in which case the slope of the stream may increase, along with the velocity. Nevertheless, the concept that width, depth, and velocity all tend to increase in the downstream direction of perennial streams and rivers is valid.

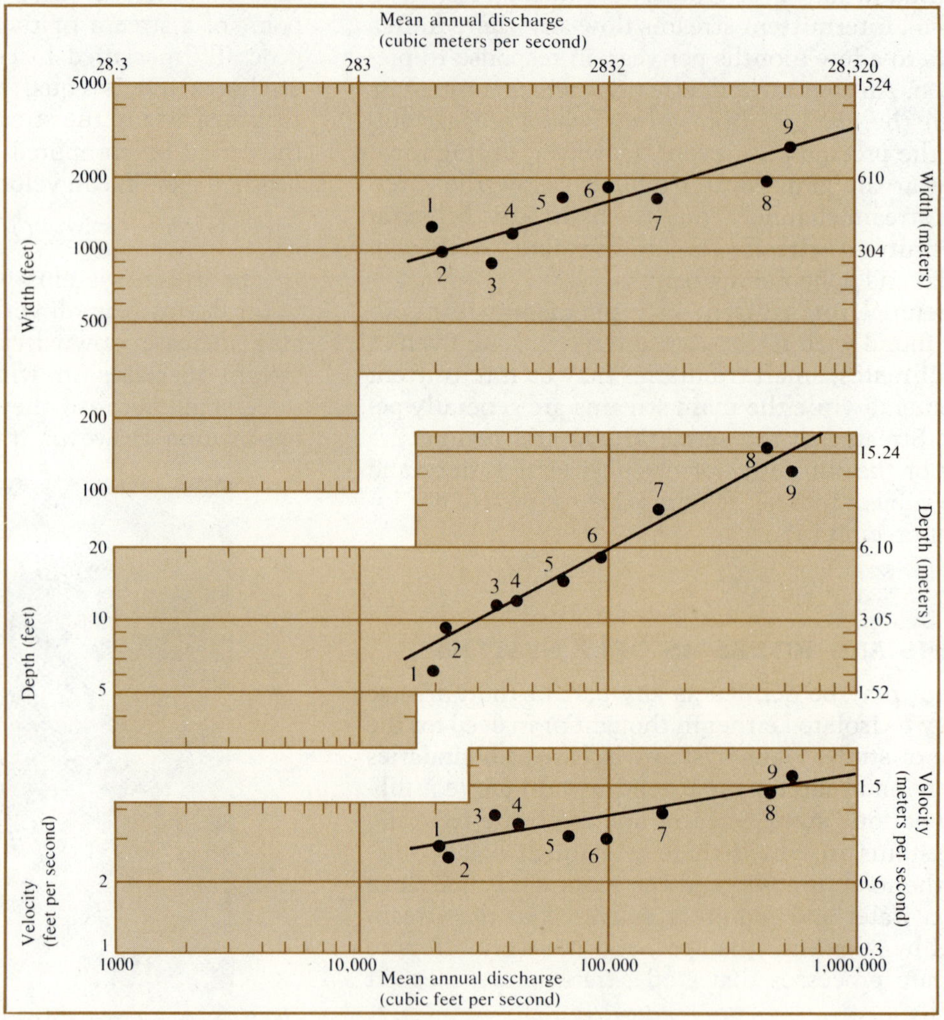

FIGURE 13–7
Downstream hydraulic geometry for the Missouri River. Notice the increase in width, depth, and velocity in the downstream direction. (After L. B. Leopold and T. Maddock, Jr., U.S. Geological Survey Professional Paper 252, 1953)

EXPLANATION

1—Missouri River at Bismarck, N.Dak.
2—Missouri River at Pierre, S.Dak.
3—Missouri River at St. Joseph, Mo.
4—Missouri River at Kansas City, Mo.
5—Missouri River at Hermann, Mo.
6—Missouri River at Alton, Ill.
7—Mississippi River at St. Louis, Mo.
8—Mississippi River at Memphis, Tenn.
9—Mississippi River near Vicksburg, Miss.

Among the basic concepts of fluvial hydrology important in understanding the behavior of streams are the following:

1. Streamflow is characteristically **turbulent**, and seldom does the flow of water in a stream parallel the stream bed or banks.
2. Areas of **convergent** streamflow are associated with **scour** (erosion), whereas areas of **divergent** flow are associated with deposition (Fig. 13–8).

Of the two concepts, the convergence and divergence of flow is probably the more significant, as it can account for many areas of observed scour and deposition in natural stream channels. The convergent-divergent relationship is fundamental to stream behavior said to be "born from the river" in that it is purely a function of the natural river processes rather than contrived by scientists to explain observations.

SEDIMENT IN STREAMS

The load that a stream carries may be derived from hillslope processes to the stream channel, or from processes within the stream itself that erode channel bed and banks. Of the two sources, hillslope processes are

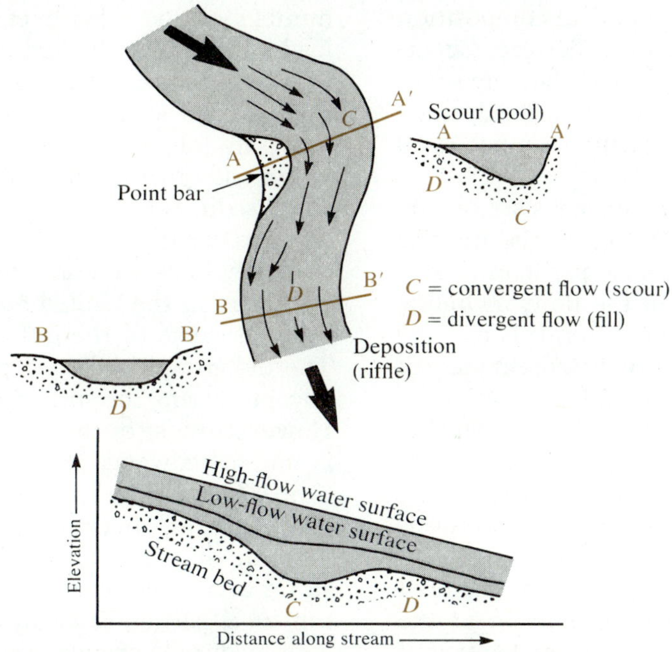

FIGURE 13–8
The convergent-divergent criterion. Convergence of flow is associated with deep areas and scour, whereas divergence of flow is associated with shallower water and deposition.

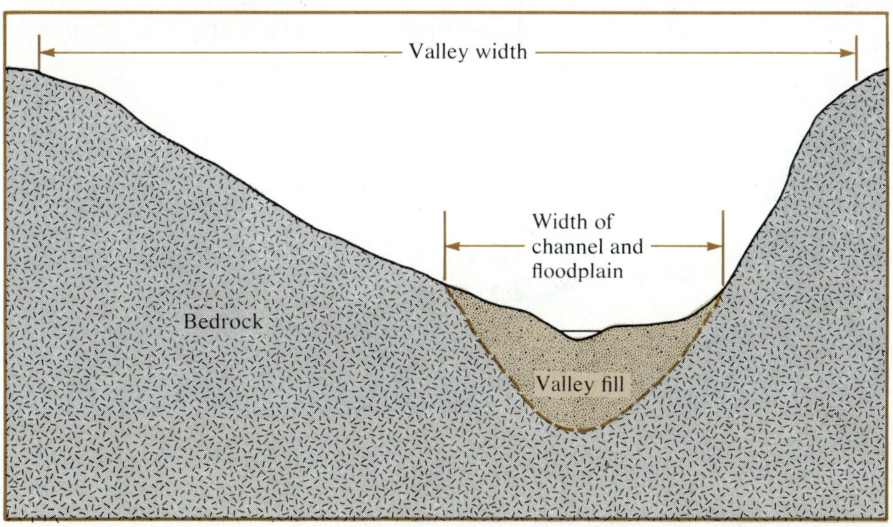

FIGURE 13–9
Cross section of a river valley. In general, the valley width is much greater than the valley depth.

more important. This is readily apparent upon examination of the cross section of the river valley shown in Fig. 13–9. In general, the width of a valley is greater than its depth, suggesting that most of the material removed from a stream valley is from the valley sides rather than directly from the bottom of the channel. Therefore, it can be concluded that valley widening is more effective than valley deepening.

The load that a stream carries can be divided into three classes: **dissolved load**, **suspended load**, and **bed load** (Fig. 13–10). The **dissolved load** of the stream consists of materials carried in chemical solution derived from chemical weathering of rocks in the river basin. Therefore, the composition of the dissolved load is somewhat dependent upon the chemical composition of the rocks from which the load was derived. Generally, the dissolved load of a stream is a relatively small portion (less than 20 percent) of the total load; however, in humid regions it may constitute as much as 60 percent of the total.

The **suspended load** of a stream consists of fine materials carried in fluid suspension by the moving water. The size of the suspended-load particles is thus dependent upon the properties of the flow, including the velocity and turbulence of the water. It is the suspended load that causes the muddy appearance of many rivers. Most of the suspended load consists of silt- and clay-sized particles that remain in suspension a relatively long time compared to larger particles such as sand and gravel. The suspended load of most rivers is usually large, often making up between two-thirds and nine-tenths of the total load by weight.

The **bed load** of a stream consists of the material that rolls, skips, or slides along the bottom of the stream channel. The amount of bed load and extent of movement depend upon hydraulic factors such as depth of flow, slope of the water surface, and discharge. In contrast to the dissolved load and suspended load which may be routinely measured in rivers, the volume and nature of bed loads are more difficult to determine; generally, however, the bed load constitutes a relatively small portion (less than 25 percent) of the total load.

Although the sediment load of a particular stream varies widely in time and space, some generalizations can be made. For instance, there is a relationship between the size of the drainage basin and sediment load such that as the basin size increases, the sediment yield per unit area decreases (Table 13–2). However, there are noticeable exceptions to this generalization. For example, in northwestern California there are a number of rivers with exceptionally high sediment loads (Table 13–3), and yet within some of these basins, as for example the Redwood Creek basin (with a suspended sediment load of about 3000 tons/km^2/year), are found a few small, steep drainage basins in virgin, old-growth redwood forest that have extremely small sediment yields (e.g., Little Lost Man Creek, 50 tons/km^2/year). Thus, in a single basin are some of the highest as well as lowest sediment yields found anywhere in the United States. These wide variations occur because of the differences in vegetative cover. The old-growth redwood provides a continuous canopy, protecting the surface of the ground from erosion. However, most of the Redwood Creek drainage basin is not in redwood, and much has been subjected to timber harvesting. As a result, erosion rates tend to be extremely high in other parts of the drainage basin.

The discharge of sediment may also vary during storm events. For example, the first storms of the season for a particular basin may give rise to relatively high loads of suspended sediment, whereas suspended loads during later storms will be much lower. This is particularly true in areas characterized by pronounced

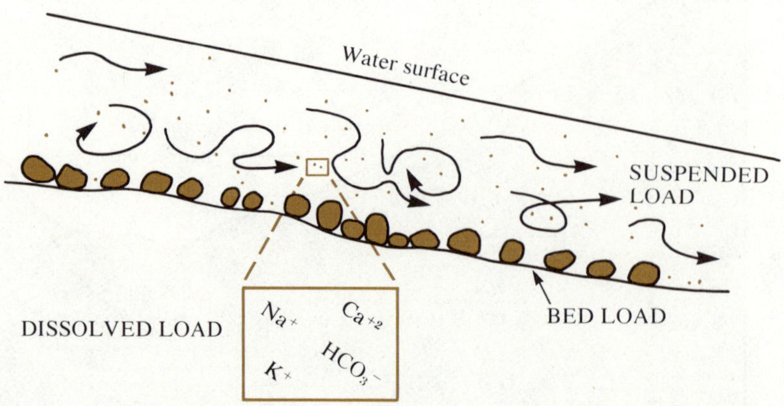

FIGURE 13–10
The three main components of the load a stream transports.

TABLE 13–2
Some major rivers of the world ranked by sediment yield per unit area.

River	Drainage Basin (thousands of km²)	Sediment Load per Year (tons/km²)
Amazon	5776	63
Mississippi	3222	97
Nile	2978	37
Yangtze	1942	257
Missouri	1370	159
Indus	969	449
Ganges	956	1518
Mekong	795	214
Yellow	673	2804
Brahmaputra	666	1090
Colorado	637	212
Irrawaddy	430	695
Red	119	1092
Kosi	62	2774
Ching	57	7158
Lo	26	7308

SOURCE: Data from Holman, 1968.

wet and dry seasons where a considerable amount of fine sediment may accumulate on hillslopes and in dry channels prior to the rainy season.

During a single storm, there might also be systematic variations in the sediment load. For example, some basins will have a sediment discharge as shown in Fig. 13–11, where there are three maxima. The first maximum is the suspended load, which peaks prior to the peak in the **hydrograph** (graph of discharge of water in cubic meters per second versus time for a particular storm). The second maximum is the bed load, and this corresponds with the peak discharge of water from the flood event. The third and final maximum is the dissolved load, which occurs sometime after the stream has returned to a low-flow condition. This results because of the time required for the storm waters from precipitation to infiltrate into rocks and perform the necessary chemical weathering to deliver the dissolved load to the stream.

Sediment yield from a river basin depends on a variety of factors including climate, vegetative cover, material type, slope, and land use. In general, the highest yields of sediment have been recorded for areas undergoing rapid urbanization or disturbance from such activities as timber harvesting or intensive off-road vehicle use. Under natural conditions, the highest sediment yields are associated with areas characterized by moderate rainfall and relatively sparse vegetation. That is, areas with the highest sediment yield are not desert regions with very low rainfall and little vegetation, but rather tend to be the semiarid to subhumid areas characterized by grasslands and brush vegetation.

Material type on hillslopes in drainage basins will also affect sediment yield. Basins underlain by weak rocks will have a higher sediment discharge than those underlain by more resistant rocks. Sandy soils are one of the easiest earth materials to erode (Fig. 13–12). Clay particles in soils tend to be held together by weak electrostatic forces and thus are more resistant to erosion than sandy soils. Soils composed of particles larger than sand are also more difficult to erode than sand because of the greater force necessary to initiate the movement.

Sediment yield in drainage basins may also fluctuate due to the nature and extent of storage compartments available for sediment being transferred through the river system. This is best examined by looking at an idealized fluvial system in terms of three major zones: the erosion zone, where sediment is produced;

TABLE 13–3
Suspended sediment yields for selected rivers.

Drainage Basin	Drainage Area (km²)	Suspended Sediment Yield (tons/km²/year)
Northern California		
Eel River	7,778	3,892
Van Duzen River	570	3,409
Redwood Creek	720	3,089
Little Lost Man Creek (tributary to Redwood Creek)	9	50
Other in United States		
Schuylkill River, Pennsylvania	4,902	202
Delaware River, New Jersey	17,560	57
Rio Grande River, New Mexico	69,153	136
Mississippi River, Louisiana	3,220,665	94

SOURCE: Data from H. Kelsey, 1977.

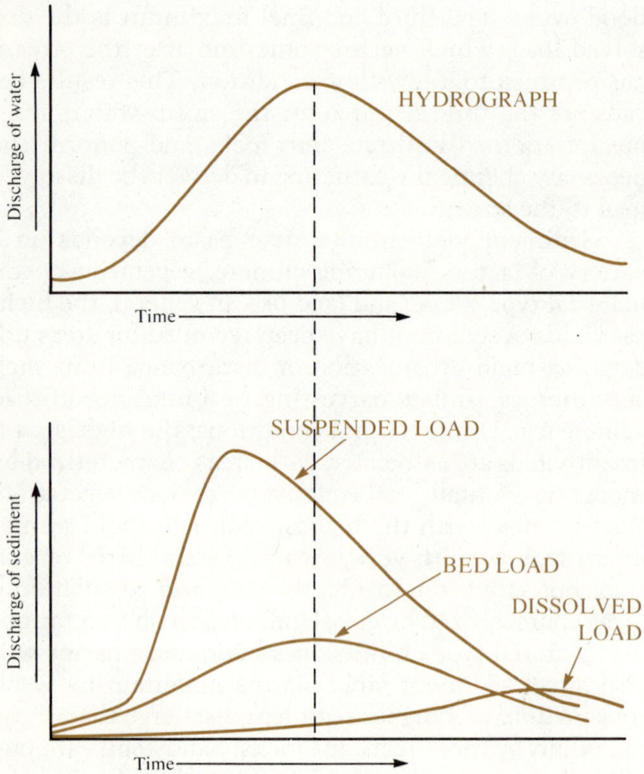

FIGURE 13–11
The relationship between the hydrograph and possible maxima for the suspended, bed, and dissolved loads associated with a flow event.

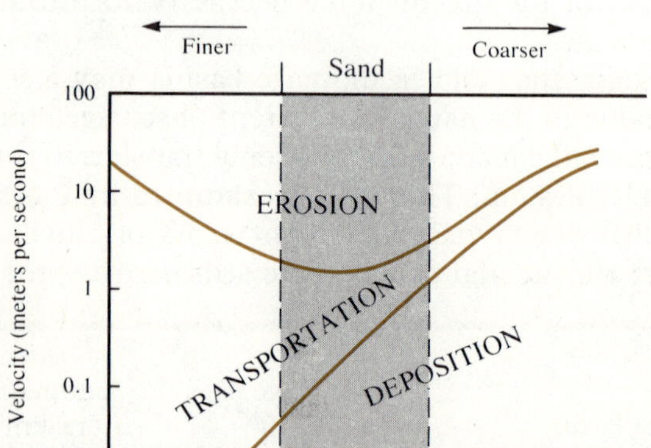

FIGURE 13–12
The water velocity necessary to move (erode) sand is less than that for finer or coarser particles. (Modified after F. Hjulstrom, 1935)

the intermediate zone, where sediment may be stored and transported; and the downstream deposition zone (Fig. 13–13). The zone of erosion in the idealized fluvial system is in the upper parts of the basin, where the small, steep streams are located. In forested regions, a great deal of sediment is added to the fluvial system in this part of the basin through landslides and debris avalanches associated with intense precipitation events. The intermediate zone is characterized by sediment storage and transport; the final zone in the fluvial system is the zone of deposition, located where the river discharges into a lake or the ocean.

Although this idealized fluvial system is valuable for making general statements concerning what happens within a stream system, there are many exceptions. For example, in old-growth redwood forests of northern California there is a tremendous amount of sediment storage in steep, small, headward basins. Most of the sediment storage sites are due to large downed redwood trees that fall into the stream channel and function as small dams. These trees may remain in the channel for several hundred years and provide storage sites for over 100 years of mean annual discharge of bed load. Another exception to the idealized model occurs in the zone of sediment transport and storage, where landsliding may intermittently add large quantities of sediment to the stream system. In spite of these and other exceptions, the general principles of the fluvial system composed of erosional, storage and transport, and depositional zones appear valid.

Other important aspects of sediment in streams and rivers are the systematic changes in the size of bed material found in the stream channel, and a stream's competence and capacity to carry sediment load. In many streams, the size of the bed material decreases systematically in the downstream direction. This results because:

1. Smaller particles are transported more easily, leaving the larger particles behind.
2. Particles are constantly being thrown together by the moving water and sediment, and thus abrasion during transport takes place.
3. Some of the minerals in the rocks of the stream bed may be soluble and slowly dissolve, reducing the size of the particles.

The maximum size of bed material a stream may carry is said to be a measure of the stream's **competence**, whereas the measure of the total load a stream may carry is termed the stream **capacity**. Both competence and capacity of a stream are very sensitive to changes in hydraulic conditions such as depth of flow, slope of the water surface, and discharge. The competence of a stream is therefore related to the velocity of flow, which is proportional to the product of the depth of flow and the slope of the water surface. Should the velocity of a stream double, then the competence will in-

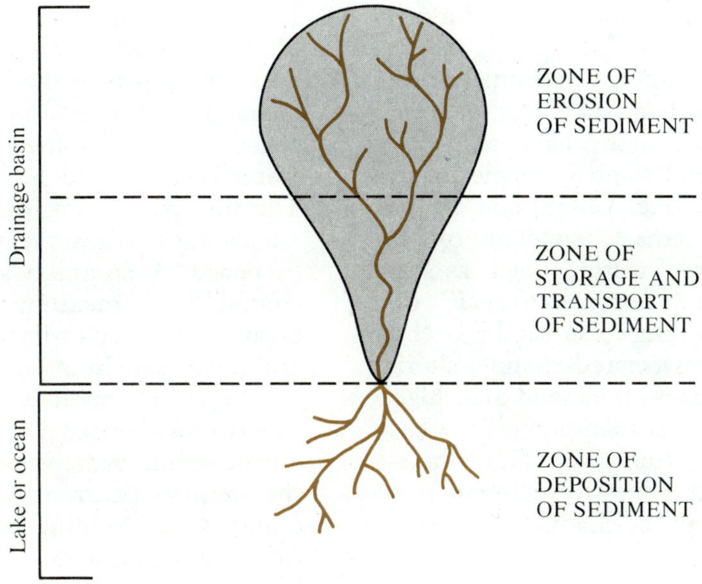

FIGURE 13–13
The zones of erosion, storage, and deposition of sediment in a fluvial system. (Modified after S. A. Schumm, 1978)

crease by a factor greater than the increase in velocity. The precise rate of increase is difficult to establish because of variations in the size and shape of the materials comprising the bed load.

CHANNEL PATTERNS

Channel pattern refers to the shape of a channel in plan view, as from the air. Three main channel patterns have been identified: **meandering**, **straight**, and **braided**. The meandering channel pattern is characterized by a sinuous channel with varying degrees of regularity of the bends in the channel. Straight channels are relatively rare and seldom exist for lengths greater than 10 times the channel width. The braided channel pattern is characterized by an abundance of midchannel bars that continuously divide and subdivide the flow of the stream. All three channel patterns may be found in any one river.

Meandering Rivers The channel of the **meandering river** develops as it continuously moves back and forth across the floodplain—creating, maintaining, and destroying reaches of channel while continually building the floodplain. The meander bends or loops (Fig. 13–14) slowly and persistently move both laterally and in the down-valley direction, while the system maintains a quasi or dynamic equilibrium in that the shape of the channel and basic channel form do not change.

There is no satisfactory explanation as to why streams should meander. However, studies of streams and rivers have established that the potential energy loss per unit mass of water may be minimized by increasing the path length of the stream (which is meandering). Minimization of potential energy loss per unit mass of water may explain why a stream should meander but this is not an explanation of how the meandering occurs. To understand how meanders develop, bed forms—topographic elements on the bottom of a

FIGURE 13–14
Meander bends of the Owens River on the east side of the Sierra Nevada, California.

stream channel produced by the interaction of moving water and sediment—must be considered.

There are two sets of bed forms: **channel-forming** bed forms and **channel-altering** bed forms. Channel-forming bed forms include **point bars**, **pools**, and **riffles**. These bed forms tend to give the meandering river its basic morphology (Fig. 13–15) and are best developed in gravel-bed streams. Channel-altering bed forms (Fig. 13–16) consist of features such as small ripples and dunes that form in response to specific flow conditions and are best developed in sand-bed channels. Ripples, dunes, and associated channel-altering forms are so designated because they tend to be highly mobile forms that mask rather than provide the basic morphology of the meandering river. Nevertheless, they are important when it comes to sediment transport and resistance to flow in the channel.

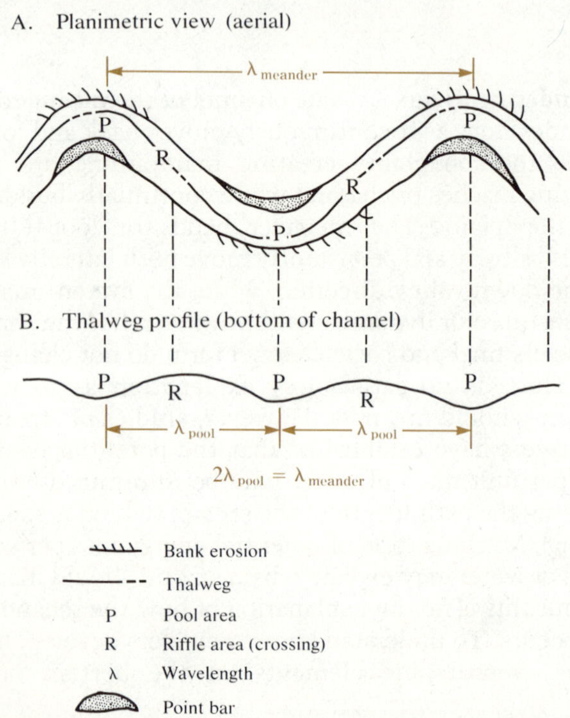

FIGURE 13–15
Pools, riffles, and point bars in a meandering stream. These forms provide the basic morphology for the meandering river. Notice that one wavelength in planimetric view for the meandering river ($\lambda_{meander}$) is equivalent to two wavelengths along the bottom of the stream channel ($2\lambda_{pool}$). Therefore, in each wavelength for a meander there exist two pools and two riffles. (Modified after E. A. Keller and W. N. Melhorn, 1978)

The bed forms shown in Fig. 13–16 comprise a sequence of forms that develop with change in the flow velocity of the water, depth of flow, and size of sediment being transported. The forms are either in the lower flow regime, characterized by relatively low velocity of flow, high resistance to flow (bed forms are out of phase with the water surface), and low rate of sediment transport; or they are in the upper flow regime, characterized by relatively high velocity of flow, low resistance to flow (bed forms are in phase with the water surface), and high rate of sediment transport. The transition from lower to upper flow regimes occurs when a critical ratio of velocity to depth of flow is exceeded. A common sequence of forms in a river channel with medium sand (0.25–0.5 mm) with increasing velocity of flow is ripples–dunes–flat beds–antidunes (shown in Fig. 13–16).

In the lower flow regime, the sediment is commonly cross-bedded (Fig. 13–16). Cross-bedding occurs during sediment transport as particles avalanche down the lee (down-current) slope of advancing ripples and dunes. Cross-bedding is common in several depositional environments, including rivers, sand dunes, beaches, and small deltas.

Consider now the channel-forming bed forms in the meandering river. Point bars are depositional features found in the inside of meander bends. Pools are deeps produced by scour at high flow and are generally found close to the outside of a meander bend and just downstream from the point of greatest curvature of the bend. The pool and point bar together produce an asymmetric cross profile (Fig. 13–8) that tends to converge the flow of water in the channel during the relatively high channel-forming flows that recur every year or so. Riffles are topographic high spots in channels located on the crossing between successive pools. Riffles are areas where the channel cross section is more symmetrical, and they tend to diverge the flow of water in the channel during the relatively high channel-forming flows, facilitating deposition. At low flow, with little sediment transport, pools are areas of slow, deep water, and riffles areas of fast, shallow water (Fig. 13–17). During flood stage, conditions are reversed and water through pools flows faster than that for adjacent riffles, even though the overall water velocities in both pools and riffles at high flow are greater than at low flow. Therefore, it can be concluded that pools scour at high flow and fill at low flow, whereas riffles fill at high flow and scour at low flow (Fig. 13–18). Pools fill and riffles scour at low flow because fine material may be moved off the riffle and deposited in the downstream pool. Riffles generally contain coarser bed material than pools, and the generalized scour-fill pattern of Fig. 13–18 provides a mechanism that maintains pools and riffles over a series of flows.

Pools and riffles have environmental significance because they provide hydrologic variability (deep-slow water in pools and fast-shallow water in riffles) at low-flow conditions, and a diversity of habitats necessary for fish and other aquatic life.

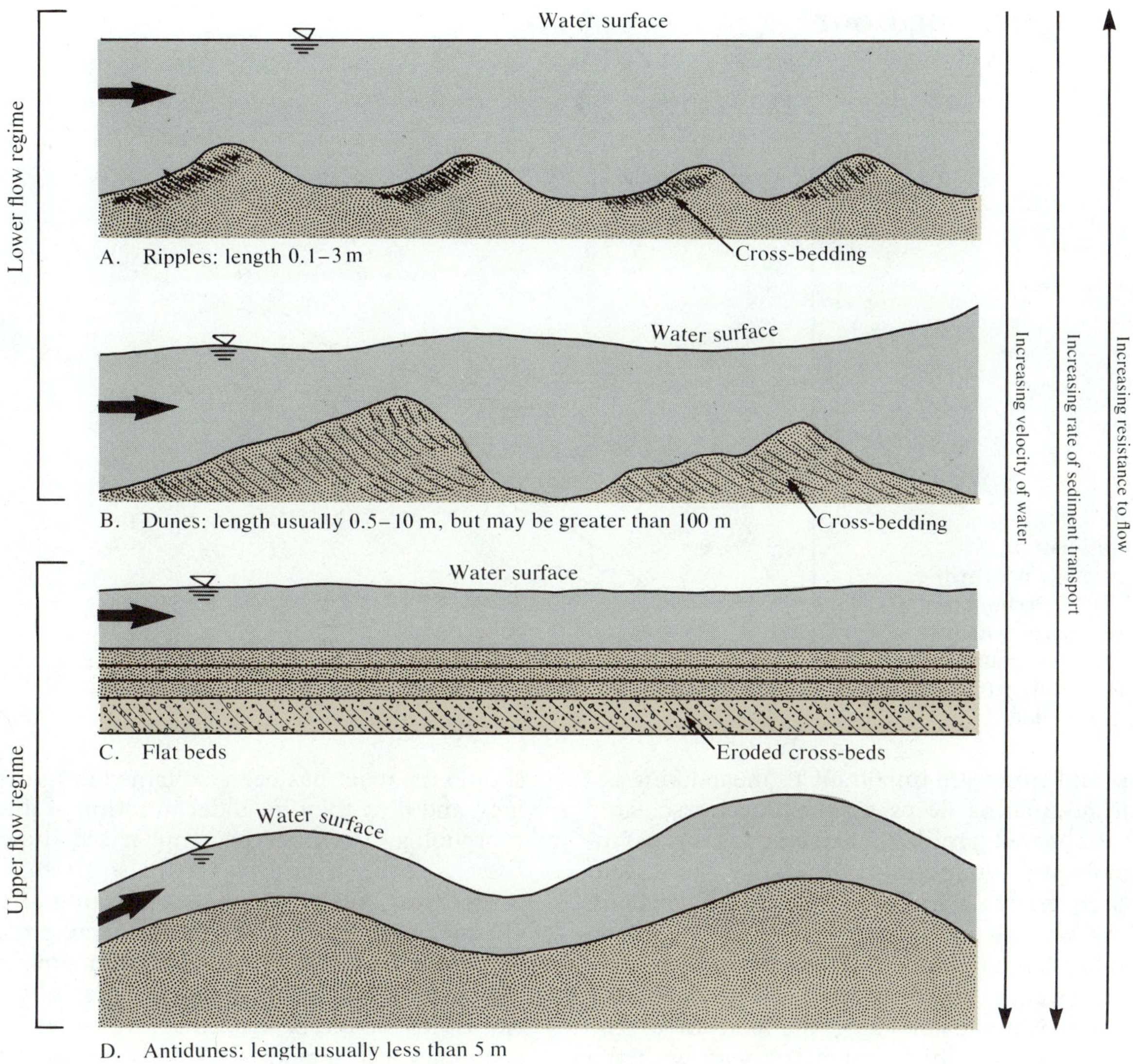

FIGURE 13–16
Channel-altering bed forms commonly associated with sand-bed streams and rivers. (Modified after D. B. Simmons and E. V. Richardson, U.S. Geological Survey Professional Paper 422J, 1966)

FIGURE 13–17
A pool–riffle–pool sequence for Big Pine Creek in northern Indiana during the late summer, low-flow condition.

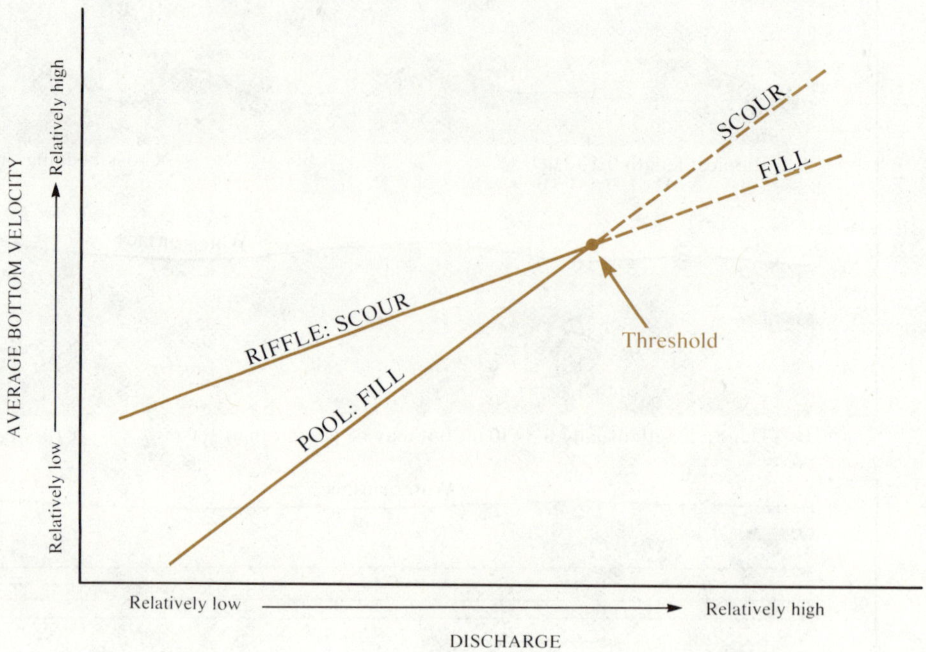

FIGURE 13–18
The scour-fill pattern characteristic of a pool–riffle sequence. The threshold is a critical discharge at which a change in process occurs from scour to fill in riffles, or fill to scour for pools.

Pools and riffles are important to meandering because the alternating deeps and shallows associated with cross-channel profiles, characterized by asymmetric pools and symmetric riffles, allow the stream to alternately erode the banks of the channel. This erosion of one bank then the other allows the stream to develop meander bends, and thus the meandering pattern. Pools and riffles therefore provide the meandering stream's third dimension along the bottom of the channel, which varies in a systematic way with the planimetric meanders visible on the surface (Fig. 13–15). One hypothesis is that the morphology of alternating pools and riffles, spaced at five to seven times the width of the channel, provides the most probable channel morphology for allowing the stream channel to increase its path length and minimize the potential energy loss per unit mass of water. Thus, why streams should meander has been explained in terms of energy loss, and how they meander in terms of their channel morphology. It must be emphasized, however, that there are other hypotheses that also attempt to explain meandering, such as random deflection of the stream channel as it erodes materials of varying resistance. It is hoped that future work will better explain why and how meanders develop and, more precisely, why pools and riffles should be spaced at five to seven times the channel width. Scientists simply do not have all the answers when it comes to the meandering problem.

Braided Channels The **braided river** is characterized by an abundance of midchannel bars that continuously divide and subdivide the flow of the stream, generally during the lower-discharge events (Fig. 13–19). Braided channels may have much in common

FIGURE 13–19
A braided river, the Nenana River in Alaska. (Photo courtesy Robert Norris)

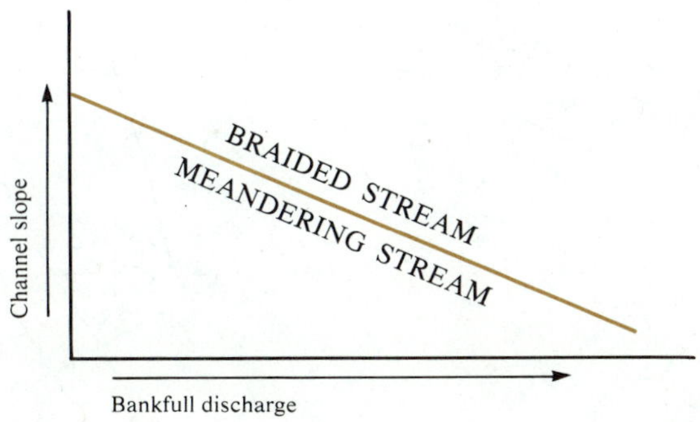

FIGURE 13–20
Generalized graph showing the relationship between meandering and braided rivers. Notice that for a given channel slope the bankfull discharge of the braided river will be greater, and for a given bankfull discharge the slope of the braided river is greater. (After L. B. Leopold and M. G. Wolman, 1957)

with meandering channels in that some streams tend to braid at low flow but resemble meandering rivers at high flow. Figure 13–20 shows the relationship between braiding and meandering in terms of channel slope and bankfull discharge (discharge that recurs approximately every 1.5 years, and theoretically fills the channel to just below the flood stage). From the graph, it can be seen that for a given channel slope the braided stream will have a greater bankfull discharge, and for a constant bankfull discharge the braided channel will be steeper.

Braided streams and rivers tend to develop because the bank materials in the stream channel are less resistant to erosion than are the midchannel bars. Thus, it is easier for the stream to widen than to deepen its channel. Most of the load of braided streams is commonly bed load, although this is not a necessary condition for braiding. Meandering rivers characteristically contain short braided reaches, which tend to form for a number of reasons. One way a short braided reach may develop in a meandering stream is if a large obstruction to flow such as a debris or log jam blocks the flow of the channel and forces the water to either side of the obstruction. The water then reunites downstream and in the process develops several bars or islands downstream from the obstruction that may become stabilized by vegetation. When the debris dam is broken up and moved downstream, the islands may remain to form a short braided reach (Fig. 13–21). Thus, large organic debris (tree trunks and large pieces of natural wood greater than 10 cm in diameter) play an important role in stream channel morphology and process. As another example, trees that fall in a stream facilitate the development of pools, and in some small, steep streams in forested areas nearly all of the pools may be formed by scour around the accumulation of organic debris.

DRAINAGE PATTERNS

Drainage pattern is the arrangement of streams in drainage basins of an area as viewed from the air. The common drainage patterns include **dendritic**, **rectangular**, **radial**, and **trellis**. Drainage patterns are important in geology because by knowing something about the drainage pattern, something may be learned about the geology of the area. Thus, analysis of drainage pattern is a useful tool in evaluating the geology from topographic maps and aerial photographs.

Dendritic Drainage The dendritic drainage pattern resembles the veins in a leaf (Fig. 13–22). Dendritic drainage tends to form on rocks that are homogeneous with respect to erosion by running water. It is commonly found on igneous and metamorphic rocks or on flat-lying sedimentary rocks. Dendritic drainage is a very common drainage pattern that has been extensively studied in attempts to explain the development of drainage basins.

Rectangular Drainage In rectangular drainage, the streams of the drainage network generally meet at nearly right angles (Fig. 13–23). The pattern develops because streams erode along fracture systems within the bedrock. Fracturing is ubiquitous in rocks, and often there are several sets of fractures along which groundwater migrates and weathering is accelerated. Thus, the drainage pattern reflects the rock fracture systems.

Radial Drainage The radial drainage pattern develops when a series of streams flow down from a central high point such as a volcanic peak. Radial drainage thus indicates the existence of a particular topographic form. Another drainage pattern that is somewhat controlled by the topography is the **centripetal** pattern, in which the streams all flow into a central low area. This pattern typically forms where internal basins exist with no through-flowing drainage to the ocean.

Trellis Drainage The trellis drainage pattern (Fig. 13–24) is commonly found on sedimentary rocks that vary in resistance to erosion by running water. The major streams of the area generally flow along valleys in the softer rocks, and the tributary streams join these master streams at nearly right angles. A trellis drainage pattern suggests the existence of folded or tilted sedimentary rocks. The longer tributary streams to the major channels flow in the direction the rocks are dipping. This results because the streams tend to migrate down the dip of the rocks. Thus, the study of the trellis drainage pattern may delineate the direction of dip of rock units as well as the existence of folds.

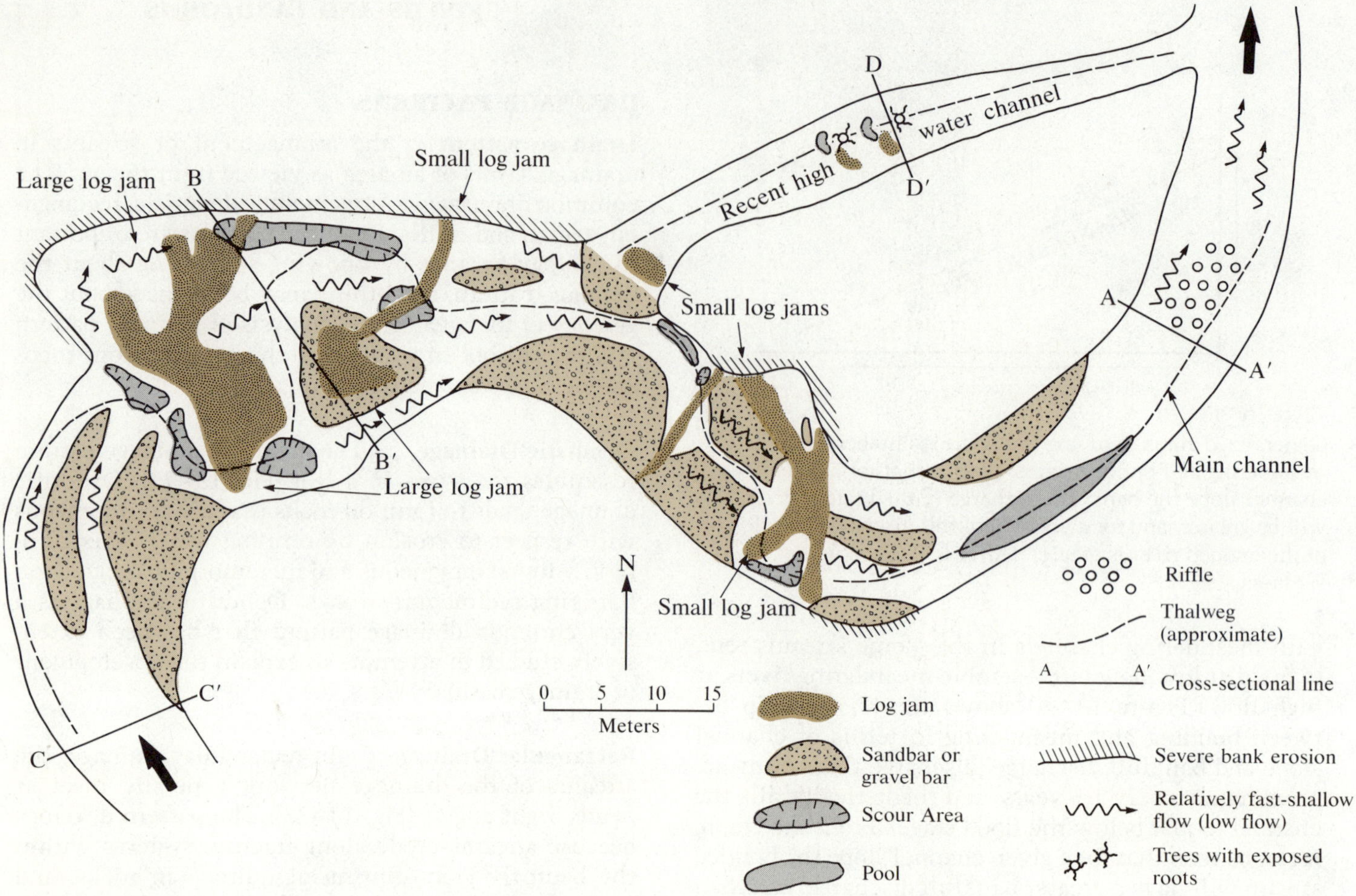

FIGURE 13–21
Morphologic map showing a large log jam in Mallard Creek near Harrisburg, North Carolina, 1976. [After E. A. Keller and F. J. Swanson, "Effects of Large Organic Material on Channel Form and Fluvial Processes," *Earth Surface Processes* (1979), Vol. 4, pp. 361–80]

FIGURE 13–22
The dendritic drainage pattern, commonly found on igneous and metamorphic rocks or on flat-lying sedimentary rocks.

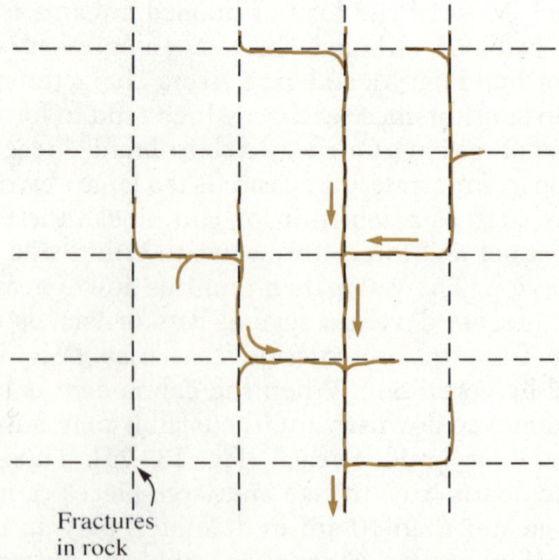

FIGURE 13–23
The rectangular drainage pattern that generally forms on rocks with pronounced sets of fractures (joints).

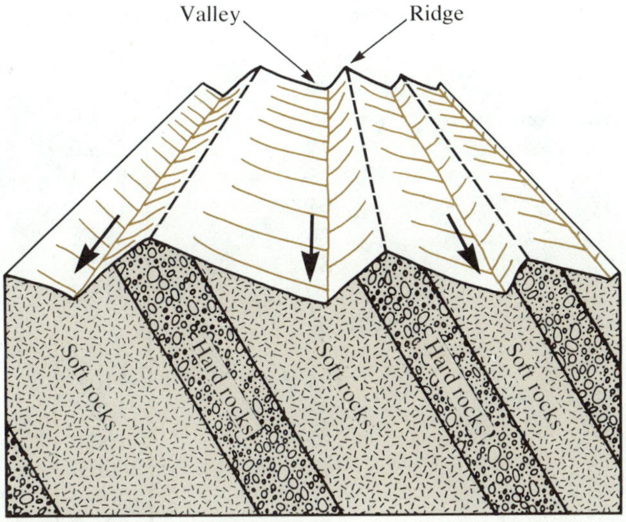

FIGURE 13–24
The trellis drainage pattern that generally forms on tilted or folded sedimentary rocks, exposing alternating layers of resistant and nonresistant rock.

RIVER FLOODING

Flooding is the natural process of overbank flow which occurs in most streams at least every year or so. The channel discharge at the point where water overflows the channel is called the **flood stage** and may or may not coincide with property damage. However, the term *flood stage* frequently denotes that the stream or river has reached a high-water condition likely to cause damage to personal property on the floodplain. Obviously, this latter use is based on human perception of the event and therefore will change or vary as human use of the floodplain changes.

A summary of flood damage in the United States is shown in Fig. 13–25, and a few of the more severe floods are listed in Table 13–4. These data emphasize the significant impact of flooding as a natural hazard. In the United States, the average loss of lives due to flooding is about 80 persons per year, with property damage averaging approximately $200 million per year.

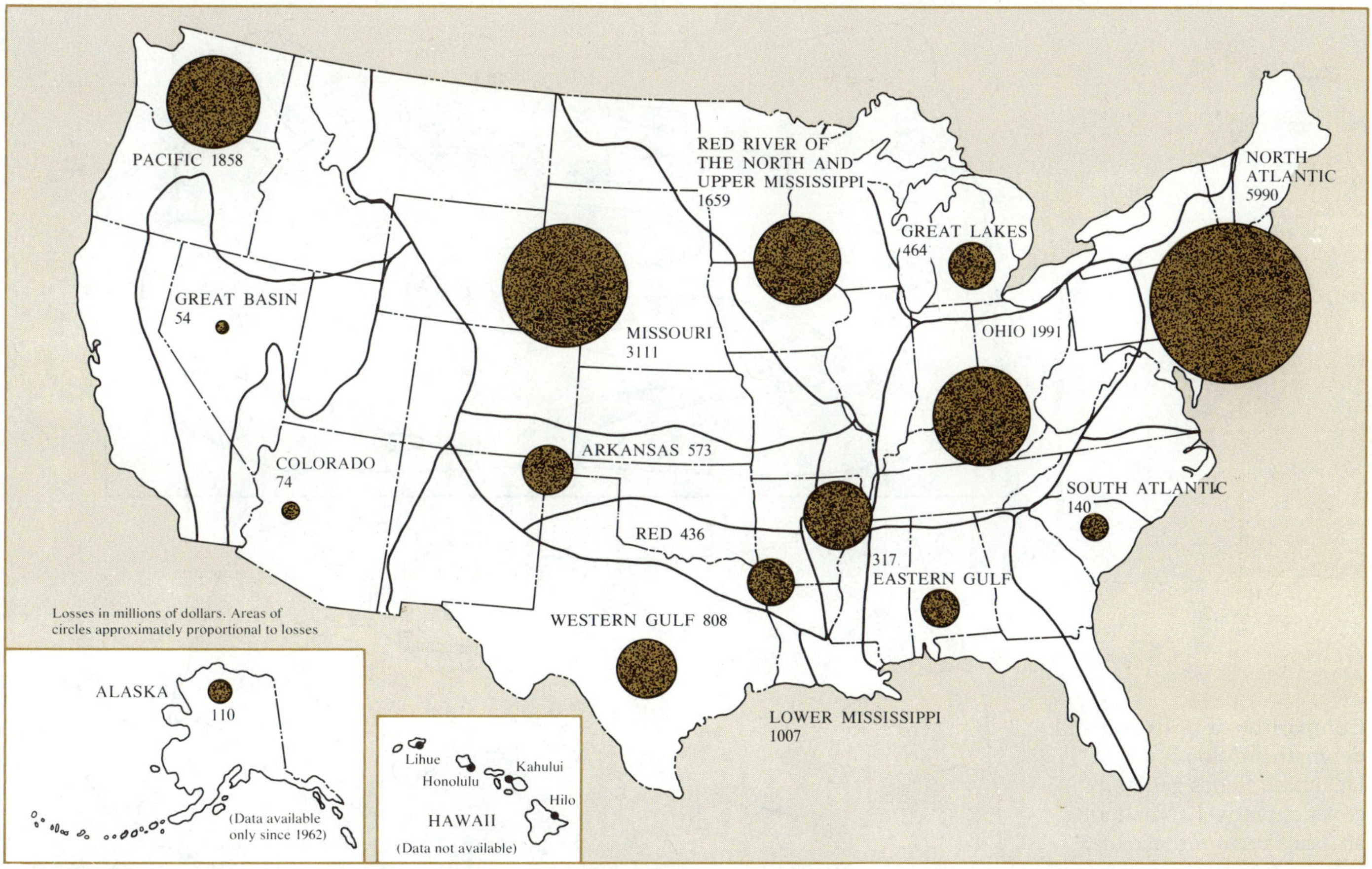

FIGURE 13–25
Summary of estimated flood losses in the United States by major river systems from 1925 through 1976. (From U.S. Department of Commerce, Climatological Data, National Summary, 1977)

TABLE 13–4
Severe river floods in the United States.

Year	Month	Location	Lives Lost	Property Damage ($ millions)
1937	Jan.–Feb.	Ohio and lower Mississippi river basins	137	417.7
1938	Mar.	Southern California	79	24.5
1940	Aug.	Southern Virginia and Carolinas, and eastern Tennessee	40	12.0
1947	May–July	Lower Missouri and middle Mississippi river basins	29	235.0
1951	June–July	Kansas and Missouri	28	923.2
1955	Dec.	West Coast	61	154.5
1963	Mar.	Ohio River basin	26	97.6
1964	June	Montana	31	54.3
1964	Dec.	California and Oregon	40	415.8
1965	June	Sanderson, Texas (flash flood)	26	2.7
1969	Jan.–Feb.	California	60	399.2
1969	Aug.	James River basin, Virginia	154	116.0
1971	Aug.	New Jersey	3	138.5
1972	June	Black Hills, South Dakota	242	163.0
1972	June	Eastern United States	113	3000.0
1973	Mar.–June	Mississippi River	—	1200.0
1976	July	Big Thompson River, Colorado	139	35.0
1977	July	Johnstown, Pennsylvania	76	330.0

SOURCES: From NOAA, Climatological Data, National Summary, 1970, 1972, 1973, and 1977; and U.S. Geological Survey.

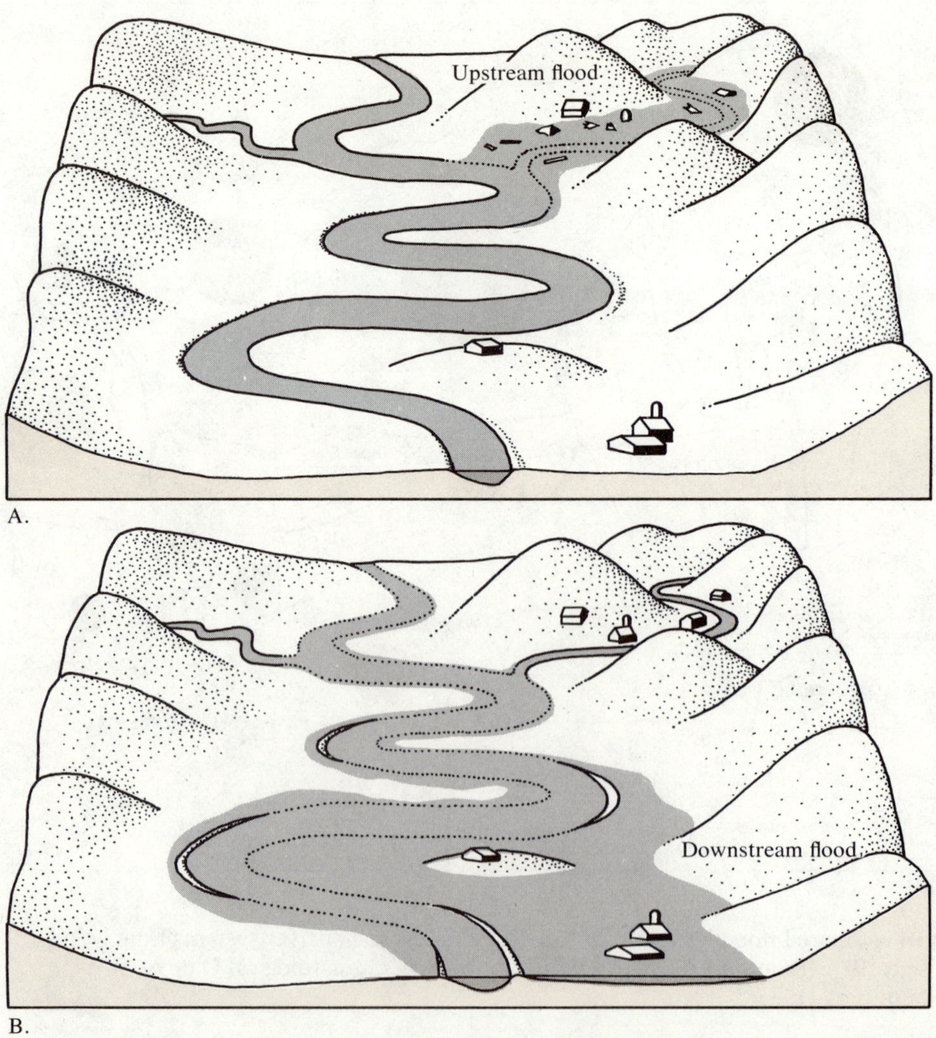

FIGURE 13–26
Comparison of upstream and downstream floods. Upstream floods generally cover a relatively small area and are due to intense local storms, whereas downstream floods cover wide areas and are caused by regional storms or spring runoff. (Modified from a drawing by U.S. Department of Agriculture)

When considering flooding, it is worthwhile to distinguish between **upstream floods** and **downstream floods** (Fig. 13–26). Upstream floods occur in the upper parts of drainage areas and are generally produced by intensive rainfall of short duration over a relatively small area. Although these floods can be severe, they do not usually cause flooding downstream in the larger rivers. The 1972 flash flood of Rapid Creek near Rapid City, South Dakota, that claimed 242 lives and inflicted $163 million in property damages is an example of an upstream flood caused by torrential rains (up to 38 cm in 5 hours). The people of Rapid City had a false sense of security owing to a reservoir built upstream. Unfortunately, the flooding occurred in a relatively small drainage basin that entered the stream downstream from the dam. Downstream floods, on the other hand, may cover a wide area, and are commonly produced by regional snowmelt or storms of long duration that saturate the soil and produce increased runoff. Flooding on smaller tributaries may be minor, but increased runoff from thousands of tributary basins causes a large downstream flood which may be recognized by the downstream migration of an ever-increasing flood wave with large rise and fall of discharge. For example, the downstream migration of a flood wave for the Chattanooga-Savannah river system is shown in Fig. 13–27A. Examination of the flood wave illustrates that a progressively longer time is necessary for the rise and fall of water as the wave proceeds downstream, and dramatically shows the tremendous increase in discharge from low-flow conditions to over 1700 m^3/s in 5 days and 257 km downstream. Figure 13–27B illustrates the same flood in terms of discharge per unit area, thus eliminating the effect of downstream increase in discharge. This shows the reduction in discharge per unit area in the

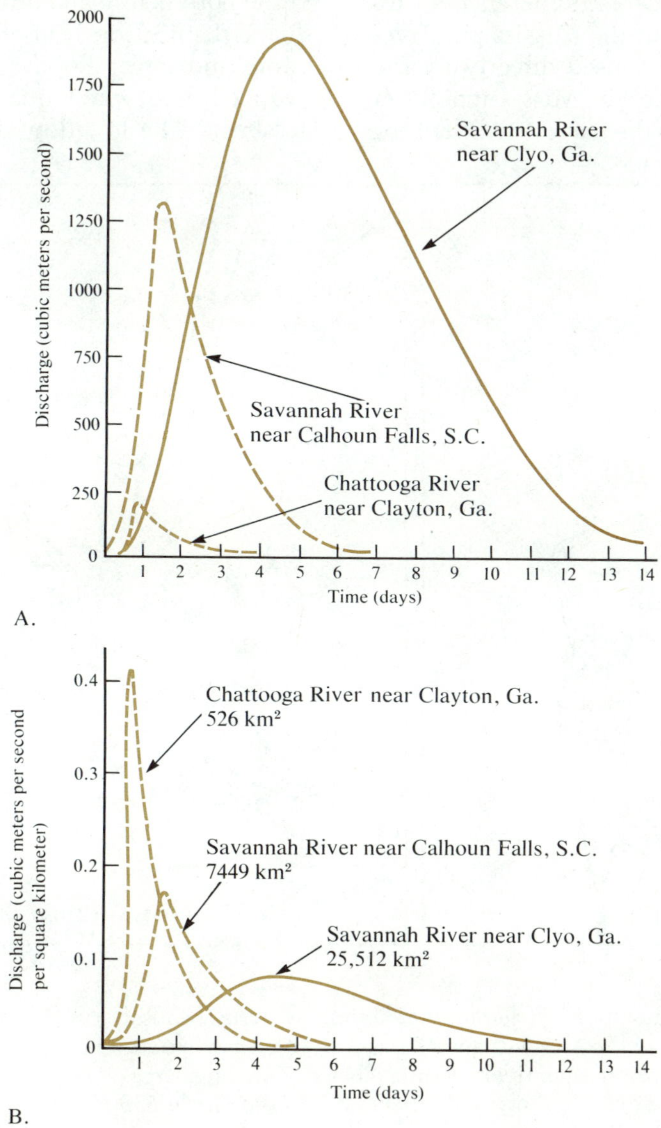

FIGURE 13–27
Downstream movement of a flood wave on the Savannah River, in South Carolina and Georgia. [After William G. Hoyt and Walter B. Langbein, *Floods* (copyright 1955 by Princeton University Press), Fig. 8, p. 39. Reprinted by permission of Princeton University Press]

downstream direction, even though the overall discharge dramatically increases. The reasons for this are that in the steeper headwater portions of the drainage the runoff is greater, and there is less storage of water on floodplains and adjacent areas. Notice in Fig. 13–27B that the shape of the hydrograph changes as the flood moves downstream. In the upper part of the basin the peak of the flood passes very quickly; however, in the downstream direction the curve becomes much flatter owing to the fact that a good deal of the floodwater is on the adjacent floodplain, which allows a fair amount of storage of floodwaters. This stored water requires a greater time to drain, and so the flood peak is flattened. A good example of a large downstream flood is the 1973 flood of the Mississippi River caused by spring runoff. This flood produced about $1.2 billion in property damage, and although tens of thousands of persons had to be evacuated and suffered appreciably as thousands of square kilometers of farmland were inundated throughout the Mississippi River valley, there were few deaths caused directly by the flooding. Part of the 1973 flood of the Mississippi River is shown in Fig. 13–28. Notice the narrow width of the river in the St. Louis, Missouri, area. This narrowing of the river was a result of the city's flood-protection works.

Urbanization and Flooding Urbanization of small drainage basins of a few square kilometers generally results in more frequent and larger floods. This occurs because urbanization means an increase in the percentage of land covered by pavement, rooftops, and other structures (impervious cover) as well as an increase in the percentage of area served by storm sewers. Both of these factors tend to increase the total runoff in addition to quickly concentrating runoff into stream channels following precipitation events. Thus, the lag time between the rainfall event and the flood event may be much reduced, and the flood peak is increased. These points are illustrated in Fig. 13–29 by the hydrographs (relationship between discharge and time) both before and after urbanization.

Urbanization leads to other effects in addition to more and larger floods. Low-flow discharges may be reduced, and water quality and aesthetic appeal are lessened. The low-flow discharge may be reduced be-

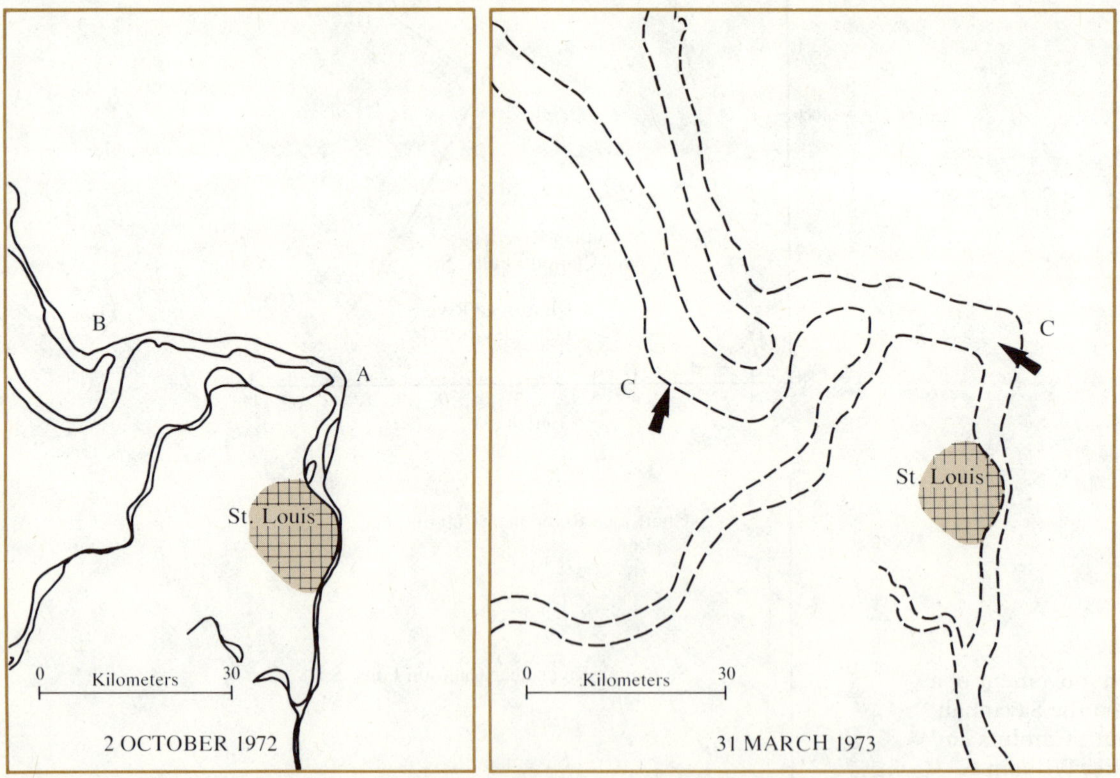

FIGURE 13–28
The spring 1973 flood of the Mississippi River as interpreted from satellite (Landsat) images. Point A is the confluence of the Missouri and Mississippi rivers, and point B is the confluence of the Illinois and Mississippi rivers. Points labeled C are large areas inundated by the floodwaters. (After N. M. Short, and others, *Mission to Earth: Landsat Views the World,* NASA, 1976)

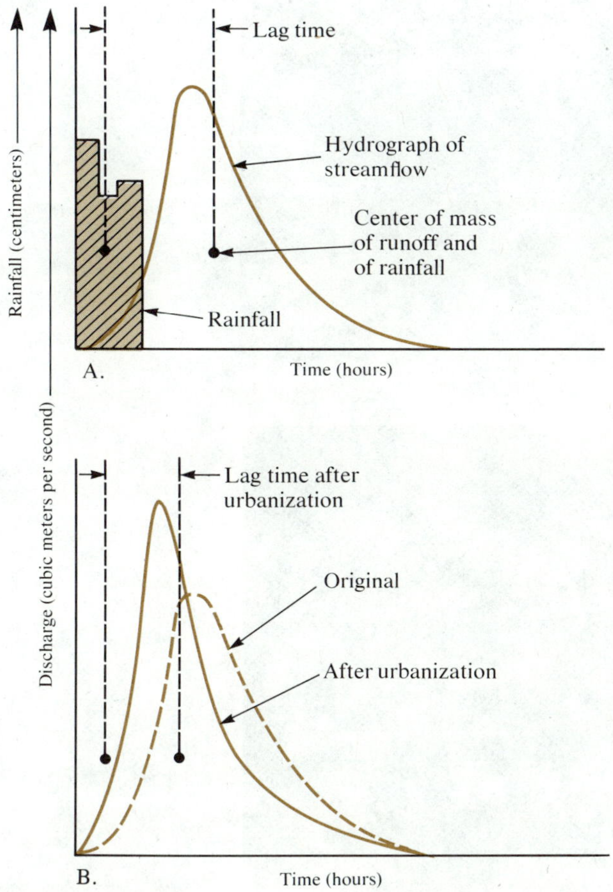

FIGURE 13–29
Generalized hydrographs for before and after urbanization. The hydrograph in part A shows the typical lag time between when most of the rainfall occurs and when the flood in the stream takes place. The hydrograph in part B shows the decrease in lag time due to urbanization. (After L. B. Leopold, U.S. Geological Survey Circular 559, 1968)

cause as the percentage of impervious cover in a small drainage basin increases, there is less infiltration, which normally restores the groundwater table which supplies the low-flow discharge to the stream. Water quality is generally lowered because any pollutants present will be concentrated during times of low flow by the lesser discharge. Also, urban streams tend to be neglected streams and are often dump sites for all sorts of urban refuse. Aesthetic degradation occurs as urban litter collects and the channel is subjected to channelization or other so-called improvements that remove trees along the banks, straighten the channel, and otherwise reduce it to an unsightly ditch. Fortunately, urban streams are not always neglected, and in several parts of the country there are programs to restore urban streams and utilize the stream channel and adjacent lands as a green belt. The extremes in dealing with urban streams are illustrated in Fig. 13–30.

Prevention of and Adjustment to Floods Different approaches are used in preventing and in adjusting to flood events. Prevention includes engineering structures and projects such as levees and floodwalls to serve as physical barriers against high water; reservoirs to store water which is later released at safe rates; on-site storm-water retention; channel improvements (channelization) to increase the channel size and quickly move the water off the land; and channel relocation or construction of a bypass to route floodwaters around flood-prone areas. Figure 13–31 shows how a floodwater retention pond in an urban area might reduce the flood peak. In other urban areas, parking lots might be planned to serve as small retention ponds also; that is, the outflow from parking lots might be regulated so that the water is released to nearby streams at a rate comparable to what the runoff was prior to urbanization.

Adjustment to the flood hazard includes floodplain regulation and flood insurance. From an environmental viewpoint, the best approach to minimizing flood damage in urban areas is floodplain regulation. This does not mean that physical barriers, reservoirs, and channel works are not desirable. In developed areas on floodplains, these are necessary to protect lives and property. However, it should be recognized that the floodplain is part of the natural river system and that encroachment increases flooding. Therefore, the ideal solution would be to discontinue development in areas subject to flooding that necessitate new physical barriers. Realistically, the most effective and practical answer is a combination of physical barriers and floodplain regulation that would result in less modification of the river system.

LANDFORMS

Speculation on the origin of landscapes including mountain ranges, plains, and river valleys is an exciting endeavor, as these are the most recent events in the history of the earth. To go beyond pure speculation in examining landforms and the development of the landscape, it is necessary to introduce several fundamental principles or concepts: base level; the graded stream; the magnitude-frequency relationship of geomorphic processes; geomorphic thresholds; erosion surfaces; and the relationship between time, space, and causality. These principles or concepts may then be applied to look at the landscape in a more critical way and evaluate possible models to explain features of the landscape.

A.

B.

FIGURE 13–30
Two extremes of channelization. A. Concrete ditch in Carpinteria Creek, California. B. Channel restoration of Briar Creek near Charlotte, North Carolina.

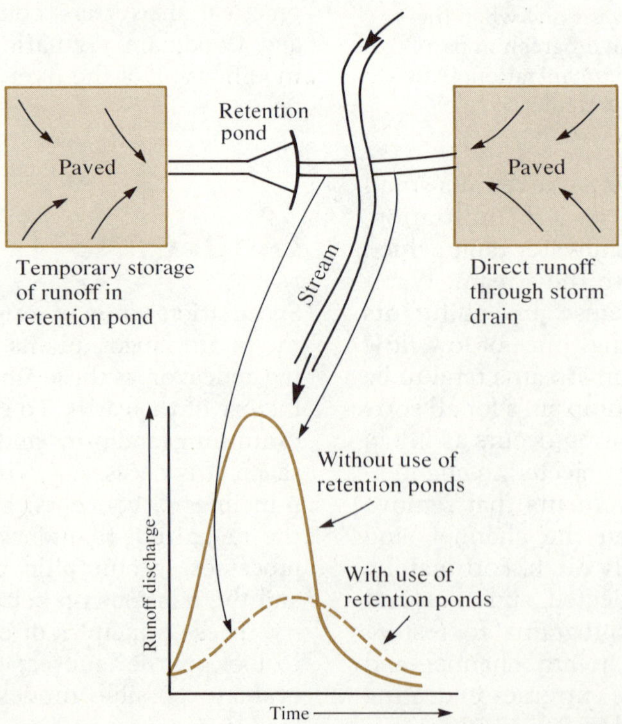

FIGURE 13–31
Idealized diagram showing how a retention pond may be used to reduce the peak runoff from a precipitation event. (Modified after U.S. Geological Survey Professional Paper 950, 1978)

PRINCIPLES AND CONCEPTS IMPORTANT TO LANDSCAPE EVOLUTION

Base Level The theoretical lowest level to which a stream or river may erode is known as the **base level**. The ultimate base level is sea level. However, there may also be local or temporary base levels produced by resistant rock layers or lakes. Even the ultimate base level (sea level) changes with time, and therefore streams and rivers adjust to these changes. When the base level lowers (**regressive sea**), streams near the coast adjust by eroding the stream bed and entrenching the channel. On the other hand, if the base level rises (**transgressive sea**), coastal valleys are drowned and the channel and valley may backfill through deposition of sediment.

Temporary or regional base levels may change with time, as for example the filling of a closed basin in the arid Southwest of the United States. With time, a basin with no outlet will slowly fill, and as base level rises, adjacent stream valleys must also fill (Fig. 13–32). Regional or local base levels may be lowered due to tectonic activity, for example when a basin marginal fault causes a downdrop of the basin (Fig. 13–32). Streams adjusting to a drop in base level will entrench their channels.

Graded Stream A **graded stream** is a stream in which there is a rough balance between the load imposed and the work done. This adjustment is such that over a period of years, the stream adjusts its slope and channel characteristics to provide just the velocity necessary to transport the load supplied to the channel. The profile of the graded stream is concave; however, there may be irregularities produced by local areas where resistance of rock is greater than in others (Fig. 13–33). Graded streams will exhibit a downstream increase in width, depth, and velocity, along with a de-

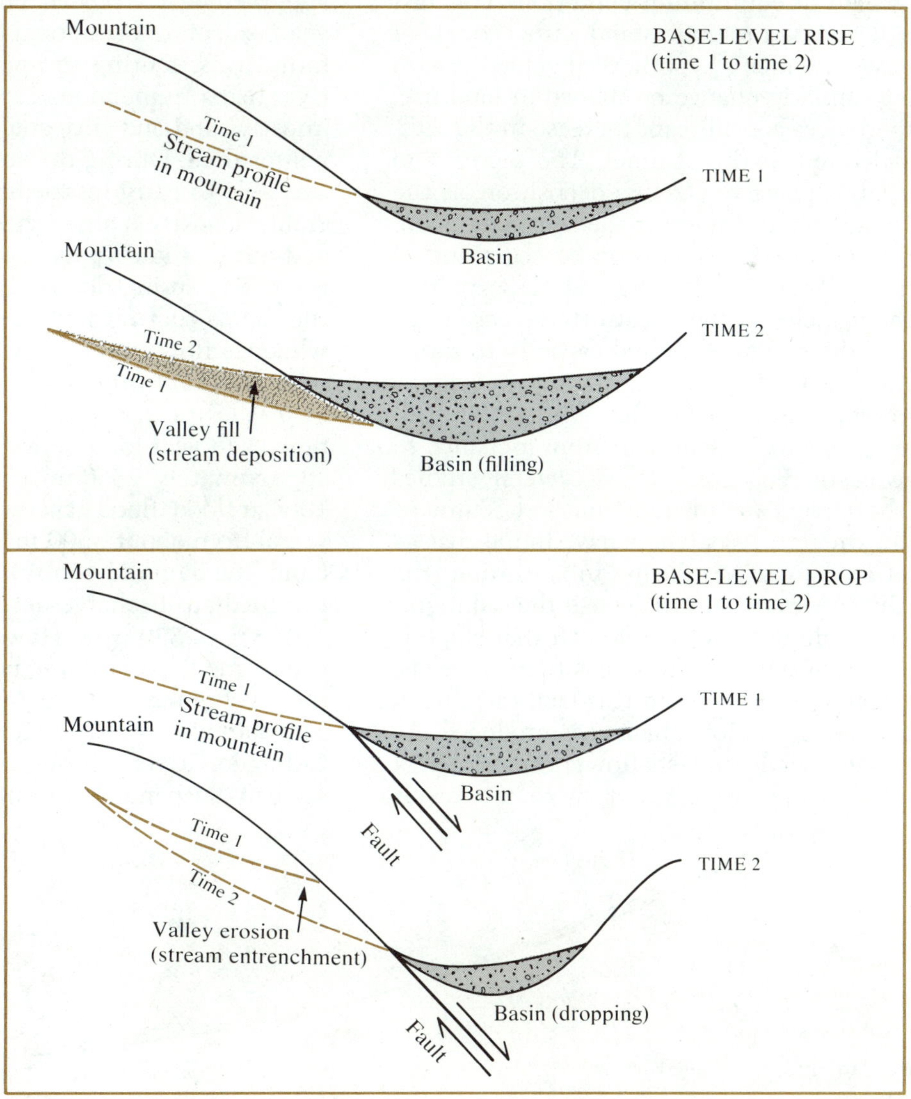

FIGURE 13–32
The response of a mountain with adjacent basin to base-level rise or base-level fall. Notice that with a rise in base level, valleys in the mountains tend to become sites of deposition, whereas with a drop in base level erosion is the dominant process, lowering the valley floor.

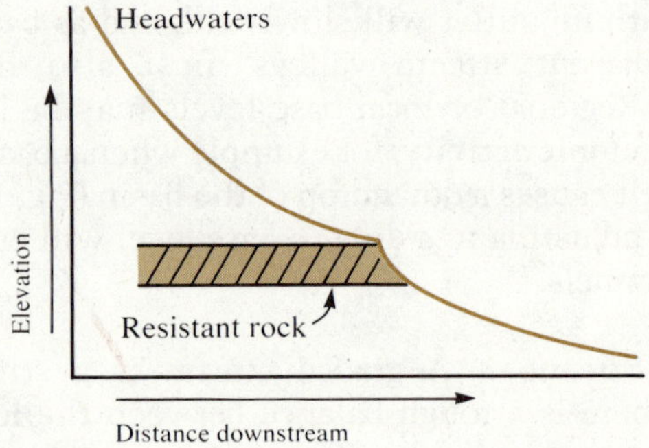

FIGURE 13–33
Longitudinal profile of a graded stream.

crease in channel slope and generally a decrease in the size of bed-load material. Channel pattern may be meandering or braided, but the graded stream resists rapid change. That is, any tendency to destroy the equilibrium will be counteracted in such a way as to restore the graded or equilibrium condition. This last statement will be easier to understand with a couple of examples. First, consider a hypothetical graded stream that, due to a climatic change or change in land use, suddenly experiences a significant increase in the sediment load delivered to the channel. The increase in load will initiate a phase of channel deposition, as the stream will be unable to transport the additional load. With time, however, the deposition of sediment in the channel will increase the channel slope, which will cause the velocity of the stream to increase until the stream has the competence and capacity to transport the increased load. It is important to recognize that channel slope may not be the only variable adjusted due to an influx of sediment from a change in the drainage basin. For example, in several studies looking at the effects of urbanization and sediment production on streams, researchers have found that as the sediment load increases during urbanization, the response of the stream may be to deposit the sediment, but more evenly along the channel such that slope is not increased. However, as the deposition proceeds, the channel may tend to fill, and this facilitates bank erosion until a wide, shallow channel is produced. As the channel gets wider and shallower, transport of bed-load sediment becomes greater, with the result that the stream may reach a new equilibrium between the load imposed and the work done. Thus, there are several ways the graded stream might adjust.

As a second example, consider what happens when a reservoir is constructed in a river valley (Fig. 13–34). Upstream from the reservoir, deposition will occur as the stream enters the lake and the sediment is deposited. Below the reservoir, water coming out of the spillway does not carry a significant load. Therefore, below the reservoir the stream erodes its bed and banks, picking up a load and lowering the stream gradient by erosion to accommodate the lower load. For example, in 1963 the Glen Canyon Dam was constructed upstream from the Grand Canyon. This structure has drastically altered the pattern of flow and channel processes downstream. From a hydrologic viewpoint, the Colorado River is tamed. Prior to construction of the dam, the river reached maximum flow in May or June during the time of spring snowmelt, and then flow receded during the remainder of the year, except for occasional flash floods due to upstream rainstorms. During the period of high discharge, the river had a tremendous capacity to transport sediment (mostly sand and silt), and vigorous scour of the channel resulted. During the summer low flow, the stream was able to carry less sediment and there was considerable deposition along the channel, forming large bars and terraces known as beaches to people who raft the river. The spring floods also moved large boulders off the rapids that formed due to shallowing of the river where it flows over alluvial deposits being delivered from tributary canyons. Following construction of the dam, the mean annual flood (average of the highest flow each year for a period of years) was reduced from approximately 2500 m^3/s to about 800 m^3/s, and the 10-year flood (flood expected on the average of every 10 years) from about 3500 m^3/s to 860 m^3/s. On the other hand, the dam did control the flow of water such that the median discharge actually increased, from about 210 m^3/s to 350 m^3/s. However, the flow is highly unstable due to fluctuating needs to generate power, and the level of the river may vary by as much as 5 m/day. The dam also greatly affected the sediment load of the Colorado River through the Grand Canyon; the median suspended sediment concentration was re-

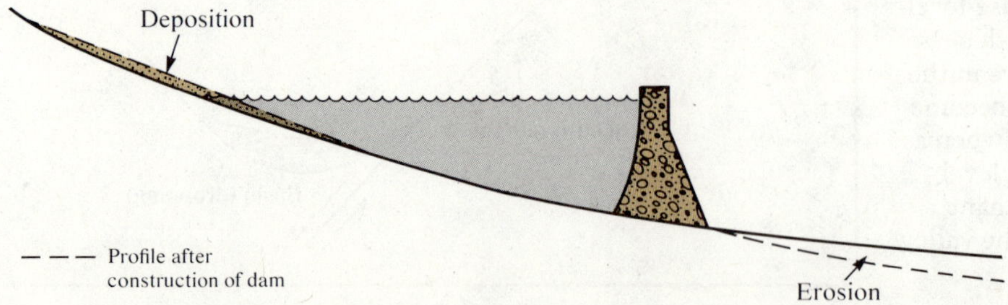

FIGURE 13–34
The effects of a dam and reservoir on a river.

duced by a factor of about 200 immediately downstream from the dam. Further downstream, there was less reduction in sediment load because tributary channels continued to add sediment to the channel.

Changes in hydrology of the Colorado River in the Grand Canyon below the Glen Canyon Dam have greatly changed the morphology of the river. The rapids may be getting more dangerous because the large floods no longer occur, and it is these flows that move some of the large boulders further downstream. In addition, some large sandbars and beaches are eroding because the river, deficient in sediment below the dam, is picking up sediment from them. The Colorado River in the 1970s and on into the 1980s is a changed river, and it can never be expected to return to what it was prior to the construction of the dam. However, eventually the stream will adjust to the new sediment load consistent with the concept of the graded stream.

Magnitude-Frequency Concept A **geomorphic process** may be defined as physical and chemical changes that effect a modification of the earth's surficial form. The **magnitude** of a process refers to the intensity or measure of energy expended, and the **frequency** refers to the recurrence of an event. In general, as the magnitude of a process increases, the frequency decreases. This holds for such processes as floods, landslides, wind storms, volcanic activity, and earthquakes, among others. Examination of the relationship between the magnitude and frequency of geomorphic processes and the work done by those processes has suggested that it is not the rare high-magnitude event or the everyday low-magnitude, high-frequency event that does most of the work in molding the landscape. Rather, the events of moderate magnitude and frequency that occur with the recurrence interval of a year or so do most of the geomorphic work in forming the landscape. However, there are exceptions to this general principle, because sometimes there are definite thresholds that must be exceeded before a change in the landscape occurs. For example, the inlets along barrier islands in the eastern United States are generally formed by hurricanes, events of high magnitude and relatively low frequency. Another example is the work done by high-magnitude, low-frequency storms that might strike mountainous areas every few tens of years. These events may trigger hundreds of landslides, introducing sediment that may take many years to work its way through the river system. Yet another example of the importance of high-magnitude, low-frequency events is the landslide-debris avalanche associated with the Mount St. Helens eruption of May 1980. That single catastrophic event will change the morphology of the Toutle River valley for the next few thousand years. In spite of these exceptions, the general tendency for most of the geomorphic work to be done by events of moderate magnitude and frequency is well founded, and in fact in many river basins 90 percent of the sediment is removed by storm discharges that recur at least once every 5 years.

Geomorphic Threshold As noted previously, changes in landforms commonly are not continuous and gradual but occur suddenly when a critical threshold is exceeded. Furthermore, when a threshold is exceeded the process usually changes, as for example from erosion to deposition. Two types of thresholds have been recognized: **extrinsic** and **geomorphic**. Extrinsic thresholds are thresholds that exist within geomorphic systems but in which change will not occur without the influence of some external force or variable. Perhaps the best-known extrinsic threshold is the water velocity necessary to move a bed-load particle of a particular size. The particle will not move until there is a sufficient velocity of water, and clearly this is an external force on the particle. Geomorphic thresholds, on the other hand, are thresholds that develop within the geomorphic system by internal change with time. For example, a failure of a slope by landsliding may occur in response to a rainstorm no greater in magnitude than rainstorms expected every year. What may have happened in the slope is a slow change in the strength of the slope materials such that a threshold eventually was crossed and failure occurred without a high-magnitude external force being applied. Other examples come from stream systems in which sediment supplied from hillslopes may be deposited in the channel until a critical threshold is exceeded, and then the sediment is eroded. The change from deposition to erosion, however, takes place without any change in the external forces acting on that river system. An example from western Colorado will illustrate the principle of the geomorphic threshold. Observation in this region showed that some small drainage basins are deeply gullied whereas others are not. Careful measurement of the slope of the stream valleys revealed that for basins greater than about 10 km^2, valley slopes for gullied valleys tend to be distinctly steeper than for ungullied valleys (Fig. 13–35). Thus, there exists a threshold slope above which trenching or valley instability is likely.

The importance of recognizing geomorphic thresholds is emphasized here because the ability to predict changes in fluvial or other systems may lead to better land management. For the example of gullies forming in western Colorado, conservation measures might be possible to reduce negative impacts such as sediment pollution and soil loss.

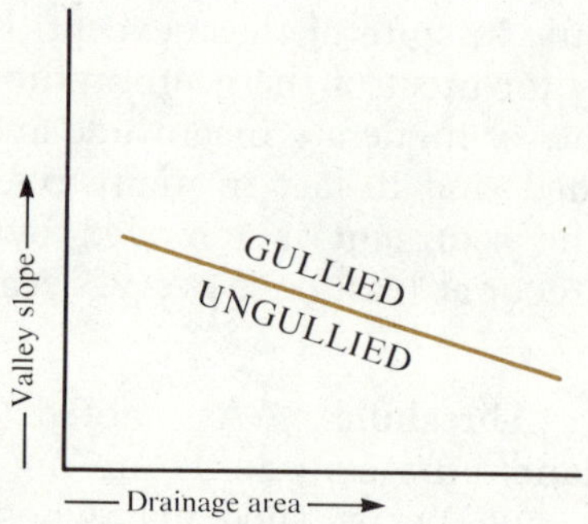

FIGURE 13–35
Generalized graph for an area in western Colorado showing a threshold valley slope above which valley instability (gulleying) is likely to occur. (After P. C. Patton and S. A. Schumm, 1975)

Geomorphic Surfaces The concept of the geomorphic surface produced either by erosion or depositional processes is significant in landform analysis. Most of the land surface of the earth has been produced by erosional and depositional processes associated with running water, moving ice, and the work of the wind. The discussion here, however, will be concerned primarily with erosion and depositional surfaces produced by running water.

Erosion surfaces may be recognized by several criteria, including the following:

1. The surface truncates rocks of varying resistance and is basically a bedrock surface of low relief (Fig. 13–36).
2. The surface may have a thick residual soil, indicating that subareal erosion has been going on for an extended period.
3. The surface may have stream gravels on the uplands between stream channels.

Of these three criteria, the most significant and the one necessary to define any erosion surface is that of truncating rocks of varying resistance. It should be recognized that older erosion surfaces may be uplifted and incised by streams. These old erosion surfaces are now relict in that the processes that originally formed them may no longer be operating or are operating at a much changed rate. In fact, many of the erosion surfaces that have been recognized by geologists appear to have been uplifted, and the surfaces now characterized by incised streams and an accordancy in elevation of hilltops. Notice that in the preceding discussion, not much has been stipulated concerning the process that formed the erosion surface. That is, nothing has been said about how the slopes form or exactly what processes are involved in the erosion. Thus, we recognize that the erosion surface may form under the influence of several different processes and modes of slope development. One characteristic common to all erosion surfaces, however, is that they develop or evolve through time. Thus, we recognize that landforms, and in fact the landscape, do change with time, and in some instances these changes may be predicted or at least expected. The existence of older, uplifted erosion surfaces attests to the dynamic nature of earth processes and to the fact that to some extent there exist cycles of erosion separated in both time and space. However, to understand how this works, the concept of time, space, and causality in geomorphic processes and landscape development must be considered.

The depositional surfaces produced by running water that are of interest here include the **floodplain**, **alluvial fan**, and **delta**. All three are dynamic, evolving surfaces, the characteristics of which may change over a short period of time due to high-magnitude events associated with running water. In spite of this, the events of moderate magnitude and frequency are the events that supply most of the sediment for these landforms.

The floodplain and stream or river channel comprise a system in which the form and processes evolved in harmony. Processes responsible for formation of the floodplain include overbank flow and lateral migration of the stream channel. Of the two processes, lateral migration is probably more important, and most floodplains appear to develop as the stream meanders back and forth over the available area in the valley. However, overbank flow during times of flood is also

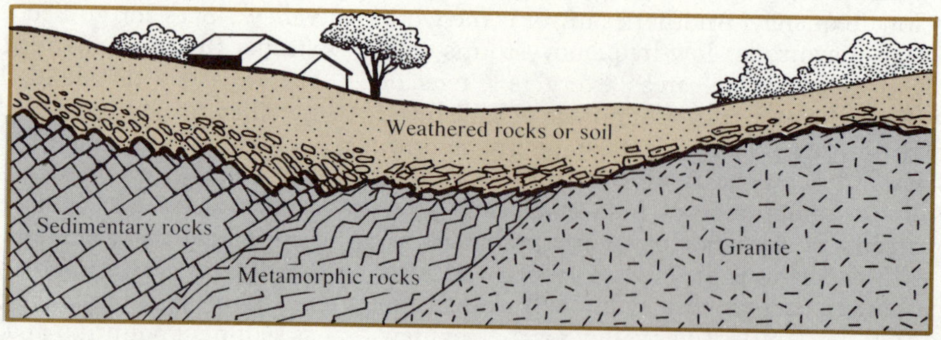

FIGURE 13–36
Diagram showing an erosion surface that truncates rocks of varying resistance.

important in adding sediment to the depositional surface. The morphology associated with the floodplain is shown in Fig. 13–37. Notice the areas of erosion, deposition, and formation of such features as **oxbows** and **meander scrolls**. Oxbows develop as meander bends are cut off by erosional processes inherent in the meandering process. The abandoned meander bend may fill with water and become an **oxbow lake**. As the stream meanders laterally, meander scrolls develop. These scrolls are prominent on large floodplains, and form as follows. As the stream moves laterally, water-loving plants such as willows develop on the inside of meander bends along the stream bank close to the low-flow water line. Roots of these plants collect sediment and so are topographically slightly higher than the surrounding floodplain material. As the stream continues to move laterally keeping a constant channel cross section, each meander scroll is the inside edge of a bend, and new scrolls form in the direction of the lateral migration. Over a period of time, numerous meander scrolls scar the floodplain, tracing the path that the meander bends took in their lateral and down-valley migration.

Alluvial fans and deltas are in the depositional zone of the fluvial system (Fig. 13–38). They are part of the erosional-depositional system in which the connecting link is the stream. Alluvial fans are composed of stream- and mudflow deposits. Commonly, the fan head is dissected by a process known as **fan-head trenching**, and therefore the active area of deposition

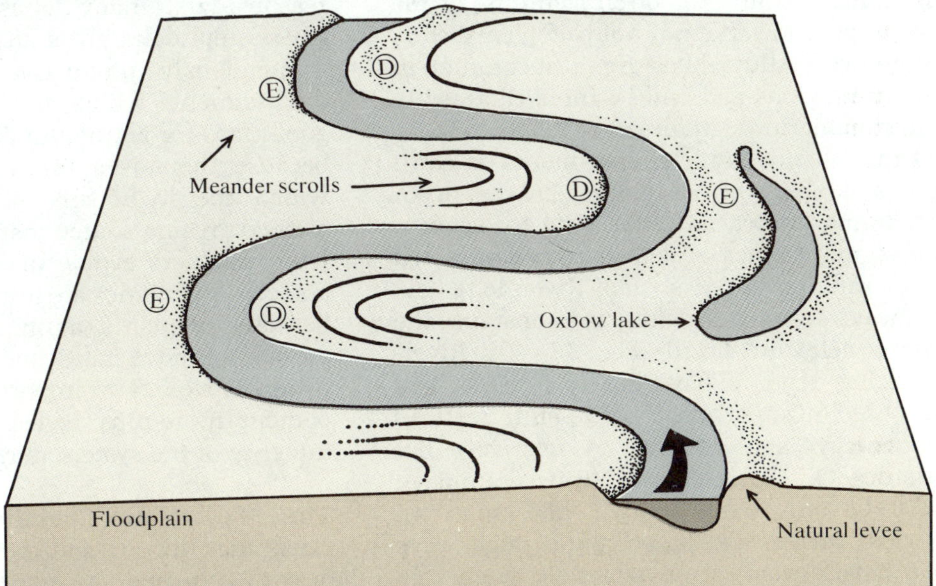

A.

Ⓔ Erosional area
Ⓓ Depositional area

B.

FIGURE 13–37
A. Diagram of a meandering river showing the areas of erosion and deposition as well as meander scrolls and an oxbow lake. B. The Missouri River showing meander scrolls. (Photo courtesy James Brice)

FIGURE 13–38
An alluvial fan in Death Valley, California. Notice the many small channels flowing across the surface of the fan. (Photo courtesy Robert Norris)

may be in a lower part of the fan. Alluvial fans are environmentally important, as they are commonly associated with large amounts of groundwater that may be tapped in semiarid areas. However, utilization of such groundwater resources is often a mining of the resource, as the groundwater may not be replenished as fast as it is removed. Alluvial fans are most common in arid to semiarid regions and will be more thoroughly discussed in conjunction with deserts (Chapter 15).

One of the most dynamic environments on earth is where the land meets the ocean. It is here that depositional landforms known as deltas tend to develop. There are two main types of deltas: **river-dominated deltas**, as for example the Mississippi River delta (Fig. 13–39), and **wave-dominated deltas**, as for instance the São Francisco delta in Brazil (Fig. 13–40). River-dominated deltas tend to have highly irregular and protruding shorelines with few wave-built features (low wave energy) and develop on relatively flat offshore profiles. On the other hand, wave-dominated deltas tend to have relatively straight shorelines with well-developed barriers and beach ridges (high wave energy) and have developed on relatively steep subaqueous offshore profiles.

The shape of deltas resembles the capital Greek letter delta (Δ) only under ideal circumstances, such as with the Nile delta (Fig. 13–41), which has some characteristics of both the river-dominated and wave-dominated types. If oceanic currents are very strong and the sediment is quickly moved off or along the shore, then deltas tend not to form off river mouths or they may be present in a subaqueous form. See Figure 13–42, which is a photo of the delta of the Ventura River in southern California. Strong longshore currents here somewhat prohibit the development of a delta, but a protrusion of the coastline is still visible. In years characterized by above-normal precipitation, the river floods, and a good deal of sediment is added to the Ventura River delta. However, in the years between storms of high magnitude the delta erodes.

River-dominated deltas such as the Mississippi River delta are dynamic, evolving systems in which the location of deposition changes often with time (Fig. 13–43). Therefore, during the life of the Mississippi River delta the location of deposition has changed even in historic time.

Deltas are environmentally important for several reasons. First, many deltas such as the Nile delta and Mississippi delta areas are extensively cultivated and tremendously productive. Second, ancient deltaic environments which are now sedimentary rocks are good traps for petroleum and natural gas. This results because there is a mixture of coarse material into which the hydrocarbons migrate along with fine-grained organic source material from which the petroleum products evolve upon burial. Finally, deltas in associated estuaries are important for providing unique habitats for plants, animals, and fish. Because deltas provide a unique landscape and environment for living things as well as an important area for resource development, more must be learned about them so that the integrity of the system may be maintained.

Time, Space, and Causality Distinction between cause and effect in landform development is a function of both time and space (area), and of the two, time is probably more significant. For example, over a long (geologic) time, such as the last 1 million years, a drainage basin is a system that, with no uplift, is progressively losing potential energy and may go through what is termed an erosion cycle. Important independent variables (causes) are time, rock type, and climate. The dependent variables (effects) are the relief and the morphology of the drainage basin, including the channel and slopes. During a moderately long interval (thousands of years), different parts of a river system may be considered. For example, the channel morphology including such factors as the width, depth, slope, and pattern is a function of the rock type, climate, and mean annual discharge of water and sediment. During this interval, time may not be particularly relevant because the stream may be in a graded or equilibrium condition; therefore, the morphology may be considered independent of time. During very short spans of time (less than 1 year), channel morphology, rather than being a dependent variable as it was with

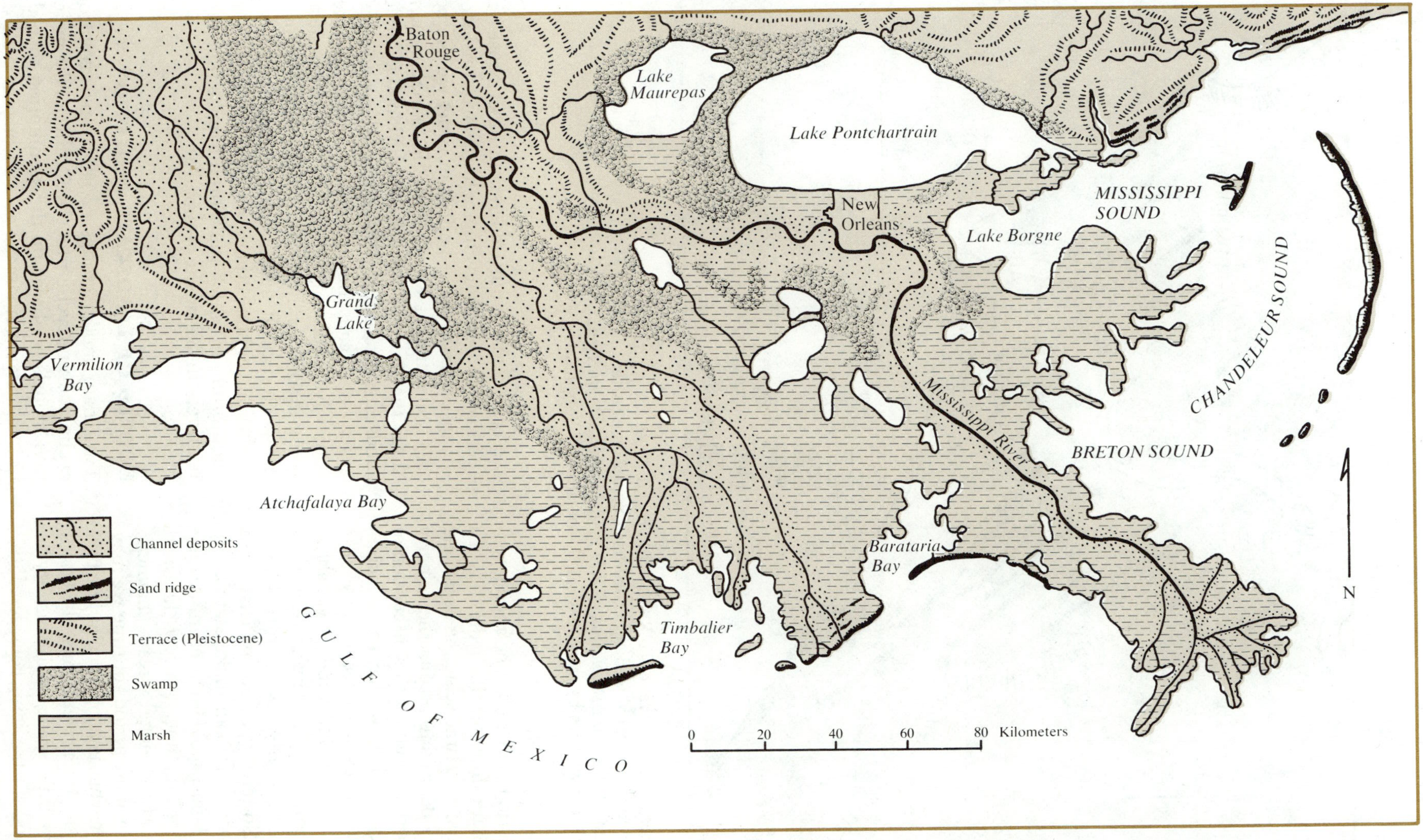

FIGURE 13–39
Major geomorphic features of the Mississippi River delta. (After L. D. Wright and J. M. Coleman, *American Association of Petroleum Geologists Bulletin,* Vol. 57, pp. 370–98, 1973, copyrighted by the American Association of Petroleum Geologists)

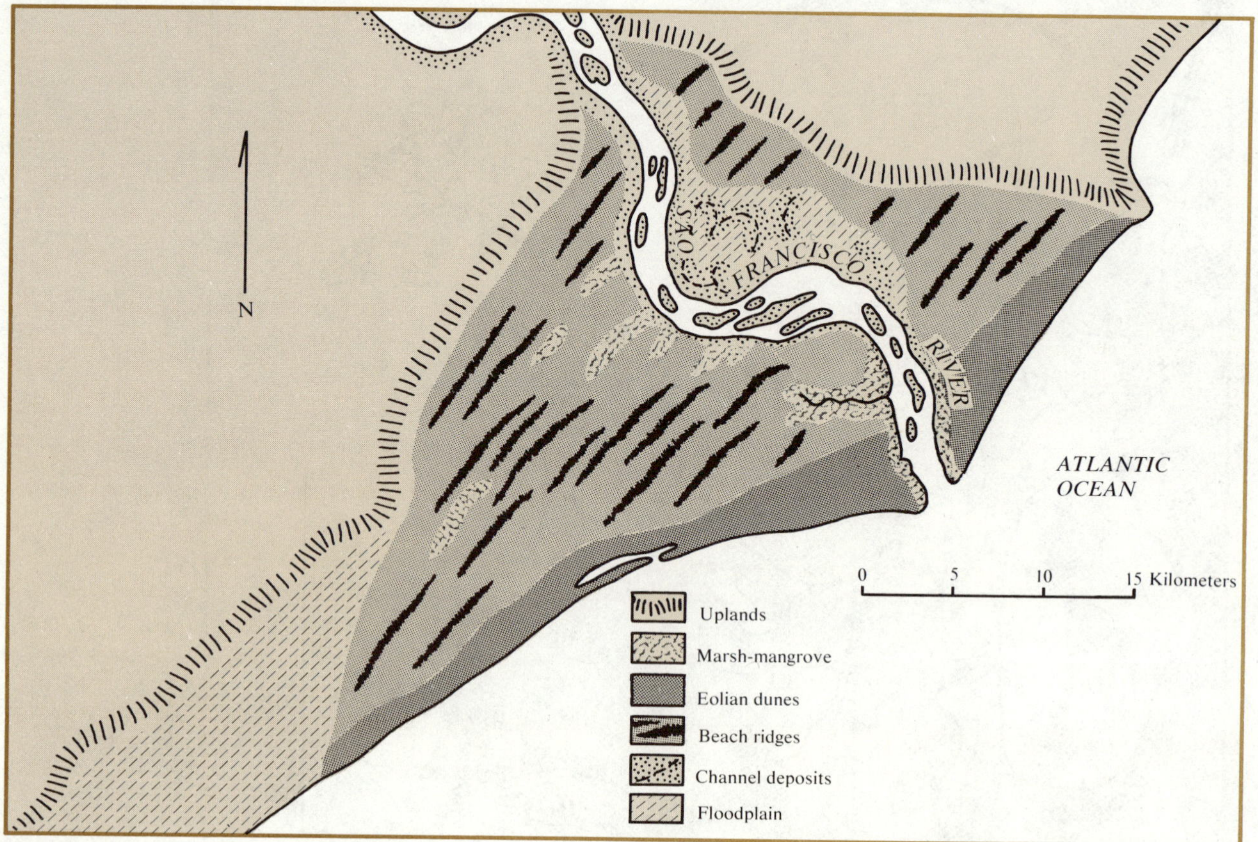

FIGURE 13–40
Major geomorphic features of the São Francisco River delta, Brazil. Note the relatively straight shoreline of this delta compared to that of the Mississippi River. (After L. D. Wright and J. M. Coleman, *American Association of Petroleum Geologists Bulletin*, Vol. 57, pp. 370–98, 1973, copyrighted by the American Association of Petroleum Geologists)

the moderately long period, may in fact be an independent variable (cause) that influences the hydraulics of a channel. For example, at this time scale the depth and slope of a channel may be used to predict the velocity of the water. The important point of this discussion of time on relative scales is to show that some factors that are dependent variables, or effects, may at a shorter period of time become independent variables, or causes. It is also important to recognize that as time was telescoped from 1 million years down to 1 year, the space (area) was also telescoped down. That is, in dealing with geologic time, we can make statements concerning the entire landscape. However, in dealing with the moderate length of time of a few thousand years we talk in terms of individual stream channels, and if we deal with very short periods of time we may only be able to consider what is happening at one particular cross-sectional site on a river. Thus, the major conclusion concerning time, space, and causality is that the longer the period of time being considered, the greater the space being considered and the less specific we can be about our understanding of causes and effects of landscape development. As we consider smaller spaces and shorter periods of time, we can become more specific but we are less able to make generalizations that may be valuable in developing models for landscape evolution. These considerations of time, space, and causality, while complex and perhaps somewhat confusing at first, are important if we are going to sort out theories of landscape evolution, some of which depend upon geologic time and thus allow evolution, while others are based on a dynamic equilibrium in which the morphology is thought to be relatively independent of time.

MODELS OF LANDSCAPE EVOLUTION

Two main model groups of landscape evolution are **cyclic** and **noncyclic** models. Cyclic landscape evolution models are based on the assumption that, during its evolution, a landscape goes through a **cycle of erosion**. Under this situation, the landscape may be con-

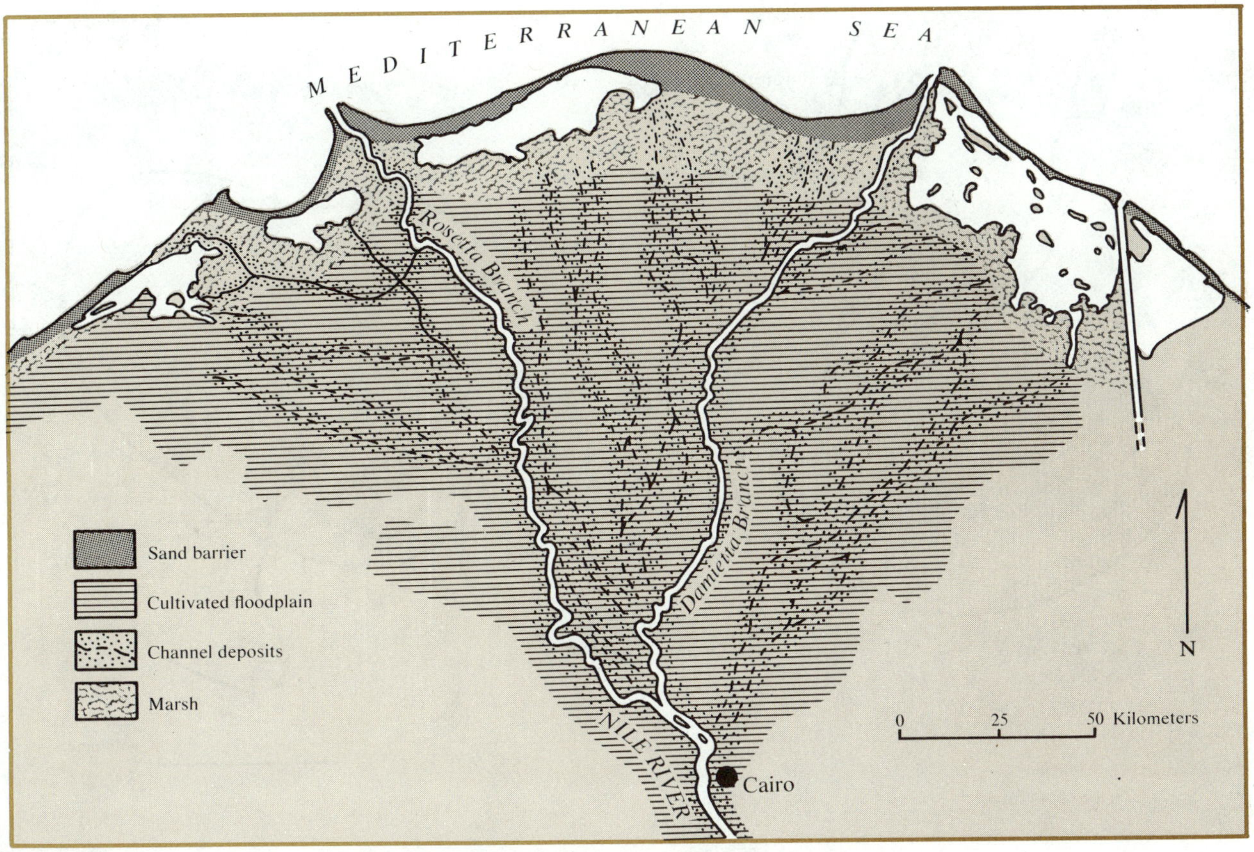

FIGURE 13–41
Major geomorphic features of the Nile River delta. The morphology of this delta is intermediate between the river-dominated Mississippi River delta and the wave-dominated São Francisco River delta. (After L. D. Wright and J. M. Coleman, *American Association of Petroleum Geologists Bulletin*, Vol. 57, pp. 370–98, 1973, copyrighted by the American Association of Petroleum Geologists)

FIGURE 13–42
The delta of the Ventura River in southern California. Strong longshore currents somewhat prohibit the development of a delta, but a protrusion of the coastline is readily visible.

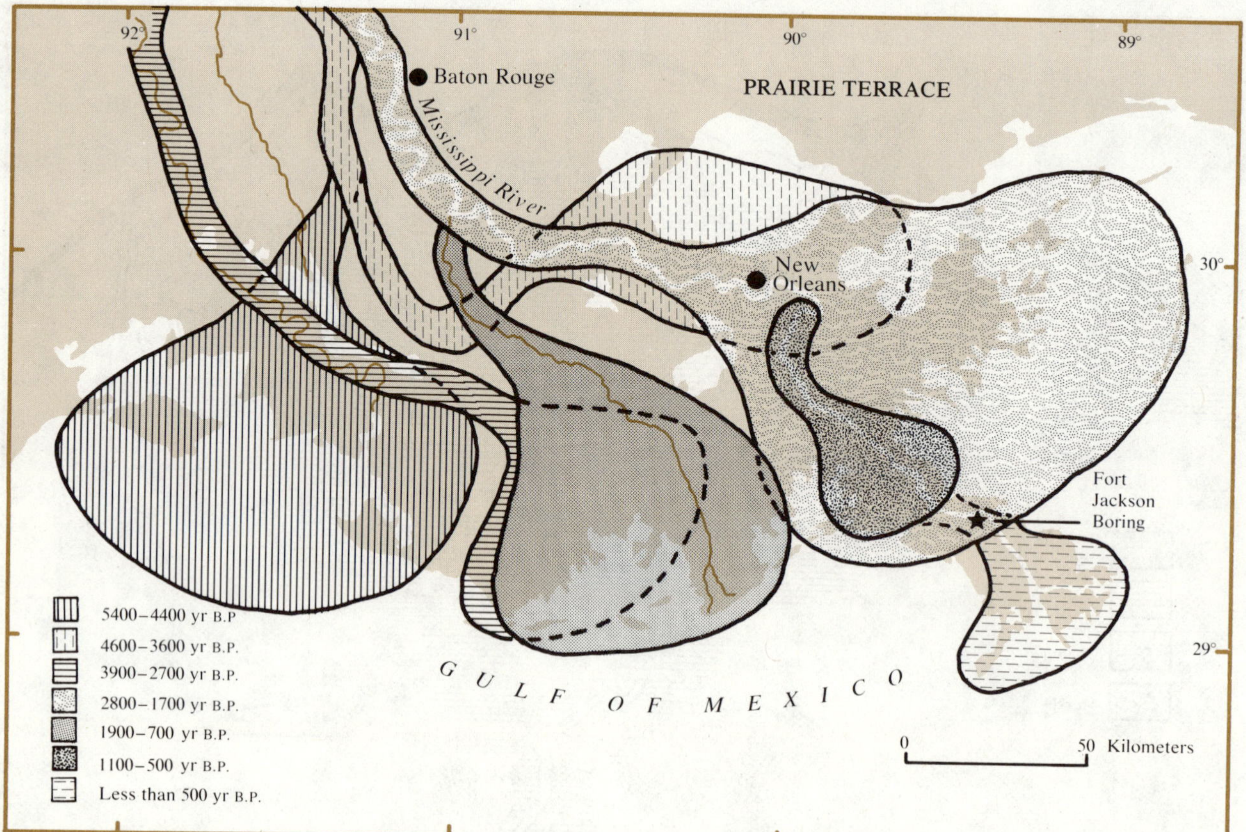

FIGURE 13–43
The change of deposition site for the Mississippi River delta during the last 5000 years. (Modified after C. R. Kolb and J. R. Van Lopik, 1958)

sidered as a special case of an open system in which a landmass is uplifted quickly with little erosion. Important here is the fact that the system of development assumes a geologic time and a broad area, and so development through time is possible. Important aspects in the development include the wearing down of slopes with time and the eventual emergence of an erosion surface of low relief known as a **peneplain**. Peneplaination was envisioned by William Morris Davis in the late nineteenth century, and the basic idea remains that landscape development begins with rapid uplift relative to other geologic processes, and then over a long period of time streams erode down to produce the erosion surface of low relief which is nearly at base level (sea level). During the development toward the peneplain, the landscape goes through a series of stages known as youth, maturity, and old age (Fig. 13–44). Youthful stream valleys are V-shaped and deeply incised, and the associated streams characteristically have an abundance of waterfalls and rapids. The youth stage is short and leads on to a long stage of maturity during which a great variety of landforms develop, including floodplains and a general, well-dissected

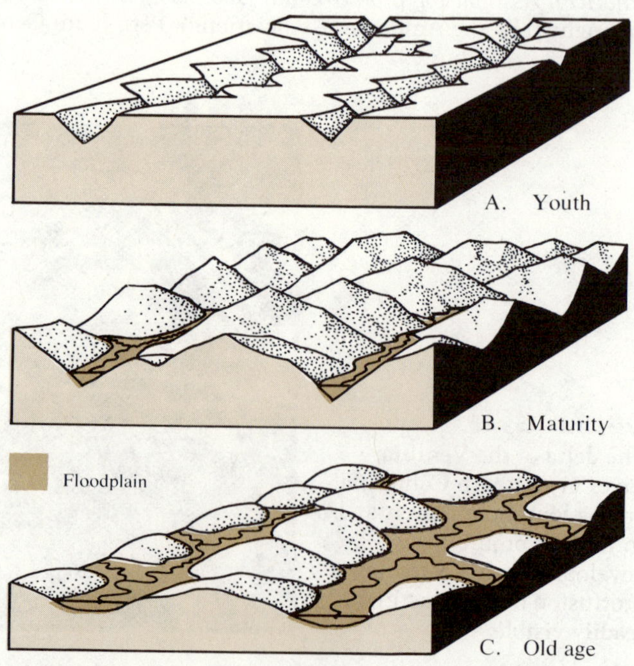

FIGURE 13–44
The cycle of erosion as envisioned by William Morris Davis.

landscape. Most of the waterfalls and rapids have disappeared, and there is a good adjustment between topography, rock type, and the streams. The final stage is very long and includes a continual reduction of hillslopes and coalescence of floodplains and slopes to form the peneplain. However, the cycle may be interrupted at any time, a phenomenon termed **rejuvenation**, and therefore any one landscape may bear the signs of several cycles of erosion (Fig. 13–45). The cycle of erosion and development of peneplains has been criticized because it requires rapid uplift, which may not be common in earth history. It also has been criticized as being too anthropocentric and rigid in requiring landscapes to go through a series of stages. Regardless of the criticism of the cycle of erosion, the model nevertheless stimulates discussion and has turned out to be a rather valuable descriptive tool for studying landscapes, if not a genetic tool. (The pediment, another product of an erosion cycle, usually in a semiarid region, will be introduced in Chapter 15.)

FIGURE 13–45
Stream terraces (central part of photograph) of the Salinas River, California. Stream terraces such as these that are cut onto bedrock have been used as evidence for rejuvenation, that is, reinitiation of a cycle of erosion.

As an analogy to the cycle of erosion, pretend you have a hot tub with a narrow slit up one side (Fig. 13–46A). Turn the water on very rapidly so that the water level rises quickly in spite of the water leaking out of the thin slit. Now, turn off the water and allow it to drain slowly out of the slit. The amount of water draining is proportional to the depth of the water in the tub, so the water level drops rapidly at first, but then gets slower and slower as there is less area in the slit available for escape. This situation is somewhat analogous to the cycle of erosion in that the water in the tub goes through a series of stages from high to low and there is no further input of energy into the system.

An important aspect of the cycle of erosion (whether or not landscapes ever tend toward the development of true peneplains) is evolution with time. It remains valid that landforms must change over long periods of time, and therefore a cyclic concept is consistent with geologic time.

Another model of landscape evolution is **dynamic equilibrium**. This model, developed by John Hack in the middle of the twentieth century, states that as a landscape develops, an equilibrium of action eventually will be reached in which every slope is adjusted to every other slope and the entire landscape will be relatively time-independent. That is, although erosion and deposition continue, the morphology will remain fairly constant. Under this model, areas containing hard, resistant rocks will remain topographically high, whereas areas with soft rocks will be topographically low. Nevertheless, the high areas will communicate with the low ones, and the morphology will not change much. This model is particularly appealing if used over a time interval intermediate between geologic time and very recent time. That is, if we are dealing with thousands of years, then indeed a landscape may approach an equilibrium of action, and there may be little change in the total morphology of the landscape.

An analogy to this model, again using the hot tub, is as follows (see Fig. 13–46B). Turn on the water in the tub, but not nearly so hard as the first time. As the tub begins to fill, water also runs out of the slit, but eventually, because the area from which water escapes increases with the water level, a dynamic equilibrium is reached in which the water going into the tub precisely equals the water going out. At this time, the level of water in the tub is a constant and will not change provided the energy (water going into the tub) does not change. Landscapes might be viewed in the same manner: eventually a dynamic equilibrium may be reached, and the form of the land will be independent of time.

An example showing hillslope processes will emphasize the dynamic equilibrium approach. Two hillslopes are shown in Fig. 13–47, one underlain by cherty limestone, and the other by pure limestone with a small amount of silt. The hilltop underlain by the silty limestone does not accumulate weathering products because they are removed by erosion as soon as they are released by chemical weathering. On the other hill, underlain by cherty rocks, bits of chert are produced in addition to fine silt. The chert fragments are too coarse to be removed by erosion on the slopes,

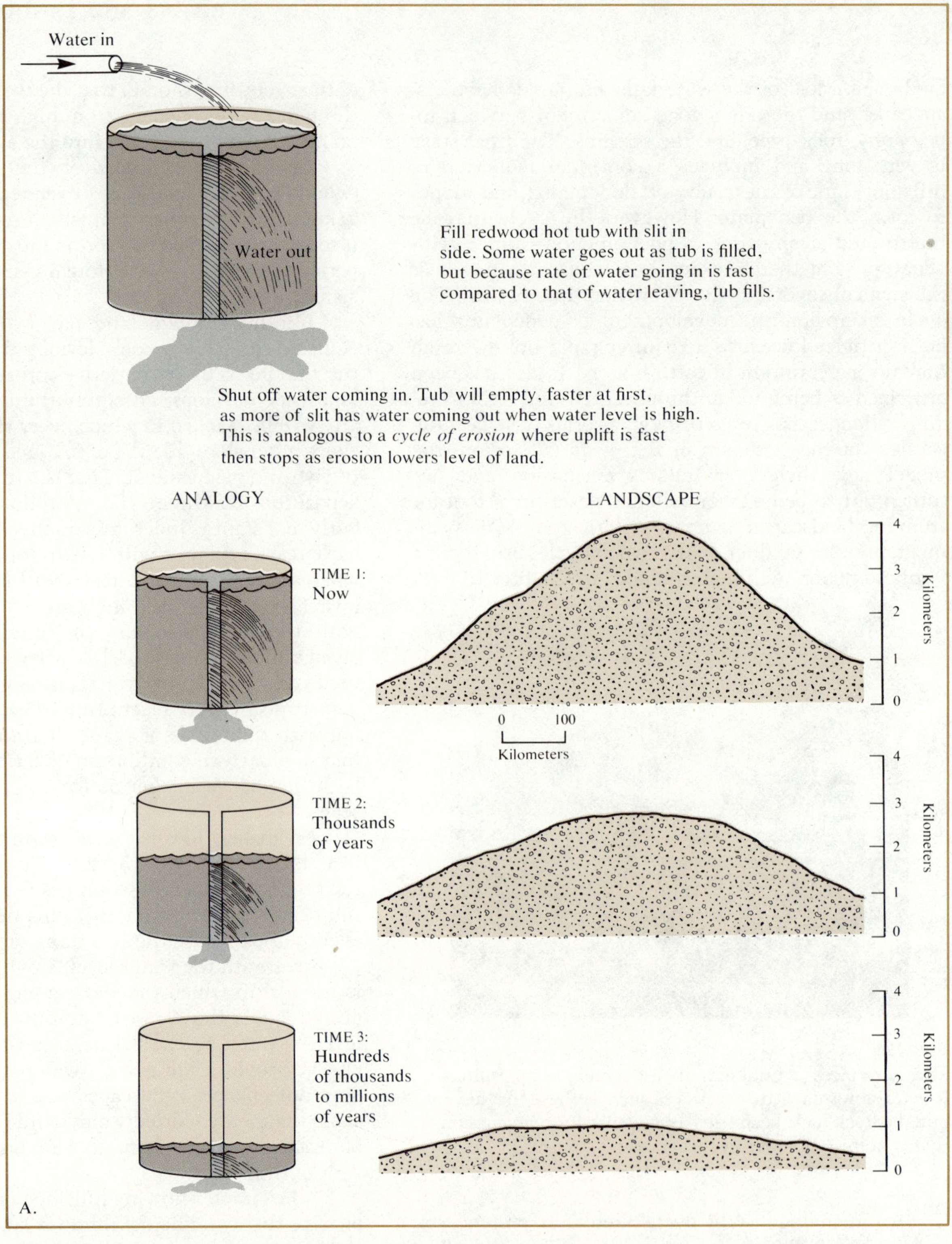

FIGURE 13–46
The difference between a system running down and a system in dynamic equilibrium. The form is independent of time.

and thus chert accumulates, forming a protective armor that prevents finer particles from being eroded. Accumulation of the deposits including the chert continues until the slopes are sufficiently steep for the chert to be removed. The topography may now be considered to be in equilibrium and the two different slopes are adjusted to each other, and continued development may be independent of time, at least over moderate time periods. This argument, which was developed by John Hack in the Appalachian Mountains, is appealing if one is looking at landscape development over a shorter period of time than geologic time.

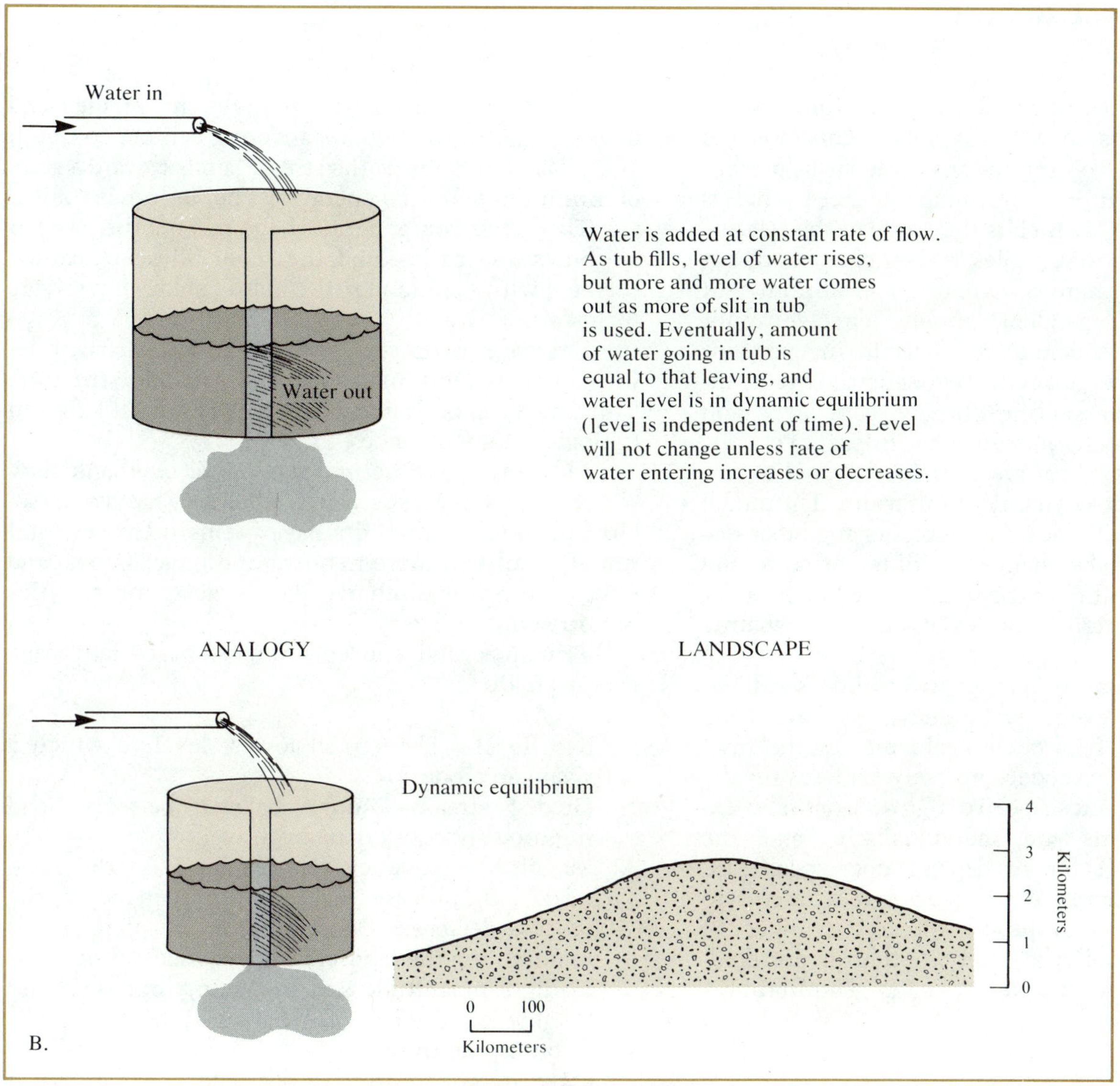

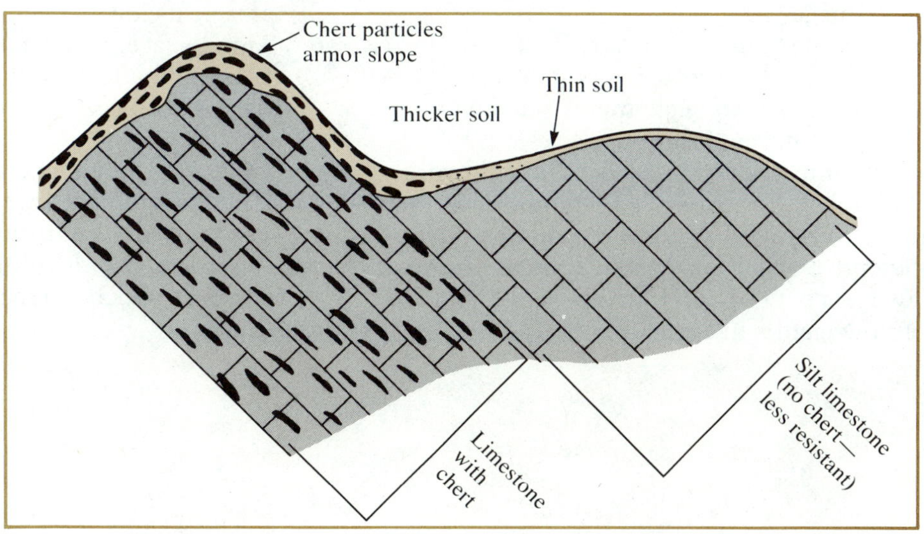

FIGURE 13–47
Two slopes in rough dynamic equilibrium. [After J. T. Hack, *American Journal of Science* (1964), Vol. 258-A, pp. 80–97]

More recently, a third model of landform evolution has been suggested by William Bull. This model is noncyclic and can be used for any time span (present through geologic time). The model utilizes what is known as **allometry**, which is the study of the relative rates of change of two variables, and emphasizes that it is unlikely that a state of dynamic equilibrium will exist because the dependent variables (effects) cannot reach a time-independent state when the independent variables (causes) do not remain constant. That is, independent variables such as climate, base level, and erodability of the rocks, as well as the impact of people, are changing at rates that are sufficiently rapid to preclude attainment of a dynamic equilibrium. The model then recognizes that landscape elements and processes may be assumed to be changing at different rates, and includes the possibility of the steady state but does not require demonstration of the existence of a dynamic equilibrium.

In the preceding paragraphs, two schools of landscape evolution—cyclic and noncyclic—have been examined, and three models evaluated. In the final analysis, none of the models precisely explains the development of the landscape. To a greater or lesser extent, all landforms are individualistic, and the evolutionary path each takes depends upon specific interactions of variables. Nevertheless, it is useful to consider landscape evolution models in light of changes in time and space and whether or not landforms may ever reach a state of dynamic equilibrium.

SUMMARY

Running water is responsible for producing most features of the landscape, and the basic unit of the landscape is the drainage basin.

In terms of the nature and extent of flow in a stream channel, there are three types of streams—perennial, intermittent, and ephemeral—and any one drainage basin may have all three types.

Among the important principles of fluvial hydrology are the following: (1) natural streamflow is neither uniform nor steady; (2) streamflow is almost exclusively turbulent; and (3) during the channel-forming flows, convergence of flow provides scour and divergence of flow results in deposition.

The two main channel patterns are braided and meandering. Of the two, meandering is more common and probably results as the stream tends toward a state of minimum potential energy loss per unit mass of water. The major components of the meandering stream are pools and riffles, and the scour-fill sequence associated with pools and riffles helps explain how rivers meander.

Drainage patterns are important because their development reflects rock type and geologic structure. Thus, analysis of drainage patterns is a useful tool in geologic reconnaissance.

Flooding is the natural process of overbank flow which occurs in most streams at least every year or so. Urbanization of small drainage basins increases total runoff, results in more frequent and larger floods, and reduces low-flow discharge and the aesthetic qualities of a stream.

Principles and concepts important to landscape evolution are:

1. Base level—Theoretical lowest level to which a river can erode.
2. Graded stream—Rough balance between load imposed and work done by a river.
3. Magnitude-frequency relationship—As the magnitude of a process increases, the frequency of the process decreases. Most work in developing and maintaining the landscape is by processes of intermediate magnitude and frequency, but there are notable exceptions.
4. Geomorphic threshold—Many changes in a landscape are not continuous but occur when a critical threshold is exceeded.
5. Geomorphic surfaces—Erosion and deposition surfaces are common landforms useful in reconstructing the history of landscape evolution in an area or region.
6. Time, space, and causality—Distinction between cause and effect in landscape evolution is a function of both time and space.

Models of landscape evolution are cyclic or noncyclic. In general, cyclic evolution of the landscape involves long periods of time (geologic) over a region, and noncyclic landscape evolution involves shorter periods of time over which a dynamic equilibrium may be established.

REVIEW EXERCISES

1 Explain how discharge is used in studying streams, and discuss how it can be measured.
2 Discuss the relative contributions of bed and suspended loads for a stream system.
3 Differentiate between stream competence and stream capacity.
4 Define *graded stream,* and discuss why the concept is important.
5 What is allometry, and how might it be important in studying the landscape?
6 Compare and contrast alluvial fans and deltas.
7 What will the existence of a trellis drainage network tell you about the geology of an area?
8 Define the *magnitude-frequency relationship,* and discuss why it is important in studying landforms.
9 Provide at least two examples of geomorphic thresholds that exist in nature.
10 Why is the concept of base level useful in studying landforms?
11 Why are streams good examples of open systems?
12 Why is rapid uplift an important assumption in the cycle of erosion?
13 Why do streams meander?
14 Why is truncation of rocks of varying resistance so important in establishing the existence of an erosion surface?
15 What can the drainage pattern tell you about the geology of an area?
16 How do meander scrolls form?
17 Why are deltas environmentally important?
18 How and why does urbanization effect flooding?
19 How are convergence and divergence of flow important in stream processes?
20 How do the nature and extent of flow determine the basic types of streams?
21 Discuss the role of time and space in evaluating causes and effects of landscape evolution.

SUGGESTED READINGS

BLOOM, A. L. *The Surface of the Earth.* Englewood Cliffs, N.J.: Prentice-Hall, 1969, 152 pp.

DOLAN, R., and HOWARD, A. "Structural Control of the Rapids and Pools of the Colorado River in the Grand Canyon." *Science* (1969), Vol. 202, No. 4365, pp. 629–31.

DUNNE, T., and LEOPOLD, L. B. *Water in Environmental Planning.* San Francisco: W. H. Freeman and Co., 1978, 818 pp.

LEOPOLD, L. B. *Water: A Primer.* San Francisco: W. H. Freeman and Co., 1974, 172 pp.

______. *Hydrology for Urban Land Planning.* U.S. Geological Survey Circular 559, 1968.

LEOPOLD, L. B.; WOLMAN, M. G.; and MILLER, J. P. *Fluvial Processes in Geomorphology.* San Francisco: W. H. Freeman and Co., 1964, 504 pp.

MORISAWA, M. *Streams: Their Dynamics and Morphology.* New York: McGraw-Hill Book Co., 1968, 175 pp.

RITTER, D. F. *Process Geomorphology.* Dubuque, Iowa: William C. Brown Publishers, 1979, 603 pp.

SCHUMM, S. A. *The Fluvial System.* New York: John Wiley and Sons, 1979, 338 pp.

WRIGHT, L. D., and COLEMAN, J. M. "Variations in Morphology of Major River Deltas as Functions of Ocean Wave and River Discharge Regimes." *The American Association of Petroleum Geologists Bulletin* (1973), Vol. 57, No. 2, pp. 370–98.

- mountain (alpine) glacier
- glacial theory
- continental glacier
- Pleistocene epoch
- firn (firn ice)
- glacial ice
- iceberg
- budget (mass balance)
- arête
- col
- U-shaped valley
- icefall
- hanging valley
- glacial trough
- tarn
- cirque lake

14

- ablation
- avalanche
- glacial surge
- headwall
- cirque
- crevasse
- bergschrund
- ice cap
- piedmont glacier
- valley glacier
- ground moraine
- proglacial deposits
- proglacial lake
- ablation moraine
- ice-walled lake plain
- drumlin
- kettle moraine
- fjord
- erratic
- terminal moraine
- end (recessional) moraine
- lateral moraine
- medial moraine
- protalus rampart
- Little Ice Age
- rock glacier
- outwash, outwash plain
- tillite
- sheet outwash
- kame
- esker
- valley train
- varve

Glaciers and Glaciation

GLACIERS ARE THE CHILDREN OF CLIMATE. THEY ARE BORN, grow and flourish, and waste away and disappear when the climates that nourished them no longer exist to provide support. Over geological time, new generations of glaciers have appeared repeatedly, to undergo the same eternal cycle of birth, growth, and death.

Glaciers, therefore, have existed for a long time. In fact, some type of glacier probably almost always has existed someplace on the earth's surface since the continental masses first congealed as the primeval planet slowly cooled and atmospheric water vapor could condense and fall to the land surface as snow. But because glacial events are transient episodes in earth history, and the record of their former presence is not always clearly preserved in the rock record, the exact time of the first glaciers cannot be determined. Almost irrefutable evidence of ancient glaciers, however, exists as peculiar, widespread assemblages of rocks accompanied by markings or grooves apparently caused by the passage of ice across an underlying surface in rocks that are dated radiometrically as 2.2 billion years old.

Three important facts of earth history should be remembered in considering the antiquity of glaciers: (1) the varying nature and distribution of planetary climates over time; (2) the relative geographic position, size, configuration, and average altitude of the continents during a given interval; and (3) the knowledge that land areas have also shifted and were not always located in relation to the poles as they are today. Let us examine these conditions.

Although planetary climates have not always been uniform, paleoclimatic evidence suggests that during most of geologic time, the typical or "normal" climates of the continents have been dominantly rather mild and uniform. Climates have been neither cold enough to generate the great ice sheets of continental proportions that engulfed the land in ice ages, nor hot and dry enough to perpetuate deserts or barren, semiarid wastelands. Thus, only rarely has climate been sufficiently cold and wet to nourish glaciers, and although such ice ages are not uncommon events, they have been the exception rather than the norm.

One of the most exciting geological developments in recent years has been the nearly universal acceptance of the concept of continental drift. In this process, internal crustal forces, over time, have constantly reshaped the surface of the planet by the splitting, movement, collision, and reassembly of individual continental masses in a manner similar to a constantly changing jigsaw puzzle. The dynamic or driving forces that initiate and determine directions and rates of movement of continents are only slightly relevant to the present discussion of glaciation; nevertheless some effects of such actions should be considered.

Continental collisions generally were accompanied by widespread, episodic mountain building and volcanism along the margins of the moving masses, so high mountain peaks often existed and rose into the colder air of the upper atmosphere, regardless of latitudinal position. Even today, glaciers exist on or near the equator on high Andean peaks in South America and the volcanic summits of Africa (Mt. Kilimanjaro and Mt. Kenya). Thus, at any given time in the past, continents could generate and support large ice sheets if the altitude was high enough and sufficient moisture was available as snow. If, however, continents were low-lying, or if the global thermal gradient diminished only slightly in a poleward direction so that subpolar regions were relatively warm, then large ice sheets either could not form or were unable to maintain themselves because of malnourishment (lack of snow accumulation).

The factors just described deal with broad-scale relations and changes of land and atmosphere in space and time. These relations are very complex and cannot be satisfactorily summarized in this brief discussion. Other difficulties lie in our uncertain knowledge about the interplay between and importance of the relative relationships between planetary geographic and climatic conditions and solar and cosmic events. However, two assumptions seem relatively sound. The first of these is that small glaciers or minor ice caps, called **mountain** or **alpine glaciers** because they originate within and are confined to mountainous terrain, have almost always existed somewhere and will continue to exist into the foreseeable future. Second, with the existing atmospheric temperature gradient from the equator toward the poles, and as long as drift of the northern continents keeps the continents at their present high-latitude positions, glaciation on some scale will continue. Those of us in central North America who have "enjoyed" the rigors of heavy snow and biting cold in recent winters have limited choices: accept the inevitability of geography or radically ameliorate global climates. The first of these is currently the only realistic option, although atmospheric scientists are studying the feasibility of large-scale climatic modification and are considering the secondary effects and problems attendant such changes. It is likely that human-induced climatic modification of limited scope will occur within our lifetime.

Studies of glaciers in the Alps of central Europe began as long ago as the fifteenth century. However, the concept that past glaciation occurred on a scale large enough to cover thousands of square kilometers was proposed much later and developed rather slowly. Even when observers began to study the plains of western Europe, which are covered by a heterogeneous veneer of sand, gravel, silt, clay, and boulders, and which contain locally the broken remains of plants and animals, at first it did not occur to anyone that these sediments were related to or could have resulted from glacial action. At that time, the powerful hold of literal belief in biblical scripture and the dogma of established religion was so pervasive that most observers simply attributed these incoherent deposits to the action of an ancient flood, perhaps that of Noah. Eventually it was recognized that many deposits were of a character or composition or lay at topographic positions totally inconsistent with deposition by floodwaters. Thus, from the decades between 1780 and 1820 issued a spate of scientific publications that presented the case for the former existence of a glacier of great extent, although some people envisioned this superglacier as merely an earlier expansion of alpine glaciers beyond their historic borders.

The **glacial theory**, in nearly modern form, was advanced more or less simultaneously by several individuals between 1838 and 1840; the Swiss scientist Louis Agassiz was the leading proponent of the theory, but others deserve nearly equal credit. Most simply stated, the glacial theory maintained that the great volumes of unconsolidated, heterogeneous debris that now mantle thousands of square kilometers of northern Europe and North America were deposited in the recent past by the action of subpolar ice sheets, or **continental glaciers**, that formed, expanded, radiated outward, and eventually melted. Subsequently, more than a century of observations and field mapping has increasingly lent support to the theory, so that it has gradually passed beyond the stage of theory and has assumed the status of proven fact. Although many things are yet unknown about the total history of this recent ice age, the physical confirmation is so overwhelming that the original theory, with some modifications, is almost universally accepted.

FORMATION AND MASS BALANCE OF GLACIERS

Glaciers, large or small, are made from snow. Although it is easiest to frame the discussion in the context of small glaciers nestled in alpine valleys, where the glaciers are most accessible and therefore most easily studied, the same basic concepts may be extended to include large modern glaciers such as the Greenland and Antarctic ice caps, and the vanished continental glaciers of the past.

SNOW

Glaciers cannot exist at any site or at any altitude unless some volume of snow remains unmelted at the end of each summer season and is carried over to the following winter season. In this manner, a **snow pack** and eventually a **perennial snowfield** may gradually accumulate. Snowfields are abundant in high mountain regions but may never become glaciers if they remain thin or small in area, or if some of the pack is continually removed by sliding or other means. However, if the pack or field is large enough, occupies a favorable topographic position, and reaches a certain critical thickness (the threshold value is unknown but is perhaps about 60–75 m), then transformation into glacial ice can occur.

GLACIAL ICE

Ice is a mass of interlocking crystals, and by geological definition is a mineral because it has a definite chemical composition (H_2O), is inorganic, and has a more or less definite crystal structure. Each ice crystal is comprised of the deformed skeletal remains of many tiny snowflakes welded together. Snowflakes first accumulate in successive layers of light, loosely packed, fluffy aggregate. Each layer differs in density, grain (crystal) size, porosity (open space between grains), and other characteristics such that the original stratification may persist and the individual layers still be distinguished long after the snow is transformed into ice. Entombed with the snow are other kinds of "rubbish": pollen grains, dust, volcanic ash, and impurities such as pollutants derived from human activities. Fortunately, the presence of these external inputs is sometimes useful, as it may give clues to the nature of climates, vegetation, or geologic events that occurred thousands of years ago.

Snow begins a change in physical characteristics almost as soon as it falls. The major changes that occur during transformation to glacial ice are: (1) an increase in density and in grain size; (2) a decrease in porosity; (3) a series of changes in crystal form (usually hexagonal); and, quite obviously, (4) a concurrent decrease in the amount of free air entrained in the original snow. This transformation may begin within a few days after the snow falls, as delicate crystals are melted, sublimed (evaporated directly), crushed, and compacted into small, nearly spherical granules, but completion of the process requires many years. As the alteration proceeds, snow first passes into "old" snow, and then gradually becomes **firn** or **firn ice**. In this latter state, at a density of about 0.55, it seems that densification rates slow, and other mechanisms take over to additionally modify the grains through deformation, edge melting, refreezing, and recrystallization; only in this way is greater densification achieved. Eventually, the process reaches a stage during which all remaining pore spaces are sealed off (though not all entrapped air has been excluded), and the mass becomes impermeable. At a density of 0.82–0.84, the mass becomes **glacial ice**. Although additional increases in density can occur—up to about 0.91 for absolutely pure ice—the ice never becomes as dense as the water from which it originated (Table 14–1). The best example is the **iceberg**, which is a broken-off portion of a glacier in a body of standing water. In the purest state, the berg, with a density of about 0.9, will have 10 percent of its mass floating above the water surface because of the density ratio (Fig. 14–1).

TABLE 14–1
Some specific gravity values for snow and ice.

Material	Specific Gravity
Dry, wind-blown snow	0.01–0.03
New snow (immediately after falling)	0.05–0.065
Settled snow	0.20–0.30
Firn ("old") snow	0.40–0.55
Firn ice	0.55–0.65
Glacial ice	>0.82
(Water)	(1.00)

The **net budget**, or **mass balance**, of a snow or ice mass is the difference between the amount accumulated and the amount lost by **ablation**, that is, the water lost either through sublimation or by direct melting at the surface or margins of the body. In addition, an ice mass may diminish by **calving** (breaking away) of icebergs at the glacier **terminus** (lower margin or extremity).

The budget may be determined on a yearly basis or averaged over a longer time period. Determined for a period of years, mass balance may be either negative or positive; obviously, if the budget is positive for a long time, the snowfield or glacier increases in size and area, whereas a long-term negative budget leads to a diminished size and the eventual disappearance of the mass. For alpine glaciers, the **firn line** (snow line) is an arbitrary contour drawn to mark the average altitude at which annual snow accumulation balances annual loss through ablation; the mass balance of this surface is therefore zero.

FIGURE 14–1
An iceberg in Wolstonholme Fjord, west Greenland. Bergs break off from a glacier where it reaches the ocean or a lake. The iceberg floats because its density is less than that of seawater. Bergs are moved by local prevailing currents and may travel hundreds of kilometers before they completely melt; many of the Greenland icebergs, like that shown, float southward in the North Atlantic Ocean and are a hazard to shipping. The dark spot seen near the center of this iceberg, which is about 20 m high, is a small cavern or *reentrant* in the ice, which may enlarge by melting until the berg splits into two or more smaller icebergs.

MELTING OF SNOW OR ICE

Snow, firn, or glacial ice melts in response to several mechanisms. Loose snow ablates mostly in place, but unstable masses may slide downslope by **avalanches** to melt at lower, warmer altitudes. Ice may melt in place either seasonally or during a period of negative budget, but it can also be moved to a lower altitude by glacial flow and then melted. Solar radiation is probably the principal melt factor, though its effectiveness varies over a wide range for a host of causes. Each glacier or snow-covered ice mass is different, as meteorological factors such as air temperature, turbulence, and wind direction and velocity play important roles in determining melt rates. Sublimation from the ice surface may not be important except at very low temperatures and extremely low atmospheric humidity.

In a moving glacier, there may be some melting because of frictional heating at the base of the ice or because of internal stress or intergranular shearing within the mass. Whether water thus liberated remains free or immediately refreezes is a function of the vertical temperature gradient of the ice. Glaciers are often called "temperate" or "cold" depending on their apparent internal temperature profiles. "Temperate" glaciers are at the pressure-melting point throughout, and free water may be present anywhere within the ice mass; meltwater may issue from below the ice, and a thin layer of free water exists at the base. Basal melting may also be increased by heat liberated by spring waters or emission of geothermal heat from underlying rocks. This type of glacier flows fairly easily and quickly because the ice can readily slide over the wet rock floor; erosive action at the base also probably is enhanced. In "cold" (polar) glaciers, ice remains well below the pressure-melting point, so meltwater is absent within the mass except near the surface in summer, there is no free water at the base, and the ice may be frozen to the bedrock on which it rests. If the ice is put in motion, some frictional melting may occur as it slides over its bed, but movement is slow and erosive power is also diminished. Many glaciers have a small amount of free water at their base, whereas others seem to be in dry contact with the ground beneath. The relations between snow and ice accumulation and the wastage or melting of the mass are shown in simple form by Fig. 14–2.

FLOW

One thing that intrigues most observers of glaciers is the rate and mode of glacial flow. This phenomenon is

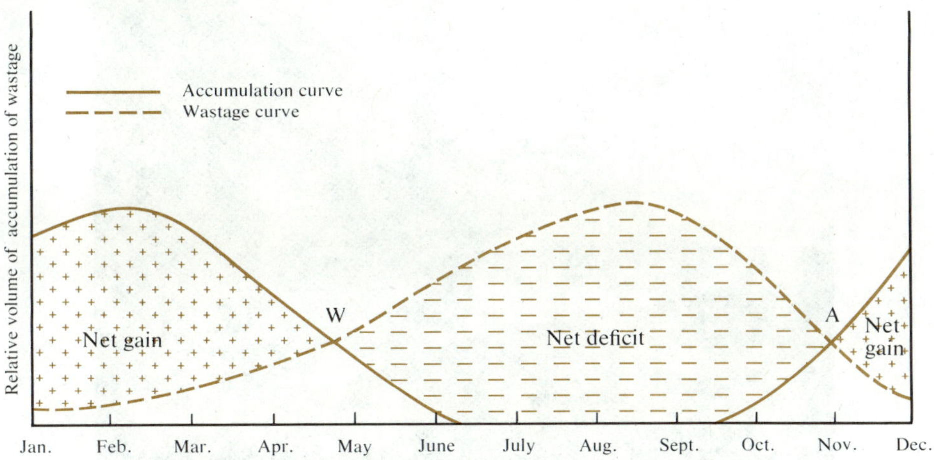

FIGURE 14–2
An idealized *budget year* of a snow or ice mass. It is readily seen that between late October and mid-April, net accumulation exceeds net wastage, and thus this is a time of positive budget; the situation reverses between mid-April and mid-October, when the budget is negative. The intersection of curves at A marks the start of the budget year, whereas intersection W defines the start of the melting season. The areas between the curves are shown as approximately equal, the snow or ice mass is in balance, and there has been no appreciable change in size or volume of the body. Budgets are meaningful only if determined over a period of years rather than for a single year as is shown. In this way, it is possible to determine the net long-term budgetary trend for a glacier.

best seen in alpine glaciers, which occupy relatively long but narrow mountain valleys where the ice is constricted by rock walls on each side (Fig. 14–3). Experiments and measurements have shown that, in general, ice moves fastest down-valley in the headward part and in and along the center of the glacier. Movement is slower at the margins and terminus, though locally this condition varies considerably because of changes in the cross-valley profile or because of differences in melt rates at the ice borders. Internally, the general tendency of ice is to have a downward component of motion in the firn region and an upward component near the glacier terminus.

Measurements made in tunnels or in boreholes drilled to determine vertical components of englacial ice flow have yielded somewhat mixed results. In general, these data show that velocity decreases downward from the glacier surface to bedrock. A few cases recorded have shown that because of local topography of the bedrock surface, movement at the **sole** (base) of the ice actually exceeds that nearer the surface. In "cold" glaciers, basal movement is either lacking or severely retarded by the roughness and dryness of the bedrock-ice contact and the tendency of the ice to adhere to the rock by freezing.

The majority of valley glaciers move at speeds that probably average not more than 0.3–0.6 m/day. Where the down-valley slope of the rock floor steepens, movement of 3–6 m/day has been noted on a seasonal basis. However, there are also shorter-term effects called **surges** that may result from abnormal accumulation of snow within the nourishment (accumulation) area at the head of the glacier, or from sudden changes in rates of internal deformation within the ice. If only ice discharge and ice thickness are considered, and if discharge remains constant while accumulation greatly thickens the ice in its headward portions, then a traveling kinematic pressure wave may be transmitted down-valley at a velocity ranging from two to six times greater than the normal mean velocity of the ice. Propagation of such kinematic waves through ice or the basal meltwater thus may transfer large volumes of ice down-valley very quickly. Surges have produced movement of as much as 8 km during a 4-year transfer episode.

Internal movement of glaciers has two components: deformation and basal sliding. It has been found that ice acts generally like a plastic solid in which no deformation occurs until a certain shear-stress level is reached; under greater stress, ice is strained and slowly

FIGURE 14–3
Athabaska Glacier, in the Canadian Rockies near Jasper, Alberta, is one of the most accessible and scenic alpine glaciers in North America. This ice mass issues as one of a series of *outlet glaciers* from the Columbia Ice Field (white area in background), and travels down-valley, first as a series of stepped *icefalls*, or *cascades*, and then as the more gently sloping Athabaska Glacier proper (center of photo). The ice surface is extensively cut by deep cracks called *crevasses.* The dark areas in the ice and paralleling the flow lines are caused by loose rock debris entrained in the glacier as *medial moraine.* The very-dark-colored embankment occupying the lower part of the photo is an ice-cored *lateral moraine,* covered with rock debris accumulating on the surface as a result of ablation.

deformed, and therefore moves by creep. The principal mechanisms accompanying creep are intergranular slip or shear and ice recrystallization. Basal sliding, probably important only in "temperate" glaciers, results from active slip or shearing between oriented ice crystals in the lower part of the ice. Sliding is enhanced by the presence of basal meltwater.

The average movement rate of unconfined glaciers of the vanished continental ice sheets is unknown, but certain precise dates or good estimates obtained from deposits left by the ice suggest that at least during their melting phase, the glacier front may have retreated of the order of 0.8–1.6 km/year. Perhaps the average rate of advance was about the same.

Glaciers resemble frozen rivers, and movement is by gravity flow in addition to the internal dynamic processes described. Glaciers may branch, or **bifurcate**, into a series of smaller, tributary ice masses just as a stream branches or divides. However, if only gravity flow were involved, glaciers would not flow uphill, spill across interfluves, and otherwise surmount higher elevations as they are seen to do. It is therefore likely that internal shearing promotes a "stacking" of ice slabs so that height disadvantages are overcome and the glacier is able to "ramp up" over obstacles that rise higher than the original ice surface.

GLACIAL SURFACE FEATURES

Where snow accumulates in high upland areas, it tends to move downslope into low spots on the existing terrain by avalanches or is blown there by the wind (Fig. 14–4). These low spots are commonly at the heads of preexisting stream valleys cut into the slopes. The net result is slow accumulation of snow to great depths in these hollows, and eventual transformation of snow to ice by the processes already described. When some critical, but unknown depth and mass are achieved, ice can escape from the basin and begin movement by gravity flow down the former stream valley. However, freeze-thaw processes on adjacent slopes and beneath the ice dislocate already-loosened rock fragments and may increase the rate of breakdown of previously unweathered bedrock. The tendency is thus to steepen the **headwall** above the glacier to a precipitous face and further deepen the basin underneath the ice by plucking or hollowing out. The front of the basin then becomes bounded by a sort of raised ramp or lip. This geometry is not clearly seen as long as ice fills the excavated basin; however, when the ice disappears, a deep, horseshoe-shaped, steep-walled **cirque** remains (Fig. 14–5).

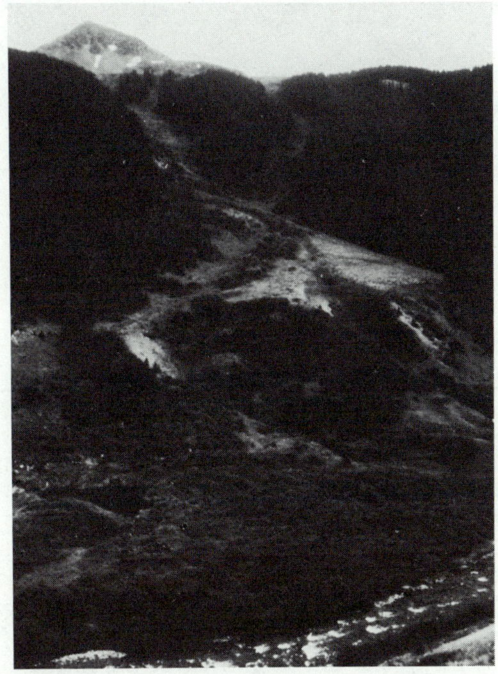

FIGURE 14–4
A snow avalanche track on the north slope of Kendall Mountain near Howardsville, Colorado. The avalanche stripe cuts vertically through the middle of the picture from the high peak in the background. Repeated snow avalanching downslope along this same "chute" drains off the accumulated snow from the higher basins and prevents formation of glacial ice there. Because timber cannot become reestablished along the avalanche chute, it is a grassy area underlain by rough, hummocky ground in which splintered trees and large boulders are embedded. The Animas River (foreground) carries away the snowmelt and much of the soil and rock debris brought down by avalanching.

As ice moves downslope, it may become cut by deep clefts called **crevasses**, which may occur in sets that resemble joint patterns in rocks. Because the uppermost part of the glacier may perform like a rigid, brittle solid rather than deform plastically, strain may produce differential compression or tensional ruptures or fractures that appear as open cracks in the ice surface. Crevasses develop primarily where the gradient of the rock floor abruptly steepens and the ice becomes concave upward. Other changes in flow lines of the ice, such as at valley bends or curves or in radially splayed areas at the glacier terminus, also produce crevasse patterns of various types.

Although crevasses may be tens to hundreds of meters long and as much as 15 m wide, they are rarely more than 30 m deep, for below this depth ice is subject to plastic flow and an opening cannot be maintained long. Especially in winter, loose snow may fill or bridge the cracks, making them difficult to see;

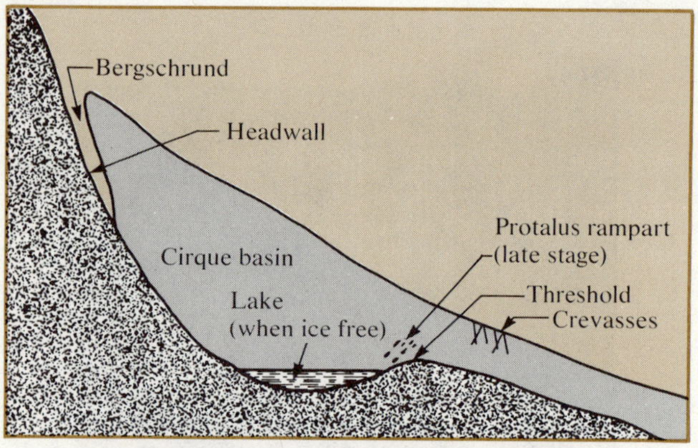

FIGURE 14–5
A simplified sketch showing cirque formation. In the *bergschrund* area, frost action on the headwall frees blocks or fragments of bedrock, which drop or slide to near the base of the ice. The remainder of the bedrock surface beneath the ice at the headwall or in the bottom of the basin is also subject to removal by plucking or freeze-thaw activity. The loose rock is frozen into the ice, and then is picked up and carried by the glacier and used to abrade the bottom and sides of the valley. The ice surface just downslope from the front or lip of the cirque is subject to crevassing because the rate of ice flow increases and the down-valley slope increases here. If the glacier melts and the cirque basin becomes ice-free, a *tarn*, or *cirque lake*, may form in the abandoned hollow. If ice re-forms, but advances only to the lip of the *threshold* of the cirque, it may build a *protalus rampart.*

travel across the glacier is dangerous, and fatalities and loss of equipment in crevasses are not unusual. Crevasses are most commonly transverse to the flow of the ice, but sets can form and trend at any angle, depending on internal stress orientation and the geometry of the valley.

Where the slope of the ice surface steepens near the terminus of the glacier, melting is vigorous, and shearing is especially active and is commonly marked by a honeycomb of deep, irregular crevasses. This part of the glacier is difficult or impossible to traverse. The cracks are also loci for the accumulation of rock debris that melts out of the ice, and are an egress for surface meltwater to discharge downward to zones either within or at the base of the glacier.

The **bergschrund** is a special type of semicircular crevasse system that marks the head of cirque glaciers. It probably forms partly by settlement, compaction, and sliding of snow and ice away from the cirque headwall; in winter it is usually bridged or filled with new snow and is invisible, but in summer it is a dangerous area of rockfalls, rock slides, and icy cascades of meltwater.

CLASSIFICATION OF ALPINE GLACIERS

More terms are used to identify and describe glacial features and landforms than perhaps any other geological process. A classification of glaciers depends primarily on their topographic position and the size of individual or compound ice masses; in a sense, there is a family of types based on shape, form, and size (Fig. 14–6). A **cirque glacier** is a glacier confined entirely to the cirque basin, and thus generally is a beginning or ending phase of glacial activity. In the United States, there are many cirque glaciers that are remnants of formerly more extensive alpine glaciers (Fig. 14–7). A **valley glacier** is any ice mass that extends down-valley from the cirque (Fig. 14–8). If a glacier succeeds in leaving a mountain valley or canyon, it is no longer confined by the valley walls and may splay outward across an adjacent lowland as a delta-shaped **piedmont glacier** (Fig. 14–9). If the valley discharges directly to the ocean, the emergent ice becomes a **tidewater glacier** (Fig. 14–10).

Because alpine glaciers generally occur in clusters, in a particular region there may be dozens or even hundreds occupying adjacent valleys. If glaciation intensifies so that the ice thickens and overtops divides between separate valley glaciers, perhaps even becoming so thick that it covers higher ground or peaks rising above the cirques, the coalescent mass becomes an **ice cap**. Such ice caps are fairly abundant in the Canadian Rockies and in the Coast Range of Alaska.

ALPINE LANDFORMS

EROSIONAL FEATURES

Alpine glaciers are surrounded by high, rocky terrain except where they fan outward as piedmont forms. Since ice is in contact with rock on three sides, freeze-thaw action together with rockfalls and avalanches contributes great quantities of broken fragments of all sizes onto the ice surface and along the glacier margins. This material can be frozen into the ice and used as an abrasive rasp in eroding additional rock from the bedrock floor or adjacent valley walls. The general result, even in areas where ice vanished during the climatic warming of the last few thousand years, is terrain characterized by angular, sharp, rough, or irregular surfaces. In fact, jagged, serrate uplands are a typical imprint of former glaciation. In the valleys, smoothed and polished rock walls may create a **U-shaped valley** cross

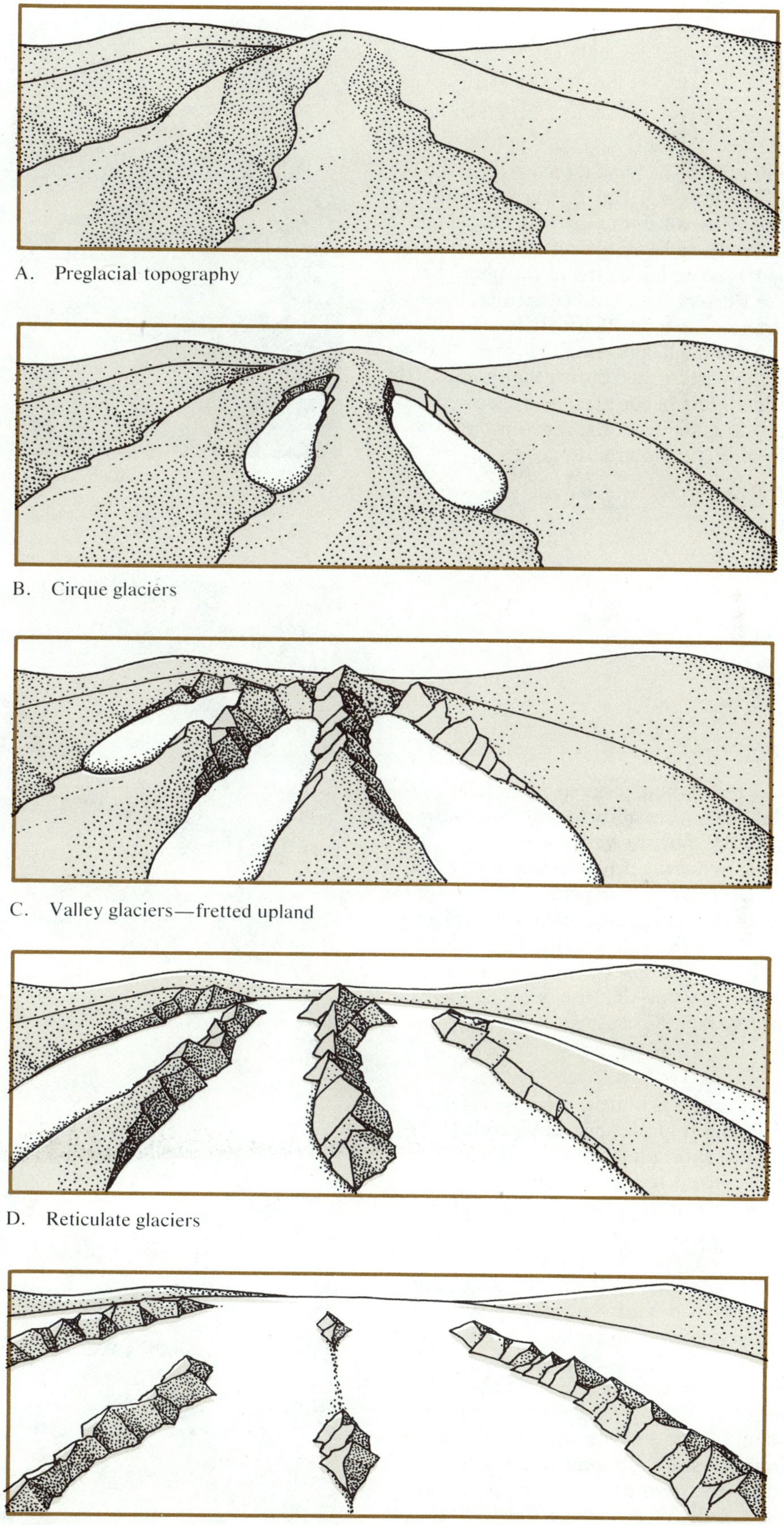

A. Preglacial topography

B. Cirque glaciers

C. Valley glaciers—fretted upland

D. Reticulate glaciers

E. Ice cap

FIGURE 14–6
This plan or bird's-eye view of a glaciated upland shows how as alpine glaciers form, different patterns and sizes of ice masses occur. A. Before glaciation. B. Initial stage during which only cirque glaciers exist on opposite sides of a mountain peak. C. As glaciation intensifies, the individual glaciers flow down-valley and erosional processes operate to steepen and sharpen the valley walls, producing a *fretted upland.* D. Adjacent glaciers have coalesced by flowing across or cutting through divides that formerly separated them. E. Most of the terrain is covered by ice to form an ice cap, and only the highest ground is exposed as isolated masses protruding above the ice surface. As glaciation wanes, the progression will be the reverse of that shown here.

FIGURE 14–7
A group of small cirque glaciers on the eastern face of the Teton Range, Wyoming. The Tetons are a fault-block range tilted westward; the eastern side is high and rises sharply above the adjacent basin of Jackson Hole. The cirque glacier in the left center of the photo has a V-shaped pile or ridge of light-colored granitic blocks at the base of the snow-/icefield. This may be an *end moraine,* but it is so fresh appearing that more likely it is a protalus rampart formed during the recent Little Ice Age. The older Pleistocene glaciers were much larger here, and extended from the canyons onto the plains at the base of the mountains.

FIGURE 14–8
McBride Glacier near where it enters Muir Inlet in the Glacier Bay area of southeastern Alaska. Note how this glacier is confined by the valley walls and meanders (bends) as did the original stream that initially carved the canyon. The points, or *spurs,* of the old meander bends are sharpened or cut off by glacially controlled abrasion of the valley walls. The dark lines are *medial moraines*; the observer stands on a prominent *lateral moraine* that lies between the valley wall and the active ice.

FIGURE 14–9
The Matanuska Glacier is not the most spectacular example of a piedmont glacier but is shown here because it is only a short distance by good road from Anchorage, Alaska, and you may be able to visit it sometime. Note how this relatively "clean" glacier splays out in a delta form as it issues from the mouth of the valley and flows a short distance onto the alluvial flats bordering the Matanuska River. A strong stand of coniferous trees exists to within a few hundred meters of the ice front, and grasses and small flowering plants encroach even closer to the ice. A relatively short but rapid advance (surge) of the glacier would block the river and pond a *proglacial lake* upstream; removal of this ice dam could cause a disastrous flood for many kilometers downstream, destroying important agricultural areas and posing a serious threat to Anchorage itself. Some piedmont glaciers are stagnant, and ablation has produced a cover of pulverized rock debris on which soils form and trees and plants grow in abundance.

FIGURE 14–10
Sawyer Glacier on Tracy Arm (inlet), Alaska, a steep-fronted snout on a tidewater glacier. The ice advances down-valley, and on some tidewater types ice may actually float on waters of an inlet, where it releases icebergs by calving, as is seen here at the left. The cruise boat provides scale. The ice front is deeply crevassed, and lateral moraines are present against the valley walls on both sides of the glacier.

section which is almost conclusive evidence of former glacial action (Fig. 14–11).

As ridges separating two immediately adjacent valleys are attacked on both sides by glaciers, the divides are narrowed, steepened, and sharpened by sapping of the valley walls. In time, a knife-edge ridge called a **serrate divide**, or **arête**, is formed (Fig. 14–12). If adjoining cirques at valley heads similarly enlarge by sapping, they may intersect, breaking through the divide to leave, on melting, a low pass, or **col**, where they meet. Continued sharpening and narrowing of high peaks or ridges above glacier heads may cut off and isolate triangular masses called **horns** (Figs. 14–12 and 14–13).

It already has been mentioned that glaciers branch and have tributaries as do streams. If in two glaciers that join together, a smaller or less active tributary ice mass does not erode its floor as much as the master or **trunk glacier** that it joins, then the tributary enters as an **icefall** or cascades onto the surface of the main ice stream (Fig. 14–14). When glaciers vanish, the valley of the former tributary is graded to a point high on the valley wall above the master valley, and is then a **hanging valley**. Excellent examples are seen in Yosemite

A.

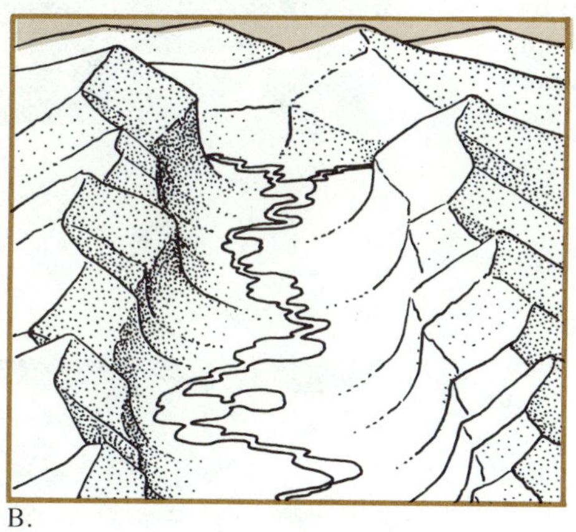

B.

FIGURE 14–11
Changes in the contour and configuration of a stream-cut mountain canyon as a result of erosion by a valley glacier. A. The V-shaped original valley of a meandering stream bordered by gently sloping rock spurs extending from the mountain front. B. The U-shaped *glacial trough* after deglaciation. Note that there are several rock basins now occupied by *tarn lakes,* but that the low point of the valley follows the same meandering course as the stream in part A. Spurs have been trimmed by glacial abrasion, and the mountain slopes are smoothed by mass wasting. The sharp breaks in the slope about two-thirds of the way up the valley walls may approximate the *trimline,* which is the height to which the glacier filled the valley. (After U.S. Geological Survey)

FIGURE 14–12
The *horn* and arêtes of Lone Walker Mountain in Glacier National Park, Montana. The arêtes are the sharp, angular, upward-pointing V-shaped ridges that are snow-free at this time of year. Note how small alpine glaciers (white areas below the ridges) are constantly sapping the divides and keep them sharpened. The weathered products are removed by avalanches and rockfalls. Some of the debris will be incorporated into the ice.

FIGURE 14–13
The Matterhorn, near Zermatt in the Swiss Alps. This is the classic form of all horns—a triangular or pyramid-shaped peak deeply eroded and truncated by glacial plucking, frost wedging, and avalanching at the head of a group of cirques that may nearly intersect. The Matterhorn is a favorite target for alpine climbers, and although it is considered a relatively easy goal by veteran mountaineers, it is also the site of numerous fatal accidents.

FIGURE 14–14
Icefall on the Mowich Glacier on Mt. Rainier, Washington. Icefalls occur wherever there is intersection of two glaciers whose floors are at different elevations, or where there is an abrupt change in the level of the ground surface of a single glacier, as shown here. The icefall is the irregular area of jumbled and crevassed ice in the center of the photo. The white areas on each side are snow-covered, relatively uncrevassed portions of the glacier.

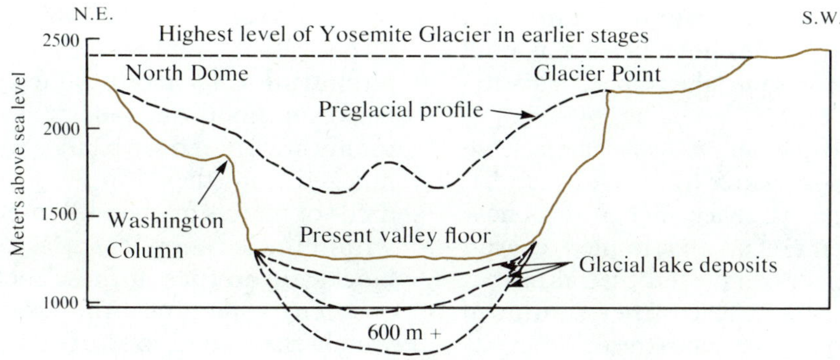

FIGURE 14–15
Glacial modification of Yosemite Valley, California. The valley was deeply eroded by a trunk glacier moving down the V-shaped Yosemite Gorge (position shown by the dotted line of the preglacial profile). Repeated glaciation deepened the gorge about 600 m, but subsequent partial backfilling by proglacial lake deposits created the present flat-floored valley. Tributary glaciers at or near Washington Column and Glacier Point joined the trunk glacier either directly or as icefalls. Because these latter glaciers were small and did not deepen their respective valleys as much as did the trunk glacier, they remain as disjunct hanging valleys several hundred meters above the floor of Yosemite Valley. (From C. Wahrhaftig, California Division of Mines and Geology Bulletin 182, 1962)

National Park, where stream water now being discharged from deglaciated hanging valleys plunges tens of meters to the floor of the old trunk glacier valley (Fig. 14–15).

The valley profile of a vanished glacier geometrically outlines, in its U-shaped cross section, a **glacial trough**. High up on the valley walls, however, there is commonly a break in slope marking the **trimline**, the height to which the former ice filled the valley. Thus, it is possible to get a fair estimate of the thickness and width of a long-vanished glacier. Also, along the down-valley profile the floor may not have been uniformly eroded by ice action. Because rock formations of differing resistance to erosion cut across the valley, softer rocks tended to more severe erosion than did harder ones. The result may be a sort of stair-step profile, or **glacial stairway**, marked by alternating reaches of valley flats and rough, steep, jagged rises (Fig. 14–16). Locally, bowl-shaped depressions in soft rocks provide basins where water collects, called **tarns**.

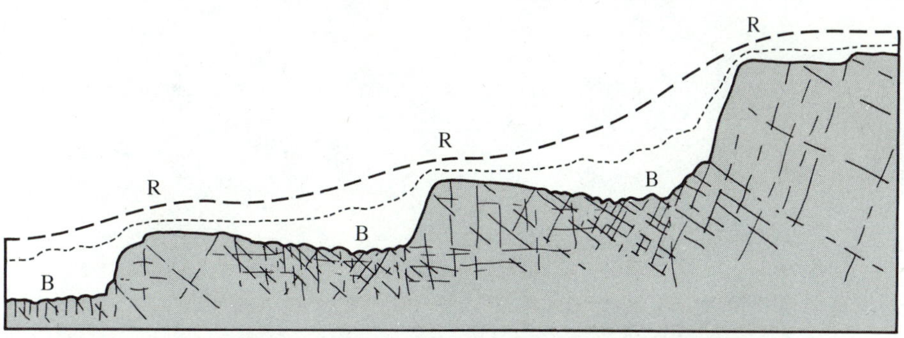

FIGURE 14–16
Longitudinal profile and section sketch of a hypothetical series of *risers* and *treads* along an eroded stairway in a glaciated valley; R is a riser, and B is a basin created at the head of a tread. The patterning shown for the rock strata suggests that the amounts of deepening of basins and erosion of the former valley floor are a function of relative hardness and intensity of fracturing of the rock. The long dashed line indicates the possible profile of the valley prior to glaciation. The short dashed line is an intermediate stage in the erosive process. The basins may collect water as *tarn lakes* (see Fig. 14–11) upon deglaciation of the area. (From F. E. Matthes, U.S. Geological Survey Professional Paper 160, 1930)

Certain valleys may contain a dozen or more tarns tied together as a string of lakes. The cirque basin may hold a particularly large tarn or **cirque lake** trapped between the headwall and the lip of the cirque (Fig. 14–17).

In the geological past, when valley glaciers were much larger, or when more water from the ocean basins was tied up on land as large ice sheets, areas now below sea level were emergent and formed coastal plains. Since valley glaciers could then extend farther outward to reach the ocean, in places they continued their troughs across land now submerged. Later, as glaciers diminished or disappeared and sea level again rose, the troughs were drowned by rising marine waters that flooded far up the old glacial valleys to produce the famous **fjords** seen in coastal Alaska, Norway, Chile, and New Zealand (Fig. 14–18).

DEPOSITIONAL FEATURES

Depositional features of alpine glaciers tend to be less prominent and picturesque than erosional forms. However, floors and sidewalls of glacial troughs commonly are littered with large boulders or blocks of rock plucked from their former positions and carried hundreds of meters or even kilometers down-valley, either within the ice or on its surface. These boulders or rock blocks are **erratics**. If these rocks are of unusual composition, sometimes the exact point or area from which they were extracted can be determined. The farthest advance of a glacier is commonly marked by a high, irregular ridge of heterogeneous rock debris called the **terminal moraine**. In places, these moraines loop around valley outlets, and some may be hundreds of meters high (Fig. 14–19). If after depositing the ter-

FIGURE 14–17
Cirque lakes nestled in the Sierra Nevada near Mt. Whitney, California. In this summertime view, no snow remains on the peaks or in the cirques, and because the area is above timber line, there are no trees and only a sparse cover of grasses or shrubs. The dozen or so dark areas are the cirque lakes. Several horns are also present. The sharp, narrow ridges between cirque basins are arêtes.

FIGURE 14–18
Tassiussaq Fjord on the west coast of Greenland. Outlet glaciers and the mass of the Greenland Ice Cap can be seen in the background. Fjords are long, linear bodies of water that came to occupy submerged glacial troughs as worldwide glacial budgets became negative and sea level rose at the end of the Ice Age. Many fjords such as the one pictured here are tens or even hundreds of kilometers long and more than 300 m deep, yet are less than 1.5 km wide.

FIGURE 14–19
A series of terminal moraines of different ages near Ridgway, Colorado. The Uncompahgre River issues from a deep canyon in the San Juan Mountains (background) but is forced to cut through the margin of a terminal moraine (light-colored ridge just left of center) about 150 m high. The grassy, rolling terrain between the large moraine and the observer is a series of older, low, arcuate terminal moraines less than 15 m high. The low, dark-colored ridge just to the right of the large moraine has outcrops of strata that are interpreted as *tillites* by some geologists, but their age is uncertain.

minal moraine the ice enters a recessive phase punctuated by minor readvances, a string of **recessional** or **end moraine** ridges may extend laterally across the valley upstream from the terminal moraine. The number and size of these vary greatly; in places, the moraines help dam tarn lakes. Loose rock cascading down the valley sidewalls accumulates along the margin of the ice as **lateral moraine**. Where two glaciers merged, debris along the margins of each glacier may become trapped between the two ice masses and moved further down-valley as **medial moraine**. After the glacier melts, these forms are usually quickly destroyed or masked by fluvial or mass-wasting processes and thus seldom remain long as prominent features, whereas terminal and end moraines may persist for centuries as recognizable landforms.

The **protalus rampart** is a special type of moraine. It is a small ridge of fresh, relatively unweathered debris confined to the area around the front or lip of a cirque. The general appearance leaves the impression that a very recent rejuvenation of glacial activity has caused cirque glaciers to expand, refill the cirque basins, and surge outward across the lip and a short distance down-valley. The hypothesis is that these ridges were deposited during the **Little Ice Age**, a short-term worldwide return to colder climates between the fifteenth and eighteenth centuries.

Other vestiges of formerly more rigorous climates are **rock glaciers**, a common landform of the mountains of western North America. These are masses of partly frozen debris containing interstitial ice between the fragments or blocks; some seem to move slowly downslope by gravity flow. The surfaces of rock glaciers are hummocky, rough, and marked by shallow hollows or basins which may represent local collapse because of differential melting of the interstitial ice. Rock glaciers may be of considerable economic importance because they collectively contain great amounts of fresh water in a frozen state; it might be possible to increase artificially the present natural rate of internal melting to provide greater streamflow in arid regions where fresh water is in chronic short supply (Fig. 14–20).

All the features just described are produced more or less directly from erosion or deposition attendant glacial activity. However, at least seasonally, a substantial amount of meltwater is discharged at the ice front and flows away from the glacier. Sediment carried in suspension by this water or as bed load is deposited at some distance downstream from the ice margin as an **outwash fan**; if confined to the course of a channel, it is a **valley train**. Often, the discharge-sediment relations are such that braided outwash channels exist (Fig. 14–21).

FIGURE 14–20
A rock glacier on the east flank of Blanca Peak in the Sangre de Cristo Mountains of southern Colorado. Note the faint series of arcuate ridges crossing the debris pile, and the rough, hummocky surface of the mass. The latter suggests that there may be differential melting of ground ice within the mass, and the ridges indicate that some sort of push or shove activity occurs periodically. Although no ice is ever seen at the surface, in the summer months water can be clearly heard rushing beneath the large, angular blocks that compose the rock glacier, suggesting that melting occurs at that time of year. Flow in the creek at the toe of the pile seems to be supported in part by this melting of interstitial ice.

FIGURE 14–21
A braided outwash channel in the valley of Nisqually Glacier on the south side of Mt. Rainier, Washington. As the ice melts, sediment and water are discharged seasonally down the valley. As water flow diminishes, the coarser-grained bed load is left as bar or channel-fill deposits. This process aggrades (builds up) the channel bottom and leaves a valley flat characterized by a network of numerous branching and joining minor channels. At times in the year when glacier melting ceases or is minor, the channel is dry, as is seen here.

CONTINENTAL GLACIERS

At the beginning of the chapter, evidence documenting the existence of long-ago glaciers was alluded to, and brief mention made of the glacial theory proposed in the mid-nineteenth century. Let us now examine in more detail how the theory has been tested, and show how it is supported by the weight of geological evidence.

Ancient glacial deposits are so deeply buried under younger rocks that they usually are seen only in deep mines. At some places, however, they have been exposed by erosion. In southern Australia, for example, rock strata called **tillites** are hundreds of meters thick and are well exposed in outcrop, so their internal structures and compositions can be determined. These originally were composed of loose sediment, which was later compacted into solid rock, but otherwise they retain all the characteristics of modern glacial deposits. Thus, tillite is comprised of heterogeneous mixtures of rock of all lithologies, shapes, and sizes embedded in a fine-grained matrix. Locally, these rocks are sandwiched between or pass laterally into stratified, cross-bedded sands and gravels that seem to be former outwash deposits. Many of the pebbles are polished or scratched as if by abrasion, as is seen in recent glacial sediments. Ancient landforms, such as moraines, cannot be observed, but the evidence is strong in support of the belief that tillites were indeed laid down by ice and not by some other geological agent.

The glacial theory began with the first systematic examinations of the action, movement, and deposits of glaciers in the Alps, where ice still existed or where there was historic documentation of the extent of former glaciers. Assuredly, it was a bold step when the theory was extended to explain similar deposits and present landforms all across northern Europe as deposits of a former glacier of continental proportions, so it is easy to see why the idea met strong resistance. Eventually, however, field evidence caused this opposition to fade; no rational alternate explanation could overcome the fact that these deposits and landforms were, except in scale, almost precise analogs of what could be seen in alpine glaciers. The final proof and greatest flowering of the theory came later, in the Great Lakes region of North America, where landscape features and deposits of glacial origin were even more extensive and better developed.

When it became evident that great areas of the northern continents were covered by sediments deposited directly by continental ice or by meltwaters flowing from it, geologists realized that this event had occurred in the very recent past. At first, there was no way to determine the time of beginning or end of this great episode. It was soon discovered, by studying the various strata, that a series of distinct and different **tills** (unconsolidated form of tillite) were locally separated by unconformities or the remnants of buried soils—evidence of former landscapes on which plants grew, animals roamed, and subaerial weathering took place. In fact, fossil proof existed locally as bits of carbonized twigs, branches and leaves of plants, pollen grains, shells of small invertebrates, and occasional vertebrate bones, all of which could be identified and compared with living species. It became obvious from this evidence that ice age continental glaciation occurred as more than a single event, and actually consisted of several events separated by at least equally long ice-free intervals (Table 14–2). From this discovery, the concept of **multiple glaciation** evolved. The land record also led to the evolution of a classification scheme that recognized four major **glacial stages** which in turn were separated by longer, warmer **interglacial stages**, the last of which we are presumably in at present. More recently, one investigator, G. J. Kukla, has claimed that as many as 17 glacial stages and an equal number of interglacials occurred in Europe (and thus presumably also in North America) in the last 1.7 million years, rather than the classical four glacials and three interglacials. His interpretations are based on the isotopic composition ratio of $^{18}O/^{16}O$ in the calcareous shells of foraminifera preserved as fossils in drill cores obtained from deep-sea sediments. As this ratio is thought to be controlled to a significant degree by global volumes of landlocked ice, the deep-sea record should be closely related to the record of Pleistocene events on land. Presumably, a high $^{18}O/^{16}O$ ratio indicates cold seawater temperatures or large volumes of land ice, whereas a low ratio indicates warm seas or reduced volumes of land ice. However, because the land record also clearly shows that smaller-scale climatic fluctuations occurred within the larger framework of the major glacial and interglacial stages, it is likely that, in part, it is these lesser fluctuations that appear in the marine sedimentary record. In time, it may be possible to reconcile the land and marine records, and this correlation will provide a better understanding of the actual scale, frequency, and magnitude of past glacial events. Undoubtedly, there will be much revision of the classical four-stage Pleistocene chronology in the next few years, accompanied by exciting research using the new methods provided by advances in paleontology and geochemistry and a reevaluation of field exposures of the glacial deposits.

The time originally assigned to the duration of the Pleistocene epoch, or Ice Age, was arbitrarily given as 1 million years because no means existed for actual dating. Subsequently, it was found that radiometric

TABLE 14–2
Time table of Pleistocene glaciation.

Years B.P.	Glacial Stage	Interglacial Stage
		Holocene (Recent)
10,000		
	Wisconsinan (4th Glacial)	
75,000		
		Sangamonian (3rd Interglacial)
200,000		
	Illinoian (3rd Glacial)	
250,000		
		Yarmouthian (2nd Interglacial)
600,000		
	Kansan (2nd Glacial)	
750,000		
		Aftonian (1st Interglacial)
900,000		
	Nebraskan (1st Glacial)	
?		

dating of the $^{14}C/^{12}C$ isotope ratio in buried, carbonized wood could yield actual dates of some accuracy for about the last 40,000 years. There have been some refinements of this radiocarbon technique that extend its application to somewhat older deposits, but no satisfactory method exists to yield precise dates in most areas for most of Pleistocene time. Clues from the Sierra Nevada, however, where radiometrically datable volcanic rocks both underlie and overlie glacial deposits, strongly suggest that the alpine glaciers in that area extended several kilometers beyond the mountain base at least 700,000 years ago, and perhaps as much as 2.5 million years ago (Fig. 14–22). Recent drilling in western Iowa has revealed similar interbedding of datable volcanic ash and glacial deposits and yields similar dates. Thus, it seems evident that the Ice Age began fully twice as long ago as was originally estimated. Based on stratigraphic, dating, and fossil evidence, the ice disappeared and the Ice Age ended in North America and Europe about 10,000 years ago, though it continues in Greenland, the Arctic islands, and Antarctica.

Climatic and oceanic changes that accompanied alternate advance and retreat of the continental glaciers spread worldwide and were of great effect. Since glaciers indirectly draw their water from the sea, through transfer of moisture to the atmosphere and reprecipitation as snow, withdrawal of vast volumes of seawater led to a global decline in sea level of perhaps

FIGURE 14–22
The classic exposure of Kansan-age glacial till overlain by volcanic air-fall and ash-flow tuffs on Sherwin Grade near Bishop, California. The till at the base, where the observer is standing, is surmounted by about 5 m of the air-fall deposits. The dark line running across the face of the exposure is the base of whitish-colored ash-flow volcanics. Glacial outwash gravels show at the very top as irregular sediments. The vertical lines are clastic dikes. The radiometric age of the basal volcanics just above the till is 710,000 years. Thus, the Sherwin till is older, and has been arbitrarily assigned to the Kansan glacial stage. The outwash gravels are thought to be of Wisconsinan age and about 60,000–75,000 years old.

as much as 100–200 m during each glacial stage. During each interglacial stage, sea level rose to near its former level. Invasion of the northern continents by ice displaced plants and animals either equatorward, onto the newly emergent continental shelves, or compressed life zones into smaller areas than before. The net result was a constant ebb and flow of living forms across the continents in search of suitable habitats. Cooling of the atmosphere and oceans was felt on a lesser scale even in tropical waters thousands of kilometers from the continental ice, as shown by changes in the rate of coral growth around atolls and reefs of the far-distant South Pacific Ocean.

CONTINENTAL GLACIER LANDFORMS

Many landforms associated with alpine glaciers also were formed by continental ice. However, the latter commonly differ in size and geometry and often are more subdued, because they formed longer ago and have been reduced and smoothed by thousands of years of normal subaerial weathering and erosion.

EROSIONAL FEATURES

Erosional features formed by continental glaciers are normally less well displayed than in their alpine counterparts. However, where bedrock is exposed in continentally glaciated areas, polished, grooved, or striated surfaces are found that are clear evidence of glacial passage. These markings occur when a sand grain, pebble, or rock fragment frozen in the ice mass is dragged across the bedrock surface at the base of the glacier; if the moving particles are harder than the bedrock, the scratches or grooves are formed. Many large erratics may litter the landscape, often possess similar markings, and are rock types exotic or foreign to the place where they now rest. Where ice flow was directed along or parallel to zones of soft, easily erodable bedrock, the glacier scoured out deep, linear troughs or valleys. Areas such as the Finger Lakes of New York State and features as large as the Great Lakes originated in this way.

DEPOSITIONAL FEATURES

Moraines of various types are conspicuous landforms in many continentally glaciated areas. Unlike moraines of mountain valleys, which may be hundreds of meters high but of limited length, some morainal ridges in the central United States and Europe extend unbroken for tens or hundreds of kilometers, though they usually rise less than 15 m and rarely more than 45 m above the adjacent plains. Each morainal surface tends to reflect the way in which it was formed. Some moraines are marked internally by abundant small, parallel ridges and troughs that give a washboard effect to the microtopography. This suggests that these moraines were built within a narrow frontal zone by a series of minor individual forward surges and retreats of active ice over a period of some years. The most spectacular variant is the **kettle moraine** characterized by deep pits, depressions, or "kettles," many still holding small **kettle lakes**. The kettles are the former positions of ice blocks that have been broken off or dislodged from the glacier and then covered with till or outwash. Slow melting of the buried, insulated blocks leaves depressions that then fill with groundwater (Fig. 14–23).

In total area, the most extensive glaciated terrain is the **ground moraine** (also called **till plain**). This is the essentially flat to gently undulating countryside that has only minor relief and lies between morainal ridges. Till plains are not very scenic but are some of the richest farmlands of the world; the fertile soils of central North America, western Europe, and Russia derive from postglacial weathering of nutrient-rich till plain sediments.

A more spectacular landform is the **drumlin** (Fig. 14–24). Drumlins are linear, streamlined ridges on ground moraine that occur as **drumlin fields**, clusters of sometimes hundreds of individual drumlins. The dynamics of their origin is not completely understood, but they seem to occur only where vigorously advancing ice moved across previously deposited ground moraine. Individual drumlins tend to have a definite width-length ratio and are rarely more than 0.8 km long and 30 m high. Drainage lines, commonly in a trellis pattern, align as narrow channels between the drumlins, imparting a roller-coaster effect to the terrain. The best examples of drumlin fields in this country are found in New York, Michigan, and Wisconsin; other good examples occur in southern Ontario, Canada, and southern Poland.

PROGLACIAL AND ICE-CONTACT DEPOSITS

Proglacial deposits are laid down by running or standing water directly adjacent to or near the ice, but are not necessarily deposited in direct contact with the glacier. Meltwaters flowing away from a lengthy reach of continental ice locally could deposit **sheet outwash**, which extends outward and gently slopes away from the former ice front. These **outwash plains** may cover many square kilometers, and are often dotted with

FIGURE 14–23
Kettle moraine in eastern North Dakota. There is a major morainal ridge on the skyline, and the photo is taken from the top of a similar moraine looking across an intervening area marked by low, irregular ridges and some rather large kettle lakes in enclosed basins. The total relief of the area is about 30 m. A moderately sized glacial erratic of white-colored quartzite rests on the surface in the foreground. This kame-kettle area is mostly the result of deposition by ablation of "dead" or stagnant ice. It is now a region of grassland, and there are few trees except along watercourses because the annual rainfall is low, winters are intensely cold, and strong, drying winds are common throughout the year.

FIGURE 14–24
At first glance, the innocent-looking linear hill in the distance looks like any other hill. Note, however, that it is smooth and elongated and slopes more steeply at the right (south) end than it does at the other end. It is a drumlin, cored with glacial till, that is about 1 km long and rises 25 m or so above the plain (foreground). This type of drumlin, steep at one end and more gentle at the other end, is characteristic of this particular class of elongate, cigar-shaped drumlins. The asymmetry in profile also confirms that the ice that formed the drumlin moved from the left (north) to the right (south). One end of a smaller drumlin is seen in the left middle of the photo just beyond the cornfield.

ice-block depressions called **pits**, whose origin is similar to that for the kettles on moraines. The pits may be filled by lakes; these areas make excellent recreational sites, as in the central United States. Elsewhere, ice locally ponded water bodies between the glacier terminus and higher terrain as short-lived **proglacial lakes**. When the lakes found outlets and could drain, or if the ice retreated and no longer supplied meltwater, the lakes vanished. However, layered sediments accumulated in the bottom while the lakes existed, so although the lakes are long extinct, the former presence is known from a flat land surface underlain by seasonally deposited, laminated or layered lake silts or clays called **varves**.

Like ice in alpine glaciers, continental ice tends to break along stress joints or cracks oriented parallel or perpendicular to the ice front, thus forming a series of intersecting crevasses. Such features develop best in the frontal zones of the continental glacier, where meltwater running across the ice surface can cascade downward and deposit entrained sediment in contact with the adjacent ice walls. If a conical-shaped pile of debris thus is deposited as the ice melts laterally or downward by ablation, a small hill, or **kame**, is left. Sometimes kames link together like beads on a string, with individual high mounds connected by a somewhat lower ridge marking the line of the old crevasse. These are **kame ridges**, or **kame eskers**. However, true **eskers** form solely by deposition in subterranean meltwater channels in or ice caves near the glacier base, and some of the largest and longest eskers of North America and Europe were probably formed in this way (Fig. 14–25). Individual eskers may extend more than 30 km.

ICE WASTAGE

Thus far, waxing and waning of continental glaciers has been discussed as if caused only by active advance or retreat of the ice front. However, evidence from landforms and glacial deposits shows that very often the ice disappeared mostly by downward melting or ablation during times when the local mass budget was negative. In this setting, the ice became essentially stagnant ("dead") and melted differentially, creating a bizarre set of landforms. **Ablation** or **dead-ice moraine** consists of a chopped-up-appearing area of small basins, irregular ridges, and knobs which have no apparent pattern. This morainal type also occurs in modern glaciers where a large ice mass splits into smaller segments, with meltwater, ablation, and ice-contact deposits of each segment interfering and interfingering with similar deposits of the other segments. A fascinating special case of inverted topography formed in a stagnant-ice area is the **ice-walled lake plains** of North Dakota (Fig. 14–26), where lake deposits that were formerly at a low elevation are now topographically higher than the ground moraine or ablation moraine that surrounds them.

REGIONAL GLACIAL LANDFORMS

Glaciers of the recent past did more than cosmetically remodel former landscapes or leave their imprint as scattered erratics, moraines, or piles of ice-contact debris. Most alpine glaciers, unless they spread some distance beyond the mountains, do little except re-

FIGURE 14–25
This oblique air view shows a complex assemblage of ice-contact deposits northwest of Christmas Bay, Kluane Lake, Yukon Territory of Canada. The bare ridges are best classified as kame eskers or esker chains, although a few individual kame forms can be seen. This esker complex was probably deposited in subice tunnels or channels that tended to bifurcate and rejoin. The ridges are barren because the coarse-grained glacial sediments drain rapidly and soil development is poor. The intervening depressions and low, marshy hollows contain organic soils and support a dense stand of conifers, in which individual trees are between 6 and 15 m tall.

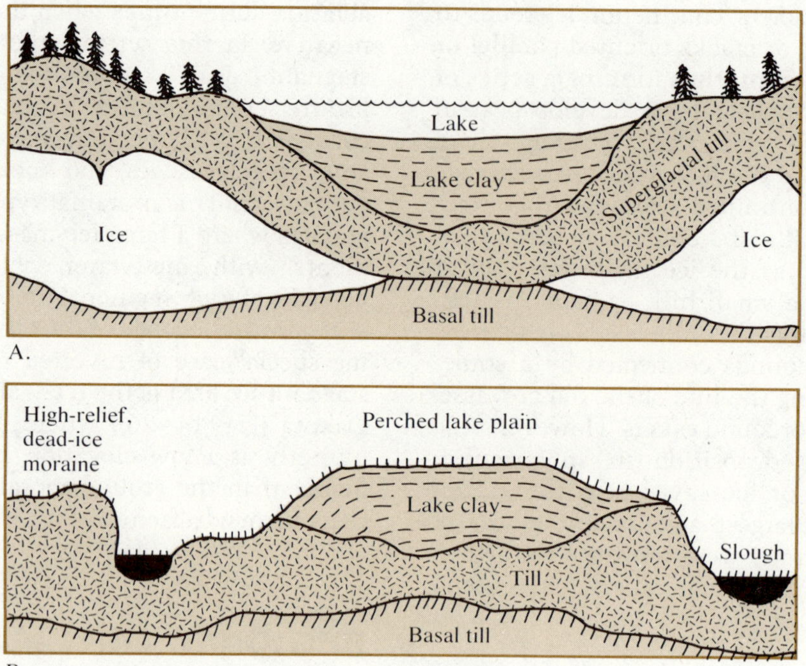

FIGURE 14–26
Stages in the development of ice-walled lake plains, common in ablation moraine of the northern Great Plains region. A. Here, stagnant or "dead" ice rests on an older *basal till.* As the ice melts, the ice surface becomes covered with an irregular mantle of till; this is *ablation till.* Local depressions or pockets accumulate water as small lakes in which additional clays or silts are deposited. B. At this time, the dead ice has totally melted, the ablation till has collapsed, and the lake has drained. The topography has now inverted, so the former lake bed is higher than (that is, it is *perched* above) the ablation moraine. Some lake beds are surrounded by *sloughs* containing moatlike water bodies or wet swamps encircling the old lake basin. The flat surfaces of ice-walled lake plains are better drained and more fertile than the surrounding areas of dead-ice moraine, and thus are local enclaves of agriculture in otherwise infertile sectors. (From L. Clayton and J. A. Cherry, North Dakota Geological Survey Miscellaneous Series 30, 1967)

sculpture the valleys to which they are confined. Continental glaciers were capable of greater and more dramatic rearrangement of the old topography. They destroyed former rivers and created new ones; formed large, fresh-water proglacial lakes in rock basins where no lakes had existed; and at times caused floods of greater intensity and magnitude than any similar event directly recorded. A few outstanding examples of these phenomena are briefly presented in the following paragraphs.

RIVERS

If one looks at the map of the maximum extent of continental glaciers in North America (Fig. 14–27), it is immediately apparent that some genetic relationship exists between the ice border and major rivers such as the Missouri and Ohio. It is now known that neither of these rivers existed in its present form prior to glaciation, and that their courses and orientations have changed since the time they were ice-border meltwater drainage channels. The Mississippi River did exist preglacially, though its course has changed considerably because of repeated diversion by different lobes of the continental ice (Fig. 14–28).

LAKES

The Great Lakes are a famous example of scoured rock basin glacial lakes. All of these lakes, except perhaps Lake Superior, occupy the general positions of preglacial river valleys cut in soft rocks. During the Ice Age, these rocks were deeply eroded by ice scour, and the former valleys were converted into rock basins in which proglacial lakes of various sizes and depths accumulated as the ice front waxed and waned. At times, these ancestral lakes were larger than their present forms, at other times much shallower and more restricted. The size, depth, and form of the present lakes are controlled by the elevations of the outlets, but over time the lakes will diminish in size and eventually be drained, as is the ultimate fate of all natural lakes unless artificially controlled.

FLOODS

Great floods occurred if proglacial lakes, blocked either by ice dams or high terrain, suddenly found an exit through newly opened discharge channels. Evidence of catastrophic floods is present in the Wabash and Illinois river valleys of the midwestern United States, but the so-called Missoula Flood event of the Pacific Northwest which occurred about 22,000 years ago may have been one of the greatest fluvial outbursts that has taken place on earth. The relict evidence consists of tremendous scour channels, dry watercourses, abandoned waterfalls, and megascale channel bed forms across the 36,000-km^2 area known as the **Channeled Scablands** (Fig. 14–30).

FLUCTUATING SEA AND LAND LEVELS

It has been noted how the formation of continental ice sheets caused the withdrawal of oceanic waters such that continental shelves became broad, subaerial platforms inhabited by plants and animals. Fresh-water bogs containing remains of land plants and vertebrates such as elephants are found well below present sea level off the Atlantic Coast. Glacial landforms such as moraines, drumlins, and outwash plains are now submerged beneath the ocean offshore from New England. The higher parts of the old moraines rise above sea level and form islands. In addition, lakes generally rose

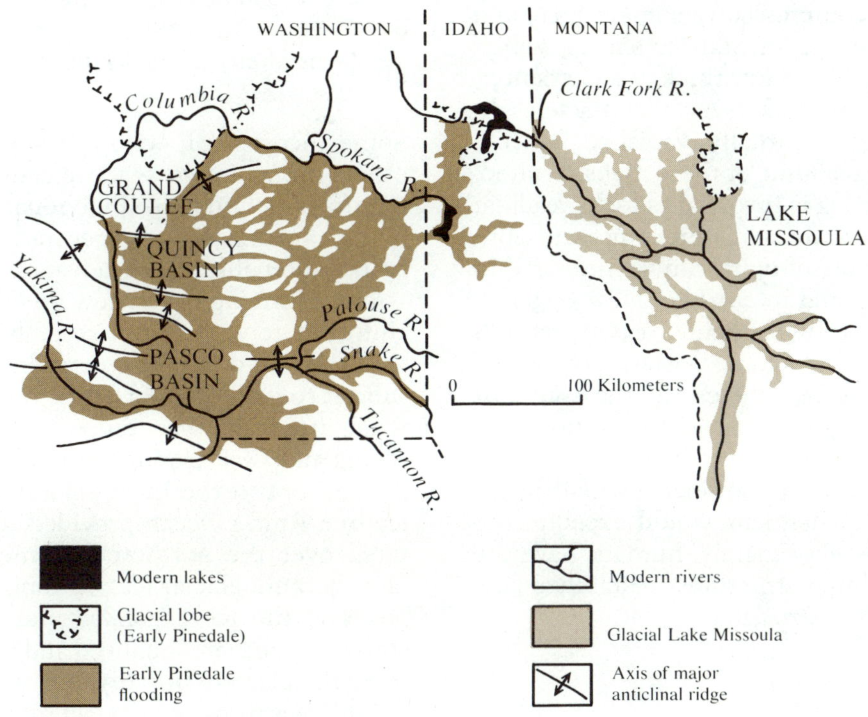

FIGURE 14–30
The setting of the Missoula Flood event in the northwestern United States. Glacial Lake Missoula (light color) was trapped behind an ice dam on Clark Fork River at the position of Lake Pend Orielle (black area). Lake Missoula may have contained 2100 km^3 of water with a hydrostatic head of 600 m. Although the exact nature of the ice-dam failure is unknown, it is speculated that the tremendous hydrostatic head lifted or "floated" the ice barrier, allowing water to undercut it and escape. The almost instantaneous discharge from the lake is estimated to have been, at maximum, nearly 27 million m^3/s. Because not all of this tremendous outflow could immediately escape through outlets farther downstream, the 36,000-km^2 area of the Channeled Scablands (dark color) was subjected to ponding, massive flood scour, and deposition of megascale channel bed forms in only a few days' to a few weeks' time, about 22,000 years ago. (After V. R. Baker, *Science,* Vol. 202, p. 1255, Dec. 22, 1978. Copyright 1978 by the American Association for the Advancement of Science)

in response to increased precipitation and decreased evaporation during glacial stages. In Africa and parts of Asia, glacial times provided a congenial habitat for a great variety of creatures, including early humans. It is argued by some that without the Ice Age, whose climates acted as a challenge and stimulus, modern humans would have evolved more slowly. In this event, our civilization might not have existed until sometime in the geological future. Elsewhere, as the ice sheets waxed and waned, animals and plants migrated in response to changing climate and in search of suitable habitats. For example, during the latest glacial maximum, the woolly mammoth lived as far south as Central America. Equatorward displacement of mean annual temperature isotherms even resulted in some cooling of oceanic waters in tropical islands of the Pacific Ocean, as evidenced by changes there in fossil marine organisms.

All rocks can be compressed to some degree, and the weight of the vast continental ice sheets, which may have been as much as 5 km thick at their center, compressed the crustal rocks. As the glaciers diminished in size and eventually vanished, a slow reelevation, or **elastic rebound**, of these rocks resulted. Because a certain time lag is involved as rocks readjust to unloading, this **isostatic adjustment** process continues today, and the rate of uplift can be measured in the Great Lakes region and in northern New England. The Great Lakes are actually being "tilted uphill"; their northern ends are undergoing isostatic compensation and are slowly rising, whereas at the southern ends compensation is completed and elevations are stable. This scenario also explains why drainage of the lakes is southward; in time, because of downcutting of the outlets, isostatic adjustment would expedite the extinction of the lakes. Fortunately, humans have the ability to design and build structures to stabilize lake levels and prevent this destruction.

CAUSES OF CONTINENTAL GLACIATION

Although inquiries about the causes of continental glaciation began about the time the glacial theory was proposed by Agassiz and others, there still are no satisfactory answers. All that is certain is that large-scale glaciation is the result of climatic change; the causes of this change are an enigma. In fact, geologists do not even know whether the remaining ice masses of Greenland and Antarctica are decaying for the last time or whether the present is merely another interglacial interlude.

Causes of climatic change may be broadly grouped into six categories, ranging from the cosmic scale to the purely terrestrial:

1. Variations in rate or amount of solar emissions (sunspot activity)
2. Clouds of interplanetary or cosmic dust
3. Geometric variation in the earth's orbital motion
4. Variation in emission-absorption rates of the earth's atmosphere over time
5. Lateral and vertical tectonic movements of the crust, and thus varying altitudes
6. Changes in oceanic-atmospheric circulation patterns of storm tracks and ocean currents

Some theories fall within a single category, whereas others draw on two or more causes. The scientific literature is rich in these individual theories. Some are of historical interest only, because they are now known to be untenable. With advancing knowledge, others will be eliminated, and new theories will be developed. Unfortunately, it is not feasible to discuss in detail here the half-dozen or so explanations still viable and subject to further proofs. One theory that has received wide attention was proposed in 1956 by Maurice Ewing and W. J. Donn, and later modified. The principal concept in the Ewing-Donn hypothesis is that an ice-free Arctic Ocean provided moisture that precipitated over the northern continents as snow, which later became glacial ice. As long as budgets remained positive, the ice expanded and produced the glacial stages. As climates changed and the Arctic Ocean froze over, the glaciers literally starved and shrank, causing oceanic warming. As deglaciation occurred, the northern continents entered an interglacial period. Although this theory provides a simple and generally satisfactory rationale to explain actual events, there has been considerable skepticism about the reality of an ice-free Arctic Ocean in the last million years or so, and whether sufficient water was available there to account for the mass of the continental ice sheets. The lack of a totally satisfactory cause-and-effect explanation of glaciation is another good example of how much exciting work remains in the geological sciences.

HUMANS AND GLACIERS

An ice sheet or even a small glacier is a cold, lonely, unfriendly place of howling winds, blowing snow, and little support of life. Nevertheless, past and present glaciers have exerted positive influences on our existence. As was noted, humans may be a product of the Ice Age, during which time they evolved from apelike brutes, through cave creatures painting pictures of hunted animals on subterranean walls, to the inhabitants of the auto-filled, skyscrapter-studded urban canyons of today.

Although the continental ice sheets of Greenland and Antarctica are now the only major vestiges of the most recent ice age, these and existing alpine glaciers help regulate planetary wind fields, exert controls on temperature patterns and ranges, and are an enormous potential source of fresh water which technology may eventually liberate for an increasingly water-deficient world. The idea of towing icebergs from Antarctica to the desert Middle East is not as absurd as it seemed a few years ago. In addition, vast reserves of oil, coal, and metallic minerals may lie either beneath glaciers or under their inhospitable peripheral regions, and these may eventually be extracted, though at great cost.

On the northern continents, glacial sediments provide natural underground reservoirs for tremendous quantities of fresh groundwater. These sediments also are the source of cheaply extracted construction materials such as sand and gravel, and nonmetallic ores such as industrial clays and peat. Perhaps most important is the fact that the fertile rolling plains and gentle hills of central North America and northern Europe are the richest farmlands on the planet. This fertility is the unique result of soil-forming processes acting over about the last 10,000 years on glacial sediments enriched with nutrients from a number of sources during mixing by the ice. Without these glacially derived natural graineries, our planet could probably not, even now, supply the food required by the billions of people that populate its surface.

SUMMARY

Glaciers have been present on the continents since early in earth history, though their former distribution in time and space is not easily defined because of gaps in the geological record. Glaciation is a transient event requiring an optimal combination of local- and planetary-scale climates and terrain in the right geographic position and of a size, shape, and altitude to retain snow for transformation into ice. The geological record shows that whereas small, local glaciers have been commonplace, global climates generally have been warm and congenial; during other periods, each perhaps a few million years long, some portion of the earth's surface has been engulfed by great ice sheets of continental proportions that covered millions of square kilometers. The most recent of these ice ages, the Pleistocene epoch, began at least 2.5 million years ago and may not yet have ended. Recognition of the physical and biological evidence for this recent event by Agassiz and others more than 100 years ago resulted in formulation of the glacial theory, a hypothesis that has become operational doctrine in geology.

Glacial ice originates from metamorphism of snow. Changes in density, porosity, crystal form, and grain size of snow crystals accompany the metamorphic transform. However, ice or snow cannot persist long unless the net budget is strongly positive; a negative budget allows an excess of melting over accumulation, and eventual disappearance of the body. Budgets that are neutral or only slightly positive produce small permanent snow- or icefields, but these do not enlarge or expand outward.

Complex classifications of glacier types exist, but they readily combine into two general classes: mountain (alpine) glaciers and continental glaciers. The first class consists of individual ice masses confined to high terrain or mountain valleys, though under some conditions these combine or coalesce into ice caps. Mountain glaciers also may extend for short distances beyond valley mouths and join on lower ground to form piedmont glaciers. With appropriate climatic controls, continental glaciers may originate not only on high ground but also on low ground or plateaus of intermediate altitude. Under strongly positive net budgets, these masses expand fanwise to cover large expanses of the continental surface.

Glaciers move or flow by gravity sliding and internal deformation, called creep. The way a given glacier moves depends on the temperature profile within the ice, how much free water exists to promote basal slippage and intergranular slip or shear between ice crystals, whether the ice is frozen to the substrate, and the slope of the underlying ground surface. Glaciers may also bifurcate into tributary ice masses. At various places in a moving glacier, ice is strained such that differential compressional or tensional ruptures occur as open, cracklike crevasses extending downward from the glacier surface.

Alpine landforms resulting from glaciation include both erosional and depositional types. Although clean glacial ice is too soft to effect much direct erosion, loose debris entrained by the glacier abrades or scours the rock surface below and along the confining valley walls. On deglaciation, such action leaves a distinctive suite of landforms—U-shaped valleys, horns, arêtes, hanging valleys, glacial stairways, and tarns. In mountainous coastal areas, fjords are drowned glacial troughs eroded below present sea level. Alpine depositional landforms are less prominent and scenic, and appear as morainal accumulations of glacial drift near the former ice terminus or along valley walls. Other features such as valley trains, outwash fans, and large erratic blocks also characterize valley deposition.

Continental glaciers differ from alpine glaciers in size and geometry, and although locally the same erosional landforms occur, they are less prominent and more widely scattered. However, glacier movement across an exposed rock surface may leave a telltale pattern of polish, scratches, or grooves; locally, alignment of these markings provides clues about direction and intensity of former ice movement. Depositional landforms, on the other hand, commonly occur on an extensive scale and are seen as broad, sweeping moraines, kame-kettle deposits, ground moraine, pitted outwash plains, and unusual linear features such as drumlins and eskers which are rarely seen in association with alpine ice deposits.

Pleistocene glaciers of Europe and North America changed the face of the land. In the interior United States, the Missouri, Ohio, and other rivers came into existence as ice-border drainage lines for the melting continental ice; other major streams such as the Mississippi, Wabash, and Columbia now have lengths or courses that reflect substantial alteration by glaciation. Deep, abrasional bedrock scour formed the Great Lakes and Finger Lakes during this time. In the western United States, a series of Ice Age lakes existed in desert regions of Nevada, Utah, Oregon, and California; glacial lakes Lahontan and Bonneville were the largest. These lakes now are dry or greatly diminished in size, but beach lines indicating former high stands are etched clearly on the sides of desert hills. Elsewhere, great meltwater floods occurred as ice-dammed lakes drained almost instantaneously as barriers broke and the lakes discharged down stream valleys to the sea.

After more than a century of scientific investigation, the ultimate cause of ice ages eludes geologists. All that is known is that they result from climatic change, which in turn results from variations in cosmic, solar, or terrestrial heat-control mechanisms. Many explanatory theories have been proposed, but no single hypothesis seems to fit all known facts and observations. Very likely, the ultimate cause is a complex combination of cause and feedback effects for which the unlocking key has not yet been found.

Glaciation has been a boon to humans. Glacial sediments are a source of extractable, nonmetallic materials such as sand, gravel, industrial clays, and peat. Buried outwash deposits and valley trains are excellent aquifers, tapped in many places for drinking and irrigation. Most importantly, soils formed on glacial deposits comprise the richest agricultural land of the Northern Hemisphere.

REVIEW EXERCISES

1 Why is glacial till heterogeneous and unstratified, whereas outwash and other deposits laid down by meltwaters are sorted and stratified?

2 Why are U-shaped valleys so convincingly diagnostic of glaciation, although no glaciers exist there now?

3 Although it is possible to radiometrically date organic matter buried between two different glacial deposits, why is it difficult or impossible to obtain precise dates for the older Pleistocene events?

4 What is the evidence for the hypothesis that glacial deposits millions of years old are still preserved in the rock record?

5 Describe the development of a cirque basin, from the first net accumulation of snow at the head of a mountain valley to the formation of a protalus rampart and cirque lake.

6 How are kames, eskers, and drumlins different? How are they similar?

7 Why are alpine glaciers of the western United States smaller than they probably were 200 years ago?

8 What are the principal changes that occur during transformation of newly fallen snow into glacier ice?

9 Explain how the Great Lakes were formed, and show why their level has fluctuated considerably during the last 10,000 years.

10 What are some of the important contributions to human economy that would not exist if the Ice Age had never occurred?

11 How do depositional landforms left by alpine glaciers differ from those left by continental glaciers?

12 The bottoms of former ice-walled lakes are commonly topographically higher than the moraine or ground moraine that surrounds them. How did this happen?

13 What evidence can be used to help determine the former ice level of a now-deglaciated valley?

14 Why are the arcuate moraines that commonly rim the outlets of mountain valleys so much higher and more massive than their morainal counterparts in regions of continental glaciation?

15 Although geological evidence of recurrent large-scale continental glaciation has existed for more than a century, why is there still no satisfactory explanation of the cause of ice ages?

SUGGESTED READINGS

ANDREWS, J. T. *Glacial Systems.* North Scituate, Mass.: Duxbury Press, 1974, 191 pp.

DYSON, J. L. *The World of Ice.* New York: Alfred A. Knopf, 1962, 292 pp.

FIELD, W. O. *Glaciers. Scientific American* (September 1955). Reprint 809, W. H. Freeman and Co., San Francisco.

FLINT, R. F. *Glacial and Quaternary Geology.* New York: John Wiley and Sons, 1971, 892 pp.

HOUGH, J. L. *Geology of the Great Lakes.* Urbana, Ill.: University of Illinois Press, 1958, 313 pp.

KUCERA, R. T. *Probing the Athabaska Glacier.* Vancouver, British Columbia: Evergreen Press, 1972, 32 pp.

KUKLA, G. J. "Pleistocene Land-Sea Correlation: I. Europe." *Earth Science Reviews* (1977), Vol. 13, No. 4, pp. 307–74.

POST, A., and LaCHAPPELLE, E. R. *Glacier Ice.* Seattle: University of Washington Press, 1971, 110 pp.

SCHULTZ, G. *Ice Age Lost.* New York: Anchor Press, 1974, 342 pp.

SHARP, R. P. *Glaciers.* Condon Lectures at the University of Oregon, 1960, 78 pp.

WOODBURY, D. O. *The Great White Mantle.* New York: Viking Press, 1962, 214 pp.

- xeric
- hot continental desert
- coastal desert
- hot steppeland
- temperate desert
- polar desert
- alluvial fan
- bajada
- claypan
- salina
- sink
- bolson
- ventifact
- desert pavement
- blowout
- deflation basin

15

- braided stream
- fan trench
- badlands
- pediment
- pediplain
- stripped surface
- pedifan (pediment veneer)
- pediment angle
- inselberg
- butte
- mesa
- ephemeral (playa) lake
- mud cracks
- fall rate
- saltation
- bed load
- clay dune
- shrub coppice dune
- sand sheet
- longitudinal dune (seif)
- transverse dune
- barchan dune
- stellate dune
- parabolic dune
- climbing and falling dunes
- relict dune

Deserts and Arid Zones

EARTH IS OFTEN CALLED THE "WATER PLANET" OF OUR SOLAR system because it may be the only planet with free water in great abundance. Even so, the present latitudinal and geographical positions of the five great continents with respect to the general circulation of the atmosphere, combined with variations in topographic relief of the land, create broad regions that are seldom reached by rain-bearing clouds. As a result, arid zones cover approximately one-third of the surface area of the continents. Furthermore, we live in a time of increasing aridity and desertification of the land; our Pleistocene ancestors prior to 10,000 years ago inhabited a somewhat colder but better-watered sphere. Certainly much of this change is due to long-term climatic trends; the effects of human activities on this progressive desertification remain unclear.

There is no universal definition of a desert. Some define it simply as a barren area capable of supporting only sparse vegetation or animal life. Following this simple definition, a desert grades into a semiarid region as the total annual rainfall increases, the terrain acquires a more complete ground cover of vegetation, and the number and diversity of animal life likewise increase. Although more precise definitions are possible, using specific climatic criteria such as the abundance and character of vegetation or measures of effective moisture availability, each individual sets his own mental image of deserts and arid zones in terms of personal experiences, water requirements, and lifestyle.

Aridity is a relative term, and is a function of rainfall and runoff, temperature, infiltration, and evapotranspiration from the ground surface and plants. There is a fair abundance of temperature and precipitation data from weather stations in arid regions, but a relative lack of evapotranspiration data. The rate of evapotranspiration is determined by a complex interaction of such factors as wind direction and velocity, ground slope, roughness of terrain, and intensity and duration of rainfall.

Although a number of climatic classification schemes have been devised, none is totally satisfactory. Clearly there is a continuum from humid conditions to the extremely dry (**xeric**) conditions of true deserts. Table 15–1

TABLE 15–1
A simple climatic classification system.

Climatic Type	Rainfall Type	Annual Rainfall (inches)	Evapotranspiration (E)/ Precipitation (P) Ratio
Arid	Scant	0–10	$E >> P$
Semiarid	Light	10–20	$E > P$
Subhumid	Moderate	20–40	$E < P$
Humid	Heavy	40–80	$E << P$

presents a simple classification based on total rainfall and relative estimates of an average precipitation-evapotranspiration ratio. Although the continuum from arid to humid conditions shown here is arbitrary, it could be used to adequately classify most points on the earth's surface.

Most people probably associate high temperatures with deserts. However, aridity may also be expressed by very cold climates with low rates of precipitation, or by a lack of hardy plant and animal life. There are, in fact, several categories of arid lands that can be separated on the bases of geographic location, biology, and climate (Fig. 15–1). The first of these, the **hot continental desert**, is typified by the Saharan, Arabian, and Great Australian deserts. These deserts are located in the continental interiors of subtropical regions and are characterized by high temperatures throughout the year. Evaporation is high, relative humidity is low, and wind speeds are moderate and usually uniform in

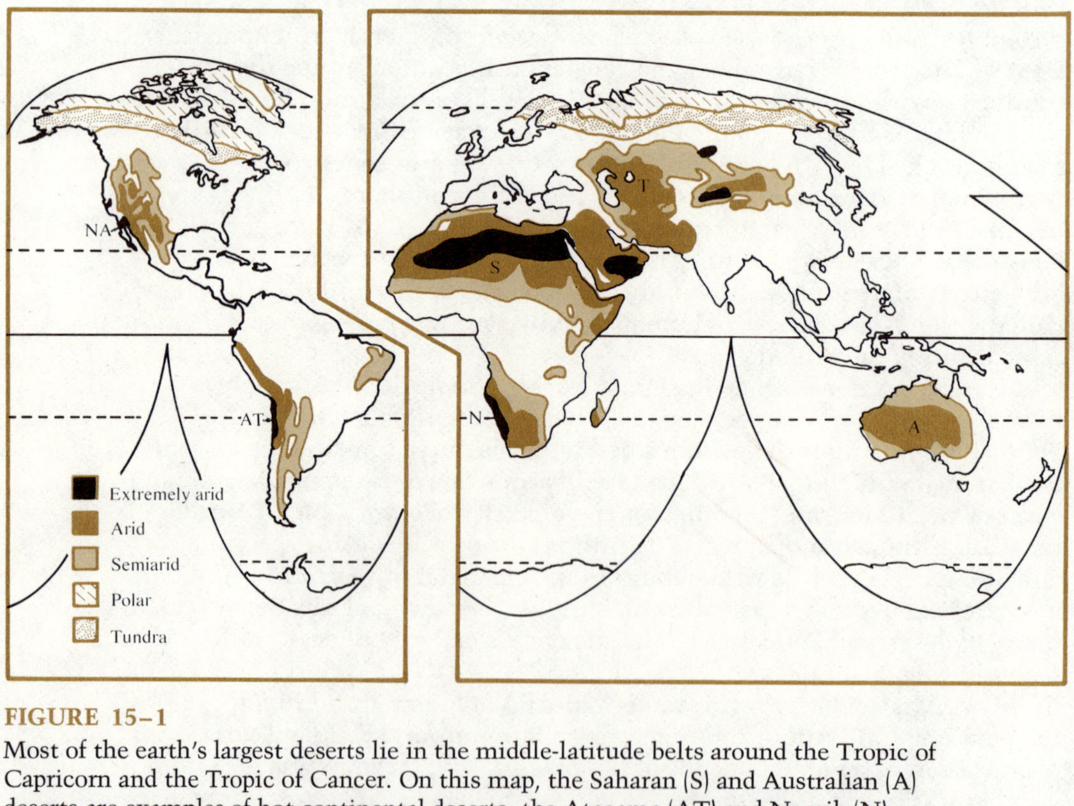

FIGURE 15–1
Most of the earth's largest deserts lie in the middle-latitude belts around the Tropic of Capricorn and the Tropic of Cancer. On this map, the Saharan (S) and Australian (A) deserts are examples of hot continental deserts; the Atacama (AT) and Namib (N) are coastal deserts; the North American (NA) and Turkestan (T) areas are temperate deserts surrounded in part by hot steppelands or prairies. The polar deserts are confined principally to the high-latitude regions of North America and Eurasia wherever annual precipitation is less than about 25 cm and much of the ground surface is permanently frozen. (Modified from P. Meigs, 1953)

direction. **Coastal deserts** are associated with unusual interactions between cool ocean currents offshore, atmospheric circulation, and latitudinal position. In these deserts, temperatures are ameliorated by winds blowing shoreward across cool offshore water. Thus, because the average climate is congenial, and though rainfall is as low as it is in hot continental deserts, coastal fogs and occasional light dew and drizzle onshore help to support at least a meager plant life and locally some belts of permanent vegetation. Examples of coastal deserts are the Atacama in Chile and the Namib of southwestern Africa. **Hot steppes** are essentially semiarid transition zones that show some characteristics of both true deserts and more humid regions. The boundary with the latter regions is usually placed where evaporation and precipitation are essentially in balance (equilibrium), as averaged over a number of years. Rainfall is commonly seasonal but is sufficient to support a cover of grasslands, low brush, and trees that can tolerate the effects of a lengthy dry season and a prevalence of saline or alkaline soils. Steppes are found around the borders of the hot deserts, for example the Sahel region in Africa marginal to the Sahara, in central Asia, and parts of California and Mexico. **Temperate deserts** lie in cooler middle-latitude zones of high atmospheric pressure beyond the range of global circulation patterns that confine the hot deserts to the subtropical regions. Local mountain barriers may impede the inland penetration of moisture-laden winds; orographic lifting wrings out most of the moisture, so very little reaches the lower terrain beyond. Examples of temperate deserts are extensive areas of interior Asia and North America. Summer temperatures are as hot as in the hot deserts, but may be followed by bitterly cold winters and substantial snowfall. Temperate deserts are characterized by high annual and diurnal (daily) temperature ranges, and by dry air, and are sites of strong, gusty winds capable of transporting fine-grained soil particles as **dust storms** for long distances (Fig. 15–2). This effect is enhanced wherever humans have subjected the transition

FIGURE 15–2
Short-lived sandstorms occur seasonally in hot deserts, and *alkali blizzards* of dust are equally common in temperate deserts such as the southwestern United States. These events have relatively little impact on human activities. Dust storms of middle-latitude prairies or steppes (as depicted here) occur naturally, but in this century have become more frequent because overgrazing and plowing of moisture-deficient, marginal agricultural lands have led to drying of the soil surface layer. Movement of strong winds across a barren ground surface, often accompanying the passage of a rapidly moving low-pressure area (cyclone), as seen in this photo, removes large amounts of fine-grained dust which may be blown many kilometers downwind before it is redeposited. (Photo courtesy NOAA)

zones between temperate deserts and steppelands to excessive agricultural development; through plowing and irrigation wells that lower local water tables, the native vegetation is removed, and the soil moisture that previously tended to aggregate soil particles and prevent movement by wind is evaporated. The classic example is the Dust Bowl of the southwestern United States of the 1930s, repeated on a smaller scale in the 1960s. A final category, **polar deserts**, seems a contradiction in terms, but these deserts have desertlike characteristics: low precipitation, relatively high evaporation, seasonally strong winds, and an absence of vegetative cover. They differ from hot deserts only in that temperatures are cold throughout the year. Polar deserts encircle the Arctic Ocean and are present in a limited way in southern Argentina and Chile. **Tundra** zones of rather abundant but stunted low brush and grasses ring the true polar deserts; in many ways, these are cold-climate analogs of the steppelands of middle latitudes.

DESERT LANDSCAPES

GEOLOGICAL PROCESSES AND LANDFORMS

Aridity leaves an impressive, often colorful set of landforms on the earth's surface. In the hot continental and coastal deserts, the general absence of true soils, a lack or scantiness of vegetal cover, and the play of light and shadow lead to a feeling of starkness and unreality. Mountains, valleys, and interstream divides appear steep and sharply defined, quite in contrast to the more smoothly eroded and rounded terrain of vegetated, temperate, humid regions. In addition, there is little surface water in lakes and streams. Nevertheless, it is apparent even to the casual observer that running water is primarily responsible for sculpturing the land, both today and in the past.

Wind action plays an important secondary role by concentrating weathered products as dunes in areas favorable for their accumulation. Other physical and chemical processes associated with rock weathering and mass wasting also are operative. Many of these geological agents have been discussed already, particularly in Chapters 4, 11, and 12. However, some repetition and expansion of information is desirable in this discourse about arid lands, for landforms mantled or obscured by vegetation in wetter, more populated areas are clearly seen, often on a grand scale, in the dry regions where vegetation is sparse and there is relatively little landscape modification by human activity.

ALLUVIAL FANS

Rainfall is scant in dry regions, but it is nearly always seasonal, and is usually precipitated as storms of great intensity but relatively short duration. Though certain areas of true deserts may receive rain only once every few years, or undergo a major storm once in a decade, the great sheets or ribbons of loose sediment that move down from every mountain canyon or valley in the arid environment attest to the overwhelming dominance of running water. Cone-shaped, concave piles of debris that are deposited downstream from canyon mouths are called **alluvial fans** (Fig. 15–3). In places, alluvial fans from a series of adjacent valleys may coalesce into continuous aprons or sheets of sediment called **bajadas** that extend along mountain fronts for many kilometers (Fig. 15–4). If sediments are deposited in interior basins where there is no outlet for running water, the basin may in time accumulate loose debris to a depth of several thousand meters. Where outlets exist, the valleys contain wide, flat, and terraced channels commonly occupied by either ephemeral streams or perennial streams that have too limited a base flow to cope with the great volumes of sediment contributed to the fluvial system. Such streams, marked by constantly shifting channels and numerous mid-channel islands and bars, are called **braided streams** (Fig. 15–5) as in more humid regions.

Some alluvial fans are broken by box-shaped discharge channels characterized by steep walls and relatively broad, flat bottoms. These are called **fan trenches** (Fig. 15–6). These trenches may be as much as 30 m deep; commonly, they are most prominent at the head of the fan and gradually die out downslope. Others extend almost the length of the fan. Some trenches obviously were cut at some time in the past but are now being backfilled with sediment. Clearly, then, channel entrenchment on a fan can result in a number of ways and be in various stages of development. Under uniform conditions, a stream will ultimately dissect the upper part of its own fan because of continued downcutting of its channel within the

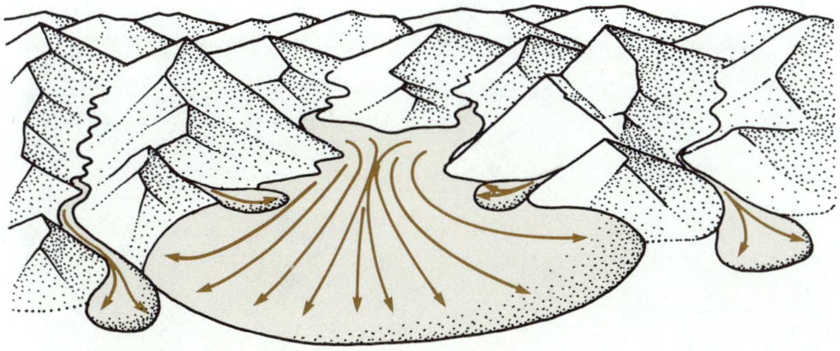

FIGURE 15–3
A schematic drawing showing an alluvial fan deposited by an intermittent stream at the mouth of a major canyon issuing from a mountainous upland. Arrows indicate how the fan builds progressively as a cone-shaped (upwardly curved), concave pile of sediment as the major flow channel shifts back and forth across the fan surface. Note that smaller fans concurrently form at the mouth of each smaller mountain-front stream, and in fact encroach at several places onto the surface of the primary fan. If the large fan ceases to grow, and smaller fans continue to expand, in time the older primary surface may be largely submerged by the younger fans. Such complex crosscutting relationships are common, and thus interpretation of the exact history of fan construction may be difficult.

FIGURE 15–4
This ground photo is taken from the Panamint Mountains bordering the west side of Death Valley, California, and looks northeastward across the floor of the valley. Large alluvial fans are prograding basinward from the eroding mountains in the foreground, but are combining to produce a large compound fan, or bajada. Alluvium is surrounding and isolating cutoff sections of the mountain front as *inselbergs* (left center). The whitish-colored streaks are distributary channels that cross the bajada and terminate in a basin or sink of interior drainage nestled against the base of the Black Mountains (background). The rather flat but intensely dissected area in the middle of the picture is a *pediment.* The pediment front ends abruptly at the basin in what is probably a fault.

FIGURE 15–5
An upstream view of the North Platte River at Ogallala, Nebraska. The Platte is a braided stream, in which numerous individual channels branch and rejoin in an intricate network. Most streams that flow east from the Rocky Mountains across the semiarid Great Plains have a similarly braided character. Because of climatic change in the last 10,000 years, the present average flow volume of these streams is not sufficient, except seasonally, to transport all of the sediment load imposed on the streams. Thus, much sediment is always in storage in the channel; braiding is a means for the stream to adjust its gradient and maintain its energy expenditure at maximum efficiency.

FIGURE 15–6
A view up the fan trench developed at the mouth of Mill Creek on the east side of the Toquima Range in central Nevada. The surface of the undissected part of the alluvial fan is about 25 m above the level of the flat-floored trench. The steep walls of the trench are marked by a sparse growth of junipers. At this site, entrenchment seems to have resulted from renewed movement along a fault at the base of the mountain front in the background, thus lowering the level of the enclosed basin to which Mill Creek drains. The sagebrush-covered area in the foreground is a small, modern fan from an adjacent canyon encroaching on the dissected Mill Creek alluvial fan, as was illustrated in Fig. 15–3.

mountains. However, any major climatic change, whether temporary or permanent, will alter the amount, duration, and intensity of rainfall and runoff from the basin. The resultant changes in amount or type of vegetative cover therefore will tend to either increase or decrease the amount and rate of erosion and sediment input into channels. Thus, permanent channel entrenchment on the fan could occur because of downcutting by accelerated erosion. In any case, over time the locus of deposition of the fan moves progressively downstream and away from the mountain front. This tends to gradually increase the size of the fan (Fig. 15–7).

In areas of active faulting such as the Great Basin region of the western United States, fan trenches are common and rather spectacular features on fan slopes. At places, alluvial fans on opposite sides of a mountain range are completely different. One flank may be bordered by a relatively smooth, rill-incised fan surface, whereas the fan on the other flank may be strongly dissected by trenches. The latter situation suggests that recent faulting has changed the former base level to which the fan was graded, and trenching alters the slope and form of the fan so as to facilitate movement of water and sediment to the new base level (Fig. 15–8).

BADLANDS

In steppelands, the landscape is less angular. Rainfall is slightly greater, and a more abundant vegetative cover on uplands and slopes tends to either diminish runoff or retard sheet flow so that the interval required for the movement downslope of direct runoff is lengthened. Mass-wasting processes also are slowed because of the stabilizing and anchoring effect of grasses or shrubs. Nevertheless, it is not uncommon to see sharp breaks between smoothly rolling or flat, grassy uplands and steep-sided, angular, narrow gullies or ravines devoid of vegetation and with normally dry channels at their bases. Where gully erosion is intense in grassland settings, **badland** topography may result (Fig. 15–9). Excellent large-scale examples of badlands are observable in western South Dakota and in the Drumheller area of eastern Alberta, Canada. Miniature badlands are widespread, and can occur almost anywhere in dry or arid regions if surface rocks are easily eroded and vegetative

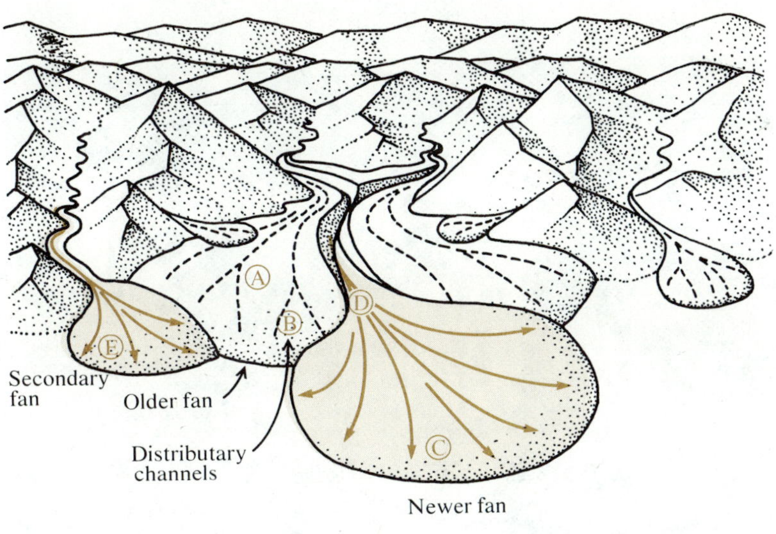

FIGURE 15–7
The development stages of an alluvial fan. A. The older fan builds outward as a cone from the mouth of a mountain canyon toward an adjoining basin. B. Numerous overflow distributary channels may cross the fan in a radial pattern. These discharge channels frequently change course because of sediment damming, mudflow events, or headward erosion of small gullies on the toe of the fan that capture and divert flow into a new channel. Such progressive, sequential diversions produce the deltalike form of typical fans, but also tend to segment the fan into a number of discrete sections.
C. Subsequently, the locus of deposition shifts downslope as fan size increases.
D. Entrenchment then may result from the changes in climatic or geologic controls described in the text. E. Young secondary fans may encroach onto the original fan.

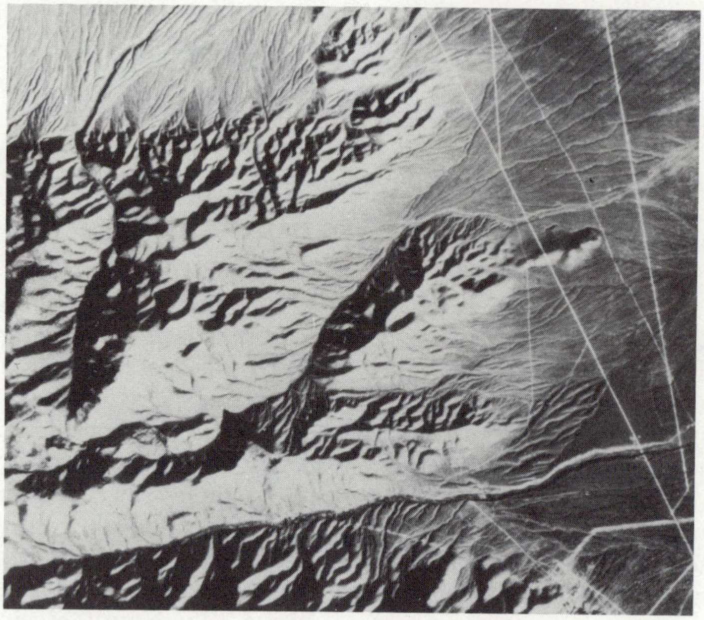

FIGURE 15–8
This photograph of the southwestern part of Battle Mountain, Nevada, shows well-developed fan trenches at Willow Canyon (upper left) and Cottonwood Canyon (lower right), yet intervening canyons display little or no evidence of entrenchment. Willow Canyon trench clearly seems controlled by the range-front fault seen in the photo. Causes for the Cottonwood Canyon trench are less clear; however, the stream originates high in the range, unlike the other streams shown, so high-altitude microclimatic controls of streamflow, and so forth, may be an important factor in the fan trench development.

FIGURE 15–9
A typical view of badland topography developing in western South Dakota. The scarp is retreating away from the viewer. The area in the foreground has partly revegetated with grass, which inhibits additional erosion. The face of the scarp is crenulated (finely scalloped or creased) with hundreds of steep-sided, narrow gullies carved in soft silts and clays that have little cementing material to bind the grains together. Heavy rains erode the scarp. Because of minor differences in the character of the sediments, the lower slopes tend to round into haystack-shaped forms, whereas the upper slopes have more angular, turreted, or castellated surfaces.

cover has never existed or has been removed. In these areas, several acres of farmland or pasture can be lost each year as gullies rapidly extend headward.

EROSION SURFACES

A visitor to deserts or steppelands is also impressed by the vastness of gently sloping plains that may extend unbroken as far as the eye can see, except where notched locally by steep, angular slopes that lead downward into canyons and ravines, or by equally sharp upward breaks into scarps that extend to the adjoining uplands. These planar surfaces generally fall into one of three classes: **pediments**, **pediplains**, or **stripped surfaces**. All are dipping or sloping surfaces of areal or regional extent that grade downward and end at some equally flat or gently dipping surface of deposition such as floodplains, interior basins, or lake plains.

The planar surfaces differ only in the way they were formed, and whether they are surfaces of erosion, transportation, or deposition. There is a tendency to subsume all of these planar surfaces under the general class **erosion surfaces**, though clearly they are not entirely the result of erosion. Geologists still disagree about their origin, and, in fact, about the proper use of the terms themselves. A **pediment** is best defined as a long, smooth surface on eroded rocks that connects the hillslope of the adjacent uplands with the depositional flats or plains below (Fig. 15–10). Coalescence of individual pediments creates a **pediplain**, analogous to the manner in which alluvial fans join to produce a bajada. The concave upward inclination or slope of a pediment varies within a somewhat limited range, from perhaps as much as 6° at the hillslope junction to as little as 0.5° near its terminus. Once beveled, the pediment becomes basically a transportational surface; it is crenulated by shallow rill channels formed by water in the process of moving additional weathered rock debris across the pediment from the uplands to a local outlet channel or depositional basin. Commonly the pediment surface is covered by fan alluvium ranging in thickness from several centimeters to 10 m or so; thus, it is difficult to see the actual slope or surface of the pediment or to determine anything about the rock types or geologic structure of strata below this **pedifan**, or **pediment veneer**. Where it is possible to study the substrate of pediments, however, there is revealed a generally planar, but locally irregular, beveled surface that cuts uniformly across the rocks regardless of their lithology or structural characteristics.

There is general agreement that pedimentation is accomplished by a slow, gradual backwasting of upland slopes; thus, over time interstream or interbasin divides slowly diminish in width and are reduced in height. Where scarp retreat and pediment development are active, weathered material tends to be removed from the foot of the slope as quickly as it is formed. Although it is not clear whether running water or downslope mass movement is primarily responsible for removing this debris, the net result is to leave the head of the pediment in juncture with the base of the scarp as a sharp obtuse break, or **pediment angle** (Fig. 15–11).

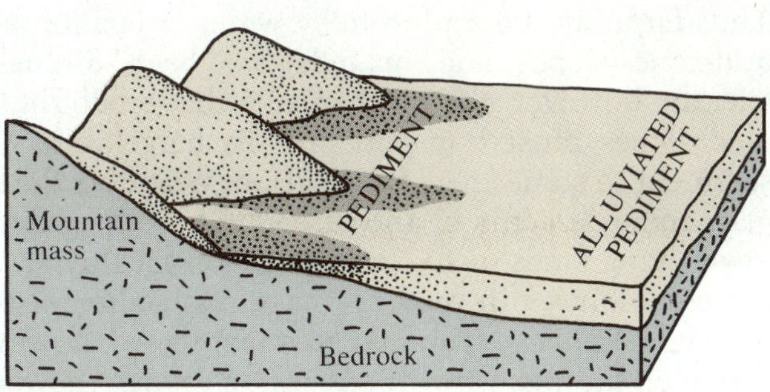

FIGURE 15–10
The development of a pediment. As the scarp at the mountain front retreats laterally and continues to degrade and backwaste by weathering and erosion, the pediment becomes a transportation surface across which sediment moves downslope to final deposition in a valley or basin. A pediment surface locally may consist of bare rock, but more commonly it is veneered with a relatively thin layer of rock fragments. This composite drawing illustrates both rock-floored and alluviated pediment surfaces; note the basinward thickening of alluvial cover.

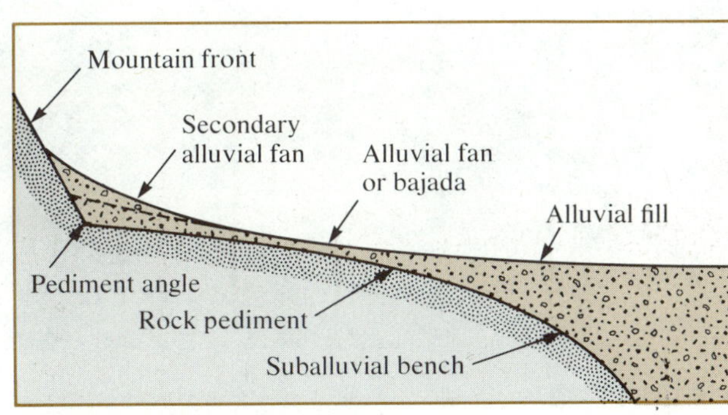

FIGURE 15–11
This sketch gives in greater detail some of the pediment relations shown in Fig. 15–10 and illustrates the pediment angle, or knickpoint, that separates the mountain-front scarp from the rock pediment or so-called suballuvial bench. As seen here, the pediment angle (sometimes called the *piedmont notch*) is somewhat obscured by encroachment of a secondary alluvial fan. However, it is not unusual to find an alluvially veneered pediment intersecting a steep, rocky mountain front at a sharp angle. The obtuse angle of juncture commonly is 25°–35° but may be greater.

As opposite slopes of an upland slowly backwaste toward each other, the upland becomes smaller, the drainage area and stream runoff decrease, and the pediments lengthen by extension. However, because of rock structure and lithology, slope retreat is not a linear or uniform process. Sections of former interstream divides between hillslope channels may be cut off and left as isolated knobs of bedrock on the pediment some distance beyond the hillslope front. These **inselbergs** ("island mountains") may be surrounded by alluvium (Fig. 15–12). However, they continue to be attacked by weathering agents, and their ultimate fate is destruction or burial beneath an ever-advancing tide of rock debris.

Pediments are best seen and perhaps achieve their maximum development in arid regions, but are not unique to these areas as was once believed. Recent investigations in humid temperate zones clearly show that pedimentation processes either operated there in the past or are occurring today.

Stripped surfaces may superficially resemble large-scale pediments, but often they extend for tens of kilometers and are continuous from upland crests or divides to the depositional basin below. They are particularly well developed in regions of nearly flat-lying or gently dipping sedimentary rocks of relatively uniform composition, such that a single rock unit or series of compositionally similar rock units is at or near the surface over broad areas. There are numerous examples of this form in the Colorado Plateau of the southwestern United States. The north flank of the Uncompahgre Plateau, south of Grand Junction, Colorado, typifies the stripped surface phenomenon. Here, fluvial erosion has removed softer rocks from above a series of coarse-grained sandstones that dip gently toward the Colorado River from the crest of the plateau (Fig. 15–13). The stripped surface declines on an almost uniform slope gradient from an elevation of 3350 m at the divide to only 1370 m at the Colorado River; this decline of 1980 m is spread over a horizontal distance of almost 55 km. However, the upland slope is not totally featureless; locally, irregular masses of unstripped rock—the uneroded remnants of older interstream divides—rise as **buttes** or **mesas** above the stripped surface. (Buttes and mesas can be more precisely defined in a geological sense, but in reality are defined by the way the terms are used by local residents.) These forms are thus somewhat analogous in origin to the inselbergs of pediment slopes. Where buttes and mesas occur, it is generally safe to assume that slow backwasting of slopes or stripping of gently dipping rock strata has occurred. Excellent examples are also seen in the colorful, castellated buttes and mesas of the Navajo country of Arizona (Fig. 15–14), the western Great Plains (Fig. 15–15), and the plains of western Morocco, North Africa.

FIGURE 15–12
One example of an inselberg has already been shown in Fig. 15–4. This example from southern Kenya, East Africa, demonstrates that inselbergs are geographically widespread in all semiarid regions. The place shown here lies approximately 1.5 km above sea level. The level area in the foreground is veneered with a thin layer of rounded, water-transported quartz pebbles. The inselberg is the last remnant on this stripped surface of formerly more extensive but now long-vanished divides between major stream valleys. It continues to undergo attack by weathering processes that are breaking the hill into a pile of loose blocks of granite. Locally, these features are called *kopjes.*

STREAM DEPOSITION

Thus far, only the sculpture by water of landforms in arid-zone slopes and uplands has been discussed. Clearly, however, the mechanisms that erode the upland slopes must transport and discharge water and sediment to some area of deposition. Wherever the few permanent streams of the drylands have a outlet to lower-lying lands or to the ocean, normal fluvial processes dominate in moving water and sediment to a final resting place. It should be remembered, however, that such through-flowing streams, although they obey the laws of fluvial dynamics, may need to adapt themselves to different sediment loading conditions than streams of more humid regions. Thus, they may assume a variety of channel configurations, as for example the braided condition already described. Other adjustments of channel shape, width, depth, and slope are common. Such channels and the adjacent floodplains commonly contain, in temporary storage, greater quantities of sediment than are found in the valleys of humid-zone streams. Unlike the fine-grained alluvium characteristic of floodplains and channels of

FIGURE 15–13
Stripped surfaces develop on gently dipping sedimentary rocks and are nowhere better displayed than here in western Colorado. This photo looks north along the stripped, sloping surface of the Uncompahgre Plateau. The level area in the lower half of the picture is now being deeply dissected by canyons, and is developed on sandstones that dip gently toward the Grand Valley of the Colorado River (dark area cutting across the middle). The difference in elevation between the place the picture was taken and Grand Valley is about 600 m. The dark line beyond Grand Valley surmounted by puffy cumulus clouds is the northward-retreating scarp of the Book Cliffs.

FIGURE 15–14
Buttes and mesas in Monument Valley, Arizona, are left isolated by backwasting of a scarp on gently dipping sedimentary rocks. The large, sandstone-capped mass at the left, surrounded by a sloping pile of weathered rock, is arbitrarily considered a mesa because the summit width is at least as great as the height of this landform. The smaller pinnacle at the right, high but with a small summit area, is a butte. Over time, mesas are reduced to a series of buttes and then progressively destroyed by lateral erosion at the base and sides.

humid regions, the sediments of arid-zone channels, terraces, and floodplains are typically composed of angular, blocky rubble or boulders. Sediment of a grain size finer than sand is winnowed out and removed during the few periods of flood flow.

INTERIOR BASINS

Some estimates show that nearly one-third of the continental surface is characterized by internal drainage, wherein streams terminate in enclosed basins totally isolated from the general base level of the oceans. The result is that streams flowing from the surrounding uplands and pediment slopes must discharge water and sediment load into these basins. Where seasonal discharge of water either exceeds the infiltration capacity of basin soils or exceeds local evaporation rates, temporary, shallow ponds or lakes form. These are called **ephemeral** or **playa lakes**. Some playa lakes may last for much of the year. Under temporary conditions in which runoff exceeds the infiltration-evaporation response of the basin for several successive years, the lakes may appear permanent. Other playa-containing basins hold water for only a few days or weeks during the spring season or for brief periods after heavy summer storms, when the playa surface may be temporarily covered with water (Fig. 15–16).

FIGURE 15–15
This photo of one of the so-called Chalk Pyramids in the Smoky Hill River valley of western Kansas shows buttes of a type different from those in Fig. 15–14. The "pyramids" are composed of soft, chalky limestone masses that were probably originally separated from the valley wall of the Smoky Hill River along a series of meander bends at a time when the climate was moister and the stream was widening its valley. In the semiarid environment of today, the buttes are wearing away under the impact of rainwater and wind erosion.

Sediment accumulated in a playa basin is dominantly composed of silt- and clay-sized particles, and on drying, the surface tends to develop a pattern of tensional desiccation markings called **mud cracks** (Fig. 15–17). Although these cracks are usually only a couple of centimeters or less deep, under some conditions the cracks enlarge into furrows or trenches about a meter wide and more than 3 m deep, especially if the water table, which is usually just below the playa surface, has been substantially lowered through pumping from nearby wells. Thus, much of the infiltrating water from the playa, which normally would be added to groundwater storage, is drained off; the playa surface may subside, and the large tension cracks form.

FIGURE 15–16
The margin of a small temporary lake, or playa, still partly filled with water after a late-afternoon summer thundershower. Because of high evapotranspiration, the remaining water will disappear in a few hours to a few days. The whitish color comes from alkali-encrusted soil particles that surround and underlie the playa basin. The puffy character of the ground around the lake is the result of swelling of certain clay minerals as moisture is added to their crystal structure. Mud cracks likely will form on the desiccated surface as the water disappears.

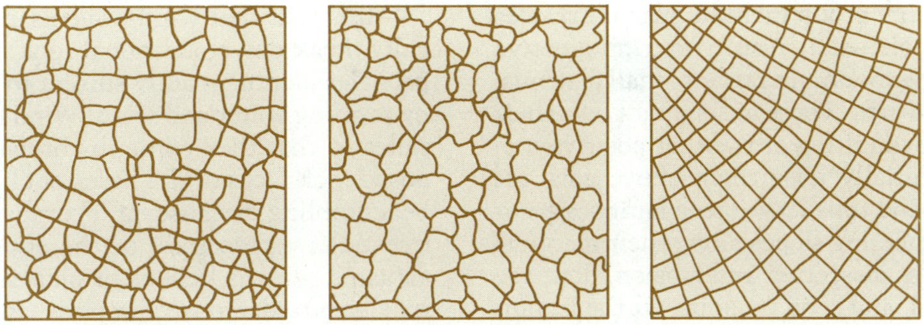

FIGURE 15–17
Some common systems of mud cracks. Cracking is caused by tensional stresses exerted as water evaporates and sediments slowly dry and contract. The rupture pattern is a function of: (1) internal conditions, such as total moisture content and sediment size and structure; and (2) external conditions, such as temperature, humidity, and drying rate. Cracks may also appear in random patterns. (From J. T. Neal, Air Force Cambridge Research Laboratories Environmental Research Paper 123, 1965)

Water often can be obtained by digging a shallow hole into or around the margins of an otherwise featureless and barren playa, but the water is usually highly mineralized with dissolved sodium, calcium or potassium chlorides, sulfides, or bromides. It is almost undrinkable by humans, though it may be used by livestock, which have a higher salt tolerance. This is the alkali water, or "alky" of cowboy movies. The excess of salts is initially derived from weathering and leaching of rocks in adjacent upland areas. The salts are then transported in solution or suspension by running water and recrystallized under the evaporative conditions of the playa. If the playa surface is dominated by clay or silt deposits, a **claypan** is formed; if it is covered by a glistening coating of whitish precipitated salts, a **saltpan**, or **salina**, is formed (Fig. 15–18). Local usage results in other interesting terms: in the western United States, interior basins are often called **sinks**. Long, linear basins may be internally divided into several subbasins, each with its own sink or playa. These are **bolsons**, or **bolson basins**.

LANDFORM ASSOCIATIONS

The landforms of arid zones just described can be fitted into two separate but characteristic assemblages, or

FIGURE 15–18
The famous Bonneville Salt Flats of western Utah exemplify the salina environment. The glistening white surface is maintained by seasonal evaporation of water that spreads across the playa surface as it runs off of adjacent highlands, such as the Silver Island Mountains in the background. If the evaporite coating is not disturbed, it provides a hard, smooth surface, which makes a good high-speed test track for wheeled vehicles. Strongly alkaline groundwater, however, is always present about a meter below the playa surface.

form families. The first group includes planar surfaces of deposition or erosion that result from progressive destruction of narrow mountain ranges, small plateaus, or uplands of limited area. These areas might owe their elevation to uplift by faulting or to other positive tectonic forces that initially raise them above adjacent lowlands or basins, and thus leave steep fault scarps or relatively steep connecting slopes along their margins. In the fluvial cycle of dissection and erosion, drainage is quickly integrated, and soon no point on the terrain would be far removed from entrenched drainage lines. Regardless of the rates of scarp retreat, the landscape would quickly be dominated by pediments, pediplains, embayed scarps, inselbergs, and alluvial fans; these would thus surpass in geographic scope the remnants of the eroded area that was originally uplifted. Furthermore, in areas of interior drainage the filling of basins with accumulated sediment would further subdue the topography.

In upland plateaus of larger extent, much of the land surface is farther from the master drainage, and because it takes longer to integrate drainage along tributary channels or canyons, there initially may be large areas of poorly drained or undrained upland. In such areas, particularly those underlain by flat-lying sedimentary rocks that differ in their relative resistance to erosion, stripped surfaces are commonplace, and are surmounted by various kinds of buttes, mesas, and erosional outliers, with local pediments or alluvial fans developed at their bases.

AEOLIAN PROCESSES AND LANDFORMS

Early-day travelers to desert regions came from better-watered countries, so it is natural that they were impressed with the immense quantity of sand piled up in giant dunes or as sand sheets extending over hundreds of square kilometers, and the fierce sandstorms that occasionally overtook their caravans. From these observations, the idea grew that deserts are dominantly sandy wastes and that wind action is the most important agent in the sculpture of arid-zone landforms. However, as has already been demonstrated, running water is the dominant force in developing the landscapes of arid regions.

There is still no accurate measure of how much of the deserts actually consists of dunes or sheets of sand thick enough to more than thinly veneer the underlying surface. A commonly accepted value, however, is 20 percent, perhaps because this is the amount crudely determined for the Sahara Desert region.

It is true that wind-driven sand helps to mechanically break down additional rock, which as weathered particles is more readily subject to additional transport by running water and wind. Wind can move small particles so that they come in corrasive contact with exposed rock faces or the bases of cliffs that may already be cumbling because of chemical attack by water, freeze-thaw processes, or the action of biological organisms such as lichens or plants. Such rock surfaces may be pitted, grooved, or polished by the sandblasting action of moving sand grains. Fine-grained, massive, loose cobbles or blocks which are subjected to constant attack by wind-driven sand moving in a constant direction may attain a faceted and highly polished surface because of continuous abrasion. These forms are **ventifacts** (Fig. 15–19).

Elsewhere, particularly on desert plains or broad alluvial fans, wind tends to blow away the finer-grained particles from the surface layer, thus leaving and concentrating the coarser-grained fragments as an armoring or protective layer above equally fine-grained materials below. Once formed, this stony or cobbly **desert pavement** may remain intact for tens or hundreds of years unless disturbed by the passage of animals or vehicles across the surface so that the pebbles are dislodged (Fig. 15–20). Though the pavement may be only a pebble or two thick, the smooth surface al-

FIGURE 15–19
Ventifacts. Ventifacts develop best from fine-grained, monomineralic intrusive igneous rocks, but other rock types such as volcanics, limestones, and sandstones may retain crude ventifactal imprinting. The examples in this photograph indicate some of the possible range of development. The large cobble in the lower right is strongly polished and striated, and the surfaces of two edges of the block intersect at an acute angle as a sharp ridge. (Photo by M. R. Campbell, U.S. Geological Survey)

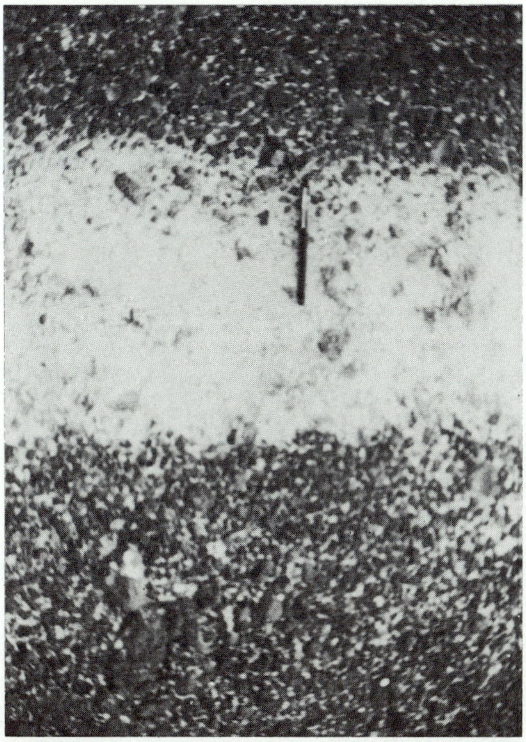

FIGURE 15–20
Desert pavement results from the removal of fine-grained particles by wind action, thus concentrating the coarser fragments as a protective surface layer that inhibits further wind erosion. However, if a vehicle is driven across the surface, the armoring layer is displaced or destroyed, and the underlying fine-grained material is again vulnerable to deflation, as seen here.

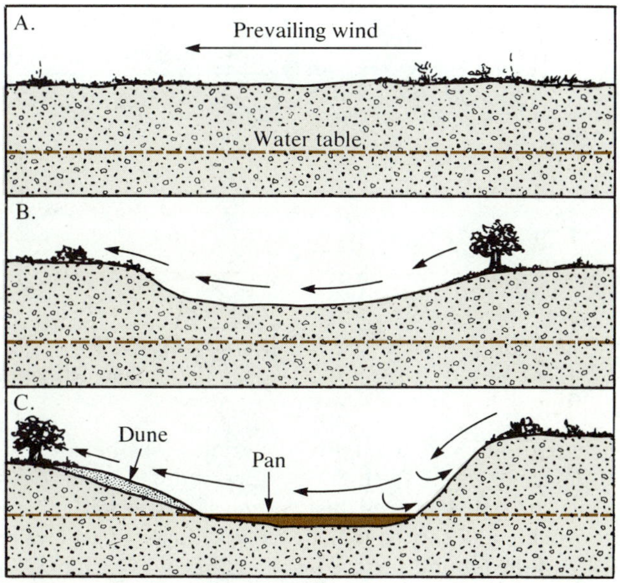

FIGURE 15–21
Deflation basins are the one desert landform produced solely by wind action. Most of these basins are rather small, though a few in the Sahara Desert are very large. These sketches illustrate how removal of an armoring layer (A) or destruction of anchoring vegetation allows progressive deflation (B) to gradually deepen and widen a deflation hollow. The water table, if it remains in a fixed position, acts as a depth limit of deflation effectiveness. Seasonal flooding or wetting of the bottom of the hollow may produce a layer of armoring salt-encrusted sediment, or *pan* (C), and thus the surface is restabilized.

lows floodwaters to pass without disturbance. However, if the pavement is dislodged, additional fine-grained material is exposed and then is subject to removal by running water or by wind transport; the latter process may create small, shallow surface depressions called **blowouts**. Unless soon stabilized, a depression may deepen and widen to become a larger **deflation basin** (Fig. 15–21). The depth of wind excavation is limited only by the depth to the water table and by the grain size. Basins resembling deflation basins may, however, be created in deserts by groundwater solution or other geological processes.

MECHANICS OF WIND DEPOSITION

Loose particles are moved by wind, either along or near the ground or in temporary suspension at some height above this surface. Whereas silt and clay particles may be picked up in the deflation process and carried for many kilometers by strong winds, sand-sized particles are too large and have a settling velocity, or **fall rate**, that prevents them from being carried for great distances except in very intense storms. Thus, sand particles remain in suspension only briefly, but if winds are strong and gusty or irregular in velocity, the particles may move by the process of **saltation** (Fig. 15–22). In saltation, the sand moves in a series of intermittent bounces or leaps of varying lengths that depend on grain size and weight, and wind velocity. Also, falling particles may strike and displace other sand grains by a sort of splash effect, thus increasing the total number of particles that are horizontally displaced. The maximum size of particles that can be displaced is limited basically by the size of a grain resting on the surface that cannot be moved, at a given velocity, either by direct pressure of the wind or by the impact of other moving grains against it. Some grains never leave the ground because they are too large, and are thus confined as **bed-load** particles. Nevertheless, these may creep along the surface, because they are driven by the impact of smaller grains.

Because of the remarkable grain-size sorting processes described, silt and clay particles may be re-

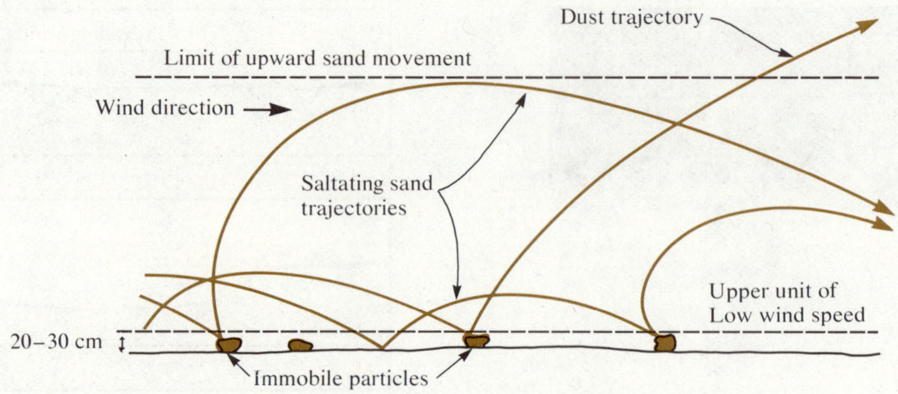

FIGURE 15–22
Movement of loose, wind-driven particles. The height and distance that particles move are functions of the settling velocities of different-sized particles, and the average wind speed. Thus, fine-grained dust particles (clay or silt) may be lifted high into the atmosphere and carried long distances. Coarser particles move in lesser trajectories closer to the ground surface by bouncing, or saltation. Creep is confined to the zone of low wind velocity (below the lower dashed line); winds are light and terrain irregularities and vegetation may break air flow into small, circular eddies. Over time, most fine-grained particles are removed.

moved entirely, blown many kilometers downwind, and redeposited by either rain or lessening of wind velocity such that they slowly settle from the air. The excess fine- and medium-grained sand is concentrated locally into dunes or sand sheets. Coarser fragments are left as a residue or protective cover of desert pavement, as already described.

DUNES

Why do sand dunes form, and what is the source of the sand? There is no clear-cut answer to either part of this question. Theoretically, all that is required to form sand dunes is an abundant supply of loose sand grains, transporting winds of sufficient velocity and persistence, and a topographic or vegetational obstacle to serve as an accumulation site. However, although it has been shown that windborne particles act as corrasive agents, creating additional particles by breaking down rocks, there is simply too much loose sand in most deserts to be accounted for by this method alone. Because most desert areas still bear the imprint of climates of hundreds or thousands of years ago, when many regions were better watered and were sites of lakes or through-flowing streams, it is thought that much of the loose material was created by normal weathering during these more humid times, and that only minor additions have come from weathering during the presently arid climatic regime.

Dunes are not exclusively made of sand. Though desert regions generally are net exporters of clay- and silt-sized particles, blown as dust to areas beyond the desert borders or to the sea, small **clay dunes** are not uncommon around the borders of former lakes or on the desiccated margins of playa basins. Here, as the playa surface dries, salt crystals in silty or clayey lake sediments crystallize, and the clays curl, crack, or blister. The clay flakes are subsequently moved a short distance by the wind or by water and piled as dunes along the margin of the lake. Rainwater then leaches the contained salts; the clay particles are dispersed and partly stabilized by plant growth; and the dune becomes a hard, immovable mass. Clay dunes, however, occur more commonly along seaside lagoons and river valleys in semiarid regions than in true desert areas.

Shrub coppice dunes are very common in desert and semiarid regions where there is a substantial cover of low brush or coarse grass, and thus especially around playa ponds or along stream courses. Mounds as much as 3.5 m high, composed of fine-grained, wind-blown sands and precipitated salts, accumulate around the bases of phreatophytes, which are plants that may put down root systems 6–9 m to reach a permanent supply of moisture. As long as the plants can extend their deep root systems to reach the water table, they flourish and the mound continues to grow; if the plants die, the aeolian deposits are soon degraded by wind and water (Fig. 15–23).

Most sand occurs in **sand sheets** rather than in discrete dunes. These bed forms tend to cover large areas of rather flat or gently rolling terrain. The sheets are not featureless; they tend to align as a series of furrows

FIGURE 15–23
Shrub coppice dunes around the margin of the famous Columbus Salt Marsh in southwestern Nevada. This marsh was the original site of mining of borax and other exotic salt deposits in the nineteenth century. The dunes are composed mostly of silt and are anchored by salt-loving plants whose roots extend deeply below the surface to reach the water table. The silts of the barren areas between the dunes are slowly deflating at present, so the mounds are getting relatively higher. If the water table declines further, the plants will die and the coppices will gradually be eroded and destroyed.

and ripples, usually only a few centimeters to a meter high, but parallel to the direction of wind flow. Sand sheets thus are the common landform in environments where there is an excess of sand available for transport but there are no local topographic or vegetative irregularities to entrap it. Sand sheets usually pass laterally into areas of fixed or stabilized dune fields.

The most spectacular dunes are mobile or moving types. These can assume almost any shape or form because of the complex interaction of factors such as shear, turbulence, and eddy effects in the near-surface winds that transport the sand; in fact, complex or compound dune forms are probably more common than dune forms characteristic of simple wind-flow patterns.

Mobile dunes can be classified into three types: **longitudinal**, **transverse**, and **barchan**. **Longitudinal dunes**, or **seifs**, are the most common dune form and occupy large areas in all deserts except in North America. Their amplitude, wavelength, and height vary; in portions of the world, they are 90–180 m high and several tens of kilometers long. Their height is probably determined by the intensity of wind shear and flow velocity in the lower atmosphere. Longitudinal dunes apparently develop where there is an abundant sand supply and a bidirectional wind regime; transverse and barchan dunes may be transformed into longitudinal dunes by crossing wind patterns (Fig. 15–24). Elsewhere, wind patterns shift sufficiently on an annual or seasonal basis so that the dunes are changed into **pyramidal** or **stellate dunes**; these types are common in the southwestern United States, where the typical seif pattern is rare (Fig. 15–25).

Transverse dunes occur in places where there is a prevalent, strong, unidirectional wind and an abundant sand supply. They orient perpendicular to prevailing wind flow, and have a gentle windward slope and a steeper lee face. If the sand supply decreases, or if the ends of the dune become permanently fixed around some obstacle such as a rock outcrop or vegetation, transverse dunes can be transformed into **parabolic dunes**, or perhaps into longitudinal forms if wind forces become bidirectional. Parabolic dunes are most common in semiarid regions, coastal deserts, and areas where geologically recent climatic change has decreased the amount of sand available for transport and dune formation.

Barchan dunes (or simply **barchans**) are a common and aesthetically appealing form. They are crescent-shaped, with the horns pointing downwind. The windward slope face is gentle, and the lee slope steep (Fig. 15–26). Because sand is continually moved from the windward slope to the lee side, these dunes retain their shape while they continue to move at 7.5–15 m/year. They develop best where wind is constant and of moderate velocity, the underlying surface is smooth, and the sand supply is adequate but not abundant. In the western United States, there are many areas where bar-

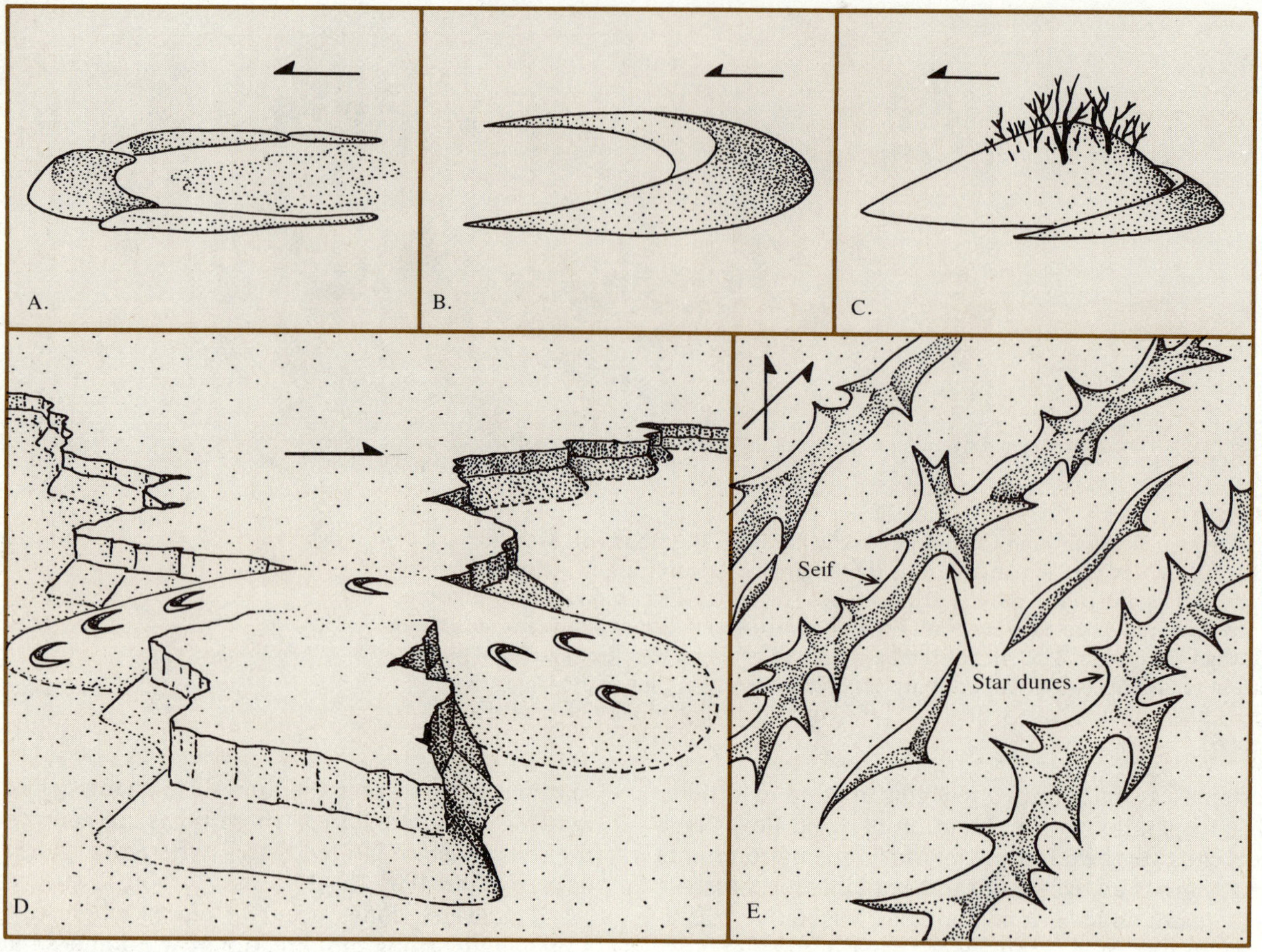

FIGURE 15–24
Major dune forms. All major aeolian landforms have elements of two basic classes: longitudinal dunes and barchans. A. This *parabolic dune* may have formed because the ends (*horns*) became anchored by some obstacle or by a rise of the water table; the perennially dry pan or playa basin on the lee side is a common associate of parabolic dunes and often is an expression of the seasonal water-table level. Some parabolic forms may be former barchans that inverted as the horns stabilized and the center "blew through." B. A typical barchan. C. A *coppice dune,* or *shrub coppice dune* if vegetated. D. *Climbing dunes* (left) and *falling dunes* (right), as pictured in Fig. 15–26. A few barchans may re-form when sand streams across the uplands, as shown here. E. A series of longitudinal dunes (seifs) depicted as sinuous, parallel ridges; crossing wind patterns may join barchanoid and longitudinal elements and form *pyramidal* or *stellate dunes* where wind shifts direction either seasonally or annually. However, the position of the stellate dune remains relatively fixed, or static. Arrows indicate prevailing wind direction.

chan dunes are being blown by deflation from one basin to another across low passes in mountain ranges. Though the barchan form is destroyed during this interbasin transfer of sand, the barchan re-forms as the sand reaches the flatter floor of the receiving basin. These are **climbing dunes** where they rise uphill and **falling dunes** where they cascade down the opposite side of the range (Fig. 15–27). Some passes through which the sand moves are nearly 600 m above the adjacent basin floors.

It should be remembered that mobile dunes exist only as long as there is a continuing source of sand, and as long as no natural or artificial stabilization process exists to fix them in place. In fact, examples may be found of geologically recent stabilized dunes of all types. Elsewhere, the stripping of anchoring vegetation

FIGURE 15–25
An oblique air view of Sand Mountain east of Fallon, Nevada. The predominant wind flow is from bottom to top; the dune rises 90 m above the surrounding surface. As shown schematically in Fig. 15–24E, the dune is a complex admixture of barchan ridges and a sinuous longitudinal ridge. Stellate dunes occur where the two ridge types intersect. The source of the sand is some distance to the southwest, but at Sand Mountain further movement is affected by mountainous terrain and strong seasonal variations in wind flow. Springs issue from the base of the dune and were once used by Indians and Pony Express riders as a fresh-water source in an area otherwise almost devoid of drinkable water.

from fixed dune or sheet sand surfaces to place land into agricultural production has reactivated formerly fixed forms. It is not generally appreciated that most sands are essentially gigantic sponges soaking up and retaining much of the limited water that reaches their surface. The digging of a shallow hole in any large dune will convince the observer that the sand is always moist at a depth not far below the surface; if true dunes did not have "wet feet" to aid in maintaining their form, they would quickly deflate away. Thus, destruction of native vegetation, lowering of the local water table by pumping, or removal of capillary water within the dune may lead to large-scale reactivation of formerly fixed surfaces (Fig. 15–28).

RELICIT DUNE FORMS

Elsewhere, **relicit** or **fossil dunes** exist. These features are usually deposits that accumulated in response to locally severe climates during the recent Ice Age. A prime example of a relict dune field is the Sand Hills region of Nebraska (Fig. 15–29). If this 57,000-km² area were active, it would be one of the largest mobile dune fields on earth. The Sand Hills were formed by deflation of sand from the floodplains of streams that flowed eastward from the Rocky Mountains at a time when the continental ice border stood several hundred kilometers north and to the northeast, so that vegetation was lacking or scanty and strong circumglacial winds could remove the sand from the valleys and redeposit it on adjacent uplands. At the same time, large quantities of outwash sand, derived from alpine glaciers in the mountains, were moved down the eastward-flowing streams. Currently, the region is a rolling grasslands (savanna); the grass stabilizes the dunes. These ancient dunes are mostly longitudinal, but are locally an intricate mixture of several dune

FIGURE 15–26
A small field of barchan dunes in southwestern Colorado. The barchan in the center of the picture is moving from right to left as indicated by the horns, which point downwind. The small depression at the foot of the *slip,* or downwind face, is covered by a sparse growth of low shrubs and grass. This is a cattle-raising area, and fence lines such as those shown crossing the dune must be constantly reset to prevent their burial by sand.

FIGURE 15–27
Sand may be deflated from one basin into another, moving across upland terrain or through low passes in bounding mountain ranges. This oblique air view, taken looking north along the axis of the Bloody Run Hills north of Winnemucca, Nevada, shows sand moving upslope and across a low pass in the hills (left to right). These climbing dunes originate in a field of barchans to the west (left) and climb 550 m above the terrain level of the adjoining basins on both sides of the pass. After crossing the divide, the sand becomes a falling dune to the east (right) before reassembling into another series of barchan and parabolic dunes.

types, indicating a complex history of changing wind-flow patterns.

In western Europe, China, and the Mississippi and Missouri river valleys of North America, glacial climates and wind flow were also responsible for extensive deflation of silts, clays, and some very-fine-grained sands from the valley trains of major streams. These airborne particles were redeposited as a thin, nonstratified veneer over much of the adjacent uplands in a downwind direction, but near the larger streams accumulated to thicknesses ranging from a few centimeters to more than 30 m. This wind-blown **loess** makes for highly fertile soils. Where cut by excavation, it tends to stand upright as bluffs or cliffs (Fig. 15–30).

FIGURE 15–28
A field of formerly stabilized barchan dunes at the south end of Silver State Valley northwest of Winnemucca, Nevada. The dark, circular areas in the distance are potato fields irrigated by center-pivot water distribution systems, the water obtained by pumping groundwater. Prior to irrigation, the water table was at or near the surface in the small depressions in the barchan field. Now, dewatering has lowered the water table; sand is starting to move and deflate from this formerly stable dune area.

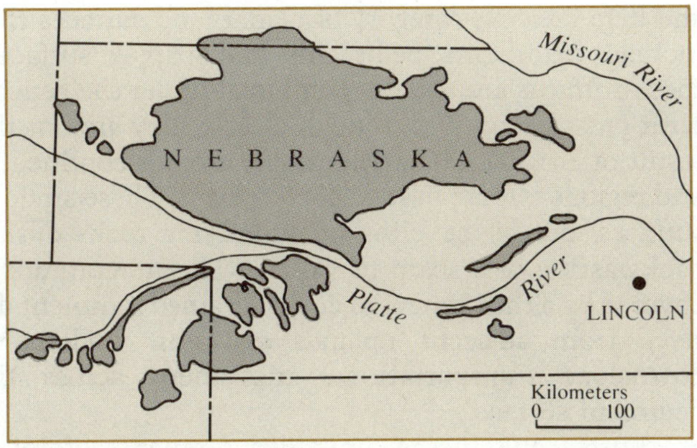

FIGURE 15–29
Map of the Sand Hills of Nebraska showing the location of one of the world's largest stabilized fields of fossil or relict dunes. The dunes probably accumulated during more than one episode of deflation of floodplain and glacial outwash deposits, because cross sections through the dunes as seen in roadcuts and stream valleys show buried soils and organic layers sandwiched between cross-bedded dune sands. (Modified after Geological Society of America, 1952)

HUMANS AND THE ARID ZONES

The world's population is growing rapidly. It is natural that the search for new sources of mineral raw materials and enlargement of agricultural lands has turned to an increased exploration of sparsely settled desert areas and an intensification of agricultural and pastoral development in semiarid regions. Any additional use of this terrain is, however, fraught with danger.

The previous discussion has shown how local vegetation of dry areas plays an important role in stabilization of climate, control of sediment erosion, and distribution and conservation of limited water resources. Deserts and their bordering semiarid lands comprise an extremely fragile ecosystem. Any poorly planned development by humans disturbs a carefully balanced system, and response to change is sudden and sometimes disastrous. History books are filled with examples of past desertification induced by overgrazing of semiarid grasslands, deforestation by cutting of timber or brush (which leads to increased slope erosion and flooding), and remobilization of stabilized dunes and sheet sands by removal of anchoring vegetation in order to place the land under the plow.

FIGURE 15–30
A loess cliff exposed along the east side of the Missouri River at the small community of Missouri Valley, Iowa. The vertical cliff face is more than 20 m high and is composed mostly of lime-cemented, wind-blown silts; however, the silts are interbedded with fine-grained sands and contain evidence of former surfaces on which plants grew and soil-forming processes operated. Thus, the total accumulation of loess is the product of repeated episodes of wind deflation of fine-grained sediments from the Missouri River floodplain when the river was a spillway that carried glacial meltwaters from the continental ice front. The organic and soil zones formed at times when there were pauses in silt sedimentation.

Clearly, water is the most precious resource of arid lands. Although through-flowing streams such as the Colorado, Nile, and Niger rivers are perennial sources, any activities in their watersheds that tend to alter streamflow, increase pollution, or result in important changes in sediment loading or water salinity may have sudden and disastrous effects on people who rely on this resource. Careful monitoring and prudent use of surface water is therefore mandatory. Groundwater is often an abundant and little-tapped resource, but withdrawal for agricultural, urban, or industrial use is commonly far in excess of natural recharge to the aquifers. By definition, this mining of water has a finite limit. Further expansions, or even the maintenance of the present consumption level by large urban areas in the industralized countries, is impossible. Unless the remaining water of such areas is carefully allocated, conserved, and recycled, history will repeat; these communities could have the same fate as certain cities of central Asia that were abandoned three centuries ago, and today are surrounded only by the shifting, whispering sands.

SUMMARY

Arid zones cover about one-third of the earth's surface. Arid lands may be categorized as hot deserts, temperate deserts, coastal deserts, steppes, tundra, or polar deserts depending on latitude, temperature regime, topographic position, and relative access to moisture. However, all share the problems of rainfall shortage, dirth or lack of vegetation, and the tendency toward an average climate that promotes an excess of evaporation over precipitation.

Although arid zones are deficient in perennial streams and lakes, running water nevertheless is the principal agent of landscape sculpture. Though rainfall is scant, a lack of continuous vegetative cover leads to rapid runoff during brief, but violent, convective storms, and flash flooding of channels and extensive bank overflow are common. Depositional landforms that result from running water, such as alluvial fans, bajadas, and braided stream channels, are found in other climatic zones, but they reach their maximum development and are best studied in arid regions.

Erosional landforms are equally well displayed. Badland topography develops in semiarid steppe regions, especially where anchoring vegetation has been removed by grazing, plowing, or deforestation. Sheet flooding and absence of a thick protective soil cover accentuate erosion, transportation, and desposition and help create a category of planar land surfaces that includes pediments, pediplains, and stripped surfaces; these surfaces are usually combined under the general term *erosion surfaces*, though clearly they are not the result of erosion alone. Pediments are not confined to arid regions but are best displayed there. These smooth surfaces, cut by beveling of underlying rocks during backwasting of a scarp or valley wall, are commonly covered by a thin veneer of coarse-grained sediment derived from adjacent uplands and now undergoing downslope transportational movement across the pediment surface.

Buttes and mesas, uneroded remnants of interstream divides or uplands, rise above the general level of stripped surfaces, which form on flat or gently dipping sedimentary strata of relatively uniform composition. Inselbergs resemble buttes but are actually circumalluviated rock spurs cut off from a mountain front by differential scarp retreat during pediplaination.

In tectonically active areas, uplift of a mountain block locally may cause a change from a depositional mode to a downcutting one as a result of base-level change. This leads to canyon incision which may leave planar rock benches high on valley walls, and is one cause for the formation of the deep, flat-bottomed fan trenches that incise the surface of alluvial fans. Climatic change without any base-level change may, however, result also in canyon cutting and fan entrenchment.

Many true desert regions are sites of interior basins. These basins have no outlet to the ocean and thus become traps, or sinks, that accumulate water and sediment from surrounding uplands. Playa lakes are ephemeral water bodies ponded in sinks or bolson basins. As water evaporates, mud cracks form in fine-grained playa sediments, and saline or alkaline salt deposits precipitate to form a glistening coating on the dry playa surface.

Wind action plays an important, but secondary, role in desert landscapes. Most windborne sand accumulates in thin, furrowed and rippled sand sheets as sand is saltated in the downwind direction. Although only about 20 percent of desert areas are covered by large dunes, sand may accumulate into giant bed-form complexes covering hundreds of square kilometers.

Dunes may be classified in a number of ways, but most fall within the general classes of transverse, longitudinal, and barchan. The longitudinal axes of transverse dunes are perpendicular to, and those of longitudinal dunes parallel to, the dominant wind-flow direction. Barchan dunes occur on smooth underlying surfaces where wind flow is moderate and constant and sand supply is adequate but not abundant. Dune sand may be deflated and moved hundreds of meters upward to spill across gaps in mountain ranges; thus, dunes re-form in adjacent basins. These are climbing dunes where they rise uphill and falling dunes where they decline down the opposite side of the range. Dunes remain mobile, however, only as long as there is an abundant sand supply, and over the long term there is a tendency toward anchoring and stabilization by vegetation. In some regions of the world, humans today are destroying stabilized dunes by removing native grasses and shrubs for grazing or plowing, and also are lowering the water table by irrigation. The latter process increases locally the amount of dry sand available for remobilization.

Fossil dunes and loess are relict aeolian deposits inherited from the Ice Age. Excellent examples of these occur in the Nebraska Sand Hills and along the bluffs of the Mississippi River valley, respectively.

REVIEW EXERCISES

1 What features do all deserts have in common, regardless of their topographic position and geographic location?

2 There are several types of dunes. How is the type influenced by wind direction and velocity, quantity of sand available, and amount and type of vegetation or other obstacles to sand movement?

3 Explain why fan trenches form in alluvial fans. Also, why are trenches in some places confined to the heads of fans, but elsewhere cross the entire fan?

4 Why is the groundwater table invariably near the ground surface in desert basins that contain salinas or playa lakes?

5 How does particle grain size determine whether wind of a certain velocity will move these particles in suspension, by saltation, or as bed-load creep?

6 Why are loess deposits thickest on uplands immediately adjacent to the river valleys from which the loess was derived?

7 The Nebraska Sand Hills are used for livestock grazing but not plowed extensively for row crop agriculture. Why is this an example of excellent land use management?

8 How is an inselberg formed during scarp retreat, and what is its ultimate fate?

9 What is the fundamental difference between an exposed pediment surface and a stripped surface, even though they strongly resemble each other?

10 Why is groundwater in deserts or semiarid zones often not fit for human consumption?

11 What is desert pavement, and how does it form?

12 Describe the geologic, climatic, and vegetative conditions necessary for the creation of badland topography.

13 What are the principal geological processes that operate in deserts? Order them according to importance.

14 Describe the formation, internal composition, and longitudinal and cross-sectional geometry of alluvial fans.

15 Compare the size, discharge, pattern, and bed form characteristics of streams in arid and humid climates.

SUGGESTED READINGS

BAGNOLD, R. A. *The Physics of Blown Sand and Desert Dunes.* London: Metheun and Co., 1941, 265 pp.

BLACKWELDER, E. "Geomorphic Processes in the Desert." In *Geomorphology of Southern California.* California Division of Mines Bulletin 170, 1954, pp. 11–20.

BULL, W. B. "Alluvial Fans." *Journal of Geological Education* (1968), Vol. 16, No. 3, pp. 101–8.

COOKE, R. U., and WARREN, A. *Geomorphology in Deserts.* Berkeley, Calif.: University of California Press, 1973, 374 pp.

DOEHRING, D O., ed. "Geomorphology in Arid Regions." *Proceedings of the Eighth Annul Geomorphology Symposium.* Binghamton, N.Y.: State University of New York, 1977, 272 pp.

HACK, J. T. "Dunes of the Western Navajo Country." *Geographical Review* (1941), Vol. 31, No. 2, pp. 240–63.

HADLEY, R. F. "Pediments and Pediment-forming Processes." *Journal of Geological Education* (1967), Vol. 15, No. 2, pp. 83–89.

MCKEE, E. D., ed. *A Study of Global Sand Seas.* U.S. Geological Survey Professional Paper 1052, 1979, 429 pp.

NORRIS, R. M. "Barchan Dunes of Imperial Valley, California." *Journal of Geology* (1966), Vol. 74, No. 3, pp. 292–306.

PETROV, M. V. *Deserts of the World.* New York: John Wiley and Sons, 1976, 447 pp.

WALTON, K. *The Arid Zones.* Chicago: Aldine Publishing Co., 1969, 175 pp.

- Coriolis effect
- boundary current
- upwelling
- deep-water mass
- wavelength
- wave height
- wave period
- wind-generated wave
- longshore drift
- rip current
- shore
- wave-cut cliff
- sea cave
- sea arch
- stack
- spit

16

- seas
- swell
- surf zone
- breaker
- beach
- tsunami
- new moon
- full moon
- tidal range
- spring tide
- neap tide
- tidal bore
- tidal current
- refraction
- longshore current
- bay barrier
- tombolo
- barrier island
- delta deposit
- marine terrace
- tectonic movement
- eustatic movement
- glacioeustatic oscillation
- Atlantic-type margin
- Pacific-type margin
- primary coast
- secondary coast
- fringing reef
- barrier reef
- atoll

Oceans and Shorelines

SINCE HERODOTUS DEDUCED THAT "EGYPT IS THE GIFT OF THE [Nile] river" and Aristotle recognized that earthquakes and volcanoes resulted from "winds" trapped within the earth rather than from supernatural manipulations, increases in knowledge of the earth's geologic condition have come primarily from study of the continents. This avoidance of the ocean floor resulted from the lack of technological capability for observing the deep-ocean floor.

It was early recognized, however, that most of the rocks exposed at the surface of continents were sedimentary rocks that had probably formed in a marine environment. In time, this tantalizing clue that there was some unrecognized relationship between the continents and the ocean floor was intensified when Alfred Wegener's theory of continental drift was presented and promptly rejected by most of the geological community.

It was not until the 1960s when technology had developed sufficiently to allow for a meaningful investigation of the geology of the deep ocean that the beautifully simple pattern of global plate tectonics discussed in Chapter 8 began to emerge. The realization of the existence of this process at work on the earth has greatly improved the understanding of such phenomena as earthquakes, volcanism, and mountain building.

In this chapter, emphasis will be placed on a brief overview of the physical motion of ocean water and a more detailed discussion of the shores of the continents modified by these motions. These motions include currents, waves generated by wind and seismic disturbances, and tides. The most significant of these phenomena in modifying the shore are wind-generated waves that erode exposed segments of the coast and deposit sediment in the protected areas.

The configuration of present shores is also determined by factors other than the shoreline processes resulting from the water motions of the present ocean. The melting of continental glaciers from about 18,000 to 3000 years ago caused a rise in sea level of some 120 m. This must have resulted in the flooding of the mouths of rivers throughout the earth, producing profound changes in the shoreline. Until wave action had a chance to modify these shorelines created by the rise in sea level, many features of the shore were primarily the product of geological phenomena such as glaciation, faulting, reef development, stream erosion, and deposition and volcanism. Still today, many stretches of shore show the combined effects of these various phenomena. Although the processes of marine erosion and deposition will be stressed in this chapter, these other processes that modify the shore must be kept in mind when trying to understand the features of any shore.

CURRENTS – THE GENERAL CIRCULATION OF THE OCEANS

Navigators have long known ocean currents to be movements of ocean water that can hinder or aid their ships on their way. Benjamin Franklin assisted in the understanding of ocean currents by publishing the first chart of the Gulf Stream in 1777 (Fig. 16–1).

In the early days of the study of marine biology, a controversy developed over the existence of life in the deep ocean. Those who thought life could not exist there based their belief on the conviction that there could be no oxygen dissolved in the deep water of the oceans. With the only sources of oxygen in the ocean being the atmosphere and plant photosynthesis confined to the top 100 m, how could free oxygen find its way into the waters at the bottom of the ocean? Further investigation proved that life did exist on the deep-ocean floor, and that oxygen was made available to the animals that lived there by currents of cold, dense water that sank into the deep ocean from subpolar surface waters.

The paths of ocean currents were first investigated using drift bottles containing requests for finders to notify the investigators of dates and positions of recovery. Current flow is now determined by **current meters** distributed throughout the water column from an anchor on the ocean floor. The data obtained are recorded electronically. In addition, **floats** designed to float at any depth in the ocean carry a "pinger" with them to tell shipboard investigators of their changing position. Floats whose positions are periodically monitored by satellites are also used in increasing our knowledge of ocean current flow.

WIND-DRIVEN CIRCULATION

The driving forces behind open-ocean circulation are the prevailing wind systems—the trade winds, westerlies, and polar easterlies (Fig. 16–2). Due to forces growing out of the fact that the earth rotates on its axis at velocities ranging from 0 km/hour at the poles to over 1600 km/hour at the equator, objects in motion in the Northern Hemisphere and Southern Hemisphere tend to veer to the right and left, respectively, of their intended paths. This phenomenon, called the **Coriolis effect** after the French physicist who first explained it, causes a clockwise rotational gyre centered in the subtropics of ocean basins in the Northern Hemisphere, and counterclockwise rotations in ocean basins of the Southern Hemisphere (Fig. 16–3). In subpolar latitudes, the polar easterlies tend to set up smaller rotational gyres in a direction opposite to that found in the adjacent subtropical gyre.

Boundary currents on the east and west sides of the subtropical gyres play an important role in the transport of excess heat energy found in the low latitudes to the heat-deficient higher latitudes. As a result of a balance reached among forces growing out of increased magnitude of the Coriolis effect with increased latitude, wind stress putting energy into the system, and bottom friction removing energy, the western boundary currents are more intense and flow at higher velocities than the eastern boundary currents. For example, the Gulf Stream flowing north along the western boundary of the North Atlantic Ocean may reach a velocity of 8 km/hour, whereas the California Current flowing south along the eastern boundary of the North Pacific Ocean usually travels at velocities below 1 km/hour.

Where ocean water is driven away from the continents and equatorial regions by winds, an **upwelling** of nutrient-rich water from depths below 100 m increases biological productivity (Fig. 16–4). A historical record of upwelling patterns is found in the sediments on the ocean floor, since many of the one-celled plants and animals that benefit from this enrichment secrete hard cell walls of calcium carbonate or silica.

DENSITY-DRIVEN VERTICAL CIRCULATION

The cold, dense subpolar water masses that sink to the ocean floor from opposite ends of the Atlantic Ocean are the North Atlantic Deep Water, which forms in surface waters of the Norwegian Sea, and the Antarctic Bottom Water, which forms primarily at the seaward margin of the Weddell Sea (Fig. 16–5). When these two water masses meet while moving across the ocean bottom just north of the equator, the North Atlantic Deep Water rises above the Antarctic Bottom Water, which is the densest water in the ocean.

Water that sinks as part of these **deep-water masses** may stay beneath the surface from 200 to 1200 years. The average age of deep-ocean water is about 1000 years. A possible pattern of deep-water circulation that indicates a more intense flow of water along the western margin of deep-ocean basins is shown in Fig. 16–6. The westward intensification, which has the same cause as the high-velocity western boundary currents at the surface, produces current velocities up to 1 km/hour. This may not seem very fast compared to the velocity of some surface currents, but it is sufficient to erode and transport the sediment on the continental rise as is discussed in Chapter 5.

FIGURE 16–1
First map of the Gulf Stream, published in 1777 by Benjamin Franklin. (Courtesy U.S. Navy)

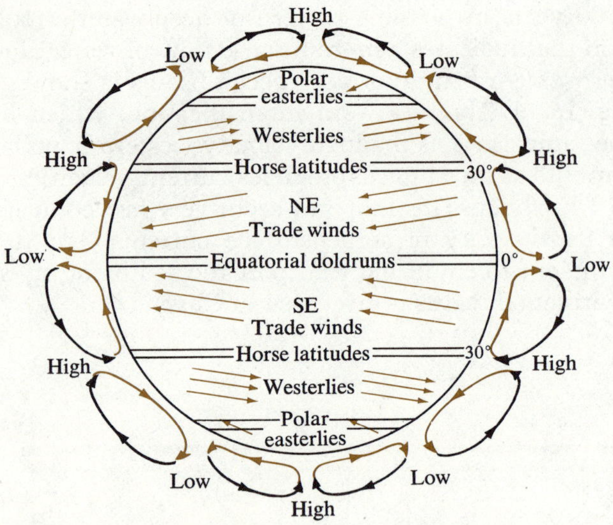

FIGURE 16–2
Prevailing wind systems. Low-pressure belts develop along the equator and the 60° latitude regions due to rising columns of warm air. Descending cool, dry air produces high-pressure belts in the 30° latitude regions. Air movements are primarily vertical in these belts. Strong lateral movements of air between the belts produce the westerlies and trade winds. Due to cold air masses overlying the polar regions, high-pressure conditions exist there also.

energy movement in **transverse waves**. Where the particles orbit, they move through the top of their orbit in the direction of energy propagation. **Orbital waves** have components of both longitudinal and transverse particle motion.

Waves on the ocean surface, like all progressive waves that transmit energy along an interface between two fluids of different density, are characterized by orbital particle motion. The fluids in contact at this interface are the low-density atmosphere and the high-density ocean water.

The terms used to describe waves are shown in Fig. 16–7. The **wavelength** is the horizontal distance between adjacent **crests** or **troughs**. **Wave height** is the vertical distance between crest and trough. The time required for the passage of one wavelength is called the **wave period**.

WIND-GENERATED WAVES

By far the most common and geologically important waves found in the ocean are **wind-generated waves**. These waves are created in storm areas called **seas**. As the wind blows over the sea area, the amount of energy put into the waves increases with increases in: (1) the wind velocity, (2) the time the wind blows in one direction, and (3) the **fetch** (distance over which the wind blows). As waves gain energy, they increase in height, length, and velocity.

When waves move out of the sea area, they become long, round, crested waves called **swell**. The swell transports the energy put into the waves in the sea across vast stretches of open ocean with very little loss of energy.

When waves eventually strike the shores of continents, they release their energy by breaking in the **surf**

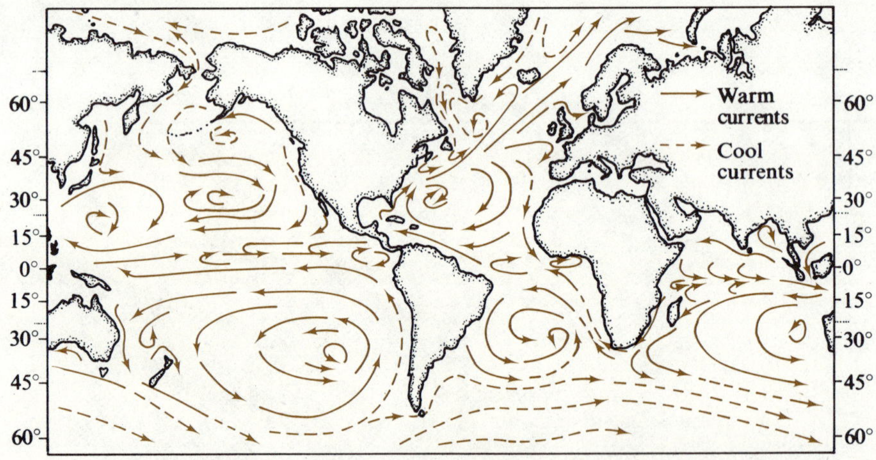

FIGURE 16–3
Circulation of surface waters of the oceans.

WAVES

Wave phenomena represent a method of translating energy from one place to another by oscillatory motion of particles. In wave manifestations where the waveform can be observed to move—that is, in **progressive waves**—the particle motion may be longitudinal, transverse, or orbital (Fig. 16–7). In **longitudinal waves**, the particles move back and forth in line with the direction of energy propagation. The particles move back and forth at right angles to the direction of

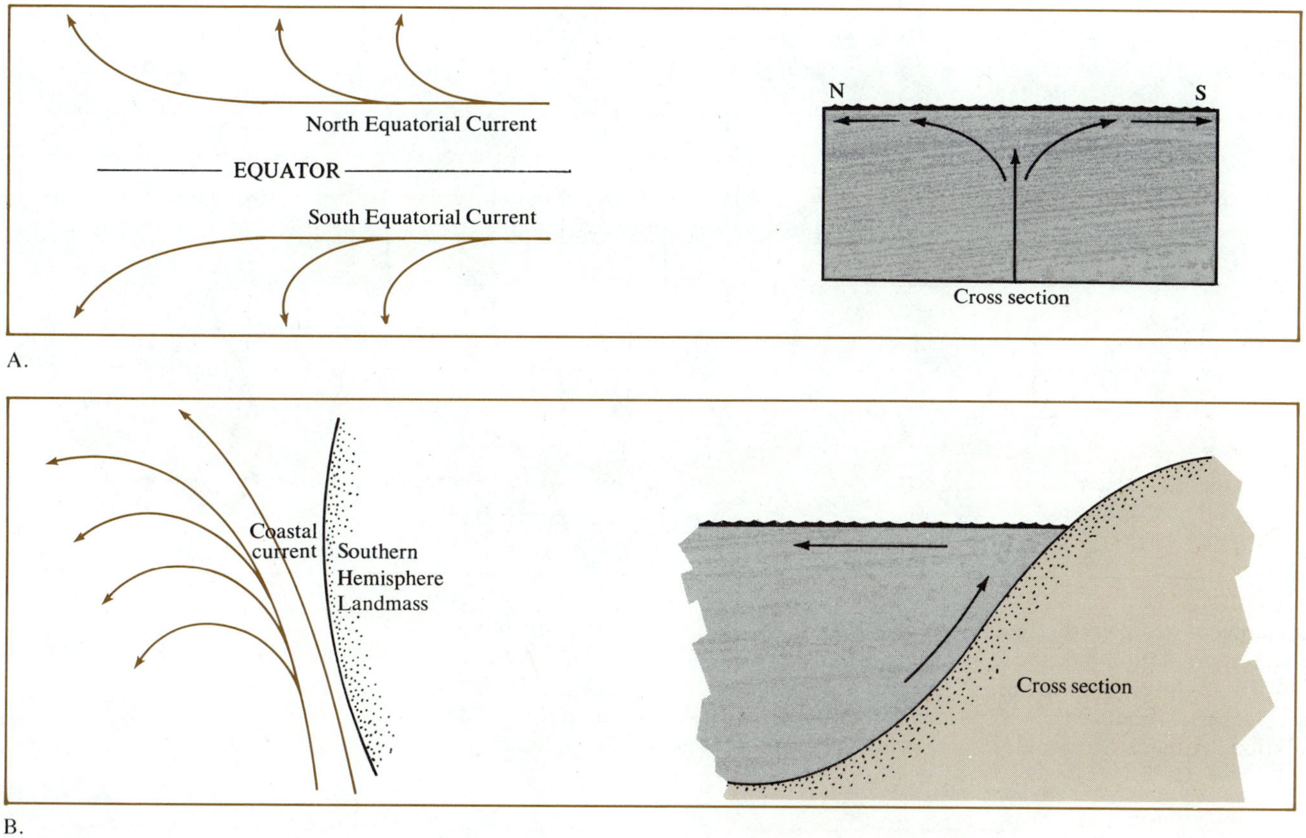

FIGURE 16–4
Upwelling. A. Equatorial upwelling. The Coriolis effect acting on westward-flowing equatorial currents pulls surface water away from the equatorial region. The water is replaced by subsurface water. B. Coastal upwelling. In areas where wind-driven coastal currents flow along the western margin of continents and toward the equator, the Coriolis effect causes surface water to flow away from the continent. An upwelling of deeper water replaces the surface water that has moved away from the coast.

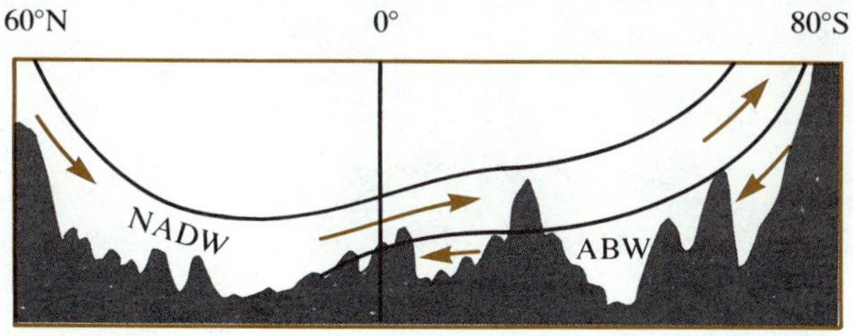

FIGURE 16–5
Deep-water masses. Deep mixing of the oceans is achieved by the sinking of the North Atlantic Deep Water (NADW) and the Antarctic Bottom Water (ABW) in the high latitudes of the North Atlantic and South Atlantic ocean basins, respectively.

zone. Particle motion associated with waves extends to a depth equal to one-half the wavelength, called the **wave base**. When water depth becomes less than the wave base, the bottom begins to interfere with the particle motion, and friction slows the wave (Fig. 16–8). Once the waves have reached shallow water, each wave is moving slower than the one behind it. This causes a decrease in wavelength and an increase in wave height. Although the waveform is slowing, the particles at the surface are not interfered with by the bottom. When the height of the wave reaches one-seventh its length, the surface particles are moving forward in their orbits faster than the waveform, and without the waveform to support them the surface particles of water crash against the shore as a **breaker**. The energy released by the breaking wave pushes the water up the shore as **swash**. The swash is a turbulent sheet of water that moves sediment up the shore slope. When conditions are right, this sediment is deposited as a **beach**. In beach areas, after the swash loses its energy against the shore, some of the swash will seep down into the beach sediment. The rest runs down the beach slope as **backwash**.

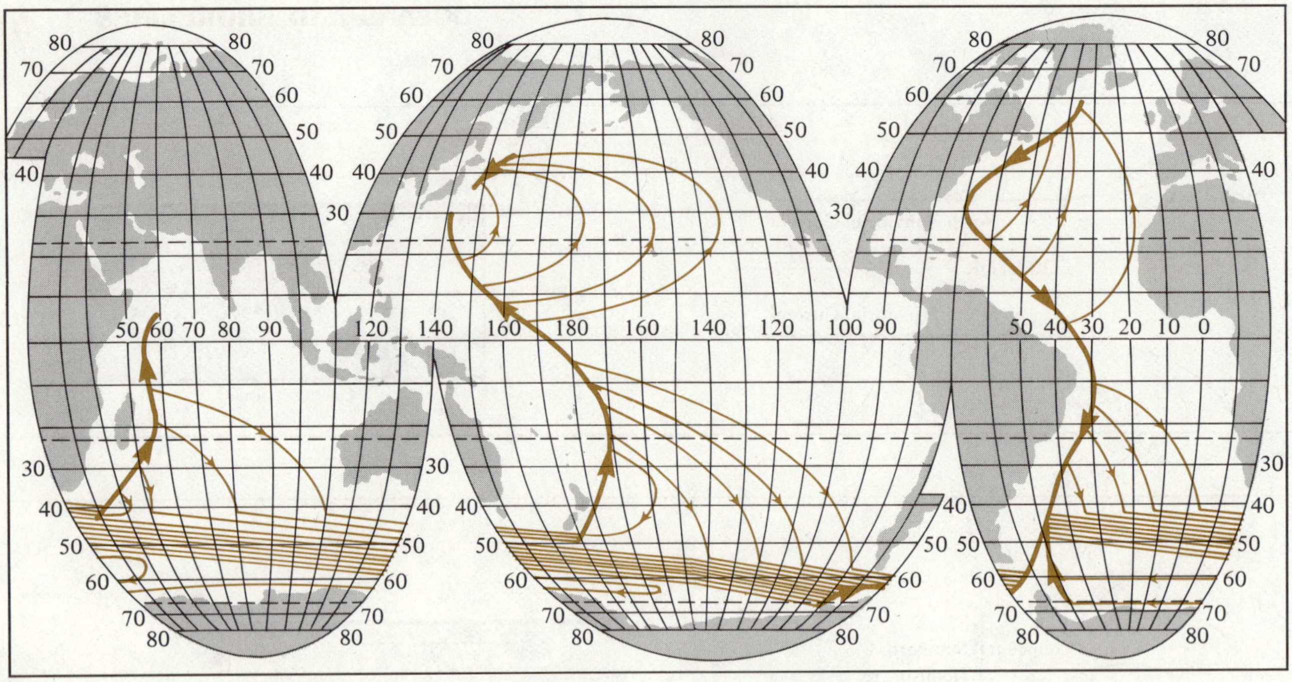

FIGURE 16–6
Model of deep-ocean circulation. Based on information obtained to date, this model of deep-water circulation developed by Henry Stommel in 1958 appears to be reasonably correct. Heavy lines mark the major western boundary currents, which result from the same forces that produce the more intense western boundary currents in the surface circulation. (After H. Stommel, 1958; base map courtesy National Ocean Survey)

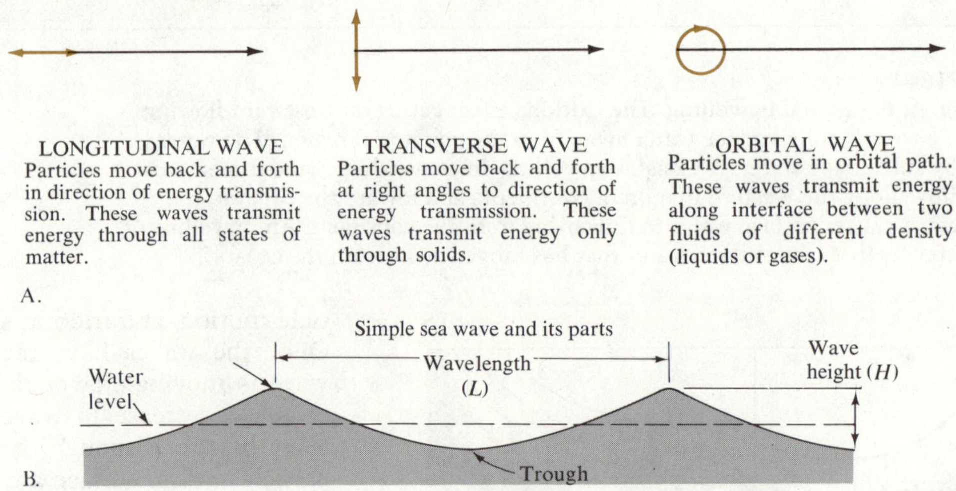

FIGURE 16–7
Types of progressive waves.

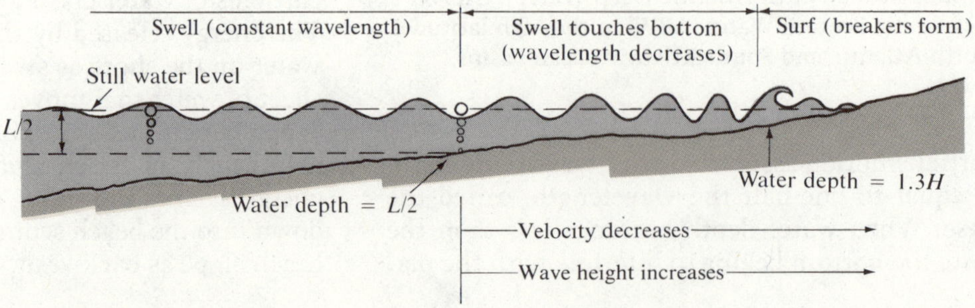

FIGURE 16–8
The surf zone. As swell approaches the shore and the depth of the ocean becomes less than one-half the wavelength, the waves begin to lose energy. The velocity and wavelength decrease while wave height increases. When the height of the wave becomes one-seventh the wavelength, the wave breaks and releases most of its energy in the surf zone.

TSUNAMI

Less common than wind-generated waves, but possessing vast quantities of energy, are **tsunami**, or seismic sea waves. Tsunami are generated by earthquakes, volcanic eruptions, and submarine landslides and can have disastrous effects on the shore.

The energy of tsunami is distributed throughout the water column as the energy is derived from disturbances on the ocean floor (Fig. 16–9). The waves move away from the point of origin in all directions at velocities ranging from 480 to 800 km/hour. The velocity increases with increasing water depth. Wavelength ranges from 55 to 200 km, and wave height from 0.3 m to 36 m. In the open ocean, wave heights rarely exceed 0.3 m, so these long, low waves can pass unnoticed. However, as they approach shallow coastal areas, they slow and increase in height. Very commonly, tsunami reach heights of about 15 m as they crash across coastal areas. There have been some instances of waves over 30 m high striking shores along the margins of the Pacific and Indian oceans.

One of the most destructive tsunami ever generated was associated with the greatest release of energy from the earth's interior observed during historical times. On 27 August 1883, the island of Krakatoa in the Sunda Strait, between Sumatra and Java, exploded and essentially disappeared. The sound of the explosion was heard 4800 km away at Madagascar, and the dust that ascended into the atmosphere circled the earth and produced unusual and beautiful sunsets for nearly a year. A tsunami that rose to heights of more than 30 m devastated the coastal region of the Sunda Strait, taking more than 36,000 lives. The energy carried by this was was still detectable after crossing the Indian Ocean, passing the southern tip of Africa, and moving north in the Atlantic Ocean into the English Channel.

The present International Tsunami Warning System was developed in response to a destructive tsunami that resulted from an earthquake in the Aleutian trench on 1 April 1946. Prior to the development of this system, the first notice that most observers had of an impending tsunami was the seaward recession of the shoreline as a trough developed ahead of the first wave crest to reach shore (Fig. 16–10). Such a recession was observed in the port of Hilo, Hawaii, when the 1 April tsunami struck. The recession was followed by a wave that carried the water to heights nearly 8 m above the normal high-tide level. The tsunami also struck Scotch Cap, Alaska, on Unimak Island. The base of the lighthouse at Scotch Cap stood 14 m above sea level and was totally destroyed. Additional evidence indicated that the wave reached a height of about 36 m.

TIDES

Ocean tides are manifested as a gradual rise and fall of the ocean surface. They are a much more subtle process of ocean motion than waves and have a less significant effect on coastal processes. Tides do, however, determine at what level wave action will strike and thereby help determine the distribution of wave energy against the shore. In restricted channels, significant tidal currents develop that will affect erosion and sediment transport.

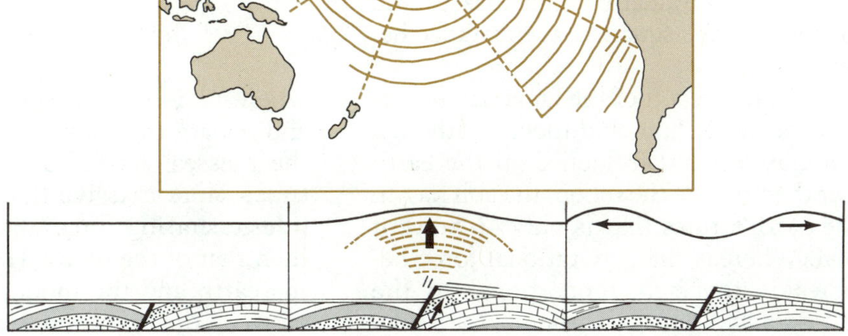

FIGURE 16–9
Movement along a fault in the earth's crust sends energy to the ocean surface. This energy then is distributed laterally along the atmosphere-ocean interface in the form of a tsunami. The energy is transmitted across the open ocean by undetectable waves up to 200 km long and about 0.3 m high. These waves, which may become more than 30 m high, release their energy as they strike the shore.

A.

B.

FIGURE 16–10
Tsunami coming ashore. A. Sandy Beach on the Island of Oahu moments before a tsunami generated by an earthquake in the Aleutian trench area struck the beach. Note that the beach has been exposed as the shoreline moves out to sea. The arrow indicates the location of the photo in part B. B. The beach a few minutes later, after the tsunami arrived. (Photos by Y. Ishii; reprinted by permission of *Honolulu Advertiser*)

Although tides were long known to have a period similar to that of lunar motions, it was not until 1686 that Sir Isaac Newton gave us the explanation of this relationship in his law of gravitation. Newton determined that two bodies attract each another with a force proportional to the product of their masses and inversely proportional to the square of the distance between their centers.

The ocean tides are a result of the gravitational attraction between the earth, sun, and moon. Although the sun exerts a gravitational influence on the earth 177 times greater than that of the moon, the sun has an influence on the earth's tides that is only 46 percent that of the moon. Whereas the gravitational force between two bodies is inversely proportional to the *square* of the distance between them, the tide-generating force is inversely proportional to the *cube* of this distance. These relationships are given in the following equations:

$$\text{Gravitational force} = G\frac{m_1 m_2}{r^2}$$

$$\text{Tide-generating force} = \propto \frac{m_1 m_2}{r^3}$$

where G is the universal gravitational constant, m_1 and m_2 are the masses, and r is the distance between the masses. As the sun is approximately 27 million times more massive than the moon, it should have a tide-generating force about 27 million times greater than that of the moon. However, the distance between the earth and the moon as compared to the distance between the earth and the sun must also be considered.

Because the sun is 390 times farther from the earth than the moon, its tide-generating force is 390^3 (about 59 million) times less than that of the moon. These conditions, as depicted in Fig. 16–11, result in the sun's tide-generating force being 27/59 (about 46 percent) that of the moon.

As a result of this dominant lunar influence on the earth's tides, two pairs of tidal bulges, a smallar solar tide and a larger lunar tide, are directed toward and away from the bodies that have produced them. As the earth rotates through the bulges, two high tides and two low tides related to the dominant moon tide are observed at most places on the earth each day (Fig. 16–12).

The lunar tide is modified by the changing position of the solar tide bulges relative to the lunar tide bulges. The path of the earth and moon as the earth-moon system revolves around the sun is shown in Fig. 16–13. It can be seen that approximately every 29.5 days (the **synodic month**), the moon is in the same phase. When the moon is between the earth and the sun, it is said to be in **conjunction**, producing a **new moon**. The moon is in **opposition** when it is on the opposite side of the earth from the sun, causing a **full moon**. A **quarter moon** results when the moon is in **quadrature**, that is, at right angles to the sun relative to the earth.

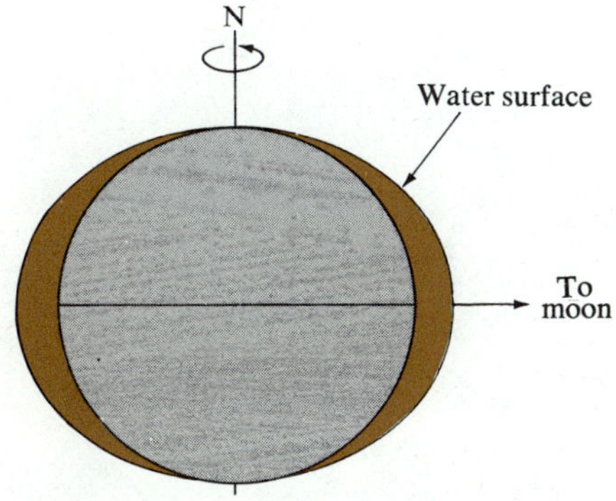

FIGURE 16–12
Tidal bulges. Because of horizontal tide-generating forces resulting from the earth-moon rotation system, water on the earth's surface is pushed into two bulges, one directly beneath the moon and the other on the opposite side of the earth from the moon. A system of solar tide bulges that are about half as large as the lunar tide bulges modify the lunar tide as they move across the earth's surface.

When the sun and moon are in opposition or conjunction, the tide-generating forces of the sun and moon are additive, and the earth experiences its greatest **tidal ranges** (vertical differences between high and low tides). During quadrature (the first or third quarter), the tide-generating force of the sun is working at right angles to the tide-generating force of the moon,

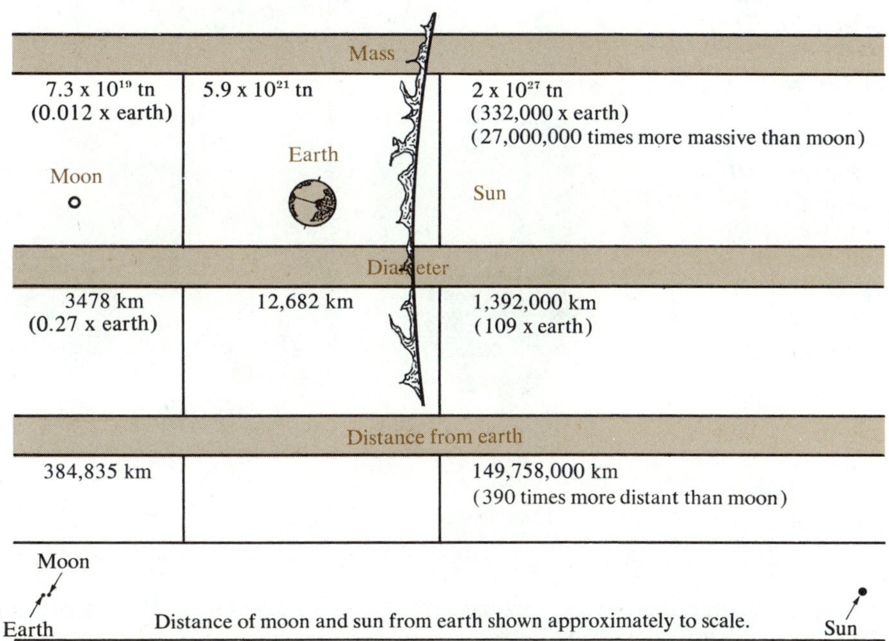

Determination of tide-generating force of sun relative to moon

$$\text{Tide-generating force} \propto \frac{\text{Mass}}{(\text{Distance})^3} \propto \frac{\text{Sun—27 million times more mass}}{(\text{Sun—390 times farther away})^3}$$

$$(390)^3 = 59{,}000{,}000 \quad \text{Thus, } \frac{27 \text{ million}}{59 \text{ million}} = 0.46 \text{ or } 46\%$$

The sun has 46% the tide-generating force of the moon.

FIGURE 16–11
A comparison of the sun and moon as tide-generating bodies.

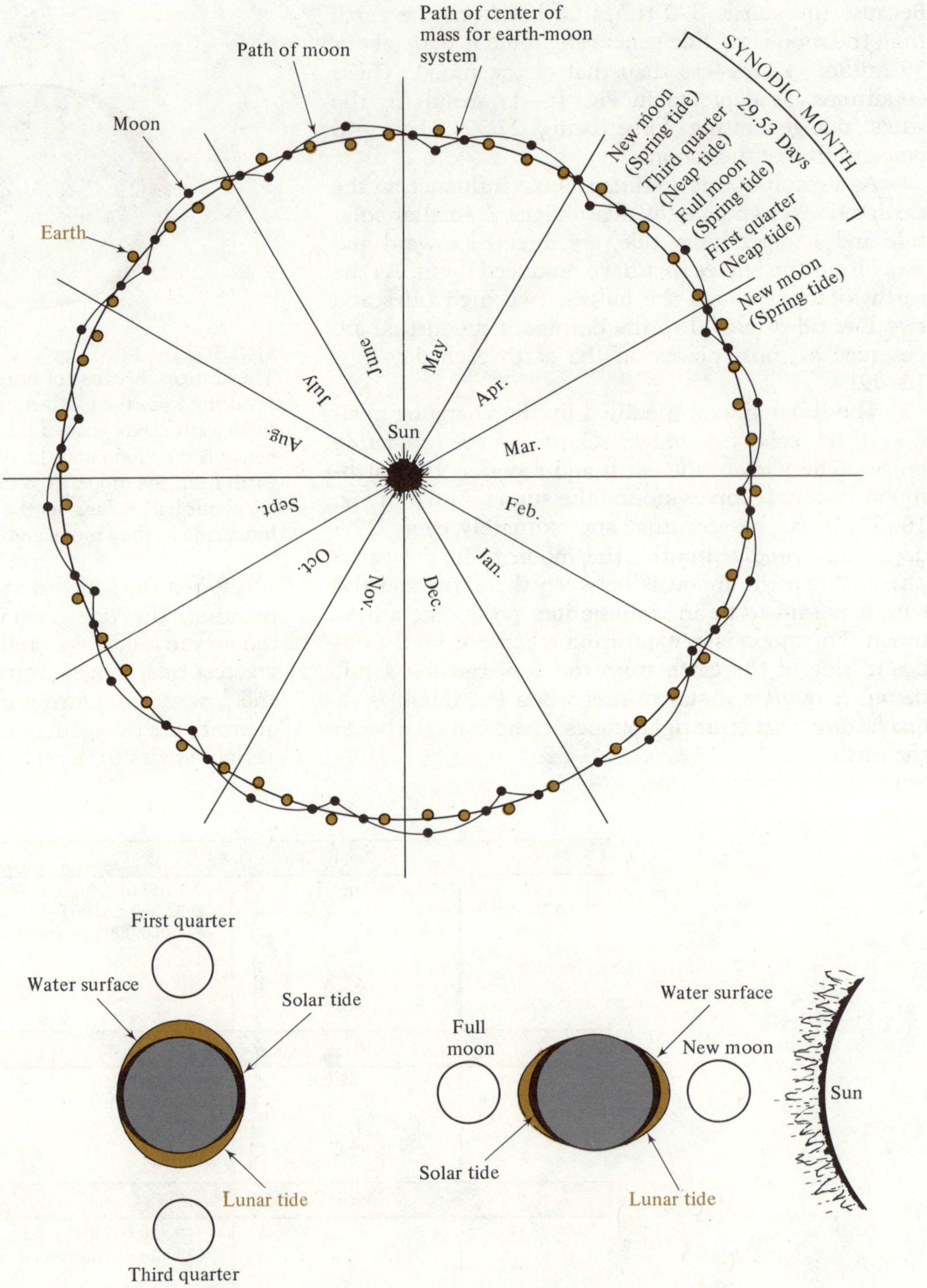

FIGURE 16–13
Sun-moon-earth motions and the tides. As the earth-moon system moves around the sun, the centers of each body follow a wavy path because they are also rotating about their center of mass. The wobble of the moon is more pronounced and is shown in exaggerated form in the figure. During a synodic month, the moon moves from a position between the sun and the earth (new moon) to a position that puts the earth between the moon and the sun (full moon) and back to its original position. When the moon is in the new or full position, the tidal bulges created by the sun and moon are aligned, producing a larger bulge. When the moon is in the positions halfway between the new and full phases—the first and third quarters—the tidal bulge produced by the moon is at right angles to the bulge created by the sun. The bulges tend to "cancel" each other, and the resulting bulge is smaller. New-moon and full-moon phases produce spring tides with maximum tidal ranges, while the first- and third-quarter phases of the moon produce neap tides with minimal tidal ranges.

and the earth's minimum tidal range occurs. The maximum tidal range condition which exists during the new- and full-moon phases is called the **spring tide**, and during the quadrature phases, **neap tides**, characterized by minimum tidal ranges, occur.

The tides are much more complicated than described in the preceding, due in part to the fact that the continents interfere with the movement of these bulges across the earth's surface. As a result of this interference, some regions on earth have only one high tide and one low tide per day. Although the difference between high and low tide throughout the earth is usually 1–2 m, it can be much greater locally. For example, as the tide is funneled into the Bay of Fundy between New Brunswick and Nova Scotia (Fig. 16–14), it rises to as much as 21 m above low-tide levels.

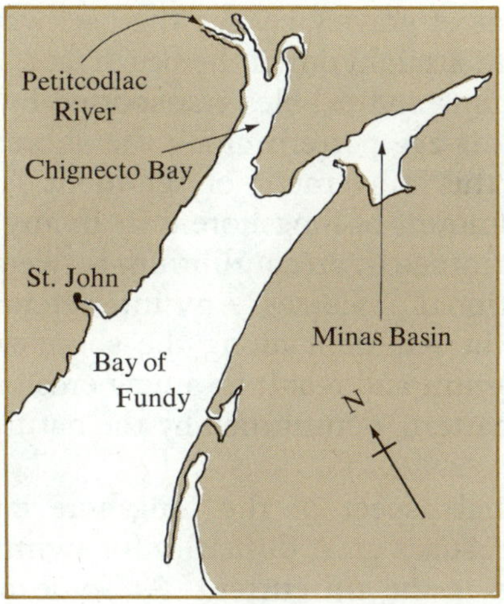

FIGURE 16–14
As the tide wave moves north into the Bay of Fundy, the "funneling" caused by decreasing depth and width of the bay produces the greatest tidal range in the world. Tidal range increases from about 2 m near the mouth of the bay to over 17 m in the Minas Basin.

Large tidal ranges are also found at the north end of the Gulf of California and along the Brittany coast of France. As tides rush up the mouths of rivers, they often reverse the flow of the river, creating steep-fronted waves called **tidal bores**. In such areas, **tidal currents** can be strong and play a significant role in eroding the ocean floor and transporting sediment.

WAVE REFRACTION

As previously discussed, waves begin to bunch up and wavelengths become shorter as swell begins to "feel bottom" on its approach to the shore. It is seldom that the swell will approach the shore at a right angle. Some segment of the wave can be expected to "feel bottom" first and therefore be slowed before the rest of the wave. This produces **refraction** (bending) of the waves as they approach the shore.

The effect of portions of a wave moving toward the shore at greater velocities than other portions of the same wave is to distribute energy unevenly along the shoreline (Fig. 16–15). Orthogonal lines constructed perpendicular to the wave front so that the amount of energy between lines is equal at all times are of great assistance in illustrating how energy is distributed along the shoreline by breaking waves. Orthogonals can be seen to converge on headlands jutting out into the ocean and to diverge in bays, indicating there is a concentration of energy released against the headlands, while energy released in the bays is spread more thinly. This condition produces a state of erosion on the headlands, while deposition may occur in the bays.

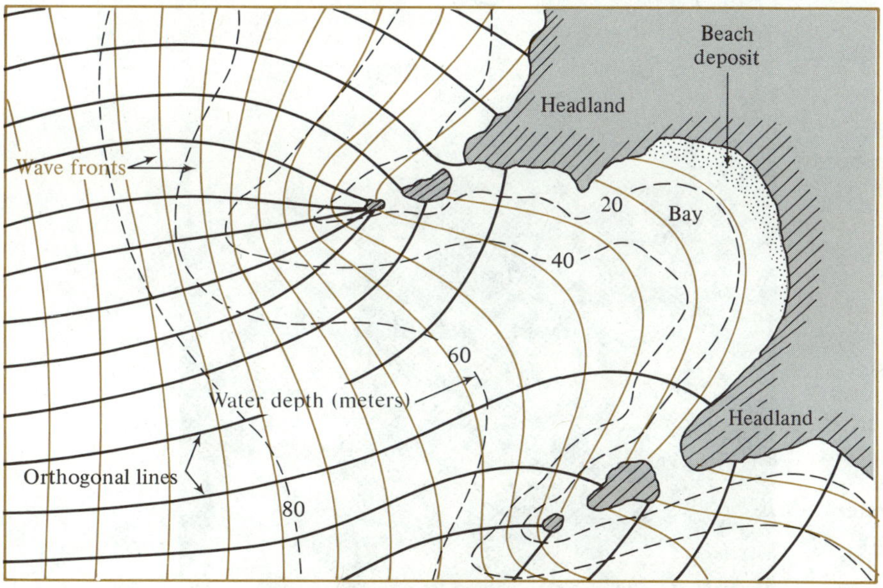

FIGURE 16–15
Refraction. As the waves "feel bottom" first in the shallow areas off the headlands, they are slowed. The segments of the waves that move through the deeper water leading into the bay are not slowed until they are well into the bay. As a result, the waves are refracted (bent), so energy is concentrated on the stacks and headlands. Erosion is active on the headlands, whereas deposition occurs in the bay where the energy level is low. Orthogonal lines spaced so that equal amounts of energy are between each line help to show the distribution of energy along the shore.

LONGSHORE CURRENT AND DRIFT

As waves strike the shore at an angle, they set up a longshore movement of water, the **longshore current**. The velocity of the longshore current increases with increasing beach slope, angle of breakers with the beach, and wave height, and with decreasing wave period. This current moves parallel to the shore between the shoreline and the breaker line, carrying with it the materials that make up the beach (Fig. 16–16A). The path of movement for water and sediment can be readily determined by observing the swash zone. The swash carries the sediment onto the beach at an angle, but the force of gravity causes the backwash to carry the sediment straight down the beach face. As a result, the sand grains and pebbles transported by the swash move in a zig-zag pattern along the shore. The term applied to this movement of sediment is **longshore drift**. The amount of longshore drift in any coastal region is determined by an equilibrium between erosional and depositional processes. Any interference with the movement of sediment along the shore will destroy this equilibrium and result in a new erosional and depositional pattern, determined by the nature of the interference.

A special aspect of the longshore current that sometimes causes great difficulty for swimmers along open coasts is the **rip current**. At some point in its

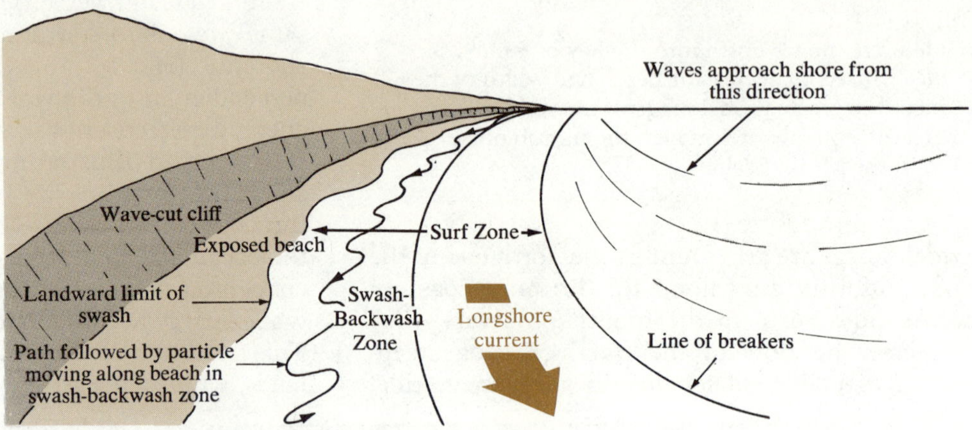

A.

B.

FIGURE 16–16
A. The longshore current extends from the line of breakers across the nearshore and the shore to the shoreline. This region is called the surf zone. The velocity of the longshore current increases with increasing beach slope, angle of the breakers with the shore, and wave height, and with decreasing wave period. The longshore drift transport of sediment follows the zig-zag pattern of the water particles in the longshore current. B. Water carried along the shore in the longshore current returns eventually to the open ocean as a rip current that flows seaward through a depression on the ocean floor. (Photo courtesy Scripps Institution of Oceanography, University of California, San Diego)

travels along the shore, the water in the longshore current moves back out to sea. The rip currents seen in Fig. 16–16B are such flows of longshore water through topographic lows on the nearshore bottom. Usually less than 25 m wide, rip currents reach velocities up to 8 km/hour. The breaking of a moderate swell will usually produce numerous rip currents of moderate velocity, and heavy swell can cause much more intense rips, although fewer in number.

GENERAL DESCRIPTION OF THE COASTAL REGION

EROSIONAL SHORE FEATURES

The landward limit of a shore dominated by erosion will normally be marked by a cliff. The **coastline**, which marks the boundary between the shore and the coast, is a line along the cliff connecting points at which the highest effective wave action takes place.

The **shore** (Fig. 16–17) is divided into the **foreshore**, that portion exposed at low tide and submerged at high tide, and the **backshore**, which extends from the normal high tide to the coastline. The **shoreline** migrates back and forth with the tide and represents the water's edge. The **nearshore** zone is that region between the low-tide shoreline and breakers. Beyond the low-tide breakers is the **offshore** zone.

The **beach** is the wave-worked sediment that moves along the **wave-cut bench** of the shore area. It may continue from the coastline across the nearshore region to the line of breakers (Fig. 16–17).

Wave Erosion Due to refraction, the wave energy is concentrated on headlands that jut out from the continent, while the amount of energy reaching the shore in bays is reduced. As the waves concentrate their energy on the headlands, erosion occurs and the shoreline retreats. The greatest concentration of wave energy on a day-to-day basis is in the foreshore region. However, during the rare periods when storm waves may batter the shore, more erosion may occur across the entire shore in one day than may be achieved by average wave conditions in a year.

The cliff shown in Fig. 16–17 is called a **wave-cut cliff**, which has been produced by wave action cutting away its base. The cliff develops as the upper portions collapse after being undermined by wave action. The undermining may be evident in the form of a notch at the base of the cliff that may be characterized by **sea caves**. Such features form as wave action erodes the softer portions of the rock outcrops. Caves may develop into openings running through the headlands, called **sea arches**. Continued erosion and crumbling of arches will produce **stacks**. Such remnants rise from the relatively smooth wave-cut bench cut from the bedrock by wave erosion (Fig. 16–18).

The rate of wave erosion will be determined by a number of variables, including the following important ones:

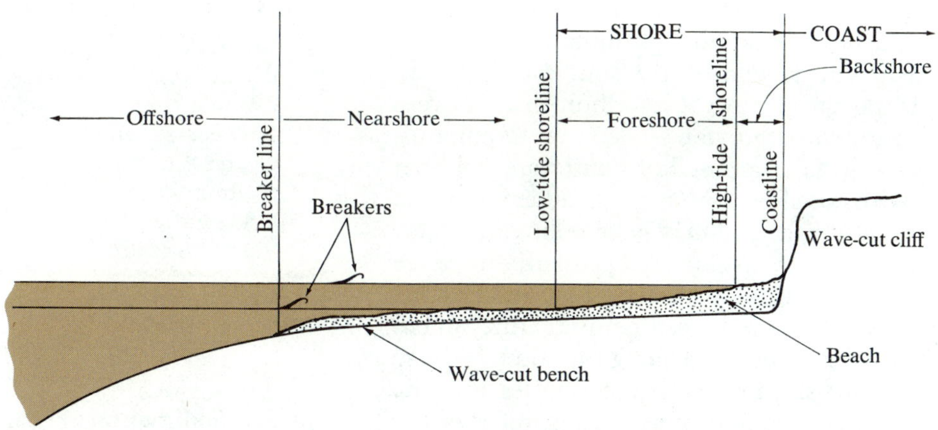

FIGURE 16–17
Landforms and terminology of coastal regions.

FIGURE 16–18
Coastal erosional features.

1. Degree of exposure of the coastal region to the open ocean—Coasts that are fully exposed receive higher energy wave action and are likely to have rugged cliffs.
2. Tidal range—Given the same amount of wave energy, a region that possesses a small tidal range will erode much more rapidly than one with a large tidal range that allows the wave energy to be spread over a much broader shore. Although high-velocity tidal currents may develop in areas where a large tidal range exists, these currents are of limited importance in eroding the coastline.
3. Composition of coastal bedrock—Crystalline igneous rocks, such as granite, are relatively resistant and generally produce rugged shoreline topography. Sedimentary rocks, such as sandstone and shale, are more easily eroded, and a gentler topography associated with more extensive beach deposits is characteristic of coasts with such bedrock.

Regardless of the rate of erosion, all coastal regions follow the same developmental path. As long as there is no change in the elevation of the landmass relative to the ocean surface, the cliffs will continue to retreat, the benches will widen, and the eroded material will be eroded from the high-energy areas and deposited on the low-energy areas.

DEPOSITIONAL SHORE FEATURES

The coastal erosion just discussed, as well as erosion being carried on by running water inland, produces large amounts of sediment that must be distributed along the continental margin. The material found in a beach deposit will depend on the source of the sediment that is transported by the longshore drift. In areas where the sediment is provided by coastal mountains, the beaches will be composed of quartz and feldspar fragments with smaller amounts of other common rock-forming minerals, and may be of relatively coarse texture. If the sediment is provided primarily by rivers that drain lowland areas, sediment that reaches the coastal regions will normally be finer in texture, and in many cases mud flats develop along the shore because only clay- and silt-sized particles are emptied into the ocean. In low-relief areas such as Florida, most of the material of the beaches is derived from the remains of the organisms that live in the coastal waters. Beaches in these areas will be predominantly composed of shell fragments and the remains of microscopic animals, particularly foraminifera. In Florida, the organic calcium carbonate content of the beach increases to the south because the bedrock underlying the state is primarily composed of limestone produced by coral reef deposits of the past. The beaches on many of the volcanic islands in the open ocean will be composed of dark-colored fragments of the basaltic lava that makes up the islands or of coarse fragments of coral debris from the coral reefs that develop around the margins of those islands that lie within the proper latitudes.

The slope of beaches is quite closely related to the size of the particles that compose the beaches. As the waves wash up onto a beach, they carry particles which are deposited, increasing the slope of the beach. When the backwash carries as much sediment back out to sea as the waves carried in, the beach has reached equilibrium and will not get any steeper. A beach composed of fine-grained sand that is relatively angular will have a gently sloping, firm surface. Because these small grains interlock closely, little of the swash sinks down among the grains. Most of it runs back down the slope to the ocean, possessing enough energy to maintain equilibrium on a gentle slope. The backshores of such beaches are usually nearly horizontal.

Beaches composed of coarse sands or pebbles will usually contain more rounded particles that are more loosely packed. The swash can quickly disappear into such deposits when it moves up the beach slope. Deposition of particles by the swash will continue until the beach slope becomes steep enough that the backwash running to the ocean has sufficient energy to maintain equilibrium. Such beaches are usually much less firm than beaches composed of finer material, and the backshores will slope significantly toward the coastline (Table 16–1).

TABLE 16–1
The relationship of particle size to beach slope.

Particle Size (Wentworth)	Mean Slope of Beach
Cobble	24°
Pebble	17°
Granule	11°
Very coarse sand	9°
Coarse sand	7°
Medium sand	5°
Fine sand	3°
Very fine sand	1°

SOURCE: Table 3 (p. 127) from *Submarine Geology,* 3rd Edition by Francis P. Shepard.

Special Sedimentary Features Numerous depositional features that are partially or wholly separated from the shore of the continent are deposited by the longshore drift and other processes that are not well understood. A **spit** is a linear ridge of sediment attached at one end to land. The other end of the deposit points in the direction of longshore drift and ends in the open water. Spits are simply extensions of beaches into the deeper water near the mouth of a bay. The open-water end of the spit will normally curve into the bay as a result of current action.

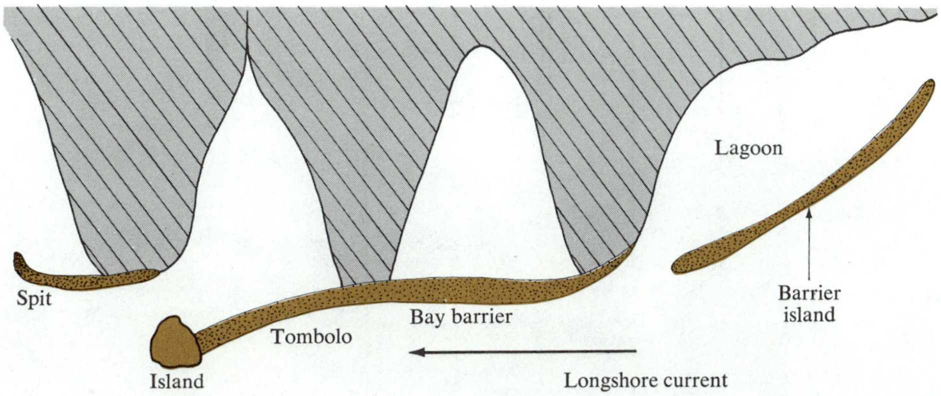

FIGURE 16–19
Coastal depositional features.

If tidal currents or the currents initiated by river runoff are too weak to keep the mouth of the bay open, the spit may eventually extend across the bay and tie to the mainland, completely separating the bay from the open ocean. The spit then has become a **bay barrier**. A **tombolo** is a sand ridge that connects an island with another island or the mainland. Such a feature probably results from the growth of a spit (Fig. 16–19).

Barrier Islands Barrier islands are long offshore deposits of sand lying parallel to the coast (Fig. 16–19). The origin of a barrier island is complex, and it may be that there are numerous explanations for their existence. However, it appears that many such structures developed during the raising of sea level that was initiated with the last melting of the glaciers some 18,000 years ago.

Barrier islands are rather continuous along the Atlantic Coast of the United States. They extend around Florida and along the Gulf of Mexico Coast, where they may be found well past the Mexican border. Barrier islands may attain lengths in excess of 100 km and have widths of several kilometers. Examples of such features are Coney Island off the New York coast and Padre Island off the coast of Texas. The seaward side of a well-developed barrier island will have a beach structure typical of the continental shore, and a dune development usually exists between the seaward shore and the barrier flat that leads into the lagoon behind the island. Barrier islands are an integral part of some estuarine environments (Fig. 16–20).

Deltas Some rivers carry more sediment than can be distributed by the longshore current. Such rivers develop a **delta deposit** as sediment settles out at the river mouth. One of the largest such features is that produced by the Mississippi River. Deltas are fertile, flat areas subject to periodic flooding.

Delta formation begins when a river has filled its estuary with sediment. Once the delta forms, it grows through the distribution of sediment by **distributaries**, branching channels that radiate out over the delta. Distributaries lengthen as they deposit sediment to produce fingerlike extensions to the delta. When the fingers get too long, they become choked with sediment. At this point, a flood may easily cause a shift in the distributary course to provide sediment to the low-lying areas between the fingers.

Where erosion and transportation processes exert a significant influence, the shoreline of the delta will be smoothed to a gentle curve like that of the Nile delta (Fig. 16–21A). In areas where depositional processes are dominant over coastal erosion and transportation processes, **bird-foot** Mississippi-type deltas result (Fig. 16–21B).

EVOLUTION OF SHORELINES

As erosion wears away headlands and deposition fills in bays, shores tend to become straight or gently curved. If the level of the ocean with respect to the shore is not changed, the shore will evolve into a state of equilibrium where energy is expended uniformly along it. Neither erosion nor sedimentation will be dominant at any point along the shore.

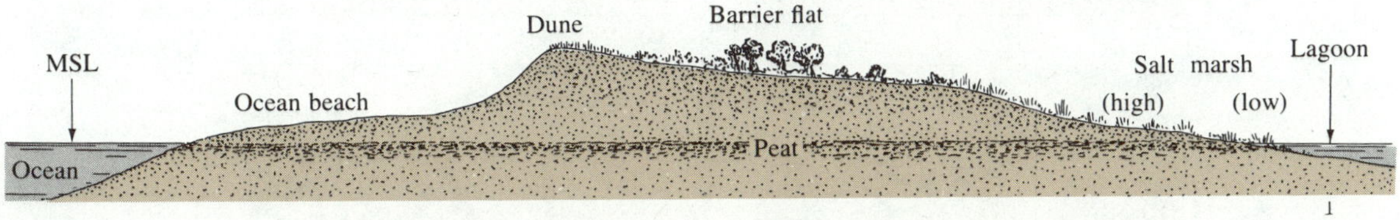

FIGURE 16–20
The major physiographic zones of a barrier island are the ocean beach, dune, barrier flat, and high and low salt marsh. MSL is mean sea level. The peat bed represents ancient marsh environments that have been covered by the island as it migrates toward the mainland as sea level rises.

A.

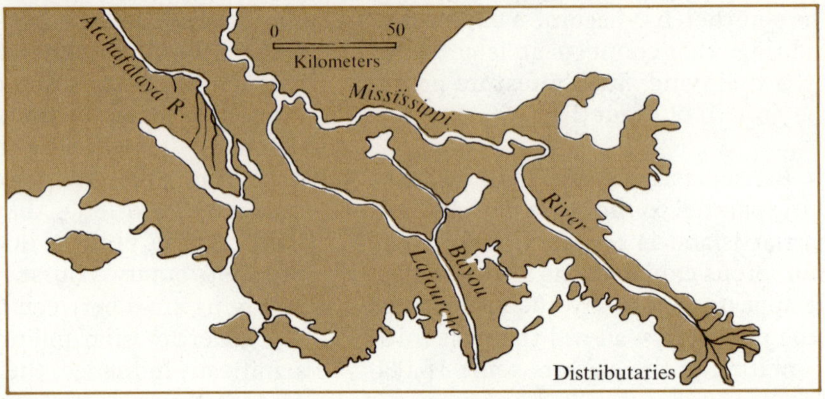

B.

FIGURE 16–21
Deltas. A. Photo taken from *Gemini IV.* The relatively smooth, curved surface of the Nile delta can be seen. The Mediterranean Sea is to the left, and the northern Red Sea is in the upper right. (Photo courtesy NASA) B. The bird-foot structure of the Mississippi River delta results from a low-energy environment. The location of the present main channel and the older main channels (Atchafalaya River and Bayou Lafourche) shows how a river shifts position with time.

This process can be followed by considering a shore that has been "drowned" by a rising level of the ocean. A shore where the dominant features are hills and valleys cut by stream erosion is illustrated in Fig. 16–22A. The shore is very irregular with the tops of submerged hills forming islands near shore. Sea cliffs form on the islands and mainland shores as they are eroded by wave action. Continued erosion wears away the islands, and deposition partially fills in the bays (Fig. 16–22B). Beaches and spits form along the straightening shore as the erosion of the sea cliffs produces a widening wave-cut bench. The wave-cut bench absorbs much of the wave energy and reduces the amount of energy available for erosion and sediment transport at the base of the cliffs (Fig. 16–22C). Finally, the sea cliffs retreat to the head of the bays, and the wave-cut bench is widened sufficiently to prevent further erosion of the cliffs (Fig. 16–22D). Essentially,

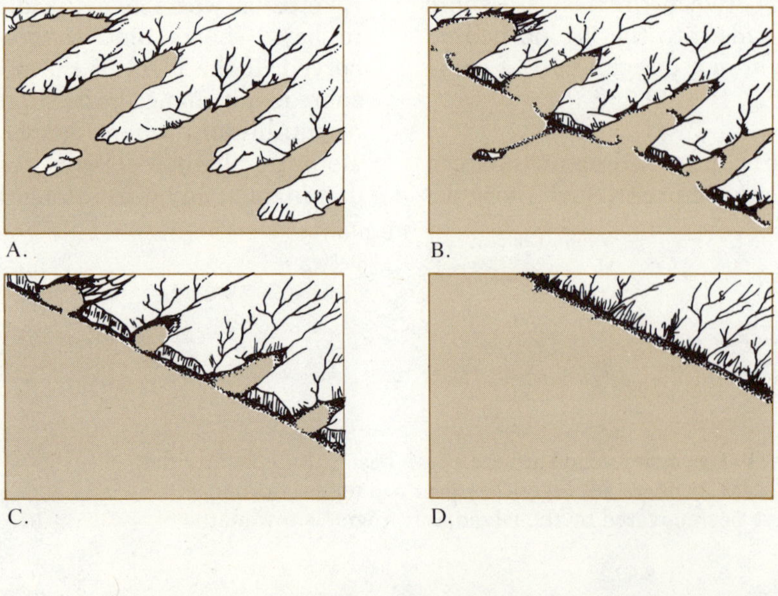

A. B. C. D.

FIGURE 16–22
The evolution of coastal regions.

all the wave energy is used up crossing the wave-cut bench and transporting sediment along the shore. Neither marine erosion nor deposition is a significant factor in the further modification of the shore, which will result primarily from surficial weathering and erosion.

Shores rarely proceed through the complete cycle leading to equilibrium because tectonic movement of the continent or changes in the volume of water in the ocean cause sea level to change before the process can be completed. However, all shorelines are moving toward such an equilibrium, and many segments of the shore of the United States have reached the stages illustrated in parts B and C in Fig. 16–22.

CHANGING LEVELS OF THE SHORELINE

Along some coasts there are flat platforms backed by cliffs called **marine terraces**, stranded beach deposits, and other evidence of marine processes many meters above the present shoreline. These features characterize shorelines of emergence that have reached their present position relative to the existing shoreline by either an uplift of the continent, the lowering of sea level, or a combination of the two.

In other areas, one may find beneath the water overlying the continental shelf relict beaches and submerged dune topography. These features along with the drowned river mouths along the present shoreline indicate that past shorelines have been submerged. This submergence must have been caused by a subsidence of the continent, a rise in sea level, or both.

Attempts to determine the causes of the change in relative level of the ocean and the continent have not met with great success. Whether the shoreline has become submerged because of rising sea level or a subsiding continent in a particular region cannot be determined by examining the coastal features in that area, since both processes produce the same end results.

TECTONIC MOVEMENTS

Changes in sea level relative to the continent may occur because of movement of the land, that is, **tectonic movement**. Such movement includes large-scale uplift or subsidence of large portions of continents or ocean basins or more localized deformation of the continental crust involving folding, faulting, and tilting. The earth's crust also responds isostatically to the accumulation of heavy loads of ice, sediment, or lava.

There is evidence that during the last 2.5–3 million years, at least four major accumulations of glacial ice developed in high latitudes. Although Antarctica is still covered by a very large glacial accumulation, much of the ice cover that once existed in northern Asia, Europe, and North America has disappeared. The most recent period of melting began about 18,000 years ago. Accumulations of ice up to 3 km in thickness have disappeared from northern Canada and Scandinavia. While these areas were beneath the thick ice sheet, they were pushed down and are still in the process of recovery after the melting of the ice. Some geologists believe that by the time the isostatic recovery is finished, the floor of Hudson Bay, which is about 150 m deep, will be above sea level. There is evidence in the Gulf of Bothnia of 275 m of isostatic recovery during the last 18,000 years. Generally, tectonic changes in the level of the shoreline are confined to a segment of the shoreline of a given continent.

EUSTATIC MOVEMENTS

Changes in the level of the shoreline that can be measured on a worldwide basis because they are caused by the increase or decrease of water volume in the ocean or the capacity of the ocean basin are termed **eustatic**. This term refers to the highly idealized situation in which all of the continents remain static while the sea rises or falls. Such changes in sea level could be created by the formation or the destruction of large inland lakes. Probably more important is the locking into or the release from continental glaciers of the earth's water during glacial and interglacial stages. It is also possible that the changes in the capacity of the ocean basin could effect such changes, but there is much less evidence pointing to this cause than there is for the former.

During the Pleistocene epoch, when the previously mentioned glacial advances were occurring, the amount of water in the ocean basin fluctuated considerably. Because the climate was colder during ice advances, some of the lowering of the shoreline might be accounted for by the contraction of the ocean volume as the ocean temperature decreased. It has been calculated that for every 1°C decrease in the mean temperature of the ocean water, sea level would drop 2 m. Temperature indications derived from study of fossils from Pleistocene ocean sediments indicate that the ocean temperature may have been as much as 5°C lower than at present. Contraction of the ocean water may have lowered sea level by about 10 m.

Although it is difficult to say with certainty what the range of shoreline fluctuation was during the Pleistocene, there is cause to believe that the shoreline was at least 120 m below the present shoreline. It is also estimated that if all the remaining glacial ice on earth were to melt, sea level would rise by another 60 m. This would give a minimum possible range of sea level during the Pleistocene of 180 m, most of which must be explained through the capture and release of the earth's water by glaciers. Such changes in sea level are termed **glacioeustatic oscillations**.

From 18,000 to 3000 years before the present, the ocean volume increased due to the expansion of the water resulting from warmer temperature and the melting of polar sea ice and glacial ice. Because the combination of tectonic and eustatic changes in sea level may be complex and difficult to identify, it is hard to classify coastal regions as purely emergent or submergent. Most coastal areas show evidence of having experienced both submergence and emergence in the recent past. It is believed, however, that sea level has not risen significantly as a result of melting glacial ice during the last 3000 years.

PLATE TECTONICS AND THE SHORE

If there has been no significant change in sea level due to glacioeustatic oscillations during the past 3000 years, tectonic processes must be looked to in order to explain the changes in the shoreline relative to the continents that can be seen to have occurred during the recent past. To understand what appears to be the underlying structural cause of emergence or submergence, we return to the concept of global plate tectonics discussed in Chapter 8.

ATLANTIC-TYPE MARGINS

Confining consideration to the Atlantic and Pacific coasts of North America, the Atlantic Coast is seen subsiding and the Pacific Coast is observed emerging. Beginning the discussion with the breakup of Pangaea and the presently forming Atlantic Ocean, this emergence-submergence pattern can be readily understood. As previously discussed, the lithosphere thickened, the ocean deepened, and heat flow decreased as the lithospheric plates cooled with age or with increasing distance from the spreading axes. Due to this process, the ocean floor of the North American Atlantic Coast is thought to have subsided about 3 km over the last 150 million years.

Erosion of North America produced sediment that has accumulated to a maximum thickness of 15 km along the Atlantic Coast. This thickness of sediment has been made possible by the plastic asthenosphere, which allows the lithosphere to flex downward as the sediment burden is increased. It is this thick sedimentary wedge underlying the continental shelf, slope, and rise that has been exploited for its large petroleum reserves in the Gulf of Mexico. Information from its exploration off the Atlantic Coast has resulted in cautiously optimistic projections of large petroleum reserves there.

When the upwelling heat initially split Pangaea, the granitic continental crustal rocks near the split were heated sufficiently to stretch plastically. This produced a gradually thinning continental or transitional crust that gives way to oceanic crust 100–150 km from the present shore beneath the continental rise (Fig. 16–23). Thus, the thick wedge of sediment underlying the continental shelf, slope, and rise is deposited on marginal continental crust, transitional crust, and oceanic crust.

The preceding summary describes the general conditions producing the subsidence of the Atlantic and Gulf coasts, although there are some relevant and more complex conditions that are not feasible to consider here. Although the rate of thermal subsidence decreases with time, and the rate of subsidence due to sediment loading depends on sediment supply, the process is never reversed.

PACIFIC-TYPE MARGINS

In great contrast to the passive and subsiding Atlantic-type margin just described, the Pacific-type margin, or active margin, is the scene of intense tectonic activity. Along such margins, the eruption of volcanoes and shocks of earthquakes are common. The thick, broad sediment wedge characteristic of the Atlantic Coast is not well developed. Instead of being the product of the creation of a new ocean basin, the Pacific-type margin is the scene of lithospheric plate destruction.

Although in the classic sense, the western coast of South America may be more characteristic of the Pacific-type margin, the western coast of North America once experienced a subduction event similar to that occurring along the western margin of South America. The characteristic alignment of mountain ranges parallel to the continental margin can be seen along both coasts. In all cases of continental margins associated with plate convergence, the compressive forces produce uplift. Along the California coast, the process has been further complicated with the development of the transform fault, the San Andreas fault. This has made possible the local subsidence of the depositional basins in the Los Angeles area. But on the broader scale, emergence along Pacific-type margins is the characteristic condition.

CLASSIFICATION OF COASTS

Attempting to classify coasts is a frustrating task because of the many variables that can contribute to the nature of any coast. However, classification can be accomplished with a reasonable degree of success using a

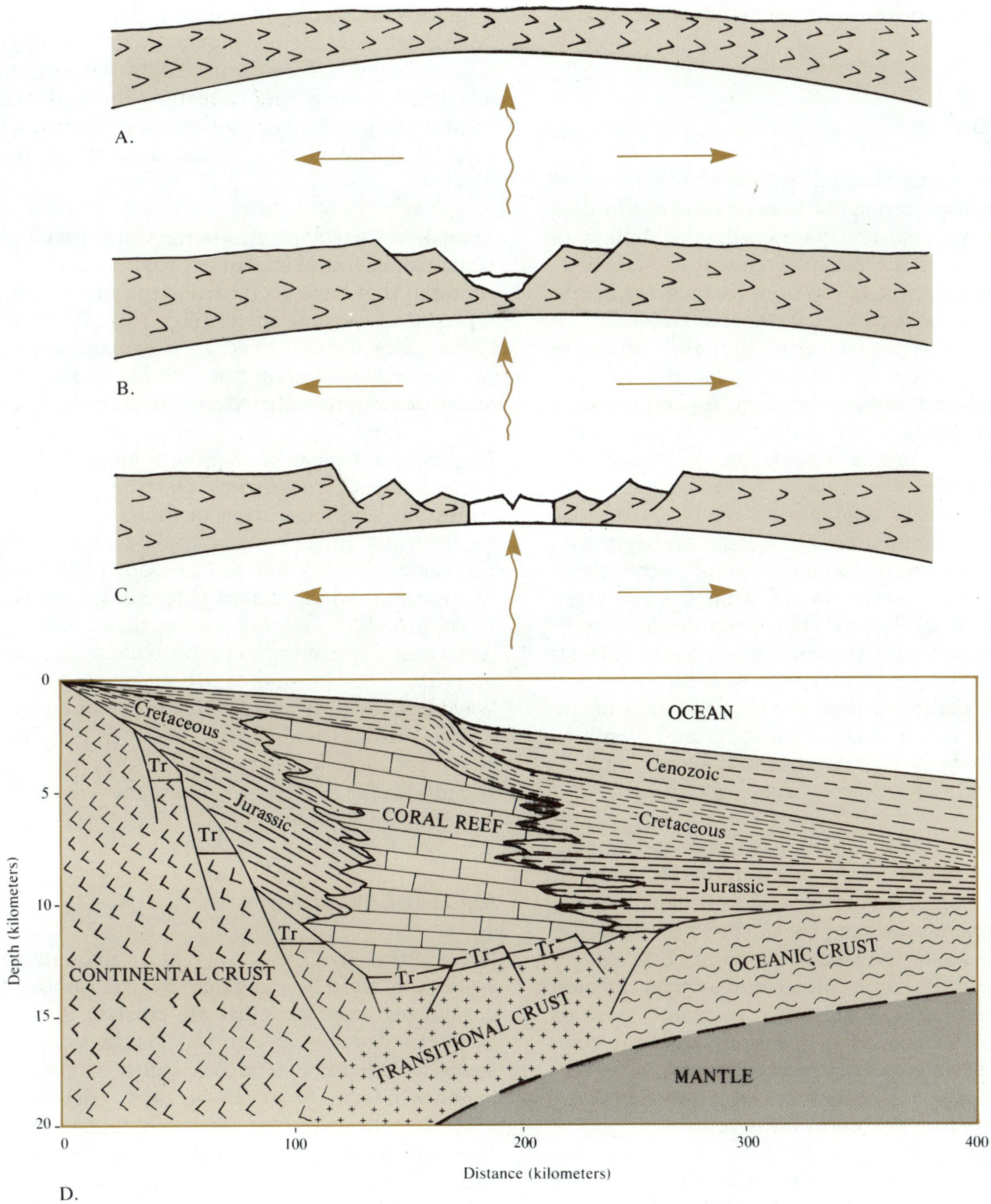

FIGURE 16–23
As Pangaea was rifted apart by upwelling heat (A), the continental crust was heated enough to flow plastically and thin at the edge of the continent, producing a faulted transitional crust (B) that eventually gives way to oceanic crust (C). The Triassic sediments that filled in the initial rift valleys are overlain by Jurassic, Cretaceous, and Cenozoic sediments that reach thicknesses of 10–15 km as the lithosphere flexes down into the asthenosphere under the increasing burden of the overlying sediment (D).

system proposed by Francis Shepard. In this system, coasts that have been modified only slightly by marine processes are called **primary coasts**. Coasts that have undergone significant modification by marine erosion, by deposition, or by growth of marine organisms are called **secondary coasts**. Some of the basic categories of this classification system are discussed in the following paragraphs.

PRIMARY COASTS

Ria Coasts Named for the Spanish word for "river," ria coasts are characterized by bays or estuaries formed when the ocean rose and "drowned" the valleys of rivers flowing into the ocean. Frequently, very little sediment will reach the ocean from rivers flowing into the embayments of such a coast, because the sediment is being deposited in embayments that are commonly less than 30 m deep. Embayments usually become gradually shallower toward the head. A good example of this type of coast, which frequently has a shoreline resembling the outline of an oak leaf, is Chesapeake Bay (Fig. 16–24A).

Fjord Coasts Named for Norwegian embayments, fjord coasts result from glacial erosion of river valleys. They usually have steep-walled, straight shorelines (Fig. 16–24B). Rock islands, remnants of glacial erosion, are common near the mouths of fjords. Glacial deposition usually produces a ridge near the lower end that makes it shallower than the upper reaches of the embayment. It is not unusual for maximum depth to exceed hundreds of meters in these scenic embayments common to the western coast of Canada.

Volcanic Coasts Most islands in the deep ocean are volcanoes that rise from the ocean floor. The slope of the ocean floor just off volcanic coasts is usually similar to the slope of the lava flows that flow into the ocean at the edge of the island. Hanauma Bay, east of Honolulu, is an example of an embayment created by the collapse of one side of a volcano. [Most of the coasts of the islands of Japan are volcanic (Fig. 16–24C)]. It has a concave character, while most volcanic coasts have lobate or convex forms resulting from lava flows moving out into the ocean.

Deltaic Coasts In some areas, sediments carried by large-volume transport rivers are being deposited at the shore to form deltas. In such areas, the shoreline is being changed rapidly. However, since these changes are primarily the result of land processes, the coasts are still considered primary. A classic example of such a coast would be that created by the Mississippi River (Fig. 16–24D).

Glacial Deposition Coasts In high latitudes where continental glaciers once extended out past the present shoreline, the shoreline is commonly the product of deposits laid down by the glaciers as they melted (Fig. 16–24E). Long Island is a glacial moraine, a depositional ridge, laid down by a retreating glacier. It was subsequently surrounded by the ocean which advanced as the glacier retreated. The oval-shaped hills in the Boston harbor are mostly drumlins, another type of glacial deposit, exposed by the retreat of the glacier.

Landslide Coasts Along the California and Oregon coast can be found irregularly shaped lobes of landslide material that have been carried into the ocean after the collapse of coastal rocks (Fig. 16–24F). The tops of these lobes usually have an irregular hummocky appearance. At the shoreline, boulders eroded from the slide material usually extend out into the ocean.

Diastrophic Coasts Where folding and faulting of coastal rocks is occurring, diastrophic coasts will be common. The west coast of the Gulf of California is such a coast. It has been created by a fault along which the seaward block has dropped down, leaving a steep escarpment. Along diastrophic coasts, the water may be thousands of meters deep less than 1 km from shore. The coastal ocean may also be made shallower by fault activity. The Alaskan earthquake of 1964 raised a wave-cut bench to the ocean surface along the coast of Prince William Sound (Fig. 16–24G). The coastal region that includes San Francisco Bay is in large part the product of movements along the San Andreas fault system.

SECONDARY COASTS

Wave Erosion Coasts When high-energy waves erode the shoreline created by the submergence of coastal features, sea cliffs are created. If the material being eroded is of uniform hardness, a relatively straight-cliffed coast with a gently sloping wave-cut bench will form. In areas where the coastal rocks are of variable hardness or are fractured, uneven rates of erosion will produce small-scale embayments where the material is more easily eroded. This can produce a rugged, cliffed shoreline like that found in Maine (Fig. 16–25A).

Marine Deposition Coasts In addition to the typical beach deposits found along many coasts, a dominant feature of marine deposition coasts is the barrier island. Barrier islands are common off the coasts of the southeastern United States and throughout the Gulf of Mexico. Between Cape Hatteras, North Carolina, and Cape Romain, South Carolina, are some interestingly shaped barrier islands called **cuspate forelands** (Fig. 16–25B). Their concave seaward shape is thought to be

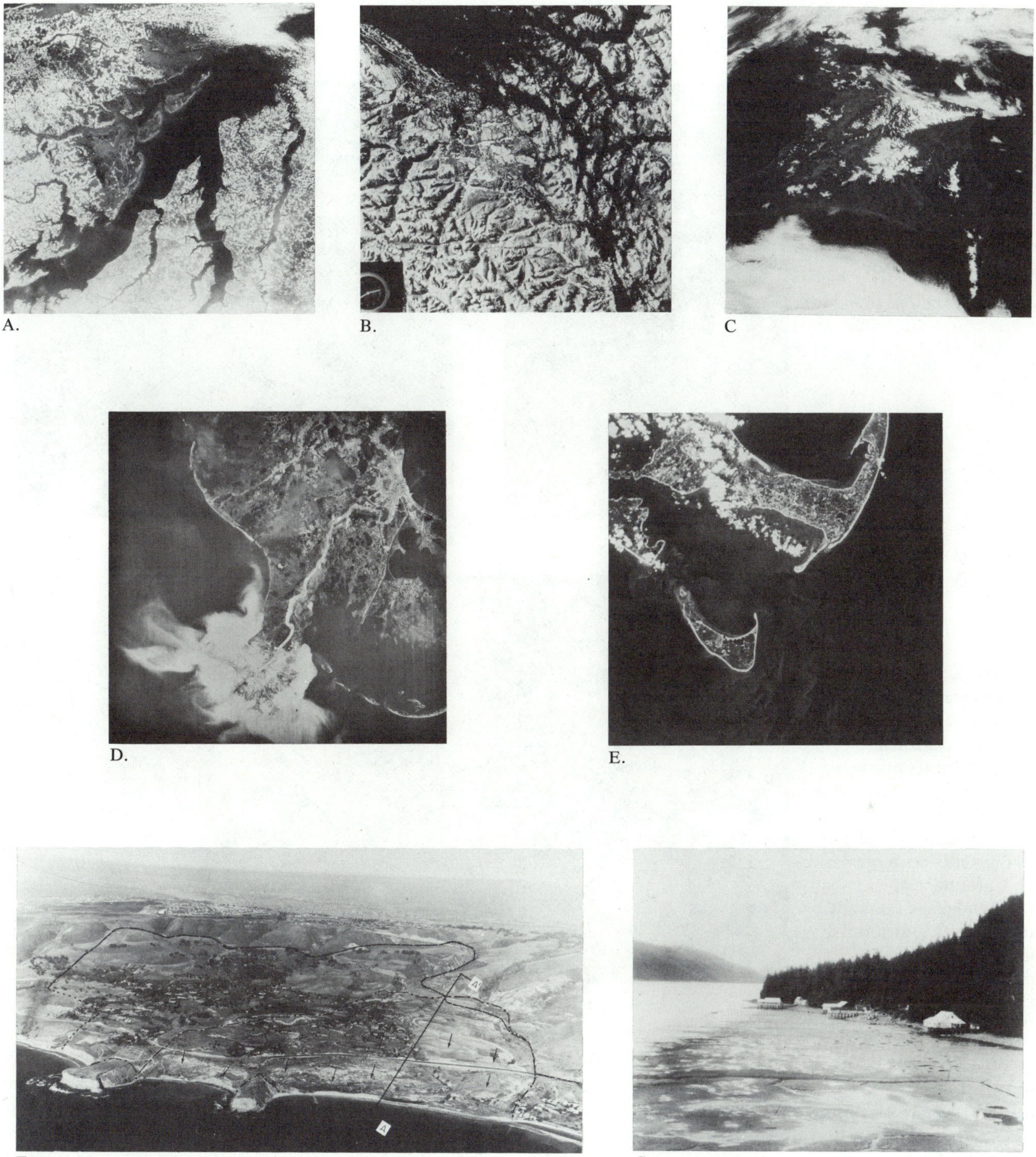

A. B. C

D. E.

F. G.

FIGURE 16–24
Primary coasts. A. Ria coast. Chesapeake Bay. B. Fjord coast. Johnstone Strait, British Columbia, Canada. Vancouver Island is in lower left half of the view. C. Volcanic coast. Hokkaido Island, Japan. D. Deltaic coast. Mississippi River. E. Glacial deposition coast. Cape Cod is composed of moraine and outwash debris. The feature has undergone some modification by marine deposition. F. Landslide coast. Palos Verdes, California. G. Diastrophic coast. Prince William Sound, Alaska, at high tide. Before uplift by a 1964 earthquake, the high-tide shoreline was beneath the docks. (Parts A–E courtesy NASA; part F courtesy Los Angeles County Department of County Engineer; part G courtesy U.S. Geological Survey)

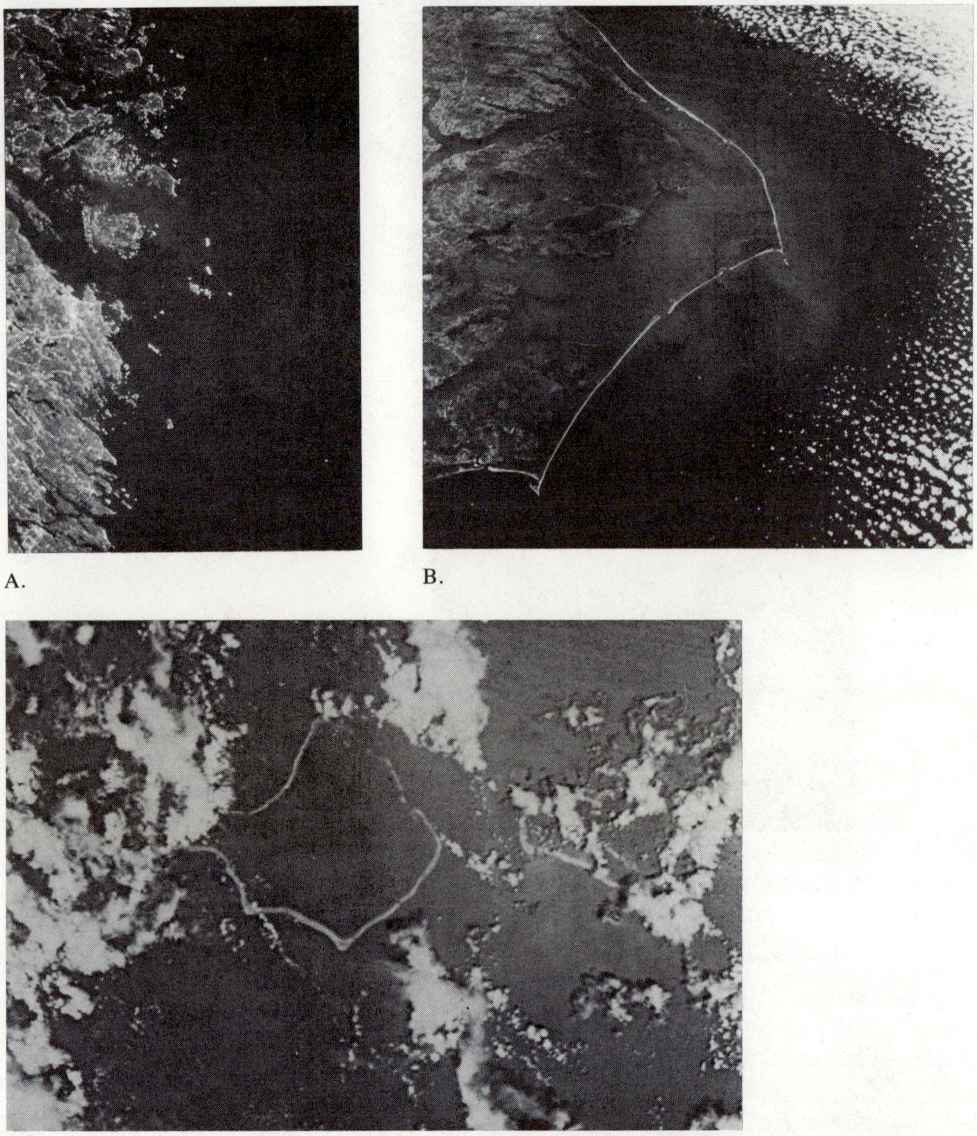

A. B. C.

FIGURE 16–25
Secondary coasts. A. Wave erosion coast. Penobscot Bay, Maine, has a coast that retains some of its primary glacial erosion character as it is being eroded by wave action. B. Marine deposition coast. Barrier islands off North Carolina. Cape Hatteras and Cape Lookout are to the north and south, respectively. C. Marshall Island atoll. (Parts A–C courtesy NASA)

the result of giant eddies of the Gulf Stream. It is believed that barrier islands form only along subsiding shores and where they and their landward lagoons migrate toward the subsiding mainland. This process has been "halted," however, in some areas where cities (e.g., Atlantic City, Miami Beach, and Galveston) have been built on them.

Coasts Built Forward by Organisms The modification of coastal areas by the growth of organisms is achieved primarily in intertropical latitudes by the development of mangrove trees and coral reefs. Mangrove coasts are composed of groves of mangrove trees that extend their roots into the ocean. Sediment is trapped in the interlocking mesh of roots and deposited, helping the mangrove barrier to grow seaward.

Although coral distribution throughout the ocean floor is rather general, coral accumulations that might be classified as reefs are restricted to the warmer water regions where the average monthly temperature exceeds 18°C throughout the year (Fig. 16–26). Such temperature conditions are found primarily between the tropics, although reefs grow to latitudes approaching 35°N and S on the western margins of ocean basins where warm water masses move into the high-latitude areas and raise average temperatures.

Not only coral, but algae, molluscs, and foraminifera, make important contributions to the reef structure. Reef corals are found only in the shallow water surrounding the continents and islands of the open ocean because they are **hermatypic**. This means that they have a symbiotic relationship with the green alga

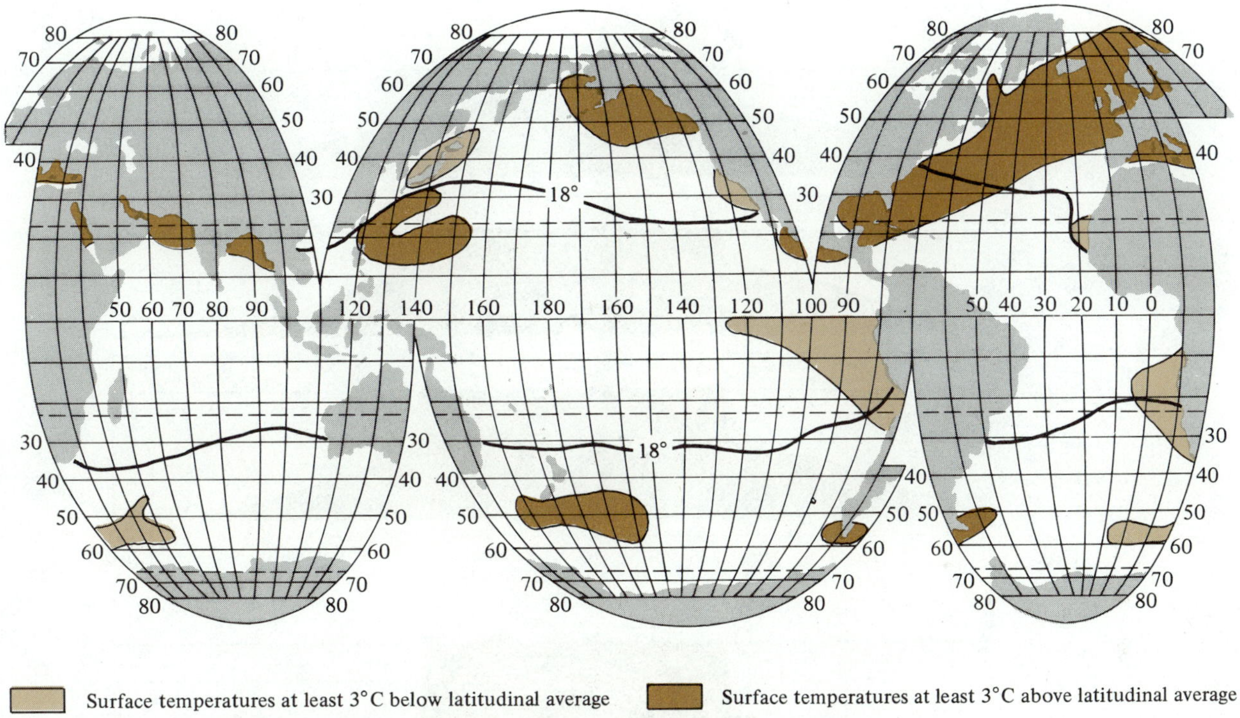

FIGURE 16–26
Areas in which coral reefs are confined. Minimum water temperatures of 18°C in the surface waters of the Northern and Southern hemispheres occur in February and August, respectively. Coral reef development is restricted to the low-latitude area between the two 18°C temperature lines shown on the map. Observing the regions where surface temperatures are 3° or more above or below latitudinal averages, the effect of surface currents on coral reef distribution can be seen. (Base map courtesy National Ocean Survey)

Zooxanthellae that lives within the tissue of the reef-building coral. The alga contributes to the calcification capability of corals by extracting carbon dioxide from the animals' body fluids, thus increasing the concentration of the carbonate ion needed for the precipitation of calcium carbonate.

Reef growth also requires that the water have a relatively normal salinity and that it be free from particulate matter. Therefore, little coral reef growth is seen near the mouths of large rivers that lower the salinity and carry large quantities of suspended material which would choke the reef colony. Maximum reef growth will occur in areas where there is a constant flow of water carrying food to the waiting tentacles of the polyps which extrude and feed primarily at night. Such conditions are optimum in areas where water circulation through current activity is relatively good.

Reefs are widely developed throughout the Pacific and Indian oceans on the flanks of the many volcanic islands rising above the ocean surface from the deep-ocean floor. Consisting primarily of the skeletal remains of hermatypic corals and calcareous algae, reefs are found around the margins of all islands and continents where the proper conditions for their existence are found.

There are three basic types of reefs: **fringing reefs**, **barrier reefs**, and **atolls** (Fig. 16–27). **Fringing reefs** develop at the margin of any landmass where the temperature, salinity, and turbidity of the water are suitable for hermatypic corals. Due to the close association between the landmass and the reef, there will be periods of time when runoff from the landmass will carry too much sediment for the coral in a particular area to survive. Very commonly, the amount of living coral in a fringing reef at any given time will be relatively small. The greatest concentration of live reef will be on the seaward margin, which is more protected from sediment and salinity changes due to runoff. Without some change in sea level relative to the landmass during the existence of the fringing reef, it is not likely that the other two types of reefs mentioned, the barrier reef and the atoll, will form.

Barrier reefs are linear or circular developments separated by a lagoon from the continent or island on whose margin they are growing. It is thought that barrier reefs form when the substratum on which a fring-

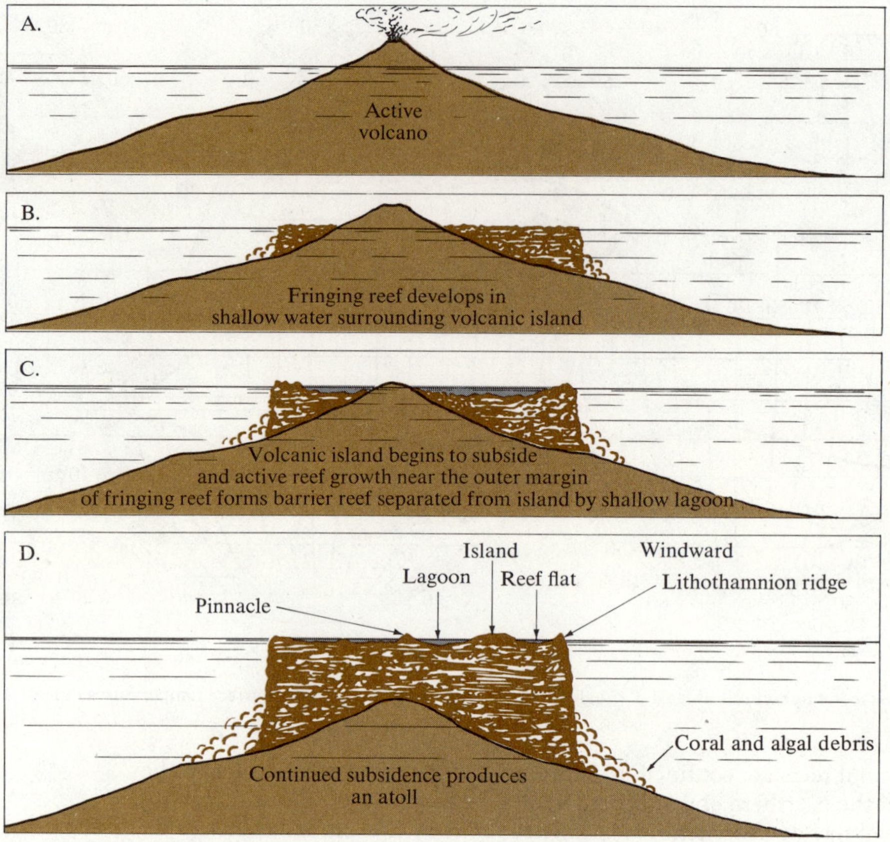

FIGURE 16–27
Darwin's theory of the development of reef types.

ing reef is built begins to subside. As the island or margin of the continent subsides, the fringe reef maintains its position at the optimum water depth by growing upward. Studies of rate of reef growth indicate that most reefs have grown at a rate of 3–5 m/1000 years during the recent geologic past. However, there is some evidence that reefs in the Caribbean have grown at rates of over 10 m/1000 years.

The most outstanding example of a barrier reef is the great Australian reef that is 150 km in width and extends for more than 2000 km along the northeast coast of Australia. Smaller barrier reefs are found around the island of Tahiti and other islands in the western Atlantic and Pacific oceans. Coral growth is much more active on the portions of the reef that are better protected from land runoff. While the barrier is actively growing, the lagoon floor may be covered with sediment derived from both the reef and the adjacent landmass.

Atolls may form either on the continental shelf or around the margin of volcanic islands in the open ocean, both of which require crustal sinking. Most atolls are of the oceanic variety, and the greatest assemblage of them occurs in the equatorial Pacific. Atolls may take a circular or irregular shape and surround a lagoon that is generally 30–50 m deep at the deepest part (Fig. 16–27C). The fringing reef will have many channels that allow circulation between the lagoon and the open ocean. The reef flat is generally broader on the windward side of the atoll, whereas channels are more numerous on the leeward side.

Islands large enough to inhabit may develop as deposits of coral debris from the windward reef flat, which is characterized by a well-developed lithothamnion ridge on the windward edge of the reef. The lithothamnion ridge, composed of calcium-secreting algae, can extend above the low-tide level, and beyond it on the reef front will be found a profuse growth of coral that must remain below the low-water line.

The floor of the lagoon will be filled with coral debris. Oval-shaped reefs called **pinnacles** or **knolls** that maintain active reef growth extend above the lagoon floor in regions of the lagoon where there is good circulation with the open ocean through the channels. Figure 16–27 shows the origin of an atoll according to the theory proposed by Charles Darwin in 1842. Investigations before and after World War II provided the evidence to prove the validity of Darwin's theory.

U.S. COASTS

THE GULF COAST

The Louisiana-Texas coastline is dominated by the Mississippi River delta, which is being deposited in a microtidal and generally low-energy environment. The tidal range is normally less than 1 m, and with the exception of the hurricane season, wave action is generally low. The sediments that have been deposited in the Gulf Coast region indicate that this area has been generally subsiding for approximately 150 million years. The sedimentary section that can be found beneath the coast is essentially continuous from the present back to the Cretaceous. For these sediments to be deposited, preserved, and overlain by younger sediment, at least a low rate of subsidence was required. Even though the Gulf Coast is subsiding, the shoreline is moving seaward in some regions. The tremendous volume of sediment that is being carried by the Mississippi River and deposited in its delta has extended the delta completely across the continental shelf (Fig. 16–28).

Along the Louisiana Gulf Coast, a nip and tuck battle is going on between the forces of deposition, attempting to advance the coastal region, and submergence, threatening to make the coastline retreat inland. Locally, one process may be winning for a short period of time. Areas some distance from the delta cease to receive new sediment and begin to subside, while others are being actively advanced by deposition. In New Orleans, it obviously would not be desirable to allow the Mississippi River to continue to add sediment. Without this deposition, the city must subside. Therefore, a constant battle must be waged to keep the area sufficiently diked to keep out the ocean and the river.

Extending along the coast east of the Mississippi delta to the Florida panhandle and west of the delta to the Mexican border are barrier islands. These long deposits, which may be over 100 km long and more than

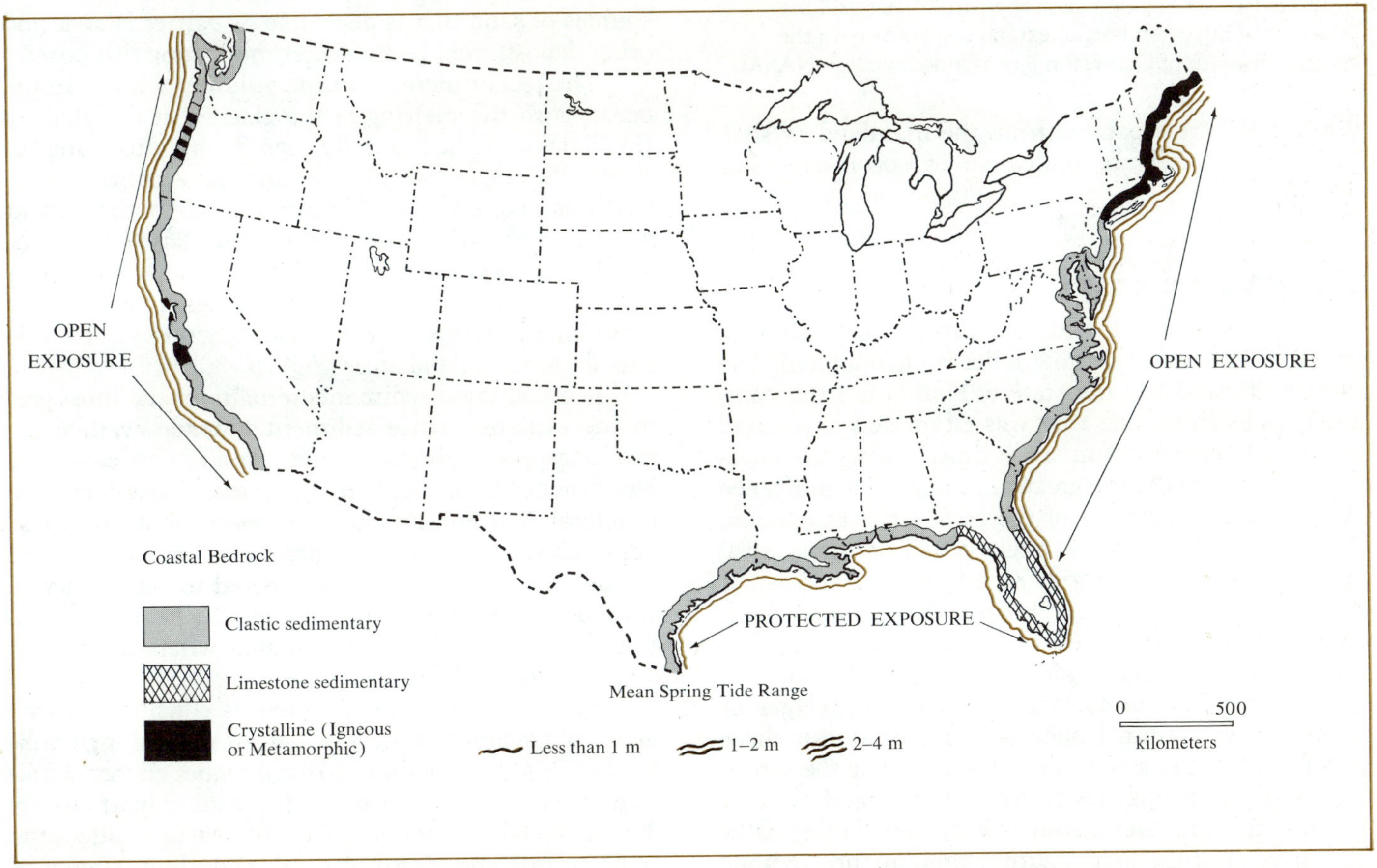

FIGURE 16–28
The distribution of conditions related to coastal erosion.

FIGURE 16–29
Apollo 9 view from an altitude of 191 km off the Texas coast from Freeport to Sabine Lake. The barrier islands, Line Island and Galveston Island, extend respectively to the north and south of Galveston Bay. (Photo courtesy NASA)

2 km wide, are separated from the mainland by shallow lagoons and are indicative of subsidence (Fig. 16–29).

THE ATLANTIC COAST

Moving east to the Florida coastline, a gradually subsiding area with a very low relief is encountered. The bedrock for most of the state of Florida is a resistant, impervious limestone that was alternately inundated and totally exposed numerous times during the Pleistocene. The area of the greatest ecological significance lies at the southern tip of the state—the Everglades. This coastal region is also a relatively low-energy area. It is protected by the Florida Keys from large waves and receives high-energy assault only during local thunderstorms and hurricanes, as is generally the case for the Mississippi delta region.

Drainage of surface water across the Everglades from northern Florida once occurred as a thin sheet less than 1 m thick that flowed south along the gentle slopes of the Everglades to the ocean. This flow was not restricted to river channels but covered essentially the breadth of the state. Within this unique environment developed an assemblage of plant and animal life. that is presently threatened.

To control flooding, the U.S. Army Corps of Engineers restricted the sheet runoff to flood-control canals that have left portions of the Everglades short of water. This shortage of water has endangered much of the plant and animal life. Also resulting from this controlled flow of water is a threat that was previously seldom known in the Everglades. Fire has become a serious problem. Since the total relief in the Everglade area is less than 10 m, it is probable that natural changes in sea level will, in time, exert stress on the organisms found in the Everglades.

Moving north along the Atlantic Coast from Florida to Maine, coastal regions are encountered where the sea level is rising an average of 0.3 m/century relative to the subsiding land. Although some rise in the continent may well have occurred along the coast from New York through Maine as a result of the melting of the continental glacier that once covered this region, evidence indicates a gradually retreating shore to the south.

The coastal rocks from Florida through New Jersey are predominantly poorly consolidated sedimentary rocks formed in the recent geologic past. These rocks do not provide great resistance to erosion and are sources of sand that is deposited as barrier islands and other depositional features common along this coast.

The recent increase in the volume of water in the ocean with the melting of the glaciers at the close of the Pleistocene has caused a significant "drowning" of river valleys that once flowed into an Atlantic Ocean that was a considerable distance seaward of the present shoreline. Delaware Bay, Chesapeake Bay, and the irregular landward margin of Pamlico Sound are of this origin. These estuaries are in the process of being filled by sediment carried to the coastal region by the rivers that drain the inland areas (Fig. 16–30).

Because these sediments remain for the most part in the estuaries, little sediment is made available to the longshore currents by river runoff. This condition has produced, in local areas, severe coastal erosion problems. The energy conditions associated with these coastal areas are greater than seen in the Gulf of Mexico, because the area is exposed to the moderate wave action of the Atlantic Ocean. High-energy conditions develop during storms and when hurricanes strike the coastal area.

From New York north, there is considerable evidence of glaciers having affected the coastal region directly. This area is subject to even higher energy conditions than the coastal region further south, particularly during the fall and winter. The "nor'easters" affect the coast from Cape Hatteras north. The high energy of these storms is manifested in up to 6-m waves with a 1-m rise in sea level that follows the low pressure as it moves northward. Such high-energy conditions will significantly change coastlines that are predominantly

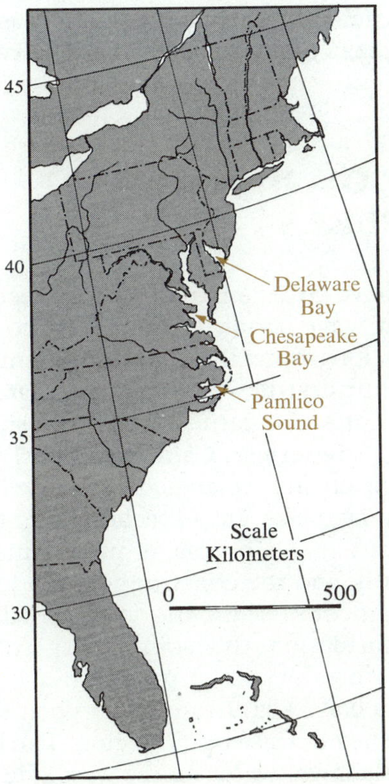

FIGURE 16–30
"Drowned" river valleys along the Atlantic Coast.

depositional, and may be expected to cause considerable erosion along the coastal region where this process is dominant. The resistant granite of the Maine coast has resisted erosion. However, zones of weakness have allowed erosion to produce a rugged, irregular coastline.

THE PACIFIC COAST

Unlike the Gulf and Atlantic coastal regions, the Pacific Coast of the United States is characterized by a continental margin that is rising rapidly relative to sea level. The folding, faulting, and volcanic activity associated with this uplift can be observed from southern California to Washington.

Throughout the Pacific Coast, the shoreline is characterized by cliffs that are being actively eroded. Old cliffs and wave-cut terraces that have been elevated above the present sea level during the past million years or so are manifested as terraces that are well developed in the Palos Verdes Hills in the Los Angeles area (Fig. 16–31). Here, 13 uplifted wave-cut terraces can be counted, and the highest one is approximately 400 m above sea level. Due to the rapid coastal uplift, it is difficult to determine the effect of sea-level fluctuation due to the melting and forming of glaciers.

Rocks outcropping along the coast are, for the most part, relatively weak marine sediments that are geologically young. Locally, more resistant granite, metamorphic, and volcanic rocks may form the bedrock along the coast. The effects of glaciation characterized by the fjords along the Canadian coast extend southward into the Washington coastal region to produce the Strait of Juan de Fuca and Puget Sound (Fig. 16–32).

The Pacific Coast of the United States is characterized by high-energy conditions in which the waves are normally about 1 m high. Frequently, the wave height will increase to 2 m, and a few times a year 6-m waves hammer the coast. In addition to the Pacific storms that generate these large waves, tsunami periodically strike the coastal region. The strong winter waves that generally strike the coast from a west-northwesterly direction produce a strong long-

FIGURE 16–31
Marine terraces of Palos Verde Hills, California. (Photo by John S. Shelton)

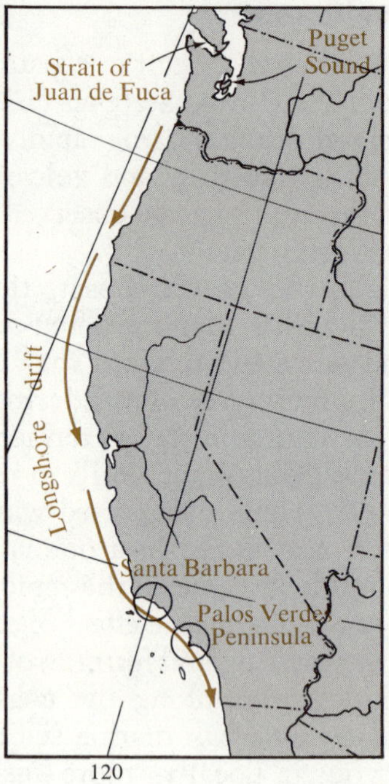

FIGURE 16–32
Features of special interest along the Pacific Coast.

shore current that carries millions of tons of sediment in the longshore drift.

During the winter, many beaches have the sand eroded from the foreshore area due to the high energy of these waves. The exposed beaches, which are composed primarily of pebbles and boulders during the winter months, regain their sand as lower-energy summer waves beat more gently against the shore. Damming of many of the rivers along the Pacific Coast for flood control has reduced the amount of sediment being provided for the longshore transport. Due to this sediment deficiency, erosion along the coast is becoming a significant problem as the sand-starved beaches disappear.

EFFECT OF HUMAN-MADE STRUCTURES

Jetties that have been constructed to protect harbor entrances from wave action are the most common barriers to the longshore transport of sediment. Many examples of the destructive results associated with the construction of such features along the shore can be cited. Along the southern California coast, cases of excessive beach erosion related to the development of local harbor facilities are especially numerous. The longshore drift in California is predominantly from north to south, and the construction of a jetty results in the entrapment of sediment on the north side with increased erosion on the south side. A jetty constructed as a breakwater to protect harbor waters at Santa Barbara had a significant effect upon the equilibrium established in this coastal region. The barrier represented by the breakwater on the west side of the harbor caused an accumulation of sand that was moving in an easterly direction along this portion of the California coast, which trends generally east-west. The beach to the west of the harbor continued to grow until finally the sand moved around the breakwater and began to fill in the harbor (Fig. 16–33).

FIGURE 16–33
Santa Barbara Harbor. Construction of a breakwater to the west of Santa Barbara Harbor interfered with the eastward-moving longshore drift, creating a broad beach. Sand being deposited against the breakwater was no longer available to replace sand being removed by wave erosion to the east. As the beach extended around the breakwater into the harbor, dredging operations were initiated to keep the harbor open and to put the sand back into the longshore drift. This helped reduce the coastal erosion east of the harbor. (From U.S. Coast and Geodetic Survey Chart 5261)

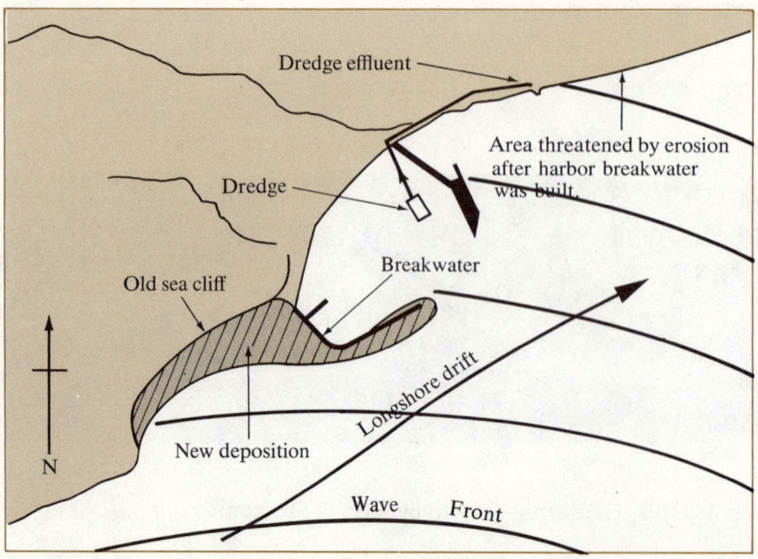

While deposition was occurring at an abnormal rate to the west, erosion was proceeding at an alarming rate east of the harbor. The energy available east of the harbor was no greater than it had previously been, but the sand that had formerly moved down the coast to replace sand removed was no longer available due to its entrapment behind the breakwater. To compensate for this deficiency of sand down-current from the harbor, a dredging activity had to be initiated within the harbor. The dredging keeps the harbor from filling in, and the sand is pumped down the coast so that it can reenter the longshore drift and replenish the eroded beach. The dredging operation has stabilized the situation, but at a considerable expense. It seems obvious that any time human beings interfere with natural processes in the coastal region, they will have to provide the energy needed to replace that which they have misdirected through modification of the shore environment.

Along the New York, New Jersey, and Florida coasts, small jettylike structures called **groins** have been constructed at intervals along the beach (Fig. 16–34). They have helped beach development by causing sand to accumulate on the north side of these structures. Being shorter than jetties built to protect harbors, the groins eventually allow the sand to migrate around their ends. An equilibrium may be reached that allows sufficient sand transport along the coast before excessive erosion occurs down-current from the last groin. Although some serious erosional problems have developed in these regions, groin construction usually results in much less beach erosion than the construction of one large jetty. Groins are, at any rate, highly effective in widening and protecting the beach in the areas where they are constructed.

FIGURE 16–34
Groins. As sand drifts along the shore of Miami Beach, it accumulates on the up-current side of the groins, increasing the width of the beach. The arrow indicates the direction of longshore transport of sediment. (Photo courtesy U.S. Army Corps of Engineers)

SUMMARY

Most of the surface of continents is covered by sedimentary rocks formed from sediment laid down in ancient oceans. These deposits tell much about the earth's history, but it was not until the 1960s, when it became possible to measure the properties of the ocean floor on a large scale, that the role of the ocean floor in the dynamics of global plate tectonics was discovered. This knowledge led to a major increase in our understanding of volcanism, earthquakes, and mountain building.

To understand the processes that affect erosion and the deposition of sediment along the margins of continents, it is necessary to have some knowledge of the physical motions of ocean water, such as currents, waves, and tides.

The horizontal surface circulation of ocean water is driven by the prevailing wind systems. The trade winds and westerlies, aided by the Coriolis effect and continents, set up subtropical gyres that rotate clockwise in the ocean basins of the Northern Hemisphere and counterclockwise in the basins of the Southern Hemisphere. Vertical circulation that carries cold, dense surface water from the Norwegian and Weddell seas to the deep-ocean floor causes the entire water column to be mixed and carries life-supporting oxygen to the deepest ocean basins. On the average, the ocean completes a mixing cycle about every 1000 years.

The most significant motion of ocean water to geology is that of wind-generated waves. Created by winds putting energy into the oceans in storm areas called seas, the waves gain energy with increases in: (1) wind velocity; (2) duration of wind blowing in one direction; and (3) fetch, the distance over which the wind blows. The energy is carried from the sea area by long, crested waves called swell that transport the energy across long distances with very little dissipation. Most of the energy is finally released along the shore of landmasses when decreasing water depth causes the waves to break in the surf zone.

Tsunami are very long waves generated by earth movements. With wavelengths of up to 200 km and maximum wave heights of only 30 cm in the open ocean, tsunami pass unnoticed. It is only when they reach shore and rise to heights of 30 m that they achieve their destructive potential.

Tides result from the gravitational attraction between the earth and sun, and the earth and moon. The dominant tide on the earth is the lunar tide because the moon is much closer than the sun. However, the solar tide significantly modifies the moon tide. Most places on the earth experience two high tides and two low tides each day. On a monthly cycle of about 29.5 days (the synodic month), the moon completes a phase cycle from new moon through first quarter, full moon, third quarter, and back to new moon. During this period of time, the earth experiences two spring tides and two neap tides. The spring tides coincide with the new- and full-moon phases, and the neap tides occur during the first- and third-quarter phases.

As waves approach a shore, one segment of each wave front will encounter shallow water first and be slowed before the rest of the wave. This causes refraction, or bending, of the wave as it approaches the shore. As a result of refraction, there is a concentration of wave energy released against headlands, while waves breaking in bays are less energetic. Breaking waves almost always strike the shore at an angle. This causes water to rush up the beach face at an angle to the shore and run straight down the slope after the wave's energy has been dissipated. This zig-zag movement of water along the shore and within the surf zone is called the longshore current. Sediment transported along the shore by the longshore current is termed longshore drift. The water that makes its way down the shore in the longshore current eventually finds its way back to the open ocean as rip currents.

The region of contact between the ocean and continent is the shore which lies between the lowest low tides and the highest elevation on the continent affected by storm waves, the coastline. The coast extends inland from the coastline as far as features related to marine processes can be found. The shore is divided into the foreshore, found between high- and low-tide extremes, and the backshore, extending from the high-tide line to the coastline. Extending seaward of the low-tide line to the line of breakers is the nearshore zone, with the offshore zone lying beyond.

Erosion along the shore produces a wave-cut cliff, sea arches, sea caves, and stacks rising above a wave-cut bench. The rate of marine erosion will increase with: (1) increasing degree of exposure to the open ocean, (2) decreasing tidal range, and (3) decreasing strength of bedrock.

The most common feature of marine deposition is the beach, which will be deposited in lower-energy environments along the shore and be composed of whatever loose material is available. The slope of the beach face increases with increased size of the particles composing it. Special beach features are spits, bay barriers, tombolos, and barrier islands. Deltas occur where rivers carry more sediment to the continental shelf than the longshore current can move away.

As shores evolve, they tend to develop into relatively straight alignments with a wave-cut cliff rising from a broad wave-cut bench that absorbs most of the wave energy in the nearshore zone before the waves reach the shore. Shores rarely attain this stage of development because the shoreline usually rises or falls before the process can be completed. Changes in the level of the shoreline relative to the continent are the result of tectonic movement that raises or lowers a segment of the margin of a continent, or of eustatic movement, which is a worldwide rise or fall of sea level caused by a change in the volume of water in the ocean or the capacity of the ocean basin. From about 18,000 to 3000 years ago, a glacioeustatic rise of 120 m in sea level resulted from the melting of large continental glaciers.

Atlantic-type margins of continents are undergoing gradual subsidence due to thermal contraction in the lithosphere and sediment loading that are occurring along the Atlantic margin of the United States. Pacific-type margins of continents are experiencing tectonic uplift due to convergent plate collisions, such as that occurring along the Pacific Coast of the United States.

Coasts are classified as primary coasts if the features observed along them are the result of nonmarine processes. If marine erosion, marine deposition, or the growth of marine organisms has greatly modified the coasts, they are called secondary coasts. Primary coasts include ria coasts, drowned river valleys whose dominant features are the result of stream erosion, fjord coasts produced by glaciation, volcanic coasts, deltaic coasts, glacial deposition coasts, landslide coasts, and diastrophic coasts resulting from faulting and folding. Secondary coasts include wave erosion coasts, marine deposition coasts and organism-built coasts produced by the growth of coral reefs or mangrove trees.

The Gulf Coast of the United States has a protected exposure and a small tidal range. It is subsiding but still advances seaward where the Mississippi River deposits its delta. Barrier islands, characteristic of subsiding coasts, and relatively soft bedrock extend along the Gulf Coast and up the Atlantic Coast to New Jersey. To the north of New Jersey, crystalline bedrock is rising out of the ocean, still responding to the melting of continental glaciers that once covered the area. The entire Atlantic Coast is openly exposed, and tidal range increases from less than 1 m to 4 m as is shown in Fig. 16–28. As a result of these factors, marine erosion is minimal along the Gulf Coast but is significant locally along the southern Atlantic Coast, particularly where human-made structures interrupt the flow of the longshore drift. The Pacific Coast has an open exposure, relatively soft bedrock, and a tidal range that averages about 2 m. Tectonic uplift and erosion are characteristic of much of this shore.

REVIEW EXERCISES

1 Discuss the driving force for open-ocean surface circulation and the reason for the different patterns of rotation in the Northern Hemisphere and Southern Hemisphere.

2 Over what portion of the ocean floor do the deep-ocean currents exert the most significant influence on the distribution of sediment?

3 What are the three fundamental types of progressive waves? Describe their characteristic pattern of particle movement, and state which is represented by ocean waves.

4 Discuss the role of the sea, swell, and the surf zone in the life cycle of wind-generated waves.

5 State the causes of tsunami, and explain why these waves, which can be so destructive when they reach shore, are undetectable in the open ocean.

6 Explain the cause of spring tides and neap tides.

7 Describe the effect of wave refraction on the distribution of wave energy released along a coast with headlands and bays.

8 Discuss the causes of and interrelationships among longshore currents, longshore drift, and rip currents.

9 What is the difference between the shore and the coast?

10 List and define four features that result from wave erosion.

11 Discuss the factors that determine the material of which a beach will be composed and how steep the beach slope will be.

12 List and define four special marine depositional features.

13 Discuss the evolution of shores where sea level does not change.

14 Compare the causes and effects of sea-level changes resulting from tectonic movement and eustatic movement.

15 Relate the characteristics of Atlantic-type margins and Pacific-type margins to global plate tectonics.

16 Give general definitions for *primary coasts* and *secondary coasts,* then list the types of coasts classified under each.

17 Describe the conditions necessary for coral reef development, and the process by which reefs evolve from fringing reefs to atolls.

18 Compare the Gulf Coast, Atlantic Coast, and Pacific Coast in terms of emergence and submergence, exposure, tidal range, bedrock type, and erosion and deposition.

19 What is the effect of constructing a barrier to longshore drift on marine erosion and deposition patterns along a shore?

SUGGESTED READINGS

BASCOM, W. *Waves and Beaches.* Garden City, N. Y.: Doubleday and Co., 1964, 267 pp.

________. "Beaches." *Scientific American* (1960), Vol. 203, No. 2, pp. 80–94.

________. "Ocean Waves." *Scientific American* (1959), Vol. 201, No. 2, pp. 89–97.

CLANCY, E. P. *The Tides: Pulse of the Earth*. Garden City, N. Y.: Doubleday and Co., 1969, 228 pp.

DEFANT, A. *Ebb and Flow: The Tides of Earth, Air, and Water.* Ann Arbor, Mich.: University of Michigan Press, 1958, 121 pp.

FAIRBRIDGE, R. W. "The Changing Level of the Sea." *Scientific American* (1960), Vol. 202, No. 5, pp. 70–79.

HAMMOND, A. L. "Oceanography: Geochemical Tracers Offer New Insight." *Science* (1971), Vol. 195, No. 4274, pp. 164–66.

HIGGINS, C. G. "Causes of Relative Sealevel Changes." *American Scientist* (1965), Vol. 53, No. 4, pp. 464–76.

HOYT, J. H. "Shoreline Processes." *Journal of Geological Education* (1972), Vol. 20, No. 1, pp. 16–22.

INMAN, D. L., and BRUSH, B. M. "The Coastal Challenge." *Science* (1973), Vol. 180, No. 4094, pp. 20–32.

KING, C. A. M. *Beaches and Coasts.* London: Edward Arnold, 1959, 437 pp.

MUNK, W. "The Circulation of the Oceans." *Scientific American* (1955), Vol. 191, No. 3, pp. 96–108.

SHEPARD, F. P. *Geological Oceanography.* New York: Crane, Russak and Co., 1977, 214 pp.

SMITH, F. G. W. "The Real Sea." *Sea Frontiers* (1971), Vol. 17, No. 5, pp. 298–311.

STOMMEL, H. "The Anatomy of the Atlantic." *Scientific American* (1955), Vol. 190, No. 1, pp. 30–35.

THURMAN, H. V. *Introductory Oceanography,* 3rd ed. Columbus, Ohio: Charles E. Merrill Publishing Co., 1981, 464 pp.

- earth resource
- mineral resource
- renewable resource
- nonrenewable resource
- resource
- reserve
- hypothetical resource
- speculative resource

17

- metal
- nonmetal
- minimum anticipated cumulative demand (MACD)
- open-pit mining
- underground mining
- in situ mining
- metallogenic province
- evaporite deposit
- bedded (banded) iron deposit
- bauxite
- manganese nodule
- placer deposit
- solubility
- Frasch process
- anion
- cation
- structural trap
- fission
- fusion
- oil shale

Earth Resources, Energy, and the Environment

EARTH RESOURCES, ENERGY, AND *ENVIRONMENT* ARE SOMETIMES called the three magic *E* terms because they are obviously interrelated and are of major significance to all people. Unfortunately, although we often believe ourselves to be knowledgeable about earth resources, energy, and environment, all too often we are really only vaguely familiar with these subjects. The purpose of this chapter is not so much to focus on environmental issues associated with resource utilization as to introduce and discuss the various categories of resources, ranging from metals, nonmetals, and fuels to earth materials used for building purposes.

An **earth resource** may be broadly defined as any useful material derived from the animal, vegetable, or mineral kingdoms on earth. The emphasis of the present discussion will be primarily on inorganic and fossilized organic materials, which may be lumped together as **mineral resources**. Such resources are concentrations of naturally occurring materials (solids, liquids, or gases) in or on the crust of the earth in a form such that economical extraction is either currently or potentially feasible. Under this broad definition, mineral resources include such diverse materials as water, crude oil, natural gas, metals, and building stone.

Mineral resources may be either **renewable** or **nonrenewable**. A renewable resource (e.g., surface-water supply) will eventually become available again following use. It is important to emphasize that all water may not be considered renewable because of the long lag time between initial use and potential reuse. This is particularly true for groundwater resources, which may include long paths between surface infiltration and eventual use of the resource. Nonrenewable resources are those materials that, once used, are essentially not available again to people. Examples include the burning of fossil fuels and the splitting of uranium atoms.

In discussing mineral resources, it is useful to distinguish between a **resource** and a **reserve** (Fig. 17–1). The reserve is that portion of the total mineral resource that is identified and from which usable materials may be economically extracted at the time of the evaluation. That is, with a reserve there is precise information as to the amount and quality of the particular mineral resource being considered. For example, if it can be stated confidently that a certain amount of rock will yield a certain amount of metal, then the deposit containing the metal is considered a reserve. Specific information as to reserves is best obtained through careful geologic evaluation. Such evaluation will vary with the resource being considered, but for metals it would probably involve drilling of rocks to obtain samples from which the amount of metal can be computed. In dealing with petroleum resources, test wells may be used to delineate the total reserve.

	Identified	Undiscovered	
		In known districts	In undiscovered districts or forms
Economic	Reserves	Hypothetical	Speculative
Subeconomic	Identified Subeconomic Resources	Resources	Resources

Increasing degree of economic feasibility (↑)

Increasing degree of geologic assurance (←)

FIGURE 17–1
Reserves are those economically recoverable materials from identified deposits (e.g., metal ore), as distinguished from resources, which are the same materials from identified but subeconomic or hypothetical deposits. Identified reserves are classified as *proved, probable,* or *possible* depending on the degree of certainty with which the amount of ore and related factors (i.e., economics of recovery) are known. For undiscovered resources, two categories can be included: (1) hypothetical resources, or undiscovered ore from known areas of ore occurrence, and (2) speculative resources, or undiscovered ore from areas where some geologic criteria indicate ore favorability but from which no ore deposits are known. Only when sufficient drilling or other techniques have been employed to indicate the presence of some ore can hypothetical ore be properly classified as a reserve. (From W. P. Pratt and D. A. Brobst, U.S. Geological Survey Circular 698, 1974)

The distinction between reserves and resources is somewhat arbitrary, as some resources will be added to or subtracted from reserves due to economic or legal conditions. Other resources in known mining areas are classified as either **hypothetical** or **speculative**, and these also may eventually be moved into the reserve category following more careful evaluation. Certainly it is more likely that additional reserves will be found in known mining areas than in locations where resources have not been discovered.

The crustal abundance of a particular element and the specific way in which that element is incorporated into minerals are the controlling factors for the identification of a reserve. That is, it is not sufficient merely to have a relatively high concentration of a particular element. It is also important that the element desired be bonded to other atoms in such a way that it can be extracted at a profit. Aluminum, the third most abundant metal by weight in the earth's crust, is a good example. Only a few minerals such as gibbsite and aluminum hydroxide can be treated for economic removal of aluminum. The aluminum in common rock-forming minerals such as the feldspars and micas is too tightly bonded to be extracted profitably under present technology.

In discussing mineral resources, it is useful to distinguish between **metals** and **nonmetals**. Metals are elements used for their metallic properties, such as conductivity and ductility. The term nonmetal refers to elements and their compounds not used for metallic properties. Thus sodium, even though it is a metal, is considered a nonmetallic mineral resource because its commercial utilization is nonmetallic. Oil, gas, and coal are other examples of nonmetallic mineral resources, as are potash (potassium chloride), gypsum (hydrated calcium sulfate), sand and gravel, cement, borax, sulfur, nitrogen, and phosphorus.

Many people are surprised to learn that the annual world consumption of nonmetals far exceeds the use of metallic elements (Table 17–1). Carbon is consumed in the greatest quantity because of the tremendous utilization of fossil fuel. Of those materials for which annual world consumption exceeds 10 million tons/year (with the exception of building materials), iron is the only metal. The remainder (carbon, sodium, nitrogen, oxygen, sulfur, potassium, and calcium) are largely consumed in agricultural and related industries. As shown in Table 17–1, some of the better-known metals such as zinc, copper, lead, and aluminum have annual world consumption rates below 10 million tons/year.

It is important for the United States to anticipate demand for mineral resources. This has been done for a number of different materials, as shown in Table 17–2. This table is expressed in terms of **minimum anticipated cumulative demand** (**MACD**) and whether a particular resource is identified or undiscovered. Those materials categorized as MACD-I are very abundant in the United States, whereas those with a higher roman numeral category are less

TABLE 17–1

Annual world consumption of some elements.

1–10 billion tons
Carbon
0.1–1 billion ton
Sodium
10–100 million tons
Iron
Nitrogen
Oxygen
Potassium
Sulfur
1–10 million tons
Aluminum
Barium
Chromium
Copper
Fluorine
Hydrogen
Lead
Magnesium
Manganese
Phosphorus
Zinc

SOURCE: After U.S. Bureau of Mines, *Minerals Yearbook, 1976.*

TABLE 17–2
MACDs for selected commodities.

Commodity	Identified Resource	Unidentified Resource
Aluminum	II	NE
Barite	II	II
Chromium	VI	V
Copper	III	III
Gold	III	NE
Iron	II	I
Nickel	III	NE
Phosphate	II	I
Sand and gravel	III	NE
Sulfur	I	I
Titanium	II	II
Uranium	II	III

SOURCE: From U.S. Geological Survey Professional Paper 820, 1973.
NOTE: Roman numerals indicate minimum anticipated cumulative demand (MACD); I = most abundant, VI = least abundant. NE = not estimated.

abundant. The NE designation in Table 17–2 denotes those materials for which no estimation has been assigned for the undiscovered resources. The assignment of NE is made for different reasons for different commodities, although the basic reason is that any MACD categorization would only be temporary. For example, in the case of aluminum there is optimism that by the year 2000 technology will have made possible extraction of the metal from clays and low-grade bauxites (an ore of aluminum). Thus, in the next decade or two the designation for undiscovered resources in aluminum may be changed from NE to I. Gold is assigned the NE category for the undiscovered resources because resource figures are simply not very meaningful for this element, which is used primarily for purposes of trade payment. Gold will continue to be recovered as a by-product of copper mining; however, note that copper falls into category III. This means that the United States may import copper to meet its needs rather than mine its own resources, thereby reducing the amount of gold produced. In the case of nickel, there is hope that technology will allow its recovery from weathered nickeliferous rocks, in which case the MACD for the undiscovered resources may change from NE to III or IV. For sand and gravel, the NE designation for the undiscovered resources reflects the lack of prospecting for these valuable building materials as well as the need to develop innovative ways to economically recover and transport huge quantities to where they can be utilized.

Natural resources are very unevenly distributed around the world, as are population and rate of industrialization. The United States, with less than 10 percent of the world's population, has a per capita consumption (annual gross consumption of commodities divided by the population) four to five times that for any other industrialized nation. Should Asia or Africa rise to our per capita consumption, then they would not be able to export mineral resources, making it difficult for the United States to import materials in short domestic supply.

A major problem related to mineral resource consumption is the threat of mass starvation and energy shortage that the world may be facing in the twenty-first century. If calculations by the National Academy of Sciences are correct, then the world's maximum capacity to produce food for its population may be reached by the mid-twenty-first century. In addition, it is probable that world oil and gas resources will have peaked early in that century, and that many mineral resources will have been exhausted. This pessimistic view is based on assumptions of continued high rates of population growth and food consumption, rapid industrialization of essentially unindustrialized countries, no new major mineral deposit discoveries, failure to use nuclear or alternative (to fossil fuel) energy sources, and no major unforeseen technical achievements. Because of these uncertainties, it is clear that trying to project MACD figures beyond the year 2000 at this time is an exercise in futility.

Where and how mineral resources are ultimately recovered depends upon legal and economic factors as well as environmental conditions at a particular site. Mining methods include **open-pit**, **underground**, and **in situ**. When relatively large amounts of a mineral resource occur close to the surface, an open-pit mine may be desirable (Fig. 17–2). If smaller amounts of high-grade ore occur at depth, then a mine shaft with lateral **drifts** (**tunnels**) from the shaft to the ore zones may be utilized. This underground method is shown in Fig. 17–3. In some cases, ore (the material containing the mineral that can be extracted at a profit) is high-grade but spotty, and is not thought to be economically recoverable by either open-pit or underground methods. In situ mining is removal of the valuable component of a mineral deposit without physical extraction of the rock. If the ore (e.g., some copper and uranium ores) can be dissolved by acid or alkaline solutions, such solutions may be injected under pressure through the ore body and the dissolved metals collected at another site. The general scheme for such in situ solution mining is shown in Fig. 17–4.

FIGURE 17–2
Part of the open pit at the Highland uranium mine in the Powder River Basin, Wyoming. Because the ore is located very near the surface, it can be economically recovered by open-pit methods.

An interesting fact about many mineral resources is that their occurrence is often related to certain types of rocks of specific ages in specific tectonic settings. Such an environment for a particular metal is called a **metallogenic province**. The porphyry copper belt, the western part of both North and South America (Fig. 17–5), is one example of a well-defined province. Another example is the tin-tungsten province of Korea-Southeast Asia (Fig. 17–5). Most metallogenic provinces have several things in common: alignment to major tectonic elements such as subduction zones as shown in Fig. 17–5, igneous rocks, and rocks of similar age and approximate equivalent chemistry.

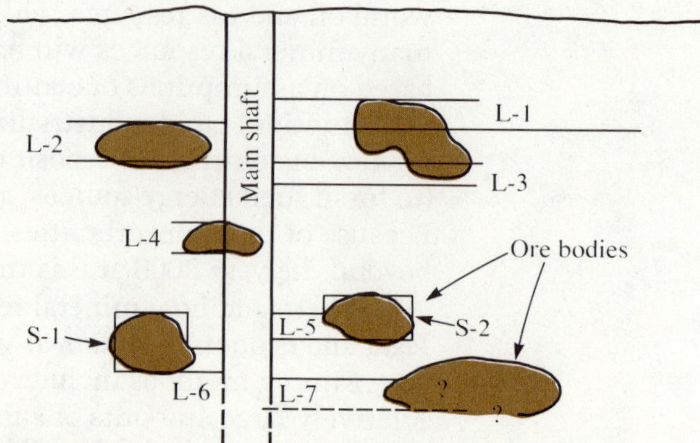

FIGURE 17–3
Underground mining. In this method, a main shaft is sunk to a predetermined depth, and tunnels (called *drifts* or *adits*) from this main shaft are used to reach the ore bodies at various levels (L-1, L-2, etc.). When the ore is thicker than the tunnels, *stopes* (S-1 and S-2) are excavated, provided the rock can withstand excavation or can be supported by other methods.

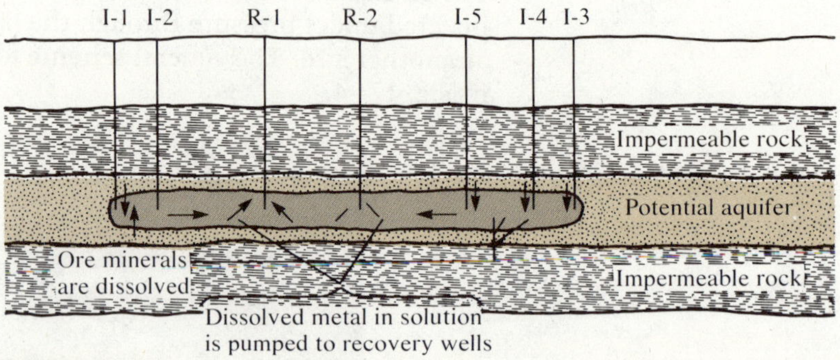

FIGURE 17–4
Solution mining. For some ore, such as copper and uranium ores, the ore minerals can be dissolved by acid or alkaline leach injected under pressure through a series of injection wells (I-1 through I-5) and recovered by pumping from recovery wells (R-1 and R-2). As shown here, the injection wells surround the recovery wells.

The origin of metallogenic provinces and ore deposits at convergent plate boundaries is hypothesized to be related to relations between seawater-saturated oceanic rocks and partial melting of rocks associated with the descent of oceanic lithosphere at the zone of convergence (subduction zone). The basic idea is that as the seawater-saturated rocks are subjected to elevated temperatures and pressures in the subduction zone, partial melting occurs. The combination of heat, pressure, and the resulting partial melting facilitates the mobilization of metals which are concentrated and ascend as components of the magma. Thus, under this hypothesis the metals are deposited in surrounding rocks as the metal-rich fluids migrate through them after escaping from the magma below. The origin of recently discovered metallic ore deposits at divergent plate boundaries (Fig. 17–7) is evidently

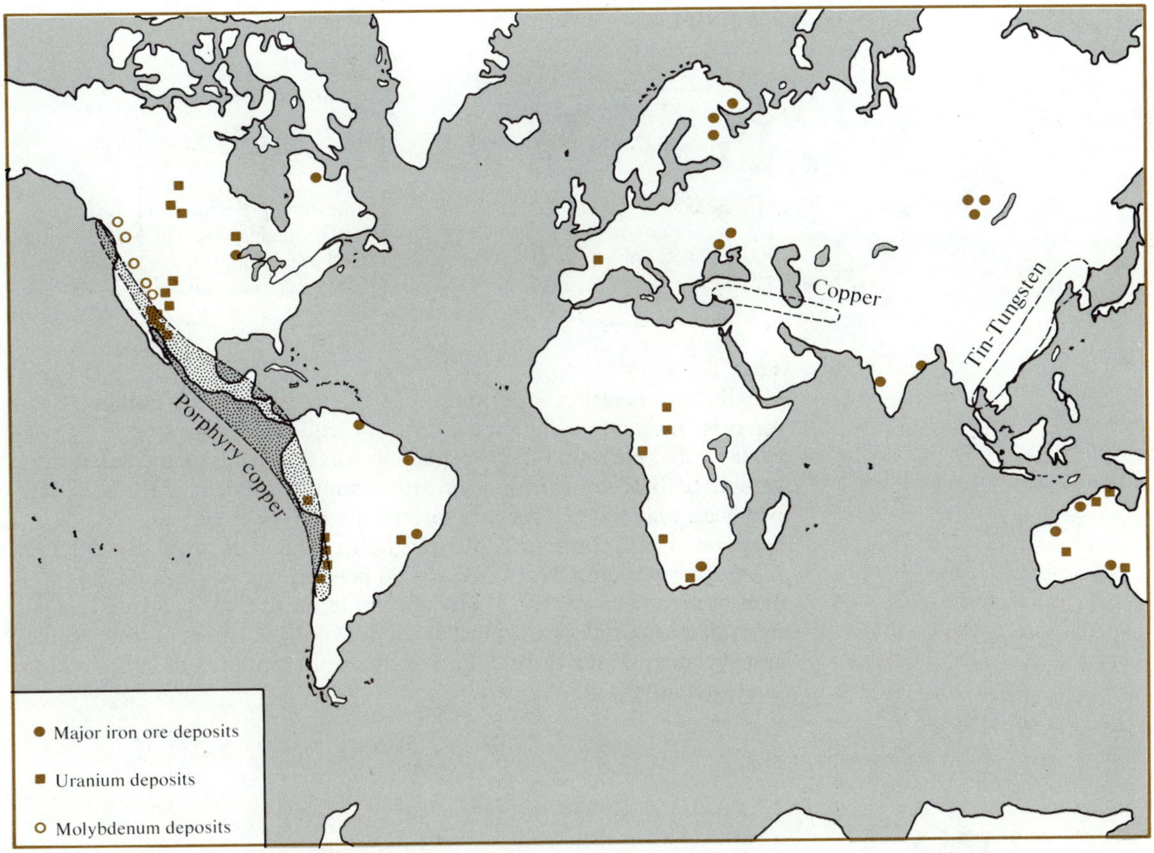

FIGURE 17–5
Worldwide metal ore distribution. The porphyry copper, molybdenum, and tin-tungsten belts are spatially associated with igneous rocks and major subduction zones. Iron and uranium do not show any regularity in their spatial relationship to orogenic belts or igneous rocks. Most of the iron deposits are Precambrian bedded sedimentary types, whereas the uranium deposits are found in sedimentary, metamorphic, or igneous rocks.

related to the migration or movement of ocean water. Basically, cold, dense ocean water moves downward through many fractures in the basaltic rocks at oceanic ridges and is heated by contact with or heat from nearby molten rocks (magma). Water so heated is lighter and chemically more active, and therefore rises (convects) up through the fractured rocks. The upward migration is accompanied by leaching of metals which are carried in solution and deposited or precipitated as metallic sulfides.

Where seawaters have evaporated, marine or nonmarine **evaporite deposits** result. Salt water contains 3.5 percent dissolved salts, and as shown by Fig. 17–6 a variety of materials will precipitate if evaporation takes place. Recoverable mineral resources include potash, gypsum, and halite (sodium chloride, or common table salt). Valuable commodities that may be extracted from nonmarine evaporite deposits include sodium carbonates, sodium sulfates, and borax. Both marine and nonmarine evaporite deposits are commonly interlayered with limestones and associated with petroliferous matter and, less commonly, with metal sulfide deposits.

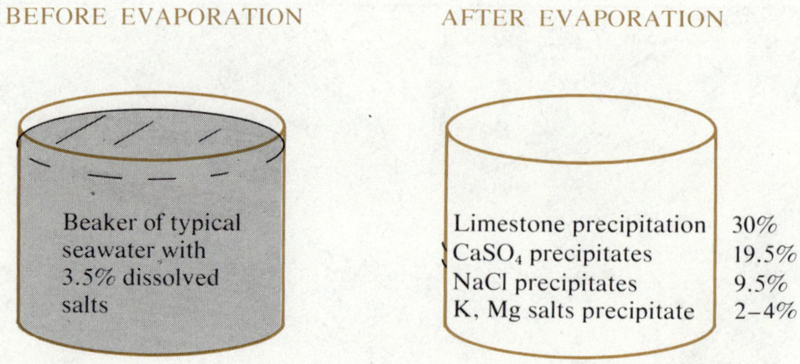

FIGURE 17–6
Usiglio's evaporation experiments. In the late 1840s, the Italian scientist J. Usiglio conducted a series of studies on seawater evaporation. He found that normal evaporation led to some calcium carbonate (limestone) formation after about 70 percent of the seawater had been evaporated. Gypsum formed after 80 percent had evaporated, and common table salt, halite (NaCl), formed after 90.5 percent evaporation. Not until 96–98 percent had evaporated did potassium and magnesium salts appear. In nature, rocks formed by evaporation follow a typical sequence of thick layers of limestone and gypsum (or anhydrite) followed by minor amounts of halite and rarely sylvite (KCl).

THE METALS

ABUNDANT METALS

Iron Iron (Fe) is the fourth most abundant element in the crust of the earth by weight percent and, more importantly, the foundation upon which industrialized nations are built. The iron industry accounts for well over 90 percent of all metals mined and is essential for the production of those elements alloyed with iron, such as cobalt, nickel, chromium, vanadium, and manganese.

Most iron production is from Precambrian banded iron formations. Principal iron minerals are magnetite, hematite, and geothite, all oxides of iron. All six major continents contain abundant reserves of iron, making the world trade in iron highly competitive. Although the United States possesses adequate reserves to meet its needs for many decades to come, it still imports roughly 30 percent of the 130 million tons it consumes each year. Every highly industrialized nation depends very heavily on iron, and there is a direct correlation between iron consumption and standard of living for these nations.

Iron ore deposits fall into four main types: (1) bedded Precambrian deposits, (2) ore associated with igneous rocks and contact metamorphic zones, (3) hydrothermal (hot-water) deposits not directly associated with magmatic processes, and (4) near-surface or surface deposits including iron laterites and other low-grade deposits enriched or formed by weathering and related processes at or near the surface of the earth. Some of the places iron is found and mined are shown in Fig. 17–5. Notice that, unlike the porphyry copper and tin-tungsten metallogenic belts discussed previously, neither the iron nor the uranium deposits show any regularity in their spatial relationship to tectonic zones. Most of the iron deposits are Precambrian bedded sedimentary deposits, and thus are related to ancient sedimentary processes much older than recent tectonic zones. Uranium deposits, on the other hand, are found in sedimentary metamorphic or igneous rocks depending upon local and regional geologic conditions. Thus, although many mineral deposits may be explained in terms of metallogenic provinces associated with the plate tectonic model as illustrated in Fig. 17–7, not all metal deposits can be so easily explained.

The Precambrian **bedded (banded) iron deposits** are so important that they merit special attention. Typically, these deposits contain approximately 20–40 percent iron deposited by sedimentary processes and are interlayered with chert with or without carbonates and other silicates. Where these deposits have been metamorphosed, the original mineral assemblages have been changed, but not the basic iron content.

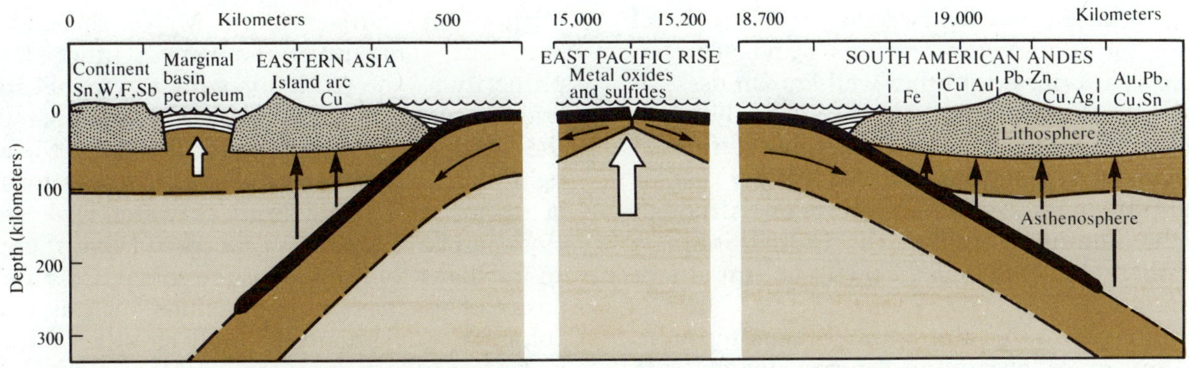

FIGURE 17–7
An idealized diagram showing the relationship between the east Pacific rise (divergent plate boundary), Pacific margins (convergent plate boundaries), and metallic ore deposits. [After NOAA, *California Geology* (1977), Vol. 30, No. 5]

The origin of these deposits is not well known. Presumably, they formed about 2–3 billion years ago when the oxygen content of the atmosphere was very low and therefore not able to oxidize ionic iron. By about 1.8 billion years ago, the oxygen content in the atmosphere had increased such that iron oxides became abundant and important in surface or near-surface processes; thus, extensive bedded iron deposits are not found in rocks younger than about this time.

Many of these iron-bearing formations are folded, and the beds dip very steeply. Iron deposits in the Lake Superior region, the Labrador deposits, and the Hammersley Range in Australia fall into this category. Mining of folded rocks requires relatively simple planning, as shown in Fig. 17–8.

Some of the banded iron deposits contain only about 15 percent iron, as for example the taconites of the Lake Superior region. Although recognized as a potential resource for many years, it was not until profitable ways of extracting the ore were found that deposits could be utilized. Once thought to be only of potential use, the taconite ore now accounts for more than 60 percent of the iron ore produced in the United States.

Aluminum Aluminum (Al) is the third most abundant element by weight in the earth's crust. However, most of it is bonded to oxygen in silicates, and is therefore extremely difficult to extract. Only iron surpasses aluminum in use as a metal, and use of aluminum is

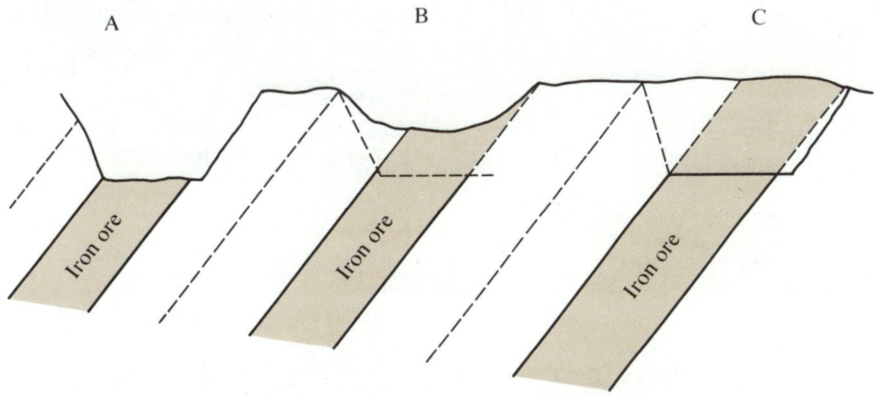

FIGURE 17–8
Banded iron ore deposits. Many of the world's iron ore deposits occur in uniform, banded, metamorphosed sedimentary rocks. Open-pit mining methods are effective because the ore is near the surface and low in grade. Ideally, as open pit A is completed, B is very nearly ready; and C will begin producing when A has been shut down and B is half finished. It is cheaper to develop pits B and C to the same depth as A than to deepen A. This type of mining is common for the extensive iron ores in Labrador.

increasing because it is a lightweight yet strong metal appropriate for construction materials and vehicles. Aluminum is also ideal for other well-known uses including electrical equipment and military hardware.

The United States in recent years has consumed between 4 and 6 million tons of aluminum per year, or about one-third of the world's consumption. Although most of the aluminum used in the United States is produced there, industry relies heavily on imports of aluminum ore.

The principal ore of aluminum is **bauxite**, more properly known as aluminum laterite, which is a hydrated oxide of aluminum. Laterite is evidently formed under conditions characterized by tropical weathering where aluminum-rich source rocks are exposed to intense and prolonged rainfall which exceeds water loss by evaporation. Under these conditions, and presuming that the rocks exposed are in a tectonically stable area, most of the common rock-forming elements such as sodium, potassium, magnesium, calcium, and silicon are removed by weathering. Aluminum remains as **gibbsite** (aluminum hydroxide), which is the principal mineral of bauxite ore. Hydrated aluminum oxide is concentrated because aluminum is extremely insoluble in natural solutions. The process by which a bauxite may form from the weathering of limestone containing clay-rich layers is illustrated in Fig. 17–9. Thus, formation of bauxite deposits depends upon climatic conditions as well as upon an aluminum-rich source rock. Therefore, clay-rich limestones as well as certain granitic rocks when exposed to tropical weathering facilitate the formation of aluminum-rich clay accumulations from which bauxite develops.

Magnesium Magnesium (Mg) is the eighth most abundant element in the earth's crust. It is used primarily as the oxide for refractory applications (refractory material can withstand high temperatures); as the carbonate for many chemicals; and as the metal for light, corrosion-resistant alloys. Methods for recovery of magnesium metal from seawater were started in 1940 and have proven so successful that there is truly an inexhaustible supply of the metal. Commercial deposits of magnesium compounds such as magnesite, dolomite, and brucite as well as other minerals are widespread throughout the world.

Titanium Titanium (Ti) is the only mineral resource listed as a metal when actually much of its use is for the chemical industry as a paint pigment. However, its increasing use for its metallic properties in the aerospace industry warrants its inclusion here; in addition, the U.S. government lists titanium as a strategic commodity and stockpiles it accordingly.

Titanium is widespread as an accessory element in many rock-forming minerals (e.g., amphiboles, pyroxenes) but fortunately it is also involved in rock-forming accessory minerals such as rutile (TiO_2) and ilmenite ($FeTiO_2$). It is from these accessory minerals that most production results. Special rocks such as anorthosites commonly contain enormous amounts of ilmenite. Lesser amounts of both ilmenite and rutile are associated with mafic and ultramafic rocks.

LESS ABUNDANT AND SCARCE METALS

Manganese (Mn) is the twelfth most abundant metal in the earth's crust—about 0.1 weight percent. Man-

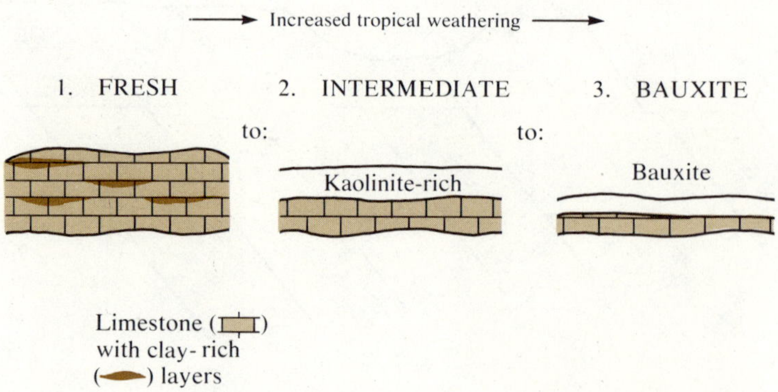

FIGURE 17–9
Formation of bauxite from limestone. Some bauxites form from rocks containing relatively aluminum-rich layers. Most limestones are impure, and the clay minerals they contain are typically aluminum-rich. When exposed to tropical weathering, the limestone itself is readily removed and kaolinite is formed or concentrated from preexisting clay minerals. Further weathering results in the formation of bauxite deposits as shown. This is typical of the deposits in the Caribbean Islands.

ganese supply is a problem for the United States; the world's supply is plentiful, yet the last mine to operate in the United States closed in 1972, and needs have been met by imports since then. The steel industry's use of manganese as a raw additive to scavenge sulfur and oxygen and as an alloy constitutes more than 90 percent of the U.S. usage of the metal. When one considers that for every ton of steel produced some 7–9 kg of manganese are needed, the importance of manganese is fully recognized.

Of special interest are the **manganese nodules** which cover wide areas of the ocean floor (Fig. 17–10). Although the nodules are extremely impure, they are iron-rich, and also enriched in such elements as copper, cobalt, chromium, and nickel, with lesser amounts of gold and other elements. Economic recovery of these nodules, which cover up to 2600 km^2 in places, is not yet possible, but there is increasing optimism among industries and agencies pursuing this matter.

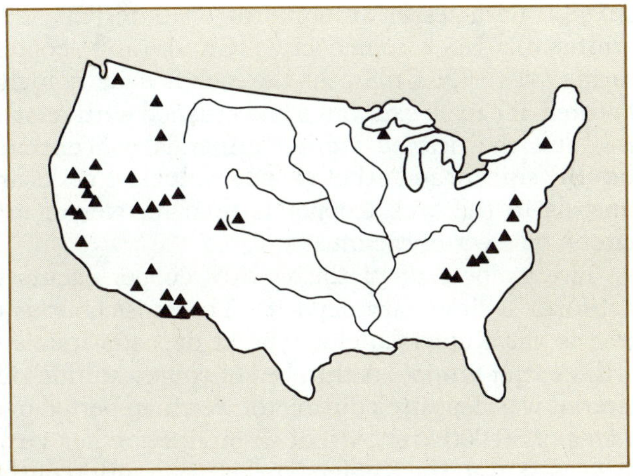

FIGURE 17–11
The occurrence of economic deposits of copper in the United States. (Data from mineral resource maps of the U.S. Geological Survey)

Scarce Metals Concentrated as Sulfides Many of the scarce metals in nature are mined from deposits of sulfides, others occur as oxides, and some occur in the native state and in other ways. In this section, we will consider the metals copper, nickel, cobalt, lead, zinc, molybdenum, and mercury, all of which are primarily concentrated in sulfides of the metal.

Copper (Cu) is perhaps the best known of these, and the copper reserves and resources of the world are apparently abundant enough for many years to come. Copper is at places concentrated as the native element, but this type of occurrence is rare. See Fig. 17–11 for the distribution of copper deposits in the United States.

Most of the world's copper comes from fractured prophyritic rocks in which copper mineralization has occurred. These so-called porphyry copper deposits contain very-low-grade ore and must be mined by open-pit methods. In several parts of the world, they occur in well-recognized belts that are classic examples of metallogenic provinces. Despite their low grade, porphyry copper deposits account for more than 50 percent of the world's copper production.

Although copper's uses are many, most is consumed for electrical applications (55 percent), and the remainder for construction, industrial machinery, transportation industry, and chemical uses. Competition is extreme, with some 65 countries producing copper, although 8 (United States, Chile, Canada, Zambia, Soviet Union, Zaire, Peru, and the Philippines, in order of production) account for more than half of the world production.

In porphyry coppers and in many other sulfide deposits, the main copper-bearing mineral is chalcopyrite

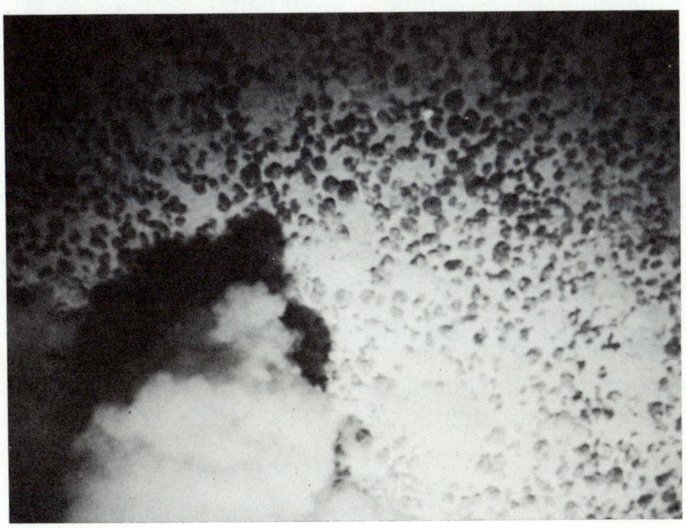

FIGURE 17–10
Manganese nodules form on the floor of the ocean as chemical precipitates. Besides manganese, the nodules contain iron, copper, nickel, and cobalt. (Photo courtesy R. F. Weiss, O. H. Kirsten, and R. Ackerman, Scripps Institute of Oceanography, University of California, San Diego)

($CuFeS_2$), with lesser amounts of enargite (Cu_3AsS_4), bornite (Cu_5FeS_4), chalcocite (Cu_2S), and copper-bearing pyrite (($(Fe,Cu)S_2$). As the rock is usually highly fractured, it can be excavated and crushed with relative ease, and the desired sulfide minerals concentrated from the silicate and other **gangue** (valueless rocks and minerals in the ore). Copper is then recovered, ultimately, by electrolytic means.

Twenty percent of the world's copper occurs in stratiform sedimentary deposits. The most famous of these is the Kupferschiefer type of deposits found in north-central Europe. A thin bed of copper-sulfide-rich material was deposited during the Permian period over an area of 50,000 km². Although production has virtually ceased from the Kupferschiefer, recognition of this type of deposit led to the discovery of the sedimentary Zambian-Zaire copper belt, which formed in a somewhat similar fashion and dominates Africa's copper production.

Nickel (Ni) and **cobalt** (Co) are discussed together here because of their very similar geochemical behavior. Both are nearly identical in their ability to substitute for each other and for other elements such as iron (Fe^{+2}), and it is for this reason that, in the absence of sulfur, nickel and cobalt will be found in the olivine and pyroxene of mafic and ultramafic rocks. When sulfur is present, both nickel and cobalt will be concentrated into the sufide minerals present to a much greater degree than to the silicate minerals. An interesting point to make here is that, should either rock just mentioned be exposed to conditions favorable for formation of laterite soils, then nickel and cobalt will behave like iron and be concentrated in the laterite.

The United States is almost completely dependent on imports of nickel from Canada, where the famous Sudbury deposit continues to account for much of the world's production.

Lead (Pb) and **zinc** (Zn) also have a similar geochemical behavior. Although they are not as close in geochemical properties as nickel and cobalt, they are concentrated together for other reasons. This is at first surprising because the lead ion Pb^{+2} is much larger than the zinc ion Zn^{+2} (1.35×10^{-7} mm versus 0.80×10^{-7} mm, respectively). Largely for this reason, lead is concentrated in galena (PbS), and zinc in sphalerite (ZnS).

Uses for lead and zinc are interesting. In addition to its well-known applications in batteries and in shielding materials, lead has been widely used as a gasoline additive. This last application is on the decline due to associated toxicity, but lead's use in industry and for shielding purposes has increased. Zinc, on the other hand, is used for alloys, brass, galvanizing iron, and for the chemical industry. It is fourth in production among all metals (behind iron, aluminum, and copper). Its use as a galvanizing agent has made the classic "tin can" almost obsolete.

Lead and zinc are found in stratiform sedimentary deposits and as contact metamorphic or hydrothermal vein types of deposits. (The leading world producers of lead and zinc are listed in Table 17–3.) The hydrothermal veins are usually located near igneous rocks, with the exception of the extensive deposits of the tristate (Missouri-Kansas-Oklahoma) district of the United States. As mentioned previously, the most famous stratiform deposit is the Kupferschiefer deposit of north-central Europe, where both lead and zinc have been recovered along with the copper.

TABLE 17–3
World leaders in zinc and lead production.

Lead
United States
Soviet Union
Australia
Canada
Peru
Zinc
Canada
Soviet Union
United States
Australia
Peru

SOURCE: U.S. Bureau of Mines, *Minerals Yearbook, 1975.*

The tristate deposits of Missouri-Kansas-Oklahoma are unique. These consist of irregular, fracture-filling veins rich in galena and sphalerite, yet no parent igneous rocks are present. The host rock is typically carbonate that has been brecciated, or else the carbonate has been replaced by the ore minerals.

Molybdenum (Mo), used in steel alloys as a toughening agent, is normally present in molybdenite (MoS_2). Much of the world's molybdenum is found in a metallogenic deposit extending from Alaska through Canada and the western United States into Mexico. Yet some 25 percent of molybdenum is produced as a by-product of porphyry copper mining.

Mercury (Hg), although extremely rare, deserves special mention for several reasons. The U.S. supplies are exhausted (the last operating mine closed in 1976), making the United States totally dependent upon imports for future supply. Also, remember that mercury is liquid at room temperatures, a property possessed by no other common metal, and substitutes for it are extremely difficult or impossible to make. In the United States (as elsewhere), the chief mercury ore mineral is cinnabar (HgS), deposits of which are located chiefly in the Coast Range of California, and in Nevada, Oregon, and Washington.

Scarce Metals Concentrated with Oxygen Many metals tend to be associated with oxygen, either in oxides or in silicates, rather than with sulfur. Consequently, many of these elements will not be concentrated but rather will be widely disseminated throughout common rocks.

Chromium (Cr) is commonly concentrated as chromite ($FeCr_2O_4$), which, because it is a heavy mineral, sinks to the bottom of the magma body, where it accumulates. Although chromite is abundant on the earth, the U.S. resources are virtually nonexistent, and thus the United States must import its supply from the Soviet Union, several countries in Africa, and the Philippines. Deposits in the Western Hemisphere are small and subeconomic except for some in Cuba.

Chromium has many uses (Table 17–4). There are no adequate substitutes for chromium in terms of a material with such high-temperature, corrosion-resistant strength.

Tin (Sn) and **tungsten** (W) are also metals for which the United States depends on imports, although the situation is somewhat better for tungsten than for tin. Tin occurs most commonly in nature as cassiterite (SnO_2), which while widely disseminated in felsic igneous rocks, is resistant to weathering and concentrated in **placer deposits**. Tin is found principally in two metallogenic belts in eastern Asia; one belt extends from the Malay Peninsula into Indonesia, the other from Bolivia into Peru. (See Table 17–4 for some common uses of tin.)

TABLE 17–4
Some uses of chromium and tin.

Chromium
Ferroalloys
Furnace linings
Pigments
Leather tanning
Tin
Tinplate
Solder
Alloys (bronze)
Protective coatings

Tungsten is found in nature as wolframite ($FeWO_4$) and scheelite ($CaWO_4$), neither of which is abundant in the United States. The major metallogenic belt for tungsten occurs in eastern Asia from Korea through China to Malaya. Conventional uses of tungsten have included cutting devices of tungsten carbide (45 percent), alloys (50 percent), and other uses (5 percent). There is now an increasing demand for tungsten in the space industry, as tungsten alloys are more suited than others for the extreme conditions to which some of the material must be subjected.

Scarce Precious Metals **Silver** (Ag), **gold** (Au), and the **platinum** (Pt) metals usually occur as the native metal, but most gold and silver production comes as a by-product of copper or lead-zinc mining operations. Porphyry coppers account for virtually all domestic gold production, and these plus lead-zinc operations constitute well over 75 percent of the domestic silver production. In the United States, gold and platinum are commonly enriched in placer deposits. The mineral is eroded from a lode that is exposed at the surface, and then mechanically transported, often by running water, to the site of the placer deposit (Fig. 17–12). Much of the gold mined in the western United States during the gold rush years came from placer deposits. Now, these deposits are too small and too sporadically distributed to be significant.

Silver more commonly occurs in sulfides than as the native metal, but it is usually lumped with gold and the platinoids when discussing production. Uses are largely in the photographic and electrical industries; the demand for silver for these industries is great and the supply is not adequate. Despite production from copper and lead-zinc operations, the United States must import most of the silver it needs.

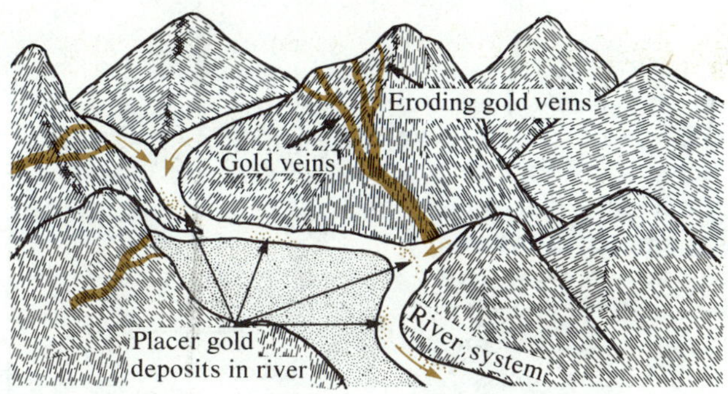

FIGURE 17–12
Placer deposits occur when a vein of mineral-bearing rock is eroded at an exposed surface, and then mechanically transported to the site of the placer deposit.

THE NONMETALS

NONMETALS FOR THE AGRICULTURAL AND CHEMICAL INDUSTRIES

Nonmetals are elements and compounds used for purposes that do not require metallic properties. Most of these elements and compounds are used in the agricultural and chemical industries.

It is necessary first to discuss the formation and distribution of certain types of sedimentary deposits. These deposits include marine and nonmarine evaporites, salt domes, and phosphorus-rich sedimentary rocks.

Seawater contains about 3.5 percent dissolved solids by weight; these solids are dominated by sodium and chlorine (Table 17–5). Different solubilities or the ability to form ions and dissolve in water accounts for the large concentrations of the ions listed in Table 17–5 in seawater rather than other abundant crustal elements. For example, silica and alumina are highly insoluble, and therefore not dominant in seawater. Instead, during weathering most of the silica and alumina is removed to form clay minerals. Similarly, iron is generally oxidized from Fe^{+2} (soluble) to Fe^{+3} (insoluble) and forms $Fe(OH)_3$. The proportion of the different cations in seawater can be explained by the preference of clay minerals for fixing one cation over another. The potassium ion is preferentially affixed to clay minerals over the sodium ion, which leaves a greater concentration of soluble sodium ions to enrich seawater (30.6 percent sodium compared to 1.1 percent potassium). In the same manner, the calcium ion is more easily fixed in clay minerals than the magnesium ion, resulting in more magnesium (3.7 percent) than calcium (1.2 percent) dissolved in seawater (Fig. 17–13).

TABLE 17–5
Percentage of dominant ions in seawater.

Cations	
Sodium	30.6%
Magnesium	3.7
Calcium	1.2
Potassium	1.1
Anions	
Chlorine	55.1
Sulfur	2.7
Bromine	0.2

Minerals have very different solubilities; calcium carbonate ($CaCO_3$) is more insoluble than calcium sulfate ($CaSO_4$), which in turn is more insoluble than halite (NaCl). Therefore, when seawater is evaporated, the dissolved salts are gradually concentrated to the point at which they precipitate. But evaporite formation involves processes more complex than mere evaporation of seawater. In the classic model, evaporite formation generally occurs under very warm conditions (between 35° north and south of the equator). Less commonly, it occurs under cold conditions where there are also very high winds and a barrier reef (or equivalent barrier) which traps the seawater. Once the seawater is trapped, evaporation begins, with new seawater occasionally overrunning the barrier, dissolving some of the more soluble salts. Eventually, however, thick accumulations of evaporite minerals result (Fig. 17–14). Due to their very great solubility, potassium (K^+)- and magnesium (Mg^{+2})-bearing salts are rare in marine evaporites, and halite (NaCl) is also relatively rare. These salts remain dissolved in the water as the less soluble salts precipitate to form the evaporites. The usual sequence for a marine evaporite is alternating layers of carbonate rocks and either anhydrite or

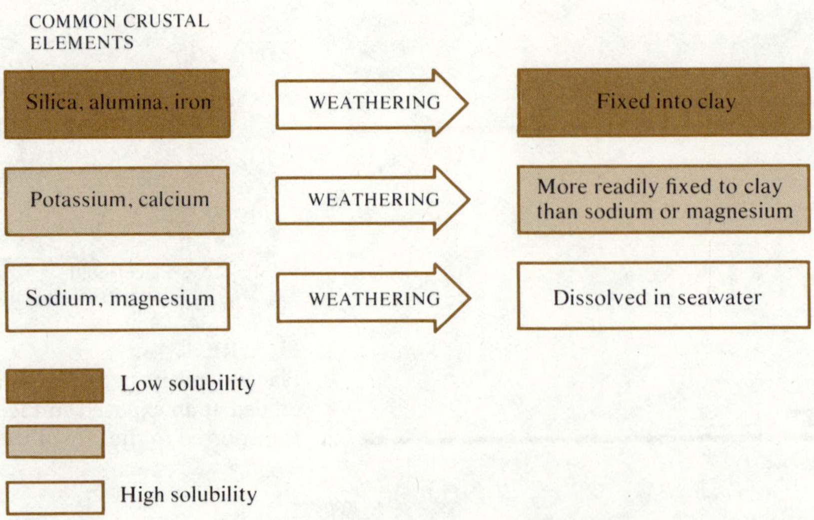

FIGURE 17–13
Solubility affects what happens to common crustal elements as they are weathered. When elements that form highly insoluble materials are weathered, they are fixed in clays. Partially soluble elements are differentially dissolved or fixed in clay depending on their solubility and the presence of other ions. Highly soluble elements are dissolved in water, and eventually make up the basic components of seawater.

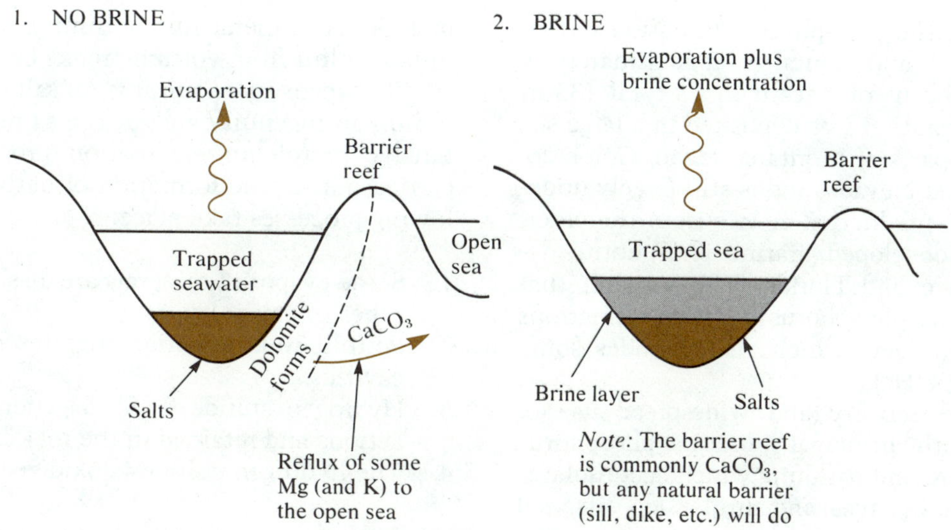

FIGURE 17–14
Marine evaporite formation. Under natural conditions, evaporites may form from either condition 1 (simple evaporation) or 2 (evaporation plus formation of a brine layer). Thick layers of limestone (at 70 percent evaporation) can accumulate before gypsum (at 80 percent evaporation) starts to form. Flooding at about the time gypsum is starting to form will precipitate more limestone. Hence, alternating layers of limestone and gypsum are the rule. Similarly, at about 90 percent evaporation, alternating layers of halite and gypsum are noted; and near total evaporation, sylvite and magnesium salts are interlayered with halite. Some of the magnesium is not incorporated into the evaporite and causes dolomitization of the barrier reef, as shown in condition 1.

gypsum. However, for an area in which evaporation has been allowed to proceed in a more uninterrupted fashion, halite will form layers, with anhydrite or gypsum forming layers in between. Finally, in almost totally uninterrupted sequences, potassium (K^-) (usually sylvite, KCl) and magnesium (Mg) salts will be found interlayered with halite.

Nonmarine evaporites are the product of the evaporation of more locally restricted seas where exchange with the open ocean is limited or nonexistent. Because nonmarine seas are small and lack exchange with normal seawater, their chemical compositions are extremely variable. Nevertheless, the dominant anion is the sulfate ion, followed by lesser amounts of borate or phosphate ions; the amount of chlorine as an anion is very small. The total amount of carbonate is smaller as well. Hence, typical minerals of nonmarine evaporites include sodium carbonates, sodium sulfates, sodium borates, some complex sodium, calcium salts, and, in general, salts that reflect the composition of the local highland areas that weathered away.

Potassium Potassium (K), largely as sylvite (KCl), is vital to the fertilizer industry. It is one of nature's quirks that, although potassium is one of the eight most abundant elements in the earth's crust by weight, it is locked up in common rock-forming silicate minerals, or else so efficiently fixed in clay minerals during weathering that essentially none is made available for the soil to use as a nutrient. When sylvite is deposited in marine evaporites, it then becomes extremely valuable because the sylvite layers are often pure and, therefore, relatively easy to mine. The sylvite can be easily treated to remove the potassium from the chlorine. In many evaporites containing sylvite, other complex potassium and magnesium salts (usually as sulfates) are also found.

Although MACD figures for potassium are not given by the U.S. Geological Survey, it is safe to assign II for both identified and hypothetical resources.

Nitrogen Prior to the development of the technology to recover nitrogen (N) from the atmosphere, most of this element was recovered from nitrate deposits. Nitrates are extremely soluble and are found only in very arid parts of the world. The largest deposits occur in the Atacama Desert of Chile, where the total precipitation between 1900 and 1970 was only 12.7 cm.

Phosphorus Phosphorus (P), while abundant, is very unevenly distributed worldwide. Typically it is concentrated in sedimentary formations close to evaporites or other marine sedimentary formations. In the United States, the Phosphoria Formation covers some 260,000 km² and contains approximately a meter-thick phosphorus-rich seam in its total 135-m thickness. This formation was deposited in a large sea that once covered parts of Montana, Idaho, Colorado, Utah, Wyoming, and Nevada, and is still largely undeveloped. Most U.S. production comes from the older, more extensively developed marine phosphorite deposits in Tennessee and Florida. The mineral that provides most of the phosphorus in these formations is apatite, $Ca_5(PO_4)_3(OH)$, which also includes some fluorine (i.e., F^- for OH^-).

Prior to the discovery and widespread use of marine evaporites, the principal sources of phosphorus were guano (sea bird and bat dung which accumulates in coastal areas and in caves) and similar deposits, and the secondary source, apatite recovery from igneous rocks. At present, about 75 percent of the world's phosphorus comes from marine phosphate rocks, 16 percent from igneous apatite (usually a by-product of other mining ventures), and the remaining 9 percent from guano deposits, mostly in newly discovered (but small) deposits in the southern Pacific Ocean region.

Sulfur Sixty percent of the sulfur (S) produced is used by agriculture either directly as a soil additive (40 percent) or indirectly as an insecticide or fungicide (20 percent). Further, because of its varied types of occurrences, the MACD figures for identified and hypothetical resources are I's, and these figures are not likely to change in the twenty-first century.

On a worldwide basis, roughly 80 percent of sulfur is made available from three prime sources, each of about equal amounts: (1) sulfur from salt domes, (2) sulfur from the petroleum industry, and (3) sulfur as a by-product of metal sulfide mining. Other sources are minor, sulfur from volcanic rocks being the largest.

The recovery of sulfur from salt domes is a tribute to human ingenuity and, at the same time, to the extensive petroleum exploration programs still being carried out. In the formation of native sulfur, the following processes take place:

1. Some evaporite minerals are dissolved by brines or other solutions.
2. Petroliferous material migrates into the solution cavities.
3. Hydrogen sulfide (H_2S) is generated by bacterial actions and retained in the rock.
4. The hydrogen sulfide is oxidized to native sulfur.

Sulfur from the salt domes is recovered by the **Frasch process**. In this process, a pipe consisting of three concentric rings is drilled into the salt dome. Hot water is forced down the outermost part of the pipe. This water melts the native sulfur, which then flows toward the next part of the pipe by a combination of gravity flow and pumping. The innermost part of the pipe is used to inject a mix of superheated air and steam into the molten sulfur; this is done not only to help the sulfur flow upward, but also to ensure its remaining in the molten state until it reaches the surface (Fig. 17–15).

Other Nonmetals The list of nonmetals used in the chemical and other industries is extremely long. Only a few will be mentioned here. **Halite** (NaCl) is best known as common table salt, yet this use accounts for very little of its total consumption in the United States and abroad. The abundant halite, available from en-

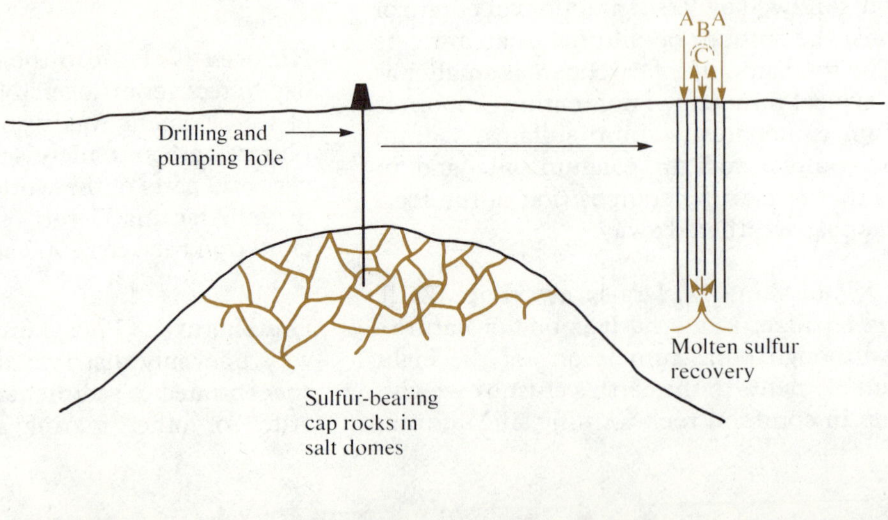

FIGURE 17–15
The Frasch process. A concentric pipe is inserted into sulfur-rich cap rocks found in many salt domes off the coasts of Louisiana, Texas, and Mississippi. Hot water injected through the outer part of the pipe (A) melts the sulfur in the fractured rock. Because of its lower density, the molten sulfur moves to the recovery part of the pipe (B), where it is mixed with superheated air from the innermost part of the pipe (C) to keep the sulfur molten until it is recovered at the surface.

trapped seawater or from evaporite deposits, is used primarily for production of chlorine gas and sodium carbonate. Although sodium carbonate is available from nonmarine evaporite deposits, it is cheaper to recover during the production of chlorine.

Bromine (Br) is used in the medical field and as a gasoline additive. It is also one of the elements that can be cheaply and efficiently recovered from seawater. Thus, despite limited use (relative to other nonmetals), the MACD figures are I and I, and no shortage is predicted in the future.

Chlorine (Cl) is the most abundant anion in seawater; estimates run as high as 250×10^{20} g of chlorine in ocean waters. Yet chlorine, unlike bromine which is present in the oceans in amounts near only 0.0065 percent, must be extracted from halite.

Barite ($BaSO_4$) deserves special mention. United States production of barite jumped from roughly 1.5 million tons in 1945 to nearly three times that amount by 1975. The U.S. Geological Survey indicates MACD figures of II and II for identified and hypothetical resources. Barite occurs primarily in bedded sedimentary rocks, where trace amounts of barium, released during mild burial and related chemical effects, are fixed by the sulfate ion (SO_4^{-2}) in groundwaters as insoluble barite. Many people think of barium's uses in the medical field, yet 80 percent of the barium produced is used as an additive to drilling mud because of its excellent sealing properties and resistance to temperature changes. Because some 600 million m of rock are drilled per year, barium is a vital commodity.

ENERGY—FUELS, ALTERNATIVES, AND PROBLEMS

As a result of war in the Mideast in 1973, oil is used as a weapon by OPEC (Overseas Petroleum Exporting Countries) against the rest of the world, with considerable success. In addition, this was the first time that many of the public became aware of energy problems and just how much we rely on oil and gas and other petroleum products. At the same time, the point was made (although not so convincingly) that even OPEC's petroleum product reserves are inadequate to meet the needs of the whole world by the early twenty-first century.

Two points also made in 1973 were the following: (1) The United States, with less than 10 percent of the world's population, consumed roughly one-third of the energy produced in the world. (2) The United States would very quickly have to develop energy alternatives, including conservation, to reduce reliance on imports of oil and gas.

OIL, GAS, AND RELATED MATERIALS

As illustrated in Fig. 17–16, most of the energy consumed in the United States (and also worldwide) is in the form of oil, gas, and liquids produced from gas. All are commonly found together in nature and will be discussed accordingly under the heading of petroliferous materials.

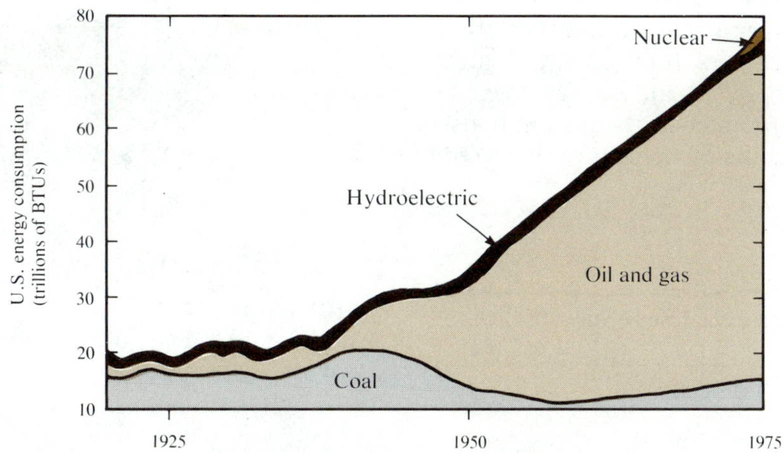

FIGURE 17–16
Trends in energy consumption in the United States for the period 1920–1975 (excluding wood). No distinction is made between domestically produced oil and gas and that provided by imports. Note that hydroelectric power has increased slightly in terms of total Btus but has decreased percentagewise in terms of total energy consumption. The graph does not adequately demonstrate the input from nuclear power because this source is too new. Note also that the total coal input has remained relatively constant. The amount of available anthracite (not shown) has decreased from about 2.5 trillion Btu (1920) to less than 0.2 trillion Btu (1975); this means reliance on lower grades (i.e., less caloric value and more impurities) of coal for the future.

Petroliferous materials form from the organic remains of animal life that have been deposited under swamplike conditions and allowed to accumulate over long periods of geologic time without destruction. In brief, the decaying hydrocarbon matter must be buried to depths of about 450 m. At these depths, the temperature and pressure are sufficient for the accumulated organic matter to be changed into new organic molecules free to migrate from their original site. In their upward migration, many of these molecules become caught in various types of **structural traps**. The classical anticline, where hydrocarbons permeate upward in sandstone overlain by impermeable shale, is the type of trap from which some 75–80 percent of the world's oil and gas has been discovered. The other traps, although often more difficult to find, may account for a proportionally higher amount of oil and gas in the future as more sophisticated exploratory techniques are developed (Fig. 17–17).

Considering world production and reserves as a function of both rock type and age, there is optimism for additional major discoveries from the tremendous accumulations of Cenozoic and especially Mesozoic rocks on the continental shelf of North America. Yet, even if significant additions to the world supply from domestic sources are made due to new discoveries, it is unlikely that world supplies can meet projected energy requirements.

How the United States uses its petroleum products is illustrated in Table 17–6. Several items from Table 17–6 are of note. First, the United States has so far been unsuccessful in its attempts to find an economic substitute for oil and gas in transportation. Whereas roughly 30 percent of total energy requirements are for transportation, oil and gas represent 77 percent of the petroleum products consumed in household and commercial use, and 69 percent of those con-

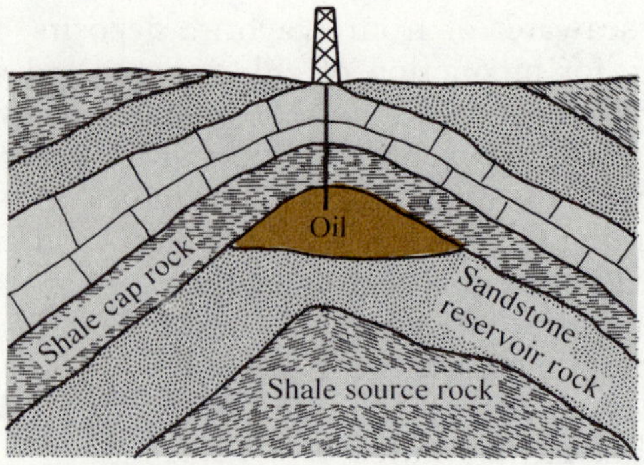

A. Anticlinal trap

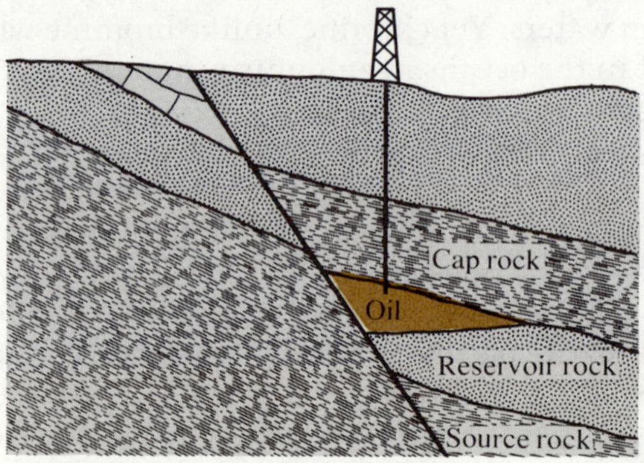

B. Fault trap

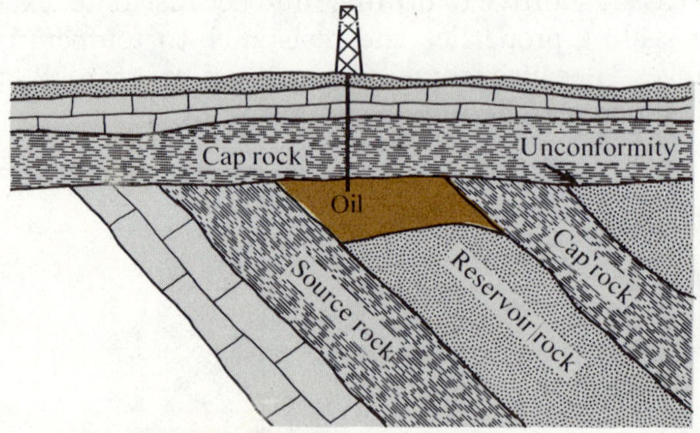

C. Unconformity (stratigraphic) trap

FIGURE 17–17
Idealized diagrams showing several types of structural oil and gas traps.

TABLE 17–6
U.S. use of petroleum products.

	Household and Commercial Use (30% of Total)	Industrial Use (40% of Total)
Natural gas	43%	44%
Oil	34	25
Electricity	21	11
Coal	2	18
Other	—	2
	Transportation Use (30% of Total)	
Oil and gas	96%	
Other	4	

SOURCE: U.S. Bureau of Mines, *Minerals Yearbook, 1974.*

sumed in industrial use. It is in these areas that substitutes for oil and gas can and must be used.

NUCLEAR ENERGY—FROM ROCKS TO REACTORS

The energy derived from nuclear processes has caused more heated debate in the last decade or so than any other form of energy. The advocates of nuclear energy point out the tremendous potential benefits of the use of energy from uranium ore. For example, because 1 g of uranium can yield as much energy as 2270 kg of coal, producing fuel for nuclear energy would mean fewer mining scars on the earth's surface. Potentially recoverable coal in the United States amounts to 1.5 trillion metric tons, from which energy could be obtained to meet domestic needs for approximately 190 years. However, to recover this amount of coal would require stripping lands (including both mined coal and overburden) roughly equivalent to the combined areas of North and South Dakota. For comparison, a 1000-MW nuclear reactor requires about 200 metric tons of uranium per year, whereas a coal-fired generating plant of the same capacity requires 2 million metric tons of coal.

Radioactive Waste In the **fission** process, during which a heavy nucleus is split to form two equal nuclei of intermediate mass, not only is intense radiation released, but many of the waste products are radioactive as well. Although these waste products initially build up in nuclear fuel rods, they are eventually removed, condensed, and stored in temporary facilities until a safe, permanent repository can be agreed to.

It has been shown that "fuel-equivalent" per "fuel-equivalent" for coal versus uranium, coal releases more radioactivity (plus large amounts of other carcinogens) to the environment. Yet the argument against possible geologic repositories is, How do we know that a repository will be safe for 250,000 years? (The figure 250,000, which is commonly used, is based on the small amount of plutonium 239 contained in the waste products. Plutonium 239 has a half-life of about 24,000 years—and the rule of thumb is that radioactive waste must be stored for 10 times its half-life to be rendered safe.)

The following geologic sites have been proposed for storage of radioactive waste:

1. Bedded salt deposits
2. Deep-sea beds
3. The Antarctic
4. Crystalline basement rocks
5. Shales
6. Others, including rocket disposal in space

Of these, storage in bedded salt deposits has received considerable attention (Fig. 17–18). Bedded salts are very dry, or else they would have dissolved by now. Thus, a site in a bedded salt in an area of low earthquake probability, great tectonic stability, and low urban development, would be almost ideal. In fact, the U.S. government plans pilot storage in such sites within the next decade. However, not all countries with a present or anticipated nuclear technology have access to bedded salt deposits; hence, research in other areas continues.

Deep-sea repositories are receiving a great deal of research, but nothing is yet known about how containers would behave in such an environment.

Waste storage in the Antarctic is at first glance somewhat attractive due to the low population density. However, the necessity to change international policy forbidding disposal of radioactive waste in the Antarctic and the unknowns about the geologic suitability of the area diminish its initial attractiveness.

Repositories in crystalline rocks are being considered by many countries, yet knowledge of basement rock technology, while increasing, is not as great as is desirable. For example, many crystalline basement rocks are fractured even in tectonically stable areas. Have these fractures resulted from relatively recent events or are they very old (i.e., safe)? Until this question can be adequately answered, crystalline rock sites will remain problematic.

To explore the feasibility of shale storage sites, let us briefly consider the lessons learned from study of the Oklo Uranium Mine in Gabon. The uranium at this site was deposited soon after sedimentation about 2 billion years ago at a time when the amount of uranium 235 in the total uranium was some 3.2 percent, just about the value to which uranium mined today is enriched. Due to a combination of high initial concentration of uranium, high uranium-235 content, proper amount of water, and lack of neutron-absorbing elements, the uranium deposited in fracture zones in shale sustained a fission chain reaction for some 500,000 years, over which time 16,500 megawatt-years of energy were produced. Further, the stable products formed after decay of the radioactive fission products are found at Oklo, as is evidence for some generation of small amounts of plutonium and neptunium. Thus, nature "beat" people in developing a nuclear reactor by some 2 billion years.

More important, however, is the fact that most of the fission products remained in place during and after the reactor's lifetime. Thus, this 2-billion-year experi-

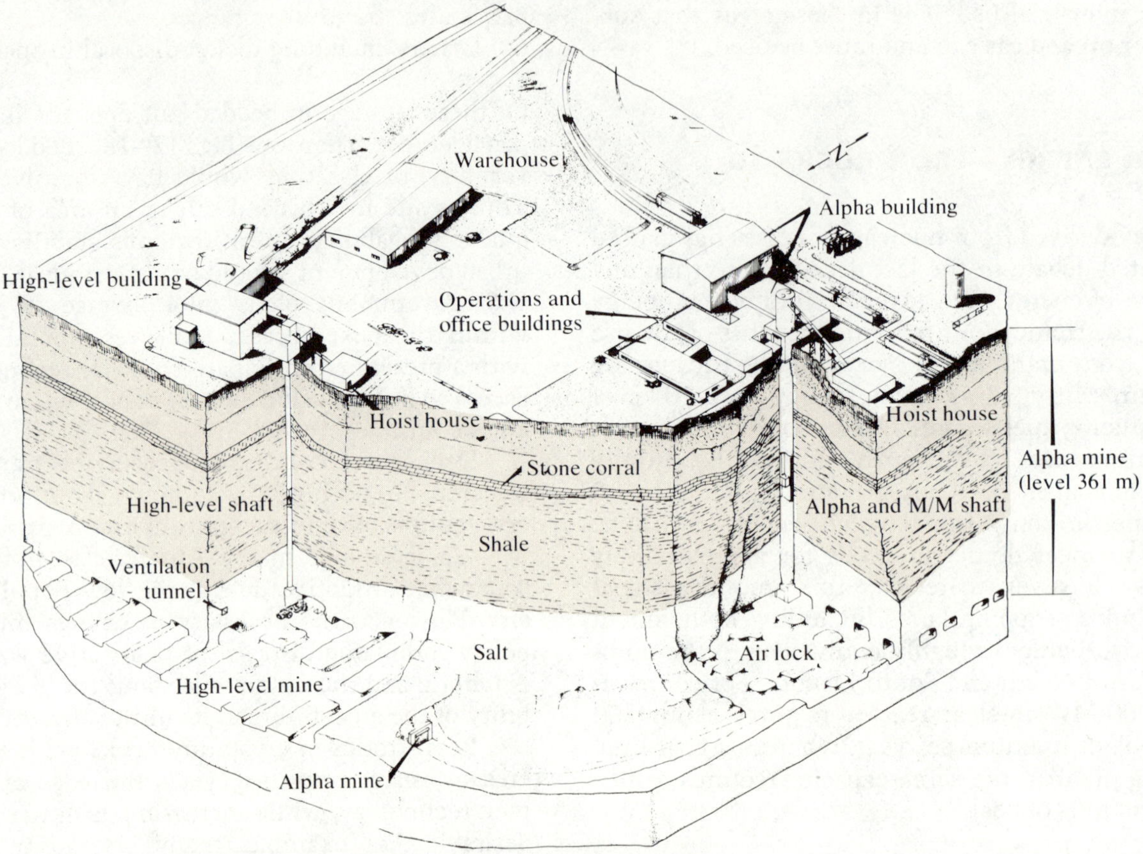

FIGURE 17–18
A schematic diagram showing how underground bedded salt may be used as a disposal site for solidified, high-level radioactive wastes. (After ERDA)

ment convincingly demonstrates that the Oklo site, an impure shale, was suitable for all forms of nuclear waste except gases (krypton and xenon). Where some migration has occurred, it is thought to be very local and, in terms of years—250,000 versus 2 billion—unimportant. Thus, Oklo also illustrates that shales in some tectonically stable environments are indeed suitable for radioactive waste disposal.

Uranium resources for the United States—uranium ore containing about 2 million tons U_3O_8—are probably adequate for planned nuclear reactors to the year 2000, assuming that 60 percent of electrical utilities needs—20 percent of our total energy consumption—is met by the nuclear industry.

Uranium deposits in the United States are found in the nonmarine sedimentary basins of Wyoming, the Colorado Plateau, and coastal Texas. Well over 50 percent of the reserves are located in New Mexico in the area of the Grants Mineral Belt. These deposits are in rocks near the surface and at depths of thousands of meters.

OTHER ENERGY SOURCES

Hydroelectric Energy This energy source continues to play an important role in the United States, but its future role will be less and less. To significantly increase hydroelectric energy in this country would require the construction of dams on an unparalleled schedule. This is not likely to occur for several reasons. First, there is some evidence that damming of our rivers is undesirable due to the problems of proper water allocations downstream from possible dam sites. Second, some of the hypothetically favorable sites are located in wilderness areas, and it is unlikely that dam construction in these regions would be approved. Third, and most important, hydroelectric energy cannot compete with the other forms of energy in an age of escalating costs.

Geothermal Energy Geothermal energy has received favorable attention in several parts of the world, including the United States. Two types of geothermal energy production are available. First, the utilization of natural steam from hot springs areas such as The Geysers in California (Fig. 17–19) has proven successful, and a significant amount of energy is provided from this field for the city of San Francisco some 120 km away. Unfortunately, many of the other natural steam fields are located in some of the more isolated parts of the western United States.

The second source of geothermal energy is derived from the high heat flow of some rocks by the so-called hot dry rock method developed at the Los Alamos Scientific Laboratory between 1973 and 1977. Here, advantage is made of areas of above-normal heat flow from within the earth (Fig. 17–20). In these areas, usually associated with young volcanism in the western United States, several wells are drilled to a depth at which high temperatures due to high heat flow are encountered. Using techniques developed by the petroleum industry, the rocks are fractured. Water is then introduced to depth by an injection well and steam is produced by heating of the water by the rocks at depth.

Solar Energy This type of energy has been attractive since early times. The source of heat from the sun is so great that figures for energy comparison are nearly meaningless. If technological breakthroughs allow the use of solar heating nationwide, some of the burden will be taken off the other sources, but probably not before the twenty-first century. Solar desalination of water could be made economical with combined solar-nuclear plants near saline lakes.

FIGURE 17–19
Geothermal steam escaping from wells at The Geysers, California. (Photo courtesy Pacific Gas and Electric Co.)

Nuclear Fusion Unlike nuclear fission, which releases energy when an atom is split, nuclear fusion releases energy when two or more atoms are joined together. Methods to utilize energy from nuclear fusion require technological advances that have not yet been made.

Oil Shale No discussion of alternate energy sources would be complete without mention of so-called **oil shale**. In this context, the word *shale* is a misnomer, as the rock is actually a kerogen-bearing marlstone formed under nonmarine, near-evaporite conditions. (Marlstone is a variety of rock rich in calcium carbonate.) This rock resembles shale because it is dark-colored (due to its kerogen content) and finely laminated. The amount of petroliferous material stored in these oil marlstones in Utah, Colorado, and Wyoming is truly impressive. The U.S. Geological Survey estimates the amount to be 2 trillion barrels. However, there is a problem with this rosy picture: the oil marlstones are located in arid to semiarid, isolated areas, and recovery of their oil is not now economical (Fig. 17–21).

Coal Whereas petroliferous products result from original accumulations of animal matter in marine or near-marine sedimentary basins, coal results primarily from the accumulation of plant material, usually in continental nonmarine basins. The world distribution of coals of various types shows them to be concentrated in the Northern Hemisphere. (The distribution of coal in the United States is shown in Fig. 17–22.) Hence, if industrialization in the Southern Hemisphere is to proceed rapidly, coal use must be carefully budgeted. For example, use in developing a steel industry may take priority over use for energy needs.

The formation of coal can be summarized as follows:

1. Accumulation of dead plant matter in swampy areas to eventually form **peat** (10 percent solids, 90 percent water)
2. Diagenetic effects (compaction, cementation, replacement, and recrystallization) due to continued burial and chemical changes resulting in the formation of **lignite**, the immediate precursor to forms of coal

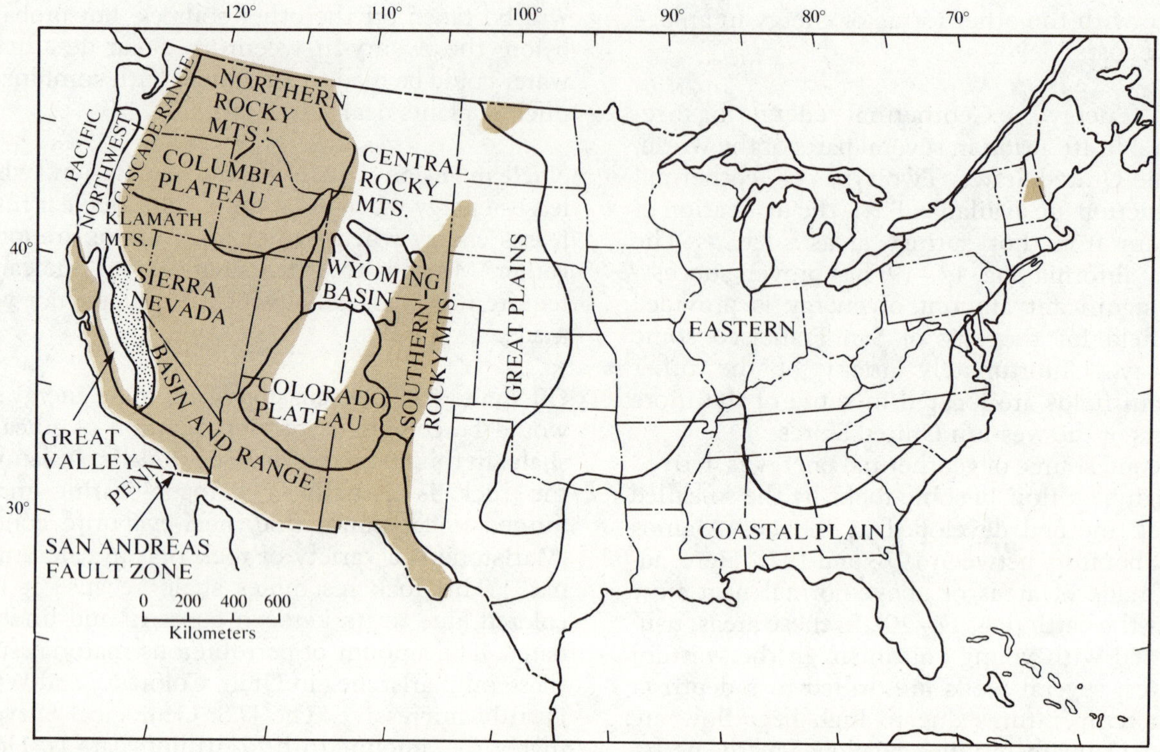

FIGURE 17–20
A map showing the generalized extent of hot (tinted), normal (white), and cold (stippled) crustal regions of the United States. Also shown are the major physiographic provinces, such as the Colorado Plateau. (After D. F. White and D. L. Williams, U.S. Geological Survey Circular 726, 1975)

3. Continued burial, causing very-low-grade metamorphism such that the lignite is converted to **subbituminous coal**
4. Continued increase in pressure and temperature to form **bituminous coal**
5. Continued increase in pressure and temperature to form **anthracite**

Once anthracite has formed, further increase in temperature and pressure so hardens the carbonaceous matter that it loses its combustible value, and eventually **graphite** will be formed. Not only is the original carbonaceous matter deposited very impure, but it will vary widely in moisture content, caloric (heat) value, and so forth, as each different type of coal is formed.

Although coal resources are enormous, coal resource development may present more problems than solutions. These problems include the high cost of land reclamation due to large-scale strip-mine operations, the fact that the industry has not kept up-to-date in technological advances, and the scarcity of clean coal (which is low in sulfur content and therefore less polluting) in the eastern United States, where most coal is burned.

BUILDING MATERIALS

In terms of dollar value, fossil fuels are our most important natural resource. Second in dollar value and first in total volume are building materials, which include stone, sand and gravel, and the lightweight aggregates. The U.S. Geological Survey (Professional Paper 820, 1973) points out that we have sufficient reserves of all of these for the next 25 or 30 years, but that supplies in the twenty-first century will depend on careful planning at all levels of industrial and urban development.

Stone Stone includes both dimension stone (natural building stone quarried and trimmed with essentially no other processing) and crushed stone (material reduced in size to meet various needs). The use of dimension and crushed stone in the United States is shown in Table 17–7. The tonnage of dimension stone has steadily decreased in the last 40 years or so due to the use of alternate building materials such as steel-reinforced concrete. The use of crushed stone has shown continued growth (from about 100 million tons in 1930 to 875 million tons in 1970) because there are

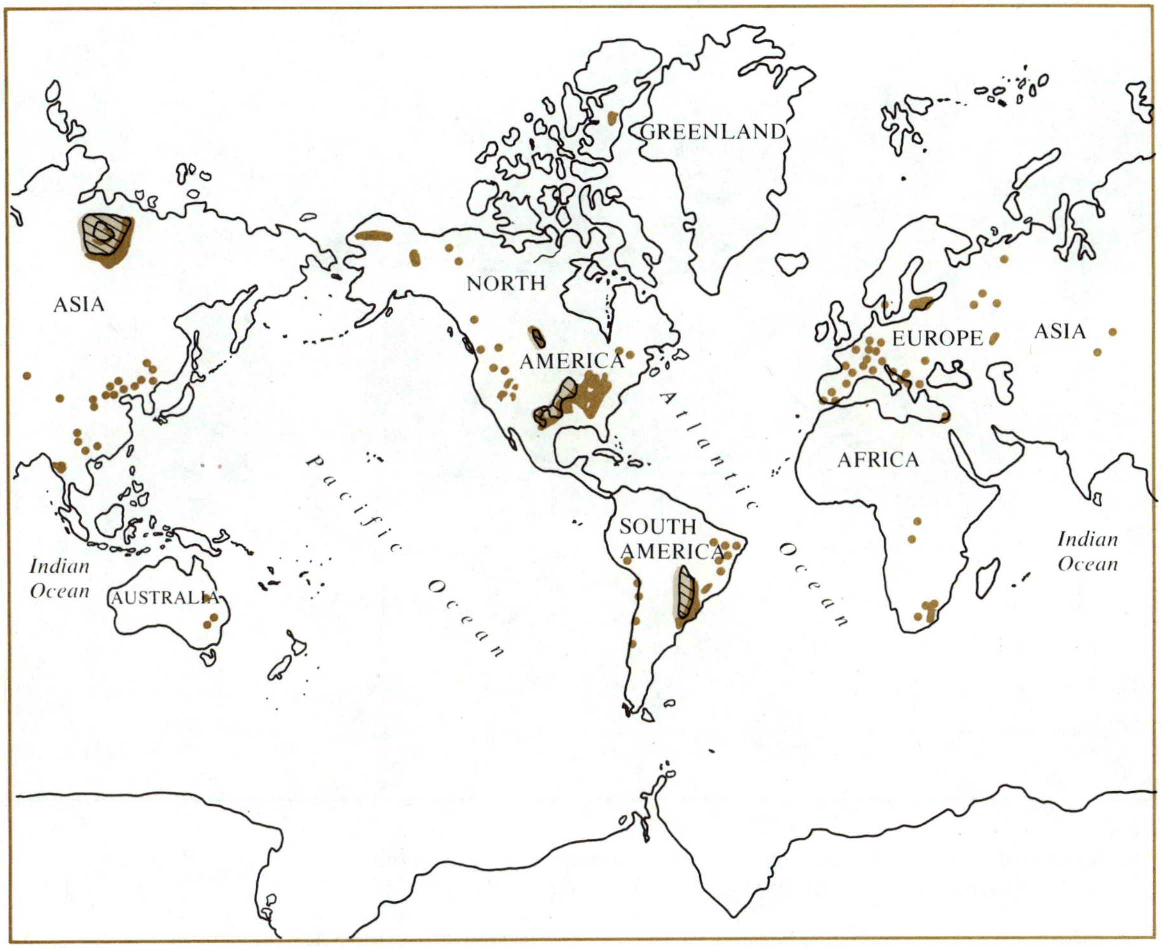

FIGURE 17–21
Oil-shale deposits of the world. Areas denoted by the diagonal lines are those where some major deposits may be extended. (From D. C. Duncan and V. E. Swanson, U.S. Geological Survey Circular 523, 1965)

TABLE 17–7
Use of stone in the United States.

Dimension Stone	Percentage of Total	Crushed Stone	Percentage of Total
Limestone	37	Limestone	73
Granite	31	Significant trap rock	10
Sandstone	17	Granite	9
Slate	8	Sandstone	3
Marble	4	Other	5

SOURCE: After U.S. Geological Survey, 1973.

no suitable substitutes in highway and railway construction and for other purposes.

Sand and Gravel The sand and gravel industry is the largest nonfuel industry in the United States in terms of volume. Traditionally, stream channels and glacial terrain have provided most of the sand and gravel. In addition, extensive sand deposits occur off the coast of northeastern United States, and these are now being dredged and will be increasingly dredged in years to come.

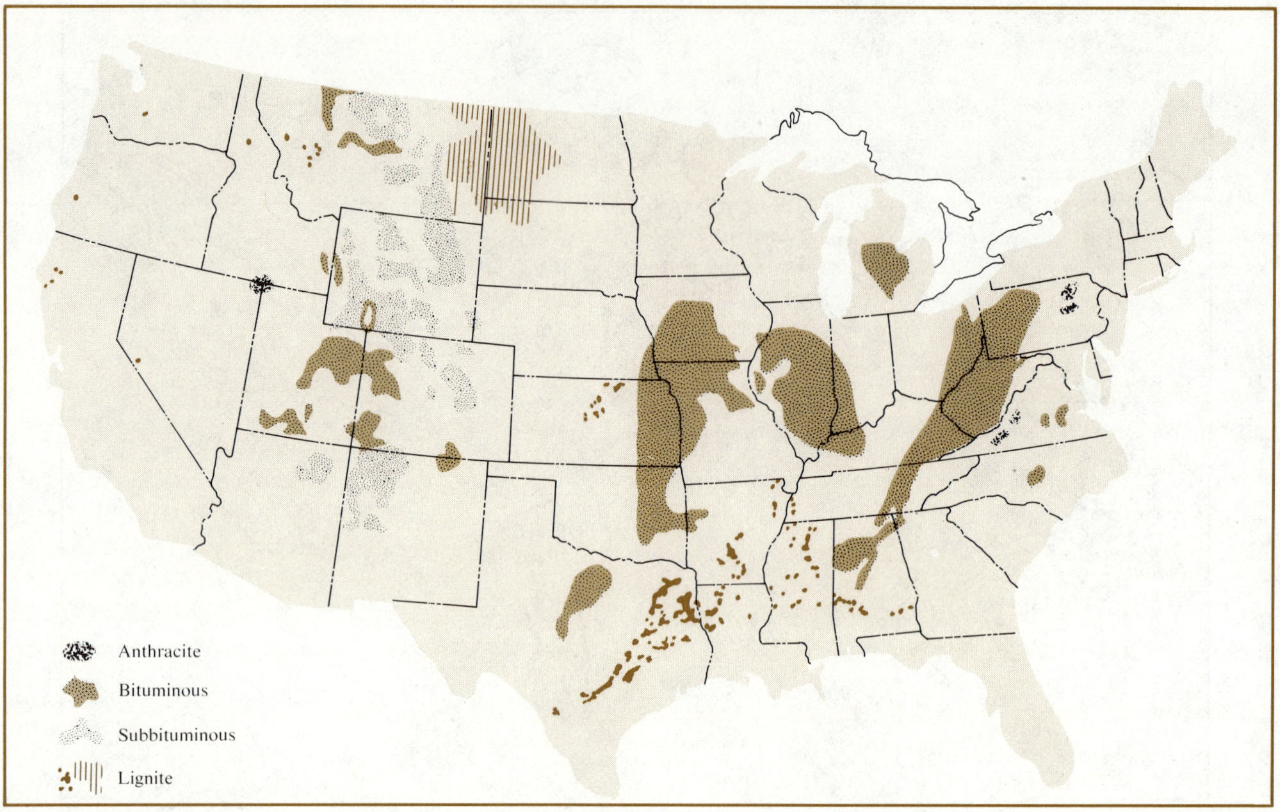

FIGURE 17–22
Coal fields of the United States. In Alaska, lignite, subbituminous, and bituminous deposits are widespread. (After U.S. Bureau of Mines Information Circular 8531, 1971)

Lightweight Aggregates Lightweight aggregates include such volcanic materials as pumice, scoria, cinder, and pumicite, as well as expandable clays, shales, some slate, and diatomite. Much of this material is used in agricultural and chemical industries.

Clays Clays, as used in the industrial sense, refer to naturally occuring material that contains only one clay material such as illite, montmorillonite, kaolinite, or halloysite. The U.S. Bureau of Mines prefers to classify clays based on the product, as follows:

Kaolin ("china clay")—essentially pure kaolinite; ideal for ceramic products
Fire clay—impure kaolinite; ideal for refractory material
Bentonite and fuller's earth—varieties of montmorillonite; widely used as drilling muds
Miscellaneous clay—illite-rich clay and shale for tiles, piping, and cement additives

Cement By itself, cement is nothing more than a fired and crushed mixture of carbonate rocks and clay minerals (or shale) moistened with water. However, once mixed with sand, gravel, or other aggregate, a myriad of construction materials emerge. Portland cement, so called because of its resemblance to the Portland Limestone in England, is the predecessor of most cements, although cements date back at least to the Roman Empire.

A typical feed to a kiln for preparing cement consists of a mixture of 1250 kg of slightly magnesian limestone and 335 kg of shale or clay. This mixture is first crushed and then fired at close to 1500°C. At this temperature, carbon dioxide and water are driven off as gases. The resultant fired material is crushed and produces about 500 kg of cement.

Abrasives and Related Materials The United States possesses adequate supplies of most abrasives used for industrial and other purposes. The abrasives include silica sand, tripoli, emery, garnet, and corundum. The one exception is diamond. Although diamonds are found in Arkansas and along the Wyoming-Colorado border in kimberlites as well as in numerous placer deposits, all naturally formed diamonds for industrial or gem uses are imported. Yet the production of artificial diamonds already accounts for more than 50 percent of our needs, and this percentage is likely to increase as threats to overseas supplies increase.

SUMMARY

Mineral resources may be divided into two categories: renewable and nonrenewable. A renewable resource is a resource that, following use, will eventually be available again. Nonrenewable resources include those materials that, once used, are for various reasons never again available. Examples of nonrenewable resources are the burning of fossil fuels and the splitting of atoms in a nuclear reactor.

A reserve is that part of the resource for which there is precise information concerning amount and quality. Some resources will be added to or subtracted from reserves due to changing economic, legal, and political conditions.

The crustal abundance of a particular element and the way that element is incorporated into minerals are important controlling factors in the formation of mineral deposits. Thus, it is not sufficient to merely have a relatively high concentration of a particular element in a rock; the element must also be bonded to other atoms in such a way that it can be extracted at a profit.

Considering mineral resources, the term *metal* is applied to elements used for their metallic properties, such as conductivity and ductility. *Nonmetal* refers to those elements and their compounds not used for metallic properties. Thus, sodium, although a metal, is used commercially as sodium compounds and is classified as a nonmetal from a resource viewpoint. Interestingly, from an annual world consumption viewpoint, the materials consumed in greatest amounts are generally the nonmetals such as carbon, nitrogen, oxygen, sulfur, potassium, and calcium. These elements are primarily consumed by the agricultural and related industries. Iron is the only metal in the top eight elements in order of consumption used for its metallic properties.

Natural resources are unevenly distributed around the world, as are population and rate of industrialization. In the United States, for example, with less than 10 percent of the world's population, the consumption of mineral resources is four to five times that of other industrialized nations, and many times that of the developing countries.

Mining methods include open-pit, underground, and in situ. Which method is used depends primarily upon the commodity being extracted, the concentration of the material, and local geologic and other environmental considerations.

Mineral resources commonly occur in specific belts, called metallogenic provinces, that have formed due to alignment of major tectonic elements, and in rocks of similar age and approximate equivalent chemistry.

The abundant metals, iron and aluminum, are the foundation upon which industrialization of nations is based. The iron industry accounts for well over 90 percent of all metals mined. Highly industrialized nations depend on iron, and there is a direct relationship between iron consumption and standard of living. Much of the iron in the world is obtained from Precambrian banded iron deposits containing between 20 and 40 percent iron deposited by sedimentary processes.

Aluminum is the third most abundant element by weight in the earth's crust and has many uses in industrialized societies. The principal ore of aluminum is bauxite, more properly known as aluminum laterite, which is formed under conditions of tropical weathering of aluminum-rich source rocks.

Other relatively abundant metals include magnesium, which may be obtained from seawater, and titanium, which may be obtained from accessory minerals in igneous rocks. Less abundant and scarce metals include copper, lead, zinc, nickel, cobalt, molybdenum, and mercury, all of which are primarily concentrated in sulfides of the metal. Many such sulfides are concentrated in metalliferous zones associated with recent tectonism. In particular, they apparently are associated with divergent and convergent plate boundaries.

Most of the nonmetals are consumed by the agricultural and chemical industries. Important in these are evaporites and phosphorus-rich sedimentary rocks. These generally form by complex physical and chemical reactions that take place in the sedimentary rock cycle.

Under the topic of energy, such diverse aspects as actual fuels, alternatives to petroliferous sources, and environmental problems associated with various energy sources must be considered. Most of the energy consumed today is in the form of oil, gas, and liquids produced from gas. These petroliferous materials are formed by complex reactions taking place in deeply buried organic remains over long periods of geologic time. Because the petroliferous resources have been extensively used in recent years, it is projected that world supplies will not meet projected energy requirements much into the twenty-first century. Therefore, analysis of alternative resources is paramount. These sources include nuclear energy, coal, hydropower, geothermal energy, solar energy, shale oil, and possibly nuclear fusion. Each of these alternative sources of energy has advantages and disadvantages in terms of availability, economic constraints, and environmental impact. It seems apparent that in the next few years, we will burn more coal than in the past, but which of the alternative energy sources will emerge as significant sources is yet to be determined.

Other very important mineral resources include those materials used for building purposes. These include stone, sand and gravel, lightweight aggregates, and clays and related materials. While fossil fuels are our most important natural resource in terms of dollar value, second in dollar value and first in total volume are the building materials.

REVIEW EXERCISES

1 What is the difference between renewable and nonrenewable resources? Are there times when a renewable resource is, for all practical purposes, a nonrenewable resource? If so, give an example.

2 What is the difference between a reserve and a resource?

3 What constitutes ore? Give examples.

4 What is meant by *concentration factor*?

5 What is the difference between metallic and nonmetallic elements (or compounds)?

6 Why are nonmetals consumed in greater abundance than metals?

7 What does *MACD* mean? Will MACD figures change in the twenty-first century?

8 Where would underground, open-pit, and solution mining techniques be used? Illustrate by use of a simple sketch.

9 What is a metallogenic province?

10 How do marine evaporites form? Do nonmarine evaporites form by a similar process? Are the mineralogies of marine and nonmarine evaporites the same? Why or why not?

11 Discuss briefly the U.S. and world supply of those elements vital to the agriculture industry: potassium, nitrogen, phosphorus, and sulfur.

12 List the most probable sources of energy for the United States for the period 2000–2025.

13 How can alternate energy sources offset current petroleum product use for industrial, transportation, and household-commercial use in the United States? Describe in some detail.

14 Compare the energy production from nuclear fuels with that from coal, solar, hydroelectric, geothermal, and oil-shale sources. Include in your discussion the effect on the environment during mining (and other excavations), total materials needed, transportation, health effects (short- and long-range), and, where applicable, disposal of waste products.

15 What geologic sites have been proposed for the storage of radioactive wastes?

16 What are some of the "good" and "bad" aspects of oil-shale use?

17 Describe the formation of coal from peat to anthracite.

18 Compare production and consumption of sand and gravel and other untreated industrial stone with treated industrial materials.

SUGGESTED READINGS

BORCHERT, H., and MUIR, R. O. *Salt Deposits: The Origin, Metamorphism, and Deformation of Evaporites.* Princeton, N.J.: Van Nostrand, 1964.

BROBST, D. A., and PRATT, W. P. *United States Mineral Resources.* U.S. Geological Survey Professional Paper 820, 1973.

BROOKINS, D. G. *Earth Resources, Energy, and the Environment.* Columbus, Ohio: Charles E. Merrill Publishing Co., 1981, 190 pp.

CAMERON, E. N., ed. *The Mineral Position of the United States, 1975–2000.* Madison, Wis.: University of Wisconsin Press, 1973, 159 pp.

FLAWN, P. T. *Environmental Geology: Conservation, Land-Use Planning, and Resource Management.* New York: McGraw-Hill Book Co., 1970, 313 pp.

FOWLER, J. M. *Energy and the Environment.* New York: McGraw-Hill Book Co., 1975.

INTERNATIONAL ATOMIC ENERGY AGENCY. *The Oklo Phenomenon.* Westview Park, N.J.: Unipub, 1976.

LAMEY, C. A. *Metallic and Industrial Mineral Deposits.* New York: McGraw-Hill Book Co., 1966.

LEGGET, R. F. *Cities and Geology.* New York: McGraw-Hill Book Co., 1973, 624 pp.

PARK, C. F., and McDIARMID, R. A. *Ore Deposits,* 3rd ed. San Francisco: W. H. Freeman and Co., 1980.

SINGER, S. F., ed. Global Effects of Environmental Pollution. Symposium of the American Association for the Advancement of Science, December 1968. New York: Springer-Verlag, 1970.

SKINNER, B. J. *Earth Resources,* 2nd ed. Englewood Cliffs, N.J.: Prentice-Hall, 1976.

APPENDIXES

APPENDIX A: Minerals
APPENDIX B: Topographic Maps
APPENDIX C: Geological Maps
APPENDIX D: Unit Measurement Systems
APPENDIX E: Rock Textures

Appendix A
MINERALS

TABLE A–1
Mineral identification key.

This table helps to identify the more common minerals. To use the chart, first decide the sort of luster (metallic or nonmetallic) the mineral has; next, its color (light or dark); third, its hardness (compared to a knife blade); fourth, whether it has cleavage; and, finally, read the brief descriptions to identify it.

Luster	Hardness	Cleavage	Description	Mineral
Light-colored nonmetallic luster	Hard—not scratched by knife	Shows cleavage	White or flesh-colored; 2 cleavage planes at nearly right angles; hardness, 6. Large crystals which show irregular veining are PERTHITE.	ORTHOCLASE (POTASSIUM FELDSPAR)
			White or green gray; 2 cleavage planes at nearly right angles; hardness, 6; striations on one cleavage.	PLAGIOCLASE
		No cleavage	White, clear, or any color; glassy luster; transparent to translucent; hexagonal (6-sided) crystals; hardness, 7; conchoidal fracture.	QUARTZ
			Various shades of green and yellow; glassy luster; granular masses and crystals in rocks; hardness, 6.5–7 (apparent hardness may be much less).	OLIVINE
			Any color or variegated; glassy luster; hardness, 5–6; conchoidal fracture.	OPAL
			Any color or variegated; waxy luster; hardness, 7; conchoidal fracture.	CHALCEDONY (AGATE)
	Soft—scratched by knife	Shows cleavage	Colorless to white; salty taste; cubic cleavage; hardness, 2.5.	HALITE
			White, yellow to colorless; rhombohedral cleavage; hardness, 3; effervesces with dilute hydrochloric acid.	CALCITE
			Pink, colorless, white, or dark; rhombohedral cleavage; hardness, 3.5–4; effervesces with dilute hydrochloric acid only if powdered.	DOLOMITE
			White to transparent; 3 unequal cleavages; hardness, 2.	GYPSUM
			Green to white; feels soapy; 1 cleavage; hardness, 1.	TALC
			Colorless to light yellow or green; transparent in thin sheets which are very elastic; 1 cleavage; hardness, 2–2.5 (white mica).	MUSCOVITE
			Green to white; fibrous cleavage; may form veins.	ASBESTOS
		No cleavage	Green to white; feels soapy; hardness, 1.	TALC
			White to transparent; hardness, 2.	GYPSUM
			Yellow to greenish; resinous luster; hardness, 1.5–2.5.	SULFUR

TABLE A–1 (cont.)
Mineral identification key.

Luster	Hardness	Cleavage	Description	Mineral
Dark-colored nonmetallic luster	Hard—not scratched by knife	Shows cleavage	Black to dark green; 2 cleavage planes at nearly right angles; hardness, 5–6.	AUGITE
Dark-colored nonmetallic luster	Hard—not scratched by knife	Shows cleavage	Black to dark green; 2 cleavage planes about 60°; hardness, 5–6.	HORNBLENDE
Dark-colored nonmetallic luster	Hard—not scratched by knife	No cleavage	Red to red brown; fracture resembles poor cleavage; brittle equidimensional crystals; hardness, 6.5–7.5.	GARNET
Dark-colored nonmetallic luster	Hard—not scratched by knife	No cleavage	Various shades of green and yellow; glassy luster; granular masses and crystals in rocks; hardness, 6.5–7 (apparent hardness may be much less).	OLIVINE
Dark-colored nonmetallic luster	Hard—not scratched by knife	No cleavage	White, clear, or any color; glassy luster; transparent to translucent; hexagonal (6-sided) crystals; hardness, 7; conchoidal fracture.	QUARTZ
Dark-colored nonmetallic luster	Hard—not scratched by knife	No cleavage	Any color or variegated; glassy luster; hardness, 5–6; conchoidal fracture.	OPAL
Dark-colored nonmetallic luster	Hard—not scratched by knife	No cleavage	Any color or variegated; waxy luster; hardness, 7; conchoidal fracture.	CHALCEDONY (AGATE)
Dark-colored nonmetallic luster	Hard—not scratched by knife	No cleavage	Red to brown; red streak; earthy appearance; hardness, 5.5–6.5 (apparent hardness may be less).	HEMATITE
Dark-colored nonmetallic luster	Hard—not scratched by knife	No cleavage	Yellow brown to dark brown; may be almost black; streak yellow brown; earthy; hardness, 5–5.5 (may have lower apparent hardness).	LIMONITE (GEOTHITE)
Dark-colored nonmetallic luster	Soft—scratched by knife	Shows cleavage	Brown to black; cleavage, 1 direction; hardness, 2.5–3 (black mica).	BIOTITE
Dark-colored nonmetallic luster	Soft—scratched by knife	Shows cleavage	Various shades of green; cleavage, 1 direction; hardness, 2–2.5 (green "mica").	CHLORITE
Dark-colored nonmetallic luster	Soft—scratched by knife	Shows cleavage	Yellow brown, dark brown, or black; streak white to pale yellow; resinous luster; cleavage, 6 directions; hardness, 3.5–4.	SPHALERITE
Dark-colored nonmetallic luster	Soft—scratched by knife	No cleavage	Red to brown; red streak; earthy appearance; hardness, 5.5–6.5 (apparent hardness may be less).	HEMATITE
Dark-colored nonmetallic luster	Soft—scratched by knife	No cleavage	Scarlet to red brown; scarlet streak; hardness, 2–2.5; high specific gravity.	CINNABAR
Dark-colored nonmetallic luster	Soft—scratched by knife	No cleavage	Lead pencil black; smudges fingers; hardness, 1; 1 cleavage that is apparent only in large crystals.	GRAPHITE
Dark-colored nonmetallic luster	Soft—scratched by knife	No cleavage	Yellow brown to dark brown; may be almost black; streak yellow brown; earthy; hardness, 5–5.5 (may have lower apparent hardness).	LIMONITE (GEOTHITE)
Dark-colored nonmetallic luster	Soft—scratched by knife	No cleavage	Dark to light green; greasy or waxy luster; some varieties are fibrous; hardness, 2–5, generally 4.	SERPENTINE
Metallic luster			Black; strongly magnetic; hardness, 6.	MAGNETITE
Metallic luster			Lead pencil black; smudges fingers; hardness, 1; 1 cleavage that is apparent only in large crystals.	GRAPHITE
Metallic luster			Brass yellow; black streak; cubic crystals, commonly with striations; hardness, 6–6.5.	PYRITE
Metallic luster			Brass yellow; may be tarnished; black streak; hardness, 3.5–4; massive.	CHALCOPYRITE
Metallic luster			Shiny gray; black streak; very heavy; cubic cleavage; hardness, 2.5.	GALENA

TABLE A–2
Summary of common minerals.

Mineral	Composition	Color	Luster	Streak	Hardness	Cleavage	Other Properties, Uses
Amphibole—*See Hornblende.*							
Andalusite	Al_2SiO_5 Aluminum silicate	Reddish brown to olive green	Vitreous to dull	White	7.5	2 poor	In metamorphic rocks
Asbestos	Hydrous magnesium silicate	Shades of green	Silky	None	Low apparent hardness	Fibrous	Fireproofing and insulation against heat and electricity
Augite	$(Ca,Na)(Mg,Fe^{II},Fe^{III},Al)(Si,Al)_2O_6$ Calcium, sodium, magnesium, iron, aluminum silicate	Black to dark green	Vitreous	None	5–6	2 at nearly right angles	Rock-forming mineral
Biotite	$K(Mg,Fe)_3(AlSi_3O_{10})(OH)_2$ Hydrous potassium, magnesium, iron, aluminum silicate	Black to dark green	Vitreous	None	2.5–3	1 perfect	Rock-forming mineral
Calcite	$CaCO_3$ Calcium carbonate	Generally white or colorless	Vitreous to earthy	None	2.5–3	Rhombohedral, 3 not at right angles	Effervesces in cold dilute hydrochloric acid. Manufacture of cement
Chalcedony	SiO_2 Silicon dioxide	Any	Waxy	None	7	None	Conchoidal fracture
Chalcopyrite	$CuFeS_2$ Copper-iron sulfide	Brass yellow—may be tarnished to bronze or iridescent	Metallic—dull	Greenish black	3.5–4	None	Brittle
Chlorite	$(Mg,Fe,Al)_6(Al,Si)_4O_{10}(OH)_8$ Hydrous magnesium, iron, aluminum silicate	Green	Vitreous	None	2–2.5	1 perfect	Rock-forming mineral
Cinnabar	HgS Mercury sulfide	Vermillion red to brownish red	Adamantine to dull earthy	Scarlet	2.5	1	High specific gravity. Only important ore of mercury
Clay	A family of hydrous aluminum silicates which may contain potassium, sodium, iron, magnesium, etc.	Generally light	Earthy	None	Very low apparent hardness	None	Rock-forming mineral
Cordierite	$(Mg,Fe)_2Al_4Si_5O_{18}$ Magnesium aluminum silicate	Blue	Vitreous	White	7–7.5	1 poor	In metamorphic rocks
Dolomite	$CaMg(CO_3)_2$ Calcium-magnesium carbonate	Generally some shade of pink, flesh color, or white	Vitreous to pearly	None	3.5–4	Rhombohedral, 3 not at right angles	Powder effervesces in cold dilute hydrochloric acid
Epidote	$Ca_2(Al,Fe)_3Si_3O_{12}(OH)$ Hydrous calcium, aluminum, iron silicate	Green to dark green	Vitreous	White to gray white	6–7	1	In metamorphic rocks
Feldspar—*See Plagioclase and Orthoclase (Potassium feldspar).*							

TABLE A–2 (cont.)
Summary of common minerals.

Mineral	Composition	Color	Luster	Streak	Hardness	Cleavage	Other Properties, Uses
Galena	PbS Lead sulfide	Lead gray	Bright metallic	Lead gray	2.5	Cubic, 3 at right angles	High specific gravity. Ore of lead
Garnet	A family of silicate minerals containing aluminum, iron, magnesium, calcium, etc.	Commonly red	Vitreous	None	6.5–7.5	None	Characteristic isometric crystals
Graphite	C Carbon	Black to steel gray	Metallic or dull earthy	Black	1–2	1	Greasy feel—marks paper
Gypsum	$CaSO_4 \cdot 2\,H_2O$ Hydrous calcium sulfate	Colorless, white, gray	Vitreous, pearly, silky to earthy	None	2	1 perfect, others poorer	Plaster of paris
Halite	NaCl Sodium chloride	Colorless to white	Vitreous	None	2.5	Cubic, 3 at right angles	Salty taste
Hematite	Fe_2O_3 Iron oxide	Reddish brown to black or gray	Earthy to bright metallic	Light to dark Indian red	5.5–6.5	None	Most important ore of iron
Hornblende	$Ca_2Na(Mg,Fe^{II})_4(Al,Fe^{III},Ti)(Al,Si)_8O_{22}(O,OH)_2$ Hydrous calcium, sodium, magnesium, iron, aluminum silicate	Black to dark green	Vitreous	None	5–6	2 at 60°	Rock-forming mineral
Kyanite	Al_2SiO_5 Aluminum silicate	Blue	Vitreous to pearly	White	4–5 along crystal, 7 across	1	In metamorphic rocks
Laumontite	$CaAl_2Si_4O_{12} \cdot 4\,H_2O$ Hydrous calcium-aluminum silicate	White	Vitreous	White	4	3	A zeolite
Lawsonite	$CaAl_2(Si_2O_7)(OH)_2 \cdot H_2O$ Calcium-aluminum silicate	Pale blue to blue gray	Vitreous to greasy	White	8	2	In metamorphic rocks
Limonite (Goethite)	$FeO(OH) \cdot n\,H_2O$ Hydrous iron oxide	Yellow brown to black	Generally earthy	Yellowish brown	5–5.5	None	Low apparent hardness
Magnetite	Fe_3O_4 Iron oxide	Black	Metallic to submetallic	Black	6	None	Strongly magnetic. Important iron ore
Mica—*See Biotite and Muscovite.*							
Muscovite	$KAl_2(AlSi_3O_{10})(OH)_2$ Hydrous potassium-aluminum silicate	Clear—light green	Vitreous	None	2–2.5	1 perfect	Rock-forming mineral

Olivine	$(Mg,Fe)_2SiO_4$ Magnesium-iron silicate	Green	Vitreous	None	6.5–7	None	Rock-forming mineral
Opal	$SiO_2 \cdot n\ H_2O$ Hydrous silicon dioxide	Any	Vitreous or resinous	None	5–6	None	Conchoidal fracture
Orthoclase (Potassium feldspar)	$KAlSi_3O_8$ Potassium-aluminum silicate	White to pink	Vitreous	None	6	2 at right angles	Rock-forming mineral
Plagioclase	$CaAl_2Si_2O_8$ and $NaAlSi_3O_8$ Calcium- and sodium-aluminum silicate	White to green	Vitreous	None	6	2 at right angles—striations on 1 cleavage direction	Rock-forming mineral
Prehnite	$Ca_2Al_2(Si_3O_{10})(OH)_2$ Hydrous calcium-aluminum silicate	Light green to white	Vitreous	White	6–6.5	1	In metamorphic rocks
Pyrite	FeS_2 Iron sulfide	Pale brass yellow	Bright metallic	Greenish or brownish black	6–6.5	None	Brittle. Cubic crystals
Pyroxene—*See Augite.*							
Quartz	SiO_2 Silicon dioxide	Any	Vitreous	None	7	None	Rock-forming mineral
Serpentine	$Mg_6Si_4O_{10}(OH)_8$ Hydrous magnesium silicate	Dark to light green	Greasy to waxy	None	2–5	None	Metamorphic rocks. Fibrous variety is asbestos
Sillimanite	Al_2SiO_5 Aluminum silicate	Brown, pale green, white	Vitreous	White	6–7	1	In metamorphic rocks
Sphalerite	ZnS Zinc sulfide	Commonly yellow brown to black	Resinous to adamantine	White to yellow and brown	3.5–4	6	Zinc ore
Staurolite	$(Fe,Mg)_2Al_9Si_4O_{23}(OH)$ Ferrous iron-aluminum silicate	Reddish brown to brown black	Vitreous to resinous	Gray	7–7.5	1 poor	In metamorphic rocks
Sulfur	S Sulfur	Yellow	Resinous	None	1.5–2.5	None	Fracture conchoidal to uneven. Brittle
Talc	$Mg_3(Si_4O_{10})(OH)_2$ Hydrous magnesium silicate	Apple green, gray, white, silver white	Pearly to greasy	None	1	1	Cosmetics

Appendix B
TOPOGRAPHIC MAPS

He had bought a large map representing the sea,
Without the least vestige of land:
And the crew were much pleased when they found it to be
A map they could all understand.

"What's the good of Mercator's North Poles and Equators,
Tropics, Zones and Meridian Lines?"
So the Bellman would cry: and the crew would reply
"They are merely conventional signs!"

Lewis Carroll
The Hunting of the Snark
Fit the Second, Stanzas 2 and 3

Maps are simply scale drawings of a part of the earth's surface. Thus, they are similar to a blueprint of an object or to a dress pattern. Most maps are drawn on sheets of paper and so show only the two horizontal dimensions. Geologists, together with civil engineers and many other users of maps, require that the third dimension, elevation, be shown on maps. Maps that indicate the shape of the land are called **topographic maps**. A number of different ways of showing topography are in current use, such as the familiar color tint that uses shades of green for low elevations and brown for higher elevations. The most accurate method, which will be described here, involves the use of contour lines.

MAP SCALES

Map scales are designated in several ways. (1) A scale can be stated as a specified unit of length on the map corresponding to a specified unit of length on the ground—for example, 1 in. equals 1 mi, meaning that 1 in. on the map represents 1 mi on the ground. (2) A scale can also be stated as a scale ratio or representative fraction, meaning that one unit (any unit) on the map equals a specified number of the same units on the ground. In the example just used, the scale would be given as 1:63,360, and 1 in. on the map would represent 63,360 in. on the ground (1 mi = 12 × 5280 = 63,360 in.). (On that same map, 1 cm would also represent 63,630 cm on the ground.) (3) Most maps contain a graphic scale, generally in the lower margin, for measuring distances.

CONTOUR LINES

A contour line is an imaginary line, every part of which is at the same elevation. Thus, a shoreline is an example of a contour. Because a single point can only have one elevation, two contour lines can never cross (except on an overhanging cliff).

By convention, contour lines on each map are a fixed interval apart. This is called the contour interval. Typical contour intervals are 5, 10, 20, 40, 80, or 100 meters or feet. The contour lines are always at an elevation that is a whole number times the contour interval. That is, the contour lines are drawn at 20, 40, 60, 80, etc., meters or feet, never at 23, 43, 63, etc.

A little imagination will reveal that if the contour lines are close together, the surface is steeper than if they are far apart (see Fig. B–1).

If we think of an island and remember that the shoreline is the zero contour, we can then imagine the successive contours, or shorelines, if the water level rises. From this, we learn that a hill is represented on a contour map by a series of concentric closed lines, and, of course, a ridge would be similar, but elongate.

If we think of a valley, we see that the contour lines form V's pointing upstream. Remember a stream flows downhill at the lowest point in a valley; hence, if we start on one of the valley sides and walk in the upstream direction at a given elevation, we will eventually reach the stream.

A few minutes' study of Fig. B–2 will reveal much information about reading contour maps.

Contour lines have the inherent limitation that the elevations of points between contours are not designated, although the approximate elevation can be determined by interpolation. In an effort to overcome this limitation, the elevation of hills, road junctions, and other features are commonly given on the maps.

CONVENTIONS USED ON TOPOGRAPHIC MAPS

Because the U.S. Geological Survey produces most of the topographic maps in use in the United States, these are described here. Until recently, these maps were made by actual field surveying, but are now made from aerial photographs with some field surveying to establish horizontal and vertical control points. The maps made since 1942 meet high accuracy standards. Ninety

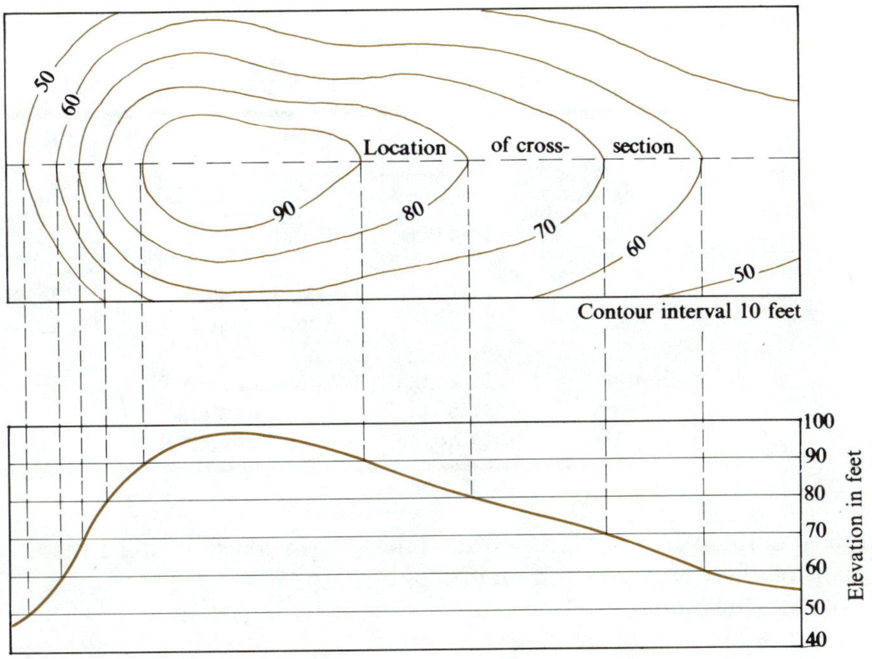

FIGURE B–1
Contour maps of a simple hill. The cross section shows that where the contour lines are closer together, the hill is steeper.

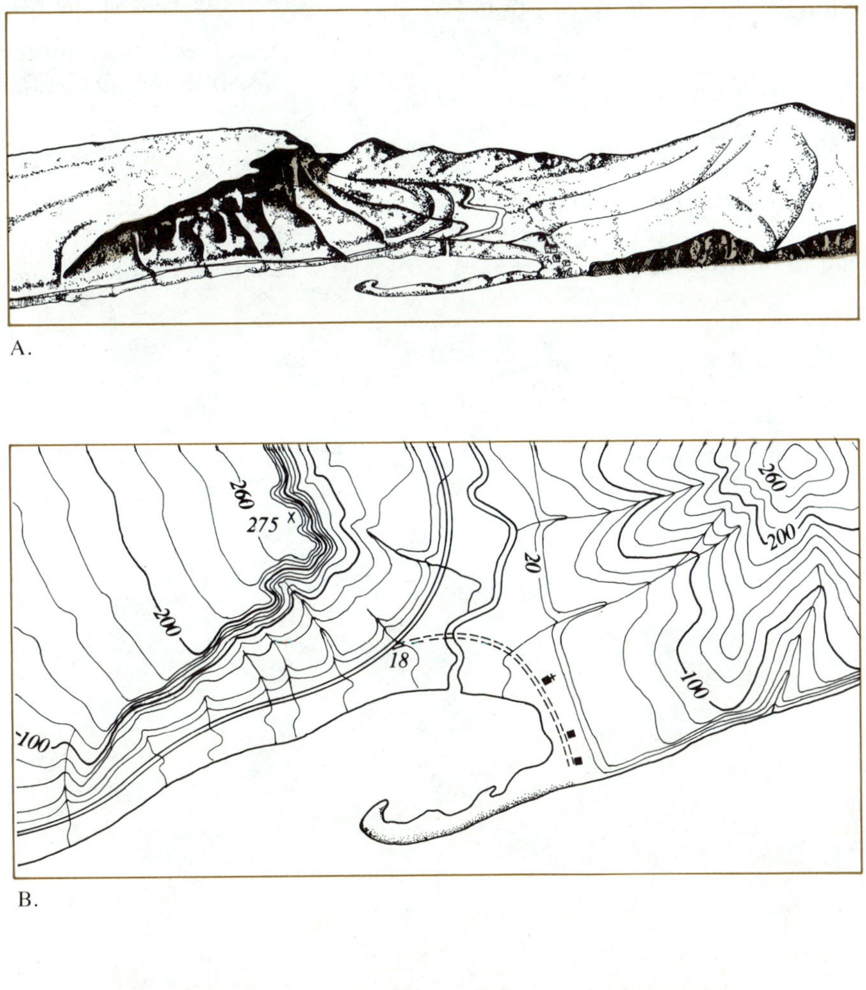

FIGURE B–2
These illustrations show how various features are depicted on a topographic map. A. A perspective view of a river valley and the adjoining hills. The river flows into a bay, which is partly enclosed by a hooked sandbar. On either side of the valley are terraces through which streams have cut gullies. The hill on the right has a smoothly eroded form and gradual slopes, whereas the one on the left rises abruptly in a sharp precipice from which it slopes gently, and forms an inclined tableland traversed by a few shallow gullies. A road provides access to a church and two houses situated across the river from a highway which follows the seacoast and curves up the river valley. B. The same features represented by symbols on a topographic map. The contour interval (the vertical distance between adjacent contours) is 20 ft. (After U.S. Geological Survey)

TABLE B–1
Sizes and scales of U.S. Geological Survey topographic maps.

Map Designation	Scale		Quadrangle Size (lat.-long.)	Quadrangle Area (mi²)	Paper Size (inches)
	Ratio	1 in. equals			
7½-min	1:24,000	Exactly 2000 ft	7½′ × 7½′	49–70	22 × 27 23 × 27
	1:31,680	Exactly 0.5 mi	7½′ × 7½′	49–68	17 × 21
15-min	1:62,500	Approximately 1 mi	15′ × 15′	197–282	17 × 21 19 × 21
30-min	1:125,000	Approximately 2 mi	30′ × 30′	789–1082	17 × 21
1-degree	1:250,000	Approximately 4 mi	1° × 1°	3173–4335	17 × 21
1:250,000	1:250,000	Approximately 4 mi	1° × 2°	6349–8669	24 × 34

percent of all well-defined features are within one-fiftieth of an inch of their true location on the published map, and the elevations of 90 percent of the features are correct within one-half of the contour interval.

The maps, which are called quadrangles, are bounded by latitude and longitude lines. They are named for features within the map area, and because the names are not duplicated within a state, some are named for less important features. The sizes and scales used are shown in Table B–1. The number of square miles in a map of a given scale varies because of the convergence of longitude lines at the North Pole.

COLORS

Water features are in blue.
Works of humans—roads, houses, and so forth—are in black.
Contour lines are in brown.

In addition, some maps show:

Important roads, urban areas, and public land subdivision lines in red.
Woodlands, orchards, and the like, in green.

If you would like to obtain topographic maps of any area, write to the addresses below for an index to topographic maps for the state desired. They are free on request and show all of the maps available.

For states east of the Mississippi River:

U.S. Geological Survey
Washington, D.C. 20242

For states west of the Mississippi River:

U.S. Geological Survey
Federal Center
Denver, Colorado 80225

Appendix C
GEOLOGICAL MAPS

Geological mapping is one of the most important methods used in the study and interpretation of the earth and earth history. Basically, a geological map is simply a map on which the various rock types that make up the earth's surface are plotted. Generally, the rock types are plotted on topographic (contour) maps because the topography is commonly related to the rock types, and knowing the topography helps another geologist to interpret the geologic map. To further aid in the interpretation of geologic maps, other symbols are used to indicate such things as the attitude or slope of planar features such as the bedding planes of sedimentary rocks. From the information on the geologic map, which is the surface geology, the underlying structure of the area is interpreted. The main difficulty in geologic mapping is that outcrops of bedrock are relatively rare in most regions because weathered material, soil, and alluvium cover much of the earth's surface.

The preparation of a geologic map begins with the field geologist locating the outcrops of the various rock types on a map or an aerial photograph of the area (Fig. C–1B). Generally, the units that are mapped are **formations**. Formations are rock units that are generally easy to recognize and thus reasonably easy to map. The geologic map should show the boundaries between formations, and these boundaries are called **contacts**. (Fig. C–1C). If the units mapped have directional properties, such as bedding in sedimentary rocks or foliation in metamorphic rocks, the **attitude** of such features should also be noted. The attitude is given by the T-shaped **strike and dip** symbol shown on the maps. The longer line shows the strike, which is the direction of the intersection of the bed and a horizontal plane. The shorter line shows the direction of the dip, and the number near it shows the dip angle (Fig. C–2). In addition, a geologic map should show the contacts of such features as intrusive rock bodies. Faults should also be shown.

Typically, different colors are used to show the various rock units on a geologic map. The ages and relationships among the rock units, and the color used for each, are shown in the legend of the map. The legend is generally at the side or bottom of a geologic map. Any other symbols used on the map will also be explained in the legend. Cross sections showing the structure may also be included (Fig. C–3).

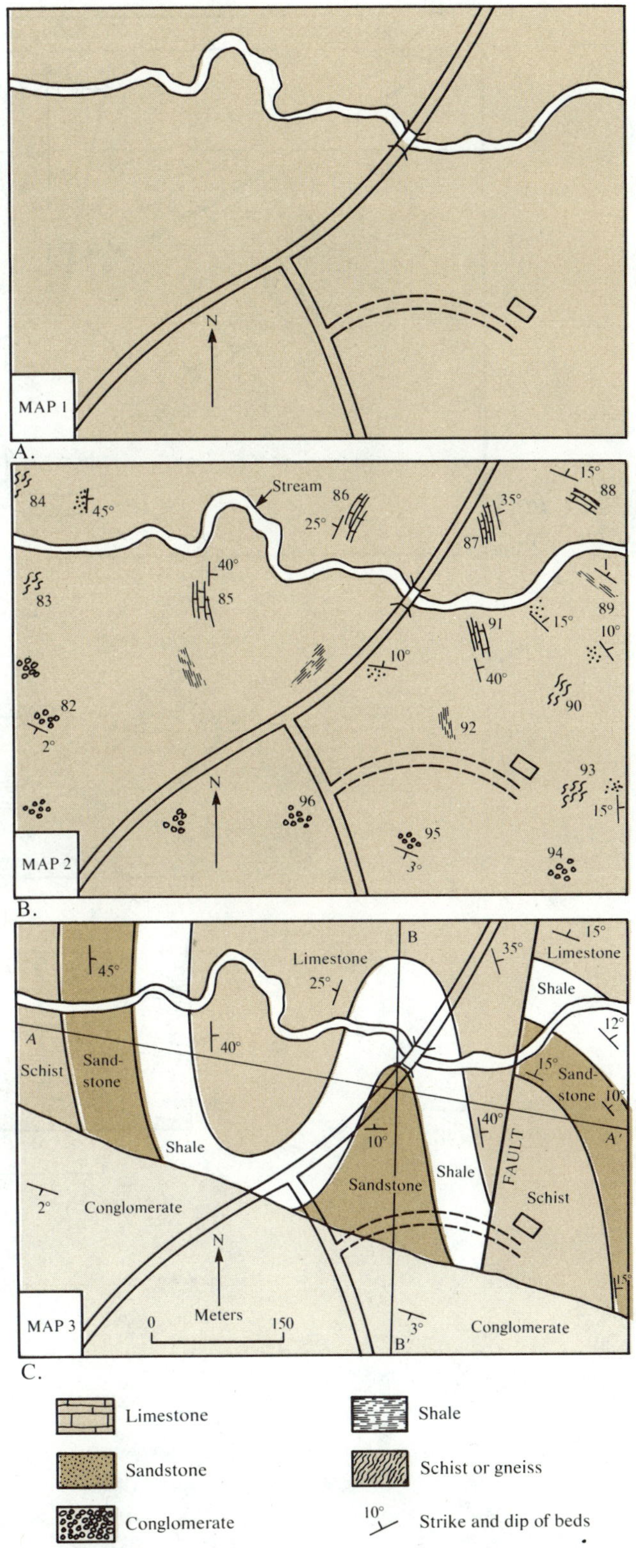

FIGURE C–1
The stages in making a geologic map.

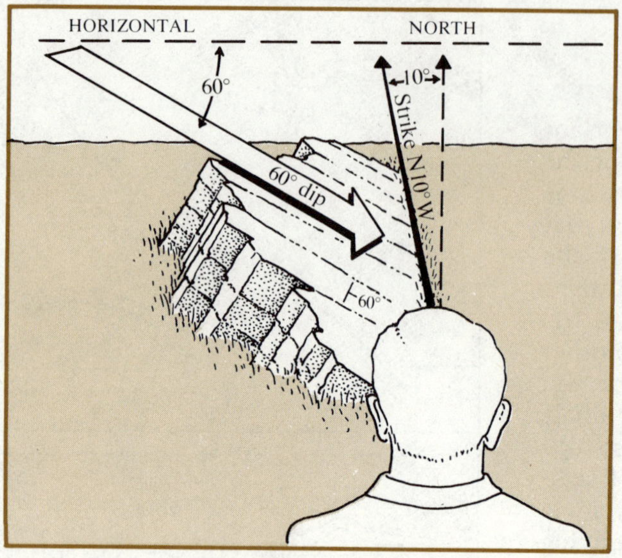

FIGURE C–2
Strike and dip.

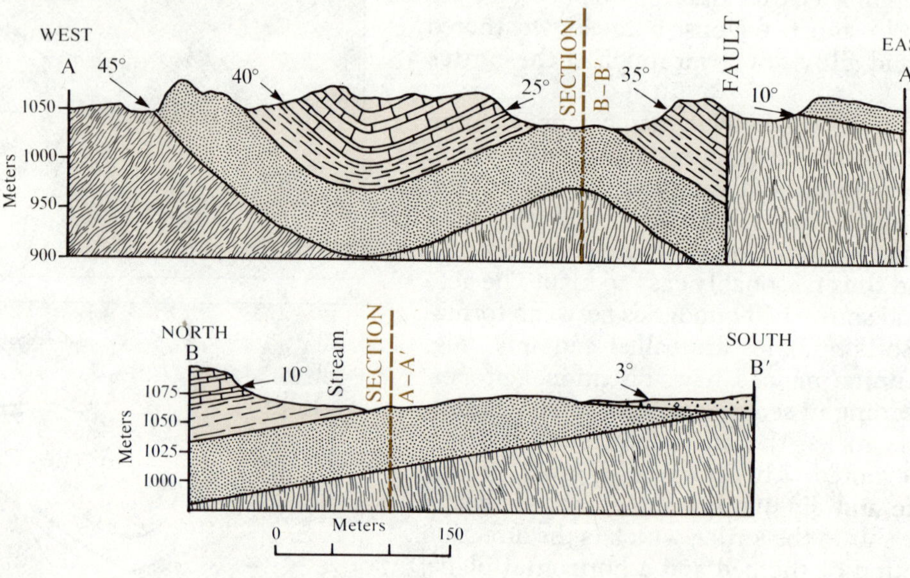

FIGURE C–3
Cross sections A–A′ and B–B′ of the map in Fig. C–1C.

Appendix D
UNIT MEASUREMENT SYSTEMS

METRIC AND ENGLISH UNITS

Length

1 millimeter (mm) = 0.001 meter (m) = 0.0394 inch
1 centimeter (cm) = 0.01 meter = 0.394 inch
1 kilometer (km) = 1000.0 meters = 0.62 mile
1 meter = 39.4 inches

Mass

1 gram (g) = 0.0022 pound = 0.0352 ounce
1 milligram (mg) = 0.001 gram
1 kilogram (kg) = 1000.0 grams = 2.2 pounds

Volume

1 milliliter (ml) = 0.001 liter (l) = (approx.) 1 cubic centimeter (cm^3) = 0.00106 quart = 0.0339 ounce

Approximate Equivalents

1 meter = 39.4 inches
1 mile = 1.6 kilometers
1 kilometer = 0.62 mile
1 inch = 2.5 centimeters
1 avoir. ounce = 28 grams
1 kilogram = 2.2 pounds
1 pound = 454 grams
1 fluid ounce = 30 milliliters
1 quart = 946 milliliters
1 liter = 1.06 quarts

TEMPERATURE SCALES

Fahrenheit (F)	Celsius (C)	
3632°	2000°	
1832°	1000°	
1000°	538.6°	
212°	100°	(boiling point of water)
32°	0°	(freezing point of water)

$$\frac{F - 32}{180} = \frac{C}{100}$$

$$C = \frac{5}{9}(F - 32)$$

$$F = \frac{9}{5}C + 32$$

Appendix E
ROCK TEXTURES

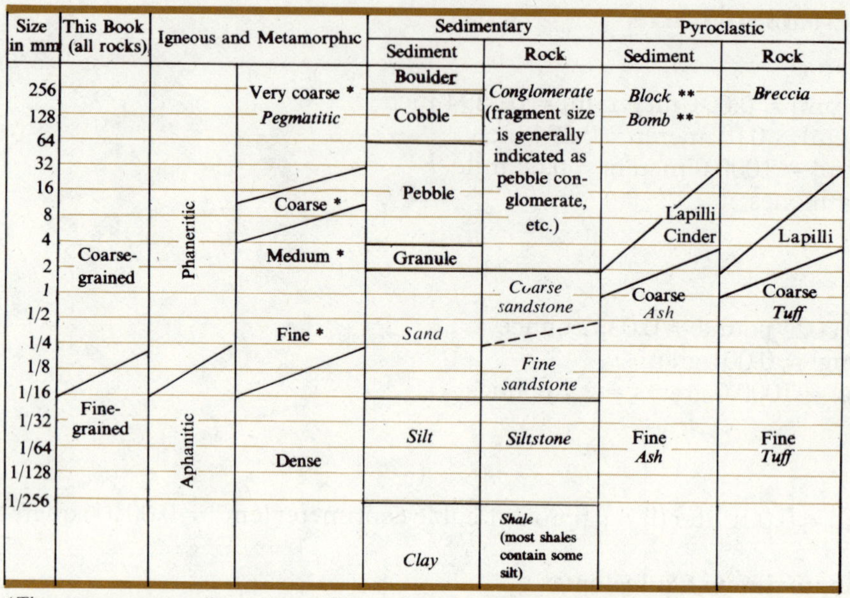

*These terms are generally used as modifiers such as medium-grained granite.
**Blocks are fragments broken from solidified rocks; bombs are molten when ejected.

FIGURE E–1
Summary of texture terms in common use and those used in this book. Terms used in this book are in italics.

GLOSSARY

GLOSSARY

A

aa lava Hawaiian term for blocky, rough-surfaced lava.

ablation All processes by which snow and ice are lost from a glacier, floating ice, or snow cover. Include melting, evaporation (sublimation), wind erosion, and calving.

ablation moraine An uneven pile or continuous layer of till or glacial debris formed when ice is lost from a glacier. May overlie inactive ice in the ablation area or rest directly on ground moraine derived from the same glacier.

absolute date A date expressed as a number of years.

abyssal plain A flat region of the ocean floor, usually at the base of a continental rise, whose slope is less than 0.1°, or essentially horizontal.

aeolian Pertaining to the wind, especially concerning such deposits as dune sand and loess and processes such as erosion caused by wind.

agglomerate Pyroclastic rock with rounded fragments over 64 mm in size.

alluvial fan A type of depositional geomorphic surface that is produced when streams emerge from a mountain front.

alpha particle Helium nucleus. Consists of two protons and two neutrons, and is positively charged.

ammonoids An extinct group of molluscs similar to the modern pearly nautilus in having a shell divided into chambers separated by partitions, which have extremely crinkly edges in contrast to the nautilus.

amphibole A ferromagnesian mineral group. The most important rock-forming amphibole is hornblende.

amphibolite A metamorphic rock composed mainly of amphibole, such as hornblende, and plagioclase feldspar.

Andean-type arc A chain of magmatic rocks that lie within or on top of continental crust whose origin is related to the subduction of oceanic lithosphere beneath continental lithosphere.

Andes A complex of mountain ranges along the western side of South America. The mountains are above a subduction zone where various eastern Pacific plates plunge beneath the continent. Structures and rocks within the Andes formed in the Paleozoic, Mesozoic, and Cenozoic eras.

andesite A fine-grained igneous rock. Intermediate in color and composition between rhyolite and basalt.

anthracite A type of coal formed by mild metamorphism of lower-grade coal.

anticline A fold, generally convex upward, whose core contains stratigraphically older rocks.

Appalachian Mountains A range of mountains in eastern North America whose structures were produced in the middle and late Paleozoic by the collision of North America, Europe, and Gondwanaland.

aquiclude Earth materials that have low permeability and therefore restrict the movement of groundwater.

aquifer Earth materials with a relatively high porosity and permeability from which groundwater may be obtained.

arc-arc collision The process whereby formerly separated island arcs come in contact by activity at a plate boundary. Commonly, the contact is caused by the subduction of intervening oceanic lithosphere.

arc-continent collision The process whereby a formerly separated island arc and continent come in contact with each other by activity at a plate boundary. Commonly, the contact is caused by the subduction of intervening oceanic lithosphere.

Archean The oldest era of geologic time, extending from the earth's origin 4.6 billion years ago down to approximately 2.5 billion years ago.

arête A narrow, serrate mountain crest or rocky sharp-edged ridge or spur, commonly present above the snowline in rugged mountains sculptured by glaciers, formed as the result of continuous erosion of the headwalls of adjacent cirques, narrowing the divide between glacial valleys.

arkosic sandstone (arkose) A sandstone containing much feldspar.

ash grain Unconsolidated pyroclastic material less than 4 mm in size.

asteroid One of a group of small planetlike bodies whose general orbit around the sun is between Mars and Jupiter.

asthenosphere A zone within the earth below the lithosphere in which the rocks are weaker (more ductile), and slow flowage of rocks takes place more rapidly than in the overlying lithosphere.

Atlantic-type margin A passive trailing edge of a continent that is subsiding due to lithospheric cooling and increasing sediment load.

atoll A ring of low coral islands surrounding a central lagoon.

atom The smallest unit of an element. Composed of protons, electrons, and neutrons.

atomic mass number The total number of protons and neutrons in the nucleus of an atom.

atomic number The number of protons, or positive charges, in the nucleus of an atom.

augite The common rock-forming pyroxene mineral. Generally black with two cleavages at right angles. $(Ca,Na)(Mg,Fe^{II},Fe^{III},Al)(Si,Al)_2O_6$.

aurora Luminous bands or areas of colored light seen in the night sky near the poles. Northern or southern lights.

avalanche A large mass of mixed snow, ice, soil, and rock, falling or sliding very rapidly under the force of gravity from a mountain slope or glacier.

B

backshore The inner portion of the shore, lying landward of the mean spring tide high-water line. Acted upon by the waves only during unusually high tides and storms.

backwash The flow back down the beach face of water that was carried onto the beach as swash.

bacteria The simplest of one-celled microorganisms; they lack a nucleus in their cells and have a fossil record back into the Archean.

badlands Topography of semiarid regions that is intricately stream-dissected on surfaces with little or no vegetative cover, and underlain by loosely consolidated clays or silts. Presence of soluble minerals such as gypsum or halite helps make this material easily erodable, and promotes a very-fine-textured drainage network of dry gulleys or ravines. *Examples:* western South Dakota; Drumheller area of eastern Alberta, Canada.

bajada A broad, continuous alluvial slope extending outward from the base of a mountain range into a basin. In contrast to pediments, bajadas are formed by coalescence of a series of separate but confluent alluvial fans, and thus are surfaces of deposition. Bajadas may merge with or pass into pediments.

barchan An isolated, crescent-shaped sand dune lying transverse to the direction of the prevailing wind, with a gently sloping convex side facing the wind, points of the crescent downwind, and a steep, concave leeward slope between the horns. One of the most common dune types, barchans form on flat, hard surfaces where sand supply is limited and wind flow is constant at a moderate velocity.

barite Barium sulfate. $BaSO_4$.

barrier island A long, narrow, wave-built island separated from the mainland by a lagoon.

barrier reef A reef whose actively growing outer edge is separated from the island or continent on which it is growing by a shallow lagoon.

basalt A fine-grained volcanic rock. Dark in color and composed mainly of plagioclase feldspar and pyroxene. One of the most common rock types.

base exchange The chemical weathering process characterized by ions in solution which exchange with ions in a mineral.

base level The theoretical lowest level to which a stream or river may erode.

batholith A very large discordant intrusive rock body, generally composed of granitic rocks. Some batholiths may be of metamorphic origin.

bauxite Aluminum ore. The term given to a weathering product composed of several aluminum-rich minerals.

bay barrier A marine deposit attached to the mainland at both ends and extending across the mouth of a bay, separating it from the open ocean.

bedding The layers, or laminae, of sedimentary rocks.

bed load The part of the total stream load that rolls, skips, or slides along the bottom of the stream channel; the larger particles that are not continuously held in solution or suspension.

belemnites An extinct group of squidlike cephalopods characterized by possession of a thick, internal, cigar-shaped calcareous shell. Most were of Mesozoic age.

Benioff zone A zone of earthquakes that may extend to depths of over 600 km. The zone of earthquakes is thought to be within the upper part of a subducting lithospheric plate.

bergschrund A deep and often wide gap or crevasse, or series of crevasses, in ice or firn at or near the head of an alpine glacier. Separates moving ice and snow from the relatively immobile ice and snow adhering to the headwall.

beta particle In radioactivity, an electron moving at high velocity.

bituminous coal Soft coal.

block A pyroclastic fragment greater than 5 cm in size.

blocky lava (aa) Lava with a blocky, rough surface.

blue-green algae The simplest organisms capable of photosynthesis; like bacteria, they lack a nucleus and have a fossil record well back into the Archean.

blueschist A metamorphic facies containing blue-colored minerals such as glaucophane.

bolson A term applied in the desert regions of southwestern United States to an extensive, flat, alluvium-floored basin or depression into which drainage from the surrounding mountains flows with gentle gradients toward a central playa. Bolsons represent interior basins, or basins with internal drainage.

bomb A pyroclastic material ejected from a volcano in a melted or plastic state. Its shape forms either in flight or upon impact.

Bouguer anomaly A gravity anomaly calculated after corrections for latitude, elevation, and terrain.

boundary current A surface or deep current that flows north or south along the east or west margin of an ocean basin.

Bowen's reaction series The sequence of reactions that take place in the crystallization of a basaltic melt.

brachiopods A major group of fossil organisms, particularly in the Paleozoic era, whose members each lived inside a bivalved calcareous shell with top and bottom portions hinged at the rear.

braided river A river with a channel pattern characterized by numerous bars and islands that continuously divide and subdivide the flow of the river. Characteristically occurs where open flow is insufficient to transport all of the sediment load of the river.

breaker The spilling, plunging, or surging pattern of energy release by swell, resulting from the interference with particle motion by the shoaling bottom.

breccia A clastic sedimentary rock with angular fragments larger than 2 mm in diameter; a sharpstone conglomerate. A volcanic breccia is composed of pyroclastic fragments greater than 2 mm in size.

brittle The behavior of a material that fractures after only a few percent elastic strain.

bryozoans A group of small colonial organisms whose fossilized remains are twiglike, lacy, massive, or sheetlike calcareous structures, consisting of numerous tiny compartments.

butte An isolated, generally flat-topped hill or mountain with steep slopes. Buttes are often capped by a layer of resistant rock and represent an erosion remnant carved from flat-lying rocks.

C

calamites Large, jointed trees characteristic of late Paleozoic coal swamps and related to the modern horsetail or scouring rush.

caldera A large volcanic depression containing volcanic vents.

Caledonian Mountains A range of mountains in the northern British Isles and Scandinavia whose structures were produced in the middle Paleozoic by the collision of North America and Europe.

cementation The process whereby binding material is precipitated or deposited between clastic grains, forming a rock.

Cenozoic era The most recent and shortest era of geologic time, encompassing the last 65 million years.

Central Province A region of the North American craton extending southwest from the Great Lakes to southern California and containing chiefly Middle Proterozoic rocks.

central vent The eruptive center of a shield volcano.

cephalopods A major mollusc group including the living squids, cuttlefish, octopus, and pearly nautilus. Are active carnivores, with tentacles for seizing prey and a funnel for propulsion. Most fossils, the nautiloids and ammonoids, resemble the living pearly nautilus in possessing a one-piece, chambered calcereous shell.

chalk A soft, usually white variety of limestone, composed of shells of microorganisms and very fine calcite.

Channeled Scablands An area of deeply eroded volcanic plateau, which was formed as the result of the erosive action of glacial meltwater released during the Missoula Flood event. This occurred about 22,000 years ago on the Columbia Plateau of eastern Washington.

chelation A type of chemical weathering associated with biological processes, such as excretion by lichens, forming agents that produce or enhance weathering.

chert An extremely fine-grained variety of quartz.

Churchill Province A region of the Canadian Shield extending in a broad arc around the west, north, and east sides of the Superior Province and containing chiefly Lower Proterozoic rocks.

cinder cone A cone-shaped, small volcano composed of pyroclastic material.

cirque A deep, steep-walled, half-bowl-like recess or hollow, which is crescent- or horseshoe-shaped in plan. Cirques are found high on the sides of mountains commonly at the heads of glacial valleys. They are formed by the erosive activity of mountain glaciers and may contain ice and snow.

cirque lake A small, deep, commonly circular lake occupying a cirque. Has no permanent inlet, being fed by runoff from surrounding slopes, and is dammed by a lip of bedrock or small moraine.

clastic A textural term for rocks or sediments composed of fragments broken from some other preexisting rock.

clay dune A dune composed of clay particles, which are formed by drying and cracking and are blown from a playa surface by wind and piled along the margin of the lake. May eventually be stabilized by vegetation and become a hard, compact mass.

claystone A rock composed largely of fine clay.

climbing dune A dune formed by the piling up of sand by wind against a cliff or mountain slope, causing it to climb over the divide from one basin to the next.

coal A sedimentary rock formed of altered plant material.

coastal desert Any desert area bordering an ocean where interactions between cool ocean currents, atmospheric circulation, and latitudinal position cause a moderation of temperature. Occasional dew and drizzle help support meager plant and animal life. *Example:* Namib Desert, southwest Africa.

coastline The landward limit of the effect of the highest storm waves on the shore.

coasts built forward by organisms Found primarily in the tropics, these coasts show seaward growth as the result of the development of coral reefs or mangrove trees.

col A high, narrow, sharp-edged pass or depression in a mountain range, generally across a ridge or through a divide between mountain peaks; especially a deep pass formed by the headward erosion and intersection of two cirques.

compaction Reduction in volume. Generally caused by pressure in sedimentary rocks.

composite volcano A volcano composed of both lava and pyroclastic material.

conchoidal fracture The rounded, shell-like fracture typical of brittle materials such as glass.

concordant intrusive An intrusive igneous body whose boundaries are parallel to the enclosing sedimentary rocks.

concurrent range zone A very-small-scale subdivision of the geologic record based upon the joint occurrence of two or more species that do not occur together in older or younger rocks.

confined groundwater system An aquifer in which the movement of water is restricted above and below by aquicludes. Thus, the water is under pressure and, if tapped, will tend to rise to the level of the recharge zone where the water enters the system from precipitation or runoff.

confining pressure A pressure equal in all directions, exerted by rocks within the earth.

conglomerate A coarse-grained clastic sedimentary rock. The fragments are larger than 2 mm.

conjunction When the sun and moon are aligned, or nearly so, on the same side of the earth, such as during the new-moon phase.

connate water Water no longer in circulation or contact with the present hydrologic cycle.

contact metamorphism Thermal metamorphism near the boundary or contact of an intrusive body such as a batholith.

continental crust The part of the earth's crust that underlies the continents and continental shelves. Ranges in thickness from about 20 km to more than 70 km under some mountain ranges. The upper part of the continental crust is less dense and chemically richer in silica than oceanic crust.

continental drift A general term used for the concept originally proposed by Wegener (1912) that continents can move relative to one another.

continental glacier A glacier of considerable thickness that covers a large part of a continent. Continental glaciers obscure the relief of the underlying landmass, such as the ice sheets covering Greenland and Antarctica.

continental rise That part of a continental margin between the continental slope and the abyssal plain, generally characterized by a gentle slope of less than 1°.

continental shelf That part of a continental margin between the shoreline and the continental slope, generally characterized by a very gentle slope of about 0.1°, and usually reaching a depth of no more than 200 m.

continental slope That part of a continental margin between the continental shelf and the continental rise, generally characterized by a relatively steep slope of 3°–6°.

continent-continent collision The process whereby two formerly separated continental masses come in contact with each other by activity at plate boundaries. Commonly, contact is caused by the subduction of intervening oceanic lithosphere.

continuous mineral series A mineral group such as plagioclase feldspar in which complete or continuous change in composition between end members is possible.

convection (tectonics) A mass movement vertically or laterally of mantle material largely in the crystalline state, mainly as the result of heat variations in the earth.

convergence and divergence of flow A fundamental concept of streamflow which states that convergence of flow is associated with scour whereas divergence of flow is associated with deposition.

convergent plate boundary The boundary that results where two plates collide. If one of the plates is oceanic crust, a subduction zone forms.

coquina A type of limestone composed largely of shells and shell fragments.

coral reef A limestone deposit as a result of secretion by corals and marine algae, found along the coasts of islands and continents located between the tropics.

cordaites Large trees characteristic of late Paleozoic coal swamps and distantly related to the conifers although their leaves are strap-shaped and their cones small.

core The central zone of the earth beginning at a depth of 2900 km and extending to the center of the earth.

Coriolis effect A fictional force resulting from the earth's rotation and manifested in the deflection of particles in motion to the right in the Northern Hemisphere and to the left in the Southern Hemisphere.

cosmic ray Very-high-speed nuclei coming from space.

cotylosaurs Primitive generalized reptiles, mostly late Paleozoic in age, that closely resemble their labyrinthodont amphibian ancestors.

covalent bond A bond formed by the sharing of electrons.

crater A volcanic depression typically developed at the top of a volcano. Also, a depression caused by impact of a falling object. Meteorites cause most large impact craters.

craton A continental area that has not been subjected to compressive mountain-building forces since the Precambrian or earliest Paleozoic.

creep The slow, nearly imperceptible flowage of rocks or soil downslope.

crevasse A deep, nearly vertical crack, fissure, or rift in a glacier or other mass of land ice or snowfield, caused by differential movement over an uneven surface.

crinoids A group of echinoderms that were very abundant in the Paleozoic era. Its members are typically stalked and have numerous bushy arms for feeding.

cross-beds Beds, or laminae, at an oblique angle to the main bedding.

crosscutting relationships The principle that a body of rock cutting across another rock body must be younger than the one it cuts.

crust The very thin outermost layer of the earth that lies above the Mohorovičić discontinuity. It generally can be divided into oceanic- and continental-type crust.

crystal A geometrical solid bounded by smooth planes that result from an orderly internal atomic arrangement.

crystalline A mineral with a crystal structure, that is, with an orderly internal atomic structure.

Curie temperature The temperature below which minerals can retain ferromagnetism.

current meter A mechanical or electronic device used to determine the direction and velocity of current flow.

cut and fill A sedimentary feature resulting from erosion of a channel that is later filled by deposition.

cycle of erosion The model of landscape evolution that states that the land surface goes through a series of erosional landforms in its development.

D

Darcy's law The law which states that the velocity of flow of groundwater is equal to the product of the coefficient of permeability and the hydraulic head, or gradient.

daughter element An element formed from another element in radioactive decay.

debris avalanche A type of mass wasting that generally occurs on steep, forested land. Involves a very rapid movement of soil and weathered debris above a bedrock surface.

debris (talus) slope The part of a slope profile found at the base of the free face. Consists of materials derived by mass-wasting and weathering processes from slope segments at higher elevations.

declination (magnetic) The horizontal angle between true north and magnetic north.

deep-water mass A subsurface mass of water that can be identified on the basis of its temperature, salinity, or chemical content.

deflation basin A topographic basin excavated and maintained by wind erosion. Unconsolidated material is removed, often leaving a rim of resistant rock surrounding the depression.

delta A depositional geomorphic surface located where a stream or river enters a lake or the ocean.

delta deposit A low-lying deposit at the mouth of a river.

deltaic coast Convex extensions of the coast into the ocean, resulting from rapid deposition of sediment at the mouth of a river.

dendritic drainage A drainage pattern which resembles the veins in a leaf. Tends to form on rocks that are homogeneous with respect to erosion by running water.

desalination A process for removing impurities from salt-bearing water.

desert pavement A natural residual concentration of wind-polished, closely packed pebbles, boulders, and other rock fragments mantling a desert surface. Wind action and sheet wash have removed the fine-grained sediments of the surface layer, whereas the pavement protects the fine-grained particles below from further erosion.

diastrophic coast A coast produced by diastrophic processes such as folding and faulting.

diatomite A rock composed largely of the siliceous shells of diatoms.

diatoms Microscopic, one-celled, floating organisms that build tiny two-valved siliceous shells and that have been one of the major photosynthetic groups in the oceans since the Cretaceous.

differentiation The process by which different rock types are derived from a single magma.

dike A tabular discordant intrusive body.

dinosaurs Two (or more) extinct groups of reptiles that dominated the land from the late Triassic through the late Cretaceous. Although some were small, most were large and probably very active, leading some experts to believe that they were warm-blooded, unlike typical reptiles.

diorite A coarse-grained igneous rock intermediate in composition between granite and gabbro.

discharge With reference to streamflow, the product of the cross-sectional area of flow times the average velocity. Units of measurement are cubic meters per second.

discontinuous series That part of the Bowen reaction series in which abrupt changes in mineral occur.

discordant intrusive An intrusive rock body that cuts across the bedding of the enclosing sedimentary rocks.

dissolved load That part of the total stream load that is carried in chemical solution.

distributary A small stream flowing away from a main stream. Such streams are characteristic of deltas.

divergence and convergence of flow A fundamental concept of streamflow which states that divergence of flow is associated with deposition whereas convergence of flow is associated with scour.

dolomite A mineral, $CaMg(CO_3)_2$, or a rock composed largely of dolomite (dolostone).

dome *See* lava plug.

Doppler effect *See* red shift.

drainage pattern The arrangement of streams in a drainage basin as viewed from the air.

drumlin A low, smoothly rounded or subrounded elongate hill, mound, or ridge of compacted glacial till or other glacial sediment. Drumlins are built under the ice and shaped by glacial flow or carved from an older moraine by readvancing ice. The long axis of a drumlin is parallel to the direction of movement of the ice; it normally has a blunt end pointing in the direction from which the ice advanced.

ductile The behavior of a material that sustains more than several percent permanent strain before fracture. Some materials may sustain hundreds of percent permanent strain without fracturing.

dunite A coarse igneous rock composed almost entirely of olivine.

dust storm Fine-grained soil particles transported by strong winds of temperate deserts or semiarid regions, causing a stormlike condition without any precipitation. Human development of transitional zones between desert and steppeland increases the mobility of dust-sized particles and adds to the intensity of duststorms.

dynamic equilibrium The model of landform evolution that states that as a landscape develops, there will eventually be an equilibrium of action in which every slope is adjusted to every other slope, and the entire landscape is then relatively time-independent.

dynamic metamorphism Metamorphism caused by deformation of the rock.

E

earthflow A common type of mass-wasting process which involves the flowage of water-saturated earth materials downslope. Earthflows differ from mudflows in that the rate of movement is often slower, and earthflows are seldom confined to channels but occur directly on slopes.

echinoderms A major group of living and fossil organisms whose members build a porous and spiny calcareous shell (test) just beneath the skin and have a unique system of pores and canals filled with seawater that is used for locomotion, feeding, breathing, and so forth. Members include sea urchins, sand dollars, sea cucumbers, starfish, and sea lilies.

echinoids A major group of spiny echinoderms characterized by a globular to flattened shell (test) containing pores for tiny feet. Living sea urchins, sand dollars, and heart urchins are examples of this group, which has left abundant Mesozoic and Cenozoic fossils.

effluent stream A stream in which the channel intersects the groundwater table. Thus, the groundwater system feeds water into the stream by seepage along the channel.

electron A fundamental subatomic particle. Has a very small mass and a negative charge.

electron capture A type of radioactive decay in which an electron from an inner shell unites with a proton to form a neutron.

element A material that cannot be changed into another element by ordinary chemical means. Just over 100 elements are known. All atoms of an element have the same atomic number (number of protons).

Elsonian Orogeny A mountain-building disturbance at the end of the Middle Proterozoic in North America.

end (recessional) moraine A ridgelike accumulation of sediment produced at the margin of an actively flowing glacier at any given time. Also, an end or lateral moraine built during a temporary pause in the final retreat (recession) of a glacier.

Eocambrian period The closing phase of Precambrian time, during which the first multicellular organisms came into existence.

ephemeral stream A stream in which the flow is only in direct response to precipitation. The channel is everywhere above the groundwater table.

epicenter The point at the earth's surface directly above the focus of an earthquake.

epoch A small-scale subdivision of relative geologic time. The major component of periods.

equilibrium assemblage A group or assemblage of minerals that formed in response to an environment.

era A large-scale subdivision of relative geologic time; groups of periods that share a fossil record of similar aspect.

erratic A rock fragment of variable size, carried by glacial ice or floating ice to some distance from its source, and which is not consistent with the lithology of rocks found in the area in which it is deposited.

esker A long, narrow, sinuous, steep-sided ridge, composed of irregularly stratified sand and gravel, which was deposited by a subglacial or englacial stream, or in an ice tunnel of a stagnant or retreating glacier. The sediment was left as a ridge, which may be branching and discontinuous, when the ice melted.

eustatic movement A rise or drop in sea level resulting from a change in the volume of water in the oceans or in the capacity of the ocean basins.

evaporite A sedimentary rock formed by the evaporation of water, leaving the dissolved material behind.

event correlation Recognition of the time equivalency of two different lithologies caused by the same physical event.

exfoliation A type of physical weathering associated with homogeneous rocks that have been deeply buried under pressure from overlying rocks. These rocks, when exposed at the surface of the earth by rapid erosion, tend to expand, and fractures form parallel to the surface of the earth.

expansive soil Soil rich in clay minerals that expand and contract upon wetting and drying.

extrusive An igneous rock erupted on the earth's surface. A volcanic rock.

F

factor of safety A concept used in evaluation of slope stability. It is the ratio of the resisting to the driving forces on a slope.

falling dune An accumulation of sand as it is blown from an upland or mesa top down into a basin. Falling dunes commonly form as sand fans extending outward from a reentrant in a mesa wall. (*See* climbing dune.)

fan trench A linear depression formed by a drainage line which is incised considerably below the surface of the alluvial fan. Characterized by steep walls and relatively broad, flat bottoms; common at the heads of alluvial fans.

fault A fracture or zone of fractures along which there has been displacement of the two sides relative to each other parallel to the fracture.

fault breccia A rock made of fragments that were broken by movement of a fault.

fault gouge The fine-grained ground rock in a fault zone.

faunal succession The occurrence of kinds of fossils in a definite and determinable sequence in the geologic record.

feldspar The most abundant mineral family. Consists of plagioclase $CaAl_2Si_2O_8$ and $NaAlSi_3O_8$, and potassium feldspar $KAlSi_3O_8$ (orthoclase and sanidine). Mixtures of potassium and sodium feldspar are termed perthite.

felsite A light-colored, fine-grained igneous rock. Most felsites have chemical compositions similar to rhyolite.

fermentation The process of converting sugar to carbon dioxide and alcohol, thus releasing energy.

fetch The distance over which the wind blows in the same direction across the ocean surface.

firn (firn ice) A material that is transitional between snow and glacier ice, and is older and denser than snow. Becomes glacier ice when compaction and recrystallization decrease its permeability to liquid water to zero while at the same time increasing its crystal grain size.

fish Aquatic vertebrates, generally with a torpedo shape and stabilizing fins. Most breathe with gills.

fissility The ability of a rock to split easily along parallel planes.

fission A reaction in which an atom is split. *Example:* the atomic bomb reaction.

fission tracks Tiny tubelike paths of high-energy particles released by some types of radioactivity.

fjord A long, narrow, winding inlet or arm of the sea, U-shaped, and steep-walled, along a mountainous coast. Typically, there is a shallow sill or rock lip submerged near its mouth. A fjord generally represents a glacial valley that was occupied when sea level was lower and later flooded when glaciation ended and sea level rose.

fjord coast A coast characterized by relatively straight-walled embayments (fjords) that were created by glacial erosion of river valleys.

flexural-slip fold A fold in which the layers have different ductility or viscosity, and the layers slip over one another during the folding.

float A device designed to float on the surface or at any depth in the ocean for the purpose of determining current direction and velocity.

floodplain A depositional geomorphic surface produced by lateral migration and overbank deposits of a stream system. Relatively flat land adjacent to a river.

flowering plants The angiosperms, the dominant plants of the Cenozoic era, which bear flowers to attract pollinating insects and have their seeds enclosed in fruits or fruitlike structures.

fluvial hydrology A type of streamflow in which the assumption of uniform and steady flow cannot always be met.

focus The initial point of rupture of an earthquake.

fold A secondary or imposed curve or bend of a rock surface such as the bedding planes in sedimentary rocks.

foliation A directional property of metamorphic rocks caused by the layering of minerals. Generally caused by crystallization under pressure. Also, the flat arrangement of features in any rock type.

footwall The rock body lying below a fault plane.

foreshore That portion of the shore lying between the normal high- and low-water marks; the intertidal zone.

fossil assemblages Groups of fossil organisms that consistently occur together, usually because in life they had similar ecologic requirements.

foraminifera (forams) A major group of microscopic organisms which build tiny calcareous shells typically consisting of several chambers. Have left abundant fossil remains since the late Paleozoic.

fossil dunes Relic dunes that accumulated in response to locally severe climates during the Ice Age. Sediment was supplied by deflation of sand from local floodplains and streams.

fossils Remains or traces of prehistoric (older than 5000 years) organisms preserved in the geologic record.

Frasch process The process for recovery of sulfur from dome salt.

free-air anomaly A gravity anomaly calculated from a theoretical model and elevation above sea level, but without accounting for the attractive effects of topography and isostatic compensation.

free face The part of a slope profile that is essentially a vertical or nearly vertical cliff, and thus probably a rock outcrop.

fringing reef A reef directly attached to the shore of an island or continent. May extend more than 1 km from shore. The outer margin is submerged and commonly consists of algal limestone, coral rock, and living coral.

frost heaving A type of physical weathering that occurs when water freezes in pore spaces of unconsolidated surficial materials. A ground-ice phenomenon involving the development of a lens or mound of ice that expands and forces the soil upward.

frost wedging A type of physical weathering that involves the process of water freezing and expanding in fractures of rocks.

full moon The phase of the moon during which it is aligned, or nearly so, with the sun and earth during opposition.

fusion A reaction in which two or more atoms are combined to form a different element. *Example:* the hydrogen bomb reaction.

G

gabbro A coarse-grained igneous rock composed mainly of plagioclase feldspar and augite.

galaxy A huge group of stars held together by gravitational attraction.

gamma rays Electromagnetic waves somewhat higher in frequency than X rays.

gastropods The largest group of molluscs, the snails. Snails have left numerous fossils in Paleozoic, Mesozoic, and Cenozoic rocks, but are most abundant in their later history. Most build a spirally coiled, unchambered, calcareous shell.

geocline *See* geosyncline.

geologic time The time interval from the earth's origin to the beginning of written human history approximately 5000 years ago.

geomorphic surfaces Landform surfaces produced by either erosion or depositional processes.

geomorphic threshold A geomorphic principle which states that many changes that occur in landforms are not continuous and gradual but occur suddenly when a critical threshold is exceeded. A threshold in which the development of the change in the system is internal rather than resulting from an external force.

geosyncline (geocline) A large, elongate, downsinking area in which many thousand meters of sedimentary rocks are deposited. Later deformed to form a mountain range.

geyser The natural discharge of hot water and steam from a geothermal system.

glacial deposition coast A coast characterized by hills and ridges of glacial deposits left behind when the glacier melted.

glacial ice Any ice that forms in or was once a part of a glacier, including land ice that is flowing or that shows evidence of having flowed, and glacier-derived ice floating in the ocean.

glacial stairway A glacial valley whose floor is shaped like a broad staircase composed of a series of irregular step-like benches (treads) separated by steep risers.

glacial surge The period of very rapid flow of a surging glacier. Also, the displacement or advance of ice resulting from very rapid flow.

glacial theory The theory, first propounded about 1840, that multiple periods of glaciation were responsible for drift deposits. These glaciers and ice sheets moved from higher to lower latitudes during the Pleistocene epoch.

glacial trough A deep, steep-sided, U-shaped valley leading down from a cirque, and excavated by a glacier that has widened, deepened, and straightened a proglacial stream valley.

glacioeustatic oscillations Eustatic changes in sea level resulting from a changing volume of water in the oceans caused by the formation or melting of glaciers.

glass A supercooled liquid. A glass does not have an orderly internal structure like a crystal.

gneiss A coarse-grained foliated metamorphic rock. Contains feldspar and is generally banded.

Gondwanaland An ancient supercontinent of the Paleozoic and early Mesozoic eras, which consisted of Africa, Antarctica, Australia, India, South America, and various smaller continental fragments (China, or some parts of it, may have been included).

graded bedding Bedding that has a gradation from coarse at the bottom to fine nearer the top.

graded stream A stream in which there is a rough balance between the load imposed and work done. Adjustment is such that over a period of years, the stream is able to adjust the slope and channel characteristics to provide just the velocity necessary to transport the load supplied to the channel from the drainage basin.

granite A coarse-grained plutonic rock containing quartz and feldspar.

graptolites An extinct group of Paleozoic organisms, most of which were attached in floating colonies and individually encased in an organic cup.

graywacke A type of sandstone, generally poorly sorted with a clay or chloritic matrix.

Great Ice Age A synonym for glacial epoch, or time of extensive glacial activity. Corresponds to the latest of the glacial epochs, also known as the Pleistocene epoch.

greenschist A schist with much chlorite and other green minerals. Generally formed by the metamorphism of basaltic rocks.

Grenville Orogeny A mountain-building disturbance near the close of the late Proterozoic in North America.

Grenville Province A region of the North American craton extending from southern Labrador through the eastern Great Lakes to the Gulf coastal plain and containing chiefly late Proterozoic rocks.

groin A low artificial structure projecting into the ocean from the shore to interfere with the longshore transportation of sediment. Usually has the purpose of trapping sand to cause the buildup of a beach deposit.

ground moraine An accumulation of till after it has been released from glacial ice during ablation. These deposits form an extensive area of low relief that has very subdued depositional features.

groundwater That part of the subsurface water within the zone of saturation.

gymnosperms The "naked seed" plants, in which the embryonic plant is not enclosed in a fruit or fruitlike structure. Most familiar are the conifers, ginkgoes, and cycads which dominated in the Mesozoic era.

gypsum A mineral. Number 2 on the hardness scale. $CaSO_4 \cdot H_2O$.

H

half-life The time required for the radioactive decay of one-half of the original amount of an isotope.

hanging valley A glacial valley in which the mouth is at a relatively high level on the steep side of a larger glacial valley. The larger valley is formed by a trunk glacier, and the hanging valley by a tributary glacier; the discordance in the level of their floors results from the greater erosive power of the trunk glacier.

hardness A mineral's resistance to scratching.

headwall A steep wall at the head of a glaciated valley, especially the rock cliff at the back of a cirque.

heat flow The amount of heat energy leaving the earth per square centimeter per second, measured in calories per square centimeter per second.

hermatypic coral Reef-building corals that have symbolic algae in their ectodermal tissue. Cannot produce a reef structure below the euphotic zone.

hexacorals The major Mesozoic-Cenozoic group of corals whose members build calcareous conical or cylindrical shells (theca) with a six-fold symmetry. Solitary and particularly colonial forms are major contributors to reefs.

Himalaya Mountains A major Asiatic mountain range formed in the middle and late Cenozoic by the collision of India with the remainder of Asia. Because they resulted from a continent-continent collision, the Himalaya Mountains are the highest mountains on earth today.

hinge The locus of maximum curvature or bending in a folded surface.

hinge line A line connecting the points of maximum curvature in a fold.

hinge surface The surface that contains all the hinge lines of a fold.

hominids The mammalian family to which the modern human species (*Homo sapiens*) and its close fossil relatives (*Australopithecus, Homo erectus*) are assigned.

hornblende A common rock-forming mineral. An amphibole. $Ca_2Na(Mg,Fe^{II})_4(Al,Fe^{III},Ti)(Al,Si)_8O_{22}(O,OH)_2$.

hornfels A metamorphic rock formed by thermal or contact metamorphism of any rock type.

hot continental desert A desert located in the continental interior of a subtropical region, and characterized by high yearly temperature, high evaporation, low relative humidity, and moderate unidirectional winds. *Examples:* Sahara Desert, Arabian Desert, Great Australian Desert.

hot spot A volcanic center persistent for at least a few million or even tens of millions of years, thought to lie above a rising plume of mantle material.

hot steppeland A semiarid transition zone which has characteristics of both deserts and humid zones. Occurs where rainfall is seasonal and supports grasslands and other vegetation that can withstand prolonged dry spells. *Example:* Sahel region in Africa.

Hudsonian Orogeny A mountain-building disturbance at the end of the early Proterozoic in North America.

hydrologic cycle The solar-driven system that constantly recycles water from the ocean to the atmosphere to the continents and back to the ocean again.

hydrolysis A type of chemical weathering which involves the universal solvent, water, plus carbon dioxide, which combine at surface conditions to form weak acids and thus attack rocks.

hypocenter (focus) The initial point of rupture of an earthquake.

I

Iapetus Ocean An ancient ocean between Europe and North America in the latest Precambrian and earliest Paleozoic. Fragments of its floor are preserved in the Appalachian and Caledonian mountains.

iceberg A large, massive piece of floating or stranded glacier ice of any shape, detached (calved) from the front of a glacier into a body of water. The greater part of its mass is below sea level.

ice cap A dome-shaped, areally extensive cover of perennial ice and snow covering the summit area of a mountain mass so that no peaks emerge through it. An ice cap may also cover a flat landmass such as an Arctic island.

icefall The part of a glacier that is highly crevassed as the result of a very steep fall in the ground surface below the ice. Icefalls commonly occur where a tributary glacier joins a trunk glacier.

ice-walled lake plain Lake deposits that accumulate within a lake formed on the surface of a glacier (surrounded by ice) and which, after the glacier melts, are topographically higher than surrounding ground moraine.

ichthyosaurs An extinct group of sea-going Mesozoic reptiles that resembled living porpoises and dolphins.

igneous rock A rock that formed by the solidification of a magma. A rock that was once melted or partially melted.

inclination (magnetic) The angle at which the magnetic field lines plunge at the earth's surface.

inclusions The principle that a body of rock that contains fragments of another rock body must be younger than the one whose fragments it includes.

index fossil A fossil organism with a brief geologic range, wide geographic range, and abundance sufficient to be useful in dating rocks.

index mineral A mineral whose first occurrence is used to determine the metamorphic grade in metamorphic rocks of a given composition.

infiltration The process whereby surface water enters into the soil and subsurface flow systems.

influent stream A stream in which the channel at all locations is above the groundwater table. Thus, the flow of water is from the channel downward through earth materials to the groundwater table, where a recharge mound may exist.

inselberg A prominent, isolated, residual knob or hill, usually smoothed or rounded, rising from and surrounded by an extensive lowland erosion surface in a hot, dry region. Generally bare and rocky but surrounded by debris derived from its slopes.

interflow The portion of direct runoff that flows beneath the surface but above the groundwater table.

interlocking texture A texture in igneous and metamorphic rocks in which the mineral grains grew together, filling all of the available space.

intermittent stream A stream that flows for a few days or weeks following precipitation. Some of the flow to the channel may be from the groundwater system.

intrusive An igneous rock that solidified below the surface.

ion An atom that has gained or lost electrons and so has an electrical charge.

ionic bond A bond in which the atoms are attracted to each other by electrical charges caused by concurrent gain and loss of electrons.

island arc A chain of islands, generally volcanic in origin, rising from deep sea floor and lying near a continent. Commonly, the island chain is convex toward the open ocean.

isograd A line on a map drawn through points of equal intensity of metamorphism.

isostasy A condition whereby the lithosphere, or some portion of it, is in equilibrium on the asthenosphere as if it were floating.

isostatic adjustment The adjustment of the crust of the earth to maintain equilibrium among units of varying mass and density. Excess mass at one spot on the crust causes compression (subsidence) which later rebounds if that mass is removed, as with a continental glacier.

isotope A form of a chemical element distinguished by the number of neutrons in its nucleus. Different isotopes of the same element have the same number of protons, but differing numbers of neutrons.

J

jetty A structure built from the shore into the ocean to protect a harbor or a navigable passage from being shoaled by deposition of longshore drift material.

juvenile water Water that is derived from the interior of the earth in rock-forming and tectonic processes but that has not previously existed as meteoric water.

K

kame A low mound, knob, or hummock, or short irregular ridge, composed of stratified sand and gravel deposited by a subglacial stream as a fan or delta at the ice margin; also deposited by a stream into a depression or hole in the glacier, and later let down onto the ground moraine.

kame esker A linear or sinuous landform consisting of kames linked by low ridges into a beadlike formation. These connecting ridges represent crevasses into which sediment was deposited.

karst topography A particular assemblage of landforms that tend to develop in soluble rocks such as limestone and marble. Landforms include sinkholes and caverns.

Kenoran Orogeny A mountain-building disturbance at the close of the Archean era in the Canadian Shield.

kettle moraine A terminal or recessional moraine whose surface is marked by steep-sided, bowl-shaped depressions called kettles that form when a block of stranded glacial ice becomes covered by till and subsequently melts.

key bed A distinctive rock layer that can be widely recognized. Some key beds, such as a distinctive bed of volcanic ash, can be useful for time correlation.

L

labyrinthodonts An extinct group of late Paleozoic-early Mesozoic amphibians that resembled large salamanders or lizards.

laccolith A concordant intrusive igneous rock body that domed up the overlying sedimentary rocks to some extent.

landslide A definite, perceptible downslope sliding or falling of soil or rock material.

landslide coast A coast resulting from landslides that carry coastal rocks into the ocean.

lapilli Pyroclastic fragments in the size range of 1–64 mm.

Laurasia An ancient supercontinent which consisted of North America, Europe, Siberia, and at least some parts of China, and which existed during the early Mesozoic era.

lava plug (dome) Viscous lava that tends to seal the vent of a volcano and so may cause violent eruptions if this causes the pressure to increase.

lignite A brown black coal that forms from the alteration of peat.

limestone A sedimentary rock composed largely of calcite.

lithification The process whereby a sediment is turned into a sedimentary rock.

lithology The type of rock based on its physical properties (granite, sandstone, schist, etc.).

lithosphere A layer of greater strength (i.e., lower ductility) relative to the underlying asthenosphere. Includes crust and a variable thickness of upper mantle and may have a thickness of about 100 km.

Little Ice Age A cool, brief interval in an otherwise warm interglacial stage. Originally employed to designate a mid-Holocene event in the Yosemite area of California about 3000 years ago, but is also widely used to describe the sixteenth and eighteenth century periods of cool climates and resurgence of mountain glaciers.

loess A widespread, homogeneous, fine-grained blanket deposit consisting predominantly of silt. Generally brown to light yellow and is wind-blown dust, generally but not exclusively of Pleistocene age, carried from desert surfaces, alluvial valleys, and glacial outwash plains or from unconsolidated glacial deposits. Characteristically stands in nearly vertical faces when exposed at the surface.

longitudinal dune (seif) A long, narrow sand dune, usually symmetrical in cross profile, oriented parallel to the direction of the prevailing wind. Longitudinal dunes form where sand is abundant and winds are strong and constant.

longitudinal waves A wave phenomenon in which particle vibration is parallel to the direction of energy propagation.

longshore current A current within the surf zone and running parallel to shore as a result of waves breaking along the shore.

longshore drift. The load of sediment transported along the shore by the longshore current.

low-velocity zone A zone in the upper mantle, generally at a depth of from 100 to 250 km but locally extending to shallower depths, where the seismic velocities are about 5 percent lower than those of the uppermost mantle.

M

magma A natural hot melt composed of a mutual solution of rock-forming materials (mainly silicates) and some volatiles (mainly steam) that are held in solution by pressure. May or may not contain suspended solids.

magmatic (volcanic) arc A linear belt of volcanic and plutonic rocks. Many of these belts have an arcuate shape.

magnetic reversal A change in the earth's magnetic field between normal and reversed polarity.

magnitude-frequency concept The geomorphic principle which states that as the magnitude of a process increases, its frequency decreases.

mammal-like reptiles An important group of later Paleozoic-early Mesozoic reptiles that had many skeletal features transitional to mammals. Many of the later ones were probably warm-blooded.

mammals The dominant group of land-dwelling vertebrates in the Cenozoic era. Have their young born alive and suckle them. Can control their own temperature internally ("warm-blooded"), typically by having a furry covering. Characterized by their highly specialized teeth.

manganese nodules Concretions rich in manganese, silicon, aluminum, chromium, cobalt, and copper that are found on the sea floor.

mantle The zone of the earth below the crust and above the core.

marble A metamorphic rock most commonly composed largely of calcite, formed by the metamorphism of limestone. There is also dolomite marble.

mare A large, low, dark, smooth area on the moon (plural: maria).

marginal sea A semienclosed sea that usually lies on the concave side of an island arc. May be underlain by oceanic or thin continental crust.

marine deposition coast A coast that has been greatly modified since the last change in sea level by the effects of marine deposition.

marine terrace An ancient wave-cut cliff and wave-cut bench that have been exposed above the present shoreline by uplift or a falling sea level.

mass wasting A term for the several types of downslope movement associated with hillslopes.

M-discontinuity *See* Mohorovičić discontinuity.

meandering river A sinuous channel with varying degrees of regularity of the bends in the channel.

medial moraine An elongate moraine in or carried upon the middle of a glacier and parallel to its sides. Normally formed as the result of the merging of a tributary glacier with a trunk glacier. Upon melting of the glacier, the moraine may be left briefly as an irregular ridge in the middle of a glacial valley.

median valley (plate tectonics) A long, narrow valley that lies along the crest of many mid-ocean ridges.

mesa An isolated, nearly level landmass standing distinctly above the surrounding country, bounded by abrupt or steeply sloping erosion escarpments on all sides. A mesa is generally capped by a layer of resistant, flat-lying rock and has a greater surface area than a butte.

Mesozoic era The middle of the three eras of geologic time in which abundant fossilizable organisms lived. Includes the time from 225 to 65 million years ago, during which the reptiles ruled the land.

metallic bond A bond in which the atoms are closely packed and the electrons move freely among the atoms.

metallogenic province A large geographic area enriched in one or more metallic elements.

metamorphic facies A group or assemblage of minerals that formed in response to an environment.

metamorphic grade A measure of the intensity of metamorphism.

metamorphic rock A rock that has undergone change in texture, mineralogy, or composition as a result of conditions below the depth affected by weathering processes.

meteor A shooting star. A meteorite burning due to friction in the atmosphere.

meteoric water Water that is in or derived from the atmosphere.

mica A mineral family. Biotite and muscovite are common rock-forming micas.

mid-ocean ridge A long, broad, continuous ridge that lies approximately in the middle of most ocean basins, except in the North Pacific Ocean. Commonly, it rises 1–3 km above the deep-ocean floor, is 2000–4000 km wide, and is completely submerged.

mineral A naturally occurring, crystalline, inorganic substance with a definite small range in chemical composition and physical properties.

Moho *See* Mohorovičić discontinuity.

Mohorovičić discontinuity (Moho, M-discontinuity) The boundary marked by a rapid change in seismic velocity that separates the earth's crust from the underlying mantle.

molluscs The animal group with the greatest variety of fossil representatives: snails, clams, cephalopods, and many smaller groups. Most have a calcareous shell covering the body, and a muscular foot.

mountain (alpine) glacier Any glacier in a mountain range except an ice cap or ice sheet. Usually originates in a cirque and may flow down into a valley previously carved by a stream.

mud cracks Irregular fractures occurring in a crudely polygonal pattern, formed by the shrinkage of clay, silt, or mud in the course of drying under atmospheric conditions.

mudflow A rapid type of mass wasting characterized by fluid motion of water-saturated debris generally confined to a channel until at a mountain front, where it may spread out over a considerable area.

mudstone *See* siltstone.

multiple glaciation The concept that repeated periods of glacial advance and retreat have occurred in the recent geologic past. This is evidenced by the layered arrangement of glacial deposits which are different from one another; the age difference may be confirmed by radiometric dating methods.

muscovite White mica. Recognized from its color and perfect cleavage. An important rock-forming mineral. $KAl_2(AlSi_3O_{10})(OH)_2$.

mylonite A fine-grained metamorphic rock formed by the milling of rocks on fault surfaces.

N

neap tide Twice each lunar month, the tidal pulls of the sun and moon on the earth are at right angles and tend to cancel each other out. During these periods, which correspond to the first- and third-quarter phases of the moon, the tidal range is minimal.

nearshore The region between the low-tide shoreline and the line of breakers.

nebula A flattened, rotating disk of interstellar matter.

net budget (mass balance) The difference between the amount of snow or ice accumulated and the amount lost by ablation in a glacier during 1 year, or averaged over a longer period.

neutrino A subatomic particle.

neutron A fundamental particle with no electrical charge found in the nuclei of atoms.

new moon The phase of the moon during which it is aligned, or nearly so, with the sun during conjunction.

nonclastic sedimentary rock A chemically or biologically precipitated sedimentary rock.

nonmarine evaporite An evaporite formed by evaporation of continental waters.

normal fault A fault on which the hanging wall has apparently moved downward relative to the footwall.

normal stress The component of stress perpendicular to a given plane.

nucleus (of an atom) The central part of an atom, composed of protons and, in most elements, neutrons.

O

obsidian Volcanic glass.

oceanic crust The part of the earth's crust that underlies the ocean basins. Ranges in thickness from 5 to 10 km, and is more dense and chemically more mafic (i.e., less rich in silica) than continental crust.

oceanic trench A long, narrow zone on the ocean floor that is deeper than the normal ocean floor and may be more than 11 km deep. Commonly, oceanic trenches lie adjacent to magmatic arcs and are thought to be the surface expression of the subduction of oceanic lithosphere.

oil The naturally occurring organic liquid from which petroleum can be recovered.

oil shale Shale with a high hydrocarbon content which can produce mineral oil through slow distillation.

olivine A common rock-forming mineral. $(Mg,Fe)_2SiO_4$.

ooze A fine-grained sediment found on deep-ocean floors, composed of more than 30 percent organic material.

ophiolite A sequence of mafic and ultramafic rocks overlain by pelagic chert. The sequence, from bottom to top, is: ultramafic rocks, gabbro, sheeted dikes, pillow lava, and radiolarian chert. Such a sequence is thought to represent oceanic crust.

opposition When the sun, moon, and earth are aligned, or nearly so, while the sun and moon are on opposite sides of the earth, such as during full moon.

ore A material that can be mined at a profit.

orthogonal lines Lines that show the paths that points on a wave front follow as it approaches shore. Are spaced along a wave front such that equal amounts of wave energy are present between any two adjacent lines.

outwash (outwash plain) Stratified sediment, chiefly sand and gravel removed from a glacier by meltwater streams and deposited beyond the end moraine. These sediments form a broad, gently sloping sheet of outwash material called an outwash plain.

overland flow The portion of direct runoff that flows directly on the surface of the earth.

overturned fold A fold whose hinge surface is inclined such that beds on one limb are inverted relative to their position before folding.

oxidation A type of chemical weathering which involves the addition of oxygen or loss of an electron to an element or compound. Oxidation products from iron-bearing minerals produce the characteristic brown, yellow, or red of many soils.

P

Pacific Ocean The earth's largest ocean, which has been in existence through much of earth history; although its present floor is no older than Jurassic, fragments of older age are incorporated in its encircling mountains.

Pacific-type margin The leading edge of a continent that undergoes tectonic uplift as a result of lithospheric plate convergence.

pahoehoe Hawaiian term for lava with a smooth, ropy surface; ropy lava.

paleontologist A scientist who studies fossil organisms.

Paleozoic era The oldest and longest era of geologic time, during which abundant fossils formed; it extends from approximately 600 to 225 million years ago.

Pangaea An ancient supercontinent which existed in the latest Paleozoic and earliest Mesozoic eras and was comprised of all the continents.

parabolic dune A sand dune with a long, scoop-shaped form, convex in the downwind direction with the horns pointing upwind. These dunes are commonly covered with sparse vegetation and are usually found along coasts where onshore winds are strong and sand is abundant.

passive fold A fold in which all layers have about the same ductility or viscosity and the layering exerts no mechanical control on the folding.

peat A dark brown material formed by the partial decomposition of plant material.

pedifan (pediment veneer) A veneer of fan alluvium from several centimeters to a meter thick that accumulates on a rock pediment surface.

pediment A broad, gently sloping, rock-floored erosion surface typically developed by weathering agents at the base of an abrupt and receding mountain front. The longitudinal profile of a pediment is generally concave upward, and the rock may be covered with a veneer of alluvium derived from upland areas and which is now transient across the pediment surface.

pediment angle The sharp, obtuse angle formed at the head of the pediment at its juncture with the base of the mountain scarp. It is the result of a rapid removal of sediment from the base of the escarpment that prevents fan formation.

pediment veneer *See* pedifan.

pediplain An extensive, thinly alluviated erosion surface formed in a desert region by the coalescence of two or more adjacent pediments.

pegmatite A very-coarse-grained igneous rock. The crystals are larger than 1 cm.

pelecypods A major mollusc group, the clams, whose members build a two-piece (bivalved) shell with right and left halves hinged along the animal's back. Most members are burrowers and thus have left a significant fossil record in Paleozoic, Mesozoic, and Cenozoic rocks, though they are most abundant in their later history.

peneplain The smooth, rolling erosion surface that develops very late in the cycle of erosion.

perennial stream A stream that flows the entire year from the headwaters to the mouth.

peridotite A coarse-grained igneous rock composed mainly of olivine and pyroxene.

period The basic subdivision of relative geologic time. Periods consist of shorter epochs and are aggregated into larger units called eras.

permeability The ability of an earth material to transmit fluids.

perthite A type of feldspar formed by the intergrowth of potassium feldspar and sodic plagioclase.

phase change When a mineral changes its internal structure without a change in chemical composition, generally in response to changes in temperature and pressure.

photosynthesis The process of converting carbon dioxide and water, using the sun's energy and the compound chlorophyll, to food (sugars) and oxygen.

piedmont glacier A thick, continuous sheet of ice at the base of a mountain range, resting on land, formed by the spreading out and coalescing of valley glaciers.

pillow lava Lava extruded underwater or in wet mud which develops a structure that resembles a pile of pillows.

pit A small depression in an outwash plain, formed by a stranded ice block that melted in place, which now contains a lake if the pit intersects the local water table.

plagioclase One of the feldspar minerals. A continuous series from $CaAl_2Si_2O_8$ to $NaAlSi_3O_8$. Recognized by its hardness, two cleavages, and striations on one of the cleavages.

plate tectonics A theory of global tectonics in which the lithosphere is divided into a number of plates whose pattern of motion is that of rigid bodies that move about various poles of rotation. Deformation and earthquake activity generally occur along the boundaries of the plates, whereas plate interiors are relatively less deformed and seismic activity is greatly reduced.

playa A dry, vegetation-free, flat area at the lowest part of an undrained desert basin. A playa is underlain by stratified sand, silt, and clay which commonly contain soluble salts. The surface seasonally may contain a playa lake formed by runoff from surrounding uplands.

Pleistocene epoch The next to the last of the seven epochs of the Cenozoic, covering from about 2.5 million years ago down to 10,000 years ago. Although originally defined on its fossil content, the epoch is best known for its extensive continental glaciations.

pleochroic halo *See* radiation halo.

plesiosaurs An extinct group of sea-going Mesozoic reptiles that resembled the mythical sea serpents with their long necks and flipperlike limbs.

plume (plate tectonics) Interpreted as a long and narrow cylindrical region of hot, rising mantle material. Plumes are thought to cause many of the volcanic centers, or hot spots, within or along the boundaries of plates.

plume track A linear zone of volcanic rocks that is progressively younger in one direction. May be manifest as a linear ridge or, less commonly, as a linear downwarp. Such tracks are thought to be the result of plate motion over a more or less fixed hot spot.

pluton Any body of igneous rock that formed below the surface.

plutonic igneous rock A plutonic rock is one that formed deep in the crust.

polar desert A desert region differing from hot deserts only in that temperatures are very cold. Characterized by low precipitation, high evaporation, seasonally strong winds, and an absence of vegetative cover except in bordering tundra regions.

polymorph Two or more minerals with the same composition but different crystal form, such as graphite and diamond (both composed of carbon).

pool A topographically low area in a stream channel produced by scour. Pools at low flow have a flat water surface.

porosity The ratio of the total volume of openings in an earth material to the total volume of the material, generally recorded as a percentage.

porphyritic An igneous rock texture in which some of the crystals are much larger than the matrix.

porphyry copper Porphyritic granitic rocks in which copper mineralization is found.

Precambrian The immense interval of geologic time from the beginning of the earth (4.6 billion years ago) down to the first appearance of abundant shelled organisms (600 million years ago).

Precambrian bedded iron deposit A stratified deposit of iron from the Precambrian era.

primary coast A coast that has not been appreciably modified by waves or other processes since the last change in sea level.

principal stress One of three mutually perpendicular normal stresses across whose planes there are no shear stresses.

principle of isostasy A condition whereby the lithosphere, or some portion of it, is in equilibrium on the asthenosphere, as if it were floating.

proglacial deposit A deposit found immediately in front of or just beyond the outer limits of a glacier or ice sheet.

proglacial lake A lake formed just beyond the frontal margin of an advancing or a retreating glacier, generally in direct contact with the glacier.

progressive metamorphism Metamorphism of increasing grade.

progressive waves Waves in which the waveform can be observed to move through a medium.

protalus rampart A small ridge of fresh glacial debris limited to the lip of a cirque. Forms as the result of a minor glacial advance that filled the cirque with ice but retreated before moving down-valley.

Proterozoic The era of Precambrian geologic time from approximately 2.5 billion to 600 million years ago.

proton A fundamental subatomic particle with a positive charge. Found in the nucleus of an atom.

pumice Very light, cellular, glassy lava.

P wave (primary wave) An elastic wave in which the particle motion is in the direction of propagation.

pyramidal dune (stellate dune, star dune) An isolated hill of sand; the base resembles a star in plan view, and sharp-crested ridges converge from basal points to a central peak that may rise as much as 90 m above the surrounding plain. These dunes tend to remain fixed for centuries at a place where the wind blows from all compass directions.

pyroclastic rock A rock formed by volcanic ejecta.

pyroxene A mineral family. Augite is the common rock-forming member.

Q

quartz A common rock-forming mineral. Recognized by its hardness (7) and lack of cleavage. SiO_2.

quartzite A metamorphic rock composed mainly of quartz. Forms from the metamorphism of sandstone or chert.

quartzose sandstone A sandstone composed largely of quartz.

R

radial drainage The drainage pattern that develops when a series of streams flow down from a central high point such as a volcanic peak.

radiation halo (pleochroic halo) A colored zone or halo surrounding a tiny radioactive mineral grain. The color of the halo changes when the grain is rotated under a polarizing microscope.

radioactivity The spontaneous breakdown of unstable isotopes by emitting particles or radiation.

radiolaria Microscopic one-celled floating organisms that build perforated spherical, conical, and cylindrical shells. Have been major contributors to oceanic oozes since the late Paleozoic.

radiometric dating The process of determining absolute dates (in years) by using the invariable decay rates of radioactive isotopes.

recessional (end) moraine A ridgelike accumulation of sediment produced at the margin of an actively flowing glacier at any given time. Also, an end or lateral moraine built during a temporary pause in the final retreat (recession) of a glacier.

rectangular drainage The drainage pattern that develops because streams erode along fracture systems within the bedrock.

recumbent fold A fold whose hinge surface is horizontal or nearly so.

red shift (Doppler effect) The apparent change in wavelength that occurs if the source and observer are moving apart.

refraction A process by which the part of a wave in shallow water is slowed down, causing the wave to bend and tend to align itself with the underwater contours.

regional metamorphism Metamorphism affecting a large region rather than a restricted basin. Typical regional metamorphic rocks are schist and gneiss.

relative dating The placement of events in their proper sequence without regard to the time of their occurrence.

reptiles A major group of terrestrial vertebrates which lay shelled eggs. Living forms have a scaly skin and cannot regulate their own temperature internally ("cold-blooded"), but not all fossil types had these traits.

respiration The process whereby food (sugars) and oxygen react to release energy, carbon dioxide, and water.

retrogressive metamorphism Metamorphism at conditions of lower grade than those under which the original rock formed.

rhyolite A fine-grained igneous rock similar in composition to granite.

ria coast An oak-leaf-shaped coast resulting from the drowning of a river valley by rising sea level.

riffle A topographic high found along a stream channel produced by deposition during relatively high flow. Riffles at low flow are characterized by relatively fast, shallow water. Riffles alternate with pools down the sinuous meandering channel.

rip current A strong, narrow current of short duration and high velocity flowing seaward through the breaker zone at nearly right angles to the shore. Represents the return to the ocean of water that has been forced onto the shore by incoming waves.

ripple mark An undulating surface on a sediment caused by water movement.

rock A natural aggregate of one or more minerals, or any essential and appreciable part of the solid portion of the earth (or any other part of the solar system).

rock glacier A mass of poorly sorted angular boulders and small rock fragments with an ice matrix, or with an ice core below the surface. Rock glaciers occur in high mountains under permafrost conditions, and move very slowly.

rock salt The mineral halite. NaCl.

Rocky Mountains The complex of mountain ranges forming the western boundary of the Great Plains and whose main structures were developed in the late Mesozoic and early Cenozoic eras and were reactivated in the late Cenozoic.

ropy lava *See* pahoehoe.

S

Saganagan Orogeny A mountain-building disturbance in the middle of the Archean era in the Canadian Shield.

salina A place where crystalline salt deposits are formed or found, such as a salt flat or salt pan, a salt lick, and, especially, a salt-encrusted playa.

saltation A mode of sediment transport in which the particles are moved by a series of short, intermittent leaps, jumps, and bounces along a substrate. Intermediate between suspension and rolling or sliding.

salt dome A structure caused by the upward movement of rock salt through overlying sedimentary rocks.

salt-water intrusion The encroachment of salt water from the ocean into fresh-water aquifers. Generally caused by overpumping or mining of groundwater resources near the coast.

sand sheet A thin, often geographically extensive accumulation of coarse sand too large to be transported by saltation. The ground surface is relatively flat and is broken only by small-scale sand ripples. Sand sheets are common where there is an excess supply of loose sand or more than can be entrapped by vegetation.

sandstone A clastic sedimentary rock with fragments between 1/16 and 2 mm in size.

satellite, natural The moon or moons of a planet.

scale trees An extinct group of large trees bearing diamond-shaped leaf scars which flourished in the late Paleozoic coal swamps. Related to the living club mosses.

schist A foliated metamorphic rock generally containing conspicuous mica.

scoria Volcanic rock with coarse bubble holes.

sea An area where wind blowing across the ocean surface puts energy into the surface water, creating wind-generated waves.

sea arch An opening through a headland cut by wave erosion.

sea cave A cavity at the base of a wave-cut cliff cut by wave erosion.

sea-floor spreading The hypothesis that oceanic crust is being generated by convective upwelling of magma along the mid-ocean ridges (i.e., spreading axes) and by laterally moving away the newly formed oceanic crust at a rate of 1–20 cm/year.

secondary coast A coast that has been appreciably modified by waves or other processes since the last change in sea level.

sedimentary rock Rocks formed near the earth's surface, generally in layers.

sediment pollution Fine-grained particles of rocks and minerals that become suspended in water, spoiling the water quality. These materials may also settle out into reservoirs or the bottom of canals, causing problems.

seif (longitudinal dune) A long, narrow sand dune, usually symmetrical in cross profile, oriented parallel to the direction of the prevailing wind. Longitudinal dunes form where sand is abundant and winds are strong and constant.

seismic waves All elastic waves generated by earthquakes or artificial explosions.

seismograph An instrument that detects and records vibrations of the earth. Commonly, these vibrations are generated by earthquakes.

serpentine A mineral family. $Mg_3Si_2O_5(OH)_4$.

serpentinite A rock composed largely of serpentine.

shale A fine-grained sedimentary rock composed largely of clay and silt and characterized by its fissility.

shear fracture A fracture that results from stresses that tend to move one part of a rock mass past another parallel to the fracture surface.

shear stress (tangential stress) The component of the stress that acts tangentially to a plane through any given point in a body.

sheeted dike complex A closely spaced sequence of mafic dikes in an ophiolite sequence that lies between pillow lavas above and gabbro below. Originally, the dikes were nearly vertical and are thought to represent intrusions at a spreading axis during the formation of oceanic crust.

sheet flow An overland flow or downslope movement of water taking the form of a thin film over relatively smooth soil and rock surfaces. The water is not concentrated into channels larger than rills.

sheet outwash Outwash deposits laid down in a layer, or sheet, by meltwater from a lengthy expanse of continental ice.

shield A portion of a craton where Precambrian or the very earliest Paleozoic rocks are exposed at the surface.

shield volcano A broad, gently sloping volcano built by flows of very fluid lava.

shore The area seaward of the coast, extending from the highest level of wave action during storms to the low-tide water line.

shrub coppice dune A small, streamlined dune that forms to the lee of brush-and-clump vegetation around playa margins and along stream courses. The dune is stable as long as the vegetation is present, but is soon degraded by wind action if the anchoring brush dies.

Sierra Nevada The highest range of the Pacific mountain systems, lying largely within California. Its eastern margin is a prominent normal fault (like those of the Basin Ranges to the east), which formed in late Cenozoic time. Many structures and rocks within the range formed in the Paleozoic and especially the Mesozoic eras.

silicate mineral A mineral containing SiO_4 tetrahedra.

sill A concordant igneous intrusive body.

siltstone A clastic sedimentary rock composed largely of silt.

sink A depression containing a central playa or saline lake with no outlet, for instance, where a desert stream comes to an end or disappears by evaporation. *Example:* Carson Sink, Nevada.

SiO_4 (silicon-oxygen) tetrahedron The basic unit of the silicate minerals. Composed of four oxygen atoms in the shape of a tetrahedron surrounding a silicon atom.

slate A fine-grained metamorphic rock characterized by well-developed foliation.

slip plane A potential or real surface in rocks or soils along which movement may take place in the downslope direction.

slump A type of mass wasting or landslide in which separate blocks of materials slide past one another with some backward rotation of these blocks.

soapstone A metamorphic rock composed of talc.

soil Earth material that has been modified by physical, chemical, and biologic processes such that it will support rooted plant life.

soil profile The organization of a soil into the A, B, and C horizons by vertical and horizontal movement of materials.

sole The lower part of basal ice of a glacier, which commonly contains rock fragments, appears dirty, and is separated from clear ice above by an abrupt boundary. The term also applies to the actual base of the ice.

solifluction A slow flowage type of mass wasting associated with permafrost.

sorting The degree to which the particles in a sediment or sedimentary rock are of the same size or shape.

specific gravity The ratio of the mass of a material to the mass of an equal volume of water.

spheroidal weathering The chemical weathering of rocks along fracture sets. Eventually, the weathered blocks may become spheroidal weathered boulders known as core stones.

sponges A group of very simple organisms with porous walls and, in many cases, a skeleton of tiny calcareous or siliceous rods and disks called spicules.

spring The natural discharge of groundwater from an aquifer.

spring tide Twice each lunar month, the tidal pulls of the sun and moon on the earth are aligned. During these alignments, which correspond to the new-moon and full-moon phases, the tidal range is maximum.

stack An isolated mass of bedrock projecting above the ocean surface off the seaward end of a headland from which it has been separated by wave erosion.

star dune *See* pyramidal dune.

stellate dune *See* pyramidal dune.

stock An igneous intrusive body smaller than a batholith.

stoping The sinking of blocks of heavier preexisting rock through molten magma.

strain The change in shape or volume of a body as a result of stress.

strain rate The velocity at which the material undergoes changes in shape or volume.

streak The color of a powdered mineral.

stream capacity A measure of the total load that a stream may carry.

stream competence The maximum size of bed material that a stream can carry.

stress In a solid, the force per unit area, acting on any surface within it.

strike-slip fault A fault in which the relative motion of the blocks on the two sides of the fault is parallel to the strike of the fault (i.e., is horizontal).

stripped surface An erosion surface of regional extent developed in an area underlain by horizontal or gently sloping rock strata of unequal resistance. An upper, softer rock layer is stripped away, exposing a smooth surface of a resistant stratum that acts as a local base level and thereby controls the depth of erosion.

stromatolite A layered mound of calcium carbonate formed by the primitive blue-green algae and bacteria; one of the dominant large fossils of the Precambrian.

subbituminous coal Coal, of a higher grade than lignite, composed of carbonaceous matter.

subduction The process whereby one lithospheric plate descends beneath another.

submarine canyon A canyon or valley on the ocean bottom, generally on the continental shelf or slope.

subsidence The downward (nearly vertical) movement of blocks of consolidated or unconsolidated earth materials. Commonly occurs over subterranean voids such as cavern systems or mines. Also forms from withdrawal of fluids such as oil and groundwater.

subsurface water The comprehensive term for all those waters that flow or are in storage in the lithosphere.

sulfide A mineral in which a metal is combined with sulfur, such as pyrite, FeS_2.

Superior Province A region of the Canadian Shield lying north of the Great Lakes and consisting mostly of Archean rocks.

superposition The principle that in an undeformed sequence of rock strata, each layer was deposited after the one below it and before the one above it.

surf A narrow zone of energy release in which swell gives up its energy, bounded on the landward side by the shoreline and on the seaward side by the line of breakers.

suspended load That part of the total stream load carried in fluid suspension by the moving water.

swash The thin layer of water that washes onto the shore at the landward edge of the surf zone.

S wave (shear wave) An elastic wave whose particle motion is at right angles to the direction of propagation.

swell Long, round, crested waves that carry energy put into the wind-generated waves of a sea across the open ocean with little loss of energy.

syncline A fold, generally concave upward, whose core contains stratigraphically younger rocks.

T

tabulate "corals" A group of fossil organisms whose remains consist of closely packed calcareous tubes containing numerous horizontal partitions (tabulae) and often having porous walls.

talc schist A foliated metamorphic rock containing the mineral talc.

talus (debris) slope The part of a slope profile found at the base of the free face. Consists of materials derived by mass-wasting and weathering processes from slope segments at higher elevations.

tangential stress (shear stress) The component of the stress that acts tangentially to a plane through any given point in a body.

tarn A relatively small and deep, steep-banked lake or pool amid high mountains, especially one occupying an ice-gouged basin or cirque in glaciated mountains.

tectonic movement The uplift or subsidence of the margin of a continent resulting from tectonic forces.

temperate desert A desert area of cooler middle latitudes. Mountain barriers may cause a rainshadow effect which prevents inland penetration of moisture-laden air, so that lower terrain to the lee of the mountains is arid. *Example:* interior Asia and North America.

tension fracture A fracture that results from stresses that tend to move one part of a rock mass away from another perpendicular to the fracture surface.

tephra Unconsolidated deposits of pyroclastic material.

terminal moraine The end moraine, extending across a glacial valley, or the outermost end moraine of a continental glacier or ice sheet which marks the farthest extent of its movement from its source. Formed at or near a stationary edge or at a place marking the cessation of an important glacial advance.

terminus The lower end or extremity of a glacier; the snout.

terracette A relatively minor slope feature formed either by animals grazing on slopes or by small failures.

Tethys Ocean An ancient ocean which separated Laurasia and Gondwanaland in the Mesozoic era.

tetracorals An extinct group of fossil corals characterized by a four-fold symmetry in their conical to cylindrical shells (thecae). Most were solitary, but some were colonial.

texture The size and arrangement of grains in a rock.

Theic Ocean An ancient ocean which existed in the Paleozoic era between Laurasia and Gondwanaland. Fragments of its floor are preserved in various low mountains of central Europe and in the southern Appalachians.

thermal gradient The rate of increase of temperature with depth in the earth.

thrust fault A fault that dips 45° or less and whose hanging wall has apparently moved up relative to its footwall.

tidal bore A steep-fronted tide crest that moves up a river in association with high tide.

tidal current A relatively rapid horizontal motion of water near shore that is associated with the rise and fall of the tide.

tidal range The vertical difference in the heights of high and low tides.

tidewater glacier A glacier that terminates in the sea, where it usually ends in an ice cliff from which icebergs are discharged.

tillite A consolidated or indurated sedimentary rock formed by the lithification of glacial till.

time correlation The process of determining that two beds of rock were formed at the same time.

tombolo A marine deposit that extends from an island to another island or the mainland.

transform fault A type of strike-slip fault in which displacement suddenly stops or changes character. Most transform faults form strike-slip plate boundaries and connect divergent or subduction boundaries.

transverse dune A strongly asymmetrical sand dune elongated perpendicularly to the direction of the prevailing winds. These dunes have a gentle windward slope and a steep leeward slope whose slipface stands at or near the angle of repose of sand.

transverse waves Waves in which particle motion is at right angles to the direction of energy propagation.

trellis drainage A drainage pattern commonly found on folded or tilted sedimentary rock units that vary in resistance to erosion.

trench *See* oceanic trench.

trilobites A major group of fossil organisms in the Paleozoic era. These extinct animals had a jointed body and appendages like the modern animals known as arthropods.

trimline A sharp boundary line that represents the upper limit of glacial ice on the sides of a vacated valley. Usually coincides with a break in slope, and the valley of a long-extinct glacier may be marked by a sharp change in the age, type, and density of vegetation.

trunk glacier A glacier that occupies a mountain valley into which tributary glaciers flow. Erodes its valley at a greater rate than the tributary glaciers.

tsunami A long-period gravity wave generated by a submarine earthquake or volcanic event. Tsunami are not noticeable on the open ocean but can build up to great heights in shallow water.

tuff A pyroclastic volcanic rock with fragments smaller than 4 mm.

tundra A treeless, level or gently undulating plain characteristic of arctic and subarctic regions. Normally has a marshy surface that supports mosses, lichens, and low shrubs, and is underlain by dark, mucky soil and permafrost.

turbidity current A current or flow caused by the movement downslope of water heavy because of its suspended material (muddy water).

turbulent flow A type of flow found in streams characterized by flow lines that are seldom, if ever, parallel to the stream bed or banks.

U

unconfined groundwater system A groundwater system in which there are no impermeable layers restricting the upper surface of the zone of saturation.

uniformitarianism The concept that ancient rocks can be understood in terms of the processes presently occurring on the earth.

upwelling The process by which deep, cold, nutrient-laden water is brought to the surface, usually by diverging equatorial currents or coastal currents that pull surface water away from the coast.

Ural Mountains A mountain range separating Europe and Siberia whose structures were formed during the late Paleozoic collision of those cratons.

Ural Ocean An ancient ocean which existed between Europe and Siberia during the Paleozoic era. Fragments of its floor are preserved in the Ural Mountains.

U-shaped valley The characteristic cross-valley profile of a glaciated valley, formed as the result of glacial erosion which reshaped a stream valley.

V

valley glacier A glacier flowing between the walls of a mountain valley throughout all or part of its length.

valley train A long, narrow body of outwash deposited by meltwater streams, far beyond the terminal moraine or front of an active glacier. These deposits are confined to the valley (which may have been preglacial) below the terminus of the glacier.

varves A sedimentary layer or sequence of layers deposited in a body of still water within 1 year's time. Glacial varves often consist of a summer layer of coarser, light-colored sediment, representing rapid summer melting, and a finer, darker layer deposited during winter.

ventifact A general term to describe any stone that has been shaped, worn, faceted, cut, or polished by the abrasive action of wind-blown sand. Normally, this occurs under desert conditions.

vertebrates A major group of living and fossil organisms whose members have a phosphatic internal skeleton, including a backbone and skull. Members include fish, amphibians, reptiles, birds, and mammals.

volcanic (magmatic) arc A linear belt of volcanic and plutonic rocks. Many of these belts have an arcuate shape.

volcanic coast A coast that consists of lava flows extending out into the ocean or concave embayments caused by explosion or collapse of one side of a volcano, opening it to the sea.

W

waning (wash) slope That part of a slope profile produced by running water which is usually concave and found at the base of the slope profile.

water table A plane that delineates the approximate elevation of the water surface in wells that penetrate the zone of saturation.

wave base The depth beneath the still-water line to which particle motion of ocean waves extends.

wave crest The highest point on a waveform.

wave-cut cliff A cliff produced by the landward cutting of wave erosion.

wave erosion coast A coast that has been greatly modified since the last change in sea level by the effects of wave erosion.

wave height The vertical distance from the trough to the crest of a waveform.

wavelength The horizontal distance between two corresponding points on successive waveforms, such as from crest to crest.

wave period The time required for the passage of one waveform past a fixed point.

wave trough The lowest point on a waveform.

waxing slope The part of a slope profile that is convex and associated with mass-wasting processes such as slow creep of the soil or rock.

weathering A group of physical, chemical, and biologic processes that operate at or near the surface of the earth to break rocks down.

welded tuff A tuff that was erupted when hot and whose fragments have fused together as a result of the action of heat and the gases present.

X

xeric A term applied to an environment characterized by a low or inadequate supply of moisture. Also applied to an organism or group of organisms existing in such a habitat.

Z

zone of aeration The zone above the water table in which some water may exist or be in storage. However, all void or empty spaces are not filled with water, and some of the water is moving downward or laterally through soil or rock.

zone of saturation The zone below the water table in which all the void or empty spaces are filled with water.

INDEX

Valarie Rolland

374-7998 Home

374-7032